de Gruyter Expositions in Mathematics 27

Editors

O. H. Kegel, Albert-Ludwigs-Universität, Freiburg
V. P. Maslov, Academy of Sciences, Moscow
W. D. Neumann, The University of Melbourne, Parkville
R. O. Wells, Jr., Rice University, Houston

de Gruyter Expositions in Mathematics

Algebra in the Stone-Čech Compactification

Theory and Applications

by

Neil Hindman
Dona Strauss

Walter de Gruyter · Berlin · New York 1998

Authors

Neil Hindman
Department of Mathematics
Howard University
Washington, DC 20059
USA

Dona Strauss
Mathematics Department
University of Hull
Hull HU6 7RX
United Kingdom

1991 Mathematics Subject Classification: 22-02; 22A15, 05D10, 54D35

Keywords: Compactification, semigroup, Stone-Čech compactification, Ramsey theory, combinatorics, right topological semigroup, idempotent, ideal minimal ideal, smallest ideal, minimal idempotent

♾ Printed on acid-free paper which falls within the guidelines of the ANSI to ensure permanence and durability.

Library of Congress — Cataloging-in-Publication Data

> Hindman, Neil, 1943—
>   Algebra in the Stone-Čech compactification : theory and applica-
> tions / by Neil Hindman, Dona Strauss.
>   p.   cm. — (De Gruyter expositions in mathematics : 27)
>   Includes bibliographical references and index.
>   ISBN 3-11-015420-X (alk. paper)
>   1. Stone-Čech compactification.   2. Topological semigroups.
>   I. Strauss, Dona, 1934—.   II. Title.   III. Series
>   QA611.23.H56   1998
>   514′.32—dc21   98-29957
>   CIP

Die Deutsche Bibliothek — Cataloging-in-Publication Data

> **Hindman, Neil:**
> Algebra in the Stone-Čech compactification : theory and applications /
> by Neil Hindman ; Dona Strauss. — Berlin ; New York : de Gruyter,
> 1998
>   (De Gruyter expositions in mathematics ; 27)
>   ISBN 3-11-015420-X

Typeset using the authors' TeX files: I. Zimmermann, Freiburg
Printing: WB-Druck GmbH & Co., Rieden/Allgäu. Binding: Lüderitz & Bauer GmbH, Berlin.
Cover design: Thomas Bonnie, Hamburg.

Errata for
"Algebra in the Stone-Čech Compactification"
by
Neil Hindman and Dona Strauss

De Gruyter Expositions in Mathematics **27** (1998)

Because of an error in the preliminary processing of the TEX files, there are **19** occurrences of the symbol "$\not\subseteq$" where the symbol "$\subsetneq$" was intended. These are as follows:

Page 49	Line −15,	Page 49	Line −13,	Page 49	Line −4,
Page 59	Line −8,	Page 123	Line 1,	Page 128	Line 1 **(twice),**
Page 131	Line 3,	Page 172	Line −14,	Page 190	**Line 18,**
Page 190	**Line −17,**	Page 190	Line −15,	Page 190	**Line −10,**
Page 190	Line −2,	Page 196	Line −16,	Page 292	Line −11,
Page 293	Line 7,	Page 293	Line 11,	Page 293	Line 12.

In addition, the following errors have been discovered:

Page 36 Line −16: "$\rho_{z|L}$ is continuous." should be "$\rho_{z|L}$ is a homeomorphism."
Page 40 Line 16: "Let $\vec{u} \in K(S)$." should be "Let $\vec{u} \in \times_{i \in I} K(S_i)$."
Page 57 Line −16: "Once we have identified $s \in S$" should be "Once we have identified $s \in D$".
Page 91 Line 16: "(a) $[A]^\kappa = \{B \subseteq A : |A| = \kappa\}$." should be
"(a) $[A]^\kappa = \{B \subseteq A : |B| = \kappa\}$."
Page 91 Line 17: "(b) $[A]^{<\kappa} = \{B \subseteq A : |A| < \kappa\}$." should be
"(b) $[A]^{<\kappa} = \{B \subseteq A : |B| < \kappa\}$."
Page 293 Line −2: "such that $f \frown x\ g$" should be "such that $f \frown x \subsetneq g$".
Page 424 Lines −4 and −3: "Theorem 20.13, one of three nonelementary results used in this book that we do not prove, is due to H. Furstenberg." should be "Theorem 20.13, one of three nonelementary results used in this book that we do not prove, is due to H. Furstenberg and Y. Katznelson in [*An ergodic Szemerédi theorem for commuting transformations*, J. Anal. Math. **34** (1978), 61–85]."

This list (possibly updated) is also available at:
`www.degruyter.com/highlights/hindman.html`.

Walter de Gruyter · Berlin · New York

Preface

The semigroup operation defined on a discrete semigroup $(S, \cdot)$ has a natural extension, also denoted by $\cdot$, to the Stone–Čech compactification βS of S. Under the extended operation, βS is a compact right topological semigroup with S contained in its topological center. That is, for each $p \in \beta S$, the function $\rho_p : \beta S \to \beta S$ is continuous and for each $s \in S$, the function $\lambda_s : \beta S \to \beta S$ is continuous, where $\rho_p(q) = q \cdot p$ and $\lambda_s(q) = s \cdot q$.

In Part I of this book, assuming only the mathematical background standardly provided in the first year of graduate school, we develop the basic background information about compact right topological semigroups, the Stone–Čech compactification of a discrete space, and the extension of the semigroup operation on S to βS. In Part II, we study in depth the algebra of the semigroup $(\beta S, \cdot)$ and in Part III present some of the powerful applications of the algebra of βS to the part of combinatorics known as *Ramsey Theory*. We conclude in Part IV with connections with Topological Dynamics, Ergodic Theory, and the general theory of semigroup compactifications.

The study of the semigroup $(\beta S, \cdot)$ has interested several mathematicians since it was first defined in the late 1950's. As a glance at the bibliography will show, a large number of research papers have been devoted to its properties.

There are several reasons for an interest in the algebra of βS.

It is intrinsically interesting as being a natural extension of S which plays a special role among semigroup compactifications of S. It is the largest possible compactification of this kind: If T is a compact right topological semigroup, φ is a continuous homomorphism from S to T, $\varphi[S]$ is dense in T, and $\lambda_{\varphi(s)}$ is continuous for each $s \in S$, then T is a quotient of βS.

We believe that $\beta \mathbb{N}$ is interesting and challenging for its own sake, as well as for its applications. Although it is a natural extension of the most familiar of all semigroups, it has an algebraic structure of extraordinary complexity, which is constantly surprising. For example, $\beta \mathbb{N}$ contains many copies of the free group on $2^{\mathfrak{c}}$ generators [152]. Algebraic questions about $\beta \mathbb{N}$ which sound deceptively simple have remained unsolved for many years. It is, for instance, not known whether $\beta \mathbb{N}$ contains any elements of finite order, other than idempotents. And the corresponding question about the existence of nontrivial finite groups was only very recently answered by E. Zelenuk. (His negative answer is presented in Chapter 7.)

The semigroup βS is also interesting because of its applications to combinatorial number theory and to topological dynamics.

Algebraic properties of βS have been a useful tool in Ramsey theory. Results in Ramsey Theory have a twin beauty. On the one hand they are representatives of pure mathematics at its purest: simple statements easy for almost anyone to understand (though not necessarily to prove). On the other hand, the area has been widely applied from its beginning. In fact a perusal of the titles of several of the original papers reveals that many of the classical results were obtained with applications in mind. (Hilbert's Theorem – Algebra; Schur's Theorem – Number Theory; Ramsey's Theorem – Logic; the Hales–Jewett Theorem – Game Theory).

The most striking example of an application of the algebraic structure of βS to Ramsey Theory is perhaps provided by the Finite Sums Theorem. This theorem says that whenever $\mathbb{N}$ is partitioned into finitely many classes (or in the terminology common within Ramsey Theory, is *finitely colored*), there is a sequence $\langle x_n \rangle_{n=1}^{\infty}$ with $\mathrm{FS}(\langle x_n \rangle_{n=1}^{\infty})$ contained in one class (or *monochrome*). (Here $\mathrm{FS}(\langle x_n \rangle_{n=1}^{\infty}) = \{\Sigma_{n \in F} \, x_n : F$ is a finite nonempty subset of $\mathbb{N}\}$.) This theorem had been an open problem for some decades, even though several mathematicians (including Hilbert) had worked on it. Although it was initially proved without using $\beta \mathbb{N}$, the first proof given was one of enormous complexity.

In 1975 F. Galvin and S. Glazer provided a brilliantly simple proof of the Finite Sums Theorem using the algebraic structure of $\beta \mathbb{N}$. Since this time numerous strong combinatorial results have been obtained using the algebraic structure of βS, where S is an arbitrary discrete semigroup. In the process, more detailed knowledge of the algebra of βS has been obtained.

Other famous combinatorial theorems, such as van der Waerden's Theorem or Rado's Theorem, have elegant proofs based on the algebraic properties of $\beta \mathbb{N}$. These proofs have in common with the Finite Sums Theorem the fact that they were initially established by combinatorial methods. A simple extension of the Finite Sums Theorem was first established using the algebra of $\beta \mathbb{N}$. This extension says that whenever $\mathbb{N}$ is finitely colored there exist sequences $\langle x_n \rangle_{n=1}^{\infty}$ and $\langle y_n \rangle_{n=1}^{\infty}$ such that $\mathrm{FS}(\langle x_n \rangle_{n=1}^{\infty}) \cup \mathrm{FP}(\langle y_n \rangle_{n=1}^{\infty})$ is monochrome, where $\mathrm{FP}(\langle y_n \rangle_{n=1}^{\infty}) = \{\Pi_{n \in F} \, y_n : F$ is a finite nonempty subset of $\mathbb{N}\}$. This combined additive and multiplicative result was first proved in 1975 using the algebraic structure of $\beta \mathbb{N}$ and it was not until 1993 that an elementary proof was found.

Other fundamental results have been established for which it seems unlikely that elementary proofs will be found. Among such results is a density version of the Finite Sums Theorem, which says roughly that the sequence $\langle x_n \rangle_{n=1}^{\infty}$ whose finite sums are monochrome can be chosen inductively in such a way that at each stage of the induction the set of choices for the next term has positive upper density. Another such result is the Central Set Theorem, which is a common generalization of many of the basic results of Ramsey Theory. Significant progress continues to be made in the combinatorial applications.

The semigroup βS also has applications in topological dynamics. A semigroup S of continuous functions acting on a compact Hausdorff space X has a closure in $^X X$ (the space of functions mapping X to itself with the product topology), which is a compact right topological semigroup. This semigroup, called the enveloping semigroup, was first studied by R. Ellis [86]. It is always a quotient of the Stone–Čech compactification

βS, as is every semigroup compactification of S, and is, in some important cases, equal to βS. In this framework, the algebraic properties of βS have implications for the dynamical behavior of the system.

The interaction with topological dynamics works both ways. Several notions which originated in topological dynamics, such as syndetic and piecewise syndetic sets, are important in describing the algebraic structure of βS. For example, a point p of βS is in the closure of the smallest ideal of βS if and only if for every neighborhood U of p, $U \cap S$ is piecewise syndetic.

This last statement can be made more concise when one notes the particular construction of βS that we use. That is, βS is the set of all ultrafilters on S, the principal ultrafilters being identified with the points of S. Under this construction, any point p of βS is precisely $\{U \cap S : U$ is a neighborhood of $p\}$. Thus p is in the closure of the smallest ideal of βS if and only if every member of p is piecewise syndetic.

In this book, we develop the algebraic theory of βS and present several of its combinatorial applications. We assume only that the reader has had graduate courses in algebra, analysis, and general topology as well as a familiarity with the basic facts about ordinal and cardinal numbers. In particular we develop the basic structure of compact right topological semigroups and provide an elementary construction of the Stone–Čech compactification of a discrete space.

With only three exceptions, this book is self contained for those with that minimal background. The three cases where we appeal to non elementary results not proved here are Theorem 6.36 (due to M. Rudin and S. Shelah) which asserts the existence of a collection of $2^{\mathfrak{c}}$ elements of $\beta \mathbb{N}$ no two of which are comparable in the Rudin–Keisler order, Theorem 12.37 (due to S. Shelah) which states that the existence of P-points in $\beta \mathbb{N} \backslash \mathbb{N}$ cannot be established in ZFC, and Theorem 20.13 (due to H. Furstenberg) which is an ergodic theoretic result that we use to derive Szemerédi's Theorem.

All of our applications involve Hausdorff spaces, so we will be assuming throughout, except in Chapter 7, that all hypothesized topological spaces are Hausdorff.

The first five chapters are meant to provide the basic preliminary material. The concepts and theorems given in the first three of these chapters are also available in other books. The remaining chapters of the book contain results which, for the most part, can only be found in research papers at present, as well as several previously unpublished results.

Notes on the historical development are given at the end of each chapter.

Let us make a few remarks about organization. Chapters are numbered consecutively throughout the book, regardless of which of the four parts of the book contains them. Lemmas, theorems, corollaries, examples, questions, comments, and remarks are numbered consecutively in one group within chapters (so that Lemma 2.4 will be found after Theorem 2.3, for example). There is no logical distinction between a theorem and a remark. The difference is that proofs are never included for remarks. Exercises come at the end of sections and are numbered consecutively within sections.

The authors would like to thank Andreas Blass, Karl Hofmann, Paul Milnes, and Igor Protasov for much helpful correspondence and discussions. Special thanks go to John Pym for a careful and critical reading of an early version of the manuscript.

The authors also wish to single out Igor Protasov for special thanks, as he has contributed several new theorems to the book. They would like to thank Arthur Grainger, Amir Maleki, Dan Tang, Elaine Terry, and Wen Jin Woan for participating in a seminar where much of the material in this book was presented, and David Gunderson for presenting lectures based on the early material in the book. Acknowledgement is also due to our collaborators whose efforts are featured in this book. These collaborators include John Baker, Vitaly Bergelson, John Berglund, Andreas Blass, Dennis Davenport, Walter Deuber, Ahmed El-Mabhouh, Hillel Furstenberg, Salvador García-Ferreira, Yitzhak Katznelson, Jimmie Lawson, Amha Lisan, Imre Leader, Hanno Lefmann, Amir Maleki, Jan van Mill, Paul Milnes, John Pym, Petr Simon, Benjamin Weiss, and Wen-jin Woan.

The authors would like to acknowledge support of a conference on the subject of this book in March of 1997 by DFG Sonderforschungsbereich 344, Diskrete Strukturen in der Mathematik, Universität Bielefeld, and they would like to thank Walter Deuber for organizing this conference. Both authors would like to thank the EPSRC (UK) for support of a visit and the first author acknowledges support received from the National Science Foundation (USA) under grant DMS 9424421 during the preparation of this book.

Finally, the authors would like to thank their spouses, Audrey and Ed, for their patience throughout the writing of this book as well as hospitality extended to each of us during visits with the other.

April 1998

Neil Hindman
Dona Strauss

Contents

Part I

Background Development

Notation

We write $\mathbb{N}$ for the set $\{1, 2, 3, \ldots\}$ of positive integers and $\omega = \{0, 1, 2, \ldots\}$ for the nonnegative integers. Also ω is the first infinite ordinal, and thus the first infinite cardinal. Each ordinal is the set of all smaller ordinals.

$\mathbb{R}^+ = \{x \in \mathbb{R} : x > 0\}$.

Given a function f and a set A contained in the domain of f, we write $f[A] = \{f(x) : x \in A\}$ and given any set B we write $f^{-1}[B] = \{x \in \mathrm{Domain}(f) : f(x) \in B\}$.

Given a set A, $\mathcal{P}_f(A) = \{F : \emptyset \neq F \subseteq A \text{ and } F \text{ is finite}\}$.

Definitions of additional unfamiliar notation can be located by way of the index.

Chapter 1

Semigroups and Their Ideals

We assume that the reader has had an introductory modern algebra course. This assumption is not explicitly used in this chapter beyond the fact that we expect a certain amount of mathematical maturity.

1.1 Semigroups

Definition 1.1. A *semigroup* is a pair $(S, *)$ where S is a nonempty set and $*$ is a binary associative operation on S.

Formally a *binary operation* on S is a function $* : S \times S \to S$ and the operation is *associative* if and only if $*(*(x, y), z) = *\big(x, *(y, z)\big)$ for all x, y, and z in S. However, we customarily write $x * y$ instead of $*(x, y)$ so the associativity requirement becomes the more familiar $(x * y) * z = x * (y * z)$. The statement that $* : S \times S \to S$, i.e., that $x * y \in S$ whenever $x, y \in S$ is commonly referred to by saying that "S is *closed* under $*$".

Example 1.2. *Each of the following is a semigroup.*

(a) $(\mathbb{N}, +)$.

(b) $(\mathbb{N}, \cdot)$.

(c) $(\mathbb{R}, +)$.

(d) $(\mathbb{R}, \cdot)$.

(e) $(\mathbb{R}\backslash\{0\}, \cdot)$.

(f) $(\mathbb{R}^+, +)$.

(g) $(\mathbb{R}^+, \cdot)$.

(h) $(\mathbb{N}, \vee)$, *where $x \vee y = \max\{x, y\}$.*

(i) $(\mathbb{N}, \wedge)$, *where $x \wedge y = \min\{x, y\}$. .*

(j) $(\mathbb{R}, \wedge)$.

(k) $(S, *)$, *where S is any nonempty set and $x * y = y$ for all $x, y \in S$.*

(l) $(S, *)$, *where S is any nonempty set and $x * y = x$ for all $x, y \in S$.*

(m) $(S, *)$, *where S is any nonempty set and $a \in S$ and $x * y = a$ for all $x, y \in S$.*

(n) $\left({}^{X}X, \circ \right)$, *where ${}^{X}X = \{f : f : X \to X\}$ and $\circ$ represents the composition of functions.*

The semigroups of Example 1.2 (k) and (l) are called respectively *right zero* and *left zero* semigroups.

An important class of semigroups are the *free* semigroups. These require a more detailed explanation.

Definition 1.3. Let A be a nonempty set. The *free semigroup on the alphabet A* is the set $S = \{f : f$ is a function and range$(f) \subseteq A$ and there is some $n \in \mathbb{N}$ such that domain$(f) = \{0, 1, \ldots, n - 1\}\}$. Given f and g in S, the operation (called *concatenation*) is defined as follows. Assume domain$(f) = \{0, 1, \ldots, n - 1\}$ and domain$(g) = \{0, 1, \ldots, m - 1\}$. Then domain$(f^\frown g) = \{0, 1, \ldots, m + n - 1\}$ and given $i \in \{0, 1, \ldots, m + n - 1\}$,

$$f^\frown g = \begin{cases} f(i) & \text{if } i < n \\ g(i - n) & \text{if } i \geq n. \end{cases}$$

The *free semigroup with identity on the alphabet A* is $S \cup \{\emptyset\}$ where S is the free semigroup on the alphabet A. Given $f \in S \cup \{\emptyset\}$ one defines $f^\frown \emptyset = \emptyset^\frown f = f$

One usually refers to the elements of a free semigroup as *words* and writes them by listing the values of the function in order. The *length* of a word is n where the domain of the word is $\{0, 1, \ldots, n - 1\}$ (and the length of $\emptyset$ is 0). Thus if $A = \{2, 4\}$ and $f = \{(0, 4), (1, 2), (2, 2)\}$ (so that the length of f is 3 and $f(0) = 4$, $f(1) = 2$, and $f(2) = 2$), then one represents f as 422. Furthermore given the "words" 422 and 24424, one has $422^\frown 24424 = 42224424$.

We leave to the reader the routine verification of the fact that concatenation is associative, so that the free semigroup *is* a semigroup.

Definition 1.4. Let $(S, *)$ and $(T, \cdot)$ be semigroups.

(a) A *homomorphism* from S to T is a function $\varphi : S \to T$ such that $\varphi(x * y) = \varphi(x) \cdot \varphi(y)$ for all $x, y \in S$.

(b) An *isomorphism* from S to T is a homomorphism from S to T which is both one-to-one and onto T.

(c) The semigroups S and T are *isomorphic* if and only if there exists an isomorphism from S to T. If S and T are isomorphic we write $S \approx T$.

(d) An *anti-homomorphism* from S to T is a function $\varphi : S \to T$ such that $\varphi(x * y) = \varphi(y) \cdot \varphi(x)$ for all $x, y \in S$.

(e) An *anti-isomorphism* from S to T is an anti-homomorphism from S to T which is both one-to-one and onto T.

(f) The semigroups S and T are *anti-isomorphic* if and only if there exists an anti-isomorphism from S to T.

Clearly, the composition of two homomorphism, if it exists, is also a homomorphism. The reader who is familiar with the concept of a category will recognize that there is a category of semigroups, in which the objects are semigroups and the morphisms are homomorphisms.

The free semigroup S on the alphabet A has the following property. Suppose that T is an arbitrary semigroup and that $g : A \to T$ is any mapping. Then there is a unique homomorphism $h : S \to T$ with the property that $h(a) = g(a)$ for every $a \in A$. (The proof of this assertion is Exercise 1.1.1.)

Definition 1.5. Let $(S, *)$ be a semigroup and let $a \in S$.

(a) The element a is a *left identity* for S if and only if $a * x = x$ for every $x \in S$.

(b) The element a is a *right identity* for S if and only if $x * a = x$ for every $x \in S$.

(c) The element a is a *two sided identity* (or simply an *identity*) for S if and only if a is both a left identity and a right identity.

Note that in a "free semigroup with identity" the element $\emptyset$ is a two sided identity (so the terminology is appropriate).

Note also that in a left zero semigroup, every element is a right identity and in a right zero semigroup, every element is a left identity. On the other hand we have the following simple fact.

Remark 1.6. *Let $(S, *)$ be a semigroup. If e is a left identity for S and f is a right identity for S, then $e = f$. In particular, a semigroup can have at most one two sided identity.*

Given a collection of semigroups $\langle (S_i, *_i) \rangle_{i \in I}$, the Cartesian product $\times_{i \in I} S_i$ is naturally a semigroup with the coordinatewise operations.

Definition 1.7. (a) Let $\langle (S_i, *_i) \rangle_{i \in I}$ be an indexed family of semigroups and let $S = \times_{i \in I} S_i$. With the operation $*$ defined by $(\vec{x} * \vec{y})_i = x_i *_i y_i$, the semigroup $(S, *)$ is called the *direct product* of the semigroups $(S_i, *_i)$.

(b) Let $\langle (S_i, *_i) \rangle_{i \in I}$ be an indexed family of semigroups where each S_i has a two sided identity e_i. Then the *direct sum* of the semigroups $(S_i, *_i)$ is $\bigoplus_{i \in I} S_i = \{\vec{x} \in \times_{i \in I} S_i : \{i \in I : x_i \neq e_i\}$ is finite$\}$.

We leave to the reader the easy verification that the direct product operation is associative as well as the verification that if $\vec{x}, \vec{y} \in \bigoplus_{i \in I} S_i$, then $\vec{x} * \vec{y} \in \bigoplus_{i \in I} S_i$.

Definition 1.8. Let $(S, *)$ be a semigroup and let $a, b, c \in S$.

(a) The element c is a *left a-inverse* for b if and only if $c * b = a$.

(b) The element c is a *right a-inverse* for b if and only if $b * c = a$.

(c) The element c is an *a-inverse* for b if and only if c is both a left a-inverse for b and a right a-inverse for b.

The terms *left a-inverse*, *right a-inverse*, and *a-inverse* are usually replaced by *left inverse*, *right inverse*, and *inverse* respectively. We introduce the more precise notions because one may have many left or right identities.

Definition 1.9. A *group* is a pair $(S, *)$ such that

(a) $(S, *)$ is a semigroup, and
(b) there is an element $e \in S$ such that
    (i) e is a left identity for S and
    (ii) for each $x \in S$ there exists $y \in S$ such that y is a left e-inverse for x.

Theorem 1.10. *Let $(S, *)$ be a semigroup. The following statements are equivalent.*

(a) *$(S, *)$ is a group.*
(b) *There is a two sided identity e for S with the property that for each $x \in S$ there is some $y \in S$ such that y is a (two sided) e-inverse for x.*
(c) *There is a left identity for S and given any left identity e for S and any $x \in S$ there is some $y \in S$ such that y is a left e-inverse for x.*
(d) *There is a right identity e for S such that for each $x \in S$ there is some $y \in S$ such that y is a right e-inverse for x.*
(e) *There is a right identity for S and given any right identity e for S and any $x \in S$ there is some $y \in S$ such that y is a right e-inverse for x.*

Proof. (a) implies (b). Pick e as guaranteed by Definition 1.9. We show first that any element has an e-inverse, so let $x \in S$ be given and let y be a left e-inverse for x. Let z be a left e-inverse for y. Then $x * y = e * (x * y) = (z * y) * (x * y) = z * \big(y * (x * y)\big) = z * \big((y * x) * y\big) = z * (e * y) = z * y = e$, so y is also a right e-inverse for x as required.

Now we show that e is a right identity for S, so let $x \in S$ be given. Pick an e-inverse y for x. Then $x * e = x * (y * x) = (x * y) * x = e * x = x$.

(b) implies (c). Pick e as guaranteed by (b). Given any left identity f for S we have by Remark 1.6 that $e = f$ so every element of S has a left f-identity.

That (c) implies (a) is trivial.

The implications (d) implies (b), (b) implies (e), and (e) implies (d) follow now by left-right switches, the details of which form Exercise 1.1.2.   □

In a right zero semigroup S (Example 1.2 (k)) every element is a left identity and given any left identity e and any $x \in S$, e is a right e-inverse for x. This is essentially the only example of this phenomenon. That is, we shall see in Theorem 1.40 that any semigroup with a left identity e such that every element has a right e-inverse is the Cartesian product of a group with a right zero semigroup. In particular we see that if a semigroup has a unique left identity e and every element has a right e-inverse, then the semigroup is a group.

In the semigroup $(\mathbb{N}, \vee)$, 1 is the unique identity and the only element with an inverse.

When dealing with arbitrary semigroups it is customary to denote the operation by $\cdot$. Furthermore, given a semigroup $(S, \cdot)$ one customarily writes xy in lieu of $x \cdot y$. We shall now adopt these conventions. Accordingly, from this point on, when we write "Let S be a semigroup" we mean "Let $(S, \cdot)$ be a semigroup" and when we write "xy" we mean "$x \cdot y$".

Definition 1.11. Let S be a semigroup.

    (a) S is *commutative* if and only if $xy = yx$ for all $x, y \in S$.

    (b) The *center* of S is $\{x \in S : \text{for all } y \in S, xy = yx\}$.

    (c) Given $x \in S, \lambda_x : S \to S$ is defined by $\lambda_x(y) = xy$.

    (d) Given $x \in S, \rho_x : S \to S$ is defined by $\rho_x(y) = yx$.

    (e) $L(S) = \{\lambda_x : x \in S\}$.

    (f) $R(S) = \{\rho_x : x \in S\}$.

Remark 1.12. *Let S be a semigroup. Then $(L(S), \circ)$ and $(R(S), \circ)$ are semigroups.*

Since our semigroups are not necessarily commutative we need to specify what we mean by $\prod_{i=1}^{n} x_i$. There are 2 reasonable interpretations (and $n! - 2$ unreasonable ones). We choose it to mean the product in increasing order of indices because that is the order that naturally arises in our applications of right topological semigroups. More formally we have the following.

Definition 1.13. Let S be a semigroup. We define $\prod_{i=1}^{n} x_i$ for $\{x_1, x_2, \ldots, x_n\} \subseteq S$ inductively on $n \in \mathbb{N}$.

    (a) $\prod_{i=1}^{1} x_i = x_1$.

    (b) Given $n \in \mathbb{N}, \prod_{i=1}^{n+1} x_i = (\prod_{i=1}^{n} x_i) \cdot x_{n+1}$.

Definition 1.14. Let S be a semigroup.

    (a) An element $x \in S$ is *right cancelable* if and only if whenever $y, z \in S$ and $yx = zx$, one has $y = z$.

    (b) An element $x \in S$ is *left cancelable* if and only if whenever $y, z \in S$ and $xy = xz$, one has $y = z$.

    (c) S is *right cancellative* if and only if every $x \in S$ is right cancelable.

    (d) S is *left cancellative* if and only if every $x \in S$ is left cancelable.

    (e) S is *cancellative* if and only if S is both left cancellative and right cancellative.

Theorem 1.15. *Let S be a semigroup.*

    (a) *The function $\lambda : S \to L(S)$ is a homomorphism onto $L(S)$.*

    (b) *The function $\rho : S \to R(S)$ is an anti-homomorphism onto $R(S)$.*

    (c) *If S is right cancellative, then S and $L(S)$ are isomorphic.*

    (d) *If S is left cancellative, then S and $R(S)$ are anti-isomorphic.*

Proof. (a) Given x, y, and z in S one has $(\lambda_x \circ \lambda_y)(z) = \lambda_x(yz) = x(yz) = (xy)z = \lambda_{xy}(z)$ so $\lambda_x \circ \lambda_y = \lambda_{xy}$.

    (c) This is part of Exercise 1.1.4.   □

Right cancellation is a far stronger requirement than is needed to have $S \approx L(S)$. See Exercise 1.1.4.

Exercise 1.1.1. Let S be the free semigroup on the alphabet A and let T be an arbitrary semigroup. Assume that $g : A \to T$ is any mapping. Prove that there is a unique homomorphism $h : S \to T$ with the property that $h(a) = g(a)$ for every $a \in A$.

Exercise 1.1.2. Prove that statements (b), (d), and (e) of Theorem 1.10 are equivalent.

Exercise 1.1.3. Prove that, in the semigroup $(^X X, \circ)$, the left cancelable elements are the injective functions and the right cancelable elements are the surjective functions.

Exercise 1.1.4. (a) Prove Theorem 1.15 (c).

(b) Give an example of a semigroup S which is not right cancellative such that $S \approx L(S)$.

Exercise 1.1.5. Let S be a right cancellative semigroup and let $a \in S$. Prove that if there is some $b \in S$ such that $ab = b$, then a is a *right* identity for S.

Exercise 1.1.6. Prove that "if S does not have an identity, one may be adjoined" (and in fact one may be adjoined even if S already has an identity). That is, Let S be a semigroup and let e be an element not in S. Define an operation $*$ on $S \cup \{e\}$ by $x * y = xy$ if $x, y \in S$ and $x * e = e * x = x$. Prove that $(S \cup \{e\}, *)$ is a semigroup with identity e. (Note that if S has an identity f, it is no longer the identity of $S \cup \{e\}$.)

Exercise 1.1.7. Suppose that S is a cancellative semigroup which does not have an identity. Prove that an identity can be adjoined to S so that the extended semigroup is also cancellative.

Exercise 1.1.8. Let S be a commutative cancellative semigroup. We define a relation $\equiv$ on $S \times S$ by stating that $(a, b) \equiv (c, d)$ if and only if $ad = bc$. Prove that this is an equivalence relation. Let $\overline{(a, b)}$ denote the equivalence class which contains the element $(a, b) \in S \times S$, and let G denote the set of all these equivalence classes. We define a binary relation $\cdot$ on G by stating that $\overline{(a, b)} \cdot \overline{(c, d)} = \overline{(ac, bd)}$. Prove that this is well defined, that $(G, \cdot)$ is a group and that it contains an isomorphic copy of S. (The group G is called the *group of quotients* of S. If $S = (\mathbb{N}, +)$, $G = (\mathbb{Z}, +)$; if $S = (\mathbb{N}, \cdot)$, $G = (\mathbb{Q}^+, \cdot)$, where $\mathbb{Q}^+ = \{x \in \mathbb{Q} : x > 0\}$.)

1.2 Idempotents and Subgroups

Our next subject is "idempotents". They will be very important to us throughout this book.

Definition 1.16. Let S be a semigroup.

(a) An element $x \in S$ is an *idempotent* if and only if $xx = x$.

(b) $E(S) = \{x \in S : x \text{ is an idempotent}\}$.

(c) T is a *subsemigroup* of S if and only if $T \subseteq S$ and T is a semigroup under the restriction of the operation of S.

(d) T is a *subgroup* of S if and only if $T \subseteq S$ and T is a group under the restriction of the operation of S.

(e) Let $e \in E(S)$. Then $H(e) = \bigcup \{G : G \text{ is a subgroup of } S \text{ and } e \in G\}$.

Lemma 1.17. *Let G be a group with identity e. Then $E(G) = \{e\}$.*

Proof. Assume $f \in E(G)$. Then $ff = f = fe$. Multiplying on the left by the inverse of f, one gets $f = e$. □

As a consequence of Lemma 1.17 the statement "$e \in G$" in the definition of $H(e)$ is synonymous with "e is the identity of G". Note that it is quite possible for $H(e)$ to equal $\{e\}$, but $H(e)$ is never empty.

Theorem 1.18. *Let S be a semigroup and let $e \in E(S)$. Then $H(e)$ is the largest subgroup of S with e as identity.*

Proof. It suffices to show that $H(e)$ is a group since e is trivially an identity for $H(e)$ and $H(e)$ contains every group with e as identity. For this it in turn suffices to show that $H(e)$ is closed. So let $x, y \in H(e)$ and pick subgroups G_1 and G_2 of S with $e \in G_1 \cap G_2$ and $x \in G_1$ and $y \in G_2$. Let $G = \{\prod_{i=1}^{n} x_i : n \in \mathbb{N}$ and $\{x_1, x_2, \ldots, x_n\} \subseteq G_1 \cup G_2\}$. Then $xy \in G$ and $e \in G$ so it suffices to show that G is a group. For this the only requirement that is not immediate is the existence of inverses. So let $\prod_{i=1}^{n} x_i \in G$. For $i \in \{1, 2, \ldots, n\}$, pick y_i such that $x_{n+1-i} y_i = e$. Then $\prod_{i=1}^{n} y_i \in G$ and $(\prod_{i=1}^{n} x_i) \cdot (\prod_{i=1}^{n} y_i) = e$. □

The groups $H(e)$ are referred to as *maximal groups*. Indeed, given any group $G \subseteq S$, G has an identity e and $G \subseteq H(e)$.

Lemma 1.19. *Let S be a semigroup, let $e \in E(S)$, and let $x \in S$. Then the following statements are equivalent.*

(a) $x \in H(e)$.

(b) $xe = x$ and there is some $y \in S$ such that $ye = y$ and $xy = yx = e$.

(c) $ex = x$ and there is some $y \in S$ such that $ey = y$ and $xy = yx = e$.

Proof. We show the equivalence of (a) and (b); the equivalence of (a) and (c) then follows by a left-right switch. The fact that (a) implies (b) is immediate.

(b) implies (a). Let $G = \{x \in S : xe = x$ and there is some $y \in S$ such that $ye = y$ and $xy = yx = e\}$. It suffices to show that G is a group with identity e. To establish closure, let $x, z \in G$. Then $xze = xz$. Pick y and w in S such that $ye = y$, $we = w$, $xy = yx = e$, and $zw = wz = e$. Then $wye = wy$ and $xzwy = xey = xy = e = wz = wez = wyxz$.

Trivially, e is a right identity for G so it suffices to show that each element of G has a right e-inverse *in G*. Let $x \in G$ and pick $y \in S$ such that $ye = y$ and $yx = xy = e$. Note that indeed y does satisfy the requirements to be in G. □

Example 1.20. *Let X be any set. Then the idempotents in $\left({}^X X, \circ\right)$ are the functions $f \in {}^X X$ with the property that $f(x) = x$ for every $x \in f[X]$.*

We next define the concept of a free group on a given set of generators. The underlying idea is simple, but the rigorous definition may seem a little troublesome. The basic idea is that we want to construct all expressions of the form $a_1^{e_1} a_2^{e_2} \ldots a_k^{e_k}$, where each $a_i \in A$ and each exponent $e_i \in \mathbb{Z}$, and to combine them in the way that we are forced to by the group axioms.

Definition 1.21. Let S be the free semigroup with identity on the alphabet $A \times \{1, -1\}$ and let

$$G = \{g \in S : \text{there do not exist } t, t+1 \in \text{domain}(g), \ a \in A \text{ and } i \in \{1, -1\}$$
$$\text{for which } g(t) = (a, i) \text{ and } g(t+1) = (a, -i)\}.$$

Given $f, g \in G \backslash \{\emptyset\}$ with

$$\text{domain}(f) = \{0, 1, \ldots, n-1\} \quad \text{and} \quad \text{domain}(g) = \{0, 1, \ldots, m-1\},$$

define $f \cdot g = f ^\frown g$ unless there exist $a \in A$ and $i \in \{1, -1\}$ with $f(n-1) = (a, i)$ and $g(0) = (a, -i)$.

In the latter case, pick the largest $k \in \mathbb{N}$ such that for all $t \in \{1, 2, \ldots, k\}$, there exist $b \in A$ and $j \in \{1, -1\}$ such that $f(n-t) = (b, j)$ and $g(t-1) = (b, -j)$. If $k = m = n$, then $f \cdot g = \emptyset$. Otherwise, domain$(f \cdot g) = \{0, 1, \ldots, n+m-2k-1\}$ and for $t \in \{1, 2, \ldots, n+m-2k-1\}$,

$$(f \cdot g)(t) = \begin{cases} f(t) & \text{if } t < n-k \\ g(t+2k-n) & \text{if } t \geq n-k. \end{cases}$$

Then $(G, \cdot)$ is the *free group generated by* A.

It is not hard to prove that, with the operation defined above, G *is* a group.

We customarily write a in lieu of $(a, 1)$ and a^{-1} in lieu of $(a, -1)$. Then in keeping with the notation to be introduced in the next section (Section 1.3) we shall write the word $ab^{-1}b^{-1}b^{-1}a^{-1}a^{-1}bb$, for example, as $ab^{-3}a^{-2}b^2$. As an illustration, we have $(ab^{-3}a^{-2}b^2) \cdot (b^{-2}a^3b^{-4}) = ab^{-3}ab^{-4}$.

We observe that the free group G generated by A has a universal property given by the following lemma.

Lemma 1.22. *Let A be a set, let G be the free group generated by A, let H be an arbitrary group, and let $\phi : A \to H$ be any mapping. There is a unique homomorphism $\widehat{\phi} : G \to H$ for which $\widehat{\phi}(g) = \phi(g)$ for every $g \in A$.*

Proof. This is Exercise 1.2.1.   □

We shall need the following result later.

Theorem 1.23. *Let A be a set, let G be the free group generated by A, and let $g \in G \backslash \{\emptyset\}$. There exist a finite group F and a homomorphism $\widehat{\phi} : G \to F$ such that $\widehat{\phi}(g)$ is not the identity of F.*

Proof. Let n be the length of g, let $X = \{0, 1, \ldots, n\}$, and let $F = \{f \in {}^X X : f$ is one-to-one and onto $X\}$. (Since X is finite, the "onto" requirement is redundant.) Then $(F, \circ)$ is a group whose identity is ι, the identity function from X to X. Given $a \in A$, let $D(a) = \{i \in \{0, 1, \ldots, n - 1\} : g(i) = a^{-1}\}$ and let $E(a) = \{i \in \{1, 2, \ldots, n\} : g(i - 1) = a\}$. Note that since $g \in G$, $D(a) \cap E(a) = \emptyset$. Define $\phi(a) : D(a) \cup E(a) \to X$ by

$$\phi(a)(i) = \begin{cases} i + 1 & \text{if } i \in D(a) \\ i - 1 & \text{if } i \in E(a), \end{cases}$$

and note that, because $g \in G$, $\phi(a)$ is one-to-one. Extend $\phi(a)$ in any way to a member of F. Let $\widehat{\phi} : G \to F$ be the homomorphism extending ϕ which was guaranteed by Lemma 1.22.

Suppose that $g = a_0{}^{i_0} a_1{}^{i_1} \ldots a_{n-1}{}^{i_{n-1}}$, where $a_r \in A$ and $i_r \in \{-1, 1\}$ for each $r \in \{0, 1, 2, \ldots, n - 1\}$. We shall show that, for each $k \in \{1, 2, \ldots, n\}$, $\widehat{\phi}(a_{k-1}{}^{i_{k-1}})(k) = k - 1$.

To see this, first suppose that $i_{k-1} = 1$. Then $k \in E(a_{k-1})$ and so $\phi(a_{k-1})(k) = k - 1$.

Now suppose that $i_{k-1} = -1$. Then $k - 1 \in D(a_{k-1})$ and so $\phi(a_{k-1})(k - 1) = k$. Thus $\widehat{\phi}(a_{k-1}{}^{-1})(k) = \phi(a_{k-1})^{-1}(k) = k - 1$.

It is now easy to see that $\widehat{\phi}(g)(n) = \widehat{\phi}(a_0{}^{i_0}) \widehat{\phi}(a_1{}^{i_1}) \ldots \widehat{\phi}(a_{n-1}{}^{i_{n-1}})(n) = 0$ and hence that $\widehat{\phi}(g)$ is not the identity map.   □

Exercise 1.2.1. Prove Lemma 1.22.

1.3 Powers of a Single Element

Suppose that x is a given element in a semigroup S. For each $n \in \mathbb{N}$, we define an element x^n in S. We do this inductively, by stating that $x^1 = x$ and that $x^{n+1} = x x^n$ if x^n has already been defined. It is then straightforward to prove by induction that $x^m x^n = x^{m+n}$ for every $m, n \in \mathbb{N}$. Thus $\{x^n : n \in \mathbb{N}\}$ is a commutative subsemigroup of S. We shall say that x has *finite order* if this subsemigroup is finite; otherwise we shall say that x has *infinite order*.

If S has an identity e, we shall define x^0 for every $x \in S$ by stating that $x^0 = e$. If x has an inverse in S, we shall denote this inverse by x^{-1}, and we shall define x^{-n} for every $n \in \mathbb{N}$ by stating that $x^{-n} = (x^{-1})^n$. If x does have an inverse, it is easy to prove that $x^m x^n = x^{m+n}$ for every $m, n \in \mathbb{Z}$. Thus $\{x^n : n \in \mathbb{Z}\}$ forms a subgroup of S.

If additive notation is being used, x^n might be denoted by nx instead. The index law mentioned above would then be written as: $mx + nx = (m + n)x$.

Theorem 1.24. *Suppose that S is a semigroup and that $x \in S$ has infinite order. Then the subsemigroup $T = \{x^n : n \in \mathbb{N}\}$ of S is isomorphic to $(\mathbb{N}, +)$.*

Proof. The mapping $n \mapsto x^n$ from $(\mathbb{N}, +)$ onto T is a surjective homomorphism, and so it will be sufficient to show that it is one-to-one. Suppose then that $x^m = x^n$ for some $m, n \in \mathbb{N}$ satisfying $m < n$. Then x^{n-m} is an identity for x^m, and the same statement holds for $x^{q(n-m)}$, where q denotes any positive integer. Suppose that s is any integer satisfying $s > m$. We can write $s - m = q(n - m) + r$ where q and m are non-negative integers and $r < (n - m)$. So $x^s = x^{s-m}x^m = x^{q(n-m)+r}x^m = x^r x^m$. It follows that $\{x^s : s > m\}$ is finite and hence that T is finite, contradicting our assumption that x has infinite order. □

Theorem 1.25. *Any finite semigroup S contains an idempotent.*

Proof. This statement is obviously true if S contains only one element. We shall prove it by induction on the number of elements in S. We make the inductive assumption that the theorem is true for all semigroups with fewer elements than S. Choose any $x \in S$. There are positive integers m and n satisfying $x^m = x^n$ and $m < n$. Then $x^{n-m}x^m = x^m$. Consider the subsemigroup $\{y \in S : x^{n-m}y = y\}$ of S. If this is the whole of S it contains x^{n-m} and so x^{n-m} is idempotent. If it is smaller than S, it contains an idempotent, by our inductive assumption. □

Exercise 1.3.1. Prove that any finite cancellative semigroup is a group.

1.4 Ideals

The terminology "ideal" is borrowed from ring theory. Given subsets A and B of a semigroup S, by AB we of course mean $\{ab : a \in A$ and $b \in B\}$.

Definition 1.26. Let S be a semigroup.
  (a) L is a *left ideal* of S if and only if $\emptyset \neq L \subseteq S$ and $SL \subseteq L$.
  (b) R is a *right ideal* of S if and only if $\emptyset \neq R \subseteq S$ and $RS \subseteq R$.
  (c) I is an *ideal* of S if and only if I is both a left ideal and a right ideal of S.

An ideal I of S satisfying $I \neq S$ is called a *proper* ideal of S.

Sometimes for emphasis an ideal is called a "two sided ideal". We often deal with semigroups in which the operation is denoted by $+$. In this case the terminology may seem awkward for someone who is accustomed to working with rings. That is, a left ideal L satisfies $S + L \subseteq L$ and a right ideal R satisfies $R + S \subseteq R$.

Of special importance for us is the notion of *minimal* left and right ideals. By this we mean simply left or right ideals which are minimal with respect to set inclusion.

Definition 1.27. Let S be a semigroup.
  (a) L is a *minimal left ideal* of S if and only if L is a left ideal of S and whenever J is a left ideal of S and $J \subseteq L$ one has $J = L$.
  (b) R is a *minimal right ideal* of S if and only if R is a right ideal of S and whenever J is a right ideal of S and $J \subseteq R$ one has $J = R$.

(c) S is *left simple* if and only if S is a minimal left ideal of S.

(d) S is *right simple* if and only if S is a minimal right ideal of S.

(e) S is *simple* if and only if the only ideal of S is S.

We do not define a minimal ideal. As a consequence of Lemma 1.29 below, we shall see that there is at most one minimal two sided ideal of a semigroup. Consequently we use the term "smallest" to refer to an ideal which does not properly contain another ideal.

Observe that S is left simple if and only if it has no proper left ideals. Similarly, S is right simple if and only if it has no proper right ideals. Whenever one has a theorem about left ideals, there is a corresponding theorem about right ideals. We shall not usually state both results.

Clearly any semigroup which is either right simple or left simple must be simple. The following simple example (pun intended) shows that the converse fails.

Example 1.28. *Let* $S = \{a, b, c, d\}$ *where* a, b, c, *and* d *are any distinct objects and let* S *have the following multiplication table. Then* S *is simple but is neither left simple nor right simple.*

$\cdot$	a	b	c	d
a	a	b	a	b
b	a	b	a	b
c	c	d	c	d
d	c	d	c	d

One can laboriously verify that the table does define an associative operation. But 128 computations $\big($of $(xy)z$ and $x(yz)\big)$ are required, somewhat fewer if one is clever. It is usually much easier to establish associativity by representing the new semigroup as a subsemigroup of one with which we are already familiar. In this case, we can represent S as a semigroup of 3×3 matrices, by putting:

$$a = \begin{pmatrix} 1 & 1 & 0 \\ 0 & 0 & 0 \\ 0 & 0 & 0 \end{pmatrix}, \quad b = \begin{pmatrix} 1 & 1 & 1 \\ 0 & 0 & 0 \\ 0 & 0 & 0 \end{pmatrix}, \quad c = \begin{pmatrix} 0 & 0 & 0 \\ 1 & 1 & 0 \\ 0 & 0 & 0 \end{pmatrix}, \quad d = \begin{pmatrix} 0 & 0 & 0 \\ 1 & 1 & 1 \\ 0 & 0 & 0 \end{pmatrix}.$$

To verify the assertions of the example, note that $\{a, b\}$ and $\{c, d\}$ are right ideals of S and $\{a, c\}$ and $\{b, d\}$ are left ideals of S.

Lemma 1.29. *Let* S *be a semigroup.*

(a) *Let* L_1 *and* L_2 *be left ideals of* S. *Then* $L_1 \cap L_2$ *is a left ideal of* S *if and only if* $L_1 \cap L_2 \neq \emptyset$.

(b) *Let* L *be a left ideal of* S *and let* R *be a right ideal of* S. *Then* $L \cap R \neq \emptyset$.

Proof. Statement (a) is immediate. To see (b), let $x \in L$ and $y \in R$. Then $yx \in L$ because $x \in L$ and $yx \in R$ because $y \in R$.   □

Lemma 1.30. *Let S be a semigroup.*

(a) *Let $x \in S$. Then xS is a right ideal, Sx is a left ideal and SxS is an ideal.*

(b) *Let $e \in E(S)$. Then e is a left identity for eS, a right identity for Se, and an identity for eSe.*

Proof. Statement (a) is immediate. For (b), let $e \in E(S)$. To see that e is a left identity for eS, let $x \in eS$ and pick $t \in S$ such that $x = et$. Then $ex = eet = et = x$. Likewise e is a right identity for Se.   □

Theorem 1.31. *Let S be a semigroup.*

(a) *If S is left simple and $e \in E(S)$, then e is a right identity for S.*

(b) *If L is a left ideal of S and $s \in L$, then $Ss \subseteq L$.*

(c) *Let $\emptyset \neq L \subseteq S$. Then L is a minimal left ideal of S if and only if for each $s \in L$, $Ss = L$.*

Proof. (a) By Lemma 1.30 (a), Se is a left ideal of S, so $Se = S$ so Lemma 1.30 (b) applies.

(b) This follows immediately from the definition of left ideal.

(c) Necessity. By Lemma 1.30 (a) Ss is a left ideal and by (b) $Ss \subseteq L$ so, since L is minimal, $Ss = L$.

Sufficiency. Since $L = Ss$ for some $s \in L$, L is a left ideal. Let J be a left ideal of S with $J \subseteq L$ and pick $s \in J$. Then by (b), $Ss \subseteq J$ so $J \subseteq L = Ss \subseteq J$.   □

We shall observe at the conclusion of the following definition that the objects defined there exist.

Definition 1.32. Let S be a semigroup.

(a) The smallest ideal of S which contains a given element $x \in S$ is called the *principal ideal generated by x*.

(b) The smallest left ideal of S which contains x is called the *principal left ideal of S generated by x*.

(c) The smallest right ideal of S which contains x is called the *principal right ideal generated by x*.

Theorem 1.33. *Let S be a semigroup and let $x \in S$.*

(a) *The principal ideal generated by x is $SxS \cup xS \cup Sx \cup \{x\}$.*

(b) *If S has an identity, then the principal ideal generated by x is SxS.*

(c) *The principal left ideal generated by x is $Sx \cup \{x\}$ and the principal right ideal generated by x is $xS \cup \{x\}$.*

Proof. This is Exercise 1.4.1.   □

Exercise 1.4.1. Prove Theorem 1.33.

Exercise 1.4.2. Describe the ideals in each of the following semigroups. Also describe the minimal left ideals and the minimal right ideals in the cases in which these exist.

  (i) $(\mathbb{N}, +)$.
 (ii) $(\mathcal{P}(X), \cup)$, where X is any set.
(iii) $(\mathcal{P}(X), \cap)$, where X is any set.
(iv) $([0, 1], \cdot)$, where $\cdot$ denotes multiplication.
 (v) The set of real-valued functions defined on a given set, with pointwise multiplication as the semigroup operation.
(vi) A left zero semigroup.
(vii) A right zero semigroup.

Exercise 1.4.3. Let X be any set. Describe the minimal left and right ideals in $^X X$.

Exercise 1.4.4. Let S be a commutative semigroup with an identity e. Prove that S has a proper ideal if and only if there is some $s \in S$ which has no e-inverse. In this case, prove that $\{s \in S : s$ has no e-inverse$\}$ is the unique maximal proper ideal of S.

1.5 Idempotents and Order

Intimately related to the notions of minimal left and minimal right ideals is the notion of minimal idempotents.

Definition 1.34. Let S be a semigroup and let $e, f \in E(S)$. Then
   (a) $e \leq_L f$ if and only if $e = ef$,
   (b) $e \leq_R f$ if and only if $e = fe$, and
   (c) $e \leq f$ if and only if $e = ef = fe$.

In the semigroup of Example 1.28, one sees that $c \leq_L a, a \leq_L c, b \leq_L d, d \leq_L b, a \leq_R b$, $b \leq_R a, c \leq_R d$, and $d \leq_R c$, while the relation $\leq$ is simply equality on this semigroup.

Remark 1.35. *Let S be a semigroup. Then $\leq_L$, $\leq_R$, and $\leq$ are transitive and reflexive relations on $E(S)$. In addition, $\leq$ is antisymmetric.*

When we say that a point e is minimal with respect to a (not necessarily antisymmetric) relation $\preceq$ on a set B, we mean that if $f \in B$ and $f \preceq e$, then $e \preceq f$ (so if $\preceq$ is antisymmetric, the conclusion becomes $e = f$).

Theorem 1.36. *Let S be a semigroup and let $e \in E(S)$. The following statements are equivalent.*

  (a) *The element e is minimal with respect to $\leq$.*
  (b) *The element e is minimal with respect to $\leq_R$.*
  (c) *The element e is minimal with respect to $\leq_L$.*

Proof. (b) implies (a). Assume that e is minimal with respect to $\leq_R$ and let $f \leq e$. Then $f = ef$ so $f \leq_R e$ so $e \leq_R f$. Then $e = fe = f$.

We show that (a) implies (b). (Then the equivalence of (a) and (c) follows by a left-right switch.) Assume that e is minimal with respect to $\leq$ and let $f \leq_R e$. Let $g = fe$. Then $gg = fefe = ffe = fe = g$ so $g \in E(S)$. Also, $g = fe = efe$ so $eg = eefe = efe = g = efee = ge$. Thus $g \leq e$ so $g = e$ by the minimality of e. That is, $e = fe$ so $e \leq_R f$ as required. □

As a consequence of Theorem 1.36, we are justified in making the following definition.

Definition 1.37. Let S be a semigroup. Then e is a *minimal idempotent* if and only if $e \in E(S)$ and e is minimal with respect to any (hence all) of the orders $\leq$, $\leq_R$, or $\leq_L$.

We see that the notions of "minimal idempotent" and "minimal left ideal" and "minimal right ideal" are intimately related. We remind the reader that there is a corresponding "right" version of the following theorem.

Theorem 1.38. *Let S be a semigroup and let $e \in E(S)$.*

(a) *If e is a member of some minimal left ideal (equivalently if Se is a minimal left ideal), then e is a minimal idempotent.*

(b) *If S is simple and e is minimal, then Se is a minimal left ideal.*

(c) *If every left ideal of S contains an idempotent and e is minimal, then Se is a minimal left ideal.*

(d) *If S is simple or every left ideal of S has an idempotent then the following statements are equivalent.*

    (i) *e is minimal.*

    (ii) *e is a member of some minimal left ideal of S.*

    (iii) *Se is a minimal left ideal of S.*

Proof. (a) Let L be a minimal left ideal with $e \in L$. (The existence of a set L with this property is equivalent to Se being minimal, by Theorem 1.31 (c).) Then $L = Se$. Let $f \in E(S)$ with $f \leq e$. Then $f = fe$ so $f \in L$ so $\left(\text{by Theorem 1.31(c)}\right)$ $L = Sf$ so $e \in Sf$ so by Lemma 1.30(b), $e = ef$ so $e = ef = f$.

(b) Let L be a left ideal with $L \subseteq Se$. We show that $Se \subseteq L$ (and hence $Se = L$). Pick some $s \in L$. Then $s \in Se$ so by Lemma 1.30(b), $s = se$. Also, since S is simple $SsS = S$, so pick u and v in S with $e = vsu$. Let $r = eue$ and $t = ev$. Then $tsr = evseue = evsue = eee = e$ and $er = eeue = eue = r$. Let $f = rts$. Then $ff = rtsrts = r(tsr)ts = rets = reevs = revs = rts = f$, so $f \in E(S)$. Also, $fe = rtse = rts = f$ and $ef = erts = rts = f$ so $f \leq e$ so $f = e$. Thus $Se = Sf = Srts \subseteq Ss \subseteq L$.

(c) Let L be a left ideal with $L \subseteq Se$. We show that $e \in L$ (so that $Se \subseteq L$ and hence $Se = L$). Pick an idempotent $t \in L$, and let $f = et$. Then $f \in L$. Since $t \in Se$, $t = te$. Thus $f = et = ete$. Therefore $ff = etet = ett = et = f$ so $f \in E(S)$. Also $ef = eete = ete = f$ and $fe = etee = ete = f$ so $f \leq e$ so $f = e$ and hence $e \in L$.

(d) This follows from (a), (b), and (c).   □

We now obtain several characterizations of a group.

Theorem 1.39. *Let S be a semigroup. The following statements are equivalent.*

(a) *S is cancellative and simple and $E(S) \neq \emptyset$.*

(b) *S is both left simple and right simple.*

(c) *For all a and b in S, the equations $ax = b$ and $ya = b$ have solutions x, y in S.*

(d) *S is a group.*

Proof. (a) implies (b). Pick an idempotent e in S. We show first that e is a (two sided) identity for S. Let $x \in S$. Then $ex = eex$ so by left cancellation $x = ex$. Similarly, $x = xe$. To see that S is left simple, let L be a left ideal of S. Then LS is an ideal of S so $LS = S$, so pick $t \in L$ and $s \in S$ such that $e = ts$. Then $sts = se = s = es$ so cancelling on the right one has $st = e$. Thus $e \in L$ so $S = SL \subseteq L$. Consequently S is left simple, and similarly S is right simple.

(b) implies (c). Let $a, b \in S$. Then $aS = S$ so there is some $x \in S$ such that $ax = b$. Similarly, since $Sa = S$, there is some $y \in S$ such that $ya = b$.

(c) implies (d). Pick $a \in S$ and pick $e \in S$ such that $ea = a$. We show that e is a left identity for S. Let $b \in S$. We show that $eb = b$. Pick some $y \in S$ such that $ay = b$. Then $eb = eay = ay = b$.

Now given any $x \in S$ there is some $y \in S$ such that $yx = e$ so every element of S has a left e-inverse.

(d) implies (a). Trivially S is cancellative and $E(S) \neq \emptyset$. To see that S is simple, let I be an ideal of S and pick $x \in I$. Let y be the inverse of x. Then $xy \in I$ so $I = S$. □

As promised earlier, we now see that any semigroup with a left identity e such that every element has a right e-inverse must be (isomorphic to) the Cartesian product of a group with a right zero semigroup.

Theorem 1.40. *Let S be a semigroup and let e be a left identity for S such that for each $x \in S$ there is some $y \in S$ with $xy = e$. Let $Y = E(S)$ and let $G = Se$. Then Y is a right zero semigroup and G is a group and $S = GY \approx G \times Y$.*

Proof. We show first that:

(∗) For all $x \in Y$ and for all $y \in S$, $xy = y$.

To establish (∗), let $x \in Y$ and $y \in S$ be given. Pick $z \in S$ such that $xz = e$. Then $xe = xxz = xz = e$. Therefore $xy = x(ey) = (xe)y = ey = y$, as required.

From (∗) it follows that for all $x, y \in Y$, $xy = y$, and $Y \neq \emptyset$ because $e \in Y$, so to see that Y is a right zero semigroup, it suffices to show that it is a semigroup, that is that Y is closed. But this also follows from (∗) since, given $x, y \in Y$ one has $xy = y \in Y$.

Now we establish that $G = Se$ is a group. By Lemma 1.30(b), e is a right identity for G. Now every element in S has a right e-inverse *in S*. So every element of G has a right e-inverse *in S*. By Theorem 1.10 we need only to show that every element of G has a right e-inverse *in G*. To this end let $x \in G$ be given and pick $y \in S$ such that $xy = e$. Then $ye \in G$ and $xye = ee = e$ so ye is as required. Since we also have $GG = SeSe \subseteq SSSe \subseteq Se = G$, it follows that G is a group.

Now define $\varphi : G \times Y \to S$ by $\varphi(g, y) = gy$. To see that φ is a homomorphism, let $(g_1, y_1), (g_2, y_2) \in G \times Y$. Then

$$
\begin{aligned}
\varphi(g_1, y_1)\varphi(g_2, y_2) &= g_1 y_1 g_2 y_2 \\
&= g_1 g_2 y_2 && \text{(by } (*) \text{)} \\
&= g_1 g_2 y_1 y_2 && \text{(by } (*) \text{)} \\
&= \varphi(g_1 g_2, y_1 y_2).
\end{aligned}
$$

To see that φ is surjective, let $s \in S$ be given. Then $se \in Se = G$, and so there exists $x \in Se$ such that $x(se) = (se)x = e$. We claim that $xs \in Y = E(S)$. Indeed,

$$
\begin{aligned}
xsxs &= xsexs && \text{(since } x \in G, \ ex = x \text{)} \\
&= xes \\
&= xs && \text{(since } x \in G, \ xe = x \text{)}.
\end{aligned}
$$

Thus $(se, xs) \in G \times Y$ and $\varphi(se, xs) = sexs = es = s$. Since φ is onto S, we have established that $S = GY$.

Finally to see that φ is one-to-one, let $(g, y) \in G \times Y$ and let $s = \varphi(g, y)$ We show that $g = se$ and $y = xs$ where x is the (unique) inverse of se in Se. Now $s = gy$ so

$$
\begin{aligned}
se &= gye \\
&= ge && \text{(by } (*) \ ye = e \text{)} \\
&= g && \text{(since } g \in Se \text{)}.
\end{aligned}
$$

Also

$$
\begin{aligned}
xs &= xgy \\
&= xgyey && (ye \in Y \text{ so by } (*) \ yey = y) \\
&= xsey \\
&= ey \\
&= y.
\end{aligned}
$$
□

We know that the existence of a left identity e for a semigroup S such that every element of S has a right e-inverse does not suffice to make S a group. A right zero semigroup is the standard example. Theorem 1.40 tells us that is essentially the only example.

Corollary 1.41. *Let S be a semigroup and assume that S has a unique left identity e and that every element of S has a right e-inverse. Then S is a group.*

Proof. This is Exercise 1.5.1.   □

Exercise 1.5.1. Prove Corollary 1.41. (Hint: Consider $|Y|$ in Theorem 1.40.)

1.6 Minimal Left Ideals

We shall see in this section that many significant consequences follow from the existence of minimal left (or right) ideals, especially those with idempotents. This is important for us, because, as we shall see in Corollary 2.6, any compact right topological semigroup has minimal left ideals with idempotents.

We begin by establishing an easy consequence of Theorem 1.40.

Theorem 1.42. *Let S be a semigroup and assume that there is a minimal left ideal L of S which has an idempotent e. Then $L = XG \approx X \times G$ where X is the (left zero) semigroup of idempotents of L, and $G = eL = eSe$ is a group. All maximal groups in L are isomorphic to G.*

Proof. Given $x \in L$, Lx is a left ideal of S and $Lx \subseteq L$ so $Lx = L$ and hence there is some $y \in L$ such that $yx = e$. By Lemma 1.30(b), e is a right identity for $Le = L$. Therefore the right-left switch of Theorem 1.40 applies (with L replacing S). It is a routine exercise to show that the maximal groups of $X \times G$ are the sets of the form $\{x\} \times G$.   □

Lemma 1.43. *Let S be a semigroup, let L be a left ideal of S, and let T be a left ideal of L.*

(a) *For all $t \in T$, Lt is a left ideal of S and $Lt \subseteq T$.*

(b) *If L is a minimal left ideal of S, then $T = L$. (So minimal left ideals are left simple.)*

(c) *If T is a minimal left ideal of L, then T is a left ideal of S.*

Proof. (a) $S(Lt) = (SL)t \subseteq Lt$ and $Lt \subseteq LT \subseteq T$.

(b) Pick any $t \in T$. By (a), Lt is a left ideal of S and $Lt \subseteq T \subseteq L$ so $Lt = L$ so $T = L$.

(c) Pick any $t \in T$. By (a), Lt is a left ideal of S, so Lt is a left ideal of L. Since $Lt \subseteq T$, $Lt = T$. Therefore, $ST = S(Lt) = (SL)t \subseteq Lt = T$.   □

As a consequence of Lemma 1.43, if L is a left ideal of S and T is a left ideal of L and either L is minimal in S or T is minimal in L, then T is a left ideal of S. Of course, the right-left switch of this statement also holds. That is, if R is a right ideal of S and T is a right ideal of R and either R is minimal in S or T is minimal in R, then T is a right ideal of S. We see now that without *some* assumptions, T need not be a right ideal of S.

Example 1.44. *Let $X = \{0, 1, 2\}$ and let $S = {}^{X}X$. Let $R = \{f \in X : \text{Range}(f) \subseteq \{0, 1\}\}$ and let $T = \{\overline{0}, a\}$ where $\overline{0}$ is the constant function and $a : \begin{matrix} 0 \to 0 \\ 1 \to 0 \\ 2 \to 1 \end{matrix}$. Then R is a right ideal of S and T is a right ideal of R, but T is not a right ideal of S.*

Lemma 1.45. *Let S be a semigroup, let I be an ideal of S and let L be a minimal left ideal of S. Then $L \subseteq I$.*

Proof. This is Exercise 1.6.1.   □

We now see that all minimal left ideals of a semigroup are intimately connected with each other.

Theorem 1.46. *Let S be a semigroup, let L be a minimal left ideal of S, and let $T \subseteq S$. Then T is a minimal left ideal of S if and only if there is some $a \in S$ such that $T = La$.*

Proof. Necessity. Pick $a \in T$. Then $SLa \subseteq La$ and $La \subseteq ST \subseteq T$ so La is a left ideal of S contained in T so $La = T$.

Sufficiency. Since $SLa \subseteq La$, La is a left ideal of S. Assume that B is a left ideal of S and $B \subseteq La$. Let $A = \{s \in L : sa \in B\}$. Then $A \subseteq L$ and $A \neq \emptyset$. We claim that A is a left ideal of S, so let $s \in A$ and let $t \in S$. Then $sa \in B$ so $tsa \in B$ and, since $s \in L$, $ts \in L$, so $ts \in A$ as required. Thus $A = L$ so $La \subseteq B$ so $La = B$.   □

Corollary 1.47. *Let S be a semigroup. If S has a minimal left ideal, then every left ideal of S contains a minimal left ideal.*

Proof. Let L be a minimal left ideal of S and let J be a left ideal of S. Pick $a \in J$. Then by Theorem 1.46, La is a minimal left ideal which is contained in J.   □

Theorem 1.48. *Let S be a semigroup and let $e \in E(S)$. Statements (a) through (f) are equivalent and imply statement (g). If either S is simple or every left ideal of S has an idempotent, then all statements are equivalent.*

(a) *Se is a minimal left ideal.*

(b) *Se is left simple.*

(c) *eSe is a group.*

(d) *$eSe = H(e)$.*

(e) *eS is a minimal right ideal.*

(f) *eS is right simple.*

(g) *e is a minimal idempotent.*

Proof. By Theorem 1.38(a), we have that (a) implies (g) and by Theorem 1.38(d), if either S is simple or every left ideal of S has an idempotent, then (g) implies (a).

$$
\begin{array}{ccccccc}
\text{(a)} & \Rightarrow & \text{(b)} & & \text{(e)} & \Rightarrow & \text{(f)} \\
\Uparrow & & \Downarrow & \text{from which} & \Uparrow & & \Downarrow \\
\text{(d)} & \Leftarrow & \text{(c)} & & \text{(d)} & \Leftarrow & \text{(c)}
\end{array}
$$

We show that the first diagram holds from which the second follows by left-right duality and the fact that (c) and (d) are two sided statements.

That (a) implies (b) follows from Lemma 1.43(b).

(b) implies (c). Trivially eSe is closed. By Lemma 1.30 e is a two sided identity for eSe. Also let $x = ese \in eSe$ be given. One has $x \in Se$ so Sx is a left ideal of Se and consequently $Sx = Se$, since Se is left simple. Thus $e \in Sx$, so pick $y \in S$ such that $e = yx$. Then $eye \in eSe$ and $eyex = eyx = ee = e$ so x has a left e-inverse in eSe.

(c) implies (d). Since eSe is a group and $e \in eSe$, one has $eSe \subseteq H(e)$. On the other hand, by Theorem 1.18, e is the identity of $H(e)$ so given $x \in H(e)$, one has that $x = exe \in eSe$, so $H(e) \subseteq eSe$

(d) implies (a). Let L be a left ideal of S with $L \subseteq Se$ and pick $t \in L$. Then $t \in Se$ so $et \in eSe$. Pick $x \in eSe$ such that $x(et) = e$. Then $xt = (xe)t = x(et) = e$ so $e \in L$ so $Se \subseteq SL \subseteq L$. □

We note that in the semigroup $(\mathbb{N}, \cdot)$, 1 is the only idempotent, and is consequently minimal, while $\mathbb{N}1$ is not a minimal left ideal. Thus Theorem 1.48(g) does not in general imply the other statements of Theorem 1.48.

We recall that in a ring there may be many minimal two sided ideals. This is because a "minimal ideal" in a ring is an ideal minimal among all ideals not equal to $\{0\}$, and one may have ideals I_1 and I_2 with $I_1 \cap I_2 = \{0\}$. By contrast, we see that a semigroup can have at most one minimal two-sided ideal.

Lemma 1.49. *Let S be a semigroup and let K be an ideal of S. If K is minimal in $\{J : J$ is an ideal of $S\}$ and I is an ideal of S, then $K \subseteq I$.*

Proof. By Lemma 1.29(b), $K \cap I \neq \emptyset$ so $K \cap I$ is an ideal contained in K so $K \cap I = K$. □

The terminology "minimal ideal" is widely used in the literature. Since, by Lemma 1.49, there can be at most one minimal ideal in a semigroup, we prefer the terminology "smallest ideal".

Definition 1.50. Let S be a semigroup. If S has a smallest ideal, then $K(S)$ is that smallest ideal.

We see that a simple condition guarantees the existence of $K(S)$.

Theorem 1.51. *Let S be a semigroup. If S has a minimal left ideal, then $K(S)$ exists and $K(S) = \bigcup\{L : L$ is a minimal left ideal of $S\}$.*

Proof. Let $I = \bigcup\{L : L$ is a minimal left ideal of $S\}$. By Lemma 1.45, if J is any ideal of S, then $I \subseteq J$, so it suffices to show that I is an ideal of S. We have that $I \neq \emptyset$ by assumption, so let $x \in I$ and let $s \in S$. Pick a minimal left ideal L of S such that $x \in L$. Then $sx \in L \subseteq I$. Also, by Theorem 1.46, Ls is a minimal left ideal of S so $Ls \subseteq I$ while $xs \in Ls$. □

Observe, however, that many common semigroups do not have a smallest ideal. This is true for example of both $(\mathbb{N}, +)$ and $(\mathbb{N}, \cdot)$.

Lemma 1.52. *Let S be a semigroup.*
(a) *Let L be a left ideal of S. Then L is minimal if and only if $Lx = L$ for every $x \in L$.*
(b) *Let I be an ideal of S. Then I is the smallest ideal if and only if $IxI = I$ for every $x \in I$.*

Proof. (a) If L is minimal and $x \in L$, then Lx is a left ideal of S and $Lx \subseteq L$ so $Lx = L$. Now assume $Lx = L$ for every $x \in L$ and let J be a left ideal of S with $J \subseteq L$. Pick $x \in J$. Then $L = Lx \subseteq LJ \subseteq J \subseteq L$.

(b) This is Exercise 1.6.2. $\square$

Theorem 1.53. *Let S be a semigroup. If L is a minimal left ideal of S and R is a minimal right ideal of S, then $K(S) = LR$.*

Proof. Clearly LR is an ideal of S. We use Lemma 1.52 to show that $K(S) = LR$. So, let $x \in LR$. Then $LRxL$ is a left ideal of S which is contained in L so $LRxL = L$ and hence $LRxLR = LR$. $\square$

Theorem 1.54. *Let S be a semigroup and assume that $K(S)$ exists and $e \in E(S)$. The following statements are equivalent and are implied by any of the equivalent statements (a) through (f) of Theorem 1.48.*

(h) $e \in K(S)$.

(i) $K(S) = SeS$.

Proof. By Theorem 1.51, it follows that Theorem 1.48(a) implies (h).

(h) implies (i). Since SeS is an ideal, we have $K(S) \subseteq SeS$. Since $e \in K(S)$, we have $SeS \subseteq K(S)$.

(i) implies (h). We have $e = eee \in SeS = K(S)$. $\square$

Two natural questions are raised by Theorems 1.51 and 1.54. First, if $K(S)$ exists, is it the union of all minimal left ideals or at least is it either the union of all minimal left ideals or be the union of all minimal right ideals? Second, given that $K(S)$ exists and e is an idempotent in $K(S)$, must Se be a minimal left ideal, or at least must e be a minimal idempotent? The following example, known as the *bicyclic* semigroup, answers the weaker versions of both of these questions in the negative. Recall that $\omega = \mathbb{N} \cup \{0\} = \{0, 1, 2, \ldots\}$.

Example 1.55. *Let $S = \omega \times \omega$ and define an operation $\cdot$ on S by*

$$(m, n) \cdot (r, s) = \begin{cases} (m, s + n - r) & \text{if } n \geq r \\ (m + r - n, s) & \text{if } n < r. \end{cases}$$

Then S is a simple semigroup (so $K(S) = S$), S has no minimal left ideals and no minimal right ideals, $E(S) = \{(n, n) : n \in \omega\}$, and for each $n \in \omega$, $(n + 1, n + 1) \leq (n, n)$.

One may verify directly that the operation in Example 1.55 is associative. It is probably easier, however, to observe that S is isomorphic to a subsemigroup of $^{\mathbb{N}}\mathbb{N}$. Specifically define $f, g \in {}^{\mathbb{N}}\mathbb{N}$ by $f(t) = t + 1$ and $g(t) = \begin{cases} t - 1 & \text{if } t > 1 \\ 1 & \text{if } t = 1 \end{cases}$. Then given $n, r \in \omega$ one has $g^n \circ f^r = \begin{cases} g^{n-r} & \text{if } n \geq r \\ f^{r-n} & \text{if } n < r \end{cases}$. Consequently, one has

$$(f^m \circ g^n) \circ (f^r \circ g^s) = \begin{cases} f^m \circ g^{s+n-r} & \text{if } n \geq r \\ f^{m+r-n} \circ g^s & \text{if } n < r. \end{cases}$$

To see that the semigroup in Example 1.55 is simple, note that given any (m, n), $(k, r) \in S$, $(k, m) \cdot (m, n) \cdot (n, r) = (k, r)$. To see that S has no minimal left ideals, let L be a left ideal of S and pick $(m, n) \in L$. Then $\{(k, r) \in S : r > n\}$ is a left ideal of S which is properly contained in L. Similarly, if R is a right ideal of S and $(m, n) \in R$ then $\{(k, r) \in S : k > m\}$ is a right ideal of S which is properly contained in R.

It is routine to verify the assertions about the idempotents in Example 1.55.

Exercise 1.6.1. Prove Lemma 1.45.

Exercise 1.6.2. Prove Lemma 1.52(b).

Exercise 1.6.3. Let $S = \{f \in {}^{\mathbb{N}}\mathbb{N} : f$ is one-to-one and $\mathbb{N} \setminus f[\mathbb{N}]$ is infinite$\}$. Prove that $(S, \circ)$ is left simple (so S is a minimal left ideal of S) and S has no idempotents.

Exercise 1.6.4. Suppose that a minimal left ideal L of a semigroup is commutative. Prove that L is a group.

Exercise 1.6.5. Let S be a semigroup and assume that there is a minimal left ideal of S. Prove that, if $K(S)$ is commutative, then it is a group.

1.7 Minimal Left Ideals with Idempotents

We present here several results that have as hypothesis "Let S be a semigroup and assume that there is a minimal left ideal of S which has an idempotent". These are important to us because, as we shall see in Corollary 2.6, this hypothesis holds in any compact right topological semigroup. (See Exercise 1.6.3 to show that the reference to the existence of an idempotent cannot be deleted from this hypothesis.)

Theorem 1.56. *Let S be a semigroup and assume that there is a minimal left ideal of S which has an idempotent. Then every minimal left ideal has an idempotent.*

Proof. Let L be a minimal left ideal with an idempotent e and let J be a minimal left ideal. By Theorem 1.46, there is some $x \in S$ such that $J = Lx$. By Theorem 1.42, $eL = eSe$ is a group, so let $y = eye$ be the inverse of exe in this group. Then $yx \in Lx = J$ and $yxyx = (ye)x(ey)x = y(exe)yx = eyx = yx$. $\square$

We shall get left and right conclusions from this one sided hypothesis. We see now that in fact the right version follows from the left.

Lemma 1.57. *Let S be a semigroup and assume that there is a minimal left ideal of S which has an idempotent. Then there is a minimal right ideal of S which has an idempotent.*

Proof. Pick a minimal left ideal L of S and an idempotent $e \in L$. By Theorem 1.31(c) Se is a minimal left ideal of S so by Theorem 1.48 eS is a minimal right ideal of S and e is an idempotent in eS.   □

Theorem 1.58. *Let S be a semigroup and assume that there is a minimal left ideal of S which has an idempotent. Let $T \subseteq S$.*

(a) T is a minimal left ideal of S if and only if there is some $e \in E(K(S))$ such that $T = Se$.

(b) T is a minimal right ideal of S if and only if there is some $e \in E(K(S))$ such that $T = eS$.

Proof. Pick a minimal left ideal L of S and an idempotent $f \in L$.

(a) Necessity. Since Sf is a left ideal contained in L, $Sf = L$. Thus by Theorem 1.48, fSf is a group. Pick any $a \in T$. Then $faf \in fSf$ so pick $x \in fSf$ such that $x(faf) = f$. Then

$$
\begin{aligned}
xaxa &= (xf)a(fx)a \\
     &= (xfaf)xa \\
     &= fxa \\
     &= xa.
\end{aligned}
$$

Consequently, xa is an idempotent. Also $xa \in T$ while $T \subseteq K(S)$ by Theorem 1.51 so $xa \in E(K(S))$. Finally, Sxa is a left ideal contained in T, so $T = Sxa$.

Sufficiency. Since $e \in K(S)$, pick by Theorem 1.51 a minimal left ideal I of S with $e \in I$. Then $Se = I$ by Theorem 1.31(c).

(b) As a consequence of Lemma 1.57 this follows by a left-right switch.   □

Theorem 1.59. *Let S be a semigroup, assume that there is a minimal left ideal of S which has an idempotent, and let $e \in E(S)$. The following statements are equivalent.*

   (a) *Se is a minimal left ideal.*
   (b) *Se is left simple.*
   (c) *eSe is a group.*
   (d) *$eSe = H(e)$.*
   (e) *eS is a minimal right ideal.*
   (f) *eS is right simple.*
   (g) *e is a minimal idempotent.*
   (h) *$e \in K(S)$.*
   (i) *$K(S) = SeS$.*

Proof. By Corollary 1.47 and Theorem 1.56 every left ideal of S contains an idempotent so by Theorem 1.48 statements (a) through (g) are equivalent. By Theorems 1.51 and 1.54 we need only show that (h) implies (a). But this follows from Theorem 1.58.   □

Theorem 1.60. *Let S be a semigroup, assume that there is a minimal left ideal of S which has an idempotent, and let e be an idempotent in S. There is a minimal idempotent f of S such that $f \le e$.*

Proof. Se is a left ideal which thus contains a minimal left ideal L with an idempotent g by Corollary 1.47 and Theorem 1.56. Now $g \in Se$ so $ge = g$ by Lemma 1.30. Let $f = eg$. Then $ff = egeg = egg = eg = f$ so f is an idempotent. Also $f \in L$ so $L = Sf$ so by Theorem 1.59 f is a minimal idempotent. Finally $ef = eeg = eg = f$ and $fe = ege = eg = f$ so $f \le e$. □

Theorem 1.61. *Let S be a semigroup and assume that there is a minimal left ideal of S which has an idempotent. Given any minimal left ideal L of S and any minimal right ideal R of S, there is an idempotent $e \in R \cap L$ such that $R \cap L = RL = eSe$ and eSe is a group.*

Proof. Let R and L be given. Pick by Theorem 1.58 an idempotent $f \in K(S)$ such that $L = Sf$. By Theorem 1.48, fSf is a group. Pick $a \in R$ and let x be the inverse of faf in fSf. Then $x \in Sf = L$ so $ax \in R \cap L$. By Theorem 1.51 $ax \in K(S)$. Also

$$\begin{aligned} axax &= a(xf)a(fx) \\ &= a(xfaf)x \\ &= afx \\ &= ax. \end{aligned}$$

Let $e = ax$. Then $eSe \subseteq Sx \subseteq L$ and $eSe \subseteq aS \subseteq R$ so $eSe \subseteq R \cap L$. To see that $R \cap L \subseteq eSe$, let $b \in R \cap L$. By Theorem 1.31 $L = Se$ and $R = eS$ so by Lemma 1.30, $b = eb = be$. Thus $b = eb = ebe \in eSe$.

Now $RL = eSSe \subseteq eSe \subseteq RL$, so $RL = eSe$.

As we have observed, $e \in K(S)$, so by Theorem 1.59 eSe is a group. □

Lemma 1.62. *Let S be a semigroup and assume that there is a minimal left ideal of S which has an idempotent. Then all minimal left ideals of S are isomorphic.*

Proof. Let L be a minimal left ideal of S with an idempotent e. Then $L = Se$ so by Theorem 1.59 eSe is a group.

We claim first that given any $s \in K(S)$ and any $t \in S$, $s(ese)^{-1} = st(este)^{-1}$, where the inverses are taken in eSe. Indeed, using the fact that $(ese)^{-1}e = e(ese)^{-1} = (ese)^{-1}$,

$$s(ese)^{-1}s(ese)^{-1} = s(ese)^{-1}ese(ese)^{-1} = s(ese)^{-1}e = s(ese)^{-1}$$

and similarly $st(este)^{-1}$ is an idempotent. By Lemma 1.57 and Theorem 1.51, $K(S) = \bigcup \{R : R$ is a minimal right ideal of $S\}$. Pick a minimal right ideal R of S such that $s \in R$. Then $s(ese)^{-1}$ and $st(este)^{-1}$ are both idempotents in $R \cap L$, which is a group by Theorem 1.61. Thus $s(ese)^{-1} = st(este)^{-1}$ as claimed.

Now let L' be any other minimal left ideal of S. By Theorem 1.59, eS is a minimal right ideal of S so by Theorem 1.61 $L' \cap eS$ is a group so pick an idempotent $d \in L' \cap eS$. Notice that $L' = Sd$ and $dS = eS$. In particular, by Lemma 1.30(b), $de = e$, $ed = d$, and for any $s \in L'$, $sd = s$.

Define $\psi : Sd \to Se$ by $\psi(s) = s(ese)^{-1}dse$, where the inverse is in the group eSe. We claim first that ψ is a homomorphism. To this end, let $s, t \in Sd$. Then

$$
\begin{aligned}
\psi(s)\psi(t) &= s(ese)^{-1}dset(ete)^{-1}dte \\
&= s(ese)^{-1}dsete(ete)^{-1}dte \quad (e(ete)^{-1} = (ete)^{-1}) \\
&= s(ese)^{-1}dsedte \\
&= s(ese)^{-1}dste \quad (ed = d \text{ and } sd = s) \\
&= st(este)^{-1}dste \quad (s(ese)^{-1} = st(este)^{-1}) \\
&= \psi(st).
\end{aligned}
$$

Now define $\gamma : Se \to Sd$ by $\gamma(t) = t(dtd)^{-1}etd$ where the inverse is in dSd, which is a group by Theorem 1.59. We claim that γ is the inverse of ψ (and hence ψ takes Sd one-to-one onto Se). To this end, let $s \in L'$. Then $ds \in L'$ so Sds is a left ideal contained in L' and thus $L' = Sds$. So pick $x \in S$ such that $s = xds$. Then

$$
\begin{aligned}
\gamma(\psi(s)) &= \psi(s)(d\psi(s)d)^{-1}e\psi(s)d \\
&= s(ese)^{-1}dse(ds(ese)^{-1}dsed)^{-1}es(ese)^{-1}dsed \\
&= xds(ese)^{-1}dsed(ds(ese)^{-1}dsed)^{-1}ese(ese)^{-1}dsed \\
&= xdedsed \\
&= xddsd \\
&= xds \\
&= s.
\end{aligned}
$$

Similarly, if $t \in L$, then $\psi(\gamma(t)) = t$. $\square$

We now analyze in some detail the structure of a particular semigroup. Our motive is that this allows us to analyze the structure of the smallest ideal of any semigroup that has a minimal left ideal with an idempotent.

Theorem 1.63. *Let X be a left zero semigroup, let Y be a right zero semigroup, and let G be a group. Let e be the identity of G, fix $u \in X$ and $v \in Y$ and let $[\ ,\ ] : Y \times X \to G$ be a function such that $[y, u] = [v, x] = e$ for all $y \in Y$ and all $x \in X$. Let $S = X \times G \times Y$ and define an operation $\cdot$ on S by $(x, g, y) \cdot (x', g', y') = (x, g[y, x']g', y')$. Then S is a simple semigroup (so that $K(S) = S = X \times G \times Y$) and each of the following statements holds.*

(a) For every $(x, y) \in X \times Y$, $(x, [y, x]^{-1}, y)$ is an idempotent (where the inverse is taken in G) and all idempotents are of this form. In particular, the idempotents in $X \times G \times \{v\}$ are of the form (x, e, v) and the idempotents in $\{u\} \times G \times Y$ are of the form (u, e, y).

(b) For every $y \in Y$, $X \times G \times \{y\}$ is a minimal left ideal of S and all minimal left ideals of S are of this form.

(c) For every $x \in X$, $\{x\} \times G \times Y$ is a minimal right ideal of S and all minimal right ideals of S are of this form.

(d) *For every* $(x, y) \in X \times Y$, $\{x\} \times G \times \{y\}$ *is a maximal group in S and all maximal groups in S are of this form.*

(e) *The minimal left ideal* $X \times G \times \{v\}$ *is the direct product of X, G, and* $\{v\}$ *and the minimal right ideal* $\{u\} \times G \times Y$ *is the direct product of* $\{u\}$, *G, and Y.*

(f) *All maximal groups in S are isomorphic to G.*

(g) *All minimal left ideals of S are isomorphic to* $X \times G$ *and all minimal right ideals of S are isomorphic to* $G \times Y$.

Proof. The associativity of $\cdot$ is immediate. To see that S is simple, let (x, g, y), $(x', g', y') \in S$. By Lemma 1.52(b), it suffices to show that $(x', g', y') \in S(x, g, y)S$. To see this, let $h = g'g^{-1}[y, x]^{-1}g^{-1}[y, x]^{-1}$. Then $(x', g', y') = (x', h, y) \cdot (x, g, y) \cdot (x, g, y')$.

(a) That $(x, [y, x]^{-1}, y)$ is an idempotent is immediate. Given an idempotent (x, g, y), one has that $g[y, x]g = g$ so $g = [y, x]^{-1}$.

(b) Let $y \in Y$. Trivially $X \times G \times \{y\}$ is a left ideal of S. To see that it is minimal, let $(x, g), (x', g') \in X \times G$. It suffices by Lemma 1.52(a) to note that $(x', g', y) = (x', g'g^{-1}[y, x]^{-1}, y) \cdot (x, g, y)$. Given any minimal left ideal L of S, pick $(x, g, y) \in L$. Then $L \cap (X \times G \times \{y\}) \neq \emptyset$ so $L = X \times G \times \{y\}$.

Statement (c) is the right-left switch of statement (b).

(d) By the equivalence of (h) and (d) in Theorem 1.59 we have that the maximal groups in S are precisely the sets of the form fSf where f is an idempotent of $K(S) = S$. That is, by (a), where $f = (x, [y, x]^{-1}, y)$. Since

$$(x, [y, x]^{-1}, y)S(x, [y, x]^{-1}, y) = \{x\} \times G \times \{y\},$$

we are done.

(e) We show that $X \times G \times \{v\}$ is a direct product, the other statement being similar. Let $(x, g), (x', g') \in X \times G$. Then

$$
\begin{aligned}
(x, g, v) \cdot (x', g', v) &= (x, g[v, x']g', v) \\
&= (x, geg', v) \\
&= (xx', gg', vv).
\end{aligned}
$$

(f) Trivially $\{u\} \times G \times \{v\}$ is isomorphic to G. Now, let $(x, y) \in X \times Y$. Then $\{x\} \times G \times \{v\}$ and $\{u\} \times G \times \{v\}$ are maximal groups in the minimal left ideal $X \times G \times \{v\}$, hence are isomorphic by Theorem 1.42. Also $\{x\} \times G \times \{v\}$ and $\{x\} \times G \times \{y\}$ are maximal groups in the minimal right ideal $\{x\} \times G \times Y$, hence are isomorphic by the left-right switch of Theorem 1.42.

(g) By Lemma 1.62 all minimal left ideals of S are isomorphic and by (e) $X \times G \times \{v\}$ is isomorphic to $X \times G$. The other conclusion follows similarly. $\square$

Note that in Theorem 1.63, the set S is the cartesian product of X, G, and Y, but is not the direct product unless $[y, x] = e$ for every $(y, x) \in Y \times X$.

Observe that as a consequence of Theorem 1.63(g) we have that for any $y \in Y$, $X \times G \times \{y\} \approx X \times G$. However, there is no transparent isomorphism unless $[y, x] = e$ for all $x \in X$, such as when $y = v$.

Theorem 1.63 spells out in detail the structure of $X \times G \times Y$. We see now that this is in fact the structure of the smallest ideal of any semigroup which has a minimal left ideal with an idempotent.

Theorem 1.64 (The Structure Theorem). *Let S be a semigroup and assume that there is a minimal left ideal of S which has an idempotent. Let R be a minimal right ideal of S, let L be a minimal left ideal of S, let $X = E(L)$, let $Y = E(R)$, and let $G = RL$. Define an operation $\cdot$ on $X \times G \times Y$ by $(x, g, y) \cdot (x', g', y') = (x, gyx'g', y')$. Then $X \times G \times Y$ satisfies the conclusions of Theorem 1.63 (where $[y, x] = yx$) and $K(S) \approx X \times G \times Y$. In particular:*

(a) The minimal right ideals of S partition $K(S)$ and the minimal left ideals of S partition $K(S)$.

(b) The maximal groups in $K(S)$ partition $K(S)$.

(c) All minimal right ideals of S are isomorphic and all minimal left ideals of S are isomorphic.

(d) All maximal groups in $K(S)$ are isomorphic.

Proof. After noting that, by Lemma 1.43 and Theorem 1.51, the minimal left ideals of S and of $K(S)$ are identical (and the minimal right ideals of S and of $K(S)$ are identical), the "in particular" conclusions follow immediately from Theorem 1.63. So it suffices to show that $X \times G \times Y$ satisfies the hypotheses of Theorem 1.63 with $[y, x] = yx$ and that $K(S) \approx X \times G \times Y$.

We know by Lemma 1.57 that S has a minimal right ideal with an idempotent (so R exists) and hence by Theorem 1.56 R has an idempotent. We know by Theorem 1.61 that RL is a group and we know by Theorem 1.42 that X is a left zero semigroup and Y is a right zero semigroup. Let e be the identity of $RL = R \cap L$ and let $u = v = e$. Given $y \in Y$ one has, since Y is a right zero semigroup, that $[y, u] = yu = u = e$. Similarly, given $x \in X$, $[v, x] = e$. Consequently the hypotheses of Theorem 1.63 are satisfied.

Define $\varphi : X \times G \times Y \to S$ by $\varphi(x, g, y) = xgy$. We claim that φ is an isomorphism onto $K(S)$. From the definition of the operation in $X \times G \times Y$ we see immediately that φ is a homomorphism. By Theorem 1.42 we have that $L = XG$ and $R = GY$. By Theorem 1.53, $K(S) = LR = XGGY = XGY = \varphi[X \times G \times Y]$. Thus it suffices to produce an inverse for φ.

For each $t \in K(S)$, let $\gamma(t)$ be the inverse of ete in $eSe = G$. Then $t\gamma(t) = t\gamma(t)e \in Se = L$ and

$$t\gamma(t)t\gamma(t) = t\gamma(t)etey(t)$$
$$= tey(t)$$
$$= t\gamma(t),$$

so $t\gamma(t) \in X$. Similarly, $\gamma(t)t \in Y$.

Define $\tau : K(S) \to X \times G \times Y$ by $\tau(t) = (t\gamma(t), ete, \gamma(t)t)$. We claim that $\tau = \varphi^{-1}$. So let $(x, g, y) \in X \times G \times Y$. Then

$$\tau\big(\varphi(x, g, y)\big) = (xgy\gamma(xgy), exgye, \gamma(xgy)xgy).$$

Now

$$xgy\gamma(xgy) = xxgy\gamma(xgy) \qquad (x = xx)$$
$$= xexgye\gamma(xgy) \qquad \big(x = xe \text{ and } \gamma(xgy) = e\gamma(xgy)\big)$$
$$= xe$$
$$= x.$$

Similarly $\gamma(xgy)xgy = y$. Since also $exgye = ege = g$, we have that $\tau = \varphi^{-1}$ as required. $\qquad\square$

The following theorem enables us to identify the smallest ideal of many semigroups that arise in topological applications.

Theorem 1.65. *Let S be a semigroup and assume that there is a minimal left ideal of S which has an idempotent. Let T be a subsemigroup of S and assume also that T has a minimal left ideal with an idempotent. If $K(S) \cap T \neq \emptyset$, then $K(T) = K(S) \cap T$.*

Proof. By Theorem 1.51, $K(T)$ exists so, since $K(S) \cap T$ is an ideal of T, $K(T) \subseteq K(S) \cap T$. For the reverse inclusion, let $x \in K(S) \cap T$ be given. Then Tx is a left ideal of T so by Corollary 1.47 and Theorem 1.56 Tx contains a minimal left ideal Te of T for some idempotent $e \in T$. Now $x \in K(S)$ so by Theorem 1.51 pick a minimal left ideal L of S with $x \in L$. Then $L = Sx$ and $e \in Tx \subseteq Sx$ so $L = Se$ so $x \in Se$ so by Lemma 1.30, $x = xe \in Te \subseteq K(T)$. $\qquad\square$

We know from the Structure Theorem (Theorem 1.64) that maximal groups in the smallest ideal are isomorphic. It will be convenient for us later to know an explicit isomorphism between them.

Theorem 1.66. *Let S be a semigroup and assume that there is a minimal left ideal of S which has an idempotent. Let $e, f \in E\big(K(S)\big)$. If g is the inverse of efe in eSe, then the function $\varphi : eSe \rightarrow fSf$ defined by $\varphi(x) = fxgf$ is an isomorphism.*

Proof. To see that φ is a homomorphism, let $x, y \in eSe$. Then

$$\varphi(x)\varphi(y) = fxgffygf$$
$$= fxgfygf$$
$$= fxgefeygf \qquad (ge = g, \ ey = y)$$
$$= fxeygf$$
$$= fxygf$$
$$= \varphi(xy).$$

To see that φ is one-to-one, let x be in the kernel of φ. Then

$$fxgf = f$$
$$efxgfe = efe$$
$$efexgefe = efe \qquad (ex = x, \ ge = g)$$
$$efexe = efe$$
$$efex = efee$$
$$x = e \qquad \text{(left cancellation in } eSe\text{)}.$$

To see that φ is onto fSf, let $y \in fSf$ and let h and k be the inverses of fgf and fef respectively in fSf. Then $ekyhe \in eSe$ and

$$
\begin{aligned}
\varphi(ekyhe) &= fekyhegf \\
&= fefkyhgf & (fk = k, \ eg = g) \\
&= fyhgf & (fefk = f) \\
&= fyhfgf & (h = hf) \\
&= fyf & (hfgf = f) \\
&= y.
\end{aligned}
$$
□

We conclude the chapter with a theorem characterizing arbitrary elements of $K(S)$.

Theorem 1.67. *Let S be a semigroup and assume that there is a minimal left ideal of S which has an idempotent. Let $s \in S$. The following statements are equivalent.*

(a) $s \in K(S)$.
(b) *For all $t \in S$, $s \in Sts$.*
(c) *For all $t \in S$, $s \in stS$.*
(d) *For all $t \in S$, $s \in stS \cap Sts$.*

Proof. (a) implies (d). Pick by Theorem 1.51 and Lemma 1.57 a minimal left ideal L of S and a minimal right ideal R of S with $s \in L \cap R$. Let $t \in S$. Then $ts \in L$ so Sts is a left ideal contained in L so $Sts = L$. Similarly $stS = R$.

The facts that (d) implies (c) and (d) implies (b) are trivial.

(b) implies (a). Pick $t \in K(S)$. Then $s \in Sts \subseteq K(S)$.

Similarly (c) implies (a). □

Exercise 1.7.1. Let S be a semigroup and assume that there is a minimal left ideal of S which has an idempotent. Prove that if $K(S) \neq S$ and $x \in K(S)$, then x is neither left nor right cancelable in S. (Hint: If x is a member of the minimal left ideal L, then $L = Sx = Lx$.)

Exercise 1.7.2. Identify $K(S)$ for those semigroups S in Exercises 1.4.2 and 1.4.3 for which the smallest ideal exists.

Exercise 1.7.3. Let S and T be semigroups and let $h : S \to T$ be a surjective homomorphism. If S has a smallest ideal, show that T does as well and that $K(T) = h[K(S)]$.

Notes

Much of the material in this chapter is based on the treatment in [39, Section II.1]. The presentation of the Structure Theorem was suggested to us by J. Pym and is based on his treatment in [202]. The Structure Theorem (Theorem 1.64) is due to A. Suschkewitsch [231] in the case of finite semigroups and to D. Rees [210] in the general case.

Chapter 2

Right Topological (and Semitopological and Topological) Semigroups

In this (and subsequent) chapters, we assume that the reader has mastered an introductory course in general topology. In particular, we expect familiarity with the notions of continuous functions, nets, and compactness.

2.1 Topological Hierarchy

Definition 2.1. (a) A *right topological semigroup* is a triple $(S, \cdot, \mathcal{T})$ where $(S, \cdot)$ is a semigroup, $(S, \mathcal{T})$ is a topological space, and for all $x \in S$, $\rho_x : S \to S$ is continuous.

(b) A *left topological semigroup* is a triple $(S, \cdot, \mathcal{T})$ where $(S, \cdot)$ is a semigroup, $(S, \mathcal{T})$ is a topological space, and for all $x \in S$, $\lambda_x : S \to S$ is continuous.

(c) A *semitopological semigroup* is a right topological semigroup which is also a left topological semigroup.

(d) A *topological semigroup* is a triple $(S, \cdot, \mathcal{T})$ where $(S, \cdot)$ is a semigroup, $(S, \mathcal{T})$ is a topological space, and $\cdot : S \times S \to S$ is continuous.

(e) A *topological group* is a triple $(S, \cdot, \mathcal{T})$ such that $(S, \cdot)$ is a group, $(S, \mathcal{T})$ is a topological space, $\cdot : S \times S \to S$ is continuous, and $\text{In} : S \to S$ is continuous (where $\text{In}(x)$ is the inverse of x in S).

We did not include any separation axioms in the definitions given above. However, all of our applications involve Hausdorff spaces. So we shall be assuming throughout, except in Chapter 7, that all hypothesized topological spaces are Hausdorff.

In a right topological semigroup we say that the operation "·" is "right continuous". We should note that many authors use the term "left topological" for what we call "right topological" and vice versa. One may reasonably ask why someone would refer to an operation for which multiplication on the right is continuous as "left continuous". The people who do so ask why we refer to an operation which is continuous in the left variable as "right continuous".

We shall customarily not mention either the operation or the topology and say something like "let S be a right topological semigroup".

Note that trivially each topological group is a topological semigroup, each topological semigroup is a semitopological semigroup and each semitopological semigroup is both a left and right topological semigroup.

Of course any semigroup which is not a group provides an example of a topological semigroup which is not a topological group simply by providing it with the discrete topology. It is the content of Exercise 2.1.1 to show that there is a topological semigroup which is a group but is not a topological group.

It is a celebrated theorem of R. Ellis [84], that if S is a locally compact *semitopological* semigroup which is a group then S is a topological group. That is, if S is locally compact and a group, then separate continuity implies joint continuity and continuity of the inverse. We shall prove this theorem in the last section of this chapter. For an example of a semitopological semigroup which is a group but is not a topological semigroup see Exercise 9.2.7.

It is the content of Exercise 2.1.2 that there is a semitopological semigroup which is not a topological semigroup.

Recall that given any topological space $(X, \mathcal{T})$, the *product topology* on ^XX is the topology with subbasis $\{\pi_x^{-1}[U] : x \in X$ and $U \in \mathcal{T}\}$, where for $f \in {}^XX$ and $x \in X$, $\pi_x(f) = f(x)$. Whenever we refer to a "basic" or "subbasic" open set in ^XX, we mean sets defined in terms of this subbasis. The product topology is also often referred to as the *topology of pointwise convergence* . The reason for this terminology is that a net $\langle f_\iota \rangle_{\iota \in I}$ converges to f in ^XX if and only if $\langle f_\iota(x) \rangle_{\iota \in I}$ converges to $f(x)$ for every $x \in X$.

Theorem 2.2. *Let $(X, \mathcal{T})$ be any topological space and let $\mathcal{V}$ be the product topology on ^XX.*

(a) $\left({}^XX, \circ, \mathcal{V} \right)$ *is a right topological semigroup.*

(b) *For each $f \in {}^XX$, λ_f is continuous if and only if f is continuous.*

Proof. Let $f \in {}^XX$. Suppose that the net $\langle g_\iota \rangle_{\iota \in I}$ converges to g in $\left({}^XX, \mathcal{V} \right)$. Then, for any $x \in X$, $\langle g_\iota\big(f(x)\big) \rangle_{\iota \in I}$ converges to $g\big(f(x)\big)$ in X. Thus $\langle g_\iota \circ f \rangle_{\iota \in I}$ converges to $g \circ f$ in $\left({}^XX, \mathcal{V} \right)$, and so ρ_f is continuous. This establishes (a).

Now λ_f is continuous if and only if $\langle f\big(g_\iota(x)\big) \rangle_{\iota \in I}$ converges to $f\big(g(x)\big)$ for every net $\langle g_\iota \rangle_{\iota \in I}$ converging to g in $\left({}^XX, \mathcal{V} \right)$ and every $x \in X$. This is obviously the case if f is continuous. Conversely, suppose that λ_f is continuous. Let $\langle x_\iota \rangle_{\iota \in I}$ be a net converging to x in X. We define $g_\iota = \overline{x_\iota}$, the function in ^XX which is constantly equal to x_ι and $g = \overline{x}$. Then $\langle g_\iota \rangle_{\iota \in I}$ converges to g in ^XX and so $\langle f \circ g_\iota \rangle_{\iota \in I}$ converges to $f \circ g$. This means that $\langle f(x_\iota) \rangle_{\iota \in I}$ converges to $f(x)$. Thus f is continuous, and we have established (b).   □

Corollary 2.3. *Let X be a topological space. The following statements are equivalent.*

(a) ^XX *is a topological semigroup.*

(b) ^XX *is a semitopological semigroup.*

(c) *For all $f \in {}^XX$, f is continuous.*

(d) *X is discrete.*

Proof. Exercise 2.1.3 □

If X is any nondiscrete space, it follows from Theorem 2.2 and Corollary 2.3 that $^X X$ is a right topological semigroup which is not left topological. Of course, reversing the order of operation yields a left topological semigroup which is not right topological.

Definition 2.4. Let S be a right topological semigroup. The *topological center* of S is the set $\Lambda(S) = \{x \in S : \lambda_x \text{ is continuous}\}$.

Thus a right topological semigroup S is a semitopological semigroup if and only if $\Lambda(S) = S$. Note that trivially the algebraic center of a right topological semigroup is contained in its topological center.

Exercise 2.1.1. Let $\mathcal{T}$ be the topology on $\mathbb{R}$ with basis $\mathcal{B} = \{(a, b] : a, b \in \mathbb{R} \text{ and } a < b\}$. Prove that $(\mathbb{R}, +, \mathcal{T})$ is a topological semigroup but not a topological group.

Exercise 2.1.2. Let $S = \mathbb{R} \cup \{\infty\}$, let S have the topology of the one point compactification of $\mathbb{R}$ (with its usual topology), and define an operation $*$ on S by

$$x * y = \begin{cases} x + y & \text{if } x, y \in \mathbb{R} \\ \infty & \text{if } x = \infty \text{ or } y = \infty. \end{cases}$$

(a) Prove that $(S, *)$ is a semitopological semigroup.
(b) Show that $* : S \times S \to S$ is not continuous at (∞, ∞).

Exercise 2.1.3. Prove Corollary 2.3.

2.2 Compact Right Topological Semigroups

We shall be concerned throughout this book with certain compact right topological semigroups. Of fundamental importance is the following theorem.

Theorem 2.5. *Let S be a compact right topological semigroup. Then $E(S) \neq \emptyset$.*

Proof. Let $\mathcal{A} = \{T \subseteq S : T \neq \emptyset, T \text{ is compact, and } T \cdot T \subseteq T\}$. That is, $\mathcal{A}$ is the set of compact subsemigroups of S. We show that $\mathcal{A}$ has a minimal member using Zorn's Lemma. Since $S \in \mathcal{A}$, $\mathcal{A} \neq \emptyset$. Let $\mathcal{C}$ be a chain in $\mathcal{A}$. Then $\mathcal{C}$ is a collection of closed subsets of the compact space S with the finite intersection property, so $\bigcap \mathcal{C} \neq \emptyset$ and $\bigcap \mathcal{C}$ is trivially compact and a semigroup. Thus $\bigcap \mathcal{C} \in \mathcal{A}$, so we may pick a minimal member A of $\mathcal{A}$.

Pick $x \in A$. We shall show that $xx = x$. (It will follow that $A = \{x\}$, but we do not need this.) We start by showing that $Ax = A$. Let $B = Ax$. Then $B \neq \emptyset$ and since $B = \rho_x[A]$, B is the continuous image of a compact space, hence compact. Also

$BB = AxAx \subseteq AAAx \subseteq Ax = B$. Thus $B \in \mathcal{A}$. Since $B = Ax \subseteq AA \subseteq A$ and A is minimal, $B = A$.

Let $C = \{y \in A : yx = x\}$. Since $x \in A = Ax$, we have $C \neq \emptyset$. Also, $C = A \cap \rho_x^{-1}[\{x\}]$, so C is closed and hence compact. Given $y, z \in C$ one has $yz \in AA \subseteq A$ and $yzx = yx = x$ so $yz \in C$. Thus $C \in \mathcal{A}$. Since $C \subseteq A$ and A is minimal, we have $C = A$ so $x \in C$ and so $xx = x$ as required. $\square$

In Section 1.7 there were several results which had as part of their hypotheses "Let S be a semigroup and assume there is a minimal left ideal of S which has an idempotent." Because of the following corollary, we are able to incorporate all of these results.

Corollary 2.6. *Let S be a compact right topological semigroup. Then every left ideal of S contains a minimal left ideal. Minimal left ideals are closed, and each minimal left ideal has an idempotent.*

Proof. If L is any left ideal L of S and $x \in L$, then Sx is a compact left ideal contained in L. (It is compact because $Sx = \rho_x[S]$.) Consequently any minimal left ideal is closed and by Theorem 2.5 any minimal left ideal contains an idempotent. Thus we need only show that any left ideal of S contains a minimal left ideal. So let L be a left ideal of S and let $\mathcal{A} = \{T : T$ is a closed left ideal of S and $T \subseteq L\}$. Applying Zorn's Lemma to $\mathcal{A}$, one gets a left ideal M minimal among all closed left ideals contained in L. But since every left ideal contains a closed left ideal, M is a minimal left ideal. $\square$

We now deduce some consequences of Corollary 2.6. Note that these consequences apply in particular to any finite semigroup S, since S is a compact topological semigroup when provided with the discrete topology.

Theorem 2.7. *Let S be a compact right topological semigroup.*

(a) *Every right ideal of S contains a minimal right ideal which has an idempotent.*

(b) *Let $T \subseteq S$. Then T is a minimal left ideal of S if and only if there is some $e \in E\big(K(S)\big)$ such that $T = Se$.*

(c) *Let $T \subseteq S$. Then T is a minimal right ideal of S if and only if there is some $e \in E\big(K(S)\big)$ such that $T = eS$.*

(d) *Given any minimal left ideal L of S and any minimal right ideal R of S, there is an idempotent $e \in R \cap L$ such that $R \cap L = eSe$ and eSe is a group.*

Proof. (a) Corollary 2.6, Lemma 1.57, Corollary 1.47, and Theorem 1.56.

(b) and (c). Corollary 2.6 and Theorem 1.58.

(d) Corollary 2.6 and Theorem 1.61. $\square$

Theorem 2.8. *Let S be a compact right topological semigroup. Then S has a smallest (two sided) ideal $K(S)$ which is the union of all minimal left ideals of S and also the union of all minimal right ideals of S. Each of $\{Se : e \in E\big(K(S)\big)\}$, $\{eS : e \in E\big(K(S)\big)\}$, and $\{eSe : e \in E\big(K(S)\big)\}$ are partitions of $K(S)$.*

Proof. Corollary 2.6 and Theorems 1.58, 1.61, and 1.64. $\square$

Theorem 2.9. *Let S be a compact right topological semigroup and let $e \in E(S)$. The following statements are equivalent.*

  (a) *Se is a minimal left ideal.*

  (b) *Se is left simple.*

  (c) *eSe is a group.*

  (d) *$eSe = H(e)$.*

  (e) *eS is a minimal right ideal.*

  (f) *eS is right simple.*

  (g) *e is a minimal idempotent.*

  (h) *$e \in K(S)$.*

  (i) *$K(S) = SeS$.*

Proof. Corollary 2.6 and Theorem 1.59.   □

Theorem 2.10. *Let S be a compact right topological semigroup. Let $s \in S$. The following statements are equivalent.*

  (a) *$s \in K(S)$.*

  (b) *For all $t \in S$, $s \in Sts$.*

  (c) *For all $t \in S$, $s \in stS$.*

  (d) *For all $t \in S$, $s \in stS \cap Sts$.*

Proof. Corollary 2.6 and Theorem 1.67.   □

The last few results have had purely algebraic conclusions. We now obtain a result with both topological and algebraic conclusions. Suppose that we have two topological spaces which are also semigroups. We say that they are *topologically and algebraically isomorphic* if there is a function from one of them onto the other which is both an isomorphism and a homeomorphism.

Theorem 2.11. *Let S be a compact right topological semigroup.*

  (a) *All maximal subgroups of $K(S)$ are (algebraically) isomorphic.*

  (b) *Maximal subgroups of $K(S)$ which lie in the same minimal right ideal are topologically and algebraically isomorphic.*

  (c) *All minimal left ideals of S are homeomorphic. In fact, if L and L' are minimal left ideals of S and $z \in L'$, then $\rho_{z|L}$ is a homeomorphism from L onto L'.*

Proof. (a) Corollary 2.6 and Theorem 1.66.

(b) Let R be a minimal right ideal of S and let $e, f \in E(R)$. Then eS and fS are right ideals contained in R and so $R = eS = fS$. Then by Lemma 1.30, $ef = f$ and $fe = e$. Let g be the inverse of efe in the group eSe and define $\varphi : eSe \to fSf$ by $\varphi(x) = fxgf$. Then by Theorem 1.66, φ is an isomorphism from eSe onto fSf. To see that φ is continuous, we show that φ is the restriction of ρ_{gf} to eSe. To this end, let $x \in eSe$. Then

$$
\begin{aligned}
\varphi(x) &= fxgf \\
&= fexgf \qquad (x = ex) \\
&= exgf \qquad (fe = e) \\
&= xgf \qquad (ex = x).
\end{aligned}
$$

Now let h and k be the inverses in fSf of fgf and fef respectively. We showed in the proof of Theorem 1.66 that if $y \in fSf$, then $\varphi^{-1}(y) = ekyhe$. Thus

$$
\begin{aligned}
\varphi^{-1}(y) &= ekyhe \\
&= fefkyhe \qquad (fk = k \text{ and } fe = e) \\
&= fyhe \qquad (fefk = f) \\
&= yhe \qquad (fy = y).
\end{aligned}
$$

So φ^{-1} is the restriction of ρ_{he} to fSf and hence is continuous.

(c) Let L and L' be minimal left ideals of S and let $z \in L'$. By Theorem 2.7(b), pick $e \in E\big(K(S)\big)$ such that $L = Se$. Then $\rho_{z|L}$ is a continuous function from Se to $Sz = L'$ and $\rho_z[Se] = L'$ because Sez is a left ideal of S which is contained in L'. To see that ρ_z is one-to-one on Se, let g be the inverse of eze in eSe. We show that for $x \in Se$, $\rho_g\big(\rho_z(x)\big) = x$, so let $x \in Se$ be given.

$$
\begin{aligned}
xzg &= xezeg \qquad (x = xe \text{ and } g = eg) \\
&= xe \\
&= x.
\end{aligned}
$$

Since $\rho_{z|L}$ is one-to-one and continuous and L is compact, $\rho_{z|L}$ is continuous. $\qquad \square$

Recall that given any idempotents e, f in a semigroup S, $e \leq_R f$ if and only if $fe = e$.

Theorem 2.12. *Let S be a compact right topological semigroup and let $e \in E(S)$. There is a $\leq_R$-maximal idempotent f in S with $e \leq_R f$.*

Proof. Let $A = \{x \in E(S) : e \leq_R x\}$. Then $A \neq \emptyset$ because $e \in A$. Let C be a $\leq_R$-chain in A. Then $\{c\ell\{r \in C : x \leq_R r\} : x \in C\}$ is a collection of closed subsets of S with the finite intersection property, so $\bigcap_{x \in C} c\ell\{r \in C : x \leq_R r\} \neq \emptyset$. Since S is Hausdorff, $\bigcap_{x \in C} c\ell\{r \in C : x \leq_R r\} \subseteq \{t \in S : \text{for all } x \in C, tx = x\}$. Consequently, $\{t \in S : \text{for all } x \in C, tx = x\}$ is a compact subsemigroup of S and hence by Theorem 2.5 there is an idempotent y such that for all $x \in C$, $yx = x$. This y is an upper bound for C, so A has a maximal member. $\qquad \square$

Given e, $f \in E\big(K(S)\big)$ and an assignment to find an isomorphism from eSe onto fSf, most of us would try first the function $\tau : eSe \to fSf$ defined by $\tau(y) = fyf$. In fact, if $eS = fS$, this works (Exercise 2.2.1). We see now that this natural function need not be a homomorphism if $eS \neq fS$ and $Se \neq Sf$

Example 2.13. *Let S be the semigroup consisting of the eight distinct elements e, f, ef, fe, efe, fef, efef, and fefe, with the following multiplication table.*

·	e	f	ef	fe	efe	fef	efef	fefe
e	e	ef	ef	efe	efe	efef	efef	e
f	fe	f	fef	fe	fefe	fef	f	fefe
ef	efe	ef	efef	efe	e	efef	ef	e
fe	fe	fef	fef	fefe	fefe	f	f	fe
efe	efe	efef	efef	e	e	ef	ef	efe
fef	fefe	fef	f	fefe	fe	f	fef	fe
efef	e	efef	ef	e	efe	ef	efef	efe
fefe	fefe	f	f	fe	fe	fef	fef	fefe

Then S with the discrete topology is a compact topological semigroup and fef is not an idempotent so the function $\tau : eSe \rightarrow fSf$ defined by $\tau(x) = fxf$ is not a homomorphism.

Once again, the simplest way to see that one has in fact defined a semigroup is to produce a concrete representation. In this case the 2×2 integer matrices $e = \begin{pmatrix} 1 & 0 \\ 0 & 0 \end{pmatrix}$ and $f = \begin{pmatrix} -1 & -2 \\ 1 & 2 \end{pmatrix}$ generate the semigroup S.

The topological center of a compact right topological semigroup is important for many applications. The following lemma will be used later in this book.

Lemma 2.14. *Let S and T be compact right topological semigroups, let D be a dense subsemigroup of S such that $D \subseteq \Lambda(S)$, and let η be a continuous function from S to T such that*

(1) $\eta[D] \subseteq \Lambda(T)$ *and*

(2) $\eta_{|D}$ *is a homomorphism.*

Then η is a homomorphism.

Proof. For each $d \in D$, $\eta \circ \lambda_d$ and $\lambda_{\eta(d)} \circ \eta$ are continuous functions agreeing on the dense subset D of S. Thus for all $d \in D$ and all $y \in S$, $\eta(dy) = \eta(d)\eta(y)$. Therefore, for all $y \in S$, $\eta \circ \rho_y$ and $\rho_{\eta(y)} \circ \eta$ are continuous functions agreeing on a dense subset of S so for all x and y in S, $\eta(xy) = \eta(x)\eta(y)$. □

Exercise 2.2.1. Let S be a compact right topological semigroup and let $e, f \in E(K(S))$ such that $eS = fS$. Let g be the inverse of efe in the group eSe and define $\varphi : eSe \rightarrow fSf$ by $\varphi(x) = fxgf$. Define $\tau : eSe \rightarrow fSf$ by $\tau(x) = fxf$. Prove that $\tau = \varphi$.

Exercise 2.2.2. Let S be a compact right topological semigroup, let T be a semigroup with topology, and let $\varphi : S \rightarrow T$ be a continuous homomorphism. Prove that $\varphi[S]$ is a compact right topological semigroup.

Exercise 2.2.3. Prove that a compact cancellative right topological semigroup is a group.

2.3 Closures and Products of Ideals

We investigate briefly the closures of right ideals, left ideals, and maximal subgroups of right topological semigroups. We also consider their Cartesian product.

Theorem 2.15. *Let S be a right topological semigroup and let R be a right ideal of S. Then $c\ell\, R$ is a right ideal of S.*

Proof. This is Exercise 2.3.1. □

On the other hand, the closure of a left ideal need not be a left ideal. We leave the verification of the details in the following example to the reader.

Example 2.16. *Let X be any compact space with a subset D such that $c\ell\, D \neq X$ and $|c\ell\, D \backslash D| \geq 2$. Define an operation · on X by*

$$x \cdot y = \begin{cases} y & \text{if } y \in D \\ x & \text{if } y \notin D. \end{cases}$$

Then X is a compact right topological semigroup, $\Lambda(X) = \emptyset$, and the set of left ideals of X is $\{X\} \cup \{B : \emptyset \neq B \subseteq D\}$. In particular, D is a left ideal of X and $c\ell\, D$ is not a left ideal.

Often, however, we see that the closure of a left ideal is an ideal.

Theorem 2.17. *Let S be a compact right topological semigroup and assume that $\Lambda(S)$ is dense in S. Let L be a left ideal of S. Then $c\ell\, L$ is a left ideal of S.*

Proof. Let $x \in c\ell\, L$ and let $y \in S$. To see that $yx \in c\ell\, L$, let U be an open neighborhood of yx. Pick a neighborhood V of y such that $Vx = \rho_x[V] \subseteq U$ and pick $z \in \Lambda(S) \cap V$. Then $zx = \lambda_z(x) \in U$ so pick a neighborhood W of x such that $zW \subseteq U$. Pick $w \in W \cap L$. Then $zw \in U \cap L$. □

In our applications we shall often be concerned with semigroups with dense center. (If S is discrete and commutative, then βS has dense center — see Theorem 4.23.)

Lemma 2.18. *Let S be a compact right topological semigroup. The following statements are equivalent.*

(a) *The (algebraic) center of S is dense in S.*

(b) *There is a dense commutative subset A of S with $A \subseteq \Lambda(S)$*

Proof. The fact that (a) implies (b) is trivial. To see that (b) implies (a), let A be a dense commutative subset of S with $A \subseteq \Lambda(S)$. For every $x \in A$, the continuous functions ρ_x and λ_x are equal on A and are therefore equal on S. So A is contained in the center of S. □

Theorem 2.19. *Let S be a compact right topological semigroup with dense center.*
(a) *If R is a right ideal of S, then $c\ell R$ is a two sided ideal of S.*
(b) *If $e \in E(K(S))$, then $c\ell(eSe) = Se$.*

Proof. Let A be the center of S.

(a) By Theorem 2.15, $c\ell R$ is a right ideal of S. To see that $S \cdot (c\ell R) \subseteq c\ell R$, let $y \in c\ell R$ be given. Then given any $x \in A$ one has $\rho_y(x) = xy = yx \in (c\ell R)x \subseteq c\ell R$. Thus $\rho_y[A] \subseteq c\ell R$, so $\rho_y[S] = \rho_y[c\ell A] \subseteq c\ell \rho_y[A] \subseteq c\ell R$. That is $Sy \subseteq c\ell R$ as required.

(b) Since Se is closed (being the continuous image of a compact space), $c\ell(eSe) \subseteq Se$. On the other hand, since ρ_e is continuous, $Se = (c\ell A)e = c\ell(Ae) = c\ell(eAe) \subseteq c\ell(eSe)$. □

Theorem 2.20. *Let S be a compact right topological semigroup with dense center. Assume that S has some minimal right ideal R which is closed. Then $R = K(S)$ and all maximal subgroups of $K(S)$ are closed and pairwise algebraically and topologically isomorphic.*

Proof. By Theorem 2.19(a), $c\ell R$ is an ideal so $K(S) \subseteq c\ell R = R \subseteq K(S)$. Given $e \in E(K(S))$, $H(e) = eSe = R \cap Se$ by Theorems 2.9 and 2.7 so $H(e)$ is closed. Any two maximal subgroups of $K(S)$ lie in the same minimal right ideal and so are algebraically and topologically isomorphic by Theorem 2.11. □

Theorem 2.21. *Let S be a compact right topological semigroup with dense center. The following statements are equivalent.*

(a) *$K(S)$ is a minimal right ideal of S.*
(b) *All maximal subgroups of $K(S)$ are closed.*
(c) *Some maximal subgroup of $K(S)$ is closed.*

Proof. (a) implies (b). Let $e \in E(K(S))$. Then by Theorem 2.9 eS is a minimal right ideal and Se is a minimal left ideal so by Theorem 2.7(d), $eSe = eS \cap Se$. Since $K(S)$ is a minimal right ideal $eS = K(S)$. Since $Se \subseteq K(S)$, $eSe = eS \cap Se = Se$, and Se is closed.

The fact that (b) implies (c) is trivial.

(c) implies (a). Pick $e \in E(K(S))$ such that eSe is closed. Let $R = eS$. By Theorem 2.19(b), $c\ell(eSe) = Se$. So $Se = eSe \subseteq eS = R$. Now any other minimal right ideal of S would be disjoint from R so would miss Se, which is impossible by Lemma 1.29(b). Thus R is the only minimal right ideal of $K(S)$, which is the union of all minimal right ideals, so $K(S) = R$. □

We have seen that the Cartesian product of semigroups is itself a semigroup under the coordinatewise operation. If the semigroups are also topological spaces the product is naturally a topological space.

Theorem 2.22. *Let $\langle S_i \rangle_{i \in I}$ be a family of right topological semigroups and let $S = \times_{i \in I} S_i$. With the product topology and coordinatewise operations, S is a right topological semigroup. If each S_i is compact, then so is S. If $\vec{x} \in S$ and for each $i \in I$, $\lambda_{x_i} : S_i \to S_i$ is continuous, then $\lambda_{\vec{x}} : S \to S$ is continuous.*

Proof. This is Exercise 2.3.5. $\square$

Note that the following theorem is entirely algebraic. We shall only need it, however, in the case in which each S_i is a compact right topological semigroup. In this case, each S_i does have a smallest ideal by Theorem 2.8.

Theorem 2.23. *Let $\langle S_i \rangle_{i \in I}$ be a family of semigroups and let $S = \times_{i \in I} S_i$. Suppose that, for each $i \in I$, S_i has a smallest ideal. Then S also has a smallest ideal and $K(S) = \times_{i \in I} K(S_i)$.*

Proof. We first note that $\times_{i \in I} K(S_i)$ is an ideal of S.

Let $\vec{u} \in K(S)$. Then

$$\left(\times_{i \in I} K(S_i) \right) \cdot \vec{u} \cdot \left(\times_{i \in I} K(S_i) \right) = \times_{i \in I} \left(K(S_i) \cdot u_i \cdot K(S_i) \right) = \times_{i \in I} K(S_i)$$

so by Lemma 1.52(b), $K(S) = \times_{i \in I} K(S_i)$. $\square$

While we are on the subject of products, we use Cartesian products to establish the following result which will be useful later.

Theorem 2.24. *Let A be a set and let G be the free group generated by A. Then G can be embedded in a compact topological group. This means that there is a compact topological group H and a one-to-one homomorphism $\varphi : G \to H$.*

Proof. Recall that $\emptyset$ is the identity of G. For each $g \in G \backslash \{\emptyset\} = G'$, pick by Theorem 1.23 a finite group F_g and a homomorphism $\phi_g : G \to F_g$ such that $\phi_g(g)$ is not the identity of F_g. Let each F_g have the discrete topology. Then $H = \times_{g \in G'} F_g$ is a compact topological group. Define a homomorphism $\varphi : G \to H$ by stating that $\varphi(h)_g = \phi_g(h)$. Then, if $g \in G'$, we know that $\varphi(g)_g$ is not the identity of F_g. So the kernel of φ is $\{\emptyset\}$ and hence φ is one-to-one as required. $\square$

Exercise 2.3.1. Prove Theorem 2.15.

Exercise 2.3.2. Let S be a right topological semigroup, and let T be a subset of the topological center of S. Prove that $c\ell\, T$ is a semigroup if T is a semigroup.

Exercise 2.3.3. Verify the assertions in Example 2.16.

Exercise 2.3.4. Let X and D be as in Example 2.16 except that D is open and $c\ell\, D \backslash D = \{z\}$. Prove that $\Lambda(X) = \{z\}$.

Exercise 2.3.5. Prove Theorem 2.22.

2.4 Semitopological and Topological Semigroups

We only scratch the surface of the theory of semitopological semigroups and the theory of topological semigroups, in order to indicate the kinds of results that hold in these settings but not in the setting of right topological semigroups.

We shall see many examples of right topological semigroups that have no closed minimal right ideals, including our favorite $(\beta \mathbb{N}, +)$. (See Theorems 6.9 and 2.20.) By way of contrast we have the following theorem.

Theorem 2.25. *Let S be a compact semitopological semigroup and let $a \in S$. Then aS, Sa, and aSa are closed.*

Proof. Exercise 2.4.1. □

Corollary 2.26. *Let S be a compact semitopological semigroup with dense center. Then S is commutative and $K(S)$ is a group.*

Proof. For any $x \in S$, λ_x and ρ_x are continuous functions agreeing on a dense subspace of S and therefore on the whole of S. So S is commutative.

By Corollary 2.6 and Exercise 1.6.5, $K(S)$ is a group. □

In fact a stronger conclusion holds. See Corollary 2.40.

Theorem 2.27. *Let S be a semitopological semigroup and let T be a subsemigroup of S. Then $c\ell\, T$ is a subsemigroup of S.*

Proof. This follows from Exercise 2.3.2. □

By way of contrast we shall now see that in a right topological semigroup, the closure of a subsemigroup need not be a semigroup. (This also holds, as we shall see in Theorem 8.21, in $(\beta \mathbb{N}, +)$.)

Theorem 2.28. *Let X be the one point compactification of $\mathbb{N}$ and let $T = \{f \in {}^X X : f$ is one-to-one$\}$. Then $c\ell\, T$ is not a semigroup.*

Proof. For each $n \in \mathbb{N}$, define $g_n : X \to X$ by

$$g_n(x) = \begin{cases} x + n & \text{if } x \in \mathbb{N} \\ \infty & \text{if } x = \infty. \end{cases}$$

Let $g : X \to X$ be the constant function $\overline{\infty}$. Then each $g_n \in T$ and $\langle g_n \rangle_{n=1}^{\infty}$ converges to g in ${}^X X$ so $g \in c\ell\, T$.

Now define $f : X \to X$ by

$$f(x) = \begin{cases} x + 1 & \text{if } x \in \mathbb{N} \\ 1 & \text{if } x = \infty. \end{cases}$$

Then $f \in T$ while $f \circ g \notin c\ell \, T$ because $\{h \in {}^X X : h(1) = h(2) = 1\}$ is a neighborhood of $f \circ g$ in ${}^X X$ which misses T. $\square$

We see, however, that many subsemigroups of ${}^X X$ do have closures that are semigroups.

Theorem 2.29. *Let X be a topological space and let $S \subseteq {}^X X$ be a semigroup such that for each $f \in S$, f is continuous. Then $c\ell \, S$ is a semigroup.*

Proof. This follows from Theorem 2.2 and Exercise 2.3.2. $\square$

The closed semigroups of Theorem 2.29 are important in topological dynamics and we will have occasion to refer to them often.

Definition 2.30. Let X be a topological space and let S be a semigroup contained in ${}^X X$ such that for all $f \in S$, f is continuous. Then $c\ell \, S$ is the *enveloping semigroup of S*.

If $T \in {}^X X$ is continuous and $S = \{T^n : n \in \mathbb{N}\}$, then $c\ell \, S$ is the *enveloping semigroup of T*.

By Theorem 2.29, if $T \in {}^X X$ is continuous, then $c\ell \{T^n : n \in \mathbb{N}\}$ is a semigroup. We see that one cannot add a single point of discontinuity and expect the same conclusion.

Example 2.31. *Let $X = [0, 1]$ with the usual topology and define $f \in {}^X X$ by*

$$f(x) = \begin{cases} 1 & \text{if } x = 0 \\ x/2 & \text{if } 0 < x \le 1. \end{cases}$$

Then $c\ell \{f^n : n \in \mathbb{N}\} = \{f^n : n \in \mathbb{N}\} \cup \{\overline{0}\}$ but $f \circ \overline{0} = \overline{1} \notin c\ell \{f^n : n \in \mathbb{N}\}$.

Theorem 2.32. *Let S be a compact topological semigroup. Then all maximal subgroups of S are closed and are topological groups.*

Proof. Let $e \in E(S)$. By Lemma 1.19, an element $x \in eSe$ is in $H(e)$ if and only if there is an element $y \in eSe$ for which $yx = xy = e$. Now eSe is closed as it is the continuous image of S under the mapping $\rho_e \circ \lambda_e$. Suppose that x is a limit point of a net $\langle x_\iota \rangle_{\iota \in I}$ in $H(e)$. For each $\iota \in I$ there is an element $y_\iota \in H(e)$ for which $x_\iota y_\iota = y_\iota x_\iota = e$. By joint continuity any limit point y of the net $\langle y_\iota \rangle_{\iota \in I}$ satisfies $xy = yx = e$ and so $x \in H(e)$.

To see that the inverse function $\text{In} : H(e) \to H(e)$ is continuous, let $x \in H(e)$ and let $\langle x_\iota \rangle_{\iota \in I}$ be a net in $H(e)$ converging to x. As observed above, any limit point y of the net $\langle x_\iota^{-1} \rangle_{\iota \in I}$ satisfies $xy = yx = e$ and so x^{-1} is the only limit point of $\langle x_\iota^{-1} \rangle_{\iota \in I}$. That is, $\langle x_\iota^{-1} \rangle_{\iota \in I}$ converges to x^{-1}. $\square$

We see that we cannot obtain the conclusion of Theorem 2.32 in an arbitrary compact semitopological semigroup.

Example 2.33. *Let* $S = \mathbb{R} \cup \{\infty\}$ *with the topology and semigroup operation given in Exercise* 2.1.2. *Then* S *is a semitopological semigroup,* $E(S) = \{0, \infty\}$ *and* $H(0) = \mathbb{R}$. *Thus* $H(0)$ *is not closed.*

Note that in Example 2.33, $H(\infty) = \{\infty\}$, which is closed. In fact, by Theorem 2.25, in any compact semitopological semigroup, all maximal groups in $K(S)$ are closed.

Exercise 2.4.1. Prove Theorem 2.25

Exercise 2.4.2. Let X be any set. We can identify $\mathcal{P}(X)$ with $^X\{0, 1\}$ by identifying each $Y \in \mathcal{P}(X)$ with its *characteristic function* χ_Y, where,

$$\chi_Y(x) = \begin{cases} 1 & \text{if } x \in Y \\ 0 & \text{if } x \in X \backslash Y. \end{cases}$$

We can use this identification to give $\mathcal{P}(X)$ a compact topology. Prove that $(\mathcal{P}(X), \cup)$ and $(\mathcal{P}(X), \cap)$ are topological semigroups. Prove also that $(\mathcal{P}(X), \Delta)$ is a topological group which can be identified topologically and algebraically with $^X\mathbb{Z}_2$ (where $Y \Delta Z = (Y \cup Z) \backslash (Y \cap Z)$, the *symmetric difference* of Y and Z).

Exercise 2.4.3. Given a set $A \subseteq \mathbb{R}^n$ and $x \in \mathbb{R}^n$ define $d(x, A) = \inf\{d(x, y) : y \in A\}$ Let $\mathcal{H}(\mathbb{R}^n)$ denote the set of non-empty compact subsets of $\mathbb{R}^n$. The *Hausdorff metric* on $\mathcal{H}(\mathbb{R}^n)$ is defined by

$$h(A, B) = \inf\{r : d(a, B) \leq r \text{ for all } a \in A \text{ and } d(b, A) \leq r \text{ for all } b \in B\}.$$

Prove that with the topology defined by this metric, $(\mathcal{H}(\mathbb{R}^n), \cup)$ and $(\mathcal{H}(\mathbb{R}^n), +)$ are topological semigroups, where $A + B = \{a + b : a \in A \text{ and } b \in B\}$.

2.5 Ellis' Theorem

We now set out to show, in Corollary 2.39, that if S is a locally compact semitopological group, then S is in fact a topological group. While this result is of fundamental importance to the theory of semitopological semigroups, it will not be used again in this book until Chapter 21. Consequently, this section may be viewed as "optional".

In a metric space (X, d), given $x \in X$ and $\epsilon > 0$ we write $N(x, \epsilon) = \{y \in X : d(x, y) < \epsilon\}$.

Lemma 2.34. *Let* X *be a compact metric space and let* $g : X \times X \to \mathbb{R}$ *be a separately continuous function. There is a dense* G_δ *subset* D *of* X *such that* g *is jointly continuous at each point of* $D \times X$.

Proof. Let d be the metric of X. For every $\epsilon > 0$ and every $\delta > 0$, let

$$E_{\delta,\epsilon} = \{x \in X : \text{whenever } y, y' \in X \text{ and } d(y, y') < \delta \text{ one has } |g(x, y) - g(x, y')| \leq \epsilon\}.$$

We first observe that each $E_{\delta,\epsilon}$ is closed.

We next observe that for each $\epsilon > 0$, $X = \bigcup_{n=1}^{\infty} E_{1/n,\epsilon}$. Indeed, for each x the function $y \mapsto g(x, y)$ is uniformly continuous, so for some n, $x \in E_{1/n,\epsilon}$.

Now let U be a nonempty open subset of X. For each $n \in \mathbb{N}$, if $U \cap \text{int}_X(E_{1/n,\epsilon}) = \emptyset$, then $U \cap E_{1/n,\epsilon}$ is nowhere dense. Thus, by the Baire Category Theorem, there exists $n \in \mathbb{N}$ such that $U \cap \text{int}_X(E_{1/n,\epsilon}) \neq \emptyset$. Let $H_\epsilon = \bigcup_{n=1}^{\infty} \text{int}_X(E_{1/n,\epsilon}) = \bigcup_{\delta>0} \text{int}_X(E_{\delta,\epsilon})$. Then H_ϵ is a dense open subset of X.

Let $D = \bigcap_{\epsilon>0} H_\epsilon = \bigcap_{m=1}^{\infty} H_{1/m}$. By the Baire Category Theorem, D is dense in X. To see that g is continuous at each point of $D \times X$, let $(x, y) \in D \times X$ and let $\epsilon > 0$. Since $x \in D \subseteq H_{\epsilon/2}$, pick $\delta > 0$ such that $x \in \text{int}_X(E_{\delta,\epsilon/2})$. Also pick a neighborhood U of x such that $|g(x, y) - g(z, y)| < \epsilon/2$ for all $z \in U$. Let $(z, w) \in (\text{int}_X(E_{\delta,\epsilon/2}) \cap U) \times N(y, \delta)$. Since $z \in E_{\delta,\epsilon/2}$ and $d(y, w) < \delta$, one has $|g(z, y) - g(z, w)| \leq \frac{\epsilon}{2}$. Since $z \in U$, one has $|g(z, y) - g(z, y)| < \frac{\epsilon}{2}$. Thus $|g(x, y) - g(z, w)| < \epsilon$. $\qquad \Box$

Lemma 2.35. *Let $(X, \cdot)$ be a compact metrizable semitopological semigroup with identity e and let x be an invertible element of X. Then $\cdot$ is jointly continuous at every point of $(\{x\} \times X) \cup (X \times \{x\})$.*

Proof. By symmetry it suffices to show that $\cdot$ is jointly continuous at every point of $\{x\} \times X$. To this end, let $y \in X$ and let $\langle x_n \rangle_{n=1}^{\infty}$ and $\langle y_n \rangle_{n=1}^{\infty}$ be sequences in X converging to x and y respectively. We shall show that $\langle x_n y_n \rangle_{n=1}^{\infty}$ converges to xy.

To see this, let z be any limit point of the sequence $\langle x_n y_n \rangle_{n=1}^{\infty}$ and choose a subsequence $\langle x_{n_r} y_{n_r} \rangle_{r=1}^{\infty}$ of $\langle x_n y_n \rangle_{n=1}^{\infty}$ which converges to z.

Suppose that $z \neq xy$. Then $x^{-1}z \neq y$ so pick a continuous function $\phi : X \to \mathbb{R}$ such that $\phi(x^{-1}z) = 0$ and $\phi(y) = 1$. Define $g : X \times X \to \mathbb{R}$ by $g(u, v) = \phi(uv)$ and note that g is separately continuous. Pick by Lemma 2.34 a dense subset D of X such that g is jointly continuous at each point of $D \times X$. Choose a sequence $\langle d_n \rangle_{n=1}^{\infty}$ in D converging to e.

For each $m \in \mathbb{N}$, $\lim_{r\to\infty} d_m x^{-1} x_{n_r} = d_m$ so, since g is jointly continuous at (d_m, y),

$$\lim_{r\to\infty} \phi(d_m x^{-1} x_{n_r} y_{n_r}) = \lim_{r\to\infty} g(d_m x^{-1} x_{n_r}, y_{n_r}) = g(d_m, y) = \phi(d_m y).$$

Also, for each m, $\lim_{r\to\infty} \phi(d_m x^{-1} x_{n_r} y_{n_r}) = \phi(d_m x^{-1} z)$. Thus

$$\phi(y) = \lim_{m\to\infty} \phi(d_m y) = \lim_{m\to\infty} \lim_{r\to\infty} \phi(d_m x^{-1} x_{n_r} y_{n_r}) = \lim_{m\to\infty} \phi(d_m x^{-1} z) = \phi(x^{-1}z),$$

a contradiction. $\qquad \Box$

We shall need the following simple lemma.

Lemma 2.36. *Let X be a semitopological semigroup, let C be a dense subset of X, and let ϕ be a continuous function from X to a topological space Z. Let $x, y \in X$. If $\phi(uxv) = \phi(uyv)$ for every $u, v \in C$, then $\phi(uxv) = \phi(uyv)$ for every $u, v \in X$.*

Proof. This is Exercise 2.5.1.   □

Lemma 2.37. *Let X be a compact semitopological semigroup and let ϕ be a continuous real valued function defined on X. We define an equivalence relation on X by stating that $x \equiv y$ if $\phi(uxv) = \phi(uyv)$ for every $u, v \in X$. Let Y denote the quotient space $X/\equiv$ and let $\pi : X \to Y$ denote the canonical projection. Then Y can be given a semigroup structure for which Y is also a compact semitopological semigroup and π is a homomorphism. Furthermore, if X is separable, Y is metrizable.*

Proof. First notice that Y is compact since it is the continuous image of a compact space. To see that Y is Hausdorff, let $x, y \in X$ and assume that $\pi(x) \neq \pi(y)$. Pick $u, v \in X$ such that $a = \phi(uxv) \neq \phi(uyv) = b$ and let $\epsilon = |a - b|$. Let $U = (\phi \circ \lambda_u \circ \rho_v)^{-1}[N(a, \frac{\epsilon}{2})]$ and let $V = (\phi \circ \lambda_u \circ \rho_v)^{-1}[N(b, \frac{\epsilon}{2})]$. Then $U = \pi^{-1}[\pi[U]]$ and $V = \pi^{-1}[\pi[V]]$ so $\pi[U]$ and $\pi[V]$ are disjoint neighborhoods of $\pi(x)$ and $\pi(y)$.

Define an operation on Y by $\pi(x)\pi(y) = \pi(xy)$. To see that the operation is well defined, assume that $x \equiv x'$ and $y \equiv y'$ and let $u, v \in X$. Then $\phi(uxyv) = \phi(ux'yv) = \phi(ux'y'v)$. The operation is clearly associative. It follows from Exercise 2.2.2 and its dual that Y is a semitopological semigroup.

Now suppose that X has a countable dense subset C. Let $\psi : X \to {}^{C \times C}\mathbb{R}$ be defined by $\psi(x)(u, v) = \phi(uxv)$. Then ψ is continuous. Define $\widehat{\psi} : Y \to {}^{C \times C}\mathbb{R}$ by $\widehat{\psi}(\pi(x)) = \psi(x)$. Then $\widehat{\psi}$ is clearly well defined. To see that $\widehat{\psi}$ is continuous, let U be an open subset of ${}^{C \times C}\mathbb{R}$ and let $V = \widehat{\psi}^{-1}[U]$. Then $\pi^{-1}[V] = \psi^{-1}[U]$ so V is open.

We claim that $\widehat{\psi}$ is one-to-one. To see this, suppose that $\widehat{\psi}(\pi(x)) = \widehat{\psi}(\pi(y))$. Then $\phi(uxv) = \phi(uyv)$ for every $u, v \in C$. It follows that $\phi(uxv) = \phi(uyv)$ for every $u, v \in X$ (by Lemma 2.36) and hence that $\pi(x) = \pi(y)$.

We can now conclude that Y is homeomorphic to $\widehat{\psi}[Y]$. This is metrizable, as it is a subspace of ${}^{C \times C}\mathbb{R}$, the product of a countable number of copies of $\mathbb{R}$.   □

Theorem 2.38. *Let $(S, \cdot)$ be a compact semitopological semigroup with identity e and let x be an invertible element of S. Then $\cdot$ is jointly continuous at every point of $(\{x\} \times S) \cup (S \times \{x\})$.*

Proof. Let $y \in S$. To show that $\cdot$ is jointly continuous at (x, y), it is sufficient to show that the mapping $(w, z) \mapsto \gamma(wz)$ is jointly continuous at (x, y) for every continuous real valued function γ.

Suppose, on the contrary, that there exists a continuous real valued function γ for which this mapping is not jointly continuous at (x, y). Then there exists $\delta > 0$ such that every neighborhood of (x, y) in $S \times S$ contains a point (w, z) for which $|\gamma(wz) - \gamma(xy)| > \delta$.

We shall inductively choose sequences $\langle x_n \rangle_{n=1}^{\infty}$ and $\langle y_n \rangle_{n=1}^{\infty}$ in S. We first choose x_1 and y_1 arbitrarily. We then assume that $m \in \mathbb{N}$ and that x_i and y_i have been chosen for every $i \in \{1, 2, \ldots, m\}$. Put

$$C_m = \left\{ \prod_{i=1}^{m} w_i : \text{for all } i, \ w_i \in \{e, x, x^{-1}, y\} \cup \{x_1, x_2, \ldots, x_m\} \cup \{y_1, y_2, \ldots, y_m\} \right\}.$$

We note that C_m is finite and that for each $u, v \in C_m$, $\gamma \circ \lambda_u \circ \rho_v$ is continuous. We can therefore choose x_{m+1} and y_{m+1} satisfying $|\gamma(ux_{m+1}v) - \gamma(uxv)| < \frac{1}{m+1}$ and $|\gamma(uy_{m+1}v) - \gamma(uyv)| < \frac{1}{m+1}$ for every $u, v \in C_m$, while $|\gamma(x_{m+1}y_{m+1}) - \gamma(xy)| > \delta$.

Let $C = \bigcup_{m=1}^{\infty} C_m$. Then C is a subsemigroup of S, because, if $u \in C_k$ and $v \in C_m$, then $uv \in C_{k+m}$. Let $X = c\ell\, C$. Then X is a subsemigroup of S by Theorem 2.27. Let $\phi = \gamma_{|X}$ and let Y be the quotient of X as described in Lemma 2.37. Then Y is a compact metrizable semitopological semigroup with identity $\pi(e)$, and $\pi(x)$ has the inverse $\pi(x^{-1})$ in Y. So, by Lemma 2.35, $\cdot$ is jointly continuous at $(\pi(x), \pi(y))$.

We claim that $\pi(x_n) \to \pi(x)$. Suppose instead that there is some $z \in X$ such that $\pi(z) \neq \pi(x)$ and $\pi(z)$ is an accumulation point of the sequence $\langle \pi(x_n) \rangle_{n=1}^{\infty}$. Choose by Lemma 2.36 some $u, v \in C$ such that $\phi(uzv) \neq \phi(uxv)$ and pick $k \in \mathbb{N}$ such that $|\phi(uzv) - \phi(uxv)| > \frac{1}{k}$. Pick $m \geq 2k$ such that $u, v \in C_m$. Let $V = \{w \in X : |\phi(uwv) - \phi(uzv)| < \frac{1}{2k}\}$. Then $V = \pi^{-1}[\pi[V]]$ so $\pi[V]$ is a neighborhood of $\pi(z)$ so choose $n > m$ such that $|\phi(ux_nv) - \phi(uzv)| < \frac{1}{2k}$. Since also $|\phi(ux_nv) - \phi(uxv)| < \frac{1}{n}$, we have that $|\phi(uzv) - \phi(uxv)| < \frac{1}{k}$, a contradiction.

Similarly, $\pi(y_n) \to \pi(y)$.

So $\pi(x_n)\pi(y_n) \to \pi(xy)$ in Y. Let $W = \{w \in X : |\phi(w) - \phi(xy)| < \delta\}$. Then $\pi[W]$ is a neighborhood of $\pi(xy)$ in Y, so pick n such that $\pi(x_ny_n) \in \pi[W]$. Then $|\gamma(x_ny_n) - \gamma(xy)| < \delta$, a contradiction.

This establishes that $\cdot$ is jointly continuous at every point of $\{x\} \times S$. By symmetry, it is also jointly continuous at every point of $S \times \{x\}$. $\square$

We can now prove Ellis' Theorem.

Corollary 2.39 (Ellis' Theorem). *Let S be a locally compact semitopological semigroup which is algebraically a group. Then S is a topological group.*

Proof. If S is compact, let $\widetilde{S} = S$. Otherwise, let $\widetilde{S} = S \cup \{\infty\}$ denote the one-point compactification of S. We extend the semigroup operation of S to $\widetilde{S}$ by putting $s \cdot \infty = \infty \cdot s = \infty$ for every $s \in \widetilde{S}$. It is then simple to check that $\widetilde{S}$ is a compact semitopological semigroup with an identity. So, by Theorem 2.38, the semigroup operation on $\widetilde{S}$ is jointly continuous at every point of $(S \times \widetilde{S}) \cup (\widetilde{S} \times S)$. In particular, S is a topological semigroup.

To see that the inverse function is continuous on S, let $x \in S$ and let $\langle x_\alpha \rangle_{\alpha \in I}$ be a net in S converging to x. We claim that $\langle x_\alpha^{-1} \rangle_{\alpha \in I}$ converges to x^{-1}. Let z be any cluster point of $\langle x_\alpha^{-1} \rangle_{\alpha \in I}$ in $\widetilde{S}$ and choose a subnet $\langle x_\delta^{-1} \rangle_{\delta \in J}$ which converges to z. Then by the continuity of $\cdot$ at (x, z), $\langle x_\delta x_\delta^{-1} \rangle_{\delta \in J}$ converges to xz. Therefore $z = x^{-1}$. $\square$

Corollary 2.40. *Let S be a compact semitopological semigroup with dense center. Then $K(S)$ is a compact topological group.*

Proof. By Corollary 2.26, $K(S)$ is a semitopological group. Thus $K(S)$ has a unique idempotent e and so $K(S) = Se$ by Theorem 2.8. Thus $K(S)$ is compact and so by Corollary 2.39, $K(S)$ is a topological group. $\square$

Exercise 2.5.1. Prove Lemma 2.36.

Notes

Theorem 2.5 was proved for topological semigroups by K. Numakura [187] and A. Wallace [240, 241, 242]. The first proof using only one sided continuity seems to be that of R. Ellis [86, Corollary 2.10].

Theorem 2.9 is due to W. Ruppert [215].

Example 2.13 is from [38], a result of collaboration with J. Berglund.

Example 2.16 is due to W. Ruppert [215], as is Theorem 2.19(a) [218].

Lemma 2.34 is a special case of a theorem of J. Christensen [63]. As we have already remarked, Corollary 2.39 is a result of R. Ellis [84]. The proof given here involves essentially the same ideas as that given by W. Ruppert in [216]. (An alternate proof of comparable length to that given here is presented by J. Auslander in [5, pp. 57–63].) Theorem 2.38 was proved by J. Lawson [173]. For additional information see [216] and [174].

Chapter 3

βD — Ultrafilters and the Stone–Čech Compactification of a Discrete Space

There are many different constructions of the Stone–Čech compactification of a topological space X. In the case in which the space is discrete, the Stone–Čech compactification of X may be viewed as the set of ultrafilters on S and it is this approach that we adopt.

3.1 Ultrafilters

Definition 3.1. Let D be any set. A *filter* on D is a non-empty set $\mathcal{U}$ of subsets of D with the following properties:

(a) If A, $B \in \mathcal{U}$, then $A \cap B \in \mathcal{U}$.
(b) If $A \in \mathcal{U}$ and $A \subseteq B \subseteq D$, then $B \in \mathcal{U}$.
(c) $\emptyset \notin \mathcal{U}$.

A classic example of a filter is the set of neighborhoods of a point in a topological space. (We remark that a *neighborhood* of a point is a set containing an open set containing that point. That is neighborhoods do not, in our view, have to be open.) Another example is the set of subsets of any infinite set whose complements are finite.

We observe that, if $\mathcal{U}$ is any filter on D, then $D \in \mathcal{U}$.

Definition 3.2. Let D be a set and let $\mathcal{U}$ be a filter on D. A family $\mathcal{A}$ is a *filter base* for $\mathcal{U}$ if and only if $\mathcal{A} \subseteq \mathcal{U}$ and for each $B \in \mathcal{U}$ there is some $A \in \mathcal{A}$ such that $A \subseteq B$.

Thus, in a topological space, the open neighborhoods of a point form a filter base for the filter of neighborhoods of that point.

Definition 3.3. An *ultrafilter* on D is a filter on D which is not properly contained in any other filter on D.

We record immediately the following very simple but also very useful fact about ultrafilters.

Remark 3.4. *Let D be a set and let $\mathcal{U}$ and $\mathcal{V}$ be ultrafilters on D. Then $\mathcal{U} = \mathcal{V}$ if and only if $\mathcal{U} \subseteq \mathcal{V}$.*

Those more familiar with measures may find it helpful to view an ultrafilter on D as a $\{0, 1\}$-valued finitely additive measure on $\mathcal{P}(D)$. (The members of the ultrafilter are the "big" sets. See Exercise 3.1.2.)

Lemma 3.5. *Let $\mathcal{U}$ be a filter on the set D and let $A \subseteq D$. Either*

(a) *there is some $B \in \mathcal{U}$ such that $A \cap B = \emptyset$ or*

(b) *$\{C \subseteq D :$ there is some $B \in \mathcal{U}$ with $A \cap B \subseteq C\}$ is a filter on D.*

Proof. This is Exercise 3.1.3. □

Recall that a set $\mathcal{A}$ of sets has the *finite intersection property* if and only if whenever $\mathcal{F}$ is a finite nonempty subset of $\mathcal{A}$, $\bigcap \mathcal{F} \neq \emptyset$.

Theorem 3.6. *Let D be a set and let $\mathcal{U} \subseteq \mathcal{P}(D)$. The following statements are equivalent.*

(a) *$\mathcal{U}$ is an ultrafilter on D.*

(b) *$\mathcal{U}$ has the finite intersection property and for each $A \in \mathcal{P}(D)\backslash\mathcal{U}$ there is some $B \in \mathcal{U}$ such that $A \cap B = \emptyset$.*

(c) *$\mathcal{U}$ is maximal with respect to the finite intersection property. (That is, $\mathcal{U}$ is a maximal member of $\{\mathcal{V} \subseteq \mathcal{P}(D) : \mathcal{V}$ has the finite intersection property$\}$.)*

(d) *$\mathcal{U}$ is a filter on D and for all $\mathcal{F} \in \mathcal{P}_f\big(\mathcal{P}(D)\big)$, if $\bigcup \mathcal{F} \in \mathcal{U}$, then $\mathcal{F} \cap \mathcal{U} \neq \emptyset$.*

(e) *$\mathcal{U}$ is a filter on D and for all $A \subseteq D$ either $A \in \mathcal{U}$ or $D\backslash A \in \mathcal{U}$.*

Proof. (a) implies (b). By conditions (a) and (c) of Definition 3.1, $\mathcal{U}$ has the finite intersection property. Let $A \in \mathcal{P}(D)\backslash\mathcal{U}$ and let $\mathcal{V} = \{C \subseteq D :$ there is some $B \in \mathcal{U}$ with $A \cap B \subseteq C\}$. Then $A \in \mathcal{V}$ so $\mathcal{U} \not\subseteq \mathcal{V}$ so $\mathcal{V}$ is not a filter on D. Thus by Lemma 3.5, there is some $B \in \mathcal{U}$ such that $A \cap B = \emptyset$.

(b) implies (c). If $\mathcal{U} \not\subseteq \mathcal{V} \subseteq \mathcal{P}(D)$, pick $A \in \mathcal{V}\backslash\mathcal{U}$ and pick $B \in \mathcal{U}$ such that $A \cap B = \emptyset$. Then $A, B \in \mathcal{V}$ so $\mathcal{V}$ does not have the finite intersection property.

(c) implies (d). Assume that $\mathcal{U}$ is maximal with respect to the finite intersection property among subsets of $\mathcal{P}(D)$. Then one has immediately that $\mathcal{U}$ is a nonempty set of subsets of D. Since $\mathcal{U} \cup \{D\}$ has the finite intersection property and $\mathcal{U} \subseteq \mathcal{U} \cup \{D\}$, one has $\mathcal{U} = \mathcal{U} \cup \{D\}$. That is, $D \in \mathcal{U}$. Given $A, B \in \mathcal{U}$, $\mathcal{U} \cup \{A \cap B\}$ has the finite intersection property so $A \cap B \in \mathcal{U}$. Given A and B with $A \in \mathcal{U}$ and $A \subseteq B \subseteq D$, $\mathcal{U} \cup \{B\}$ has the finite intersection property so $B \in \mathcal{U}$. Thus $\mathcal{U}$ is a filter.

Now let $\mathcal{F} \in \mathcal{P}_f\big(\mathcal{P}(D)\big)$ with $\bigcup \mathcal{F} \in \mathcal{U}$, and suppose that for each $A \in \mathcal{F}, A \notin \mathcal{U}$. Then given $A \in \mathcal{F}$, $\mathcal{U} \not\subseteq \mathcal{U} \cup \{A\}$ so $\mathcal{U} \cup \{A\}$ does not have the finite intersection property so there exists $\mathcal{G}_A \in \mathcal{P}_f(\mathcal{U})$ such that $A \cap \bigcap \mathcal{G}_A = \emptyset$. Let $\mathcal{H} = \bigcup_{A \in \mathcal{F}} \mathcal{G}_A$. Then $\mathcal{H} \cup \{\bigcup \mathcal{F}\} \subseteq \mathcal{U}$ while $(\bigcup \mathcal{F}) \cap \bigcap \mathcal{H} = \emptyset$, a contradiction.

(d) implies (e). Let $\mathcal{F} = \{A, D\backslash A\}$.

(e) implies (a). Assume that $\mathcal{U}$ is a filter on D and for all $A \subseteq D$ either $A \in \mathcal{U}$ or $D \backslash A \in \mathcal{U}$. Let $\mathcal{V}$ be a filter with $\mathcal{U} \subseteq \mathcal{V}$ and suppose that $\mathcal{U} \neq \mathcal{V}$. Pick $A \in \mathcal{V} \backslash \mathcal{U}$. Then $D \backslash A \in \mathcal{U} \subseteq \mathcal{V}$ while $A \cap (D \backslash A) = \emptyset$, a contradiction. $\square$

If $a \in D$, $\{A \in \mathcal{P}(D) : a \in A\}$ is easily seen to be an ultrafilter on D. This ultrafilter is called the *principal ultrafilter* defined by a.

Theorem 3.7. *Let D be a set and let $\mathcal{U}$ be an ultrafilter on D. The following statements are equivalent.*

(a) *$\mathcal{U}$ is a principal ultrafilter.*

(b) *There is some $F \in \mathcal{P}_f(D)$ such that $F \in \mathcal{U}$.*

(c) *The set $\{A \subseteq D : D \backslash A \text{ is finite}\}$ is not contained in $\mathcal{U}$.*

(d) *$\bigcap \mathcal{U} \neq \emptyset$.*

(e) *There is some $x \in D$ such that $\bigcap \mathcal{U} = \{x\}$.*

Proof. (a) implies (b). Pick $x \in D$ such that $\mathcal{U} = \{A \subseteq D : x \in A\}$. Let $F = \{x\}$.

(b) implies (c). Given $F \in \mathcal{P}_f(D) \cap \mathcal{U}$ one has $D \backslash F \notin \mathcal{U}$.

(c) implies (d). Pick $A \subseteq D$ such that $D \backslash A$ is finite and $A \notin \mathcal{U}$. Let $F = D \backslash A$. Then $F \in \mathcal{U}$ and $F = \bigcup \{\{x\} : x \in F\}$ so by Theorem 3.6, we may pick $x \in F$ such that $\{x\} \in \mathcal{U}$. Then for each $B \in \mathcal{U}$, $B \cap \{x\} \neq \emptyset$ so $x \in \bigcap \mathcal{U}$.

(d) implies (e). Assume that $\bigcap \mathcal{U} \neq \emptyset$ and pick $x \in \bigcap \mathcal{U}$. Then $D \backslash \{x\} \notin \mathcal{U}$ so $\{x\} \in \mathcal{U}$ so $\bigcap \mathcal{U} \subseteq \{x\}$.

(e) implies (a). Pick $x \in D$ such that $\bigcap \mathcal{U} = \{x\}$. Then $\mathcal{U}$ and $\{A \subseteq D : x \in A\}$ are both ultrafilters so by Remark 3.4 it suffices to note that $\mathcal{U} \subseteq \{A \subseteq D : x \in A\}$. $\square$

It is a fact that principal ultrafilters are the only ones whose members can be explicitly defined. There are no others which can be defined within Zermelo–Fraenkel set theory. (See the notes to this chapter.) However, the axiom of choice produces a rich set of nonprincipal ultrafilters on any infinite set.

Theorem 3.8. *Let D be a set and let $\mathcal{A}$ be a subset of $\mathcal{P}(D)$ which has the finite intersection property. Then there is an ultrafilter $\mathcal{U}$ on D such that $\mathcal{A} \subseteq \mathcal{U}$.*

Proof. Let $\Gamma = \{\mathcal{B} \subseteq \mathcal{P}(D) : \mathcal{A} \subseteq \mathcal{B}$ and $\mathcal{B}$ has the finite intersection property$\}$. Then $\mathcal{A} \in \Gamma$ so $\Gamma \neq \emptyset$. Given a chain $\mathcal{C}$ in Γ one has immediately that $\mathcal{A} \subseteq \bigcup \mathcal{C}$. Given $\mathcal{F} \in \mathcal{P}_f(\bigcup \mathcal{C})$ there is some $\mathcal{B} \in \mathcal{C}$ with $\mathcal{F} \subseteq \mathcal{B}$ so $\bigcap \mathcal{F} \neq \emptyset$. Thus by Zorn's Lemma we may pick a maximal member $\mathcal{U}$ of Γ. Trivially $\mathcal{U}$ is not only maximal in Γ, but in fact $\mathcal{U}$ is maximal with respect to the finite intersection property. By Theorem 3.6, $\mathcal{U}$ is an ultrafilter on D. $\square$

Corollary 3.9. *Let D be a set, let $\mathcal{A}$ be a filter on D, and let $A \subseteq D$. Then $A \notin \mathcal{A}$ if and only if there is some ultrafilter $\mathcal{U}$ with $\mathcal{A} \cup \{D \backslash A\} \subseteq \mathcal{U}$.*

Proof. Since one cannot have a filter $\mathcal{U}$ with $A \in \mathcal{U}$ and $D \backslash A \in \mathcal{U}$, the sufficiency is trivial.

For the necessity it suffices by Theorem 3.8 to show that $\mathcal{A} \cup \{D\backslash A\}$ has the finite intersection property. So suppose instead that there is some finite nonempty $\mathcal{F} \subseteq \mathcal{A}$ such that $(D\backslash A) \cap \bigcap \mathcal{F} = \emptyset$. Then $\bigcap \mathcal{F} \subseteq A$ so $A \in \mathcal{A}$. $\qquad\square$

The following concept will be important in the combinatorial applications of ultra-filters.

Definition 3.10. Let $\mathcal{R}$ be a nonempty set of sets. We say that $\mathcal{R}$ is *partition regular* if and only if whenever $\mathcal{F}$ is a finite set of sets and $\bigcup \mathcal{F} \in \mathcal{R}$, there exist $A \in \mathcal{F}$ and $B \in \mathcal{R}$ such that $B \subseteq A$.

Given a property Ψ of subsets of some set D, (i.e. a statement about these sets), we say the property is *partition regular* provided $\{A \subseteq D : \Psi(A)\}$ is partition regular.

Notice that we do not require that a partition regular family be closed under supersets (in some set D). (The reason is that there are combinatorial families, such as the sets of finite products from infinite sequences, that are partition regular under our requirement (by Corollary 5.15) but not under the stronger requirement.) See however Exercise 3.1.6

Theorem 3.11. *Let D be a set and let $\mathcal{R} \subseteq \mathcal{P}(D)$ be nonempty and assume that $\emptyset \notin \mathcal{R}$. Let $\mathcal{R}^\uparrow = \{B \in \mathcal{P}(D) : A \subseteq B \text{ for some } A \in \mathcal{R}\}$. The following statements are equivalent.*

(a) *$\mathcal{R}$ is partition regular.*

(b) *Whenever $\mathcal{A} \subseteq \mathcal{P}(D)$ has the property that every finite nonempty subfamily of $\mathcal{A}$ has an intersection which is in $\mathcal{R}^\uparrow$, there is an ultrafilter $\mathcal{U}$ on D such that $\mathcal{A} \subseteq \mathcal{U} \subseteq \mathcal{R}^\uparrow$.*

(c) *Whenever $A \in \mathcal{R}$, there is some ultrafilter $\mathcal{U}$ on D such that $A \in \mathcal{U} \subseteq \mathcal{R}^\uparrow$.*

Proof. (a) implies (b). Let $\mathcal{B} = \{A \subseteq D : \text{for all } B \in \mathcal{R}, A \cap B \neq \emptyset\}$ and note that $\mathcal{B} \neq \emptyset$ since $D \in \mathcal{B}$. Note also that we may assume that $\mathcal{A} \neq \emptyset$, since $\{D\}$ has the hypothesized property. Let $\mathcal{C} = \mathcal{A} \cup \mathcal{B}$. We claim that $\mathcal{C}$ has the finite intersection property. To see this it suffices (since $\mathcal{A}$ and $\mathcal{B}$ are nonempty) to let $\mathcal{F} \in \mathcal{P}_f(\mathcal{A})$ and $\mathcal{G} \in \mathcal{P}_f(\mathcal{B})$ and show that $\bigcap \mathcal{F} \cap \bigcap \mathcal{G} \neq \emptyset$. So suppose instead that we have such $\mathcal{F}$ and $\mathcal{G}$ with $\bigcap \mathcal{F} \cap \bigcap \mathcal{G} = \emptyset$. Pick $B \in \mathcal{R}$ such that $B \subseteq \bigcap \mathcal{F}$. Then $B \cap \bigcap \mathcal{G} = \emptyset$ and so $B = \bigcup_{A \in \mathcal{G}}(B\backslash A)$. Pick $A \in \mathcal{G}$ and $C \in \mathcal{R}$ such that $C \subseteq B\backslash A$. Then $A \cap C = \emptyset$, contradicting the fact that $A \in \mathcal{B}$.

By Theorem 3.8 there is an ultrafilter $\mathcal{U}$ on D such that $\mathcal{C} \subseteq \mathcal{U}$. Given $C \in \mathcal{U}$, $D\backslash C \notin \mathcal{B}$ (since $C \cap (D\backslash C) = \emptyset \notin \mathcal{U}$). So pick some $B \in \mathcal{R}$ such that $B \cap (D\backslash C) = \emptyset$. That is, $B \subseteq C$.

(b) implies (c). Let $\mathcal{A} = \{A\}$.

(c) implies (a). Let $\mathcal{F}$ be a finite set of sets with $\bigcup \mathcal{F} \in \mathcal{R}$ and let $\mathcal{U}$ be an ultrafilter on D such that $\bigcup \mathcal{F} \in \mathcal{U}$ and for each $C \in \mathcal{U}$ there is some $B \in \mathcal{R}$ such that $B \subseteq C$. Pick by Theorem 3.6 some $A \in \mathcal{F} \cap \mathcal{U}$. $\qquad\square$

Definition 3.12. Let D be a set and let $\mathcal{U}$ be an ultrafilter on D. The *norm* of $\mathcal{U}$ is $\|\mathcal{U}\| = \min\{|A| : A \in \mathcal{U}\}$.

Note that, by Theorem 3.7, if $\mathcal{U}$ is an ultrafilter, then $\|\mathcal{U}\|$ is either 1 or infinite.

Definition 3.13. Let D be a set and let κ be an infinite cardinal. A κ-*uniform ultrafilter on D* is an ultrafilter $\mathcal{U}$ on D such that $\|\mathcal{U}\| \geq \kappa$. The set $U_\kappa(D) = \{\mathcal{U} : \mathcal{U} \text{ is a } \kappa\text{-uniform ultrafilter on } D\}$. A *uniform ultrafilter on D* is a κ-uniform ultrafilter on D, where $\kappa = |D|$.

Corollary 3.14. *Let D be any set and let A be a family of subsets of D. If the intersection of every finite subfamily of A is infinite, then A is contained in an ultrafilter on D all of whose members are infinite. More generally, if κ is an infinite cardinal and if the intersection of every finite subfamily of A has cardinality at least κ, then A is contained in a κ-uniform ultrafilter on D.*

Proof. The property of being infinite is partition regular, and so is the property of having cardinality at least κ.   □

If the intersection of every finite subfamily of $\mathcal{A}$ is infinite, $\mathcal{A}$ is said to have the *infinite finite intersection property*. Thus, Corollary 3.14 says that if $\mathcal{A}$ has the infinite finite intersection property, then there is a nonprincipal ultrafilter $\mathcal{U}$ on D with $\mathcal{A} \subseteq \mathcal{U}$.

Exercise 3.1.1. Let $\mathcal{A}$ be a set of sets. Prove that there is some filter $\mathcal{U}$ on $D = \bigcup \mathcal{A}$ such that $\mathcal{A}$ is a filter base for $\mathcal{U}$ if and only if $\mathcal{A} \neq \emptyset$, $\emptyset \notin \mathcal{A}$, and for every finite nonempty subset $\mathcal{F}$ of $\mathcal{A}$, there is some $A \in \mathcal{A}$ such that $A \subseteq \bigcap \mathcal{F}$.

Exercise 3.1.2. Let D be any set. Show that the ultrafilters on D are in one-to-one correspondence with the finitely additive measures defined on $\mathcal{P}(D)$ which take values in $\{0, 1\}$ and are not identically zero.

Exercise 3.1.3. Prove Lemma 3.5.

Exercise 3.1.4. Let D be any set. Show that the ultrafilters on D are in one-to-one correspondence with the Boolean algebra homomorphisms mapping $(\mathcal{P}(D), \cup, \cap)$ onto the Boolean algebra $(\{0, 1\}, \vee, \wedge)$.

Exercise 3.1.5. Let D be a set with cardinality $\mathfrak{c}$. Show that there is no nonprincipal ultrafilter $\mathcal{U}$ on D with the property that every countable subfamily of $\mathcal{U}$ has an intersection which belongs to $\mathcal{U}$. (Hint: D can be taken to be the interval $[0, 1]$ in $\mathbb{R}$. If an ultrafilter $\mathcal{U}$ with these properties did exist, every $a \in [0, 1]$ would have a neighborhood which did not belong to $\mathcal{U}$.)

Exercise 3.1.6. Let $\mathcal{R}$ be a set of sets and let D be a set such that $\bigcup \mathcal{R} \subseteq D$. Let $\mathcal{R}^\uparrow = \{B \subseteq D : \text{there is some } A \in \mathcal{R} \text{ with } A \subseteq B\}$. Prove that the following statements are equivalent.

(a) $\mathcal{R}$ is partition regular.
(b) If $\mathcal{F}$ is a finite set of sets and there is some $C \in \mathcal{R}$ with $C \subseteq \bigcup \mathcal{F}$, then there exist $A \in \mathcal{F}$ and $B \in \mathcal{R}$ with $B \subseteq A$.
(c) $\mathcal{R}^\uparrow$ is partition regular.
(d) If $\mathcal{F}$ is a finite set of sets and $\bigcup \mathcal{F} \in \mathcal{R}^\uparrow$, then $\mathcal{F} \cap \mathcal{R}^\uparrow \neq \emptyset$.

3.2 The Topological Space βD

In this section we define a topology on the set of all ultrafilters on a set D, and establish some of the properties of the resulting space.

Definition 3.15. Let D be a discrete topological space.
     (a) $\beta D = \{p : p \text{ is an ultrafilter on } D\}$.
     (b) Given $A \subseteq D$, $\widehat{A} = \{p \in \beta D : A \in p\}$.

The reason for the notation βD will become clear in the next section. We shall henceforward use lower case letters to denote ultrafilters on D, since we shall be thinking of ultrafilters as points in a topological space.

Definition 3.16. Let D be a set and let $a \in D$. Then $e(a) = \{A \subseteq D : a \in A\}$.

Thus for each $a \in D$, $e(a)$ is the principal ultrafilter corresponding to a.

Lemma 3.17. *Let D be a set and let $A, B \subseteq D$.*
     (a) $\widehat{(A \cap B)} = \widehat{A} \cap \widehat{B}$;
     (b) $\widehat{(A \cup B)} = \widehat{A} \cup \widehat{B}$;
     (c) $\widehat{(D \backslash A)} = \beta D \backslash \widehat{A}$;
     (d) $\widehat{A} = \emptyset$ if and only if $A = \emptyset$;
     (e) $\widehat{A} = \beta D$ if and only if $A = D$;
     (f) $\widehat{A} = \widehat{B}$ if and only if $A = B$.

Proof. This is Exercise 3.2.1. $\square$

We observe that the sets of the form $\widehat{A}$ are closed under finite intersections, because $\widehat{A} \cap \widehat{B} = \widehat{(A \cap B)}$. Consequently, $\{\widehat{A} : A \subseteq D\}$ forms a basis for a topology on βD. We define the topology of βD to be the topology which has these sets as a basis.

The following theorem describes some of the basic topological properties of βD.

Theorem 3.18. *Let D be any set.*
     (a) βD is a compact Hausdorff space.
     (b) The sets of the form $\widehat{A}$ are the clopen subsets of βD.
     (c) For every $A \subseteq D$, $\widehat{A} = c\ell_{\beta D}\, e[A]$.
     (d) For any $A \subseteq D$ and any $p \in D$, $p \in c\ell_{\beta D}\, e[A]$ if and only if $A \in p$.
     (e) The mapping e is injective and $e[D]$ is a dense subset of βD whose points are precisely the isolated points of βD.
     (f) If U is an open subset of βD, $c\ell_{\beta D}\, U$ is also open.

Proof. (a) Suppose that p and q are distinct elements of βD. If $A \in p \backslash q$, then $D \backslash A \in q$. So $\widehat{A}$ and $\widehat{(D \backslash A)}$ are disjoint open subsets of βD containing p and q respectively. Thus βD is Hausdorff.

We observe that the sets of the form $\widehat{A}$ are also a base for the closed sets, because $\beta D \backslash \widehat{A} = \widehat{(D \backslash A)}$. Thus, to show that βD is compact, we shall consider a family $\mathcal{A}$

of sets of the form $\widehat{A}$ with the finite intersection property and show that $\mathcal{A}$ has a non-empty intersection. Let $\mathcal{B} = \{A \subseteq D : \widehat{A} \in \mathcal{A}\}$. If $\mathcal{F} \in \mathcal{P}_f(\mathcal{B})$, then there is some $p \in \bigcap_{A \in \mathcal{F}} \widehat{A}$ and so $\bigcap \mathcal{F} \in p$ and thus $\bigcap \mathcal{F} \neq \emptyset$. That is, $\mathcal{B}$ has the finite intersection property, so by Theorem 3.8 pick $q \in \beta D$ with $\mathcal{B} \subseteq q$. Then $q \in \bigcap \mathcal{A}$.

(b) We pointed out in the proof of (a) that each set $\widehat{A}$ was closed as well as open. Suppose that C is any clopen subset of βD. Let $\mathcal{A} = \{\widehat{A} : A \subseteq D$ and $\widehat{A} \subseteq C\}$. Since C is open, $\mathcal{A}$ is an open cover of C. Since C is closed, it is compact by (a) so pick a finite subfamily $\mathcal{F}$ of $\mathcal{P}(D)$ such that $C = \bigcup_{A \in \mathcal{F}} \widehat{A}$. Then by Lemma 3.17(b), $C = \widehat{\bigcup \mathcal{F}}$.

(c) Clearly, for each $a \in A$, $e(a) \in \widehat{A}$ and therefore $c\ell_{\beta D} e[A] \subseteq \widehat{A}$. To prove the reverse inclusion, let $p \in \widehat{A}$. If $\widehat{B}$ denotes a basic neighborhood of p, then $A \in p$ and $B \in p$ and so $A \cap B \neq \emptyset$. Choose any $a \in A \cap B$. Since $e(a) \in e[A] \cap \widehat{B}$, $e[A] \cap \widehat{B} \neq \emptyset$ and thus $p \in c\ell_{\beta D} e[A]$.

(d) By (c) and the definition of $\widehat{A}$,

$$p \in c\ell_{\beta D} e[A] \Leftrightarrow p \in \widehat{A}$$
$$\Leftrightarrow A \in p.$$

(e) If $a, b \in D$ are distinct, $D \setminus \{a\} \in e(b) \setminus e(a)$ and so $e(a) \neq e(b)$.

If $\widehat{A}$ is a non-empty basic open subset of βD, then $A \neq \emptyset$. Any $a \in A$ satisfies $e(a) \in e[D] \cap \widehat{A}$ and so $e[D] \cap \widehat{A} \neq \emptyset$. Thus $e[D]$ is dense in βD.

For any $a \in D$, $e(a)$ is isolated in βD because $\widehat{\{a\}}$ is an open subset of βD whose only member is $e(a)$. Conversely if p is an isolated point of βD, then $\{p\} \cap e[D] \neq \emptyset$ and so $p \in e[D]$.

(f) If $U = \emptyset$, the conclusion is trivial and so we assume that $U \neq \emptyset$. Put $A = e^{-1}[U]$. We claim first that $U \subseteq c\ell_{\beta D} e[A]$. So let $p \in U$ and let $\widehat{B}$ be a basic neighborhood of p. Then $U \cap \widehat{B}$ is a nonempty open set and so by (e), $U \cap \widehat{B} \cap e[D] \neq \emptyset$. So pick $b \in B$ with $e(b) \in U$. Then $e(b) \in \widehat{B} \cap e[A]$ and so $\widehat{B} \cap e[A] \neq \emptyset$.

Also $e[A] \subseteq U$ and hence $U \subseteq c\ell_{\beta D} e[A] \subseteq c\ell_{\beta D} U$. Thus $c\ell_{\beta D} U = c\ell_{\beta D} e[A] = \widehat{A}$ (by (c)), and so $c\ell_{\beta D} U$ is open in βD. $\square$

We next establish a characterization of the closed subsets of βD which will be useful later.

Definition 3.19. Let D be a set and let $\mathcal{A}$ be a filter on D. Then $\widehat{\mathcal{A}} = \{p \in \beta D : \mathcal{A} \subseteq p\}$.

Theorem 3.20. *Let D be a set.*
(a) *If $\mathcal{A}$ is a filter on D, then $\widehat{\mathcal{A}}$ is a closed subset of βD.*
(b) *If $A \subseteq \beta D$ and $\mathcal{A} = \bigcap A$, then $\mathcal{A}$ is a filter on D and $\widehat{\mathcal{A}} = c\ell A$.*

Proof. (a) Let $p \in \beta D \setminus \widehat{\mathcal{A}}$. Pick $B \in \mathcal{A} \setminus p$. Then $\widehat{D \setminus B}$ is a neighborhood of p which misses $\widehat{\mathcal{A}}$.

(b) $\mathcal{A}$ is the intersection of a set of filters, so $\mathcal{A}$ is a filter. Further, for each $p \in A$, $\mathcal{A} \subseteq p$ so $A \subseteq \widehat{\mathcal{A}}$ and thus by (a), $c\ell A \subseteq \widehat{\mathcal{A}}$. To see that $\widehat{\mathcal{A}} \subseteq c\ell A$, let $p \in \widehat{\mathcal{A}}$ and let $B \in p$. Suppose $\widehat{B} \cap A = \emptyset$. Then for each $q \in A$, $D \setminus B \in q$ so $D \setminus B \in \mathcal{A} \subseteq p$, a contradiction. $\square$

It is customary to identify a principal ultrafilter $e(x)$ with the point x, and we shall adopt that practice ourselves after we have proved that βD is the Stone–Čech compactification of the discrete space D in Section 3.3. Once this is done, we can write the following definition as $A^* = \widehat{A} \backslash A$. For the moment, we shall continue to maintain the distinction between x and $e(x)$.

Definition 3.21. Let D be a set and let $A \subseteq D$. Then $A^* = \widehat{A} \backslash e[A]$.

The following theorem is simple but useful. Once $e(x)$ is identified with x the conclusion becomes "$U \cap D \in p$".

Theorem 3.22. *Let $p \in \beta D$ and let U be a subset of βD. If U is a neighborhood of p in βD, then $e^{-1}[U] \in p$.*

Proof. If U is a neighborhood of p, there is a basic open subset $\widehat{A}$ of βD for which $p \in \widehat{A} \subseteq U$. This implies that $A \in p$ and so $e^{-1}[U] \in p$, because $A \subseteq e^{-1}[U]$. □

Recall that a space is *zero dimensional* if and only if it has a basis of clopen sets.

Theorem 3.23. *Let X be a zero dimensional space and let Y be a compact subset of X. The clopen subsets of Y are the sets of the form $C \cap Y$ where C is clopen in X. In particular, if D is an infinite set, then the nonempty clopen subsets of D^* are the sets of the form A^* where A is an infinite subset of D.*

Proof. Trivially, if C is clopen in X, then $C \cap Y$ is clopen in Y. For the converse, let B be clopen in Y and let $\mathcal{A} = \{A \cap Y : A \text{ is clopen in } X \text{ and } A \cap Y \subseteq B\}$. Since X is zero dimensional, $\mathcal{A}$ is an open cover of B. Since Y is compact and hence B is compact, pick a finite set $\mathcal{F}$ of clopen subsets of X such that $B = \bigcup_{A \in \mathcal{F}} (A \cap Y)$ and let $C = \bigcup \mathcal{F}$.

The "in particular" conclusion follows from Corollary 3.14 and Theorem 3.18(b). □

Exercise 3.2.1. Prove Lemma 3.17.

3.3 Stone–Čech Compactification

In this section we show that βD is the Stone–Čech compactification of the discrete space D.

Recall that by an *embedding* of a topological space X into a topological space Z, one means a function $\varphi : X \to Z$ which defines a homeomorphism from X onto $\varphi[X]$.

We remind the reader that we are assuming that all *hypothesized* topological spaces are Hausdorff.

Definition 3.24. Let X be a topological space. A *compactification* of X is a pair (φ, C) such that C is a compact space, φ is an embedding of X into C, and $\varphi[X]$ is dense in C.

Any completely regular space X has a largest compactification called its Stone–Čech compactification.

Definition 3.25. Let X be a completely regular topological space. A *Stone–Čech compactification of X* is a pair (φ, Z) such that

(a) Z is a compact space,
(b) φ is an embedding of X into Z,
(c) $\varphi[X]$ is dense in Z, and
(d) given any compact space Y and any continuous function $f : X \to Y$ there exists a continuous function $g : Z \to Y$ such that $g \circ \varphi = f$. (That is the diagram

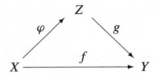

commutes.)

One customarily refers to *the* Stone–Čech compactification of a space X rather than *a* Stone–Čech compactification of X. The reason is made clear by the following remark. If one views φ as an inclusion map, then Remark 3.26 may be viewed as saying: "The Stone–Čech compactification of X is unique up to a homeomorphism leaving X pointwise fixed."

Remark 3.26. *Let X be a completely regular space and let (φ, Z) and (τ, W) be Stone–Čech compactifications of X. Then there is a homeomorphism $\gamma : Z \to W$ such that $\gamma \circ \varphi = \tau$.*

Theorem 3.27. *Let D be a discrete space. Then $(e, \beta D)$ is a Stone–Čech compactification of D.*

Proof. Conditions (a), (b), and (c) of Definition 3.25 hold by Theorem 3.18. It remains for us to verify condition (d).

Let Y be a compact space and let $f : D \to Y$. For each $p \in \beta D$ let $\mathcal{A}_p = \{c\ell_Y f[A] : A \in p\}$. Then for each $p \in \beta D$, $\mathcal{A}_p$ has the finite intersection property (Exercise 3.3.1) and so has a nonempty intersection. Choose $g(p) \in \bigcap \mathcal{A}_p$. Then we have the following diagram.

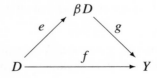

We need to show that the diagram commutes and that g is continuous.

For the first assertion, let $x \in D$. Then $\{x\} \in e(x)$ so $g\bigl(e(x)\bigr) \in c\ell_Y f[\{x\}] = c\ell_Y[\{f(x)\}] = \{f(x)\}$ so $g \circ e = f$ as required.

To see that g is continuous, let $p \in \beta D$ and let U be a neighborhood of $g(p)$ in Y. Since Y is regular, pick a neighborhood V of $g(p)$ with $c\ell_Y V \subseteq U$ and let $A = f^{-1}[V]$. We claim that $A \in p$ so suppose instead that $D \backslash A \in p$. Then $g(p) \in c\ell_Y f[D \backslash A]$ and V is a neighborhood of $g(p)$ so $V \cap f[D \backslash A] \neq \emptyset$, contradicting the fact that $A = f^{-1}[V]$. Thus $\widehat{A}$ is a neighborhood of p. We claim that $g[\widehat{A}] \subseteq U$, so let $q \in \widehat{A}$ and suppose that $g(q) \notin U$. Then $Y \backslash c\ell_Y V$ is a neighborhood of $g(q)$ and $g(q) \in c\ell_Y f[A]$ so $(Y \backslash c\ell_Y V) \cap f[A] \neq \emptyset$, again contradicting the fact that $A = f^{-1}[V]$. □

Although we have not used that fact, each of the sets $\bigcap \mathcal{A}_p$ is a singleton. (See Exercise 3.3.2.)

We have shown in Theorem 3.27 that βD is the Stone–Čech compactification of D. This explains the reason for using the notation βD for this space: if X is any completely regular space, βX is the standard notation for its Stone–Čech compactification. X is usually regarded as being a subspace of βX.

Identifying D with $e[D]$

It is common practice in dealing with βD to identify the points of D with the principal ultrafilters generated by those points, and we shall adopt this practice from this point on. Only rarely will it be necessary to remind the reader that when we write s we sometimes mean $e(s)$.

Once we have identified $s \in S$ with $e(s) \in \beta D$, we shall suppose that $D \subseteq \beta D$ and shall write $D^* = \beta D \backslash D$, rather than $D^* = \beta D \backslash e[D]$. Further, with this identification, $\widehat{A} = c\ell_{\beta D} A$ for every $A \in \mathcal{P}D$, by Theorem 3.18. So the notations $\widehat{A}$ and $\overline{A}$ become interchangeable.

We illustrate the conversion process by restating Theorem 3.27.

Theorem 3.28 (Stone–Čech Compactification — restated). *Let D be an infinite discrete space. Then*

(a) *βD is a compact space,*

(b) *$D \subseteq \beta D$,*

(c) *D is dense in βD, and*

(d) *given any compact space Y and any function $f : D \to Y$ there exists a continuous function $g : \beta D \to Y$ such that $g_{|D} = f$.*

Identifying βT with $\widehat{T}$ for $T \subseteq S$

In a similar fashion, if $T \subseteq S$, we shall identify $p \in \widehat{T}$ (which is an ultrafilter on S) with the ultrafilter $\{A \cap T : A \in p\}$ (which is an ultrafilter on T) and thus we shall pretend that $\beta T \subseteq \beta S$.

Exercise 3.3.1. Let D be a discrete space, let Y be a compact space, and let $f : D \to Y$. For each $p \in \beta D$ let $\mathcal{A}_p = \{cl_Y f[A] : A \in p\}$. Prove that for each $p \in \beta D$, $\mathcal{A}_p$ has the finite intersection property.

Exercise 3.3.2. Let the sets $\mathcal{A}_p$ be as defined in the proof of Theorem 3.27. Prove that for each $p \in \beta D$, $\bigcap \mathcal{A}_p$ is a singleton. (Hint: Consider the fact that two continuous functions agreeing on $e[D]$ must be equal.)

Exercise 3.3.3. Let D be any discrete space and let $A \subseteq D$. Prove that $cl_{\beta D} A$ can be identified with βA.

3.4 More Topology of βD

If f is a continuous mapping from a completely regular space X into a compact space Y, we shall often use $\tilde{f}$ to denote the continuous mapping from βX to Y which extends f, although in some cases we may use the same notation for a function and its extension. (Notice that there can be only one continuous extension, since any two extensions agree on a dense subspace.)

Definition 3.29. Let D be a discrete space, let Y be a compact space, and let $f : D \to Y$. Then $\tilde{f}$ is the continuous function from βD to Y such that $\tilde{f}_{|D} = f$.

If $f : X \to Y$ is a continuous function between completely regular spaces, it has a continuous extension $f^\beta : \beta X \to \beta Y$. The reader with an interest in category theory might like to know that this defines a functor from the category of completely regular spaces to that of compact Hausdorff spaces, and that this is an adjoint to the inclusion functor embedding the second category in the first.

Lemma 3.30. *Let D and E be discrete spaces and let $f : D \to E \subseteq \beta E$. For each $p \in \beta D$, $\tilde{f}(p) = \{A \subseteq E : f^{-1}[A] \in p\}$. In particular, if $A \in p$, then $f[A] \in \tilde{f}(p)$; and if $B \in \tilde{f}(p)$, then $f^{-1}[B] \in p$.*

Proof. It is routine to verify that $\{A \subseteq E : f^{-1}[A] \in p\}$ is an ultrafilter on E. For each $p \in \beta D$, let $g(p) = \{A \subseteq E : f^{-1}[A] \in p\}$. Now, given $x \in D$ we have $g(x) = \{A \subseteq E : f(x) \in A\}$. Recalling that we are identifying x with $e(x)$ (and $f(x)$ with $e(f(x))$) we have $g(x) = f(x)$. To see that g is continuous, let $\widehat{A}$ be a basic open set in βE. Then $g^{-1}[\widehat{A}] = \widehat{f^{-1}[A]}$. Since g is a continuous extension of f, we have $\tilde{f} = g$.   □

Lemma 3.31. *Let D and E be discrete spaces, let $f, g : D \to E$, and let $p \in \beta D$ satisfy $\tilde{f}(p) = \tilde{g}(p)$. Then for each $A \in p$, $\{x \in D : f(x) \in g[A]\} \in p$.*

Proof. By Lemma 3.30, $g[A] \in \tilde{g}(p) = \tilde{f}(p)$ so, again applying Lemma 3.30, $f^{-1}[g[A]] \in p$.   □

Lemma 3.32. *Suppose that D is any discrete space and that f is a mapping from D to itself. The mapping $\widetilde{f} : \beta D \to \beta D$ has a fixed point if and only if every finite partition of D has a cell C for which $C \cap f[C] \neq \emptyset$.*

Proof. Suppose first that D has a finite partition in which every cell C satisfies $C \cap f[C] = \emptyset$. Let $p \in \beta D$ and pick a cell C satisfying $C \in p$. This implies by Lemma 3.30 that $f[C] \in \widetilde{f}(p)$ and hence that $\widetilde{f}(p) \neq p$.

Conversely, suppose that $\widetilde{f}$ has no fixed points. For each $p \in \beta D$ pick $A_p \in p \setminus \widetilde{f}(p)$, pick $B_p \in p$ such that $\widetilde{f}[\widehat{B_p}] \cap \widehat{A_p} = \emptyset$, and let $C_p = A_p \cap B_p$. Then $\{\widehat{C_p} : p \in \beta D\}$ is an open cover of βD so pick finite $F \subseteq \beta D$ such that $\{\widehat{C_p} : p \in F\}$ covers βD. Then $\{C_p : p \in F\}$ covers D and can thus be refined to a finite partition $\mathcal{F}$ of D such that each $C \in \mathcal{F}$ satisfies $C \cap f[C] = \emptyset$. $\square$

Lemma 3.33. *Suppose that D is a set and that $f : D \to D$ is a function with no fixed points. Then D can be partitioned into three sets A_0, A_1, and A_2 with the property that $A_i \cap f[A_i] = \emptyset$ for each $i \in \{0, 1, 2\}$.*

Proof. We consider the set

$$\mathcal{G} = \big\{ g : \text{(i) } g \text{ is a function,}$$
$$\text{(ii) } \mathrm{dom}(g) \subseteq D,$$
$$\text{(iii) } \mathrm{ran}(g) \subseteq \{0, 1, 2\},$$
$$\text{(iv) } f[\mathrm{dom}(g)] \subseteq \mathrm{dom}(g), \text{ and}$$
$$\text{(v) for each } a \in \mathrm{dom}(g), \ g(a) \neq g\big(f(a)\big) \big\}.$$

We observe that $\mathcal{G}$ is non-empty, because the function $\emptyset$ is a member of $\mathcal{G}$.

Then $\mathcal{G}$ is partially ordered by set inclusion and, if $\mathcal{C}$ is a chain in $\mathcal{G}$, then $\bigcup \mathcal{C} \in \mathcal{G}$, so Zorn's Lemma implies that $\mathcal{G}$ has a maximal element g. We shall show that $\mathrm{dom}(g) = D$.

We assume the contrary and suppose that $b \in D \setminus \mathrm{dom}(g)$. We then define a function h extending g. To do so, we put $h(a) = g(a)$ for every $a \in \mathrm{dom}(g)$. We choose $h(b)$ to be a value in $\{0, 1, 2\}$, choosing it so that $h(b) \neq h\big(f(b)\big)$ if $h(f(b))$ has already been defined. Now suppose that $h\big(f^m(b)\big)$ has been defined for each $m \in \{0, 1, 2, \ldots, n\}$. (Where $f^0(b) = b$ and $f^{t+1}(b) = f\big(f^t(b)\big)$.) If $h\big(f^{n+1}(b)\big)$ has not yet been defined, we choose $h\big(f^{n+1}(b)\big)$ to be a value in $\{0, 1, 2\}$ different from $h\big(f^n(b)\big)$ and from $h\big(f^{n+2}(b)\big)$, if the latter has already been defined. In this way, we can inductively define a function $h \in \mathcal{G}$ with $\mathrm{dom}(h) = \mathrm{dom}(g) \cup \{f^n(b) : n \in \omega\}$. Since $g \not\subseteq h$, we have contradicted our choice of g as being maximal in $\mathcal{G}$. Thus $\mathrm{dom}(g) = D$.

We now define the sets A_i by putting $A_i = g^{-1}[\{i\}]$ for each $i \in \{0, 1, 2\}$. $\square$

Theorem 3.34. *Let D be a discrete space and let $f : D \to D$. If f has no fixed points, neither does $\widetilde{f} : \beta D \to \beta D$.*

Proof. This follows from Lemmas 3.32 and 3.33. $\square$

Theorem 3.35. *Let D be a set and let $f : D \to D$. If $\widetilde{f} : \beta D \to \beta D$ and if $p \in \beta D$, then $\widetilde{f}(p) = p$ if and only if $\{a \in D : f(a) = a\} \in p$.*

Proof. Let E denote $\{a \in D : f(a) = a\}$. We first assume that $\widetilde{f}(p) = p$. We want to show that $E \in p$ so suppose instead that $D \backslash E \in p$. We choose any $b \in D \backslash E$, and define $g : D \to D$ by stating that $g(a) = f(a)$ if $a \in D \backslash E$ and $g(a) = b$ if $a \in E$. Then g has no fixed points. However, $f = g$ on $D \backslash E$ and so $\widetilde{f} = \widetilde{g}$ on $cl(D \backslash E)$. Since $p \in cl(D \backslash E)$, $\widetilde{g}(p) = \widetilde{f}(p) = p$. This contradicts Theorem 3.34 and hence $E \in p$.

Conversely, suppose that $E \in p$. Let $\iota : D \to D$ be the identity function. Since $f = \iota$ on E, $\widetilde{f} = \widetilde{\iota}$ on $cl\, E$. Since $p \in cl\, E$, $\widetilde{f}(p) = \widetilde{\iota}(p) = p$. $\qquad\square$

Theorem 3.36. *If D is any infinite set, every non-empty G_δ-subset of D^* has a non-empty interior in D^*.*

Proof. Suppose that, for each $n \in \mathbb{N}$, U_n is an open subset of βD and that $\bigcap_{n=1}^{\infty}(U_n \cap D^*) \neq \emptyset$. Choose any $p \in \bigcap_{n=1}^{\infty}(U_n \cap D^*)$. For each $n \in \mathbb{N}$, we can choose a subset A_n of D for which $p \in A_n{}^* \subseteq U_n$. We may assume that these sets are decreasing, because we can replace each A_n by $\bigcap_{i=1}^{n} A_i$. Observe that each A_n is infinite, for otherwise we would have $A_n{}^* = \emptyset$. We can thus choose an infinite sequence $\langle a_n \rangle_{n=1}^{\infty}$ of distinct points of D for which $a_n \in A_n$ for each $n \in \mathbb{N}$. Put $A = \{a_n : n \in \mathbb{N}\}$. If $q \in A^*$, then $q \in \widehat{A_n}$ for every $n \in \mathbb{N}$, because $A \backslash A_n$ is finite. Thus the non-empty open subset A^* of D^* is contained in $\bigcap_{n=1}^{\infty}(U_n \cap D^*)$. $\qquad\square$

Corollary 3.37. *Let D be any set. Any countable union of nowhere dense subsets of D^* is again nowhere dense in D^*.*

Proof. For each $n \in \mathbb{N}$, let A_n be a nowhere dense subset of D^*. To see that $\bigcup_{n=1}^{\infty} A_n$ is nowhere dense, it suffices to show that $B = \bigcup_{n=1}^{\infty} cl\, A_n$ is nowhere dense. Suppose instead one has $U = \mathrm{int}_{D^*}\, cl\, B \neq \emptyset$. By the Baire Category Theorem $D^* \backslash B$ is dense in D^* so $U \backslash B \neq \emptyset$. Thus $U \backslash B = \bigcap_{n=1}^{\infty}(U \backslash cl\, A_n)$ is a nonempty G_δ which thus, by Theorem 3.36, has nonempty interior, say V. But then $V \cap cl\, B = \emptyset$ while $V \subseteq U$, a contradiction. $\qquad\square$

A point p in a topological space is said to be a *P-point* if the intersection of any countable family of neighborhoods of p is also a neighborhood of p. Thus an ultrafilter $p \in \mathbb{N}^*$ is a P-point of $\mathbb{N}^*$ if and only if whenever $\langle A_n \rangle_{n=1}^{\infty}$ is a sequence of members of p, there is some $B \in p$ such that $|B \backslash A_n| < \omega$ for all $n \in \mathbb{N}$.

We see now that the Continuum Hypothesis implies that P-points exist in $\mathbb{N}^*$. It is a fact that their existence cannot be proved in ZFC. (See the notes to this chapter.)

Theorem 3.38. *The Continuum Hypothesis implies that P-points exist in $\mathbb{N}^*$.*

Proof. Assume the Continuum Hypothesis and enumerate $\mathcal{P}(\mathbb{N})$ as $\langle C_\alpha \rangle_{\alpha < \omega_1}$ with $C_0 = \mathbb{N}$. Let $\mathcal{A}_0 = \{C_0\}$. Inductively let $0 < \sigma < \omega_1$ and assume that we have chosen $\mathcal{A}_\alpha$ for all $\alpha < \sigma$ such that

(1) $\mathcal{A}_\alpha$ has the infinite finite intersection property,

(2) either $C_\alpha \in \mathcal{A}_\alpha$ or $\mathbb{N} \backslash C_\alpha \in \mathcal{A}_\alpha$,

(3) if $\delta < \alpha$, then $\mathcal{A}_\delta \subseteq \mathcal{A}_\alpha$,

(4) $|\mathcal{A}_\alpha| \le \omega$, and

(5) there exists $A \in \mathcal{A}_\alpha$ such that for each $\mathcal{F} \in \mathcal{P}_f(\mathcal{A}_\alpha)$, $|A \setminus \bigcap \mathcal{F}| < \omega$.

Let $\langle B_n \rangle_{n=1}^\infty$ list the elements of $\bigcup_{\alpha < \sigma} \mathcal{A}_\alpha$ (with repetition if necessary). If there is some n such that $|C_\sigma \cap \bigcap_{m=1}^n B_m| < \omega$, let $D = \mathbb{N} \setminus C_\sigma$ and otherwise let $D = C_\sigma$. Then for each $n \in \mathbb{N}$, $|D \cap \bigcap_{m=1}^n B_m| = \omega$ so we may choose a one-to-one sequence $\langle x_n \rangle_{n=1}^\infty$ with each $x_n \in D \cap \bigcap_{m=1}^n B_m$. Let $A = \{x_n : n \in \mathbb{N}\}$ and let $\mathcal{A}_\sigma = \{A, D\} \cup \bigcup_{\alpha < \sigma} \mathcal{A}_\alpha$. Then the induction hypotheses are satisfied.

Let $p = \bigcup_{\sigma < \omega_1} \mathcal{A}_\sigma$. Then by hypotheses (1) and (2), $p \in \mathbb{N}^*$. Given a sequence $\langle E_n \rangle_{n=1}^\infty$ of members of p, pick $\sigma < \omega_1$ such that $\{E_n : n \in \mathbb{N}\} \subseteq \{C_\alpha : \alpha < \sigma\}$. By hypothesis (5), there is some $A \in \mathcal{A}_\sigma \subseteq p$ such that $|A \setminus E_n| < \omega$ for each $n \in \mathbb{N}$. $\quad\square$

Definition 3.39. A topological space is said to be *extremally disconnected* if the closure of every open subset is open.

We showed in Theorem 3.18(f) that βD is an extremally disconnected space. Since we often work with D^*, the reader should be cautioned that D^* is *not* extremally disconnected [104, Exercise 6W].

The following theorem will be very useful in our algebraic investigations of countable semigroups.

Theorem 3.40. *Let D be a discrete space and let A and B be σ-compact subsets of βD. If $A \cap cl\,B = cl\,A \cap B = \emptyset$, then $cl\,A \cap cl\,B = \emptyset$.*

Proof. Write $A = \bigcup_{n=1}^\infty A_n$ and $B = \bigcup_{n=1}^\infty B_n$ where A_n and B_n are compact for each n. Since βD is a compact (Hausdorff) space, it is normal. For each $n \in \mathbb{N}$, A_n and $cl\,B$ are disjoint closed sets and $cl\,A$ and B_n are disjoint closed sets so pick open sets T_n, U_n, V_n, and W_n such that $T_n \cap U_n = V_n \cap W_n = \emptyset$, $A_n \subseteq T_n$, $cl\,B \subseteq U_n$, $cl\,A \subseteq V_n$, and $B_n \subseteq W_n$. For each $n \in \mathbb{N}$, let $G_n = T_n \cap \bigcap_{k=1}^n V_k$ and let $H_n = W_n \cap \bigcap_{k=1}^n U_k$. Then for each n, one has $A_n \subseteq G_n$ and $B_n \subseteq H_n$. Further, given any $n, m \in \mathbb{N}$, $G_n \cap H_m = \emptyset$.

Let $C = \bigcup_{n=1}^\infty G_n$ and let $D = \bigcup_{n=1}^\infty H_n$. Then C and D are disjoint open sets (so $D \cap cl\,C = \emptyset$), $A \subseteq C$, and $B \subseteq D$. By Theorem 3.18 (f) $cl\,C$ is open, so $cl\,B \cap cl\,C = \emptyset$. Since $cl\,A \subseteq cl\,C$ we have $cl\,A \cap cl\,B = \emptyset$ as required. $\quad\square$

The following equivalent (see Exercise 3.4.2) version of Theorem 3.40 is also useful.

Corollary 3.41. *Let D be a discrete space and let A and B be σ-compact subsets of βD. Then $cl\,A \cap cl\,B \subseteq cl(A \cap cl\,B) \cup cl(B \cap cl\,A)$.*

Proof. Let $p \in cl\,A \cap cl\,B$ and suppose that $p \notin cl(A \cap cl\,B) \cup cl(B \cap cl\,A)$. Pick $H \in p$ such that $\widehat{H} \cap (A \cap cl\,B) = \emptyset$ and $\widehat{H} \cap (B \cap cl\,A) = \emptyset$. Let $A' = \widehat{H} \cap A$ and $B' = \widehat{H} \cap B$. Then A' and B' are σ-compact subsets of βD and $p \in cl\,A' \cap cl\,B'$ so by Theorem 3.40 we may assume without loss of generality that $A' \cap cl\,B' \ne \emptyset$ so pick $q \in A' \cap cl\,B'$. Then $q \in \widehat{H} \cap (A \cap cl\,B)$, a contradiction. $\quad\square$

We isolate some specific instances of Theorem 3.40 that we shall often use.

Corollary 3.42. *Let D be a discrete space.*

(a) *Let A and B be σ-compact subsets of D^*. If $A \cap c\ell\, B = c\ell\, A \cap B = \emptyset$, then $c\ell\, A \cap c\ell\, B = \emptyset$.*

(b) *Let A and B be countable subsets of βD. If $A \cap c\ell\, B = c\ell\, A \cap B = \emptyset$, then $c\ell\, A \cap c\ell\, B = \emptyset$.*

(c) *Let A and B be countable subsets of D^*. If $A \cap c\ell\, B = c\ell\, A \cap B = \emptyset$, then $c\ell\, A \cap c\ell\, B = \emptyset$.*

Proof. Countable sets are σ-compact. By Theorem 3.18 (e) $D^* = \beta D \backslash e[D]$ is a closed subset of βD so σ-compact subsets of D^* are σ-compact in βD. $\square$

The following notion will be used in the exercises and later.

Definition 3.43. Let X be a topological space and let $A \subseteq X$. Then A is *strongly discrete* if and only if there is a family $\langle U_x \rangle_{x \in A}$ of open subsets of X such that $x \in U_x$ for each $x \in A$ and $U_x \cap U_y = \emptyset$ whenever x and y are distinct members of A.

Exercise 3.4.1. Suppose that $f : D \to E$ is a mapping from a discrete space D to a discrete space E. Prove that $\tilde{f} : \beta D \to \beta E$ is injective if f is injective, surjective if f is surjective and a homeomorphism if f is bijective.

Exercise 3.4.2. Derive Theorem 3.40 from Corollary 3.41.

Exercise 3.4.3. Prove that any infinite regular space has an infinite strongly discrete subset.

Exercise 3.4.4. Let X be an extremally disconnected regular space. Prove that no sequence can converge in X unless it is eventually constant.

Exercise 3.4.5. Let D be any set. Prove that no proper F_σ-subset of D^* can be dense in D^*.

Exercise 3.4.6. Let D be any set. Prove that no zero set in D^* can be a singleton. (A subset Z of a topological space X is said to be a *zero set* if $Z = f^{-1}[\{0\}]$ for some continuous function $f : X \to [0, 1]$.) (Hint: Use Theorem 3.36.)

Exercise 3.4.7. Let D be any discrete space. Prove that every separable subspace of βD is extremally disconnected. (Hint: Use Theorem 3.40.)

3.5 Uniform Limits via Ultrafilters

We now introduce the notion of p-limit. The definition of p-limit is very natural to anyone familiar with nets. We would like $p\text{-}\lim_{s \in D} x_s = y$ to mean that x_s is "often" "close to" y. Closeness is of course determined by neighborhoods of y while "often" is determined by members of p. (Recall that an ultrafilter can be thought of as a $\{0, 1\}$-valued measure.)

As we shall see, the notion is as versatile as the notion of nets, and has two significant advantages: (1) in a compact space a p-limit always converges and (2) it provides a "uniform" way of taking limits, as opposed to randomly choosing from among many possible limit points of a net.

Definition 3.44. Let D be a discrete space, let $p \in \beta D$, let $\langle x_s \rangle_{s \in D}$ be an indexed family in a topological space X, and let $y \in X$. Then $p\text{-}\lim_{s \in D} x_s = y$ if and only if for every neighborhood U of y, $\{s \in D : x_s \in U\} \in p$.

Recall that we have identified $\widehat{A}$ with βA for $A \subseteq D$. Consequently, if $A \in p$ and x_s is defined for $s \in A$ we write $p\text{-}\lim_{s \in A} x_s$ without worrying about possible values of x_s for $s \in D \backslash A$.

There is a more general concept of limit in a topological space, which may well be familiar to the reader. We shall show that p-limits and limits coincide for functions defined on βD.

Definition 3.45. Suppose that X and Y are topological spaces, that $A \subseteq X$ and that $f : A \to Y$. Let $x \in c\ell_X A$ and $y \in Y$. We shall write $\lim_{a \to x} f(a) = y$ if and only if, for every neighborhood V of y, there is a neighborhood U of x such that $f[A \cap U] \subseteq V$.

We observe that $\lim_{a \to x} f(a)$, if it exists, is obviously unique.

Theorem 3.46. *Let D be a discrete space, let Y be a topological space, and let $p \in \beta D$ and $y \in Y$. If $A \in p$ and $f : A \to Y$, then $p\text{-}\lim_{a \in A} f(a) = y$ if and only if $\lim_{a \to p} f(a) = y$.*

Proof. Suppose that $p\text{-}\lim_{a \in A} f(a) = y$. Then, if V is a neighborhood of y, $f^{-1}[V] \in p$. Let $B = f^{-1}[V]$. Then $\widehat{B}$ is a neighborhood of p by Theorem 3.18 and $f[\widehat{B} \cap A] = f[B] \subseteq V$.

Conversely, suppose that $\lim_{a \to p} f(a) = y$. Then, if V is a neighborhood of y, there is a neighborhood U of p in βD for which $f[A \cap U] \subseteq V$. Now $U \cap A \in p$ and so, since $U \cap A \subseteq f^{-1}[V]$, it follows that $f^{-1}[V] \in p$. Thus $p\text{-}\lim_{a \in A} f(a) = y$. $\square$

Remark 3.47. *Let D be a discrete space and let $p \in \beta D$. Viewing $\langle s \rangle_{s \in D}$ as an indexed family in βD one has $p\text{-}\lim_{s \in D} s = p$.*

Theorem 3.48. *Let D be a discrete space, let $p \in \beta D$, and let $\langle x_s \rangle_{s \in D}$ be an indexed family in a topological space X.*

(a) *If $p\text{-}\lim_{s \in D} x_s$ exists, then it is unique.*

(b) *If X is a compact space, then $p\text{-}\lim_{s \in D} x_s$ exists.*

Proof. Statement (a) is obvious.

(b) Suppose that $p\text{-}\lim \langle x_s \rangle_{s \in D}$ does not exist and for each $y \in X$, pick an open neighborhood U_y of y such that $\{s \in D : x_s \in U_y\} \notin p$. Then $\{U_y : y \in X\}$ is an open cover of X so pick finite $F \subseteq X$ such that $X = \bigcup_{y \in F} U_y$. Then $D = \bigcup_{y \in F} \{s \in D : x_s \in U_y\}$ so pick $y \in F$ such that $\{s \in D : x_s \in U_y\} \in p$. This contradiction completes the proof. $\square$

Theorem 3.49. *Let D be a discrete space, let $p \in \beta D$, let X and Y be topological spaces, let $\langle x_s \rangle_{s \in D}$ be an indexed family in X, and let $f : X \to Y$. If f is continuous and $p\text{-}\lim_{s \in D} x_s$ exists, then $p\text{-}\lim_{s \in D} f(x_s) = f(p\text{-}\lim_{s \in D} x_s)$.*

Proof. Let U be a neighborhood of $f(p\text{-}\lim_{s \in D} x_s)$ and pick a neighborhood V of $p\text{-}\lim_{s \in D} x_s$ such that $f[V] \subseteq U$. Let $A = \{s \in D : x_s \in V\}$. Then $A \in p$ and $A \subseteq \{s \in D : f(x_s) \in U\}$. $\square$

We pause to verify that Theorems 3.48 and 3.49 in fact provide a characterization of ultrafilters.

Definition 3.50. Let D be a discrete space. A *uniform operator on functions from D to compact spaces* is an operator $\mathcal{O}$ which assigns to each function f from D to a compact space Y some point $\mathcal{O}(f)$ of Y such that whenever Y and Z are compact spaces, g is a function from D to Y, and f is a continuous function from Y to Z, one has $f(\mathcal{O}(g)) = \mathcal{O}(f \circ g)$.

Theorems 3.48 and 3.49 tell us that given a discrete space D and $p \in \beta D$, "p-lim" is a uniform operator on functions from D to compact spaces. The following theorem provides the converse.

Theorem 3.51. *Let D be a discrete space and let $\mathcal{O}$ be a uniform operator on functions from D to compact spaces. There is a unique $p \in \beta D$ such that for every compact space Y and every function $f : D \to Y$, $\mathcal{O}(f) = p\text{-}\lim_{s \in D} f(s)$.*

Proof. Let $\iota : D \to D \subseteq \beta D$ be the inclusion map. (Before we had identified the points of D with principal ultrafilters, we would have said $\iota = e$.) Now the choice of $p \in \beta D$ is forced, since we must have $\mathcal{O}(\iota) = p\text{-}\lim_{s \in D} s = p$, so let $p = \mathcal{O}(\iota)$. Now let Y be a compact space and let $f : D \to Y$. Let $\tilde{f} : \beta D \to Y$ be the continuous extension of f.

Then

$$\begin{aligned}
p\text{-}\lim_{s \in D} f(s) &= p\text{-}\lim_{s \in D} \tilde{f}(s) \\
&= \tilde{f}(p\text{-}\lim_{s \in D} s) \\
&= \tilde{f}(p) \\
&= \tilde{f}(\mathcal{O}(\iota)) \\
&= \mathcal{O}(\tilde{f} \circ \iota) \\
&= \mathcal{O}(f).
\end{aligned}$$

$\square$

We now verify the assertion that the notion of p-limit is as versatile as the notion of nets by proving some of the standard theorems about nets in the context of p-limits.

Theorem 3.52. *Let X be a topological space. Then X is compact if and only if whenever $\langle x_s \rangle_{s \in D}$ is an indexed family in X and p is an ultrafilter on D, $p\text{-}\lim\limits_{s \in D} x_s$ exists.*

Proof. The necessity is Theorem 3.48(b).

Sufficiency. Let $\mathcal{A}$ be a collection of closed subsets of X with the finite intersection property. Let $D = \{\bigcap \mathcal{F} : \mathcal{F}$ is a finite nonempty subset of $\mathcal{A}\}$. For each $A \in D$ let $\mathcal{B}_A = \{B \in D : B \subseteq A\}$. Then $\{\mathcal{B}_A : A \in D\}$ has the finite intersection property. (The proof of this assertion is Exercise 3.5.1.) So by Theorem 3.8, pick an ultrafilter p on D such that $\{\mathcal{B}_A : A \in D\} \subseteq p$. For each $A \in D$, pick $x_A \in A$. Let $y = p\text{-}\lim\limits_{A \in D} x_A$. We claim that $y \in \bigcap \mathcal{A}$. To see this let $A \in \mathcal{A}$ and suppose that $y \notin A$. Then $X \backslash A$ is a neighborhood of y so $\{B \in D : x_B \in X \backslash A\} \in p$. Also $\mathcal{B}_A \in p$ so pick $B \in \mathcal{B}_A$ such that $x_B \in X \backslash A$. Then $(X \backslash A) \cap B \neq \emptyset$ while $B \subseteq A$, a contradiction. $\square$

Theorem 3.53. *Let X be a topological space, let $A \subseteq X$ and let $y \in X$. Then $y \in c\ell\, A$ if and only if there exists an indexed family $\langle x_s \rangle_{s \in D}$ in A such that $p\text{-}\lim\limits_{s \in D} x_s = y$.*

Proof. Sufficiency. Let U be a neighborhood of y. Then $\{s \in D : x_s \in U\} \in p$, so $\{s \in D : x_s \in U\} \neq \emptyset$.

Necessity. Let D be the set of neighborhoods of y. For each $U \in D$ pick $x_U \in U \cap A$ and let $\mathcal{B}_U = \{V \in D : V \subseteq U\}$. Then $\{\mathcal{B}_U : U \in D\}$ has the finite intersection property so pick an ultrafilter p on D with $\{\mathcal{B}_U : U \in D\} \subseteq p$. To see that $p\text{-}\lim\limits_{U \in D} x_U = y$, let U be a neighborhood of y. Then $\mathcal{B}_U \subseteq \{V \in D : x_V \in U\}$ so $\{V \in D : x_V \in U\} \in p$. $\square$

Theorem 3.54. *Let X and Y be topological spaces and let $f : X \to Y$. Then f is continuous if and only if whenever $\langle x_s \rangle_{s \in D}$ is an indexed family in X, $p \in \beta D$, and $p\text{-}\lim\limits_{s \in D} x_s$ exists, one has $f(p\text{-}\lim\limits_{s \in D} x_s) = p\text{-}\lim\limits_{s \in D} f(x_s)$.*

Proof. The necessity is Theorem 3.49

For the sufficiency, let $a \in X$ and let W be a neighborhood of $f(a)$. Suppose that no neighborhood U of a has $f[U] \subseteq W$. Then $a \in c\ell\, f^{-1}[Y \backslash W]$ so pick by Theorem 3.53 an indexed family $\langle x_s \rangle_{s \in D}$ in $f^{-1}[Y \backslash W]$ such that $p\text{-}\lim\limits_{s \in D} x_s = a$. Then $p\text{-}\lim\limits_{s \in D} f(x_s) = f(a)$ so $\{s \in D : f(x_s) \in W\}$ is in p and is hence nonempty, a contradiction. $\square$

Comment 3.55. The concept of limit is closely related to that of continuity. Suppose that X, Y are topological spaces, that $A \subseteq X$ and that $x \in A$. If $f : A \to Y$, then f is continuous at x if and only if for every neighborhood V of $f(x)$, there is a neighborhood U of x for which $f[U] \subseteq V$. So f is continuous at x if and only if $\lim\limits_{a \to x} f(a) = f(x)$.

Also, if $x \in c\ell\, A$, then f has a continuous extension to $A \cup \{x\}$ if and only if $\lim_{a \to x} f(a)$ exists.

Exercise 3.5.1. Let $\mathcal{A}$ be a collection of closed subsets of the topological space X with the finite intersection property. Let $D = \{\bigcap \mathcal{F} : \mathcal{F}$ is a finite nonempty subset of $\mathcal{A}\}$. For each $A \in D$ let $\mathcal{B}_A = \{B \in D : B \subseteq A\}$. Prove that $\{\mathcal{B}_A : A \in D\}$ has the finite intersection property.

Exercise 3.5.2. Let $\langle x_n \rangle_{n=1}^{\infty}$ be a sequence in a topological space X and assume that $\lim_{n \to \infty} x_n = a$. Prove that for each $p \in \beta\mathbb{N} \backslash \mathbb{N}$, $p\text{-}\lim_{n \in \mathbb{N}} x_n = a$.

3.6 The Cardinality of βD

We show here that for any infinite discrete space D, $|\beta D| = 2^{2^{|D|}}$. We do the proof for the case in which D is countable first. The proof uses the following surprising theorem which is of significant interest in its own right.

Recall that $\mathfrak{c}$ is the cardinality of the continuum. That is $\mathfrak{c} = 2^\omega = |\mathbb{R}| = |\mathcal{P}(\mathbb{N})|$.

Theorem 3.56. *Let D be a countably infinite set. There is a $\mathfrak{c} \times \mathfrak{c}$ matrix $\langle\langle A_{\sigma,\delta}\rangle_{\sigma < \mathfrak{c}}\rangle_{\delta < \mathfrak{c}}$ of subsets of D satisfying the following two statements.*

(a) *Given $\sigma, \delta, \tau < \mathfrak{c}$ with $\delta \neq \tau$, $|A_{\sigma,\delta} \cap A_{\sigma,\tau}| < \omega$.*
(b) *Given any $F \in \mathcal{P}_f(\mathfrak{c})$ and any $g : F \to \mathfrak{c}$, $|\bigcap_{\sigma \in F} A_{\sigma,g(\sigma)}| = \omega$.*

Proof. Let

$$S = \{(k, f) : k \in \mathbb{N} \text{ and } f : \mathcal{P}(\{1, 2, \dots, k\}) \to \mathcal{P}(\{1, 2, \dots, k\})\}.$$

Then S is countable so it suffices to produce a $\mathfrak{c} \times \mathfrak{c}$ matrix of subsets of S satisfying statements (a) and (b).

Enumerate $\mathcal{P}(\mathbb{N})$ as $\langle X_\sigma \rangle_{\sigma < \mathfrak{c}}$. Given $(\sigma, \delta) \in \mathfrak{c} \times \mathfrak{c}$, let

$$A_{\sigma,\delta} = \big\{(k, f) \in S : f(X_\sigma \cap \{1, 2, \dots, k\}) = X_\delta \cap \{1, 2, \dots, k\}\big\}.$$

To verify statement (a), let $\sigma, \delta, \tau < \mathfrak{c}$ with $\delta \neq \tau$. Pick $m \in \mathbb{N}$ such that $X_\delta \cap \{1, 2, \dots, m\} \neq X_\tau \cap \{1, 2, \dots, m\}$. Then $X_{\sigma,\delta} \cap X_{\sigma,\tau} \subseteq \{(k, f) \in S : k < m\}$, a finite set.

To verify statement (b), let $F \in \mathcal{P}_f(\mathfrak{c})$ and $g : F \to \mathfrak{c}$ be given. Pick $m \in \mathbb{N}$ such that given any $\sigma \neq \tau$ in F, $X_\sigma \cap \{1, 2, \dots, m\} \neq X_\tau \cap \{1, 2, \dots, m\}$. Then given any $k > m$, there exists

$$f : \mathcal{P}(\{1, 2, \dots, k\}) \to \mathcal{P}(\{1, 2, \dots, k\})$$

such that $(k, f) \in \bigcap_{\sigma \in F} A_{\sigma,g(\sigma)}$. Consequently $\bigcap_{\sigma \in F} A_{\sigma,g(\sigma)}$ is infinite. $\square$

Corollary 3.57. *Let D be a countably infinite discrete space. Then* $|\beta D| = 2^{\mathfrak{c}}$.

Proof. Since $\beta D \subseteq \mathcal{P}\big(\mathcal{P}(D)\big)$, one has that $|\beta D| \leq 2^{\mathfrak{c}}$. Let $\big\langle \langle A_{\sigma,\delta} \rangle_{\sigma < \mathfrak{c}} \big\rangle_{\delta < \mathfrak{c}}$ be as in Theorem 3.56. For each $f : \mathfrak{c} \to \{0, 1\}$, let $\mathcal{A}_f = \{A_{\sigma, f(\sigma)} : \sigma < \mathfrak{c}\}$. Then by Theorem 3.56(b), for each f, $\mathcal{A}_f$ has the property that all finite intersections are infinite, so pick by Corollary 3.14 a nonprincipal ultrafilter p_f on D such that $\mathcal{A}_f \subseteq p_f$. If f and g are distinct functions from $\mathfrak{c}$ to $\{0, 1\}$, then since p_f and p_g are nonprincipal, one has by Theorem 3.56(a) that $p_f \neq p_g$. □

We now prove the general result, using a modification of the construction in Theorem 3.56.

Theorem 3.58. *Let D be an infinite set with cardinality* κ. *Then* $|U_\kappa(D)| = |\beta D| = 2^{2^\kappa}$.

Proof. Again we note that $\beta D \subseteq \mathcal{P}\big(\mathcal{P}(D)\big)$ and so $|U_\kappa(D)| \leq |\beta D| \leq 2^{2^\kappa}$.

We define $S = \{(F, f) : F \in \mathcal{P}_f(D) \text{ and } f : \mathcal{P}(F) \to \mathcal{P}(F)\}$. Note that $|S| = \kappa$, and so it will be sufficient to prove that $|U_\kappa(S)| \geq 2^{2^\kappa}$.

For each $X \subseteq D$, we put $A_{X,0} = \{(F, f) \in S : f(F \cap X) = F \cap X\}$ and $A_{X,1} = \{(F, f) \in S : f(F \cap X) = F \backslash X\}$.

Then $A_{X,0} \cap A_{X,1} = \emptyset$ for every X. Let $g : \mathcal{P}(D) \to \{0, 1\}$. We shall show that, for every finite subset $\mathcal{C}$ of $\mathcal{P}(D)$, $\big|\bigcap_{X \in \mathcal{C}} A_{X,g(X)}\big| = \kappa$.

We can choose $B \in \mathcal{P}_f(D)$ for which the sets of the form $X \cap B$ with $X \in \mathcal{C}$, are all distinct. (For each pair (X, Y) of distinct elements of $\mathcal{C}$, pick a point $b(X, Y)$ in the symmetric difference $X \triangle Y$. Let $B = \{b(X, Y) : X, Y \in \mathcal{C} \text{ and } X \neq Y\}$.) Then, whenever $F \in \mathcal{P}_f(D)$ and $B \subseteq F$, the sets $X \cap F$ for $X \in \mathcal{C}$ are all distinct. For each $F \in \mathcal{P}_f(D)$ such that $B \subseteq F$, we can define $f : \mathcal{P}(F) \to \mathcal{P}(F)$ such that, for each $X \in \mathcal{C}$, $f(X \cap F) = X \cap F$ if $g(X) = 0$ and $f(X \cap F) = F \backslash X$ if $g(X) = 1$. Then $(F, f) \in A_{X,g(X)}$ for every $X \in \mathcal{C}$. Since there are κ possible choices for F, $\big|\bigcap_{X \in \mathcal{C}} A_{X,g(X)}\big| = \kappa$.

It follows from Corollary 3.14 that, for every function $g : \mathcal{P}(D) \to \{0, 1\}$, there is an ultrafilter $p \in \mathcal{U}_\kappa(D)$ such that $A_{X,g(X)} \in p$ for every $X \in \mathcal{P}(D)$. Since $A_{X,0} \cap A_{X,1} = \emptyset$ for every X, different functions correspond to different ultrafilters in $\mathcal{U}_\kappa(D)$. Since there are 2^{2^κ} functions $g : \mathcal{P}(D) \to \{0, 1\}$, it follows that $|\mathcal{U}_\kappa(D)| \geq 2^{2^\kappa}$. □

We now see that infinite closed subsets of βD must be reasonably large.

Theorem 3.59. *Let D be an infinite discrete space and let A be an infinite closed subset of* βD. *Then A contains a topological copy of* $\beta \mathbb{N}$. *In particular* $|A| \geq 2^{\mathfrak{c}}$.

Proof. Choose an infinite strongly discrete subset B of A. (The fact that one can do this is Exercise 3.4.3.) Let $\langle p_n \rangle_{n=1}^\infty$ be a one-to-one sequence in B. Since $\{p_n : n \in \mathbb{N}\}$ is strongly discrete choose for each n some $C_n \in p_n$ such that $C_n \cap C_m = \emptyset$ whenever $n \neq m$.

Define $f : \mathbb{N} \to \beta D$ by $f(n) = p_n$. Then $\tilde{f} : \beta \mathbb{N} \to A$ is a continuous function with compact domain, which is one-to-one and so $\tilde{f}[\beta \mathbb{N}]$ is homeomorphic to $\beta \mathbb{N}$. (The proof that $\tilde{f}$ is one-to-one is Exercise 3.6.1.)   □

We conclude by showing in Theorem 3.62 that sets which are not too large, but have finite intersections that are large, extend to many κ-uniform ultrafilters.

Definition 3.60. Let $\mathcal{A}$ be a set of sets and let κ be an infinite cardinal. Then $\mathcal{A}$ has the κ-*uniform finite intersection property* if and only if whenever $\mathcal{F} \in \mathcal{P}_f(\mathcal{A})$, one has $|\bigcap \mathcal{F}| \geq \kappa$.

Thus the "infinite finite intersection property" is the same as the "ω-uniform finite intersection property".

Lemma 3.61. *Let D be an infinite set with cardinality κ and let $\mathcal{A}$ be a set of at most κ subsets of D with the κ-uniform finite intersection property. There is a set $\mathcal{B}$ of κ pairwise disjoint subsets of D such that for each $B \in \mathcal{B}$, $\mathcal{A} \cup \{B\}$ has the κ-uniform finite intersection property.*

Proof. Since $|\mathcal{P}_f(\mathcal{A})| \leq \kappa$ we may presume that $\mathcal{A}$ is closed under finite intersections. Since $\kappa \cdot |\mathcal{A}| = \kappa$, we may choose a κ-sequence $\langle A_\sigma \rangle_{\sigma < \kappa}$ of members of $\mathcal{A}$ such that, for each $C \in \mathcal{A}$,

$$|\{\sigma < \kappa : A_\sigma = C\}| = \kappa.$$

Let $Y_0 = \emptyset$ and inductively let $\tau < \kappa$ and assume that we have chosen $\langle Y_\gamma \rangle_{\gamma < \tau}$ such that:

(1) for all $\gamma < \tau$, $|Y_\gamma| = |\gamma|$,
(2) for all $\gamma < \tau$, $Y_\gamma \subseteq A_\gamma$, and
(3) for all $\gamma < \sigma < \tau$, $Y_\gamma \cap Y_\sigma = \emptyset$.

Now $|A_\tau| = \kappa$ and $\left| \bigcup_{\gamma < \tau} Y_\gamma \right| = \Sigma_{\gamma < \tau} |\gamma| < \kappa$. Thus we may pick $Y_\tau \subseteq A_\tau$ such that $|Y_\tau| = |\tau|$ and $Y_\tau \cap \left(\bigcup_{\gamma < \tau} Y_\gamma \right) = \emptyset$.

Having chosen $\langle Y_\sigma \rangle_{\sigma < \kappa}$, enumerate each Y_σ as $\langle y_{\sigma,\eta} \rangle_{\eta < \sigma}$. For each $\eta < \kappa$, let $B_\eta = \{ y_{\sigma,\eta} : \eta < \sigma < \kappa \}$. Since $Y_\sigma \cap Y_\gamma = \emptyset$ whenever $\sigma < \gamma < \kappa$, we have that each $|B_\eta| = \kappa$. Further, if $\delta < \eta < \kappa$, then $B_\delta \cap B_\eta = \emptyset$. (Since $\langle y_{\sigma,\eta} \rangle_{\eta < \sigma}$ is an enumeration of Y_σ we can't have $y_{\sigma,\delta} = y_{\sigma,\eta}$ and since $Y_\sigma \cap Y_\gamma = \emptyset$ for $\sigma \neq \gamma$, we can't have $y_{\sigma,\eta} = y_{\gamma,\delta}$.)

Finally, let $\eta < \kappa$ and $\mathcal{F} \in \mathcal{P}_f(\mathcal{A})$ be given. Let

$$C = \{\sigma : \eta < \sigma < \kappa \quad \text{and} \quad A_\sigma = \bigcap \mathcal{F}\}.$$

By assumption $|C| = \kappa$ and $\{ y_{\sigma,\eta} : \sigma \in C \} \subseteq B_\eta \cap \bigcap \mathcal{F}$.   □

Theorem 3.62. *Let D be an infinite set with cardinality κ and let $\mathcal{A}$ be a set of at most κ subsets of D with the κ-uniform finite intersection property. Then*

$$|\{p \in U_\kappa(D) : \mathcal{A} \subseteq p\}| = 2^{2^\kappa}.$$

Proof. Choose by Lemma 3.61 a family $\mathcal{B}$ of κ pairwise disjoint subsets of D such that for each $B \in \mathcal{B}$, $\mathcal{A} \cup \{B\}$ has the κ-uniform finite intersection property.

Choose a one-to-one function $\theta : D \to \mathcal{B}$ and for each $x \in D$, pick by Corollary 3.14 some $f(x) \in U_\kappa(D)$ such that $\mathcal{A} \cup \{\theta(x)\} \subseteq f(x)$.

Notice that $U_\kappa(D)$ is a closed, hence compact, subset of βD. Thus, by Theorem 3.28, there is a continuous extension $\widetilde{f} : \beta D \to U_\kappa(D)$. Since for each $x \in D$, $\mathcal{A} \subseteq f(x)$, we have for each $q \in \beta D$ that $\mathcal{A} \subseteq \widetilde{f}(q)$.

To complete the proof, it suffices to show by Theorem 3.58 that $\widetilde{f}$ is one-to-one. For this it in turn suffices to show that for each $q \in \beta D$ and each $A \in q$, $\bigcup_{x \in A} \theta(x) \in \widetilde{f}(q)$. (For then, given $q \neq r$ and $A \in q$ and $C \in r$ with $A \cap C = \emptyset$ one has that $\left(\bigcup_{x \in A} \theta(x)\right) \cap \left(\bigcup_{x \in C} \theta(x)\right) = \emptyset$.) The proof of this assertion is Exercise 3.6.2. $\square$

Exercise 3.6.1. Prove that the function $\widetilde{f}$ in the proof of Theorem 3.59 is one-to-one.

Exercise 3.6.2. Let D and E be discrete spaces, let

$$\theta : D \to \mathcal{P}(E) \quad \text{and} \quad f : D \to \beta E$$

be functions such that for each $x \in D$, $\theta(x) \in f(x)$, and let $\widetilde{f} : \beta D \to \beta E$ be the continuous extension of f. Prove that for each $q \in \beta D$ and each $A \in q$, $\bigcup_{x \in A} \theta(x) \in \widetilde{f}(q)$.

Exercise 3.6.3. Prove that $\mathbb{N}^*$ contains a collection of $\mathfrak{c}$ pairwise disjoint clopen sets.

Exercise 3.6.4. Prove that $\mathbb{N}^*$ contains exactly $\mathfrak{c}$ clopen subsets.

Exercise 3.6.5. Prove that every clopen subset of $\mathbb{N}^*$ is homeomorphic to $\mathbb{N}^*$.

Notes

The standard reference for detailed information about ultrafilters is [69].

We used the Axiom of Choice (or to be precise, Zorn's Lemma) in our proof of Theorem 3.8, which we used to deduce that nonprincipal ultrafilters exist. Such an appeal is unavoidable as there are models of set theory (without choice) in which no nonprincipal ultrafilters on $\mathbb{N}$ exist. (See the discussion of this point in [69, p.162].)

The Stone–Čech compactification was produced independently by M. Stone [227] and E. Čech [60]. The approach used by Čech was to embed a space X in a product of lines. A method that is close to the one that we chose in defining βD, is due to H. Wallman [244]. Suppose that X is a normal space. Let $\mathcal{Z}$ denote the set of closed subsets of X. The points of βX are defined to be the ultrafilters in the lattice of closed subsets of X, and the topology of βX is defined by taking the sets of the form $\{p \in \beta X : Z \in p\}$, where $Z \in \mathcal{Z}$, as a base for the closed sets. The space X is embedded in βX by mapping each $x \in X$ to $\{Z \in \mathcal{Z} : x \in Z\}$, this being an ultrafilter in the lattice of closed sets. The approach used here for discrete spaces is based on the treatment in [104]. See the notes in [104, p. 269] for a more detailed discussion of the development of this approach.

Lemma 3.33 is due to M. Katetǒv [167].

P-points were invented (and shown to exist in $\mathbb{N}^*$ under the assumption of the continuum hypothesis) by W. Rudin [214] who used them to show that $\mathbb{N}^*$ is not homogeneous. It is a result of S. Shelah [222, VI, §4] that it is consistent relative to ZFC that there are no P-points in $\mathbb{N}^*$.

The reader may be interested to know that extremally disconnected spaces have remarkable properties. For example, any continuous function mapping a dense subspace of an extremally disconnected space into a compact space, has a continuous extension to the entire space. [104, Exercise 6M2]

Extremally disconnected spaces can be characterized among completely regular spaces by the following property: suppose that X is a completely regular space and that $C_{\mathbb{R}}(X)$ denotes the set of bounded continuous real-valued functions defined on X. Then X is extremally disconnected if and only if every bounded subset of $C_{\mathbb{R}}(X)$ has a least upper bound in $C_{\mathbb{R}}(X)$. [104, Exercise 3N6]

For the reader with some acquaintance with category theory, we might mention that it is the extremally disconnected compact Hausdorff spaces which are the projective objects in the category of compact Hausdorff spaces. [107]

The notion of p-limit is apparently due originally to Z. Frolík [95]

Theorem 3.56 is due to K. Kunen [170], and answered a question of F. Galvin. The proof given here is a simplification of Kunen's original proof due to P. Simon.

Closing Remarks

More general constructions of compact spaces. Our construction of βD is a special case of more general constructions in which compact Hausdorff spaces are obtained by using sets which are maximal subject to having certain algebraic properties. We shall briefly mention two of these constructions, because they are relevant to the Stone–Čech compactification.

(i) Any unital commutative C*-algebra is associated with a compact Hausdorff space, called its spectrum, whose points could be described as maximal ideals of the algebra or, alternatively, as homomorphisms from the algebra to the complex numbers.

Let X be a completely regular space and let $C(X)$ denote the C*-algebra formed by the continuous bounded complex-valued functions defined on X, with each $f \in C(X)$ having the uniform norm $\|f\| = \sup\{|f(X)| : x \in X\}$. Then βX can be identified with the spectrum of $C(X)$.

In this framework, if D denotes a discrete space, βD is the spectrum of the Banach algebra $l^{\infty}(D) \simeq C(D)$. See [166].

(ii) Any Boolean algebra B is associated with a totally disconnected compact Hausdorff space, called its Stone space, whose points could be described as the ultrafilters of B. All totally disconnected compact Hausdorff spaces arise in this way, as any such space can be identified with the Stone space associated with the Boolean algebra formed by its clopen subsets. This theory displays the category of totally disconnected compact Hausdorff spaces as being the dual of the category of Boolean algebras. The extremally disconnected compact spaces are those corresponding to complete Boolean algebras. If D is a discrete space, βD could be described as the Stone space of the Boolean algebra $\mathcal{P}(D)$, while D^* could be described as the Stone space of the quotient of $\mathcal{P}(D)$ by the ideal of finite subsets of D. See [165].

Compactifications and subalgebras of $C_{\mathbb{R}}(X)$. Suppose that X is a completely regular space and that $C_{\mathbb{R}}(X)$ denotes the subspace of $C(X)$ which consists of the real-valued functions. A subalgebra A of $C_{\mathbb{R}}(X)$ is said to separate points and closed sets if, for each closed subset C of X and each $x \in X \backslash C$, there is a function $f \in A$ for which $f(x) \notin c\ell\, f[C]$. The compact spaces in which X can be densely embedded correspond to the closed subalgebras of $C_{\mathbb{R}}(X)$ which contain the constant functions and separate points and closed sets. For any such algebra A, there is a compactification (Y, φ) of X which has the following property: a function $f \in C_{\mathbb{R}}(X)$ can be expressed as $f = g \circ \varphi$ for some $g \in C_{\mathbb{R}}(Y)$ if and only if $f \in A$. Y could be defined as the spectrum of A, and, for each $x \in X$, $\varphi(x)$ could be defined as the homomorphism from A to the complex numbers for which $\varphi(x)(f) = f(x)$.

Conversely, if (Y, φ) is a compactification of X, it can be identified with the compactification arising in this way from the subalgebra $A = \{g \circ \varphi : g \in C_{\mathbb{R}}(Y)\}$ of $C_{\mathbb{R}}(X)$.

The Stone–Čech compactification of X corresponds to the largest possible subalgebra, namely $C_{\mathbb{R}}(X)$ itself. Any other compactification is a quotient of βX. If Y is such a compactification, corresponding to the subalgebra A of $C_{\mathbb{R}}(X)$, Y can be identified with the quotient of βX by the equivalence relation $p \simeq q$ if and only if $\widetilde{f}(p) = \widetilde{f}(q)$ for every $f \in A$.

We shall address this issue again in Chapter 21.

F-Spaces. Extremally disconnected spaces are examples of spaces called *F-spaces*. A completely regular space X is said to be an F-space if any two disjoint cozero subsets A and B of X are completely separated. This means that there is a continuous function $f : X \to [0, 1]$ for which $f[A] = \{0\}$ and $f[B] = \{1\}$. These spaces are significant in the theory of the Stone–Čech compactification, because, if X is locally compact, $\beta X \backslash X$ is an F-space (See [104].)

It is quite easy to see that a compact Hausdorff space X satisfies Theorem 3.40 if and only if it is an F-space. (See for example [154, Lemma 1.1].) In particular, if D is a discrete space, every compact subset of βD is an F-space. Thus D^* is an F-space, although — unlike βD — it need not be extremally disconnected.

Chapter 4

βS — The Stone–Čech Compactification of a Discrete Semigroup

4.1 Extending the Operation to βS

We shall see in this chapter that we can extend the operation of a discrete semigroup to its Stone–Čech compactification. When we say that $(S, \cdot)$ is a semigroup we intend also that it have the discrete topology.

We remind the reader that we are pretending that $S \subseteq \beta S$. Without the identification of the point s of S with the principal ultrafilter $e(s)$, conclusions (a) and (c) of Theorem 4.1 would read as follows:

(a) *The embedding e is a homomorphism.*

(c) *For all $s \in S$, $\lambda_{e(s)}$ is continuous.*

Theorem 4.1. *Let S be a discrete space and let $\cdot$ be a binary operation defined on S. There is a unique binary operation $* : \beta S \times \beta S \to \beta S$ satisfying the following three conditions:*

(a) *For every $s, t \in S$, $s * t = s \cdot t$.*

(b) *For each $q \in \beta S$, the function $\rho_q : \beta S \to \beta S$ is continuous, where $\rho_q(p) = p * q$.*

(c) *For each $s \in S$, the function $\lambda_s : \beta S \to \beta S$ is continuous, where $\lambda_s(q) = s * q$.*

Proof. We establish uniqueness and existence at the same time, defining $*$ as we are forced to define it. We first define $*$ on $S \times \beta S$. Given any $s \in S$, define $\ell_s : S \to S \subseteq \beta S$ by $\ell_s(t) = s \cdot t$. Then by Theorem 3.28, there is a continuous function $\lambda_s : \beta S \to \beta S$ such that $\lambda_{s|S} = \ell_s$. If $s \in S$ and $q \in \beta S$, we define $s * q = \lambda_s(q)$. Then (c) holds and so does (a), because λ_s extends ℓ_s. Furthermore, the extension λ_s is unique (because continuous functions agreeing on a dense subspace are equal), and so this is the only possible definition of $*$ satisfying (a) and (c).

Now we extend $*$ to the rest of $\beta S \times \beta S$. Given $q \in \beta S$, define $r_q : S \to \beta S$ by $r_q(s) = s * q$. Then there is a continuous function $\rho_q : \beta S \to \beta S$ such that $\rho_{q|S} = r_q$. For $p \in \beta S \setminus S$, we define $p * q = \rho_q(p)$ and note that if $s \in S$, $\rho_q(s) = r_q(s) = s * q$. So for all $p \in \beta S$, $\rho_q(p) = p * q$. We observe that (b) holds. Again, by the uniqueness of continuous extensions, this is the only possible definition which satisfies the required conditions. $\qquad\square$

It is customary to denote the operation on βS by the same symbol as that used for the operation on S and we shall adopt that practice. Thus if the operation on S is denoted by $\cdot$, we shall talk about the operation $\cdot$ on βS. If the operation is denoted by $+$, we shall talk about the operation $+$ on βS. (There will be exceptions. If we are working with the semigroup $(\mathcal{P}_f(\mathbb{N}), \cup)$ we shall *not* talk about the semigroup $(\beta\mathcal{P}_f(\mathbb{N}), \cup)$, since the union of two ultrafilters already means something else, because ultrafilters are sets.)

We do not yet know that $(\beta S, \cdot)$ is a semigroup if $(S, \cdot)$ is a semigroup, as we have not yet verified associativity. We shall do this shortly, but first we want to get a handle on the operations. The operation on βS has a characterization in terms of limits. The statements in the following remark follow immediately from the fact that λ_s is continuous for every $s \in S$ and ρ_q is continuous for every $q \in \beta S$. (Here s and t represent members of S.)

Remark 4.2. *Let $\cdot$ be a binary operation on a discrete space S.*
   (a) *If $s \in S$ and $q \in \beta S$, then $s \cdot q = \lim\limits_{t \to q} s \cdot t$.*
   (b) *If $p, q \in \beta S$, then $p \cdot q = \lim\limits_{s \to p} (\lim\limits_{t \to q} s \cdot t)$.*

The following version of Remark 4.2(b) will often be useful.

Remark 4.3. *Let $\cdot$ be a binary operation on a discrete space S, let $p, q \in \beta S$, let $P \in p$, and let $Q \in q$. Then $p \cdot q = p\text{-}\lim\limits_{s \in P}(q\text{-}\lim\limits_{t \in Q} s \cdot t)$.*

We remind the reader that if $f : X \to Y$ is a continuous function and $\lim\limits_{s \to p} f(x_s)$ and $\lim\limits_{s \to p} x_s$ exist, then $\lim\limits_{s \to p} f(x_s) = f(\lim\limits_{s \to p} x_s)$, by Theorem 3.49. We shall regularly use this fact without mentioning it explicitly.

Theorem 4.4. *Let $(S, \cdot)$ be a semigroup. Then the extended operation on βS is associative.*

Proof. Let $p, q, r \in \beta S$. We consider $\lim\limits_{a \to p} \lim\limits_{b \to q} \lim\limits_{c \to r} (a \cdot b) \cdot c$, where a, b and c denote elements of S. We have:

$$
\begin{aligned}
\lim_{a \to p} \lim_{b \to q} \lim_{c \to r} (a \cdot b) \cdot c &= \lim_{a \to p} \lim_{b \to q} (a \cdot b) \cdot r && \text{(because } \lambda_{a \cdot b} \text{ is continuous)} \\
&= \lim_{a \to p} (a \cdot q) \cdot r && \text{(because } \rho_r \circ \lambda_a \text{ is continuous)} \\
&= (p \cdot q) \cdot r && \text{(because } \rho_r \circ \rho_q \text{ is continuous).}
\end{aligned}
$$

Also:

$$
\begin{aligned}
\lim_{a \to p} \lim_{b \to q} \lim_{c \to r} a \cdot (b \cdot c) &= \lim_{a \to p} \lim_{b \to q} a \cdot (b \cdot r) && \text{(because } \lambda_a \circ \lambda_b \text{ is continuous).} \\
&= \lim_{a \to p} a \cdot (q \cdot r) && \text{(because } \lambda_a \circ \rho_r \text{ is continuous)} \\
&= p \cdot (q \cdot r) && \text{(because } \rho_{q \cdot r} \text{ is continuous).}
\end{aligned}
$$

So $(p \cdot q) \cdot r = p \cdot (q \cdot r)$.   □

The following theorem illustrates again the virtue of the uniformity of the process of passing to limits using p-limit.

Theorem 4.5. *Let $(S, \cdot)$ be a semigroup, let X be a topological space, let $\langle x_s \rangle_{s \in S}$ be an indexed family in X, and let $p, q \in \beta S$. If all limits involved exist, then $(p \cdot q)\text{-}\lim_{v \in S} x_v = p\text{-}\lim_{s \in S} q\text{-}\lim_{t \in S} x_{st}$.*

Proof. Let $z = (p \cdot q)\text{-}\lim_{v \in S} x_v$ and, for each $s \in S$, let $y_s = q\text{-}\lim_{t \in S} x_{st}$. Suppose that $p\text{-}\lim_{s \in S} y_s \neq z$ and pick disjoint open neighborhoods U and V of $p\text{-}\lim_{s \in S} y_s$ and z respectively. Let $A = \{v \in S : x_v \in V\}$ and let $B = \{s \in S : y_s \in U\}$. Then $A \in p \cdot q$ and $B \in p$. Since $\widehat{A}$ is a neighborhood of $\rho_q(p)$, pick $C \in p$ (so that $\widehat{C}$ is a basic neighborhood of p) such that $\rho_q[\widehat{C}] \subseteq \widehat{A}$. Then $B \cap C \in p$ so pick $s \in B \cap C$. Since $s \in B, q\text{-}\lim_{t \in S} x_{st} \in U$. Let $D = \{t \in S : x_{st} \in U\}$. Then $D \in q$. Since $s \in C, s \cdot q \in \widehat{A}$ so $\widehat{A}$ is a neighborhood of $\lambda_s(q)$. Pick $E \in q$ such that $\lambda_s[\widehat{E}] \subseteq \widehat{A}$. Then $D \cap E \in q$ so pick $t \in D \cap E$. Since $t \in D, x_{st} \in U$. Since $t \in E, st \in A$ so $x_{st} \in V$ and hence $U \cap V \neq \emptyset$, a contradiction.   □

Observe that the hypotheses in Theorem 4.5 regarding the existence of limits can be dispensed with if X is compact by Theorem 3.48. Observe also, that if X is contained in a compact space Y (i.e. if X is a completely regular space) the proof is much shorter. For then let $f : S \to X$ be defined by $f(s) = x_s$ and let $\widetilde{f}$ be the continuous extension of f to βS. Then one has

$$\begin{aligned}
(p \cdot q)\text{-}\lim_{v \in S} x_v &= (p \cdot q)\text{-}\lim_{v \in S} \widetilde{f}(v) \\
&= \widetilde{f}\big((p \cdot q)\text{-}\lim_{v \in S} v\big) \\
&= \widetilde{f}(p \cdot q) \\
&= \widetilde{f}(p\text{-}\lim_{s \in S} q\text{-}\lim_{t \in S} st) \\
&= p\text{-}\lim_{s \in S} q\text{-}\lim_{t \in S} f(st) \\
&= p\text{-}\lim_{s \in S} q\text{-}\lim_{t \in S} x_{st}.
\end{aligned}$$

In the case in which $(S, \cdot)$ is a semigroup, conclusion (c) of Theorem 4.1 may seem extraneous. Indeed it would appear more natural to replace (c) by the requirement that the extended operation be associative. We see in fact that we lose the uniqueness of the extension by such a replacement.

Example 4.6. *Let $+$ be the extension of addition on $\mathbb{N}$ to $\beta \mathbb{N}$ satisfying* (a), (b), *and* (c) *of Theorem* 4.1. *Define an operation $*$ on $\beta \mathbb{N}$ by*

$$p * q = \begin{cases} q & \text{if } q \in \beta \mathbb{N} \backslash \mathbb{N} \\ p + q & \text{if } q \in \mathbb{N}. \end{cases}$$

Then $*$ is associative and satisfies statements (a) and (b) of Theorem 4.1, but if $q \in \beta\mathbb{N}\backslash\mathbb{N}$, then $1 * q \neq 1 + q$.

The verification of the assertions in Example 4.6 is Exercise 4.1.1.

As a consequence of Theorems 4.1 and 4.4 we see that $(\beta S, \cdot)$ is a compact right topological semigroup, so that all of the results of Chapter 2 apply. In addition, we shall see that $(\beta S, \cdot)$ is maximal among semigroup compactifications of S in a sense similar to that in which βS is maximal among topological compactifications.

Recall that, given a right topological semigroup T, $\Lambda(T) = \{x \in T : \lambda_x$ is continuous$\}$.

Definition 4.7. Let S be a semigroup which is also a topological space. A *semigroup compactification* of S is a pair (φ, T) where T is a compact right topological semigroup, $\varphi : S \to T$ is a continuous homomorphism, $\varphi[S] \subseteq \Lambda(T)$, and $\varphi[S]$ is dense in T.

It is customary to assume in the definition of a semigroup compactification that S is a semitopological semigroup. However, as we shall see in Chapter 21, this restriction is not essential.

Note that a semigroup compactification need not be a (topological) compactification because the homomorphism φ need not be one-to-one.

Theorem 4.8. *Let $(S, \cdot)$ be a discrete semigroup, and let $\iota : S \to \beta S$ be the inclusion map.*

(a) $(\iota, \beta S)$ is a semigroup compactification of S.

(b) If T is a compact right topological semigroup and $\varphi : S \to T$ is a continuous homomorphism with $\varphi[S] \subseteq \Lambda(T)$ (in particular, if (φ, T) is a semigroup compactification of S), then there is a continuous homomorphism $\eta : \beta S \to T$ such that $\eta_{|S} = \varphi$. (That is the diagram

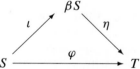

commutes.)

Proof. Conclusion (a) follows from Theorems 3.28, 4.1, and 4.4.

(b) Denote the operation of T by $*$. By Theorem 3.28 choose a continuous function $\eta : \beta S \to T$ such that $\eta_{|S} = \varphi$. We need only show that η is a homomorphism. By Theorems 3.28 and 4.1, S is dense in βS and $S \subseteq \Lambda(\beta S)$, so Lemma 2.14 applies. $\square$

Comment 4.9. We remark that the characterization of the semigroup operation in βS in terms of limits is valid in all semigroup compactifications. Let S be any semigroup with topology and let (φ, T) be any semigroup compactification of S. For every $p, q \in T$, we have:

$$pq = \lim_{\varphi(s) \to p} \lim_{\varphi(t) \to q} \varphi(s)\varphi(t)$$

where s and t denote elements of S. (The notation $\lim\limits_{\varphi(s)\to p} x_s = y$ means that for any neighborhood U of y, there is a neighborhood V of p such that whenever $s \in S$ and $\varphi(s) \in V$, one has $x_s \in U$.)

Since the points of βS are ultrafilters, we want to know which subsets of S are members of $p \cdot q$.

Definition 4.10. Let $(S, \cdot)$ be a semigroup, let $A \subseteq S$, and let $s \in S$.
    (a) $s^{-1}A = \{t \in S : st \in A\}$.
    (b) $As^{-1} = \{t \in S : ts \in A\}$.

Note that $s^{-1}A$ is simply an alternative notation for $\lambda_s^{-1}[A]$, and its use does not imply that s has an inverse in A. If $t \in S$, we may use tA to denote $\{ta : a \in A\}$. This does not introduce any conflict, because, if s does have an inverse s^{-1} in A, then $\lambda_s^{-1}[A] = \{s^{-1}a : a \in A\}$. However, even if S can be embedded in a group G (so that s^{-1} is defined in G), $s^{-1}A$ need not equal $\{s^{-1}a : a \in A\}$. For example, in $(\mathbb{N}, \cdot)$, let $A = \mathbb{N}2 + 1 = \{2n + 1 : n \in \mathbb{N}\}$. Then $2^{-1}A = \{t \in \mathbb{N} : 2t \in A\} = \emptyset$.

We shall often deal with semigroups where the operation is denoted by $+$, and so we introduce the appropriate notation.

Definition 4.11. Let $(S, +)$ be a semigroup, let $A \subseteq S$, and let $s \in S$.
    (a) $-s + A = \{t \in S : s + t \in A\}$.
    (b) $A - s = \{t \in S : t + s \in A\}$.

Theorem 4.12. *Let $(S, \cdot)$ be a semigroup and let $A \subseteq S$.*
    (a) *For any $s \in S$ and $q \in \beta S$, $A \in s \cdot q$ if and only if $s^{-1}A \in q$.*
    (b) *For any $p, q \in \beta S$, $A \in p \cdot q$ if and only if $\{s \in S : s^{-1}A \in q\} \in p$.*

Proof. (a) Necessity. Let $A \in s \cdot q$. Then $\widehat{A}$ is a neighborhood of $\lambda_s(q)$ so pick $B \in q$ such that $\lambda_s[\widehat{B}] \subseteq \widehat{A}$. Since $B \subseteq s^{-1}A$, we are done.

Sufficiency. Assume $s^{-1}A \in q$ and suppose that $A \notin s \cdot q$. Then $S\backslash A \in s \cdot q$ so, by the already established necessity, $s^{-1}(S\backslash A) \in q$. This is a contradiction since $s^{-1}A \cap s^{-1}(S\backslash A) = \emptyset$.

    (b) This is Exercise 4.1.3. $\square$

The following notion will be of significant interest to us in applications involving idempotents in βS.

Definition 4.13. Let $(S, \cdot)$ be a semigroup, let $A \subseteq S$, and let $p \in \beta S$. Then $A^{\star}(p) = \{s \in A : s^{-1}A \in p\}$. .

We shall frequently write $A^{\star}$ rather than $A^{\star}(p)$.

As an immediate consequence of Theorem 4.12(b), one has that a point p of βS is an idempotent if and only if for every $A \in p$, $A^{\star}(p) \in p$. And, of course, if $A \in p$ and $s \in A^{\star}(p)$, then $s^{-1}A \in p$. In fact, one has the following stronger result.

Lemma 4.14. *Let $(S, \cdot)$ be a semigroup, let $p \cdot p = p \in \beta S$, and let $A \subseteq S$. For each $s \in A^{\star}(p)$, $s^{-1}\big(A^{\star}(p)\big) \in p$.*

Proof. Let $s \in A^{\star}(p)$, and let $B = s^{-1}A$. Then $B \in p$ and, since p is an idempotent, $B^{\star}(p) \in p$. We claim that $B^{\star}(p) \subseteq s^{-1}\big(A^{\star}(p)\big)$. So let $t \in B^{\star}(p)$. Then $t \in B$ so $st \in A$. Also $t^{-1}B \in p$. That is, $(st)^{-1}A \in p$. (See Exercise 4.1.4.) Since $st \in A$ and $(st)^{-1}A \in p$, one has that $st \in A^{\star}(p)$ as required. $\qquad\square$

Theorem 4.15. *Let $(S, \cdot)$ be a semigroup, let $p, q \in \beta S$, and let $A \subseteq S$. Then $A \in p \cdot q$ if and only if there exists $B \in p$ and an indexed family $\langle C_s \rangle_{s \in B}$ in q such that $\bigcup_{s \in B} sC_s \subseteq A$.*

Proof. This is Exercise 4.1.5. $\qquad\square$

Lemma 4.16. *Let $(S, \cdot)$ be a semigroup, let $s \in S$, let $q \in \beta S$ and let $A \subseteq S$.*
    (a) *If $A \in q$, then $sA \in s \cdot q$.*
    (b) *If S is left cancellative and $sA \in s \cdot q$ then $A \in q$.*

Proof. (a) We have $A \subseteq s^{-1}(sA)$ so Theorem 4.12(a) applies.
    (b) Since S is left cancellative, $s^{-1}(sA) = A$. $\qquad\square$

The following results will frequently be useful to us later.

Theorem 4.17. *Let S be a discrete semigroup, let (φ, T) be a semigroup compactification of S, and let $A \subseteq B \subseteq S$. Suppose that B is a subsemigroup of S.*
    (a) $c\ell(\varphi[B])$ *is a subsemigroup of T.*
    (b) *If A is a left ideal of B, then $c\ell(\varphi[A])$ is a left ideal of $c\ell(\varphi[B])$.*
    (c) *If A is a right ideal of B, then $c\ell(\varphi[A])$ is a right ideal of $c\ell(\varphi[B])$.*

Proof. (a) This is a consequence of Exercise 2.3.2.
    (b) Suppose that A is a left ideal of B. Let $x \in c\ell(\varphi[B])$ and $y \in c\ell(\varphi[A])$. Then $xy = \lim_{\varphi(s) \to x} \lim_{\varphi(t) \to y} \varphi(s)\varphi(t)$, where s denotes an element of B and t denotes an element of A. If $s \in B$ and $t \in A$, we have $\varphi(s)\varphi(t) = \varphi(st) \in \varphi[A]$ and so $xy \in c\ell\,\varphi[A]$.
    (c) The proof of (c) is similar to that of (b). $\qquad\square$

Corollary 4.18. *Let S be a subsemigroup of the discrete semigroup T. Then $c\ell\, S$ is a subsemigroup of βT. If S is a right or left ideal of T, then $c\ell\, S$ is respectively a right or left ideal of βT.*

Proof. By Theorem 4.8, $(\iota, \beta T)$ is a semigroup compactification of T, so Theorem 4.17 applies. $\qquad\square$

Remark 4.19. *Suppose that S is a subsemigroup of a discrete semigroup T. In the light of Exercise 4.1.10 (below), we can regard βS as being a subsemigroup of βT. In particular, it will often be convenient to regard $\beta \mathbb{N}$ as embedded in $\beta \mathbb{Z}$.*

Theorem 4.20. *Let $(S, \cdot)$ be a semigroup and let $\mathcal{A} \subseteq \mathcal{P}(S)$ have the finite intersection property. If for each $A \in \mathcal{A}$ and each $x \in A$, there exists $B \in \mathcal{A}$ such that $xB \subseteq A$, then $\bigcap_{A \in \mathcal{A}} \widehat{A}$ is a subsemigroup of βS.*

Proof. Let $T = \bigcap_{A \in \mathcal{A}} \widehat{A}$. Since $\mathcal{A}$ has the finite intersection property, $T \neq \emptyset$. Let $p, q \in T$ and let $A \in \mathcal{A}$. Given $x \in A$, there is some $B \in \mathcal{A}$ such that $xB \subseteq A$ and hence $x^{-1}A \in q$. Thus $A \subseteq \{x \in S : x^{-1}A \in q\}$ so $\{x \in S : x^{-1}A \in q\} \in p$. By Theorem 4.12(b), $A \in p \cdot q$. $\qquad\square$

Theorem 4.21. *Let $(S, \cdot)$ be a semigroup and let $\mathcal{A} \subseteq \mathcal{P}(S)$ have the finite intersection property. Let $(T, \cdot)$ be a compact right topological semigroup and let $\varphi : S \to T$ satisfy $\varphi[S] \subseteq \Lambda(T)$. Assume that there is some $A \in \mathcal{A}$ such that for each $x \in A$, there exists $B \in \mathcal{A}$ for which $\varphi(x \cdot y) = \varphi(x) \cdot \varphi(y)$ for every $y \in B$. Then for all $p, q \in \bigcap_{A \in \mathcal{A}} \widehat{A}$, $\widetilde{\varphi}(p \cdot q) = \widetilde{\varphi}(p) \cdot \widetilde{\varphi}(q)$.*

Proof. Let $p, q \in \bigcap_{A \in \mathcal{A}} \widehat{A}$. For each $x \in A$, one has

$$
\begin{aligned}
\widetilde{\varphi}(x \cdot q) &= \widetilde{\varphi}(x \cdot \lim_{y \to q} y) \\
&= \lim_{y \to q} \varphi(x \cdot y) && \text{because } \widetilde{\varphi} \circ \lambda_x \text{ is continuous} \\
&= \lim_{y \to q} \varphi(x) \cdot \varphi(y) && \text{because } \varphi(x \cdot y) = \varphi(x) \cdot \varphi(y) \text{ on a member of } q \\
&= \varphi(x) \cdot \lim_{y \to q} \varphi(y) && \text{because } \varphi(x) \in \Lambda(T) \\
&= \varphi(x) \cdot \widetilde{\varphi}(q).
\end{aligned}
$$

Since $A \in p$ one then has

$$
\begin{aligned}
\widetilde{\varphi}(p \cdot q) &= \widetilde{\varphi}\big((\lim_{x \to p} x) \cdot q\big) \\
&= \lim_{x \to p} \widetilde{\varphi}(x \cdot q) && \text{because } \widetilde{\varphi} \circ \rho_q \text{ is continuous} \\
&= \lim_{x \to p} \big(\varphi(x) \cdot \widetilde{\varphi}(q)\big) \\
&= \big(\lim_{x \to p} \varphi(x)\big) \cdot \widetilde{\varphi}(q) && \text{by the continuity of } \rho_{\widetilde{\varphi}(q)} \\
&= \widetilde{\varphi}(p) \cdot \widetilde{\varphi}(q). && \square
\end{aligned}
$$

Corollary 4.22. *Let $(S, \cdot)$ be a semigroup and let $\varphi : S \to T$ be a homomorphism to a compact right topological semigroup $(T, \cdot)$ such that $\varphi[S] \subseteq \Lambda(T)$. Then $\widetilde{\varphi}$ is a homomorphism from βS to T.*

Proof. Let $\mathcal{A} = \{S\}$. $\qquad\square$

Exercise 4.1.1. (a) Prove that if $q \in \beta\mathbb{N}\setminus\mathbb{N}$ and $r \in \mathbb{N}$, then $q + r \in \beta\mathbb{N}\setminus\mathbb{N}$.

(b) Verify that the operation $*$ in Example 4.6 is associative.

(c) Verify that the operation $*$ in Example 4.6 satisfies conclusions (a) and (b) of Theorem 4.1.

(d) Let $q \in \beta\mathbb{N}\setminus\mathbb{N}$ and let $A = \mathbb{N}2 = \{2n : n \in \mathbb{N}\}$. Prove that $A \in q = 1 * q$ if and only if $A \notin 1 + q$.

Exercise 4.1.2. Show that one cannot dispense with the assumption that $\varphi[S] \subseteq \Lambda(T)$ in Theorem 4.8. (Hint: consider Example 4.6.)

Exercise 4.1.3. Prove Theorem 4.12(b).

Exercise 4.1.4. Let S be a semigroup, let $s, t \in S$, and let $A \subseteq S$. Prove that $s^{-1}(t^{-1}A) = (ts)^{-1}A$. (Caution: This is very easy, but the reason is *not* that $s^{-1}t^{-1} = (ts)^{-1}$, for no such objects need exist.)

Exercise 4.1.5. Prove Theorem 4.15.

We can generalize Theorem 4.15 to the product of any finite number of ultrafilters. The following exercise shows that we obtain a set in $p_1 \cdot p_2 \cdot \ldots \cdot p_k$ by choosing all products of the form $a_1 \cdot a_2 \cdot \ldots \cdot a_k$, where each a_i is chosen to lie in a member of p_i which depends on $a_1, a_2, \ldots, a_{i-1}$.

Exercise 4.1.6. Let $(S, \cdot)$ be a semigroup, let $k \in \mathbb{N}\setminus\{1\}$, and let $p_1, p_2, \ldots, p_k \in \beta S$. Suppose that $D \subseteq \{\emptyset\} \cup \bigcup_{i=1}^{k-1} S^i$, with $\emptyset \in D$. Suppose also that, for each $\sigma \in D$, there is a subset A_σ of S such that the following conditions hold:

(1) $A_\emptyset \in p_1$,

(2) $A_\emptyset \subseteq D$,

(3) $A_\sigma \in p_{i+1}$ if $\sigma \in D \cap S^i$, and

(4) if $k > 2$, then for every $i \in \{1, 2, \ldots, k-2\}$ and every $\sigma = (a_1, a_2, \ldots, a_i) \in D$, $a \in A_\sigma$ implies that $(a_1, a_2, \ldots, a_i, a) \in D$.

Let $P = \{a_1 \cdot a_2 \cdot \ldots \cdot a_k : a_1 \in A_\emptyset \text{ and } (a_1, a_2, \ldots, a_{i-1}) \in D \text{ and } a_i \in A_{(a_1,a_2,\ldots,a_{i-1})}$ for every $i \in \{2, 3, \ldots, k\}\}$. Prove that $P \in p_1 \cdot p_2 \cdot \ldots \cdot p_k$. (Hint: This can be done by induction on k, using Theorem 4.12 or 4.15).

Exercise 4.1.7. Let $p \in \mathbb{N}^* + \mathbb{N}^*$ and let $A \in p$. Prove that there exists $k \in \mathbb{N}$ such that $\{a \in A : a + k \in A\}$ is infinite. Deduce that A cannot be arranged as a sequence $\langle x_n \rangle_{n=1}^\infty$ for which $x_{n+1} - x_n \to \infty$ as $n \to \infty$.

Exercise 4.1.8. Let $(S, \cdot)$ be a semigroup, let $A \subseteq S$, let $s \in S$ and let $q \in \beta S$.
(a) Prove that $A \in q \cdot s$ if and only if $As^{-1} \in q$.
(b) Prove that if $A \in q$, then $As \in q \cdot s$.
(c) Prove that if $As \in q \cdot s$ and S is right cancellative, then $A \in q$.

Exercise 4.1.9. Let $(S, \cdot)$ be a semigroup and let $s \in S$. Prove that the following statements are equivalent.

(a) For every $p \in \beta S$, $s \cdot p = \{sA : A \in p\}$.
(b) There is some $p \in \beta S$ such that $s \cdot p = \{sA : A \in p\}$.
(c) The function $\lambda_s : S \to S$ is surjective.

Exercise 4.1.10. Recall that if $S \subseteq T$, we have identified βS with the subset $\widehat{S}$ of βT. Show that if $(S, \cdot)$ is a subsemigroup of $(T, \cdot)$ and $p, q \in \beta S$, then the product $p \cdot q$ is the same whether it is computed in βS or in βT. That is, if $r = \{A \subseteq S : \{x \in S : x^{-1}A \in q\} \in p\}$ (the product $p \cdot q$ computed in βS), then $\{B \subseteq T : B \cap S \in r\} = \{A \subseteq T : \{x \in T : x^{-1}A \in q\} \in p\}$ (the product $p \cdot q$ computed in βT). (The notation $x^{-1}A$ is ambiguous. In the first case it should be $\{y \in S : xy \in A\}$ and in the second case it should be $\{y \in T : xy \in A\}$.)

Exercise 4.1.11. We recall that the semigroup operations $\vee$ and $\wedge$ are defined on $\mathbb{N}$ by $n \vee m = \max\{n, m\}$ and $n \wedge m = \min\{n, m\}$. As customary, we use the same notation for the extensions of these operations to $\beta\mathbb{N}$.

(a) Prove that for $p, q \in \beta\mathbb{N}$,

$$p \vee q = \begin{cases} q & \text{if } q \in \beta\mathbb{N}\backslash\mathbb{N} \\ p & \text{if } q \in \mathbb{N} \text{ and } p \in \beta\mathbb{N}\backslash\mathbb{N}. \end{cases}$$

(b) Derive and verify a similar characterization of $p \wedge q$.

4.2 Commutativity in βS

We provide some elementary results about commutativity in βS here. We shall study this subject more deeply in Chapter 6.

Theorem 4.23. *If $(S, \cdot)$ is a commutative semigroup, then S is contained in the center of $(\beta S, \cdot)$.*

Proof. Let $s \in S$ and $p \in \beta S$. Then

$$\begin{aligned} s \cdot q &= \lim_{t \to q} st & \text{by Remark 4.2(a)} \\ &= \lim_{t \to q} ts & \\ &= (\lim_{t \to q} t) \cdot s & \text{since } \rho_s \text{ is continuous} \\ &= q \cdot s. & \qquad\square \end{aligned}$$

Theorem 4.24. *Let S be a discrete commutative semigroup. Then the topological center of βS coincides with its algebraic center.*

Proof. Suppose that $p \in \Lambda(\beta S)$ and that $q \in \beta S$. Since λ_p is continuous, it follows that

$$\begin{aligned} p \cdot q &= q\text{-}\lim_{t \in S}(p \cdot t) \\ &= q\text{-}\lim_{t \in S}(t \cdot p) & \text{(by Theorem 4.23)} \\ &= q \cdot p. \end{aligned}$$

Thus p is also in the algebraic center of βS.

Conversely, if p is in the algebraic center of βS, λ_p is continuous because $\lambda_p = \rho_p$. Thus $p \in \Lambda(\beta S)$. $\qquad\square$

We have *defined* the operation on βS in such a way as to make βS right topological. It is not clear whether this somehow forces it to be semitopological. In the case in which S is commutative, we see that this question is equivalent to asking whether βS is commutative.

Theorem 4.25. *Let $(S, \cdot)$ be a commutative semigroup. The following statements are equivalent.*

(a) *$(\beta S, \cdot)$ is commutative.*

(b) *$(\beta S, \cdot)$ is a left topological semigroup.*

(c) *$(\beta S, \cdot)$ is a semitopological semigroup.*

Proof. The fact that (a) implies (b) is trivial as is the fact that (b) implies (c). It follows from Theorem 4.24 that (c) implies (a). $\qquad\square$

Lemma 4.26. *Let $(S, \cdot)$ be a semigroup, let $\langle x_n \rangle_{n=1}^{\infty}$ and $\langle y_n \rangle_{n=1}^{\infty}$ be sequences in S, and let $p, q \in \beta S$. If $\big\{ \{x_n : n > k\} : k \in \mathbb{N} \big\} \subseteq q$ and $\{y_k : k \in \mathbb{N}\} \in p$, then $\{y_k \cdot x_n : k, n \in \mathbb{N} \text{ and } k < n\} \in p \cdot q$.*

Proof. This follows from Theorem 4.15. $\qquad\square$

We are now able to characterize when βS is commutative. (A moment's thought will tell you that this characterization says "hardly ever".)

Theorem 4.27. *Let $(S, \cdot)$ be a semigroup. Then $(\beta S, \cdot)$ is not commutative if and only if there exist sequences $\langle x_n \rangle_{n=1}^{\infty}$ and $\langle y_n \rangle_{n=1}^{\infty}$ such that*

$$\{y_k \cdot x_n : k, n \in \mathbb{N} \text{ and } k < n\} \cap \{x_k \cdot y_n : k, n \in \mathbb{N} \text{ and } k < n\} = \emptyset.$$

Proof. Necessity. Pick p and q in βS such that $p \cdot q \neq q \cdot p$. Pick $A \subseteq S$ such that $A \in p \cdot q$ and $S \setminus A \in q \cdot p$. Let $B = \{s \in S : s^{-1}A \in q\}$ and let $C = \{s \in S : s^{-1}(S \setminus A) \in p\}$. Then $B \in p$ and $C \in q$. Choose $x_1 \in B$ and $y_1 \in C$. Inductively given $x_1, x_2, \ldots, x_n$ and $y_1, y_2, \ldots, y_n$, choose $x_{n+1} \in B \cap \bigcap_{k=1}^{n} y_k^{-1}(S \setminus A)$ and $y_{n+1} \in C \cap \bigcap_{k=1}^{n} x_k^{-1}A$. Then $\{y_k \cdot x_n : k, n \in \mathbb{N} \text{ and } k < n\} \subseteq S \setminus A$ and $\{x_k \cdot y_n : k, n \in \mathbb{N} \text{ and } k < n\} \subseteq A$.

Sufficiency. Now $\big\{ \{x_n : n > k\} : k \in \mathbb{N} \big\}$ has the finite intersection property so choose $p \in \beta S$ such that $\big\{ \{x_n : n > k\} : k \in \mathbb{N} \big\} \subseteq p$. Similarly, choose $q \in \beta S$ such that $\big\{ \{y_n : n > k\} : k \in \mathbb{N} \big\} \subseteq q$. Then by Lemma 4.26 $\{y_k \cdot x_n : k, n \in \mathbb{N} \text{ and } k < n\} \in q \cdot p$ and $\{x_k \cdot y_n : k, n \in \mathbb{N} \text{ and } k < n\} \in p \cdot q$. $\qquad\square$

As a consequence of Theorem 4.27, we see that neither $(\beta \mathbb{N}, +)$ nor $(\beta \mathbb{N}, \cdot)$ is commutative, and hence by Theorem 4.25 neither is a left topological semigroup. (As we shall see in Chapter 6, in fact the center of each is $\mathbb{N}$.)

Exercise 4.2.1. Prove that if S is a left zero semigroup, so is βS. In this case show that βS is not only a semitopological semigroup, but in fact a topological semigroup.

Exercise 4.2.2. Prove that if S is a right zero semigroup, so is βS. In this case show that βS is not only a semitopological semigroup, but in fact a topological semigroup. (Note that because of our lack of symmetry in the definition of the operation on βS this is *not* a "dual" of Exercise 4.2.1.)

4.3 S^*

For many reasons we are interested in the semigroup $S^* = \beta S \backslash S$. In the first place, it is the algebra of S^* that is the "new" material to study. In the second place it turns out that the structure of S^* provides most of the combinatorial applications that are a large part of our motivation for studying this subject.

One of the first things we want to know about the "semigroup S^*" is whether it *is* a semigroup.

Theorem 4.28. *Let S be a semigroup. Then S^* is a subsemigroup of βS if and only if for any $A \in \mathcal{P}_f(S)$ and for any infinite subset B of S there exists $F \in \mathcal{P}_f(B)$ such that $\bigcap_{x \in F} x^{-1} A$ is finite.*

Proof. Necessity. Let a finite nonempty subset A and an infinite subset B of S be given. Suppose that for each $F \in \mathcal{P}_f(B)$, $\bigcap_{x \in F} x^{-1} A$ is infinite. Then $\{x^{-1} A : x \in B\}$ has the property that all of its finite intersections are infinite so by Corollary 3.14 we may pick $p \in S^*$ such that $\{x^{-1} A : x \in B\} \subseteq p$. Pick $q \in S^*$ such that $B \in q$. Then $A \in q \cdot p$ and A is finite so by Theorem 3.7, $q \cdot p \in S$, a contradiction. (Recall once again that we have identified the principal ultrafilters with the points of S.)

Sufficiency. Let $p, q \in S^*$ be given and suppose that $q \cdot p = y \in S$, (that is, precisely, that $q \cdot p$ is the principal ultrafilter generated by y). Let $A = \{y\}$ and let $B = \{x \in S : x^{-1} A \in p\}$. Then $B \in q$ while for each $F \in \mathcal{P}_f(B)$, one has $\bigcap_{x \in F} x^{-1} A \in p$ so that $\bigcap_{x \in F} x^{-1} A$ is infinite, a contradiction.   □

Corollary 4.29. *Let S be a semigroup. If S is either right or left cancellative then S^* is a subsemigroup of βS.*

Proof. This is Exercise 4.3.1.   □

We can give simple conditions characterizing when S^* is a left ideal of βS.

Definition 4.30. A semigroup S is *weakly left cancellative* if and only if for all $u, v \in S$, $\{x \in S : ux = v\}$ is finite. Similarly, S is *weakly right cancellative* if and only if for all $u, v \in S$, $\{x \in S : xu = v\}$ is finite.

Of course a left cancellative semigroup is weakly left cancellative. On the other hand the semigroup $(\mathbb{N}, \vee)$ is weakly left (and right) cancellative but is far from being cancellative.

When we say that a function f is *finite-to-one*, we mean that for each x in the range of f, $f^{-1}[\{x\}]$ is finite. Thus a semigroup S is weakly left cancellative if and only if for each $x \in S$, λ_x is finite-to-one.

Theorem 4.31. *Let S be an infinite semigroup. Then S^* is a left ideal of βS if and only if S is weakly left cancellative.*

Proof. Necessity. Let $x, y \in S$ be given, let $A = \lambda_y^{-1}[\{x\}]$ and suppose that A is infinite. Pick $p \in S^* \cap \widehat{A}$. Then $y \cdot p = x$, a contradiction.

Sufficiency. Since S is infinite, $S^* \neq \emptyset$. Let $p \in S^*$, let $q \in \beta S$ and suppose that $q \cdot p = x \in S$. Then $\{x\} \in q \cdot p$ so $\{y \in S : y^{-1}\{x\} \in p\} \in q$ and is hence nonempty. So pick $y \in S$ such that $y^{-1}\{x\} \in p$. But $y^{-1}\{x\} = \lambda_y^{-1}[\{x\}]$ so $\lambda_y^{-1}[\{x\}]$ is infinite, a contradiction. □

The characterization of S^* as a right ideal is considerably more complicated.

Theorem 4.32. *Let S be an infinite semigroup. The following statements are equivalent.*
(a) *S^* is a right ideal of βS.*
(b) *Given any finite subset A of S, any sequence $\langle z_n \rangle_{n=1}^{\infty}$ in S, and any one-to-one sequence $\langle x_n \rangle_{n=1}^{\infty}$ in S, there exist $n < m$ in $\mathbb{N}$ such that $x_n \cdot z_m \notin A$.*
(c) *Given any $a \in S$, any sequence $\langle z_n \rangle_{n=1}^{\infty}$ in S, and any one-to-one sequence $\langle x_n \rangle_{n=1}^{\infty}$ in S, there exist $n < m$ in $\mathbb{N}$ such that $x_n \cdot z_m \neq a$.*

Proof. (a) implies (b). Suppose $\{x_n \cdot z_m : n, m \in \mathbb{N} \text{ and } n < m\} \subseteq A$. Pick $p \in \beta S$ such that $\big\{\{z_m : m > n\} : n \in \mathbb{N}\big\} \subseteq p$ and pick $q \in S^*$ such that $\{x_n : n \in \mathbb{N}\} \in q$, which one can do, since $\{x_n : n \in \mathbb{N}\}$ is infinite. Then by Lemma 4.26, $A \in q \cdot p$ so by Theorem 3.7, $q \cdot p \in S$, a contradiction.

The fact that (b) implies (c) is trivial.

(c) implies (a). Since S is infinite, $S^* \neq \emptyset$. Let $p \in \beta S$, let $q \in S^*$, and suppose that $q \cdot p = a \in S$. Then $\{s \in S : s^{-1}\{a\} \in p\} \in q$ so choose a one-to-one sequence $\langle x_n \rangle_{n=1}^{\infty}$ such that $\{x_n : n \in \mathbb{N}\} \subseteq \{s \in S : s^{-1}\{a\} \in p\}$. Inductively choose a sequence $\langle z_n \rangle_{n=1}^{\infty}$ in S such that for each $m \in \mathbb{N}$, $z_m \in \bigcap_{n=1}^{m} x_n^{-1}\{a\}$ (which one can do since $\bigcap_{n=1}^{m} x_n^{-1}\{a\} \in p$). Then for each $n < m$ in $\mathbb{N}$, $x_n \cdot z_m = a$, a contradiction. □

We see that cancellation on the appropriate side guarantees that S^* is a left or right ideal of βS. In particular, if S is cancellative, then S^* is a (two sided) ideal of βS.

Corollary 4.33. *Let S be an infinite semigroup.*
(a) *If S is left cancellative, then S^* is a left ideal of βS.*
(b) *If S is right cancellative, then S^* is a right ideal of βS.*

Proof. This is Exercise 4.3.2. □

Corollary 4.34. *Let S be an infinite semigroup. If S^* is a right ideal of βS, then S is weakly right cancellative.*

Proof. Let $a, y \in S$ and suppose that $\rho_y^{-1}[\{a\}]$ is infinite. Choose a one-to-one sequence $\langle x_n \rangle_{n=1}^{\infty}$ in $\rho_y^{-1}[\{a\}]$ and for each $m \in \mathbb{N}$ let $z_m = y$. □

Corollary 4.35. *Let S be an infinite semigroup. If S is weakly right cancellative and for all but finitely many $y \in S$, λ_y is finite-to-one, then S^* is a right ideal of βS.*

Proof. Suppose S^* is not a right ideal of βS and pick by Theorem 4.32(c) some $a \in S$, a one-to-one sequence $\langle x_n \rangle_{n=1}^{\infty}$ in S, and a sequence $\langle z_n \rangle_{n=1}^{\infty}$ in S such that $x_n \cdot z_m = a$ for all $n < m$ in $\mathbb{N}$. Now if $\{z_m : m \in \mathbb{N}\}$ is infinite, then for all $n \in \mathbb{N}$, $\lambda_{x_n}^{-1}[\{a\}]$ is infinite, a contradiction. Thus $\{z_m : m \in \mathbb{N}\}$ is finite so we may pick b such that $\{m \in \mathbb{N} : z_m = b\}$ is infinite. Then, given $n \in \mathbb{N}$ one may pick $m > n$ such that $z_m = b$ and conclude that $x_n \cdot b = a$. Consequently, $\rho_b^{-1}[\{a\}]$ is infinite, a contradiction. □

Somewhat surprisingly, the situation with respect to S^* as a two sided ideal is considerably simpler than the situation with respect to S^* as a right ideal.

Theorem 4.36. *Let S be an infinite semigroup. Then S^* is an ideal of βS if and only if S is both weakly left cancellative and weakly right cancellative.*

Proof. Necessity. Theorem 4.31 and Corollary 4.34.
Sufficiency. Theorem 4.31 and Corollary 4.35. □

The following simple fact is of considerable importance, since it is frequently easier to work with S^* than with βS.

Theorem 4.37. *Let S be an infinite semigroup. If S^* is an ideal of βS, then the minimal left ideals, the minimal right ideals, and the smallest ideal of S^* and of βS are the same.*

Proof. For this it suffices to establish the assertions about minimal left ideals and minimal right ideals, since the smallest ideal is the union of all minimal left ideals (and of all minimal right ideals). Further, the proofs are completely algebraic, so it suffices to establish the result for minimal left ideals.

First assume L is a minimal left ideal of βS. Since S^* is an ideal of βS we have $L \subseteq S^*$ and hence L is a left ideal of S^*. To see that L is a minimal left ideal of S^*, let L' be a left ideal of S^* with $L' \subseteq L$. Then by Lemma 1.43(b), $L' = L$.

Now assume L is a minimal left ideal of S^*. Then by Lemma 1.43(c), L is a left ideal of βS. If L' is a left ideal of βS with $L' \subseteq L$, then L' is a left ideal of S^* and consequently $L' = L$. □

Exercise 4.3.1. Prove Corollary 4.29.

Exercise 4.3.2. Prove Corollary 4.33.

Exercise 4.3.3. Give an example of a semigroup S which is not left cancellative such that S^* is a left ideal of βS.

Exercise 4.3.4. Give an example of a semigroup S which is not right cancellative such that S^* is a right ideal of βS.

Exercise 4.3.5. Prove that $(\mathbb{N}^*, +)$ and $(-\mathbb{N}^*, +)$ are both left ideals of $(\beta \mathbb{Z}, +)$. (Here $-\mathbb{N}^* = \{-p : p \in \mathbb{N}^*\}$ and $-p$ is the ultrafilter on $\mathbb{Z}$ generated by $\{-A : A \in p\}$.)

Exercise 4.3.6. Let $T = \bigcap_{n \in \mathbb{N}} c\ell_{\beta \mathbb{Z}} \, n\mathbb{Z}$.
  (a) Prove that T is a subsemigroup of $(\beta \mathbb{Z}, +)$ which contains all of the idempotents. (Hint: Use the canonical homomorphisms from $\mathbb{Z}$ onto $\mathbb{Z}_n$.)
  (b) Prove that T is an ideal of $(\beta \mathbb{Z}, \cdot)$.

4.4 $K(\beta S)$ and Its Closure

In this section we determine precisely which ultrafilters are in the smallest ideal $K(\beta S)$ of βS and which are in its closure.

We borrow some terminology from topological dynamics. The terms *syndetic* and *piecewise syndetic* originated in the context of $(\mathbb{N}, +)$. In $(\mathbb{N}, +)$, a set A is syndetic if and only if it has bounded gaps and a set is piecewise syndetic if and only if there exist a fixed bound b and arbitrarily long intervals in which the gaps of A are bounded by b.

Definition 4.38. Let S be a semigroup.
  (a) A set $A \subseteq S$ is *syndetic* if and only if there exists some $G \in \mathcal{P}_f(S)$ such that $S = \bigcup_{t \in G} t^{-1}A$.
  (b) A set $A \subseteq S$ is *piecewise syndetic* if and only if there is some $G \in \mathcal{P}_f(S)$ such that $\{a^{-1}(\bigcup_{t \in G} t^{-1}A) : a \in S\}$ has the finite intersection property.

Equivalently, a set $A \subseteq S$ is piecewise syndetic if and only if there is some $G \in \mathcal{P}_f(S)$ such that for every $F \in \mathcal{P}_f(S)$ there is some $x \in S$ with $F \cdot x \subseteq \bigcup_{t \in G} t^{-1}A$.

We should really call the notions defined above *right syndetic* and *right piecewise syndetic*. If we were taking βS to be left topological we would have replaced "$S = \bigcup_{t \in G} t^{-1}A$" in the definition of syndetic by "$S = \bigcup_{t \in G} At^{-1}$". We also would have replaced "$a^{-1}(\bigcup_{t \in G} t^{-1}A)$" in the definition of piecewise syndetic by "$(\bigcup_{t \in G} At^{-1})a^{-1}$". We shall see in Lemma 13.39 that the notions of left and right piecewise syndetic are different.

The equivalence of statements (a) and (c) in the following theorem follows from Theorem 2.10. However, it requires no extra effort to provide a complete proof here, so we do.

Theorem 4.39. *Let S be a semigroup and let $p \in \beta S$. The following statements are equivalent.*

  (a) *$p \in K(\beta S)$.*
  (b) *For all $A \in p$, $\{x \in S : x^{-1}A \in p\}$ is syndetic.*
  (c) *For all $q \in \beta S$, $p \in \beta S \cdot q \cdot p$.*

Proof. (a) implies (b). Let $A \in p$ and let $B = \{x \in S : x^{-1}A \in p\}$. Let L be the minimal left ideal of βS for which $p \in L$. For every $q \in L$, we have $p \in \beta S \cdot q = c\ell_{\beta S}(S \cdot q)$. Since $\overline{A}$ is a neighbourhood of p in βS, $t \cdot q \in \overline{A}$ for some $t \in S$ and so $q \in \overline{t^{-1}A}$. Thus the sets of the form $\overline{t^{-1}A}$ cover the compact set L and hence $L \subseteq \bigcup_{t \in G} \overline{t^{-1}A}$ for some finite subset G of S.

To see that $S \subseteq \bigcup_{t \in G} t^{-1}B$, let $a \in S$. Then $a \cdot p \in L$ so pick $t \in G$ such that $a \cdot p \in \overline{t^{-1}A}$. Then $t^{-1}A \in a \cdot p$ so that $(ta)^{-1}A = a^{-1}(t^{-1}A) \in p$ and so $ta \in B$ and thus $a \in t^{-1}B$.

(b) implies (c). Let $q \in \beta S$ and suppose that $p \notin \beta S \cdot q \cdot p$. Pick $A \in p$ such that $\overline{A} \cap \beta S \cdot q \cdot p = \emptyset$. Let $B = \{x \in S : x^{-1}A \in p\}$ and pick $G \in \mathcal{P}_f(S)$ such that $S = \bigcup_{t \in G} t^{-1}B$. Pick $t \in G$ such that $t^{-1}B \in q$. Then $B \in tq$. That is, $\{x \in S : x^{-1}A \in p\} \in tq$ so $A \in tqp$, a contradiction.

(c) implies (a). Pick $q \in K(\beta S)$. $\square$

Theorem 4.40. *Let S be a semigroup and let $A \subseteq S$. Then $\overline{A} \cap K(\beta S) \neq \emptyset$ if and only if A is piecewise syndetic.*

Proof. Necessity. Let $p \in K(\beta S) \cap \overline{A}$ and let $B = \{x \in S : x^{-1}A \in p\}$. Then by Theorem 4.39, B is syndetic and so $S = \bigcup_{t \in G} t^{-1}B$ for some $G \in \mathcal{P}_f(S)$. For each $a \in S$, $a \in t^{-1}B$ for some $t \in G$ and so $a^{-1}(t^{-1}A) = (ta)^{-1}A \in p$. It follows that $a^{-1}(\bigcup_{t \in G} t^{-1}A) \in p$ and hence that $\{a^{-1}(\bigcup_{t \in G} t^{-1}A) : a \in S\}$ has the finite intersection property.

Sufficiency. Assume that A is piecewise syndetic and pick $G \in \mathcal{P}_f(S)$ such that $\{a^{-1}(\bigcup_{t \in G} t^{-1}A) : a \in S\}$ has the finite intersection property. Pick $q \in \beta S$ such that $\{a^{-1}(\bigcup_{t \in G} t^{-1}A) : a \in S\} \subseteq q$. Then $S \cdot q \subseteq \overline{\bigcup_{t \in G} t^{-1}A}$. This implies that $(\beta S) \cdot q \subseteq \overline{\bigcup_{t \in G} t^{-1}A}$. We can choose $y \in K(\beta S) \cap (\beta S \cdot q)$. We then have $y \in \overline{t^{-1}A}$ for some $t \in G$ and so $t \cdot y \in \overline{A} \cap K(\beta S)$. $\square$

Corollary 4.41. *Let S be a semigroup and let $p \in \beta S$. Then $p \in c\ell K(\beta S)$ if and only if every $A \in p$ is piecewise syndetic.*

Proof. This is an immediate consequence of Theorem 4.40. $\square$

The members of idempotents in $K(\beta S)$ are of particular combinatorial interest. (See Chapter 14.)

Definition 4.42. *Let S be a semigroup and let $A \subseteq S$. Then A is central in S if and only if there is some idempotent $p \in K(\beta S)$ such that $A \in p$.*

Theorem 4.43. *Let S be an infinite semigroup and let $A \subseteq S$. The following statements are equivalent.*

(a) *A is piecewise syndetic.*

(b) *The set $\{x \in S : x^{-1}A$ is central$\}$ is syndetic.*

(c) *There is some $x \in S$ such that $x^{-1}A$ is central.*

Proof. (a) implies (b). Pick by Theorem 4.40 some $p \in K(\beta S)$ with $A \in p$. Now $K(\beta S)$ is the union of all minimal left ideals of βS by Theorem 2.8. So pick a minimal left ideal L of βS with $p \in L$ and pick an idempotent $e \in L$. Then $p = p \cdot e$ so pick $y \in S$ such that $y^{-1}A \in e$.

Now by Theorem 4.39 $B = \{z \in S : z^{-1}(y^{-1}A) \in e\}$ is syndetic, so pick finite $G \subseteq S$ such that $S = \bigcup_{t \in G} t^{-1}B$. Let $D = \{x \in S : x^{-1}A$ is central$\}$. We claim that $S = \bigcup_{t \in (y \cdot G)} t^{-1}D$. Indeed, let $x \in S$ be given and pick $t \in G$ such that $t \cdot x \in B$. Then $(t \cdot x)^{-1}(y^{-1}A) \in e$ so $(t \cdot x)^{-1}(y^{-1}A)$ is central. But $(t \cdot x)^{-1}(y^{-1}A) = (y \cdot t \cdot x)^{-1}A$. Thus $y \cdot t \cdot x \in D$ so $x \in (y \cdot t)^{-1}D$ as required.

(b) implies (c). This is trivial.

(c) implies (a). Pick $x \in S$ such that $x^{-1}A$ is central and pick an idempotent $p \in K(\beta S)$ such that $x^{-1}A \in p$. Then $A \in x \cdot p$ and $x \cdot p \in K(\beta S)$ so by Theorem 4.40, A is piecewise syndetic. □

Recall from Theorem 2.15 and Example 2.16 that the closure of any right ideal in a right topological semigroup is again a right ideal but the closure of a left ideal need not be a left ideal.

Theorem 4.44. *Let S be a semigroup. Then* $c\ell\, K(\beta S)$ *is an ideal of βS.*

Proof. This is an immediate consequence of Theorems 2.15, 2.17, and 4.1. □

Exercise 4.4.1. Let S be a discrete semigroup and let $A \subseteq S$. Show that A is piecewise syndetic if and only if there is a finite subset F of S for which $c\ell_{\beta S}\left(\bigcup_{t \in F} t^{-1}A\right)$ contains a left ideal of βS.

Exercise 4.4.2. Let $A \subseteq \mathbb{N}$. Show that A is piecewise syndetic in $(\mathbb{N}, +)$ if and only if there exists $k \in \mathbb{N}$ such that, for every $n \in \mathbb{N}$, there is a set J of n consecutive positive integers with the property that A intersects every subset of J containing k consecutive integers.

Exercise 4.4.3. Prove that $\{\sum_{n \in F} 2^{2n} : F \in \mathcal{P}_f(\omega)\}$ is not piecewise syndetic in $(\mathbb{N}, +)$.

Exercise 4.4.4. Let $A \subseteq \mathbb{N}$. Show that $K(\beta \mathbb{N}, +) \subseteq \overline{A}$ if and only if, for every $k \in \mathbb{N}$, there exists $n_k \in \mathbb{N}$ such that every subset of $\mathbb{N}$ which contains n_k consecutive integers must contain k consecutive integers belonging to A.

Exercise 4.4.5. Let G be a group and let H be a subgroup of G. Prove that H is syndetic if and only if H is piecewise syndetic. (Hint: By Theorem 4.40, pick $p \in K(\beta G)$ such that $H \in p$ and consider $\{x \in G : x^{-1}H \in p\}$.)

Exercise 4.4.6. Let G be a group and let H be a subgroup of G. Show that the index of H is finite if and only if $c\ell_{\beta G}(H) \cap K(\beta G) \neq \emptyset$. (By the index of H, we mean the number of left cosets of the form aH).

Exercise 4.4.7. Let G be a group. Suppose that G can be expressed as the union of a finite number of subgroups, $H_1, H_2, \ldots, H_n$. Show that, for some $i \in \{1, 2, \ldots, n\}$, H_i has finite index.

Exercise 4.4.8. Let S be a discrete semigroup and let T be a subsemigroup of S. Show that T is central if $(c\ell_{\beta S}\, T) \cap K(\beta S) \neq \emptyset$. (Apply Corollary 4.18 and Theorem 1.65.)

Exercise 4.4.9. Show that if S is commutative, then the closure of any right ideal of βS is a two sided ideal of βS. (Hint: See Theorems 2.19 and 4.23.)

Notes

We have chosen to extend the operation in such a way as to make $(\beta S, \cdot)$ a right topological semigroup. One can equally well extend the operation so as to make $(\beta S, \cdot)$ a left topological semigroup and in fact this choice is often made in the literature. (It used to be the customary choice of the first author of this book.) It would seem then that one could find two situations in the literature. But no, one instead finds four! The reader will recall that in Section 2.1 we remarked that what we refer to as "right topological" is called by some authors "left topological" and *vice versa*. In the following table we include one citation from our Bibliography, where the particular combination of choices is made. Obviously, when referring to the literature one must be careful to determine what the author means.

	Called Right Topological	Called Left Topological
ρ_p Continuous	[192]	[45]
λ_p Continuous	[215]	[178]

Example 4.6 is due to J. Baker and R. Butcher [6]. The existence of a compactification with the properties of βS given by Theorem 4.8 is [40, Theorem 4.5.3], where it is called the $\mathcal{LMC}$-compactification of S. The fact that an operation on a discrete semigroup S can be extended to βS was first implicitly established by M. Day in [75] using methods of R. Arens [3]. The first explicit statement seems to have been made by P. Civin and B. Yood [64]. These mathematicians tended to view βS as a subspace of the dual of the real or complex valued functions on S. The extension to βS as a space of ultrafilters is done by R. Ellis [86, Chapter 8] in the case that S is a group.

Corollary 4.22 is due to P. Milnes [184].

The results in Section 4.3 are special cases of results from [130] which are in turn special cases of results due to D. Davenport in [72].

The notion of "central" has its origins in topological dynamics. See Chapter 19 for the dynamical definition and a proof of the equivalence of the notions.

Exercise 4.4.7 was suggested by I. Protasov. This result is due to B. Neumann [186], who proved something stronger: if the union of the subgroups H_i is irredundant, then every H_i has finite index.

Chapter 5

βS and Ramsey Theory — Some Easy Applications

We describe in this chapter some easy applications of the algebraic structure of βS to the branch of combinatorics known as "Ramsey Theory".

5.1 Ramsey Theory

We begin our discussion of the area of mathematics known as *Ramsey Theory* by illustrating the subject area by example. We cite here several of the classic theorems of the field. They will all be proved in this book. The oldest is the 1892 result of Hilbert. It involves the notion of *finite sums*, which will be of continuing interest to us in this book, so we pause to introduce some notation. Recall that in a noncommutative semigroup we have defined $\prod_{i=1}^{n} x_i$ to be the product in increasing order of indices. Similarly, we take $\prod_{n \in F} x_n$ to be the product in increasing order of indices. So, for example, if $F = \{2, 5, 6, 9\}$, then $\prod_{n \in F} x_n = x_2 \cdot x_5 \cdot x_6 \cdot x_9$.

We introduce separate notation ("FP" for "finite products" and "FS" for "finite sums") depending on whether the operation of the semigroup is denoted by $\cdot$ or $+$.

Definition 5.1. (a) Let $(S, \cdot)$ be a semigroup. Given an infinite sequence $\langle x_n \rangle_{n=1}^{\infty}$ in S, $\mathrm{FP}(\langle x_n \rangle_{n=1}^{\infty}) = \{\prod_{n \in F} x_n : F \in \mathcal{P}_f(\mathbb{N})\}$. Given a finite sequence $\langle x_n \rangle_{n=1}^{m}$ in S, $\mathrm{FP}(\langle x_n \rangle_{n=1}^{m}) = \{\prod_{n \in F} x_n : F \in \mathcal{P}_f(\{1, 2, \ldots, m\})\}$.

(b) Let $(S, +)$ be a semigroup. Given an infinite sequence $\langle x_n \rangle_{n=1}^{\infty}$ in S, $\mathrm{FS}(\langle x_n \rangle_{n=1}^{\infty}) = \{\sum_{n \in F} x_n : F \in \mathcal{P}_f(\mathbb{N})\}$. Given a finite sequence $\langle x_n \rangle_{n=1}^{m}$ in S, $\mathrm{FS}(\langle x_n \rangle_{n=1}^{m}) = \{\sum_{n \in F} x_n : F \in \mathcal{P}_f(\{1, 2, \ldots, m\})\}$.

Theorem 5.2 (Hilbert [116]). *Let $r \in \mathbb{N}$ and let $\mathbb{N} = \bigcup_{i=1}^{r} A_i$. For each $m \in \mathbb{N}$ there exist $i \in \{1, 2, \ldots, r\}$, a sequence $\langle x_n \rangle_{n=1}^{m}$ in $\mathbb{N}$, and an infinite set $B \subseteq \mathbb{N}$ such that for each $a \in B$, $a + \mathrm{FS}(\langle x_n \rangle_{n=1}^{m}) \subseteq A_i$.*

Proof. This is a consequence of Corollary 5.10. $\qquad\square$

The next classical result is the 1916 result of Schur, which allows one to omit the translates on the finite sums when $m = 2$.

Theorem 5.3 (Schur [221]). *Let $r \in \mathbb{N}$ and let $\mathbb{N} = \bigcup_{i=1}^{r} A_i$. There exist $i \in \{1, 2, \ldots, r\}$ and x and y in $\mathbb{N}$ with $\{x, y, x + y\} \subseteq A_i$.*

Proof. This is a consequence of Corollary 5.10. □

One of the most famous results of the field is the 1927 result of van der Waerden guaranteeing "monochrome" arithmetic progressions. (The statement "let $r \in \mathbb{N}$ and let $\mathbb{N} = \bigcup_{i=1}^{r} A_i$" is often replaced by "let $r \in \mathbb{N}$ and let $\mathbb{N}$ be r-colored", in which case the conclusion "$\ldots \subseteq A_i$" is replaced by "$\ldots$ is monochrome".)

Theorem 5.4 (van der Waerden [238]). *Let $r \in \mathbb{N}$ and let $\mathbb{N} = \bigcup_{i=1}^{r} A_i$. For each $\ell \in \mathbb{N}$ there exist $i \in \{1, 2, \ldots, r\}$ and $a, d \in \mathbb{N}$ such that $\{a, a + d, \ldots, a + \ell d\} \subseteq A_i$.*

Proof. This is Corollary 14.2. □

Since the first of the classical results in this area are due to Hilbert, Schur, and van der Waerden, one may wonder why it is called "Ramsey Theory". The reason lies in the kind of result proved by Ramsey in 1930. It is a more general structural result not dependent on the arithmetic structure of $\mathbb{N}$.

Definition 5.5. Let A be a set and let κ be a cardinal number.
(a) $[A]^{\kappa} = \{B \subseteq A : |A| = \kappa\}$.
(b) $[A]^{<\kappa} = \{B \subseteq A : |A| < \kappa\}$.

Theorem 5.6 (Ramsey [209]). *Let Y be an infinite set and let $k, r \in \mathbb{N}$. If $[Y]^{k} = \bigcup_{i=1}^{r} A_i$, then there exist $i \in \{1, 2, \ldots, r\}$ and an infinite subset B of Y with $[B]^{k} \subseteq A_i$.*

Proof. This is Theorem 18.2. □

Notice that the case $k = 1$ of Ramsey's Theorem is the pigeon hole principle.

We mention also that another fundamental early result of Ramsey Theory is Rado's Theorem [206]. It involves the introduction of several new notions, so we shall not state it here. We shall however present and prove Rado's Theorem in Chapter 15.

Notice that all of these theorems are instances of the following general statement: One has a set X and a collection of "good" subsets $\mathcal{G}$ of X. One then asserts that whenever $r \in \mathbb{N}$ and $X = \bigcup_{i=1}^{r} A_i$ there will exist $i \in \{1, 2, \ldots, r\}$ and $G \in \mathcal{G}$ with $G \subseteq A_i$. (Consider the following table.)

Theorem	X	$G \in \mathcal{G}$
Hilbert	$\mathbb{N}$	$\bigcup_{a \in B} \left(a + \mathrm{FS}(\langle x_n \rangle_{n=1}^{m}) \right)$
Schur	$\mathbb{N}$	$\{x, y, x + y\}$
van der Waerden	$\mathbb{N}$	$\{a, a + d, \ldots, a + \ell d\}$
Ramsey	$[Y]^{k}$	$[B]^{k}$

It would not be entirely accurate, but certainly not far off the mark, to define *Ramsey Theory* as the classification of pairs $(X, \mathcal{G})$ for which the above statement is true. We see now that under this definition, any question in Ramsey Theory is a question about ultrafilters.

Theorem 5.7. *Let X be a set and let $\mathcal{G} \subseteq \mathcal{P}(X)$. The following statements are equivalent.*

(a) *Whenever $r \in \mathbb{N}$ and $X = \bigcup_{i=1}^{r} A_i$, there exist $i \in \{1, 2, \ldots, r\}$ and $G \in \mathcal{G}$ such that $G \subseteq A_i$.*

(b) *There is an ultrafilter p on X such that for every member A of p, there exists $G \in \mathcal{G}$ with $G \subseteq A$.*

Proof. (a) implies (b). If $\emptyset \in \mathcal{G}$, then any ultrafilter on X will do, so assume $\emptyset \notin \mathcal{G}$. Let $\mathcal{A} = \{B \subseteq X :$ for every $G \in \mathcal{G}, B \cap G \neq \emptyset\}$. Then $\mathcal{A}$ has the finite intersection property. (For suppose one has $\{B_1, B_2, \ldots, B_r\} \subseteq \mathcal{A}$ with $\bigcap_{i=1}^{r} B_i = \emptyset$. Then $X = \bigcup_{i=1}^{r}(X \setminus B_i)$ so there exist some $i \in \{1, 2, \ldots, r\}$ and some $G \in \mathcal{G}$ with $G \cap B_i = \emptyset$, a contradiction.) Pick by Theorem 3.8 some ultrafilter p on X with $\mathcal{A} \subseteq p$. Then given any $A \in p$, $X \setminus A \notin \mathcal{A}$ so there is some $G \in \mathcal{G}$ with $G \cap (X \setminus A) = \emptyset$.

(b) implies (a). For some $i \in \{1, 2, \ldots, r\}$, $A_i \in p$. $\square$

Exercise 5.1.1. Prove that Theorem 5.4 implies the following superficially stronger statement.

Let $r \in \mathbb{N}$ and let $\mathbb{N} = \bigcup_{i=1}^{r} A_i$. There exists $i \in \{1, 2, \ldots, r\}$ such that for each $\ell \in \mathbb{N}$, there exist $a, d \in \mathbb{N}$ such that $\{a, a + d, \ldots, a + \ell d\} \subseteq A_i$.

5.2 Idempotents and Finite Products

The first application of the algebraic structure of βS to Ramsey Theory was the Galvin–Glazer proof of the Finite Sums Theorem (Corollary 5.10). It established an intimate relationship between finite sums (or products) in S and idempotents in βS. We present two proofs. The first of these is the original and is given for historic reasons as well as the fact that variations on this proof will be used later. The second, simpler, proof is of recent origin.

Theorem 5.8. *Let S be a semigroup, let p be an idempotent in βS, and let $A \in p$. There is a sequence $\langle x_n \rangle_{n=1}^{\infty}$ in S such that $\mathrm{FP}(\langle x_n \rangle_{n=1}^{\infty}) \subseteq A$.*

First Proof. Let $A_1 = A$ and let $B_1 = \{x \in S : x^{-1}A_1 \in p\}$. Since $A_1 \in p = p \cdot p$, $B_1 \in p$. Pick $x_1 \in B_1 \cap A_1$, let $A_2 = A_1 \cap (x_1^{-1}A_1)$, and note that $A_2 \in p$. Inductively given $A_n \in p$, let $B_n = \{x \in S : x^{-1}A_n \in p\}$. Since $A_n \in p = p \cdot p$, $B_n \in p$. Pick $x_n \in B_n \cap A_n$ and let $A_{n+1} = A_n \cap (x_n^{-1}A_n)$.

To see for example why $x_2 \cdot x_4 \cdot x_5 \cdot x_7 \in A$, note that $x_7 \in A_7 \subseteq A_6 \subseteq x_5^{-1}A_5$ so that $x_5 \cdot x_7 \in A_5 \subseteq x_4^{-1}A_4$. Thus $x_4 \cdot x_5 \cdot x_7 \in A_4 \subseteq A_3 \subseteq x_2^{-1}A_2$ so that $x_2 \cdot x_4 \cdot x_5 \cdot x_7 \in A_2 \subseteq A_1 = A$.

More formally, we show by induction on $|F|$ that if $F \in \mathcal{P}_f(\mathbb{N})$ and $m = \min F$ then $\prod_{n \in F} x_n \in A_m$. If $|F| = 1$, then $\prod_{n \in F} x_n = x_m \in A_m$. Assume $|F| > 1$, let $G = F \setminus \{m\}$, and let $k = \min G$. Note that since $k > m$, $A_k \subseteq A_{m+1}$. Then by the induction hypothesis, $\prod_{n \in G} x_n \in A_k \subseteq A_{m+1} \subseteq x_m^{-1} A_m$ so $\prod_{n \in F} x_n = x_m \cdot \prod_{n \in G} x_n \in A_m$.

Second Proof. Recall that $A^\star(p) = \{x \in A : x^{-1} A \in p\}$ and write $A^\star$ for $A^\star(p)$. Pick $x_1 \in A^\star$. Let $n \in \mathbb{N}$ and assume we have chosen $\langle x_t \rangle_{t=1}^n$ such that $\mathrm{FP}(\langle x_t \rangle_{t=1}^n) \subseteq A^\star$. Let $E = \mathrm{FP}(\langle x_t \rangle_{t=1}^n)$. Then E is finite and for each $a \in E$ we have by Lemma 4.14 that $a^{-1} A^\star \in p$ and hence $\bigcap_{a \in E} a^{-1} A^\star \in p$. Pick $x_{n+1} \in A^\star \cap \bigcap_{a \in E} a^{-1} A^\star$. Then $x_{n+1} \in A^\star$ and given $a \in E$, $a \cdot x_{n+1} \in A^\star$ and hence $\mathrm{FP}(\langle x_t \rangle_{t=1}^{n+1}) \subseteq A^\star$. $\square$

Corollary 5.9. *Let S be a semigroup, let $r \in \mathbb{N}$ and let $S = \bigcup_{i=1}^r A_i$. There exist $i \in \{1, 2, \ldots, r\}$ and a sequence $\langle x_n \rangle_{n=1}^\infty$ in S such that $\mathrm{FP}(\langle x_n \rangle_{n=1}^\infty) \subseteq A_i$.*

Proof. Pick by Theorem 2.5 an idempotent $p \in \beta S$ and pick $i \in \{1, 2, \ldots, r\}$ such that $A_i \in p$. Apply Theorem 5.8. $\square$

Note that in the proof of Theorem 5.8, the fact that we have chosen to take our products in increasing order of indices is important. However, the corresponding version of Corollary 5.9 taking products in decreasing order of indices remains true.

The hard way to see this is to redefine the operation on βS so that it is a left topological semigroup. In this case one gets that $A \in p \cdot q$ if and only if $\{x \in S : Ax^{-1} \in p\} \in q$, and mimicking the proof of Theorem 5.8 produces sequences with products in decreasing order of indices contained in any member of any idempotent.

The easy way to see this is to consider the semigroup $(S, *)$ where $x * y = y \cdot x$.

We isolate the following corollary for historical reasons. We also point out that this corollary is strong enough to derive all of the combinatorial results of this section. (See the exercises at the end of this section.)

Corollary 5.10 (Finite Sums Theorem). *Let $r \in \mathbb{N}$ and let $\mathbb{N} = \bigcup_{i=1}^r A_i$. There exist $i \in \{1, 2, \ldots, r\}$ and a sequence $\langle x_n \rangle_{n=1}^\infty$ in $\mathbb{N}$ such that $\mathrm{FS}(\langle x_n \rangle_{n=1}^\infty) \subseteq A_i$.*

Proof. This is a special case of Corollary 5.9. $\square$

It turns out that the relationship between idempotents and finite products is even more intimate than indicated by Theorem 5.8.

Lemma 5.11. *Let S be a semigroup and let $\langle x_n \rangle_{n=1}^\infty$ be a sequence in S. Then $\bigcap_{m=1}^\infty \overline{\mathrm{FP}(\langle x_n \rangle_{n=m}^\infty)}$ is a subsemigroup of βS. In particular, there is an idempotent p in βS such that for each $m \in \mathbb{N}$, $\mathrm{FP}(\langle x_n \rangle_{n=m}^\infty) \in p$.*

Proof. Let $T = \bigcap_{m=1}^\infty \overline{\mathrm{FP}(\langle x_n \rangle_{n=m}^\infty)}$. To see that T is a semigroup, we use Theorem 4.20. Trivially, $\{\mathrm{FP}(\langle x_n \rangle_{n=m}^\infty) : m \in \mathbb{N}\}$ has the finite intersection property. Let $m \in \mathbb{N}$ and let $s \in \mathrm{FP}(\langle x_n \rangle_{n=m}^\infty)$ be given. Pick $F \in \mathcal{P}_f(\mathbb{N})$ with $\min F \geq m$ such that $s = \prod_{n \in F} x_n$. Let $k = \max F + 1$. To see that $s \cdot \mathrm{FP}(\langle x_n \rangle_{n=k}^\infty) \subseteq \mathrm{FP}(\langle x_n \rangle_{n=m}^\infty)$, let

$t \in \mathrm{FP}(\langle x_n \rangle_{n=k}^{\infty})$ be given and pick $G \in \mathcal{P}_f(\mathbb{N})$ with min $G \geq k$ such that $t = \prod_{n \in G} x_t$. Then max $F < \min G$ so $st = \prod_{n \in F \cup G} x_n \in \mathrm{FP}(\langle x_n \rangle_{n=m}^{\infty})$.

For the "in particular" conclusion note that by Theorem 2.5, $E(T) \neq \emptyset$. $\square$

Theorem 5.12. *Let S be a semigroup and let $A \subseteq S$. There exists an idempotent p of βS with $A \in p$ if and only if there exists a sequence $\langle x_n \rangle_{n=1}^{\infty}$ in S with $\mathrm{FP}(\langle x_n \rangle_{n=1}^{\infty}) \subseteq A$.*

Proof. The necessity is Theorem 5.8 and the sufficiency follows from Lemma 5.11. $\square$

One should note what is *not* guaranteed by Theorem 5.8 and Lemma 5.11. That is, given an idempotent p and $A \in p$ one can obtain a sequence $\langle x_n \rangle_{n=1}^{\infty}$ with $\mathrm{FP}(\langle x_n \rangle_{n=1}^{\infty}) \subseteq A$. One can then in turn obtain an idempotent q with $\mathrm{FP}(\langle x_n \rangle_{n=1}^{\infty}) \in q$, but one is not guaranteed that $p = q$ or even that $\mathrm{FP}(\langle x_n \rangle_{n=1}^{\infty}) \in p$. (See Chapter 12 for more information on this point.)

We now show that sets of finite products are themselves partition regular.

Definition 5.13. (a) Let $(S, \cdot)$ be a semigroup and let $\langle x_n \rangle_{n=1}^{\infty}$ be a sequence in S. The sequence $\langle y_n \rangle_{n=1}^{\infty}$ is a *product subsystem* of $\langle x_n \rangle_{n=1}^{\infty}$ if and only if there is a sequence $\langle H_n \rangle_{n=1}^{\infty}$ in $\mathcal{P}_f(\mathbb{N})$ such that for every $n \in \mathbb{N}$, max $H_n < \min H_{n+1}$ and $y_n = \prod_{t \in H_n} x_t$.

(b) Let $(S, +)$ be a semigroup and let $\langle x_n \rangle_{n=1}^{\infty}$ be a sequence in S. The sequence $\langle y_n \rangle_{n=1}^{\infty}$ is a *sum subsystem* of $\langle x_n \rangle_{n=1}^{\infty}$ if and only if there is a sequence $\langle H_n \rangle_{n=1}^{\infty}$ in $\mathcal{P}_f(\mathbb{N})$ such that for every $n \in \mathbb{N}$, max $H_n < \min H_{n+1}$ and $y_n = \sum_{t \in H_n} x_t$.

Theorem 5.14. *Let S be a semigroup, let $\langle x_n \rangle_{n=1}^{\infty}$ be a sequence in S and let p be an idempotent in βS such that for every $m \in \mathbb{N}$, $\mathrm{FP}(\langle x_n \rangle_{n=m}^{\infty}) \in p$. Let $A \in p$. There is a product subsystem $\langle y_n \rangle_{n=1}^{\infty}$ of $\langle x_n \rangle_{n=1}^{\infty}$ such that $\mathrm{FP}(\langle y_n \rangle_{n=1}^{\infty}) \subseteq A$.*

Proof. Let $A^{\star} = A^{\star}(p)$ and pick $y_1 \in A^* \cap \mathrm{FP}(\langle x_n \rangle_{n=1}^{\infty})$. Pick $H_1 \in \mathcal{P}_f(\mathbb{N})$ such that $y_1 = \prod_{t \in H_1} x_t$.

Inductively, let $n \in \mathbb{N}$ and assume that we have chosen $\langle y_i \rangle_{i=1}^{n}$ and $\langle H_i \rangle_{i=1}^{n}$ such that:

(1) each $y_i = \prod_{t \in H_i} x_t$,
(2) if $i < n$, then max $H_i < \min H_{i+1}$, and
(3) $\mathrm{FP}(\langle y_i \rangle_{i=1}^{n}) \subseteq A^{\star}$.

Let $E = \mathrm{FP}(\langle y_i \rangle_{i=1}^{n})$ and let $k = \max H_n + 1$. Let

$$B = \mathrm{FP}(\langle x_i \rangle_{i=k}^{\infty}) \cap A^{\star} \cap \bigcap_{a \in E} a^{-1} A^{\star}.$$

By Lemma 4.14 $a^{-1} A^{\star} \in p$ for each $a \in E$ so $B \in p$. Pick $y_{n+1} \in B$ and H_{n+1} with min $H_{n+1} \geq k$ such that $y_{n+1} = \prod_{t \in H_{n+1}} x_t$.

As in the second proof of Theorem 5.8 we see that $\mathrm{FP}(\langle y_i \rangle_{i=1}^{n+1}) \subseteq A^{\star}$. $\square$

Corollary 5.15. *Let S be a semigroup, let $\langle x_n \rangle_{n=1}^{\infty}$ be a sequence in S, let $r \in \mathbb{N}$, and let $\mathrm{FP}(\langle x_n \rangle_{n=1}^{\infty}) = \bigcup_{i=1}^{r} A_i$. There exist $i \in \{1, 2, \ldots, r\}$ and a product subsystem $\langle y_n \rangle_{n=1}^{\infty}$ of $\langle x_n \rangle_{n=1}^{\infty}$ such that $\mathrm{FP}(\langle y_n \rangle_{n=1}^{\infty}) \subseteq A_i$.*

Proof. Pick by Lemma 5.11 an idempotent $p \in \beta S$ such that for each $m \in \mathbb{N}$, $FP(\langle x_n \rangle_{n=m}^{\infty}) \in p$. Then $\bigcup_{i=1}^{r} A_i \in p$ so pick $i \in \{1, 2, \ldots, r\}$ such that $A_i \in p$. Now Theorem 5.14 applies. $\qquad \square$

The semigroup $(\mathcal{P}_f(\mathbb{N}), \cup)$ is of sufficient importance to introduce special notation for it, and to make special mention of its partition regularity.

Definition 5.16. Let $\langle F_n \rangle_{n=1}^{\infty}$ be a sequence in $\mathcal{P}_f(\mathbb{N})$.

(a) $FU(\langle F_n \rangle_{n=1}^{\infty}) = \{\bigcup_{n \in G} F_n : G \in \mathcal{P}_f(\mathbb{N})\}$.

(b) The sequence $\langle G_n \rangle_{n=1}^{\infty}$ is a *union subsystem* of $\langle F_n \rangle_{n=1}^{\infty}$ if and only if there is a sequence $\langle H_n \rangle_{n=1}^{\infty}$ in $\mathcal{P}_f(\mathbb{N})$ such that for every $n \in \mathbb{N}$, $\max H_n < \min H_{n+1}$ and $G_n = \bigcup_{t \in H_n} F_t$.

Corollary 5.17. *Let $\langle F_n \rangle_{n=1}^{\infty}$ be a sequence in $\mathcal{P}_f(\mathbb{N})$, let $r \in \mathbb{N}$ and let $FU(\langle F_n \rangle_{n=1}^{\infty}) = \bigcup_{i=1}^{r} A_i$. There exist $i \in \{1, 2, \ldots, r\}$ and a union subsystem $\langle G_n \rangle_{n=1}^{\infty}$ of $\langle F_n \rangle_{n=1}^{\infty}$ such that $FU(\langle G_n \rangle_{n=1}^{\infty}) \subseteq A_i$. In particular, if $r \in \mathbb{N}$ and $\mathcal{P}_f(\mathbb{N}) = \bigcup_{i=1}^{r} A_i$, then there exist $i \in \{1, 2, \ldots, r\}$ and a sequence $\langle G_n \rangle_{n=1}^{\infty}$ in $\mathcal{P}_f(\mathbb{N})$ such that $FU(\langle G_n \rangle_{n=1}^{\infty}) \subseteq A_i$ and for each $n \in \mathbb{N}$, $\max G_n < \min G_{n+1}$.*

Proof. The first conclusion is an immediate consequence of Corollary 5.15. To see the "in particular" conclusion, let $F_n = \{n\}$ for each $n \in \mathbb{N}$, so that $FU(\langle F_n \rangle_{n=1}^{\infty}) = \mathcal{P}_f(\mathbb{N})$. $\qquad \square$

We conclude this section with a sequence of exercises designed to show that the Finite Sums Theorem (Corollary 5.10) can be used to derive the combinatorial conclusions of this section. (That is, Corollaries 5.9, 5.15, and 5.17.) Since Corollary 5.9 is a consequence of Corollary 5.15, it suffices to establish the last two of these results. The intent in all of these exercises is that one should *not* use the algebraic structure of βS.

Exercise 5.2.1. Let $\langle x_n \rangle_{n=1}^{\infty}$ be a sequence in $\mathbb{N}$ and let $t \in \mathbb{N}$. Prove that for each $m \in \mathbb{N}$ there exists $H \in \mathcal{P}_f(\{n \in \mathbb{N} : n > m\})$ such that $2^t | \sum_{n \in H} x_n$.

Exercise 5.2.2. Let $\langle x_n \rangle_{n=1}^{\infty}$ be a sequence in $\mathbb{N}$. Prove that there is a sum subsystem $\langle y_n \rangle_{n=1}^{\infty}$ of $\langle x_n \rangle_{n=1}^{\infty}$ such that for each n and t in $\mathbb{N}$, if $2^t \leq y_n$, then $2^{t+1} | y_{n+1}$.

Exercise 5.2.3. Prove, using Corollary 5.10, that if $r \in \mathbb{N}$ and $\mathcal{P}_f(\mathbb{N}) = \bigcup_{i=1}^{r} A_i$, then there exist $i \in \{1, 2, \ldots, r\}$ and a sequence $\langle G_n \rangle_{n=1}^{\infty}$ in $\mathcal{P}_f(\mathbb{N})$ such that $FU(\langle G_n \rangle_{n=1}^{\infty}) \subseteq A_i$ and for each $n \in \mathbb{N}$, $\max G_n < \min G_{n+1}$. (Hint: Consider the function $\tau : \mathcal{P}_f(\mathbb{N}) \to \mathbb{N}$ defined by $\tau(F) = \sum_{n \in F} 2^{n-1}$.)

Exercise 5.2.4. Prove Corollary 5.15.

5.3 Sums and Products in $\mathbb{N}$

The results of the previous section are powerful and yet easily proved using the algebraic structure of βS. But their first proofs were combinatorial. We present in this section a simple result whose first proof was found using the algebraic structure of $\beta \mathbb{N}$ and for which an elementary proof was only found much later. (In subsequent chapters we shall encounter many examples of combinatorial results whose only known proofs utilize the algebraic structure of βS.)

One knows from the Finite Sums Theorem that whenever $r \in \mathbb{N}$ and $\mathbb{N} = \bigcup_{i=1}^{r} A_i$, there exist $i \in \{1, 2, \ldots, r\}$ and a sequence $\langle x_n \rangle_{n=1}^{\infty}$ in $\mathbb{N}$ such that $FS(\langle x_n \rangle_{n=1}^{\infty}) \subseteq A_i$. It is an easy consequence of this fact that whenever $r \in \mathbb{N}$ and $\mathbb{N} = \bigcup_{i=1}^{r} A_i$, there exist $j \in \{1, 2, \ldots, r\}$ and a sequence $\langle y_n \rangle_{n=1}^{\infty}$ in $\mathbb{N}$ such that $FP(\langle y_n \rangle_{n=1}^{\infty}) \subseteq A_j$. (Of course the second statement follows from Corollary 5.9. It is also an elementary consequence of the Finite Sums Theorem in the fashion outlined in the exercises at the end of the previous section. However, it follows much more quickly from the Finite Sums Theorem by a consideration of the homomorphism $\varphi : (\mathbb{N}, +) \to (\mathbb{N}, \cdot)$ defined by $\varphi(n) = 2^n$.)

Two questions naturally arise. First can one choose $i = j$ in the above statements? Second, if so, can one choose $\langle x_n \rangle_{n=1}^{\infty} = \langle y_n \rangle_{n=1}^{\infty}$? We answer the first of these questions affirmatively in this section. The negative answer to the second question is a consequence of Theorem 17.16.

Definition 5.18. $\Gamma = c\ell\{p \in \beta\mathbb{N} : p = p + p\}$

We have the following simple combinatorial characterization of Γ.

Lemma 5.19. *Let $p \in \beta\mathbb{N}$. Then $p \in \Gamma$ if and only if for every $A \in p$ there exists a sequence $\langle x_n \rangle_{n=1}^{\infty}$ such that $FS(\langle x_n \rangle_{n=1}^{\infty}) \subseteq A$.*

Proof. Necessity. Let $A \in p$ be given. Then $\overline{A}$ is a neighborhood of p so pick $q = q + q$ in $\beta\mathbb{N}$ with $q \in \overline{A}$. Then $A \in q$ so by Theorem 5.12 there is a sequence $\langle x_n \rangle_{n=1}^{\infty}$ in $\mathbb{N}$ with $FS(\langle x_n \rangle_{n=1}^{\infty}) \subseteq A$.

Sufficiency. Let a basic neighborhood $\overline{A}$ of p be given. Then there is a sequence $\langle x_n \rangle_{n=1}^{\infty}$ in $\mathbb{N}$ with $FS(\langle x_n \rangle_{n=1}^{\infty}) \subseteq A$, so by Theorem 5.12 there is some $q = q + q$ in $\beta\mathbb{N}$ with $A \in q$. □

We shall see as a consequence of Exercise 6.1.4 that Γ is not a subsemigroup of $(\beta\mathbb{N}, +)$. However, we see that Γ does have significant multiplicative structure.

Theorem 5.20. Γ *is a left ideal of* $(\beta\mathbb{N}, \cdot)$.

Proof. Let $p \in \Gamma$ and let $q \in \beta\mathbb{N}$. To see that $q \cdot p \in \Gamma$, let $A \in q \cdot p$ be given. Then $\{z \in \mathbb{N} : z^{-1}A \in p\} \in q$ so pick $z \in \mathbb{N}$ such that $z^{-1}A \in p$. Then by Lemma 5.19 we may pick a sequence $\langle x_n \rangle_{n=1}^{\infty}$ in $\mathbb{N}$ such that $FS(\langle x_n \rangle_{n=1}^{\infty}) \subseteq z^{-1}A$. For each $n \in \mathbb{N}$, let $y_n = zx_n$. Then $FS(\langle y_n \rangle_{n=1}^{\infty}) \subseteq A$ so by Lemma 5.19, $q \cdot p \in \Gamma$. □

We shall see in Corollary 17.17 that there is no $p \in \beta\mathbb{N}$ with $p+p = p \cdot p$. However, we see that there are multiplicative idempotents close to the additive idempotents.

Corollary 5.21. *There exists $p = p \cdot p$ in Γ.*

Proof. By Theorem 5.20, $(\Gamma, \cdot)$ is a compact right topological semigroup so Theorem 2.5 applies.   □

Corollary 5.22. *Let $r \in \mathbb{N}$ and let $\mathbb{N} = \bigcup_{i=1}^{r} A_i$. There exist $i \in \{1, 2, \ldots, r\}$ and sequences $\langle x_n \rangle_{n=1}^{\infty}$ and $\langle y_n \rangle_{n=1}^{\infty}$ in $\mathbb{N}$ with $\mathrm{FS}(\langle x_n \rangle_{n=1}^{\infty}) \cup \mathrm{FP}(\langle y_n \rangle_{n=1}^{\infty}) \subseteq A_i$.*

Proof. Pick by Corollary 5.21 some $p = p \cdot p$ in Γ. Pick $i \in \{1, 2, \ldots, r\}$ such that $A_i \in p$. Since $p = p \cdot p$ there is by Theorem 5.8 a sequence $\langle y_n \rangle_{n=1}^{\infty}$ in $\mathbb{N}$ with $\mathrm{FP}(\langle y_n \rangle_{n=1}^{\infty}) \subseteq A_i$. Since $p \in \Gamma$ there is by Lemma 5.19 a sequence $\langle x_n \rangle_{n=1}^{\infty}$ in $\mathbb{N}$ with $\mathrm{FS}(\langle x_n \rangle_{n=1}^{\infty}) \subseteq A_i$.   □

5.4 Adjacent Finite Unions

In Corollary 5.17, there is no reason to expect the terms of the sequence $\langle G_n \rangle_{n=1}^{\infty}$ to be close to each other. In fact, if as in Corollary 5.17 one has $\max G_n < \min G_{n+1}$ for all n, we can absolutely guarantee that they are not.

Theorem 5.23. *There is a partition of $\mathcal{P}_f(\mathbb{N})$ into two cells such that, if $\langle G_n \rangle_{n=1}^{\infty}$ is any sequence in $\mathcal{P}_f(\mathbb{N})$ such that $\mathrm{FU}(\langle G_n \rangle_{n=1}^{\infty})$ is contained in one cell of the partition and $\max G_n < \min G_{n+1}$ for all n, then $\{\min G_{n+1} - \max G_n : n \in \mathbb{N}\}$ is unbounded.*

Proof. Given $a \in \mathbb{N}$ and $F \in \mathcal{P}_f(\mathbb{N})$, define

$$\varphi(a, F) = |\{x \in F : \{x+1, x+2, \ldots, x+a\} \cap F \neq \emptyset\}|.$$

For $i \in \{0, 1\}$, let

$$\mathcal{A}_i = \{F \in \mathcal{P}_f(\mathbb{N}) : \varphi(\min F, F) \equiv i \pmod{2}\}.$$

Suppose that we have a $i \in \{0, 1\}$ and a sequence $\langle G_n \rangle_{n=1}^{\infty}$ such that

(1) $\max G_n < \min G_{n+1}$ for all $n \in \mathbb{N}$,
(2) $\mathrm{FU}(\langle G_n \rangle_{n=1}^{\infty}) \subseteq \mathcal{A}_i$, and
(3) $\{\min G_{n+1} - \max G_n : n \in \mathbb{N}\}$ is bounded.

Pick $b \in \mathbb{N}$ such that for all $n \in \mathbb{N}$, $\min G_{n+1} - \max G_n \leq b$, and pick $k \in \mathbb{N}$ such that $\min G_k > b$. Let $a = \min G_k$ and pick ℓ such that $\min G_\ell - \max G_k > a$, noticing that $\ell > k + 1$. Let $F = \bigcup_{t=k+1}^{\ell-1} G_t$. Then each of G_k, $G_k \cup F$, $G_k \cup G_\ell$, and $G_k \cup F \cup G_\ell$ is in $\mathcal{A}_i$ and

$$\min G_k = \min(G_k \cup F) = \min(G_k \cup G_\ell) = \min(G_k \cup F \cup G_\ell) = a$$

so

$$\varphi(a, G_k) \equiv \varphi(a, G_k \cup F) \equiv \varphi(a, G_k \cup G_\ell) \equiv \varphi(a, G_k \cup F \cup G_\ell) \pmod 2.$$

Now

$$\varphi(a, G_k \cup F) = \varphi(a, G_k) + \varphi(a, F) + 1,$$
$$\varphi(a, G_k \cup G_\ell) = \varphi(a, G_k) + \varphi(a, G_\ell),$$

and

$$\varphi(a, G_k \cup F \cup G_\ell) = \varphi(a, G_k) + \varphi(a, F) + \varphi(a, G_\ell) + 2$$

so

$$\begin{aligned}
\varphi(a, G_k) &\equiv \varphi(a, G_k) + \varphi(a, F) + 1 \\
&\equiv \varphi(a, G_k) + \varphi(a, G_\ell) \\
&\equiv \varphi(a, G_k) + \varphi(a, F) + \varphi(a, G_\ell) + 2 \pmod 2,
\end{aligned}$$

which is impossible.   □

We show in the remainder of this section that, given any finite partition of $\mathcal{P}_f(\mathbb{N})$, we can get a sequence $\langle H_n \rangle_{n=1}^\infty$ with any specified separation between max H_n and min H_{n+1} (including min $H_{n+1} = \max H_n + 1$) and all finite unions *that do not use adjacent terms* in one cell of the partition. In the following definition "naFU" is intended to represent "non adjacent finite unions".

Definition 5.24. Let $\langle H_k \rangle_{k=1}^\infty$ be a sequence in $\mathcal{P}_f(\mathbb{N})$.

(a) $\text{naFU}(\langle H_k \rangle_{k=1}^\infty) = \big\{ \bigcup_{t \in F} H_t : F \in \mathcal{P}_f(\mathbb{N}) \text{ and for all } t \in F, \ t + 1 \notin F \big\}$.

(b) If $n \in \mathbb{N}$, then $\text{naFU}(\langle H_k \rangle_{k=1}^n) = \big\{ \bigcup_{t \in F} H_t : \emptyset \neq F \subseteq \{1, 2, \ldots, n\} \text{ and for all } t \in F, \ t + 1 \notin F \big\}$.

(c) $\text{naFU}(\langle H_k \rangle_{k=1}^0) = \emptyset$.

We shall be using the semigroup $(\mathcal{P}_f(\mathbb{N}), \cup)$ and the customary extension of the operation to $\beta(\mathcal{P}_f(\mathbb{N}))$. However, we cannot follow our usual practice of denoting the extension of the operation to $\beta(\mathcal{P}_f(\mathbb{N}))$ by the same symbol used to denote the operation in $\mathcal{P}_f(\mathbb{N})$. (If $p, q \in \beta(\mathcal{P}_f(\mathbb{N}))$, then $p \cup q$ already means something.) So, we shall denote the extension of the operation $\cup$ to $\beta(\mathcal{P}_f(\mathbb{N}))$ by $\uplus$.

Lemma 5.25. *Let $f : \mathbb{N} \to \mathbb{N}$ be a nondecreasing function. Define $M : \mathcal{P}_f(\mathbb{N}) \to \mathbb{N}$ and $m : \mathcal{P}_f(\mathbb{N}) \to \mathbb{N}$ by $M(F) = \max F + f(\max F)$ and $m(F) = \min F$. Then M is a homomorphism from $(\mathcal{P}_f(\mathbb{N}), \cup)$ to $(\mathbb{N}, \vee)$ and m is a homomorphism from $(\mathcal{P}_f(\mathbb{N}), \cup)$ to $(\mathbb{N}, \wedge)$. Consequently, $\widetilde{M}$ is a homomorphism from $(\beta(\mathcal{P}_f(\mathbb{N})), \uplus)$ to $(\beta\mathbb{N}, \vee)$ and $\widetilde{m}$ is a homomorphism from $(\beta(\mathcal{P}_f(\mathbb{N})), \uplus)$ to $(\beta\mathbb{N}, \wedge)$.*

Proof. That m is a homomorphism is trivial. Given $F, G \in \mathcal{P}_f(\mathbb{N})$, one has

$$f\big(\max(F \cup G)\big) = f(\max F) \vee f(\max G)$$

since f is a nondecreasing function, so that $M(F \cup G) = M(F) \vee M(G)$. The fact that $\widetilde{M}$ and $\widetilde{m}$ are homomorphisms then follows from Corollary 4.22 and Theorem 4.24.   □

Lemma 5.26. *Let* $f : \mathbb{N} \to \mathbb{N}$ *be a nondecreasing function, define* M *and* m *as in Lemma 5.25, and let* $Y = \{a + f(a) : a \in \mathbb{N}\}$. *If* $x \in \overline{Y} \backslash \mathbb{N}$ *and*

$$T = \{p \in \beta(\mathcal{P}_f(\mathbb{N})) : \widetilde{M}(p) = \widetilde{m}(p) = x\},$$

then T *is a compact subsemigroup of* $(\beta(\mathcal{P}_f(\mathbb{N})), \uplus)$.

Proof. Trivially T is compact. Further, by Exercise 4.1.11, $x \vee x = x \wedge x = x$, so by Lemma 5.25, if $p, q \in T$, then $p \uplus q \in T$. Thus it suffices to show that $T \neq \emptyset$.

Let $\mathcal{A} = \{M^{-1}[X] \cap m^{-1}[X] : X \in x\}$. We claim that $\mathcal{A}$ has the finite intersection property, for which it suffices to show that $M^{-1}[X] \cap m^{-1}[X] \neq \emptyset$ for every $X \in x$. So let $X \in x$ be given, pick $a \in X$, and pick $c \in X \cap Y$ such that $c > a + f(a)$. Since $c \in Y$, pick $b \in \mathbb{N}$ such that $c = b + f(b)$. Let $F = \{a, b\}$. If one had $b < a$, then one would have $c = b + f(b) < a + f(a)$, so $\max F = b$. Then $m(F) = a \in X$ and $M(F) = b + f(b) = c \in X$, so $F \in M^{-1}[X] \cap m^{-1}[X]$.

Pick $p \in \beta(\mathcal{P}_f(\mathbb{N}))$ such that $\mathcal{A} \subseteq p$. By Lemma 3.30 $p \in T$. □

We can now prove the promised theorem yielding adjacent sequences with non adjacent unions in one cell.

Theorem 5.27. *Let* $f : \mathbb{N} \to \mathbb{N}$ *be a nondecreasing function and let* $\mathcal{P}_f(\mathbb{N}) = \bigcup_{i=1}^{r} \mathcal{A}_i$. *Then there exist some* $i \in \{1, 2, \ldots, r\}$ *and a sequence* $\langle H_n \rangle_{n=1}^{\infty}$ *in* $\mathcal{P}_f(\mathbb{N})$ *such that*

(i) *for each* n, $\min H_{n+1} = \max H_n + f(\max H_n)$ *and*
(ii) $\mathrm{naFU}(\langle H_n \rangle_{n=1}^{\infty}) \subseteq \mathcal{A}_i$.

Proof. Let M and m be as defined in Lemma 5.25 and let $Y = \{a + f(a) : a \in \mathbb{N}\}$. Pick $x \in \overline{Y} \backslash \mathbb{N}$ and by Lemma 5.26 and Theorem 2.5 pick an idempotent p in $(\beta(\mathcal{P}_f(\mathbb{N})), \uplus)$ such that $\widetilde{M}(p) = \widetilde{m}(p) = x$. We show that for each $\mathcal{U} \in p$, there is a sequence $\langle H_n \rangle_{n=1}^{\infty}$ in $\mathcal{P}_f(\mathbb{N})$ such that

(a) *for each* n, $m(H_{n+1}) = M(H_n)$ *and*
(b) $\mathrm{naFU}(\langle H_n \rangle_{n=1}^{\infty}) \subseteq \mathcal{U}$.

Then choosing $i \in \{1, 2, \ldots, r\}$ such that $\mathcal{A}_i \in p$ completes the proof.

So let $\mathcal{U} \in p$ be given. For each $G \in \mathcal{P}_f(\mathbb{N})$ and any $\mathcal{V} \subseteq \mathcal{P}_f(\mathbb{N})$ let $G^{-1}\mathcal{V} = \{F \in \mathcal{P}_f(\mathbb{N}) : G \cup F \in \mathcal{V}\}$ and let $\mathcal{U}^{\star} = \{G \in \mathcal{U} : G^{-1}\mathcal{U} \in p\}$. Then by Lemma 4.14, $\mathcal{U}^{\star} \in p$ and for each $G \in \mathcal{U}^{\star}$, $G^{-1}\mathcal{U}^{\star} \in p$.

We claim now that, given any $H \in \mathcal{P}_f(\mathbb{N})$, $\{J \in \mathcal{P}_f(\mathbb{N}) : m(J) > M(H)\} \in p$. Indeed, otherwise we have some $t \leq M(H)$ such that $\{J \in \mathcal{P}_f(\mathbb{N}) : m(J) = t\} \in p$ so that $\widetilde{m}(p) = t$, a contradiction.

Now, by Lemma 3.31, $\{H \in \mathcal{P}_f(\mathbb{N}) : M(H) \in m[\mathcal{U}^{\star}]\} \in p$ so pick $H_1 \in \mathcal{U}^{\star}$ such that $M(H_1) \in m[\mathcal{U}^{\star}]$. Inductively, let $n \in \mathbb{N}$ and assume that we have chosen $\langle H_j \rangle_{j=1}^{n}$ in $\mathcal{P}_f(\mathbb{N})$ such that for each $k \in \{1, 2, \ldots, n\}$,

(1) $\mathrm{naFU}(\langle H_j \rangle_{j=1}^{k}) \subseteq \mathcal{U}^{\star}$,

(2) if $k > 1$, then $M(H_k) \in m[\mathcal{U}^\star \cap \bigcap\{G^{-1}\mathcal{U}^\star : G \in \text{naFU}(\langle H_j\rangle_{j=1}^{k-1})\}]$, and

(3) if $k > 1$, then $m(H_k) = M(H_{k-1})$.

Hypothesis (1) holds at $n = 1$ and hypotheses (2) and (3) are vacuous there.

If $n = 1$, pick $H \in \mathcal{U}^\star$ such that $m(H) = M(H_1)$. Otherwise, by hypothesis (2), pick

$$H \in \mathcal{U}^\star \cap \bigcap\{G^{-1}\mathcal{U}^\star : G \in \text{naFU}(\langle H_j\rangle_{j=1}^{n-1})\}$$

such that $m(H) = M(H_n)$. Let $\mathcal{V}$ be the family of sets $J \in \mathcal{P}_f(\mathbb{N})$ satisfying

(i) $m(J) > M(H)$,

(ii) $H \cup J \in \mathcal{U}^\star$,

(iii) $G \cup H \cup J \in \mathcal{U}^\star$ for all $G \in \text{naFU}(\langle H_j\rangle_{j=1}^{n-1})$, and

(iv) $M(J) \in m[\mathcal{U}^\star \cap \bigcap\{G^{-1}\mathcal{U}^\star : G \in \text{naFU}(\langle H_j\rangle_{j=1}^{n})\}]$.

We have seen that $\{J \in \mathcal{P}_f(\mathbb{N}) : m(J) > M(H)\} \in p$ and $H^{-1}\mathcal{U}^\star \in p$ because $H \in \mathcal{U}^\star$. If $n > 1$ and $G \in \text{naFU}(\langle H_j\rangle_{j=1}^{n-1})$, then $H \in G^{-1}\mathcal{U}^\star$ so $G \cup H \in \mathcal{U}^\star$ so $(G \cup H)^{-1}\mathcal{U}^\star \in p$. Thus the family of sets satisfying (i), (ii), and (iii) is in p.

Finally, if $G \in \text{naFU}(\langle H_j\rangle_{j=1}^{n})$, then $G \in \mathcal{U}^\star$ by hypothesis (1), so $G^{-1}\mathcal{U}^\star \in p$. Thus $\mathcal{U}^\star \cap \bigcap\{G^{-1}\mathcal{U}^\star : G \in \text{naFU}(\langle H_j\rangle_{j=1}^{n})\} \in p$ so

$$\{J \in \mathcal{P}_f(\mathbb{N}) : M(J) \in m[\mathcal{U}^\star \cap \bigcap\{G^{-1}\mathcal{U}^\star : G \in \text{naFU}(\langle H_j\rangle_{j=1}^{n})\}]\} \in p$$

by Lemma 3.31. Thus $\mathcal{V} \in p$ and so is nonempty.

Pick $J \in \mathcal{V}$ and let $H_{n+1} = H \cup J$. Then $M(H_{n+1}) = M(J)$ and also $m(H_{n+1}) = m(H) = M(H_n)$ so hypothesis (3) holds.

Further, $J \in \mathcal{V}$ so

$$M(H_{n+1}) = M(J) \in m[\mathcal{U}^\star \cap \bigcap\{G^{-1}\mathcal{U}^\star : G \in \text{naFU}(\langle H_j\rangle_{j=1}^{n})\}]$$

so hypothesis (2) holds.

To complete the proof, we show that $\text{naFU}(\langle H_j\rangle_{j=1}^{n+1}) \subseteq \mathcal{U}^\star$. So let $\emptyset \neq F \subseteq \{1, 2, \ldots, n+1\}$ such that for each $t \in F$, $t+1 \notin F$. If $n+1 \notin F$, then $\bigcup_{t\in F} H_t \in \mathcal{U}^\star$ by hypothesis (1) at n, so assume that $n + 1 \in F$.

If $F = \{n+1\}$, then we have $H_{n+1} = H \cup J \in \mathcal{U}^\star$. Thus, assume that $F \neq \{n+1\}$, let $K = F\backslash\{n + 1\}$, and let $G = \bigcup_{t\in K} H_t$. Then $\bigcup_{t\in F} H_t = G \cup H \cup J \in \mathcal{U}^\star$. $\square$

Exercise 5.4.1. Let $f : \mathbb{N} \to \mathbb{N}$ be a nondecreasing function and let Z be any infinite subset of $\{a + f(a) : a \in \mathbb{N}\}$. Show that the conclusion of Theorem 5.27 can be strengthened to require that $\min H_n \in Z$ for each $n \in \mathbb{N}$. (Hint: In Lemma 5.26, choose $x \in \overline{Z}\backslash\mathbb{N}$.)

5.5 Compactness

Most of the combinatorial results that we shall prove in this book are infinite in nature. We deal here with a general method used to derive finite analogues from the infinite versions. This version is commonly referred to as "compactness". (One will read in the literature, "by a standard compactness argument one sees…".) We illustrate two forms of this method of proof in this section, first deriving a finite version of the Finite Products Theorem (Corollary 5.9).

For either of the common forms of compactness arguments it is more convenient to work with the "coloring" method of stating results in Ramsey Theory.

Definition 5.28. Let X be a set and let $r \in \mathbb{N}$.

(a) An *r-coloring* of X is a function $\varphi : X \to \{1, 2, \ldots, r\}$.

(b) Given an r-coloring φ of X, a subset B of X is said to be *monochrome* (with respect to φ) provided φ is constant on B.

Theorem 5.29. *Let S be a countable semigroup and enumerate S as $\langle s_n \rangle_{n=1}^{\infty}$. For each r and m in $\mathbb{N}$, there exists $n \in \mathbb{N}$ such that whenever $\{s_t : t \in \{1, 2, \ldots, n\}\}$ is r-colored, there is a sequence $\langle x_t \rangle_{t=1}^{m}$ in S such that $\mathrm{FP}(\langle x_t \rangle_{t=1}^{m})$ is monochrome.*

Proof. Let $r, m \in \mathbb{N}$ be given and suppose the conclusion fails. For each $n \in \mathbb{N}$ choose an r-coloring φ_n of $\{s_t : t \in \{1, 2, \ldots, n\}\}$ such that for no sequence $\langle x_t \rangle_{t=1}^{m}$ in S is $\mathrm{FP}(\langle x_t \rangle_{t=1}^{m})$ monochrome.

Choose by the pigeon hole principle an infinite subset B_1 of $\mathbb{N}$ and an element $\sigma(1) \in \{1, 2, \ldots, r\}$ such that for all $k \in B_1$, $\varphi_k(s_1) = \sigma(1)$. Inductively, given $\ell \in \mathbb{N}$ with $\ell > 1$ and an infinite subset $B_{\ell-1}$ of $\mathbb{N}$ choose an infinite subset B_ℓ of $B_{\ell-1}$ and an element $\sigma(\ell) \in \{1, 2, \ldots, r\}$ such that $\min B_\ell \geq \ell$ and for all $k \in B_\ell$, $\varphi_k(s_\ell) = \sigma(\ell)$.

Define an r-coloring τ of S by $\tau(s_\ell) = \sigma(\ell)$. Then $S = \bigcup_{i=1}^{r} \tau^{-1}[\{i\}]$ so pick by Corollary 5.9 some $i \in \{1, 2, \ldots, r\}$ and an infinite sequence $\langle x_t \rangle_{t=1}^{\infty}$ in S such that $\mathrm{FP}(\langle x_t \rangle_{t=1}^{\infty}) \subseteq \tau^{-1}[\{i\}]$. Pick $k \in \mathbb{N}$ such that $\mathrm{FP}(\langle x_t \rangle_{t=1}^{m}) \subseteq \{s_1, s_2, \ldots, s_k\}$ and pick $n \in B_k$. We claim that $\mathrm{FP}(\langle x_t \rangle_{t=1}^{m})$ is monochrome with respect to φ_n, a contradiction. To see this, let $a \in \mathrm{FS}(\langle x_t \rangle_{t=1}^{m})$ be given and pick $\ell \in \{1, 2, \ldots, k\}$ such that $a = s_\ell$. Then $n \in B_k \subseteq B_\ell$ so $\varphi_n(a) = \sigma(\ell)$. Since $a \in \mathrm{FP}(\langle x_t \rangle_{t=1}^{\infty}) \subseteq \tau^{-1}[\{i\}]$, $i = \tau(a) = \tau(s_\ell) = \sigma(\ell)$. Thus φ_n is constantly equal to i on $\mathrm{FP}(\langle x_t \rangle_{t=1}^{m})$ as claimed.□

The reader may wonder why the term "compactness" is applied to the proof of Theorem 5.29. One answer is that such results can be proved using the compactness theorem of logic. Another interpretation is provided by the proof of the following theorem which utilizes topological compactness.

Theorem 5.30. *For each r and m in $\mathbb{N}$ there exists $n \in \mathbb{N}$ such that whenever $\{1, 2, \ldots, n\}$ is r-colored, there exist sequences $\langle x_t \rangle_{t=1}^{m}$ and $\langle y_t \rangle_{t=1}^{m}$ in $\mathbb{N}$ such that $\mathrm{FS}(\langle x_t \rangle_{t=1}^{m})$ and $\mathrm{FP}(\langle y_t \rangle_{t=1}^{m})$ are contained in $\{1, 2, \ldots, n\}$ and $\mathrm{FS}(\langle x_t \rangle_{t=1}^{m}) \cup \mathrm{FP}(\langle y_t \rangle_{t=1}^{m})$ is monochrome.*

Proof. Let $r, m \in \mathbb{N}$ be given and suppose the conclusion fails. For each $n \in \mathbb{N}$ choose an r-coloring φ_n of $\{1, 2, \ldots, n\}$ such that there are no sequences $\langle x_t \rangle_{t=1}^m$ and $\langle y_t \rangle_{t=1}^m$ in $\mathbb{N}$ for which $\mathrm{FS}(\langle x_t \rangle_{t=1}^m)$ and $\mathrm{FP}(\langle y_t \rangle_{t=1}^m)$ are contained in $\{1, 2, \ldots, n\}$ and $\mathrm{FS}(\langle x_t \rangle_{t=1}^m) \cup \mathrm{FP}(\langle y_t \rangle_{t=1}^m)$ is monochrome.

Let $Y = \bigtimes_{n=1}^\infty \{1, 2, \ldots, r\}$, where $\{1, 2, \ldots, r\}$ is viewed as a topological space with the discrete topology. For each $n \in \mathbb{N}$ define $\mu_n \in Y$ by

$$\mu_n(t) = \begin{cases} \varphi_n(t) & \text{if } t \leq n \\ 1 & \text{if } t > n. \end{cases}$$

Then $\langle \mu_n \rangle_{n=1}^\infty$ is a sequence in the compact space Y and so we can choose a cluster point τ of $\langle \mu_n \rangle_{n=1}^\infty$ in Y.

Then $\mathbb{N} = \bigcup_{i=1}^r \tau^{-1}[\{i\}]$. So by Corollary 5.22 there exist $i \in \{1, 2, \ldots, r\}$ and sequences $\langle x_n \rangle_{n=1}^\infty$ and $\langle y_n \rangle_{n=1}^\infty$ in $\mathbb{N}$ such that $\mathrm{FS}(\langle x_n \rangle_{n=1}^\infty) \cup \mathrm{FP}(\langle y_n \rangle_{n=1}^\infty) \subseteq \tau^{-1}[\{i\}]$. Let $k = \max\big(\mathrm{FS}(\langle x_t \rangle_{t=1}^m) \cup \mathrm{FP}(\langle y_t \rangle_{t=1}^m)\big)$ and let $U = \{g \in Y : \text{for all } t \in \{1, 2, \ldots, k\}, g(t) = \tau(t)\}$. Then U is a neighborhood of τ in Y, so pick $n > k$ such that $\mu_n \in U$. But then μ_n agrees with τ on $\mathrm{FS}(\langle x_t \rangle_{t=1}^m) \cup \mathrm{FP}(\langle y_t \rangle_{t=1}^m)$. Also μ_n agrees with φ_n on $\{1, 2, \ldots, n\}$ so $\mathrm{FS}(\langle x_t \rangle_{t=1}^m) \cup \mathrm{FP}(\langle y_t \rangle_{t=1}^m)$ is monochrome with respect to φ_n, a contradiction. $\square$

Exercise 5.5.1. Prove, using compactness and Theorem 5.6, the following version of Ramsey's Theorem. Let $k, r, m \in \mathbb{N}$. There exists $n \in \mathbb{N}$ such that whenever Y is a set with $|Y| = n$ and $[Y]^k = \bigcup_{i=1}^r A_i$, there exist $i \in \{1, 2, \ldots, r\}$ and $B \in [Y]^m$ with $[B]^k \subseteq A_i$.

Notes

The basic reference for *Ramsey Theory* is the book by that title [111].

Corollary 5.10 was originally proved in [118], with a purely elementary (but very complicated) combinatorial proof. A simplified combinatorial proof was given by J. Baumgartner [15], and a proof using tools from Topological Dynamics has been given by H. Furstenberg and B. Weiss [100].

The first proof of the Finite Sums Theorem given here is due to F. Galvin and S. Glazer. This proof was never published by the originators, although it has appeared in several surveys, the first of which was [66]. The idea for the construction occurred to Galvin around 1970. At that time the Finite Sums Theorem was a conjecture of R. Graham and B. Rothschild [110], as yet unproved. Galvin asked whether an "almost translation invariant ultrafilter" existed. That is, is there an ultrafilter p on $\mathbb{N}$ such that whenever $A \in p$, one has $\{x \in S : A - x \in p\} \in p$? (In terms of the measure μ corresponding to p which was introduced in Exercise 3.1.2, this is asking that any set of measure 1 should almost always translate to a set of measure 1.) Galvin had invented the construction used in the first proof of Theorem 5.8, and knew that an affirmative answer to his question would provide a proof of the conjecture of Graham and Rothschild. One of the current authors tried to answer this question and succeeded only in showing that, under the assumption of the continuum hypothesis, the validity of the conjecture of Graham and Rothschild implied an affirmative answer to Galvin's question. With the subsequent elementary proof of the Finite Sums Theorem [118], Galvin's almost translation invariant ultrafilters became a figment of the continuum hypothesis. Galvin was interested in establishing their existence in ZFC, and one day in 1975 he asked Glazer whether such ultrafilters existed. When Glazer quickly answered "yes", Galvin tried to explain that he must be missing something because it couldn't be that easy. In fact it was that easy, because Glazer (1) knew that $\beta\mathbb{N}$ could be made into a right topological semigroup with an operation extending ordinary addition and (2) knew the characterization of that operation in terms of ultrafilters. (Most of the mathematicians who were aware of the algebraic structure of $\beta\mathbb{N}$ did not think of $\beta\mathbb{N}$ as a space of ultrafilters.) In terms of that characterization, it was immediate that Galvin's almost translation invariant ultrafilters were simply idempotents.

Theorem 5.12 is a result of F. Galvin (also not published by him).

An elementary proof of Corollary 5.22 can be found in [30], a result of collaboration with V. Bergelson.

Theorem 5.27 is due to A. Blass in [44] where it was proved using Martin's Axiom, followed by an absoluteness argument showing that it is a theorem of ZFC.

Part II

Algebra of βS

Chapter 6

Ideals and Commutativity in βS

A very striking fact about $\beta\mathbb{N}$ is how far it is from being commutative. Although $(\beta\mathbb{N}, +)$ is a natural extension of the semigroup $(\mathbb{N}, +)$, which is the most familiar of all semigroups, its algebraic structure is amazingly complicated. For example, as we shall show in Corollary 7.36, it contains many copies of the free group on $2^{\mathfrak{c}}$ generators.

It is a simple observation that a semigroup $(S, \cdot)$ which contains two disjoint left ideals or two disjoint right ideals cannot be commutative. If L_1 and L_2, say, are two disjoint left ideals of S and if $s_1 \in L_1$ and $s_2 \in L_2$, then $s_1 \cdot s_2 \in L_2$ and $s_2 \cdot s_1 \in L_1$. So $s_1 \cdot s_2 \neq s_2 \cdot s_1$. We shall show that $(\beta\mathbb{N}, +)$ contains $2^{\mathfrak{c}}$ mutually disjoint left ideals and $2^{\mathfrak{c}}$ mutually disjoint right ideals, and that this is a property shared by a large class of semigroups of the form βS.

Using this observation, it is fairly easy to see that $\beta\mathbb{Z}$ is not commutative. In fact, no element of $\mathbb{Z}^*$ can be in the center of $\beta\mathbb{Z}$. We first observe that $\mathbb{N}^*$ and $(-\mathbb{N})^*$ are both left ideals of $\beta\mathbb{Z}$, as shown in Exercise 4.3.5. Thus, if $p \in \mathbb{N}^*$ and $q \in (-\mathbb{N})^*$, $q + p \in \mathbb{N}^*$ and $p + q \in (-\mathbb{N})^*$. So $q + p \neq p + q$ and neither p nor q is in the center of $\beta\mathbb{Z}$. (Here, as elsewhere, if we mention "the semigroup $\beta\mathbb{Z}$" or "the semigroup $\beta\mathbb{N}$" without mention of the operation, we assume that the operation is addition.)

We shall show in Theorem 6.10 that $\mathbb{N}$ is the center of $(\beta\mathbb{N}, +)$ and of $(\beta\mathbb{N}, \cdot)$, and we shall show in Theorem 6.54 that these facts follow from a more general theorem.

6.1 The Semigroup $\mathbb{H}$

We shall use the binary expansion of positive integers to show how rich the algebraic structure of $(\beta\mathbb{N}, +)$ is. This tool will yield information, not only about $\beta\mathbb{N}$, but about all semigroups βS which arise from infinite, discrete, cancellative semigroups S.

Definition 6.1. $\mathbb{H} = \bigcap_{n \in \mathbb{N}} c\ell_{\beta\mathbb{N}}(2^n\mathbb{N})$.

Recall that each $n \in \mathbb{N}$ can be expressed uniquely as $n = \sum_{i \in F} 2^i$, where $F \in \mathcal{P}_f(\omega)$. (Recall that $\omega = \mathbb{N} \cup \{0\}$.)

Definition 6.2. Given $n \in \mathbb{N}$, $\mathrm{supp}(n) \in \mathcal{P}_f(\omega)$ is defined by $n = \sum_{i \in \mathrm{supp}(n)} 2^i$.

In our study of $\beta \mathbb{N}$ it will often be the case that functions which are not homomorphisms on $\mathbb{N}$ nonetheless extend to functions whose restrictions to $\mathbb{H}$ are homomorphisms. We remind the reader that the topological center $\Lambda(T)$ of a right topological semigroup T is the set of elements $t \in T$ for which λ_t is continuous.

Lemma 6.3. *Let $(T, \cdot)$ be a compact right topological semigroup and let $\varphi : \mathbb{N} \to T$ with $\varphi[\mathbb{N}] \subseteq \Lambda(T)$. Assume that there is some $k \in \omega$ such that whenever $x, y \in \mathbb{N}$ and $\max \operatorname{supp}(x) + k < \min \operatorname{supp}(y)$ one has $\varphi(x + y) = \varphi(x) \cdot \varphi(y)$. Then for each $p, q \in \mathbb{H}$, $\widetilde{\varphi}(p + q) = \widetilde{\varphi}(p) \cdot \widetilde{\varphi}(q)$.*

Proof. This follows from Theorem 4.21 with $\mathcal{A} = \{2^n \mathbb{N} : n \in \mathbb{N}\}$. $\square$

The following theorem shows that homomorphic images of $\mathbb{H}$ occur very widely. In the following theorem, the possibility of a finite dense topological center can only occur if T itself is finite, since we are assuming that all hypothesized topological spaces are Hausdorff.

Theorem 6.4. *Let T be a compact right topological semigroup with a countable dense topological center. Then T is the image of $\mathbb{H}$ under a continuous homomorphism.*

Proof. We enumerate $\Lambda(T)$ as $\{t_i : i \in I\}$, where I is either ω or $\{0, 1, 2, \ldots, k\}$ for some $k \in \omega$. We then choose a disjoint partition $\{A_i : i \in I\}$ of $\{2^n : n \in \omega\}$, with each A_i being infinite.

We define a mapping $\tau : \mathbb{N} \to T$ by first stating that $\tau(2^n) = t_i$ if $2^n \in A_i$. We then extend τ to $\mathbb{N}$ by putting $\tau(n) = \prod_{i \in \operatorname{supp}(n)} \tau(2^i)$, with the terms in this product occurring in the order of increasing i.

It follows from Lemma 6.3, that $\widetilde{\tau} : \beta \mathbb{N} \to T$ is a homomorphism on $\mathbb{H}$. We must show that $\widetilde{\tau}[\mathbb{H}] = T$. To see this, let $i \in I$ and let $x \in A_i^*$. Then $x \in \mathbb{H}$ and $\widetilde{\tau}(x) = t_i$. So $\widetilde{\tau}[\mathbb{H}] \supseteq \Lambda(T)$ and thus $\widetilde{\tau}[\mathbb{H}] \supseteq c\ell(\Lambda(T)) = T$. $\square$

Corollary 6.5. *Every finite (discrete) semigroup is the image of $\mathbb{H}$ under a continuous homomorphism.*

Proof. A finite discrete semigroup is a compact right topological semigroup which is equal to its own topological center. $\square$

The following simple result will frequently be useful.

Lemma 6.6. *Let p be an idempotent in $(\beta \mathbb{N}, +)$. Then for every $n \in \mathbb{N}$, $n\mathbb{N} \in p$.*

Proof. Let $\gamma : \mathbb{N} \to \mathbb{Z}_n$ denote the canonical homomorphism. Then $\widetilde{\gamma} : \beta \mathbb{N} \to \mathbb{Z}_n$ is also a homomorphism, by Corollary 4.22. Thus $\widetilde{\gamma}(p) = \widetilde{\gamma}(p) + \widetilde{\gamma}(p)$ and so $\widetilde{\gamma}(p) = 0$. It follows that $\widetilde{\gamma}^{-1}[\{0\}]$ is a neighborhood of p, and hence that $\widetilde{\gamma}^{-1}[\{0\}] \cap \mathbb{N} = n\mathbb{N} \in p$. $\square$

Definition 6.7. We define $\phi : \mathbb{N} \to \omega$ and $\theta : \mathbb{N} \to \omega$ by stating that $\phi(n) = \max(\operatorname{supp}(n))$ and $\theta(n) = \min(\operatorname{supp}(n))$.

Lemma 6.8. *The set $\mathbb{H}$ is a compact subsemigroup of $(\beta\mathbb{N}, +)$, which contains all the idempotents of $(\beta\mathbb{N}, +)$. Furthermore, for any $p \in \beta\mathbb{N}$ and any $q \in \mathbb{H}$, $\widetilde{\phi}(p+q) = \widetilde{\phi}(q)$ and $\widetilde{\theta}(p+q) = \widetilde{\theta}(p)$.*

Proof. $\mathbb{H}$ is obviously compact. Now, if $n \in 2^k\mathbb{N}$ and if $r > \phi(n)$, $n + 2^r\mathbb{N} \subseteq 2^k\mathbb{N}$. Thus it follows from Theorem 4.20 that $\mathbb{H}$ is a semigroup.

By Lemma 6.6, $\mathbb{H}$ contains all of the idempotents of $(\beta\mathbb{N}, +)$.

Now suppose that $p \in \beta\mathbb{N}$ and $q \in \mathbb{H}$. Given any $m \in \mathbb{N}$, choose $r \in \mathbb{N}$ satisfying $r > \phi(m)$. Then, if $n \in 2^r\mathbb{N}$, $\phi(m + n) = \phi(n)$ and $\theta(m + n) = \theta(m)$. Hence since $2^r\mathbb{N} \in q$,

$$
\begin{aligned}
\widetilde{\phi}(m + q) &= q\text{-}\lim_{n \in 2^r\mathbb{N}} \phi(m + n) \\
&= q\text{-}\lim_{n \in 2^r\mathbb{N}} \phi(n) \\
&= \widetilde{\phi}(q).
\end{aligned}
$$

It follows that $\widetilde{\phi}(p + q) = \lim_{m \to p} \widetilde{\phi}(m + q) = \widetilde{\phi}(q)$. Similarly, $\widetilde{\theta}(p + q) = \widetilde{\theta}(p)$. $\qquad\square$

Theorem 6.9. *$(\beta\mathbb{N}, +)$ contains $2^{\mathfrak{c}}$ minimal left ideals and $2^{\mathfrak{c}}$ minimal right ideals. Each of these contains $2^{\mathfrak{c}}$ idempotents.*

Proof. Let $A = \{2^n : n \in \mathbb{N}\}$. Since $\phi(2^n) = \theta(2^n) = n$ for each $n \in \mathbb{N}$, $\phi_{|A} = \theta_{|A} : A \to \mathbb{N}$ is bijective and so $\widetilde{\phi}_{|\overline{A}} = \widetilde{\theta}_{|\overline{A}} : \overline{A} \to \beta\mathbb{N}$ is bijective as well. Now if $q_1, q_2 \in A^*$ and $(\beta\mathbb{N} + q_1) \cap (\beta\mathbb{N} + q_2) \neq \emptyset$, then by Lemma 6.8, $q_1 = q_2$. (For if $r + q_1 = s + q_2$, then $\widetilde{\phi}(q_1) = \widetilde{\phi}(r + q_1) = \widetilde{\phi}(s + q_2) = \widetilde{\phi}(q_2)$.) By Corollary 3.57 $|A^*| = 2^{\mathfrak{c}}$, so $\beta\mathbb{N}$ has $2^{\mathfrak{c}}$ pairwise disjoint left ideals, and each contains a minimal left ideal, by Corollary 2.6.

Now observe that any idempotent e which is minimal (with respect to the ordering of idempotents where $e \leq f$ if and only if $e = e + f = f + e$) in $\mathbb{H}$ is in fact minimal in $\beta\mathbb{N}$. For otherwise there is an idempotent f of $\beta\mathbb{N}$ with $f < e$. But every idempotent of $\beta\mathbb{N}$ is in $\mathbb{H}$ and so $f \in \mathbb{H}$, contradicting the minimality of e in $\mathbb{H}$.

Given any $q \in A^*$, $q + \mathbb{H}$ contains an idempotent $e(q)$ which is minimal in $\mathbb{H}$ by Theorems 2.7 and 2.9, and is hence minimal in $\beta\mathbb{N}$. Thus by Theorem 2.9, $e(q) + \beta\mathbb{N}$ is a minimal right ideal of $\beta\mathbb{N}$. We claim that if $q_1 \neq q_2$ in A^*, then $e(q_1) + \beta\mathbb{N} \neq e(q_2) + \beta\mathbb{N}$. For otherwise, $e(q_2) \in e(q_1) + \beta\mathbb{N}$ and so $e(q_2) = e(q_1) + e(q_2)$ by Lemma 1.30. However, $\widetilde{\theta}(e(q_2)) \in \widetilde{\theta}[q_2 + \mathbb{H}] = \{q_2\}$ (by Lemma 6.8), while $\widetilde{\theta}(e(q_1) + e(q_2)) = \widetilde{\theta}(e(q_1)) = q_1$, a contradiction.

If L is any minimal left ideal and R is any minimal right ideal of $(\beta\mathbb{N}, +)$, $L \cap R$ contains an idempotent, by Theorem 1.61. It follows that L and R both contain $2^{\mathfrak{c}}$ idempotents. $\qquad\square$

As we shall see in Theorem 6.54, the following theorem is a special case of a theorem that is far more general. We include it, however, because the special case is important and has a simpler proof.

Theorem 6.10. *$\mathbb{N}$ is the center of $(\beta\mathbb{N}, +)$ and of $(\beta\mathbb{N}, \cdot)$.*

Proof. By Theorem 4.23 $\mathbb{N}$ is contained in the centers of $(\beta\mathbb{N}, +)$ and of $(\beta\mathbb{N}, \cdot)$.

Let $A = \{2^n : n \in \mathbb{N}\}$. Suppose that $p \in \mathbb{N}^*$ and $q \in A^*$. By Lemma 6.8, $\widetilde{\phi}(p + q) = \widetilde{\phi}(q)$. On the other hand, suppose that $m, n \in \mathbb{N}$ and that $n > m$. Then $\phi(m+n) = \phi(n)$ or $\phi(m+n) = \phi(n)+1$. For each m, $\{n \in \mathbb{N} : \phi(m+n) = \phi(n)\} \in p$ or $\{n \in \mathbb{N} : \phi(m + n) = \phi(n) + 1\} \in p$. In the first case,

$$p \in c\ell\{n \in \mathbb{N} : \phi(m + n) = \phi(n)\} \quad \text{and so} \quad \widetilde{\phi}(m + p) = \widetilde{\phi}(p).$$

In the second case, $\widetilde{\phi}(m + p) = \widetilde{\phi}(p) + 1$. Now $\{m \in \mathbb{N} : \widetilde{\phi}(m + p) = \widetilde{\phi}(p)\} \in q$ or $\{m \in \mathbb{N} : \widetilde{\phi}(m + p) = \widetilde{\phi}(p) + 1\} \in q$. So $\widetilde{\phi}(q + p) = \widetilde{\phi}(p)$ or $\widetilde{\phi}(q + p) = \widetilde{\phi}(p) + 1$. Since there are $2^{\mathfrak{c}}$ different values of $\widetilde{\phi}(q)$, we can choose $q \in A^*$ satisfying $\widetilde{\phi}(q) \notin \{\widetilde{\phi}(p), \widetilde{\phi}(p) + 1\}$. Then $q \in \mathbb{H}$ so $\widetilde{\phi}(p + q) = \widetilde{\phi}(q) \neq \widetilde{\phi}(q + p)$ so $p + q \neq q + p$, and thus p cannot be in the center of $(\beta\mathbb{N}, +)$.

Now for any $m, n \in \mathbb{N}$, $\phi(m2^n) = \phi(m) + n = \phi(m) + \phi(2^n)$. Now given $p \in \mathbb{N}^*$ and $q \in A^*$, one has

$$\begin{aligned}
\widetilde{\phi}(p \cdot q) &= \lim_{m \to p} \lim_{2^n \to q} \phi(m \cdot 2^n) \\
&= \lim_{m \to p} \lim_{2^n \to q} \big(\phi(m) + \phi(2^n)\big) \\
&= \lim_{m \to p} \big(\phi(m) + \lim_{2^n \to q} \phi(2^n)\big) \\
&= \lim_{m \to p} \big(\phi(m) + \widetilde{\phi}(q)\big) \\
&= \widetilde{\phi}(p) + \widetilde{\phi}(q).
\end{aligned}$$

And similarly, $\widetilde{\phi}(q \cdot p) = \widetilde{\phi}(q) + \widetilde{\phi}(p)$.

Now let $p \in \mathbb{N}^*$. We show that p is not in the center of $(\beta\mathbb{N}, \cdot)$. Since ϕ is finite-to-one, $\widetilde{\phi}(p) \in \mathbb{N}^*$ so pick $r \in \mathbb{N}^*$ such that $\widetilde{\phi}(p) + r \neq r + \widetilde{\phi}(p)$. Since $\widetilde{\phi}_{|\overline{A}}$ maps $\overline{A}$ bijectively to $\beta\mathbb{N}$, pick $q \in A^*$ such that $\widetilde{\phi}(q) = r$. Then $\widetilde{\phi}(p \cdot q) = \widetilde{\phi}(p) + r \neq r + \widetilde{\phi}(p) = \widetilde{\phi}(q \cdot p)$ so $p \cdot q \neq q \cdot p$. $\square$

Remark 6.11. *It follows from Theorem 4.24 that $\mathbb{N}$ is the topological center of $(\beta\mathbb{N}, +)$ and of $(\beta\mathbb{N}, \cdot)$.*

We shall in fact see in Theorems 6.79 and 6.80 that there is a subset Δ of $\mathbb{N}^*$ such that Δ and $\mathbb{N}^* \backslash \Delta$ are both left ideals of $(\beta\mathbb{N}, +)$ and of $(\beta\mathbb{N}, \cdot)$.

Theorem 6.12. $\mathbb{H}$ *contains an infinite decreasing sequence of idempotents.*

Proof. Let $\langle A_n \rangle_{n=1}^{\infty}$ be an infinite increasing sequence of subsets of $\mathbb{N}$ such that $A_{n+1} \backslash A_n$ is infinite for every n. Let $S_n = \{m \in \mathbb{N} : \text{supp}(m) \subseteq A_n\}$. We observe that, for every $n, r \in \mathbb{N}$, we have $k+m \in 2^r \mathbb{N} \cap S_n$ whenever $k, m \in 2^r \mathbb{N} \cap S_n$ and $\text{supp}(k) \cap \text{supp}(m) = \emptyset$. Thus it follows from Theorem 4.20 (with $\mathcal{A} = \{S_n \cap 2^r \mathbb{N} : r \in \mathbb{N}\}$) that $T_n = \overline{S_n} \cap \mathbb{H}$ is a subsemigroup of $\beta\mathbb{N}$.

We shall inductively construct a sequence $\langle e_n \rangle_{n=1}^{\infty}$ of distinct idempotents in $\mathbb{N}$ satisfying $e_{n+1} \leq e_n$ and $e_n \in K(T_n)$ for every n.

We first choose e_1 to be any minimal idempotent in T_1. We then assume that e_i has been chosen for each $i \in \{1, 2, \ldots, m\}$.

By Theorem 1.60, we can choose an idempotent $e_{m+1} \in K(T_{m+1})$ for which $e_{m+1} \leq e_m$. We shall show that $e_{m+1} \neq e_m$ by showing that $e_m \notin K(T_{m+1})$.

To see this, we choose any $x \in \mathbb{N}^* \cap \overline{\{2^n : n \in A_{m+1} \backslash A_m\}}$. Let

$$M = \{r + 2^n + s : r, s \in \mathbb{N}, n \in A_{m+1} \backslash A_m \text{ and } \max \operatorname{supp}(r) < n < \min \operatorname{supp}(s)\}.$$

Since $M \cap S_m = \emptyset$, $e_m \notin \overline{M}$. However, it follows from Exercise 4.1.6, that $y + x + z \in \overline{M}$ for every $y, z \in \mathbb{H}$. So $\overline{M}$ contains the ideal $T_{m+1} + x + T_{m+1}$ of T_{m+1} and therefore contains $K(T_{m+1})$. Hence $e_m \notin K(T_{m+1})$. $\square$

We have seen that it is remarkably easy to produce homomorphisms on $\mathbb{H}$. This is related to the fact that $\mathbb{H}$ can be defined in terms of the concept of oids, and the algebraic structure of an oid is very minimal. Theorem 6.15 shows that the only algebraic information needed to define $\mathbb{H}$ is the knowledge of how to multiply by 1.

Definition 6.13. (a) A set A is called an *id* if A has a distinguished element 1 and a multiplication mapping $(\{1\} \times A) \cup (A \times \{1\})$ to A with the property that $1a = a1 = a$ for every $a \in A$.

(b) If for each i in some index set I, A_i is an id, then

$$\bigoplus_{i \in I} A_i = \{x \in \times_{i \in I} A_i : \{i \in I : x_i \neq 1\} \text{ is finite }\}$$

is an *oid*.

(c) If $S = \bigoplus_{i \in I} A_i$ is an oid and $x \in S$, then $\operatorname{supp}(x) = \{i \in I : x_i \neq 1\}$.

Notice that we have already defined the notation "$\operatorname{supp}(x)$" when $x \in \mathbb{N}$. The correspondence between the two versions will become apparent in the proof of Theorem 6.15 below.

Definition 6.14. Let $S = \bigoplus_{i \in I} A_i$ be an oid. If $x, y \in S$ and $\operatorname{supp}(x) \cap \operatorname{supp}(y) = \emptyset$, then $x \cdot y$ is defined by $(x \cdot y)_i = x_i \cdot y_i$ for all $i \in I$.

Notice that in an oid, one does not require $x \cdot y$ to be defined if $\operatorname{supp}(x) \cap \operatorname{supp}(y) \neq \emptyset$.

Theorem 6.15. *Let $A = \{a, 1\}$ be an id with two elements and, for each $i \in \omega$, let $A_i = A$. Let $S = \bigoplus_{i \in \omega} A_i$ and let $H = \bigcap_{n \in \omega} c\ell_{\beta S}\{x \in S : \min(\operatorname{supp}(x)) \geq n\}$. Extend the operation $\cdot$ to all of $S \times S$ arbitrarily and then extend the operation $\cdot$ to βS as in Theorem 4.1. Then H is topologically and algebraically isomorphic to $\mathbb{H}$.*

Proof. Define $\varphi : \mathbb{N} \to S$ by

$$\varphi(x)_i = \begin{cases} 1 & \text{if } i \notin \operatorname{supp}(x) \\ a & \text{if } i \in \operatorname{supp}(x) \end{cases}$$

and notice that φ is one-to-one, $\varphi[\mathbb{N}] = S \backslash \{\overline{1}\}$, and for all $x \in \mathbb{N}$, $\operatorname{supp}(\varphi(x)) = \operatorname{supp}(x)$. In particular, for all $n \in \mathbb{N}$, $\varphi[2^n \mathbb{N}] = \{x \in S : \min(\operatorname{supp}(x)) \geq n\}$ so that $\widetilde{\varphi}[\mathbb{H}] = H$.

Since φ is one-to-one, so is $\widetilde{\varphi}$ by Exercise 3.4.1. Thus $\widetilde{\varphi}_{|\mathbb{H}}$ is a homeomorphism onto H.

To see that $\widetilde{\varphi}_{|\mathbb{H}}$ is a homomorphism, let $p, q \in \mathbb{H}$. Then by Remark 4.2,

$$\widetilde{\varphi}(p + q) = \widetilde{\varphi}(p\text{-}\lim_{x \in \mathbb{N}} q\text{-}\lim_{y \in \mathbb{N}} x + y) = p\text{-}\lim_{x \in \mathbb{N}} q\text{-}\lim_{y \in \mathbb{N}} \varphi(x + y).$$

Now, given any $x \in \mathbb{N}$, if $n = \max(\text{supp}(x)) + 1$, then for all $y \in 2^n \mathbb{N}$, $\varphi(x + y) = \varphi(x) \cdot \varphi(y)$, so that, since $q \in \mathbb{H}$ and $\lambda_{\varphi(x)}$ and $\widetilde{\varphi}$ are continuous,

$$q\text{-}\lim_{y \in \mathbb{N}} \varphi(x + y) = q\text{-}\lim_{y \in 2^n \mathbb{N}} \varphi(x + y) = q\text{-}\lim_{y \in 2^n \mathbb{N}} \big(\varphi(x) \cdot \varphi(y)\big) = \varphi(x) \cdot \widetilde{\varphi}(q).$$

Thus $\widetilde{\varphi}(p + q) = p\text{-}\lim_{x \in \mathbb{N}} \big(\varphi(x) \cdot \widetilde{\varphi}(q)\big) = \widetilde{\varphi}(p) \cdot \widetilde{\varphi}(q)$ as required. $\square$

Notice that in the statement of Theorem 6.15, we did not require the arbitrary extension of the operation to be associative on S. However, the theorem says that it nonetheless induces an associative operation on H.

Exercise 6.1.1. Show that $\mathbb{H}$ has $2^{\mathfrak{c}}$ pairwise disjoint closed right ideals. (Hint: Consider $\{\widetilde{\theta}(p) : p \in A^*\}$ where $A = \{2^n : n \in \mathbb{N}\}$.)

Exercise 6.1.2. Show that no two closed right ideals of $\beta \mathbb{N}$ can be disjoint. (Hint: Use Theorem 2.19.)

Exercise 6.1.3. Show that no two closed right ideals of $\mathbb{N}^*$ can be disjoint. (Hint: Consider Theorem 4.37 and Exercise 6.1.2.)

Exercise 6.1.4. Show that there are two minimal idempotents in $(\beta \mathbb{N}, +)$ whose sum is not in Γ, the closure of the set of idempotents. (Hint: Let S denote the semigroup described in Example 2.13. By Corollary 6.5 there is a continuous homomorphism h mapping $\mathbb{H}$ onto S. Show that there are minimal idempotents u and v in $\mathbb{H}$ for which $h(u) = e$ and $h(v) = f$. Observe that idempotents minimal in $\mathbb{H}$ are also minimal in $\beta \mathbb{N}$.)

6.2 Intersecting Left Ideals

In this section we consider left ideals of βS of the form $\beta S \cdot p$.

Definition 6.16. Let S be a semigroup and let $s \in S$. The left ideal Ss of S is called the *semiprincipal left ideal of S generated by s*.

Note that the semiprincipal left ideal generated by s is equal to the principal left ideal generated by s if and only if $s \in Ss$.

Definition 6.17. Let S be a semigroup and let $p \in \beta S$. Then $C(p) = \{A \subseteq S : \text{for all } x \in S, x^{-1}A \in p\}$.

Theorem 6.18. *Let S be a semigroup and let $p \in \beta S$. Then $\beta S \cdot p = \{q \in \beta S : C(p) \subseteq q\}$.*

Proof. Let $r \in \beta S$. Then given $A \in C(p)$ one has $S = \{x \in S : x^{-1}A \in p\}$ so $A \in r \cdot p$.

For the other inclusion, let $q \in \beta S$ such that $C(p) \subseteq q$. For $A \in q$, let $B(A) = \{x \in S : x^{-1}A \in p\}$. We claim that $\{B(A) : A \in q\}$ has the finite intersection property. To see this observe that $\{B(A) : A \in q\}$ is closed under finite intersections and that if $B(A) = \emptyset$, then $S \backslash A \in C(p) \subseteq q$. Pick $r \in \beta S$ such that $\{B(A) : A \in q\} \subseteq r$. Then $q = r \cdot p$. □

Theorem 6.19. *Suppose that S is a countable discrete semigroup and that $p, q \in \beta S$. If $\beta S \cdot p \cap \beta S \cdot q \neq \emptyset$, then $sp = xq$ for some $s \in S$ and some $x \in \beta S$, or $yp = tq$ for some $t \in S$ and some $y \in \beta S$.*

Proof. Since $\beta S \cdot p = \overline{Sp}$ and $\beta S \cdot q = \overline{Sq}$ we can apply Corollary 3.42 and deduce that $sp \in \overline{Sq}$ for some $s \in S$, or else $tq \in \overline{Sp}$ for some $t \in S$. In the first case, $sp = xq$ for some $x \in \beta S$. In the second case, $tq = yp$ for some $y \in \beta S$. □

Corollary 6.20. *Let S be a countable group. Suppose that p and q are elements of βS for which $\beta S \cdot p \cap \beta S \cdot q \neq \emptyset$. Then $p \in \beta S \cdot q$ or $q \in \beta S \cdot p$.*

Proof. By Theorem 6.19, $sp = xq$ for some $s \in S$ and some $x \in \beta S$, or $tq = yp$ for some $t \in S$ and some $y \in \beta S$. In the first case, $p = s^{-1}xq$. In the second, $q = t^{-1}yp$. □

Corollary 6.21. *Let p and q be distinct elements of $\beta \mathbb{N}$. If the left ideals $\beta \mathbb{N} + p$ and $\beta \mathbb{N} + q$ are not disjoint, $p \in \beta \mathbb{N} + q$ or $q \in \beta \mathbb{N} + p$.*

Proof. Suppose that $(\beta \mathbb{N} + p) \cap (\beta \mathbb{N} + q) \neq \emptyset$. It then follows by Theorem 6.19, that $m + p = x + q$ for some $m \in \mathbb{N}$ and some $x \in \beta \mathbb{N}$, or else $n + q = y + p$ for some $n \in \mathbb{N}$ and some $y \in \beta \mathbb{N}$. Assume without loss of generality that $m + p = x + q$ for some $m \in \mathbb{N}$ and some $x \in \beta \mathbb{N}$. If $x \in \mathbb{N}^*$, then $-m + x \in \mathbb{N}^*$, because $\mathbb{N}^*$ is a left ideal of $\mathbb{Z}^*$, as shown in Exercise 4.3.5. So $p = -m + x + q \in \beta \mathbb{N} + q$. Suppose then that $x \in \mathbb{N}$. Then $x \neq m$, for otherwise we should have $p = -m + x + q = q$. If $x > m$, $-m + x \in \mathbb{N}$ and again $p = -m + x + q \in \beta \mathbb{N} + q$. If $x < m$, $-x + m \in \mathbb{N}$ and so $q = -x + m + p \in \beta \mathbb{N} + p$. □

By Corollaries 6.20 and 6.21, we have the following.

Remark 6.22. *If S is a countable group or if $S = \mathbb{N}$, any two semiprincipal left ideals of βS are either disjoint or comparable for the relationship of inclusion.*

Recall from Chapter 3 that if $T \subseteq S$ we have identified βT with the subset $c\ell\, T$ of βS.

Corollary 6.23. *Let S be a countable semigroup which can be embedded in a group G, and let $e, f \in E(S)$. If $\beta S \cdot e \cap \beta S \cdot f \neq \emptyset$, $ef = e$ or $fe = f$.*

Proof. By Theorem 6.19, we may assume that $se = xf$ for some $s \in S$ and some $x \in \beta S$. This implies that $e = s^{-1}xf$ in βG, and hence that $ef = s^{-1}xff = s^{-1}xf = e$. $\square$

We can now show that, for many semigroups S, the study of commutativity for idempotents in βS is equivalent to the study of the order relation $\leq$.

Corollary 6.24. *Suppose that S is a countable discrete semigroup which can be embedded in a group, and that $e, f \in E(\beta S)$. Then e and f commute if and only if $e \leq f$ or $f \leq e$.*

Proof. By definition of the relation $\leq$, if say $e \leq f$, then $e = ef = fe$.
     Conversely, suppose that $ef = fe$. Then $\beta S \cdot e \cap \beta S \cdot f \neq \emptyset$, and so, by Corollary 6.23, we may suppose that $ef = e$. Since $ef = fe$, this implies that $e \leq f$. $\square$

Corollary 6.25. *Suppose that S is a countable discrete semigroup which can be embedded in a group, and that C is a subsemigroup of βS. If $e, f \in E(C)$ satisfy $Ce \cap Cf = \emptyset$, then $\beta S \cdot e \cap \beta S \cdot f = \emptyset$.*

Proof. If $\beta S \cdot e \cap \beta S \cdot f \neq \emptyset$, we may suppose by Corollary 6.23, that $ef = e$. But then $e \in Ce \cap Cf$, a contradiction. $\square$

Exercise 6.2.1. Let S be a semigroup and let $p \in \beta S$. Prove that $C(p)$ is a filter on S and $C(p) = \bigcap(\beta S \cdot p)$.

Exercise 6.2.2. Let S be an infinite semigroup and let $p \in S^*$. Let $C_\omega(p) = \{A \subseteq S : \{x \in S : x^{-1}A \notin p\}$ is finite$\}$. Prove that $C_\omega(p)$ is a filter on S, $C_\omega(p) = \bigcap(S^* \cdot p)$, and $S^* \cdot p = \{q \in \beta S : C_\omega(p) \subseteq q\}$.

6.3 Numbers of Idempotents and Ideals — Copies of $\mathbb{H}$

The semigroup $\mathbb{H}$ arises in $\beta \mathbb{N}$. We shall see, however, that copies of $\mathbb{H}$ can be found in S^* for any infinite cancellative semigroup S.
     We introduced the notation $\mathrm{FP}(\langle x_n \rangle_{n=1}^\infty)$ in Definition 5.1. We extend it now to apply to sequences indexed by linearly ordered sets other than $\mathbb{N}$.

Definition 6.26. Let S be a semigroup, let $(D, <)$ be a linearly ordered set, and let $\langle x_s \rangle_{s \in D}$ be a D-sequence (i.e., a set indexed by D) in S.

(a) For $F \in \mathcal{P}_f(D)$, $\prod_{s \in F} x_s$ is the product in increasing order of indices.

(b) $\text{FP}(\langle x_s \rangle_{s \in D}) = \{\prod_{s \in F} x_s : F \in \mathcal{P}_f(D)\}$.

(c) The D-sequence $\langle x_s \rangle_{s \in D}$ has *distinct finite products* if and only if whenever $F, G \in \mathcal{P}_f(D)$ and $\prod_{s \in F} x_s = \prod_{s \in G} x_s$, one has $F = G$.

Thus, if $F = \{s_1, s_2, s_3, s_4\}$ where $s_1 < s_2 < s_3 < s_4$, then $\prod_{s \in F} x_s = x_{s_1} \cdot x_{s_2} \cdot x_{s_3} \cdot x_{s_4}$.

If additive notation is being used we define $\Sigma_{s \in F} x_s$, $\text{FS}(\langle x_s \rangle_{s \in D})$, and "distinct finite sums" analogously.

Theorem 6.27. *Let S be a discrete semigroup and let $\langle x_n \rangle_{n=1}^{\infty}$ be a sequence in S with distinct finite products. Let $T = \bigcap_{m=1}^{\infty} c\ell\big(\text{FP}(\langle x_n \rangle_{n=m}^{\infty})\big)$. Then T is a subsemigroup of βS which is algebraically and topologically isomorphic to $\mathbb{H}$.*

Proof. It was shown in Lemma 5.11 that T is a subsemigroup of βS. We define a mapping f from $\text{FP}(\langle x_n \rangle_{n=1}^{\infty})$ to $2\mathbb{N}$ by stating that $f(\prod_{n \in F} x_n) = \Sigma_{n \in F} 2^n$ for every $F \in \mathcal{P}_f(\mathbb{N})$. Since f is a bijection (because of the distinct finite products assumption), $\widetilde{f} : c\ell(\text{FP}(\langle x_n \rangle_{n=1}^{\infty})) \to \beta(2\mathbb{N})$ is a homeomorphism. We shall show that $\widetilde{f}_{|T}$ is a homomorphism from T onto $\mathbb{H}$.

Let $T_m = \text{FP}(\langle x_n \rangle_{n=m}^{\infty})$. Then $f[T_m] = 2^m \mathbb{N}$ and so

$$\widetilde{f}[T] = \widetilde{f}[\bigcap_{m=1}^{\infty} \overline{T_m}] = \bigcap_{m=1}^{\infty} \widetilde{f}[\overline{T_m}] = \bigcap_{m=1}^{\infty} \overline{f[T_m]} = \bigcap_{m=1}^{\infty} \overline{2^m \mathbb{N}} = \mathbb{H}.$$

To see that $\widetilde{f}_{|T}$ is a homomorphism, it suffices by Theorem 4.21 to show that for each $s \in \text{FP}(\langle x_n \rangle_{n=1}^{\infty})$ there is some $m \in \mathbb{N}$ such that $f(st) = f(s) + f(t)$ for all $t \in \text{FP}(\langle x_n \rangle_{n=m}^{\infty})$. So let $s \in \text{FP}(\langle x_n \rangle_{n=1}^{\infty})$ and pick $F \in \mathcal{P}_f(\mathbb{N})$ such that $s = \prod_{n \in F} x_n$. Let $m = \max F + 1$. $\qquad \square$

We now turn our attention to the effect of certain cancellation assumptions. Much more will be said on this subject in Chapter 8.

Lemma 6.28. *Let S be a discrete semigroup and let s be a cancelable element of S. For every $t \in S$ and $p \in \beta S$*

(i) *if S is right cancellative and $sp = tp$, then $s = t$, and*

(ii) *if S is left cancellative and $ps = pt$, then $s = t$.*

Proof. (i) Suppose that S is right cancellative. Let $t \in S$, let $p \in \beta S$, and suppose that $s \neq t$. We shall define a function $f : S \to S$ by stating that $f(su) = tu$ for every $u \in S$ and that $f(v) = s^2$ for every $v \in S \setminus sS$. We note that f is well-defined because every element in sS has a unique expression of the form su, and that f has no fixed points. We also note that since $\widetilde{f} \circ \lambda_s$ and λ_t agree on S, one has that $\widetilde{f}(sp) = tp$. Now by Lemma 3.33 we can partition S into three sets A_0, A_1, A_2 such that $A_i \cap f[A_i] = \emptyset$ for each $i \in \{0, 1, 2\}$. Choose i such that $sp \in A_i$. Then by Lemma 3.30, $f[A_i] \in \widetilde{f}(sp) = tp$, so $sp \neq tp$.

(ii) This is proved in essentially the same way.   □

Recall that a semigroup S is weakly left cancellative if and only if for all $u, v \in S$, $\{x \in S : ux = v\}$ is finite, and that S is weakly right cancellative if and only if for all $u, v \in S$, $\{x \in S : xu = v\}$ is finite.

Lemma 6.29. *Let S be an infinite weakly left cancellative semigroup and let I denote the set of right identities of S. Let κ be an infinite cardinal with $\kappa \leq |S|$, and let $\langle s_\lambda \rangle_{\lambda < \kappa}$ be a one-to-one κ-sequence in S. If T is a subset of S with cardinality κ, then there exists a one-to-one κ-sequence $\langle t_\lambda \rangle_{\lambda < \kappa}$ in T such that*

(a) *for every $\mu < \kappa$, $t_\mu \notin \mathrm{FP}(\langle t_\lambda \rangle_{\lambda < \mu})$,*

(b) *for every $\lambda < \mu < \kappa$ and every $u, v \in I \cup \{s_\iota : \iota < \mu\} \cup \mathrm{FP}(\langle t_\iota \rangle_{\iota < \mu})$, $u \neq vt_\mu$ and $ut_\lambda \neq vt_\mu$, and*

(c) *$I \cap \mathrm{FP}(\langle t_\lambda \rangle_{\lambda < \kappa}) = \emptyset$.*

Proof. For each $u, v \in S$, let $A_{u,v} = \{s \in S : u = vs\}$. Since S is weakly left cancellative, $A_{u,v}$ is finite. We note that this implies that I is finite, because $I \subseteq A_{u,u}$ for every $u \in S$.

We construct $\langle t_\lambda \rangle_{\lambda < \kappa}$ inductively. We choose t_0 to be any element of $T \backslash I$. We then assume that $\nu < \kappa$ and that t_ι has been chosen for every $\iota < \nu$ so that conditions (a), (b), and (c) hold with ν in place of κ.

Let $G = I \cup \{s_\iota : \iota < \nu\} \cup \mathrm{FP}(\langle t_\iota \rangle_{\iota < \nu})$, let $H = G \cup \{ut_\lambda : u \in G$ and $\lambda < \nu\}$, and let $K = \bigcup\{A_{u,v} : u, v \in H\}$. If $\nu < \omega$, then H and K are finite so we can choose $t_\nu \in T \backslash (H \cup K)$. If $\nu \geq \omega$, then $|H| = |\nu| < \kappa$ and $|K| \leq |\nu| < \kappa$ so we can again choose $t_\nu \in T \backslash (H \cup K)$.

Since $t_\nu \notin G$, hypothesis (a) holds. To verify hypothesis (b), let $\lambda < \nu$ and let $u, v \in I \cup \{s_\iota : \iota < \nu\} \cup \mathrm{FP}(\langle t_\iota \rangle_{\iota < \nu})$. Then $u, v \in G \subseteq H$ so $t_\nu \notin A_{u,v}$ so $u \neq vt_\nu$. Also $ut_\lambda \in H$ so $t_\nu \notin A_{ut_\lambda, v}$ and so $ut_\lambda \neq vt_\nu$. To verify hypothesis (c), note that $t_\nu \notin G$ so $t_\nu \notin I$. Also, given $v \in \mathrm{FP}(\langle t_\iota \rangle_{\iota < \nu})$ and $u \in I$, $t_\nu \notin A_{u,v}$ so $vt_\nu \notin I$.   □

Theorem 6.30. *Let S be an infinite discrete semigroup which is weakly left cancellative. Let κ be an infinite cardinal with $\kappa \leq |S|$ and let R and T be subsets of S of cardinality κ. Then there is a subset V of T with cardinality κ such that the set P of uniform ultrafilters on V has the following properties:*

(1) *$|P| = 2^{2^\kappa}$,*

(2) *for each pair of distinct elements $p, q \in P$, $c\ell_{\beta S}(Rp)$ and $c\ell_{\beta S}(Rq)$ are disjoint, and*

(3) *if S is also right cancellative, then for each $p \in P$, $ap \neq bp$ whenever $a \neq b$ in R and Rp is strongly discrete in βS.*

Proof. We apply Lemma 6.29, taking $\langle s_\lambda \rangle_{\lambda < \kappa}$ to be a one-to-one enumeration of all the elements of R. We put $V = \{t_\lambda : \lambda < \kappa\}$, where $\langle t_\lambda \rangle_{\lambda < \kappa}$ is the κ-sequence guaranteed by Lemma 6.29.

We have $|P| = 2^{2^\kappa}$, by Theorem 3.58, so (1) holds.

Suppose that $\alpha, \beta, \lambda, \mu \in \kappa$ satisfy $\alpha < \lambda$ and $\beta < \mu$, and that $s_\alpha t_\lambda = s_\beta t_\mu$. We claim that $\lambda = \mu$. To see this, suppose instead without loss of generality that $\lambda < \mu$. Then the equation $s_\alpha t_\lambda = s_\beta t_\mu$ contradicts condition (b) of Lemma 6.29.

For each $\mu < \kappa$, let $V_\mu = \{t_\nu : \mu < \nu < \kappa\}$. Then $V_\mu \in p$ for every $p \in P$. Let p and q be distinct elements of P. We can choose disjoint subsets A and B of V satisfying $A \in p$ and $B \in q$. For any $\alpha, \beta \in \kappa$, $s_\alpha(V_\alpha \cap A) \in s_\alpha p$ and $s_\beta(V_\beta \cap B) \in s_\beta q$ and we have seen that these two sets are disjoint. Thus, if $X = \bigcup_{\alpha \leq \kappa} s_\alpha(V_\alpha \cap A)$ and $Y = \bigcup_{\beta < \kappa} s_\beta(V_\beta \cap B)$, we have $X \cap Y = \emptyset$, $Rp \subseteq \overline{X}$ and $Rq \subseteq \overline{Y}$. So $\overline{X} \cap \overline{Y} = \emptyset$, $c\ell_{\beta S}(Rp) \subseteq \overline{X}$ and $c\ell_{\beta S}(Rq) \subseteq \overline{Y}$ so (2) holds.

Now suppose that S is right cancellative and $p \in P$. If $\alpha < \lambda < \kappa$ and $\beta < \mu < \kappa$, the equation $s_\alpha t_\lambda = s_\beta t_\mu$ implies that $\lambda = \mu$ and therefore that $\alpha = \beta$. For any $\alpha \in \kappa$, we have $s_\alpha V_\alpha \in s_\alpha p$. Now $s_\alpha V_\alpha \cap s_\beta V_\beta = \emptyset$ if $\alpha \neq \beta$, and so Rp is strongly discrete. $\square$

Lemma 6.31. *Let S be an infinite discrete semigroup, which is right cancellative and weakly left cancellative. Let T be an infinite subset of S, and let $\kappa = |T|$. There is a κ-sequence $\langle t_\lambda \rangle_{\lambda < \kappa}$ in T which has distinct finite products.*

Proof. Let $\langle s_\lambda \rangle_{\lambda < \kappa}$ be any one-to-one κ-sequence in S and choose a one-to-one κ-sequence $\langle t_\lambda \rangle_{\lambda < \kappa}$ in T with the properties guaranteed by Lemma 6.29.

To see that $\langle t_\lambda \rangle_{\lambda < \kappa}$ has distinct finite products, suppose instead that there exist $F \neq G$ in $\mathcal{P}_f(\kappa)$ with $\prod_{\lambda \in F} t_\lambda = \prod_{\lambda \in G} t_\lambda$, and choose F and G with $|F \cup G|$ as small as possible. We assume without loss of generality that $\max F \leq \max G = \mu$. The assumption that $\max F < \mu$ contradicts condition (a) of Lemma 6.29 if $G = \{\mu\}$ and contradicts condition (b) of Lemma 6.29 otherwise. So we also have $\max F = \mu$.

If $|F| > 1$ and $|G| > 1$, we contradict the assumption that $|F \cup G|$ is small as possible by cancelling t_μ from our equation and deducing that $\prod_{\lambda \in F \setminus \{\mu\}} t_\lambda = \prod_{\lambda \in G \setminus \{\mu\}} t_\lambda$. Thus we assume without loss of generality that $|F| = 1$ and $|G| > 1$. Let $x = \prod_{\lambda \in G \setminus \{\mu\}} t_\lambda$. Then $t_\mu = x t_\mu$ and so $s t_\mu = s x t_\mu$ for every $s \in S$. Hence, by our right cancellative assumption, $s = sx$ for every $s \in S$ and so $x \in I$. This contradicts condition (c) of Lemma 6.29. $\square$

Theorem 6.32. *Let S be an infinite discrete right cancellative and weakly left cancellative semigroup. Then every G_δ subset of S^* which contains an idempotent contains a copy of $\mathbb{H}$.*

Proof. Let $p \in E(S^*)$ and let A be a G_δ-subset of S^* for which $p \in A$. We may suppose that $A = \bigcap_{n=1}^\infty \overline{A_n}$, where $\langle A_n \rangle_{n=1}^\infty$ is a decreasing sequence of subsets of S.

We shall inductively construct a sequence $\langle x_n \rangle_{n=1}^\infty$ of elements of S with the property that, for every $m \in \mathbb{N}$, $\mathrm{FP}(\langle x_n \rangle_{n=m}^\infty) \subseteq A_m{}^\star$.

We first choose x_1 to be any element of $A_1{}^\star$. We then suppose that $k \in \mathbb{N}$ and that $x_1, x_2, \ldots, x_k$ have been chosen so that $\mathrm{FP}(\langle x_n \rangle_{n=i}^k) \subseteq A_i{}^\star$ for every $i \in \{1, 2, \ldots, k\}$.

Now $A_{k+1}{}^{\star} \in p$ and, for each $i \in \{1, 2, \ldots, k\}$ and each $y \in \mathrm{FP}(\langle x_n \rangle_{n=i}^{k})$, we have $y^{-1}A_i{}^{\star} \in p$ (by Lemma 4.14). We can therefore continue the induction by choosing

$$x_{k+1} \in A_{k+1}{}^{\star} \cap \bigcap\{y^{-1}A_i{}^{\star} : i \in \{1, 2, \ldots, k\} \text{ and } y \in \mathrm{FP}(\langle x_n \rangle_{n=i}^{k})\} \backslash \{x_1, x_2, \ldots, x_k\}.$$

By Lemma 6.31, we may suppose that $\langle x_n \rangle_{n=1}^{\infty}$ has distinct finite products, because we can replace $\langle x_n \rangle_{n=1}^{\infty}$ by a subsequence with this property. (Any sequence with distinct finite products is necessarily one-to-one. A one-to-one sequence in $\{x_n : n \in \mathbb{N}\}$ can be thinned to be a subsequence of $\langle x_n \rangle_{n=1}^{\infty}$.) Then, by Theorem 6.27, $\bigcap_{m=1}^{\infty} c\ell\big(\mathrm{FP}(\langle x_n \rangle_{n=m}^{\infty})\big)$ is a copy of $\mathbb{H}$. Since $c\ell\big(\mathrm{FP}\langle x_n \rangle_{n=m}^{\infty}\big) \subseteq \overline{A_m}$ for every $m \in \mathbb{N}$, $\bigcap_{m=1}^{\infty} c\ell\big(\mathrm{FP}\langle x_n \rangle_{n=m}^{\infty}\big) \subseteq A$. $\square$

Corollary 6.33. *Let S be an infinite discrete right cancellative and weakly left cancellative semigroup. Then every G_δ-subset of S^* which contains an idempotent contains $2^{\mathfrak{c}}$ non-minimal idempotents.*

Proof. Let A be a G_δ subset of S^*. By Theorem 6.32, A contains a copy of $\mathbb{H}$. So it suffices to show that $\mathbb{H}$ contains $2^{\mathfrak{c}}$ non-minimal idempotents.

Let $T = \bigcap_{m=1}^{\infty} c\ell_{\beta\mathbb{N}}\big(\mathrm{FS}(\langle 2^{2n} \rangle_{n=m}^{\infty})\big)$. Trivially $T \subseteq \mathbb{H}$ and by Theorem 6.27 T is a copy of $\mathbb{H}$. By Exercise 4.4.3 and Theorem 4.40, $T \cap K(\beta\mathbb{N}) = \emptyset$. Thus by Lemma 6.8 and Theorem 1.65, $T \cap K(\mathbb{H}) = \emptyset$. By Theorem 6.32, T contains a copy of $\mathbb{H}$ so by Lemma 6.8 and Theorem 6.9 T has $2^{\mathfrak{c}}$ idempotents. $\square$

Corollary 6.34. *Let S be an infinite discrete semigroup which is right cancellative and weakly left cancellative. Then every G_δ subset of S^* which contains an idempotent, contains an infinite decreasing chain of idempotents.*

Proof. This follows immediately from Theorems 6.32 and 6.12. $\square$

We now see that, for many familiar semigroups S, S^*S^* is nowhere dense in S^*.

Theorem 6.35. *Let S be a countable semigroup which is weakly left cancellative and right cancellative. Then S^*S^* is nowhere dense in S^*.*

Proof. We enumerate S as a sequence $\langle s_n \rangle_{n=1}^{\infty}$ of distinct elements. Let T denote any infinite subset of S. We shall produce an infinite subset V of T such that $V^* \cap S^*S^* = \emptyset$. To this end we shall inductively choose a sequence $\langle t_n \rangle_{n=1}^{\infty}$ of distinct elements of T with the property that the equations $s_m x = t_n$ and $s_{m'} x = t_{n'}$ have no simultaneous solution in S if $m < n$, $m' < n'$ and $n < n'$.

We choose t_1 to be any element of T. We then assume that $t_1, t_2, \ldots, t_k$ have been chosen and and we put $X = \{x \in S : s_m x = t_n \text{ for some } m, n \le k\}$. Since S is weakly left cancellative, X is finite. So we can choose $t_{k+1} \in T \backslash \big(\bigcup_{m=1}^{k} s_m X \cup \{t_1, t_2, \ldots, t_k\}\big)$. This shows that the sequence $\langle t_n \rangle_{n=1}^{\infty}$ can be constructed as claimed.

Let $V = \{t_n : n \in \mathbb{N}\}$. We claim that, for any $p, q \in S^*$, $pq \notin V^*$. Suppose, on the contrary, that $pq \in V^*$. Then, if $P = \{s \in S : sq \in \overline{V}\}$, $P \in p$. If $s \in P$, it follows from Theorem 4.31 that $sq \in V^*$. Pick $m < m'$ with $s_m, s_{m'} \in P$. If

$Q = \{x \in S : s_m x \text{ and } s_{m'} x \text{ are in } \{t_i : i > m'\}\}$, then $Q \in q$. So we can choose $x \in Q$ and will then have $s_m x = t_n$ and $s_{m'} x = t_{n'}$ for some $n > m$ and some $n' > m'$. Since S is right cancellable, $n \neq n'$. But then we have a contradiction to the choice of the sequence $\langle t_n \rangle_{n=1}^{\infty}$. $\square$

Notice that the hypotheses of Corollary 6.34 and Theorem 6.35 cannot be significantly weakened. The semigroup $(\mathbb{N}, \vee)$ is weakly right and weakly left cancellative, but every member of $\mathbb{N}^*$ is a minimal idempotent in $(\beta\mathbb{N}, \vee)$ and $\mathbb{N}^* \vee \mathbb{N}^* = \mathbb{N}^*$.

We have need of the following topological fact which it would take us too far afield to prove.

Theorem 6.36. *There is a subset L of $\mathbb{N}^*$ such that $|L| = 2^{\mathfrak{c}}$ and for all $p \neq q$ in L and all $f : \mathbb{N} \to \mathbb{N}$, $\widetilde{f}(p) \neq q$.*

Proof. This is a special case of [213]. $\square$

Lemma 6.37. *Let $p, q \in \mathbb{N}^*$. If there is a one-to-one function $f : \mathbb{N} \to \beta\mathbb{N}$ such that $f[\mathbb{N}]$ is discrete and $\widetilde{f}(p) = q$, then there is a function $g : \mathbb{N} \to \mathbb{N}$ such that $\widetilde{g}(q) = p$.*

Proof. For each $n \in \mathbb{N}$ choose $A_n \in f(n)$ such that $\overline{A_n} \cap f[\mathbb{N}\setminus\{n\}] = \emptyset$. For each $n > 1$ in $\mathbb{N}$, let $B_n = A_n \setminus \bigcup_{k=1}^{n-1} A_k$ and let $B_1 = \mathbb{N} \setminus \bigcup_{n=2}^{\infty} B_n$. Then $\{B_n : n \in \mathbb{N}\}$ is a partition of $\mathbb{N}$ and for each $n \in \mathbb{N}$, $B_n \in f(n)$. Define $g : \mathbb{N} \to \mathbb{N}$ by $g(k) = n$ if and only if $k \in B_n$. Then for each $n \in \mathbb{N}$ one has g is constantly equal to n on a member of $f(n)$ so $\widetilde{g}(f(n)) = n$. Therefore $\widetilde{g} \circ \widetilde{f} : \beta\mathbb{N} \to \beta\mathbb{N}$ is the identity on $\beta\mathbb{N}$. Thus $\widetilde{g}(q) = \widetilde{g}(\widetilde{f}(p)) = p$. $\square$

Recall that two points x and y in a topological space X have the same *homeomorphism type* if and only if there is a homeomorphism f from X onto X such that $f(x) = y$. See [69, Chapter 9].

Theorem 6.38. *Let D be a discrete space and let X be an infinite compact subset of βD. Then X has at least $2^{\mathfrak{c}}$ distinct homeomorphism types.*

Proof. By Theorem 3.59, X contains a copy of $\beta\mathbb{N}$, so we shall assume that $\beta\mathbb{N} \subseteq X$. Let $L \subseteq \mathbb{N}^*$ be as guaranteed by Theorem 6.36. We claim that the elements of L all have different homeomorphism types in X.

Suppose instead that some two elements of L have the same homeomorphism type in X. We claim that in fact

(*) there exist some homeomorphism h from X onto X and some $p \neq q$ in L such that $h(p) = q$ and $q \in c\ell(h[\mathbb{N}] \cap c\ell\,\mathbb{N})$.

To verify (*), pick a homeomorphism h from X onto X and $p \neq q$ in L such that $h(p) = q$. Now $q \in c\ell\,\mathbb{N} \cap c\ell\,h[\mathbb{N}]$ so by Corollary 3.41, $q \in c\ell(\mathbb{N} \cap c\ell\,h[\mathbb{N}]) \cup c\ell(h[\mathbb{N}] \cap c\ell\,\mathbb{N})$. If $q \in c\ell(h[\mathbb{N}] \cap c\ell\,\mathbb{N})$, then we have that (*) holds directly, so assume that $q \in c\ell(\mathbb{N} \cap c\ell\,h[\mathbb{N}])$. Then $p = h^{-1}(q) \in h^{-1}[c\ell(\mathbb{N} \cap c\ell\,h[\mathbb{N}])] =$

$c\ell(h^{-1}[\mathbb{N}] \cap c\ell\,\mathbb{N})$. Thus (*) holds with p and q interchanged and with h^{-1} replacing h.

Thus (*) holds and we have $q \in c\ell(h[\mathbb{N}] \cap \beta\mathbb{N})$. Let $A = h[\mathbb{N}] \cap \mathbb{N}$ and let $B = h[\mathbb{N}] \cap \mathbb{N}^*$. Then $q \in c\ell\,A \cup c\ell\,B$.

Assume first that $q \in c\ell\,A$. Define $f : \mathbb{N} \to \mathbb{N}$ by

$$f(n) = \begin{cases} h(n) & \text{if } n \in h^{-1}[A] \\ 1 & \text{if } n \in \mathbb{N}\backslash h^{-1}[A]. \end{cases}$$

Then $p = h^{-1}(q) \in h^{-1}[c\ell\,A] = c\ell\,h^{-1}[A]$ and f and h agree on $h^{-1}[A]$ so $\widetilde{f}(p) = h(p) = q$, contradicting the choice of L.

Thus we must have $q \in c\ell\,B$. Since $h[\mathbb{N}]$ is discrete, and no countable subset of $\mathbb{N}^*$ is dense in $\mathbb{N}^*$ by Corollary 3.37, we may pick a one-to-one function $f : \mathbb{N} \to \mathbb{N}^*$ such that $f(n) = h(n)$ for all $n \in h^{-1}[B]$ and $f[\mathbb{N}]$ is discrete. Then $p = h^{-1}(q) \in h^{-1}[c\ell\,B] = c\ell\,h^{-1}[B]$ and f and h agree on $h^{-1}[B]$. So $\widetilde{f}(p) = q$. But then by Lemma 6.37 there is a function $g : \mathbb{N} \to \mathbb{N}$ such that $\widetilde{g}(q) = p$, again contradicting the choice of L. $\qquad\square$

Recall from Theorem 2.11 that the minimal left ideals of βS are homeomorphic. Thus the hypothesis of the following theorem is the same as stating that some minimal left ideal of βS is infinite.

Theorem 6.39. *Let S be a discrete semigroup. If the minimal left ideals of βS are infinite, βS contains at least $2^{\mathfrak{c}}$ minimal right ideals.*

Proof. Suppose that L is an infinite minimal left ideal of βS. Then L is compact. Now two points x and y of L which belong to the same minimal right ideal R belong to the same homeomorphism type in L. To see this, observe that $R \cap L$ is a group, by Theorem 1.61. Let x^{-1} and y^{-1} be the inverses of x and y in $R \cap L$. The mapping $\rho_{x^{-1}y}$ is a homeomorphism from L to itself by Theorem 2.11(c) and $\rho_{x^{-1}y}(x) = y$.

By Theorem 6.38 the points of L belong to at least $2^{\mathfrak{c}}$ different homeomorphism classes. So βS contains at least $2^{\mathfrak{c}}$ minimal right ideals. $\qquad\square$

Lemma 6.40. *Let S be an infinite discrete cancellative semigroup. Then every minimal left ideal of βS is infinite.*

Proof. Let L be a left ideal of βS and let $p \in L$. If s, t are distinct elements of S, then $sp \neq tp$, by Lemma 6.28. $\qquad\square$

Corollary 6.41. *Let S be an infinite discrete cancellative semigroup. Then βS contains at least $2^{\mathfrak{c}}$ minimal right ideals.*

Proof. This follows from Theorem 6.39 and Lemma 6.40. $\qquad\square$

Theorem 6.42. *Let S be an infinite discrete semigroup which is weakly left cancellative. If $|S| = \kappa$, then there are $2^{2^{\kappa}}$ pairwise disjoint left ideals of βS.*

Proof. This is an immediate consequence of Theorem 6.30, with $R = S$. $\qquad\square$

Corollary 6.43. *If S is a weakly left cancellative infinite discrete semigroup with cardinality κ, then βS contains $2^{2^{\kappa}}$ minimal idempotents.*

Proof. Each left ideal of βS contains a minimal idempotent. $\qquad\square$

Theorem 6.44. *Let S be an infinite cancellative discrete semigroup. Then βS contains at least 2^{c} minimal left ideals and at least 2^{c} minimal right ideals. Each minimal right ideal and each minimal left ideal contains at least 2^{c} idempotents.*

Proof. By Theorem 6.42 and Corollary 6.41, βS contains at least 2^{c} minimal left ideal and at least 2^{c} minimal right ideals. Now each minimal right ideal and each minimal left ideal have an intersection which contains an idempotent, by Theorem 1.61. So each minimal right ideal and each minimal left ideal contains at least 2^{c} idempotents. $\qquad\square$

Since minimal left ideals of βS are closed, it is immediate that, if S is weakly left cancellative, there are many pairwise disjoint closed left ideals of βS. We see that it is more unusual for S^* to be the disjoint union of finitely many closed left ideals of βS.

Definition 6.45. *Let S be a semigroup and let $B \subseteq S$. Then B is almost left invariant if and only if for each $s \in S$, $sB \backslash B$ is finite.*

Theorem 6.46. *Let S be a semigroup and let $n \in \mathbb{N}$. Then S^* is the union of n pairwise disjoint closed left ideals of βS if and only if*

(a) S is weakly left cancellative and

(b) S is the union of n pairwise disjoint infinite almost left invariant subsets.

Proof. Necessity. Assume that $S^* = \bigcup_{i=1}^{n} L_i$, where each L_i is a closed left ideal of βS and $L_i \cap L_j = \emptyset$ for $i \neq j$. Then S^* is a left ideal of βS so by Theorem 4.31, S is weakly left cancellative.

Since $L_i \cap L_j = \emptyset$ when $i \neq j$, each L_i is clopen so by Theorem 3.23, there is some $C_i \subseteq S$ such that $L_i = C_i^*$. Let $B_1 = C_1$. For $i \in \{2, 3, 4, \ldots, n-1\}$ (if any) let $B_i = C_i \backslash \bigcup_{j=1}^{i-1} C_j$, and let $C_n = S \backslash \bigcup_{j=1}^{n-1} C_j$. Then for each i, $B_i^* = C_i^* = L_i$ and $B_i \cap B_j = \emptyset$ when $i \neq j$.

Finally, let $i \in \{1, 2, \ldots, n\}$, let $s \in S$, and suppose that $sB_i \backslash B_i$ is infinite. Since $sB_i \backslash B_i = s \cdot (B_i \backslash s^{-1}B_i)$, it follows that $B_i \backslash s^{-1}B_i$ is infinite and so there exists $p \in (B_i \backslash s^{-1}B_i)^*$. Then $p \in L_i$ while $sp \notin L_i$, a contradiction.

Sufficiency. Let $S = \bigcup_{i=1}^{n} B_i$ where each B_i is infinite and almost invariant and $B_i \cap B_j = \emptyset$ when $i \neq j$. For each $i \in \{1, 2, \ldots, n\}$, let $L_i = B_i^*$. By Theorem 4.31, S^* is a left ideal of βS, so it suffices to show that for each $i \in \{1, 2, \ldots, n\}$, each $p \in L_i$, and each $s \in S$, $B_i \in sp$. To this end, let $i \in \{1, 2, \ldots, n\}$, $p \in L_i$, and $s \in S$ be given. Then $sB_i \backslash B_i$ is finite and $B_i \backslash s^{-1}B_i \subseteq \bigcup_{x \in sB_i \backslash B_i} \{y \in S : sy = x\}$ so, since S is weakly left cancellative, $B_i \backslash s^{-1}B_i$ is finite. Thus $s^{-1}B_i \in p$ so $B_i \in sp$ as required. $\qquad\square$

Exercise 6.3.1. Show that, for any discrete semigroup S, the number of minimal right ideals of βS is either finite or at least 2^c.

Exercise 6.3.2. Suppose that S is a commutative discrete semigroup. Show that, for any minimal right ideal R and any minimal left ideal L of βS, $R \cap L$ is dense in L. Deduce that the number of minimal right ideals of βS is either one or at least 2^c.

Exercise 6.3.3. Several of the results of this section refer to a right cancellative and weakly left cancellative semigroup. Give an example of an infinite right cancellative and weakly left cancellative semigroup which is not cancellative. (Hint: Such examples can be found among the subsemigroups of $(^{\mathbb{N}}\mathbb{N}, \circ)$.)

Exercise 6.3.4. Show that there is a semigroup S which is both weakly left cancellative and weakly right cancellative but βS has only one minimal right ideal.

Exercise 6.3.5. Show that $\mathbb{N}^*$ is not the union of two disjoint closed left ideals of $(\beta\mathbb{N}, +)$.

Exercise 6.3.6. Show that $\mathbb{Z}^*$ is the union of two disjoint closed left ideals of $(\beta\mathbb{Z}, +)$ but is not the union of three such left ideals.

6.4 Weakly Left Cancellative Semigroups

We have shown that the center of $(\beta\mathbb{N}, +)$ is disjoint from $\mathbb{N}^*$, and so is the center of $(\beta\mathbb{N}, \cdot)$. In this section, we shall show that this is a property shared by a large class of semigroups.

We shall use κ to denote an infinite cardinal and shall regard κ as being a discrete space.

Lemma 6.47. *Let S be an infinite weakly left cancellative semigroup with cardinality κ and let T be a given subset of S with cardinality κ. Then there exists a function $f : S \to \kappa$ such that*

(1) *$f[T] = \kappa$,*
(2) *for every $\lambda < \kappa$, $f^{-1}[\lambda]$ is finite if λ is finite and $|f^{-1}[\lambda]| \leq |\lambda|$ if λ is infinite, and*
(3) *for every $s, s' \in S$, if $f(s)+1 < f(s')$, then $f(ss') \in \{f(s')-1, f(s'), f(s')+1\}$ if $f(s')$ is not a limit ordinal and $f(ss') \in \{f(s'), f(s') + 1\}$ otherwise.*

Proof. We assume that S has been arranged as a κ-sequence $\langle s_\alpha \rangle_{\alpha < \kappa}$. We shall construct an increasing κ-sequence $\langle E_\alpha \rangle_{\alpha < \kappa}$ of subsets of S satisfying the following conditions for $\gamma < \kappa$:

(i) $s_\gamma \in E_\gamma$;

  (ii) $T \cap \bigcup_{\alpha < \gamma} E_\alpha \not\subseteq T \cap E_\gamma$;

  (iii) if $\alpha < \gamma$, then $E_\alpha E_\alpha \subseteq E_\gamma$;

  (iv) if $\gamma = \alpha + 1$, $s \in E_\alpha$, and $ss' \in E_\alpha$, then $s' \in E_\gamma$;

  (v) if $\gamma < \omega$, then $|E_\alpha| < \omega$; and

  (vi) if $\gamma \geq \omega$, then $|E_\gamma| \leq |\gamma|$.

We choose any $t \in T$ and put $E_0 = \{s_0, t\}$. We then assume that $0 < \gamma < \kappa$ and that we have defined E_α for every $\alpha < \gamma$.

Let $U = \bigcup_{\alpha < \gamma} E_\alpha$ and notice that $|U| < \omega$ if $\gamma < \omega$ and that $|U| \leq |\gamma|$ if $\gamma \geq \omega$. If γ is a limit ordinal let $V = \emptyset$. If $\gamma = \alpha + 1$, let $V = E_\alpha E_\alpha \cup \{t \in S : E_\alpha t \cap E_\alpha \neq \emptyset\}$ and note that, since S is weakly left cancellative, $|V| < \omega$ if $\gamma < \omega$ and $|V| \leq |\gamma|$ if $\gamma \geq \omega$. Pick $t \in T \setminus U$ and let $E_\gamma = U \cup V \cup \{s_\gamma, t\}$. It is easy to verify that the induction hypotheses are satisfied.

We now define $f : S \to \kappa$ by putting $f(s) = \min\{\alpha < \kappa : s \in E_\alpha\}$.

For every $\beta < \kappa$ there exists $t \in T \cap E_\beta \setminus \bigcup_{\alpha < \beta} E_\alpha$. Since $f(t) = \beta$, it follows that $f[T] = \kappa$. For every $\lambda < \kappa$, $f^{-1}[\lambda] \subseteq E_\lambda$. So $f^{-1}[\lambda]$ is finite if λ is finite and $|f^{-1}[\lambda]| \leq |\lambda|$ if λ is infinite.

Finally, suppose that $f(s) = \alpha$, $f(s') = \beta$ and $\alpha + 1 < \beta$. Let $\gamma = f(ss')$. Since $s, s' \in E_\beta$, we have $ss' \in E_{\beta+1}$ and so $\gamma \leq \beta + 1$. We now claim that $\gamma + 1 \geq \beta$. So suppose instead that $\gamma + 1 < \beta$ and let $\mu = \max\{\alpha, \gamma\}$. Then $s, ss' \in E_\mu$ and so $s' \in E_{\mu+1}$. Thus $\beta \leq \mu + 1 < \beta$, a contradiction. $\qquad\square$

Definition 6.48. Let S be a discrete semigroup. We define a binary relation R on $U_\kappa(S)$ by stating that pRq if $(\beta S)p \cap (\beta S)q \neq \emptyset$. We extend this to an equivalence relation $\sim$ on $U_\kappa(S)$ by stating that $p \sim q$ if $pR^n q$ for some $n \in \mathbb{N}$.

By R^n we mean the composition of R with itself n times. Thus, given $p, q \in U_\kappa(S)$, $p \sim q$ if and only if there exist elements $x_0, x_1, x_2, \ldots, x_n \in U_\kappa(S)$ such that $p = x_0$, $q = x_n$ and $x_i R x_{i+1}$ for every $i \in \{0, 1, 2, \ldots, n-1\}$.

Lemma 6.49. *Let S be a discrete infinite weakly left cancellative semigroup with cardinality κ and let T be any subset of S with cardinality κ. Let f be the function guaranteed by Lemma 6.47 and let $\widetilde{f} : \beta S \to \beta \kappa$ be its continuous extension.*

  (i) *If $\kappa = \omega$, we put $A = \{2^m : m \in \mathbb{N}\}$.*

  (ii) *If $\kappa > \omega$, we put A equal to the set of limit ordinals in κ.*

If B and C are disjoint subsets of A with cardinality κ and if $B \in \widetilde{f}(p)$ and $C \in \widetilde{f}(q)$ for some $p, q \in U_\kappa(S)$, then $p \not\sim q$.

Proof. (i) We first suppose that $\kappa = \omega$ and that $A = \{2^m : m \in \mathbb{N}\}$. We shall show by induction that, for every $n \in \mathbb{N}$, every $x, y \in S^*$ and every $X \subseteq \mathbb{N}$, if $X \in \widetilde{f}(x)$ and $x R^n y$, then $\bigcup_{i=-2n}^{2n} (i + X) \in \widetilde{f}(y)$.

Suppose first that $n = 1$. If xRy then $ux = vy$ for some $u, v \in \beta S$. Let $U = \{ss' : s, s' \in S, f(s) + 1 < f(s'), \text{ and } f(s') \in X\}$. Then $U \in ux$ by Theorem 4.15, and

$f[U] \subseteq (X - 1) \cup X \cup (X + 1)$ by condition (3) of Lemma 6.47. So $(X - 1) \cup X \cup (X + 1) \in \widetilde{f}(ux)$. Similarly, if $Y \in \widetilde{f}(y)$, then $(Y - 1) \cup Y \cup (Y + 1) \in \widetilde{f}(vy)$. So $Y \cap \bigcup_{i=-2}^{2}(i + X) \neq \emptyset$. Thus $\bigcup_{i=-2}^{2}(i + X) \in \widetilde{f}(y)$.

Now assume that our claim holds for $n \in \mathbb{N}$. Suppose that $x R^{n+1} y$. Then $x R^n z$ and $z R y$ for some $z \in S^*$. Let $Z = \bigcup_{i=-2n}^{2n}(i+X)$. By our inductive assumption, $Z \in \widetilde{f}(z)$. By what we have just proved with $n = 1$, $\bigcup_{i=-2}^{2}(i + Z) = \bigcup_{i=-2(n+1)}^{2(n+1)}(i + X) \in \widetilde{f}(y)$. So we have established our claim.

We now observe that, for every $n \in \mathbb{N}$, $B \cap \left(\bigcup_{i=-2n}^{2n}(i + C)\right)$ is finite and so $\bigcup_{i=-2n}^{2n}(i + C) \notin \widetilde{f}(p)$. Thus we cannot have $p R^n q$.

(ii) Suppose that $\kappa > \omega$ and that A is the set of limit ordinals in κ.

Let $B' = \{\lambda + n : \lambda \in B \text{ and } n \in \omega\}$. We shall show that, if $x, y \in U_\kappa(S)$, $B' \in \widetilde{f}(x)$, and $x R^n y$, then $B' \in \widetilde{f}(y)$.

Suppose first that $x R y$. Then $ux = vy$ for some $u, v \in \beta S$. Now, if $U = \{ss' : s, s' \in S, f(s) + 1 < f(s'), \text{ and } f(s') \in B'\}$, then $U \in ux$ (by Theorem 4.15). We also have $f[U] \subseteq B'$, by condition (3) of Lemma 6.47. So $B' \in \widetilde{f}(ux)$. Similarly, if $\kappa \setminus B' \in \widetilde{f}(y)$, then $\kappa \setminus B' \in \widetilde{f}(vy)$. This contradicts our assumption that $ux = vy$ and so $B' \in \widetilde{f}(y)$.

It follows immediately, by induction, that $x R^n y$ implies that $B' \in \widetilde{f}(y)$.

Since $B' \in \widetilde{f}(p)$ and $B' \notin \widetilde{f}(q)$, it follows that $p \not\sim q$.   □

Lemma 6.50. *Let S be a discrete infinite weakly left cancellative semigroup with cardinality κ. For every $p \in U_\kappa(S)$, $\{q \in U_\kappa(S) : q \sim p\}$ is a nowhere dense subset of $U_\kappa(S)$.*

Proof. Suppose instead that we have some subset T of cardinality κ such that $U_\kappa(T) \subseteq c\ell\{q \in U_\kappa(S) : q \sim p\}$. Let f be the function guaranteed by Lemma 6.47.

If $\kappa = \omega$, we put $A = \{2^m : m \in \mathbb{N}\}$. If $\kappa > \omega$, we put A equal to the set of limit ordinals in κ.

Let B and C be disjoint subsets of A with cardinality κ. It follows from Lemma 6.49 that $\{q \in U_\kappa(S) : q \sim p\}$ is disjoint from $c\ell(f^{-1}[B]) \cap U_\kappa(S)$ or from $c\ell(f^{-1}[C]) \cap U_\kappa(S)$. So either $c\ell(f^{-1}[B]) \cap U_\kappa(T)$ or $c\ell(f^{-1}[C]) \cap U_\kappa(T)$ is a non-empty open subset of $U_\kappa(T)$ disjoint from $\{q \in S^* : q \sim p\}$, a contradiction.   □

Corollary 6.51. *Let S be a discrete infinite weakly left cancellative semigroup with cardinality κ. For every $p \in U_\kappa(S)$, $\{q \in U_\kappa(S) : qp = pq\}$ is nowhere dense in $U_\kappa(S)$.*

Proof. If $qp = pq$, $(\beta S)p \cap (\beta S)q \neq \emptyset$ and so $q R p$.   □

Corollary 6.52. *Let S be a discrete infinite weakly left cancellative semigroup with cardinality κ. Then the center of $U_\kappa(S)$ is empty.*

Proof. This follows immediately from Corollary 6.51.   □

Theorem 6.53. *Let S be a discrete infinite weakly left cancellative semigroup with car-dinality κ. Then there exists a decomposition $\mathcal{I}$ of $U_\kappa(S)$ with the following properties:*

(1) *Each $I \in \mathcal{I}$ is a left ideal of βS.*

(2) *Each $I \in \mathcal{I}$ is nowhere dense in $U_\kappa(S)$.*

(3) $|\mathcal{I}| = 2^{2^\kappa}$.

Proof. We use the equivalence relation described in Definition 6.48 and denote the equivalence class of an element $p \in U_\kappa(S)$ by $[p]$. We put $\mathcal{I} = \{[p] : p \in U_\kappa(S)\}$.

(1) By Exercise 6.4.1, $U_\kappa(S)$ is a left ideal of βS. So, for each $p \in U_\kappa(S)$ and each $x \in \beta S$, $pRxp$. Thus $[p]$ is a left ideal of βS.

(2) This is Lemma 6.50.

(3) Let f denote the function guaranteed by Lemma 6.47, with $T = S$, and let $\widetilde{f} : \beta S \to \beta \kappa$ denote its continuous extension. We note that $\widetilde{f}$ is surjective because f is surjective. Furthermore, $\widetilde{f}^{-1}[U_\kappa(\kappa)] \subseteq U_\kappa(S)$.

If $\kappa = \omega$, we put $A = \{2^m : m \in \mathbb{N}\}$. If $\kappa > \omega$, we put A equal to the set of limit ordinals in κ. For each $u \in U_\kappa(A)$, we can choose $p_u \in U_\kappa(S)$ for which $\widetilde{f}(p_u) = u$. If u and v are distinct elements of $U_\kappa(A)$, there exist disjoint subsets B and C of A for which $B \in u$ and $C \in v$. Since $B \in \widetilde{f}(p_u)$ and $C \in \widetilde{f}(p_v)$, it follows from Lemma 6.49 that $[p_u] \neq [p_v]$.

Now $|A| = \kappa$ and so $|U_\kappa(A)| = 2^{2^\kappa}$ (by Theorem 3.58), and so $|\mathcal{I}| = 2^{2^\kappa}$. $\square$

Theorem 6.54. *Let S be a weakly left cancellative semigroup. Then the center of βS is equal to the center of S. The center of S^* is empty.*

Proof. First, if s is in the center of S, then s is also in the center of βS. Indeed, for every $p \in \beta S$, we have $sp = p\text{-}\lim_{t \in S} st = p\text{-}\lim_{t \in S} ts = ps$.

To see the reverse inclusion, let p be in the center of βS and suppose $p \notin S$. (Of course, if $p \in S$, then p is in the center of S.) Let $\kappa = \|p\|$. We choose any $B \in p$ with $|B| = \kappa$ and let T denote the subsemigroup of S generated by B. Since $|T| = \kappa$ and since we can regard p as being in $U_\kappa(T)$, it follows from Corollary 6.52 that p is not in the center of βT. So p is not in the center of βS. $\square$

Exercise 6.4.1. Let S be a discrete infinite weakly left cancellative semigroup with cardinality κ. Prove that $U_\kappa(S)$ is a left ideal of βS.

6.5 Semiprincipal Left Ideals and the Center of $p(\beta S)p$

We show in this section that in $\mathbb{N}^*$ there are infinite decreasing chains of semiprincipal left ideals. We also study the center of $p(\beta S)p$ for nonminimal idempotents p.

If S is embedded in a group G, βS is also embedded (topologically and algebraically) in βG by Remark 4.19. We shall assume, whenever it is convenient to do so, that $\beta S \subseteq \beta G$.

Theorem 6.55. *Let S be a countably infinite discrete semigroup embedded in a countable discrete group G. Let $q \in K(\beta S)$ and let $A \in q$. Let $n \in \mathbb{N}$ and let $p_1, p_2, \ldots, p_n \in S^* \backslash K(\beta S)$. Then there is an infinite subset B of A such that, for every $r \in B^*$ and every $i, j \in \{1, 2, \ldots, n\}$, $p_i \notin (\beta G) r p_j$.*

Proof. We claim that, for every $i, j \in \{1, 2, \ldots, n\}$, $p_i \notin (\beta G) q p_j$. To see this, we note that $q p_j \in K(\beta S)$ and hence that $q p_j = q p_j e$ for some idempotent $e \in K(\beta S)$ (by Theorem 2.8). So if $p_i \in (\beta G) q p_j$, we have $p_i = x q p_j$ for some $x \in \beta G$ and so $p_i e = x q p_j e = x q p_j = p_i$ and thus $p_i = p_i e \in K(\beta S)$, a contradiction.

Thus there is a subset D of G such that

$$\{p_1, p_2, \ldots, p_n\} \subseteq \overline{D} \quad \text{and} \quad \overline{D} \cap (\bigcup_{i=1}^{n} (\beta G) q p_i) = \emptyset.$$

For each $a \in G$, let $E_a = \{x \in A^* : \text{there exists } i \in \{1, 2, \ldots, n\} \text{ such that } a x p_i \in \overline{D}\}$. Notice that $E_a = A^* \cap \bigcup_{i=1}^{n} (\lambda_a \circ \rho_{p_i})^{-1}[\overline{D}]$ and is therefore closed. Now $q \in S^*$, by Theorem 4.36. Since $q \in A^* \backslash \bigcup_{a \in G} E_a$, it follows that $\bigcap_{a \in G} (A^* \backslash E_a)$ is a nonempty G_δ set in S^*. So by Theorem 3.36 there is an infinite subset B of A such that $B^* \subseteq \bigcap_{a \in G} (A^* \backslash E_a)$.

Let $r \in B^*$ and let $i, j \in \{1, 2, \ldots, n\}$. If $p_i \in (\beta G) r p_j = cl_{\beta G}(G r p_j)$, then $a r p_j \in \overline{D}$ for some $a \in G$ and hence $r \in B^* \cap E_a$, contradicting our choice of B^*. $\square$

Theorem 6.56. *Let S be a countably infinite discrete semigroup embedded in a countable discrete group G. Let $p \in S^* \backslash K(\beta S)$, let $q \in K(\beta S)$ and let $A \in q$. There is an infinite subset B of A with the property that, for every $r \in B^*$, $p \notin (\beta G) r p$, and, whenever r_1 and r_2 are distinct members of B^*, we have $(\beta G) r_1 p \cap (\beta G) r_2 p = \emptyset$. Furthermore, for every $r \in B^*$, $r p$ is right cancelable in βG.*

Proof. By Theorem 6.55, there is an infinite subset C of A with the property that $r \in C^*$ implies that $p \notin (\beta G) r p$.

We choose a sequential ordering for G and write $a < b$ if a precedes b in this ordering.

We can choose a sequence $\langle b_n \rangle_{n=1}^{\infty}$ in C with the property that $a b_m \neq b_n$ whenever $m < n$ and $a \in G$ satisfies $a < b_m$. We then let $B = \{b_n : n \in \mathbb{N}\}$.

We shall now show that whenever $r_1, r_2 \in B^*$, $v \in \beta G$, and $r_1 p = v r_2 p$, we have $r_1 = r_2$ and $v = 1$, the identity of G. We choose subsets B_1 and B_2 of B which satisfy $B_1 \in r_1$ and $B_2 \in r_2$, choosing them to be disjoint if $r_1 \neq r_2$.

Now $r_1 p \in cl(B_1 p)$ and $v r_2 p \in cl(G r_2 p)$. Furthermore, $v r_2 p \in cl((G \backslash \{1\}) r_2 p)$ in the case in which $v \neq 1$. By applying Corollary 3.42, we can deduce that $b p = v' r_2 p$ for some $b \in B_1$ and some $v' \in \beta G$, or else $r' p = a r_2 p$ for some $r' \in B_1^*$ and some $a \in G$. We can strengthen the second statement in the case in which $v \neq 1$, by asserting that $a \in G \backslash \{1\}$. The first possibility contradicts the fact that $p \notin (\beta G) r_2 p$, because it implies that $p = b^{-1} v' r_2 p$, and so we assume the second.

Put $B_1' = B_1 \cap \{s \in S : s > a^{-1}\}$ and $B_2' = B_2 \cap \{s \in S : s > a\}$. Then $B_1' \in r'$ and $B_2' \in r_2$. Since $r' p \in cl(B_1' p)$ and $a r_2 p \in cl(a B_2' p)$, another application of Corollary 3.42 shows that $c p = a w p$ for some $c \in B_1'$ and some $w \in cl\, B_2'$, or else $z p = a d p$

for some $z \in c\ell B_1'$ and some $d \in B_2'$. Now, in the first case, $w \notin (B_2')^*$, because $p \notin (\beta G)wp$ if $w \in (B_2')^*$. Similarly, in the second case, $z \notin (B_1')^*$. Thus we can deduce that $b_m p = ab_n p$ for some $b_m \in B_1'$ and some $b_n \in B_2'$. This equation implies that $b_m = ab_n$ by Lemma 6.28. By our choice of the sequence $\langle b_n \rangle_{n=1}^\infty$, $a^{-1}b_m \neq b_n$ if $n > m$ and $ab_n \neq b_m$ if $m > n$. Thus the equation $b_m = ab_n$ can only hold if $m = n$, and this implies that $a = 1$. We can thus deduce that $r_1 = r_2$ and that $v = 1$, as claimed.

Now if $(\beta G)r_1 p \cap (\beta G)r_2 p \neq \emptyset$, then $r_1 p \in (\beta G)r_2 p$ or $r_2 p \in (\beta G)r_1 p$, by Corollary 6.20. We have seen that this implies that $r_1 = r_2$.

Finally, let $r \in B^*$. If rp is not right cancelable in βG, there are distinct elements $w, z \in \beta G$ for which $wrp = zrp$. We can choose disjoint subsets W and Z of G satisfying $W \in w$ and $Z \in z$. Since $wrp \in c\ell(Wrp)$ and $zrp \in c\ell(Zrp)$, another application of Corollary 3.42 allows us to deduce that $drp = z'rp$ for some $d \in W$ and some $z' \in c\ell Z$, or else $w'rp = d'rp$ for some $d' \in Z$ and some $w' \in c\ell W$. In either case, it follows that $rp \in (\beta G\backslash\{1\})rp$, because $d^{-1}z' \in \beta G\backslash\{1\}$ in the first case, and $(d')^{-1}w' \in \beta G\backslash\{1\}$ in the second case. (If $z' \in G$, then $d^{-1}z' \neq 1$ because $z \notin W$. If $z' \in G^*$, then $d^{-1}z' \in G^*$ by Theorem 4.36.) We have seen that this is not possible. $\square$

Theorem 6.57. *Suppose that S is a countable discrete semigroup embedded in a countable discrete group G. Let $n \in \mathbb{N}$ and let $p_1, p_2, \ldots, p_n \in S^*\backslash K(\beta S)$. Suppose that, in addition, $p_i \notin Gp_j$ whenever i and j are distinct members of $\{1, 2, \ldots, n\}$. Let B be an infinite subset of S with the property that $p_i \notin (\beta G)rp_j$ whenever $r \in B^*$ and $i, j \in \{1, 2, \ldots n\}$. Then $(\beta G)rp_i \cap (\beta G)rp_j = \emptyset$ for every $r \in B^*$ and every pair of distinct members i, j of $\{1, 2, \ldots, n\}$.*

Proof. Suppose, on the contrary, that there is an element $r \in B^*$ for which $(\beta G)rp_i \cap (\beta G)rp_j \neq \emptyset$ for some pair i, j of distinct members of $\{1, 2, \ldots, n\}$. We may then suppose that $rp_i = xrp_j$ for some $x \in \beta G$, by Corollary 6.20. Now $rp_i \in c\ell(Bp_i)$ and $xrp_j \in c\ell(Grp_j)$. Thus, by Corollary 3.42, $r'p_i = arp_j$ for some $r' \in c\ell(B)$ and some $a \in G$, or $bp_i = x'rp_j$ for some $b \in B$ and some $x' \in \beta G$. The second possibility can be ruled out because $p_i \notin (\beta G)rp_j$, and so the first must hold.

We now observe that $r'p_i \in c\ell(Bp_i)$ and $arp_j \in c\ell(aBp_j)$. So another application of Corollary 3.42 shows that $cp_i = asp_j$ for some $c \in B$ and some $s \in c\ell(B)$, or else $tp_i = adp_j$ for some $t \in B^*$ and some $d \in B$. The first possibility cannot hold if $s \in B^*$, because then $p_i \notin (\beta G)sp_j$. Neither can it hold if $s \in B$, because $p_i \notin Gp_j$. So the first possibility can be ruled out. The second possibility cannot hold since $p_j \notin (\beta G)tp_i$. $\square$

Theorem 6.58. *Let G be a countably infinite discrete group and let $p \in G^*\backslash K(\beta G)$. Suppose that $B \subseteq G$ is an infinite set with the property that, for every $r \in B^*$, $p \notin (\beta G)rp$, and that $(\beta G)rp \cap (\beta G)r'p = \emptyset$ for every pair of distinct elements r and r' in B^*. (The existence of such a set follows from Theorem 6.56 with $S = G$.) Then, for every $r \in B^*$, $(\beta G)rp$ is maximal subject to being a principal left ideal of βG strictly contained in $(\beta G)p$.*

Proof. We note that $(\beta G)rp$ is strictly contained in $(\beta G)p$, because $p \in (\beta G)p$ and $p \notin (\beta G)rp$.

Suppose that we have $(\beta G)rp \not\subseteq (\beta G)x \not\subseteq (\beta G)p$ for some $x \in \beta G$.

Then $rp = yx$ for some $y \in \beta G$ and thus $\overline{Bp} \cap \overline{Gx} \neq \emptyset$. So by Corollary 3.42, either $bp = zx$ for some $b \in B$ and some $z \in \beta G$ or else $r'p = ax$ for some $r' \in B^*$ and some $a \in G$. The first possibility can be ruled out since $p = b^{-1}zx \in (\beta G)x$ implies that $(\beta G)p \subseteq (\beta G)x$.

Therefore assume that $r'p = ax$ for some $r' \in B^*$ and some $a \in G$. Then $x = a^{-1}r'p \in (\beta G)r'p$ and so $(\beta G)rp \subseteq (\beta G)x \subseteq (\beta G)r'p$. In particular $(\beta G)rp \cap (\beta G)r'p \neq \emptyset$ and so $r = r'$. But then $(\beta G)x = (\beta G)rp$, a contradiction.   □

We now prove a theorem about semiprincipal left ideals in $\mathbb{N}^*$. An analogous theorem is true if $\mathbb{N}$ is replaced by any countably infinite commutative cancellative discrete semigroup. However, we shall confine ourselves to the special case of $\mathbb{N}$, because it is the most important and because the more general theorem is somewhat more complicated to formulate.

Theorem 6.59. *Let $p \in \mathbb{N}^* \backslash K(\beta \mathbb{N})$. Suppose that $B \subseteq \mathbb{N}$ is an infinite set with the property that, for every $x \in B^*$, $p \notin \beta \mathbb{Z} + x + p$, and that for every distinct pair of elements x, x' in B^*, $(\beta \mathbb{Z} + x + p) \cap (\beta \mathbb{Z} + x' + p) = \emptyset$. (The existence of such a set follows from Theorem 6.56). Then, for every $x \in B^*$, $\mathbb{N}^* + x + p$ is maximal subject to being a semiprincipal left ideal of $\mathbb{N}^*$ strictly contained in $\mathbb{N}^* + p$.*

Proof. We first note that $\mathbb{N}^* + x + p$ is strictly contained in $\mathbb{N}^* + p$. To see this, pick $x' \in B^* \backslash \{x\}$. Then $x' + p \in \mathbb{N}^* + p$ and $x' + p \notin \mathbb{N}^* + x + p$ because $(\mathbb{N}^* + x' + p) \cap (\mathbb{N}^* + x + p) = \emptyset$.

Suppose that $\mathbb{N}^* + x + p \subseteq \mathbb{N}^* + y \subseteq \mathbb{N}^* + p$, where $y \in \mathbb{N}^*$ and $x \in B^*$. We shall show that $\mathbb{N}^* + y$ is equal to $\mathbb{N}^* + x + p$ or $\mathbb{N}^* + p$.

By Corollary 6.21, $x + p \in \beta \mathbb{N} + y$ or $y \in \beta \mathbb{N} + x + p$. The second possibility implies that $\mathbb{N}^* + y = \mathbb{N}^* + x + p$, and so we assume the first.

Then $x + p = z + y$ for some $z \in \beta \mathbb{N}$. Now $x + p \in c\ell(B + p)$ and $z + y \in c\ell(\mathbb{N} + y)$. We can therefore deduce from Corollary 3.42, that $n + p = z' + y$ for some $n \in B$ and some $z' \in \beta \mathbb{N}$ or else $x' + p = m + y$ for some $x' \in B^*$ and some $m \in \mathbb{N}$. In the first case $p = -n + z' + y$ and so $\mathbb{N}^* + p = \mathbb{N}^* + y$. Thus we assume that the second case holds.

The left ideals $\beta \mathbb{Z} + x + p$ and $\beta \mathbb{Z} + x' + p$ intersect, because, for any $u \in \mathbb{N}^*$, $u + x + p \in \mathbb{N}^* + y = \mathbb{N}^* + (-m) + x' + p$. It follows that $x' = x$. Then $y = -m + x + p$ and this implies that $\mathbb{N}^* + y = \mathbb{N}^* + x + p$.   □

Corollary 6.60. *If $p \in \mathbb{N}^* \backslash K(\beta \mathbb{N})$, the semiprincipal left ideal $\mathbb{N}^* + p$ of $\mathbb{N}^*$ contains 2^c disjoint semiprincipal left ideals, each maximal subject to being a semiprincipal left ideal strictly contained in $\mathbb{N}^* + p$.*

Proof. We choose a set B satisfying the hypotheses of Theorem 6.59. The corollary then follows from the fact that $|B^*| = 2^c$. (See Theorem 3.59.)   □

Corollary 6.60 does not hold as it stands if $\mathbb{N}^*$ is replaced by $\beta\mathbb{N}$. For example, if p is a right cancelable element of $\beta\mathbb{N}$, there is precisely one semiprincipal left ideal of $\beta\mathbb{N}$ maximal subject to being strictly contained in $\beta\mathbb{N} + p$, namely $\beta\mathbb{N} + 1 + p$.

Corollary 6.61. *Let $p \in \mathbb{N}^* \backslash K(\beta\mathbb{N})$. Then $\mathbb{N}^* + p$ belongs to an infinite decreasing sequence of semiprincipal left ideals of $\mathbb{N}^*$, each maximal subject to being strictly contained in its predecessor.*

Proof. By Theorem 6.56 choose an infinite set $B \subseteq \mathbb{N}$ such that for every $x \in B^*$, $p \notin \beta\mathbb{Z} + x + p$ and $x + p$ is right cancelable in $\beta\mathbb{Z}$ and such that whenever x and x' are distinct members of B^*, $\beta\mathbb{Z} + x + p \cap \beta\mathbb{Z} + x' + p = \emptyset$. Pick any $x \in B^*$. Then by Theorem 6.59 $\mathbb{N}^* + x + p$ is maximal among all semiprincipal left ideals that are properly contained in $\mathbb{N}^* + p$. Since $x + p$ is right cancelable, $x + p \notin K(\beta\mathbb{N})$ by Exercise 1.7.1, so one may repeat the process with $x + p$ in place of p. $\square$

It is possible to prove [180] — although we shall not do so here — that the statement of Corollary 6.61 can be strengthened by replacing the word "sequence" by "ω_1-sequence". Thus $\mathbb{N}^*$ certainly contains many reverse well-ordered chains of semiprincipal left ideals. It is not known whether every totally ordered chain of semiprincipal left ideals of $\mathbb{N}^*$ is reverse well-ordered.

Theorem 6.62. *Let S be a countably infinite semigroup, embedded in a countable group G. If p is a nonminimal idempotent in S^*, the center of the semigroup $p(\beta S)p$ is contained in Gp.*

Proof. Suppose that x is an element of the center of $p(\beta S)p$. Then, for every $r \in \beta S$, we have $x(prp) = (prp)x$ and hence $xrp = prx$, because $xp = px = x$.

We first show that $x \notin K(\beta S)$. Suppose instead that $x \in K(\beta S)$ and pick an idempotent $e \in K(\beta S)$ such that $x = xe$. By Theorem 6.56 there is an infinite subset B of S such that $(\beta G)r_1 p \cap (\beta G)r_2 p = \emptyset$ whenever r_1 and r_2 are distinct elements of B^* and rp is right cancelable in βG for every $r \in B^*$. We can choose $r \in B^*$ such that $x \notin (\beta G)rp$, because this must hold for every $r \in B^*$ with at most one exception. Now $rp \notin (\beta G)x$ because otherwise $rp = rpe \in K(\beta S)$, while rp is right cancelable. (By Theorem 4.36 $K(\beta S) \subseteq S^*$, so by Exercise 1.7.1 no member of $K(\beta S)$ is right cancelable.) Since $x \notin (\beta G)rp$ and $rp \notin (\beta G)x$ we have by Corollary 6.20 that $(\beta G)rp \cap (\beta G)x = \emptyset$ and hence that $xrp \neq prx$, a contradiction.

Now suppose that $x \notin Gp$. Then also $p \notin Gx$ and so by Theorems 6.55 and 6.57 with $n = 2$, $p_1 = p$, and $p_2 = x$, there is an infinite subset C of S such that $(\beta G)rp \cap (\beta G)rx = \emptyset$ for every $r \in C^*$. This implies that $xrp \neq prx$, again contradicting our assumption that x is in the center of $p(\beta S)p$. $\square$

Recall that if S is an infinite, commutative, and cancellative semigroup, then S has a "group of quotients", G. This means that G is a group in which S can be embedded in such a way that each element of G has the form $s^{-1}t$ for some $s, t \in S$. The group G is also commutative. If S is countable, then so is G.

Theorem 6.63. *Let S be a countable commutative cancellative semigroup and let G be its group of quotients. If p is a nonminimal idempotent in βS the center of $p(\beta S)p$ is equal to $Gp \cap p(\beta S)p$.*

Proof. By Theorem 6.62 the center of $p(\beta S)p$ is contained in Gp. Conversely, let $a \in G$ such that $ap \in p(\beta S)p$. Now a is in the center of βG by Theorem 4.23. Thus, if $y \in p(\beta S)p$, then $apy = ay = ya = ypa = yap$.   □

Corollary 6.64. *If p is a nonminimal idempotent in $\mathbb{N}^*$, the center of $p + \beta\mathbb{N} + p$ is equal to $\mathbb{Z} + p$.*

Proof. This follows from Theorem 6.63 and the fact that $\mathbb{Z} + \mathbb{N}^* \subseteq \mathbb{N}^*$ (by Exercise 4.3.5).   □

Notice that the requirement that p be nonminimal in Theorem 6.63 is important. It is not even known whether $\mathbb{Z} + p$ is the center of $p + \beta\mathbb{N} + p$ for an idempotent which is minimal in $(\beta\mathbb{N}, +)$.

Lemma 6.65. *Let S be a countably infinite, commutative, and cancellative semigroup, and let G be its group of quotients. Then $K(\beta S) = K(\beta G) \cap \beta S$.*

Proof. By Theorem 1.65, it is sufficient to show that $K(\beta G) \cap \beta S \neq \emptyset$.

Let q be a minimal idempotent in βS. We claim that $Gq \subseteq \beta S$. To see this, choose any $r \in G$ and put $r = s^{-1}t$ for some $s, t \in S$. Now sS is an ideal of S and so $c\ell(sS) = s\beta S$ is an ideal of βS, by Corollary 4.18. It follows that $K(\beta S) \subseteq s\beta S$. Hence $q \in s\beta S$ so $s^{-1}q \in \beta S$ and so $rq = s^{-1}tq = ts^{-1}q \in t\beta S \subseteq \beta S$.

Now there is a minimal idempotent p in $K(\beta G)$ for which $pq = p$, by Theorem 1.60. Thus $p \in (\beta G)q = c\ell(Gq) \subseteq \beta S$.   □

Exercise 6.5.1. Suppose that $p \in \mathbb{N}^* \backslash (\mathbb{N}^* + \mathbb{N}^*)$. Prove that $\mathbb{N}^* + p$ is a maximal semiprincipal left ideal of $\mathbb{N}^*$. (Hint: Use Corollary 6.21.)

Exercise 6.5.2. Let F denote the free semigroup on two generators, a and b, and let G denote the free group on these generators. Prove that $K(\beta G) \cap \beta F = \emptyset$. (Hint: Suppose that $p \in \beta F$ and that $q \in c\ell_{\beta G}\{b^{-n}a : n \in \mathbb{N}\} \cap G^*$. Each $x \in G$ can be expressed uniquely in the form $a^{n_1}b^{n'_1}a^{n_2}b^{n'_2}\ldots a^{n_k}b^{n'_k}$, where all the exponents are in $\mathbb{Z}$ and all, except possibly n_1 and n'_k, are in $\mathbb{Z}\backslash\{0\}$. Define $f(x) = \Sigma_{i=1}^{k}|n'_i|$. Let $E = \{xb^{-n}ay : x \in G, y \in F, \text{ and } n > f(x)\}$. Show that $E \cap F = \emptyset$, that $(\beta G)qp \subseteq c\ell_{\beta G} E$, and hence that $p \notin (\beta G)qp$.)

Exercise 6.5.3. Let S be a countably infinite, commutative, and cancellative semigroup, and let G be its group of quotients. Show that if p is a minimal idempotent in βS, then $Gp \subseteq p(\beta S)p$. (Hint: Consider the proof of Lemma 6.65.)

Exercise 6.5.4. Let $S = (\mathbb{N}, \cdot)$, so that the group of quotients G of S is $(\mathbb{Q}^+, \cdot)$. Show that there are idempotents $p \in S^*$ for which $Gp \nsubseteq S^*$.

6.6 Principal Ideals in $\beta\mathbb{Z}$

It is a consequence of Theorems 6.56 and 6.58 that, given any $p \in \mathbb{Z}^* \setminus K(\beta\mathbb{Z})$ there is an infinite sequence $\langle p_n \rangle_{n=1}^{\infty}$ with $p_1 = p$ such that $\beta\mathbb{Z} + p_{n+1} \not\subseteq \beta\mathbb{Z} + p_n$ for each n. Whether there is an infinite strictly increasing chain of principal left ideals of $\beta\mathbb{Z}$ is an old and notoriously difficult problem. (See the notes to this chapter for a discussion of the history of this problem.) We do not address this problem here, but instead solve the corresponding problem for principal right ideals and principal closed ideals (defining the latter term below).

Recall from Exercise 4.4.9 that if S is a commutative semigroup, then the closure of any right ideal of βS is a two sided ideal of βS.

Definition 6.66. Let G be a commutative group and let $p \in \beta G$. Then $c\ell(p + \beta G)$ is the *principal closed ideal generated by* p.

Notice that the terms "principal closed ideal" and "closed principal ideal" mean two different things. Indeed the later can only rarely be found in βS where S is a semigroup.

The reasons for defining the term "principal closed ideal" only for groups are first, that this guarantees that p is a member, and second, because we are only going to use the notion in $\mathbb{Z}$.

We now introduce some special sets needed for our construction. Recall that $\bigoplus_{i=1}^{\infty} \omega = \{\vec{a} \in \bigtimes_{i=1}^{\infty} : \{i : a_i \neq 0\} \text{ is finite}\}$.

Definition 6.67. (a) For $m, n \in \mathbb{N}$,

$$\theta_n^m = \{\vec{a} \in \bigoplus_{i=1}^{\infty} \omega : \Sigma_{i=1}^{\infty} \frac{a_i}{2^i} = \frac{1}{2^n} \text{ and } \min\{i \in \mathbb{N} : a_i \neq 0\} \geq m + n\}.$$

(b) Fix a sequence $\langle z_k \rangle_{k=1}^{\infty}$ such that for each k, $z_{k+1} > \Sigma_{i=1}^{k} 2^{i+1} z_i$.
(c) For $m, n \in \mathbb{N}$, $A_n^m = \{\Sigma_{i=1}^{\infty} a_i z_i : \vec{a} \in \theta_n^m\}$.

Lemma 6.68. *For each* $m, n \in \mathbb{N}$, $A_n^{m+1} \subseteq A_n^m$ *and* $|A_n^m| = \omega$. *Further, for each* $\vec{a} \in \theta_n^m$ *and each* $i \in \mathbb{N}$, $a_i < 2^i$.

Proof. That $A_n^{m+1} \subseteq A_n^m$ is trivial. For the second assertion notice that $\{2^{k-n} z_k : k \geq m + n\} \subseteq A_n^m$. For the third assertion notice that if $a_i \geq 2^i$, then $\frac{a_i}{2^i} > \frac{1}{2^n}$. $\square$

Lemma 6.69. *Let* $m, n \in \mathbb{N}$. *Then* $A_{n+1}^m + A_{n+1}^m \subseteq A_n^{m+1}$.

Proof. Let $x, y \in A_{n+1}^m$ and pick $\vec{a}, \vec{b} \in \theta_{n+1}^m$ such that $x = \Sigma_{i=1}^{\infty} a_i z_i$ and $y = \Sigma_{i=1}^{\infty} b_i z_i$. Let $\vec{c} = \vec{a} + \vec{b}$. Then $x + y = \Sigma_{i=1}^{\infty} c_i z_i$ so it suffices to show that $\vec{c} \in \theta_n^{m+1}$. First $\Sigma_{i=1}^{\infty} \frac{c_i}{2^i} = \Sigma_{i=1}^{\infty} \frac{a_i}{2^i} + \Sigma_{i=1}^{\infty} \frac{b_i}{2^i} = \frac{1}{2^{n+1}} + \frac{1}{2^{n+1}} = \frac{1}{2^n}$. Also, $\min\{i \in \mathbb{N} : c_i \neq 0\} = \min(\{i \in \mathbb{N} : a_i \neq 0\} \cup \{i \in \mathbb{N} : b_i \neq 0\}) \geq n + 1 + m$. $\square$

Lemma 6.70. *If* $\vec{a}, \vec{b} \in \bigoplus_{i=1}^{\infty} \mathbb{Z}$, $\Sigma_{i=1}^{\infty} a_i z_i = \Sigma_{i=1}^{\infty} b_i z_i$, *and for each* $i \in \mathbb{N}$, $|a_i| \leq 2^i$ *and* $|b_i| \leq 2^i$, *then* $\vec{a} = \vec{b}$.

Proof. Suppose $\vec{a} \neq \vec{b}$. Since $\vec{a}$ and $\vec{b}$ each have only finitely many nonzero coordinates, we may pick the largest n such that $a_n \neq b_n$ and assume without loss of generality that $a_n > b_n$. Then $\Sigma_{i=1}^{n} a_i z_i = \Sigma_{i=1}^{n} b_i z_i$ so

$$z_n \leq (a_n - b_n)z_n = \Sigma_{i=1}^{n-1}(b_i - a_i)z_i \leq \Sigma_{i=1}^{n-1} 2^{i+1} z_i < z_n,$$

a contradiction. $\qquad\square$

Definition 6.71. For each $n \in \mathbb{N}$, $B_n = \bigcap_{m=1}^{\infty} c\ell\, A_n^m$.

Lemma 6.72. *For each* $n \in \mathbb{N}$, B_n *is a nonempty* G_δ *subset of* $\mathbb{N}^*$ *such that*

  (a) $B_{n+1} + B_{n+1} \subseteq B_n$ *and*
  (b) *for each* $p \in B_n$, $c\ell(p + \beta\mathbb{Z}) \cap B_{n+1} = \emptyset$.

Proof. By Lemma 6.68, we have that $B_n \neq \emptyset$. Since for each m, $\min A_n^m = 2^m \cdot z_{n+m}$, we have $B_n \subseteq \mathbb{N}^*$.

To verify (a), let $p, q \in B_{n+1}$. To see that $p + q \in B_n$, let $m \in \mathbb{N}$ be given. Then by Lemma 6.69, for each $x \in A_{n+1}^m$, $A_{n+1}^m \subseteq -x + A_n^{m+1}$ so

$$A_{n+1}^m \subseteq \{x \in \mathbb{Z} : -x + A_n^{m+1} \in q\}$$

and consequently $A_n^{m+1} \in p+q$. Since $A_n^{m+1} \subseteq A_n^m$, we have $A_n^m \in p+q$ as required.

To verify (b), let $p \in B_n$ and suppose that we have some $r \in B_{n+1} \cap c\ell(p + \beta\mathbb{Z})$. Then $A_{n+1}^1 \in r$ so $A_{n+1}^1 \cap (p + \beta\mathbb{Z}) \neq \emptyset$. Pick $q \in \beta\mathbb{Z}$ such that $A_{n+1}^1 \in p + q$ and let $C = \{x \in \mathbb{Z} : -x + A_{n+1}^1 \in q\}$. Then $C \in p$ so pick $x \in C \cap A_n^1$ and pick $\vec{a} \in \theta_n^1$ such that $x = \Sigma_{i=1}^{\infty} a_i z_i$. Let $m = \max\{i \in \mathbb{N} : a_i \neq 0\}$. Since $A_n^m \in p$, pick $y \in A_n^m \cap C$ and pick $\vec{b} \in \theta_n^m$ such that $y = \Sigma_{i=1}^{\infty} b_i z_i$. Since x and y are in C, pick $w \in (-x + A_{n+1}^1) \cap (-y + A_{n+1}^1)$. Then $x + w \in A_{n+1}^1$ and $y + w \in A_{n+1}^1$ so pick $\vec{c}$ and $\vec{d}$ in θ_{n+1}^1 such that $x + w = \Sigma_{i=1}^{\infty} c_i z_i$ and $y + w = \Sigma_{i=1}^{\infty} d_i z_i$. Then $w = \Sigma_{i=1}^{\infty}(c_i - a_i)z_i = \Sigma_{i=1}^{\infty}(d_i - b_i)z_i$. By Lemma 6.68, for each i, $|c_i - a_i| < 2^i$ and $|d_i - b_i| < 2^i$ so by Lemma 6.70 $\vec{c} - \vec{a} = \vec{d} - \vec{b}$. Now $\vec{b} \in \theta_n^m$ so $b_i = 0$ for $i < n + m$ and thus $\Sigma_{i=n+m}^{\infty} \dfrac{b_i}{2^i} = \dfrac{1}{2^n}$. Since $a_i = 0$ for $i > m$ we have that for $i \geq n + m$, $d_i = b_i + c_i$. But then

$$\dfrac{1}{2^{n+1}} = \Sigma_{i=1}^{\infty} \dfrac{d_i}{2^i} \geq \Sigma_{i=n+m}^{\infty} \dfrac{d_i}{2^i} = \Sigma_{i=n+m}^{\infty} \dfrac{b_i + c_i}{2^i} \geq \Sigma_{i=n+m}^{\infty} \dfrac{b_i}{2^i} = \dfrac{1}{2^n},$$

a contradiction. $\qquad\square$

The following lemma is quite general. We state it for semigroups written additively because we intend to use it with $(\beta\mathbb{Z}, +)$.

Lemma 6.73. *Let $(S, +)$ be a compact right topological semigroup and for each $n \in \mathbb{N}$, let D_n be a nonempty closed subset of S such that for each n, $D_{n+1} + D_{n+1} \subseteq D_n$. Given any sequence $\langle q_n \rangle_{n=1}^{\infty}$ in S with each $q_n \in D_n$, there is a sequence $\langle p_n \rangle_{n=1}^{\infty}$ with each $p_n \in D_n$, such that, for each n, $p_{n+1} + q_{n+1} = p_n$.*

Proof. Pick some $r \in \mathbb{N}^*$. For n and m in $\mathbb{N}$ with $n \geq m$, let $t_{n,m} = q_n$. By downward induction on n, for $n < m$, let $t_{n,m} = t_{n+1,m} + q_{n+1}$ and notice that each $t_{n,m} \in D_n$. For each $n \in \mathbb{N}$, let $p_n = r\text{-}\lim_{m \in S} t_{n,m}$ and notice that, since D_n is closed, $p_n \in D_n$. Since for $m > n$, $t_{n,m} = t_{n+1,m} + q_{n+1}$ one has that

$$p_n = r\text{-}\lim_{m > n} (t_{n+1,m} + q_{n+1}) = (r\text{-}\lim_{m > n} t_{n+1,m}) + q_{n+1} = p_{n+1} + q_{n+1}. \qquad \square$$

Theorem 6.74. *There is a sequence $\langle p_n \rangle_{n=1}^{\infty}$ in $\mathbb{N}^*$ such that $\langle p_n + \beta\mathbb{Z} \rangle_{n=1}^{\infty}$ is a strictly increasing chain of principal right ideals of $\beta\mathbb{Z}$ and $\langle c\ell(p_n + \beta\mathbb{Z}) \rangle_{n=1}^{\infty}$ is a strictly increasing chain of principal closed ideals of $\beta\mathbb{Z}$.*

Proof. Pick some $q_n \in B_n$ for each $n \in \mathbb{N}$. By Lemma 6.72, we have the sequences $\langle B_n \rangle_{n=1}^{\infty}$ and $\langle q_n \rangle_{n=1}^{\infty}$ satisfy the hypotheses of Lemma 6.73 with $S = \beta\mathbb{Z}$ so pick a sequence $\langle p_n \rangle_{n=1}^{\infty}$ with each $p_n \in B_n$ and each $p_n = p_{n+1} + q_{n+1}$. Then one has immediately that for each n, $p_n + \beta\mathbb{Z} \subseteq p_{n+1} + \beta\mathbb{Z}$ and so, of course, $c\ell(p_n + \beta\mathbb{Z}) \subseteq c\ell(p_{n+1} + \beta\mathbb{Z})$. Further, by Lemma 6.72(b), for each n,

$$p_{n+1} = p_{n+1} + 0 \in (p_{n+1} + \beta\mathbb{Z}) \setminus c\ell(p_n + \beta\mathbb{Z})$$

so both chains are strictly increasing. $\square$

6.7 Ideals and Density

The concept of density for a set of positive integers has interesting algebraic implications in $\beta\mathbb{N}$. We shall show that the set of ultrafilters in $\beta\mathbb{N}$ all of whose members have positive upper density is a left ideal of $(\beta\mathbb{N}, +)$ as well as of $(\beta\mathbb{N}, \cdot)$, and that the same statement holds for the complement in $\mathbb{N}^*$ of this set.

Lemma 6.75. *Let S be an arbitrary semigroup and let $\mathcal{R}$ be a partition regular set of subsets of S. Let $\Delta_{\mathcal{R}}$ denote the set of ultrafilters in βS all of whose members are supersets of sets in $\mathcal{R}$. If $\mathcal{R}$ has the property that $sA \in \mathcal{R}$ for every $A \in \mathcal{R}$ and every $s \in S$, then $\Delta_{\mathcal{R}}$ is a closed left ideal of βS.*

Proof. It follows from Theorem 3.11 that $\Delta_{\mathcal{R}}$ is non-empty and it is immediate that $\Delta_{\mathcal{R}}$ is closed.

To see that $\Delta_{\mathcal{R}}$ is a left ideal, let $p \in \Delta_{\mathcal{R}}$. It suffices to show that $Sp \subseteq \Delta_{\mathcal{R}}$, for then $(\beta S)p = c\ell_{\beta S}(Sp) \subseteq \Delta_{\mathcal{R}}$. So let $s \in S$ and let $B \in sp$. Then $s^{-1}B \in p$ so pick $A \in \mathcal{R}$ such that $A \subseteq s^{-1}B$. Then $sA \in \mathcal{R}$ and $sA \subseteq B$. $\square$

Definition 6.76. Let $A \subseteq \mathbb{N}$. We define the *upper density* $\overline{d}(A)$ of A by

$$\overline{d}(A) = \lim_{n \to \infty} \sup \frac{|A \cap \{1, 2, \ldots, n\}|}{n}.$$

Remark 6.77. *Let $A \subseteq \mathbb{N}$ and let $m \in \mathbb{N}$. Then*

$$\overline{d}(m + A) = \overline{d}(A) \quad and \quad \overline{d}(mA) = \frac{1}{m}\overline{d}(A).$$

Definition 6.78. We define $\Delta \subseteq \beta\mathbb{N}$ by

$$\Delta = \{p \in \beta\mathbb{N} : \overline{d}(A) > 0 \text{ for all } A \in p\}.$$

Theorem 6.79. Δ *is a closed left ideal of $(\beta\mathbb{N}, +)$ and of $(\beta\mathbb{N}, \cdot)$.*

Proof. We apply Lemma 6.75 with $\mathcal{R} = \{A \subseteq \mathbb{N} : \overline{d}(A) > 0\}$. It is clear that $\mathcal{R}$ is partition regular. By Remark 6.77 we have $m + A \in \mathcal{R}$ and $mA \in \mathcal{R}$ for every $A \in \mathcal{R}$ and every $m \in \mathbb{N}$. Thus our conclusion follows. $\square$

Theorem 6.80. $\mathbb{N}^* \backslash \Delta$ *is a left ideal of $(\beta\mathbb{N}, +)$ and of $(\beta\mathbb{N}, \cdot)$.*

Proof. Let $p \in \mathbb{N}^* \backslash \Delta$ and let $q \in \beta\mathbb{N}$. There is a set $B \in p$ such that $\overline{d}(B) = 0$. For each $n \in \mathbb{N}$ we put $f(n) = \frac{|B \cap \{1, 2, \ldots, n\}|}{n}$. So $f(n) \to 0$ as $n \to \infty$. For each $m \in \mathbb{N}$, we choose $n_m \in \mathbb{N}$ so that $f(n) < \frac{1}{m^2}$ whenever $n > n_m$. We put $B_m = \{n \in B : n > n_m\}$ and observe that $B_m \in p$. It follows from Theorem 4.15 that $S = \bigcup_{m \in \mathbb{N}}(m + B_m) \in q + p$ and $P = \bigcup_{m \in \mathbb{N}} mB_m \in qp$. We shall show that these sets both have zero upper density. It will follow that $p + q$ and pq are both in $\mathbb{N}^* \backslash \Delta$.

Suppose then that $m \in \mathbb{N}$ and that $n \in B_m$ and that $m + n < r$ for some $r \in \mathbb{N}$. This implies that $n_m < r$ because $n > n_m$, and hence that $f(r) < \frac{1}{m^2}$. So, for a given value of r, if $f(r) \neq 0$, the number of possible choices of m is at most $\sqrt{1/f(r)}$. Since we also have $n \in B \cap \{1, 2, \ldots, r\}$, the number of possible choices of n is at most $f(r)r$. Thus the number of possible choices of $m + n$ is at most $r\sqrt{f(r)}$. It follows that

$$\frac{|S \cap \{1, 2, \ldots, r\}|}{r} \leq \sqrt{f(r)}.$$

The same argument shows that

$$\frac{|P \cap \{1, 2, \ldots r\}|}{r} \leq \sqrt{f(r)}.$$

This establishes the claim that $\overline{d}(S) = \overline{d}(P) = 0$ and it follows that $\beta\mathbb{N}^* \backslash \Delta$ is a left ideal of $(\beta\mathbb{N}, +)$ and of $(\beta\mathbb{N}, \cdot)$. $\square$

Notice that it did not suffice to show in the proof of Theorem 6.80 that $n + p$ and np are in $\mathbb{N}^* \backslash \Delta$ for every $p \in \mathbb{N}^* \backslash \Delta$ because $\mathbb{N}^* \backslash \Delta$ is not closed. Notice also that Theorems 6.79 and 6.80 provide an alternative proof that the centers of $(\beta\mathbb{N}, +)$ and of $(\beta\mathbb{N}, \cdot)$ are both equal to $\mathbb{N}$.

Notes

The fact in Theorem 6.9 that $\beta\mathbb{N}$ has 2^c minimal right ideals is due to J. Baker and P. Milnes [10], while the fact that $\beta\mathbb{N}$ has 2^c minimal left ideals is due to C. Chou [61].

The concept of an oid (" '$\oplus$ id' being unpronounceable") is due to J. Pym [204], who showed that the semigroup structure $\mathbb{H}$ depended only on the oid structure of $FS(\langle 2^n \rangle_{n=1}^{\infty})$. Theorem 6.27 is due to A. Lisan [178]. Theorem 6.32 is due to T. Budak (nee Papazyan) [192].

Exercise 6.1.4 is from [38], a result of collaboration with J. Berglund.

Theorem 6.36, one of three nonelementary results used in this book that we do not prove, is due to M. Rudin and S. Shelah. Theorem 6.36 and Lemma 6.37 deal (without ever explicitly defining them) with the Rudin–Keisler and Rudin Frolík orderings of ultrafilters. See Chapter 11 for more information about these orderings as well as a discussion of their origins. For a simpler proof that there exist *two* points in $\mathbb{N}^*$ that are not Rudin–Keisler comparable, see [182, Section 3.4].

The proof of Theorem 6.38 is from [66].

Theorem 6.46 is due to D. Parsons in [193].

Theorem 6.53 is, except for its weaker hypotheses, a special case of the following theorem which is due to E. van Douwen (in a letter to the first author). A proof, obtained in collaboration with D. Davenport, can be found in [74].

Theorem *Let S be an infinite cancellative semigroup with cardinality κ. There exists a decomposition $\mathcal{L}$ of $U_\kappa(S)$ with the following properties:*

(1) $|\mathcal{L}| = 2^{2^\kappa}$.

(2) *Each $I \in \mathcal{L}$ is a left ideal of βS.*

(3) *For each $I \in \mathcal{L}$ and each $p \in I$, $c\ell(pS) \subseteq I$.*

(4) *Each $I \in \mathcal{L}$ is nowhere dense in $U_\kappa(S)$.*

Theorem 6.63 is from [180], a result of collaboration with A. Maleki.

The results of Section 6.6 are from [149], a result of collaboration with J. van Mill and P. Simon. Sometime in the 1970's or 1980's M. Rudin was asked by some, now anonymous, analysts whether every point of $\mathbb{Z}^*$ is a member of a maximal orbit closure under the continuous extension $\tilde{\sigma}$ of the shift function σ, where $\sigma(n) = n + 1$. This question was not initially recognized as a question about the algebra of $\beta\mathbb{Z}$. However, if $p \in \beta\mathbb{Z}$, then $\tilde{\sigma}(p) = 1 + p$ and so, for all $n \in \mathbb{Z}$, $\tilde{\sigma}^n(p) = n + p$. Thus the orbit closure of p is $\beta\mathbb{Z} + p$ and so the question can be rephrased as asking whether every point of $\beta\mathbb{Z}$ lies in some maximal (proper) principal left ideal of $\beta\mathbb{Z}$. One could answer this question in the affirmative by determining that there is no strictly increasing sequence of principal left ideals of $\beta\mathbb{Z}$.

The results of Section 6.7 are due to E. van Douwen in [80].

Groups in βS

If S is a discrete semigroup, it is often quite easy to find large groups contained in βS. For example, the maximal groups in the smallest ideal of $\beta \mathbb{N}$ contain $2^{\mathfrak{c}}$ elements. More generally, as we shall show in this chapter, if S is infinite and cancellative with cardinality κ, βS contains algebraic copies of the free group on $2^{2^{\kappa}}$ generators. This provides a remarkable illustration of how far βS is from being commutative.

However, it can be tantalizingly difficult to find nontrivial small groups in βS. For many years, one of the difficult open questions about the algebra of $\beta \mathbb{N}$ was whether or not $\beta \mathbb{N}$ contained any nontrivial finite groups. This question has now been answered by E. Zelenuk, and we give the proof of his theorem in Section 1 below. Whether or not $\beta \mathbb{N}$ contains any elements of finite order which are not idempotent still remains a challenging open question.

Of course, there are many copies of $\mathbb{Z}$ in $\mathbb{N}^*$, since $\mathbb{Z} + p$ provides a copy of $\mathbb{Z}$ if p denotes any idempotent in $\mathbb{N}^*$. It is consistent with ZFC that there are maximal groups in $\mathbb{N}^*$ isomorphic to $\mathbb{Z}$, since Martin's Axiom can be used to show that there are idempotents p in $\mathbb{N}^*$ for which $H(p)$, the largest group with p as identity, is just $\mathbb{Z} + p$. (See Theorem 12.42). It is not known whether the existence of such an idempotent can be proved within ZFC.

Because we shall be constructing several topological spaces in this chapter, some of which are not necessarily Hausdorff, we depart for this chapter only from our standing assumption that all hypothesized spaces are Hausdorff.

7.1 Zelenuk's Theorem

The proof of Zelenuk's Theorem uses the notion of a *left invariant topology* on a group.

Definition 7.1. Let G be a group. A topology $\mathcal{T}$ on G is *left invariant* if and only if for every $U \in \mathcal{T}$ and every $a \in G$, $aU \in \mathcal{T}$.

Notice that a topology on G is left invariant if and only if for every $a \in G$, λ_a is a homeomorphism. Notice also that to say that $(G, \cdot)$ is a group with a left invariant topology $\mathcal{T}$ is the same as saying that $(G, \cdot, \mathcal{T})$ is a left topological group, i.e., a group which is a left topological semigroup.

The next result, Lemma 7.4, is unfortunately rather lengthy and involves a good deal of notation. We shall see after the proof of this lemma how a topology on G and a set X satisfying the hypotheses of this lemma arise naturally from the assumption that βG has a nontrivial finite subgroup while G does not.

Definition 7.2. (a) F will denote the free semigroup on the letters 0 and 1 with identity $\emptyset$.

(b) If $m \in \omega$ and $i \in \{0, 1, 2, \ldots, m\}$, s_i^m will denote the element of F consisting of i 0's followed by $m - i$ 1's. We also write $u_m = s_m^m$ (so that $u_0 = \emptyset$).

(c) If $s \in F$, $l(s)$ will denote the length of s and supp $s = \{i \in \{1, 2, \ldots, l(s)\} : s_i = 1\}$ where s_i is the ith letter of s.

(d) If $s, t \in F$, we write $s << t$ if max supp $s + 1 <$ min supp t.

(e) If $s, t \in F$, we define $s + t$ to be the element of F for which $l(s + t) = \max\{l(s), l(t)\}$ and $(s + t)_i = 1$ if and only if $s_i = 1$ or $t_i = 1$.

Given any $t \in F$, t has a unique representation in the form $t = s_{i_0}^{m_0} + s_{i_1}^{m_1} + \cdots + s_{i_k}^{m_k}$ where $0 \leq i_0 < m_0 < i_1 < m_1 < \cdots < i_k \leq m_k$ (except that, if $k = 0$, the requirement is $0 \leq i_0 \leq m_0$). We shall call this the *canonical representation* of t. When we write $t = s_{i_0}^{m_0} + s_{i_1}^{m_1} + \cdots + s_{i_k}^{m_k}$ we shall assume that this is the canonical representation.

Definition 7.3. (a) Given $t \in F$, if $t = s_i^m$ for some $i, m \in \omega$, then $t' = \emptyset$ and $t^* = t$. Otherwise, if $t = s_{i_0}^{m_0} + s_{i_1}^{m_1} + \cdots + s_{i_{k+1}}^{m_{k+1}}$, then $t' = s_{i_0}^{m_0} + s_{i_1}^{m_1} + \cdots + s_{i_k}^{m_k}$ and $t^* = s_{i_{k+1}}^{m_{k+1}}$.

(b) The mapping $c : F \to \omega$ is defined by stating that $c(t) = |$ supp $t|$.

(c) For each $p \in \mathbb{N}$, define $g_p : F \to \mathbb{Z}_p$ by $g_p(t) \equiv c(t) \pmod{p}$.

While the proof of Lemma 7.4 is long, because there are a large number of statements to be checked, the reader will see that checking any one of them is not difficult.

Recall that a topology is *zero dimensional* if it has a basis of clopen sets.

Lemma 7.4. *Suppose that G is a group with identity e, and that X is a countable subset of G containing e. Suppose also that G has a left invariant Hausdorff zero dimensional topology and that X has no isolated points in the relative topology. We also suppose that, for every $a \in X$, $aX \cap X$ is a neighborhood of a in X. We suppose in addition that $p \in \mathbb{N}$ and that there is a mapping $h : X \to \mathbb{Z}_p$ such that, for each $a \in X$, there is a neighborhood $V(a)$ of e in X satisfying $aV(a) \subseteq X$ and $h(ab) = h(a) + h(b)$ for every $b \in V(a)$. Furthermore, we suppose that $h[Y] = \mathbb{Z}_p$ for every non-empty open subset Y of X and that $V(e) = X$.*

Then we can define $x(t) \in X$ and $X(t) \subseteq X$ for every $t \in F$ so that $x(\emptyset) = e$, $X(\emptyset) = X$, and the following conditions are all satisfied:

(1) *$X(t)$ is clopen in X.*

(2) *$x(t) \in X(t)$.*

(3) $X(t^\frown 0) \cup X(t^\frown 1) = X(t)$ and $X(t^\frown 0) \cap X(t^\frown 1) = \emptyset$.

(4) $x(t^\frown 0) = x(t)$.

(5) $x(t) = x(t')x(t^*)$.

(6) $X(t) = x(t')X(t^*)$.

(7) $X(t^*) \subseteq V(x(t'))$.

(8) $h(x(t)) = g_p(t)$.

(9) For any $s, t \in F$, $x(s) = x(t)$ if and only if s is equal to t followed by 0's or vice-versa.

(10) If $s, t \in F$ and $s \ll t$, then $x(s + t) = x(s)x(t)$.

(11) If $s, t \in F$ and $s \ll t$, then $h(x(s + t)) = h(x(s)) + h(x(t))$.

(12) For every $n \in \omega$, $x[u_n F] = X(u_n)$, so that in particular $x[F] = X$.

(13) If $\langle W_n \rangle_{n=1}^\infty$ is any preassigned sequence of neighborhoods of e in X, we can choose $X(u_n)$ to satisfy $X(u_{n+1}) \subseteq W_n$ for every $n \in \mathbb{N}$.

Proof. We assume that we have chosen a sequential ordering for X. We define $x(t)$ and $X(t)$ by induction on $l(t)$.

We start by stating that $x(\emptyset) = e$ and $X(\emptyset) = X$. We then make the inductive assumption that, for some $n \in \omega$, $x(t)$ and $X(t)$ have been defined for every $t \in F$ with $l(t) \leq n$ so that conditions (1)–(9) and (13) are satisfied and further if $n \geq 1$, sequences $\langle a_k \rangle_{k=0}^{n-1}$ in X and $\langle t_k \rangle_{k=0}^{n-1}$ in F have been chosen satisfying the following additional hypotheses for each k:

(i) $a_k = \min X \setminus \{x(t) : t \in F$ and $l(t) \leq k\}$,

(ii) $a_k \in X(t_k^\frown 1)$, and

(iii) if $h(a_k) = g_p(t_k^\frown 1)$, then $a_k = x(t_k^\frown 1)$.

Of these hypotheses, only (8) requires any effort to verify at $n = 0$. For this, note that $h(e) = h(ee) = h(e) + h(e)$ so $h(e) = 0 = g_p(\emptyset)$.

Let a_n be the first element of X which is not in $\{x(t) : t \in F$ and $l(t) \leq n\}$. Now the sets $X(t)$ with $l(t) = n$ form a disjoint partition of X by condition (3) and the choice of $X(\emptyset)$ and so $a_n \in X(t_n)$ for a unique $t_n \in F$ with $l(t_n) = n$. Since $X(t_n) = x(t_n')X(t_n^*)$, it follows that $a_n = x(t_n')c_n$ for some $c_n \in X(t_n^*)$.

For each $s \in \{s_i^n : i \in \{0, 1, \ldots, n\}\}$ choose $b_s \in X(s)$ such that $b_s \neq x(s)$ and $h(b_s) = g_p(s^\frown 1)$. (To see that we can do this, note that, by conditions (1) and (2), $X(s)$ is a nonempty open subset of X so, since X has no isolated points, $X(s) \setminus \{x(s)\}$ is a nonempty open subset of X and hence $h[X(s) \setminus \{x(s)\}] = \mathbb{Z}_p$.) In the case in which $s = t_n^*$, we choose $b_{t_n^*} = c_n$ if $h(c_n) = g_p(t_n^{*\frown} 1)$. To see that this can be done, notice that if $c_n = x(s)$, then by condition (5), $a_n = x(t_n')c_n = x(t_n')x(t_n^*) = x(t_n)$. For each $s \in \{s_i^n : i \in \{0, 1, \ldots, n\}\}$ define $x(s^\frown 0) = x(s)$ and $x(s^\frown 1) = b_s$. Notice that $x(s^\frown 1) = x(s')x(s^{*\frown} 1)$.

Now, given any other $t \in F$ with $l(t) = n$, notice that we have already defined $x(t^{*\frown} 1) \in X(t^*)$. We define $x(t^\frown 0) = x(t)$ and $x(t^\frown 1) = x(t')x(t^{*\frown} 1)$. Notice that

$$x(t) = x(t')x(t^*) \neq x(t')x(t^{*\frown} 1) = x(t^\frown 1)$$

and $x(t\frown 1) \in x(t')X(t^*) = X(t)$.

We can choose a clopen neighborhood U_n of e in G such that $U_n \cap X \subseteq V(x(v))$ and $x(v)U_n \cap X \subseteq x(v)X$ for every $v \in F$ with $l(v) \leq n$. (We can get the latter inclusion because $x(v)X \cap X$ is a neighborhood of $x(v)$ in X.) We can also require that $a_n \notin x(t_n)U_n$ and that U_n satisfies both of the following conditions for every $t \in F$ with $l(t) = n$:

$$x(t)U_n \cap X \subseteq X(t) \quad \text{and} \quad x(t\frown 1) \notin x(t)U_n.$$

Furthermore, in the case in which there is an assigned sequence $\langle W_n \rangle_{n=1}^{\infty}$ of neighborhoods of e in X, we choose U_n to satisfy $U_n \cap X \subseteq W_n$.

Let us note now that for any $v \in F$ with $l(v) \leq n$, we have $x(v)U_n \cap X = x(v)(U_n \cap X)$. To see this, notice that $x(v)U_n \cap X \subseteq x(v)X$ so that

$$x(v)U_n \cap X \subseteq x(v)U_n \cap x(v)X = x(v)(U_n \cap X).$$

Also $U_n \cap X \subseteq V(x(v))$ and so $x(v)(U_n \cap X) \subseteq x(v)V(x(v)) \subseteq X$. Hence $x(v)(U_n \cap X) \subseteq x(v)U_n \cap X$.

We put $X(t\frown 0) = x(t)U_n \cap X$ and $X(t\frown 1) = X(t) \backslash X(t\frown 0)$.

We need to check that conditions (1) - (9) and (13) and hypotheses (i), (ii), and (iii) are satisfied for elements of F of length $n + 1$. Suppose then that $t \in F$ and that $l(t) = n$. We put $v = t\frown 1$ and $w = t\frown 0$. Notice that $v' = t'$ and $v^* = t^*\frown 1$. Notice also that t is equal to w' followed by a certain number (possibly zero) of 0's so that $x(t) = x(w')$ and that $w^* = u_{n+1}$.

Conditions (1), (2), (3) and (4) are immediate. In particular, notice that $x(u_{n+1}) = x(u_n) = e$.

We observe that (5) holds for v because $x(v) = x(t')x(t^*\frown 1) = x(v')x(v^*)$. It holds for w because $x(w) = x(t) = x(w') = x(w')e = x(w')x(w^*)$.

We now check condition (6). By the definition of $X(w)$, we have

$$
\begin{aligned}
X(w) &= x(t)U_n \cap X \\
&= x(t)(U_n \cap X) \\
&= x(w')X(u_{n+1}) \\
&= x(w')X(w^*).
\end{aligned}
$$

We also have

$$
\begin{aligned}
X(v) &= X(t) \backslash (x(t)U_n \cap X) \\
&= X(t) \backslash (x(t)(U_n \cap X)) \\
&= x(t')X(t^*) \backslash (x(t')x(t^*)(U_n \cap X)) \\
&= x(t')(X(t^*) \backslash x(t^*)(U_n \cap X)) \\
&= x(t')(X(t^*) \backslash (x(t^*)U_n \cap X)) \\
&= x(t')X(t^*\frown 1) \\
&= x(v')X(v^*).
\end{aligned}
$$

To verify condition (7) for v, we note that $X(v^*) = X(t^*\frown 1) \subseteq X(t^*) \subseteq V(x(t')) = V(x(v'))$. It also holds for w, because $X(w^*) = X(u_{n+1}) = U_n \cap X \subseteq V(x(t)) = V(x(w'))$.

Condition (8) for w is immediate because $h(x(w)) = h(x(t)) = g_p(t) = g_p(w)$. To verify condition (8) for v, we have $h(x(v)) = h(x(v')x(v^*)) = h(x(v')) + h(x(v^*))$ because $x(v^*) \in X(v^*) \subseteq V(x(v'))$. Also $x(v^*) = x(t^{*}{}^\frown 1) = b_{t^*}$ and so

$$\begin{aligned}
h(x(v)) &= g_p(v') + h(b_{t^*}) \\
&= g_p(v') + g_p(t^{*}{}^\frown 1) \\
&= g_p(v') + g_p(v^*) \\
&= g_p(v' + v^*) \\
&= g_p(v)
\end{aligned}$$

because the supports of v' and v^* are disjoint.

To see that (9) holds, we first note that s and t can be assumed to have the same length, because we can add 0's to s or t to achieve this. By condition (4), this will not change the value of $x(s)$ or $x(t)$. Then $x(t)$ and $x(s)$ belong to disjoint clopen sets if one of the elements s or t has a 1 in a position where the other has a 0. Also, if there is an assigned sequence $\langle W_n \rangle_{n=1}^{\infty}$, then $X(u_{n+1}) = U_n \cap X \subseteq W_n$, so (13) holds.

Now a_n was chosen to satisfy (i). Further, since $a_n \notin x(t_n)U_n$, one has $a_n \notin X(t_n{}^\frown 0)$ so, since $a_n \in X(t_n)$, one has $a_n \in X(t_n{}^\frown 1)$ and thus (ii) holds. To verify (iii), assume that $h(a_n) = g_p(t_n{}^\frown 1)$. Observe that $t_n{}^\frown 1 = t_n' + t_n^{*}{}^\frown 1$ so $g_p(t_n{}^\frown 1) = g_p(t_n') + g_p(t_n^{*}{}^\frown 1)$. Also, $a_n = x(t_n')c_n$ and $c_n \in X(t_n^*) \subseteq V(x(t_n'))$ so $h(a_n) = h(x(t_n')) + h(c_n)$. Thus $h(c_n) = g_p(t_n^{*}{}^\frown 1)$ so $x(t_n^{*}{}^\frown 1) = b_{t_n^*} = c_n$. Therefore $a_n = x(t_n')c_n = x(t_n')x(t_n^{*}{}^\frown 1) = x(t_n{}^\frown 1)$.

Thus we can extend our definition of x and X to sequences in F of length $n+1$ so that conditions (1)–(9) and (13), as well as hypotheses (i), (ii), and (iii) remain true. This shows that these functions can be defined on the whole of F with these conditions remaining valid.

It is now easy to show by induction on k using condition (5) that if the canonical representation of t is

$$t = s_{i_0}^{m_0} + s_{i_1}^{m_1} + \cdots + s_{i_k}^{m_k},$$

then $x(t) = x(s_{i_0}^{m_0})x(s_{i_1}^{m_1})\ldots x(s_{i_k}^{m_k})$, so that condition (10) holds.

To verify condition (11) notice that for any $t \in F$ we have

$$h(x(t)) = h(x(t')x(t^*)) = h(x(t')) + h(x(t^*))$$

because $x(t^*) \in X(t^*) \subseteq V(x(t'))$. Thus one can show by induction on k that if the canonical representation of t is

$$t = s_{i_0}^{m_0} + s_{i_1}^{m_1} + \cdots + s_{i_k}^{m_k},$$

then $h(x(t)) = h(x(s_{i_0}^{m_0})) + h(x(s_{i_1}^{m_1})) + \cdots + h(x(s_{i_k}^{m_k}))$.

To check condition (12), we show first that each $a \in X$ eventually occurs as a value of x. Suppose instead that $X \setminus x[F] \neq \emptyset$ and let $a = \min X \setminus x[F]$. Then a has only finitely many predecessors so $a = a_n$ for some n. Pick $m \in \mathbb{Z}_p$ such that $h(a_n) = g_p(t_n{}^\frown 1) + m$. If one had $m = 0$, then one would have by (iii) that $a_n = x(t_n{}^\frown 1)$ so $m \geq 1$. By (i) and the assumption that $a_n \notin x[F]$, one has $a_{n+1} = a_n$. Also by (ii) $a_n \in X(t_n{}^\frown 1)$ and

$a_{n+1} \in X(t_{n+1}\frown 1) \subseteq X(t_{n+1})$ so $t_{n+1} = t_n \frown 1$. Then $g_p(t_{n+1}\frown 1) = g_p(t_n\frown 1) + 1$ so $h(a_{n+1}) = h(a_n) = g_p(t_n\frown 1) + m = g_p(t_{n+1}\frown 1) + (m-1)$. Repeating this argument m times, one has $h(a_{n+m}) = g_p(t_{n+m}\frown 1)$ so by (iii), $a_n = a_{n+m} = x(t_{n+m}\frown 1)$, a contradiction.

To complete the verification of condition (12), we note that, for every $n \in \omega$ and every $t \in F$, we have $x(u_n t) \in X(u_n t) \subseteq X(u_n)$ and so $x[u_n F] \subseteq X(u_n)$. On the other hand, suppose that $a \in X \setminus x[u_n F]$. We have already established that $a = x(v)$ for some $v \in F \setminus u_n F$. Since $e \in x[u_n F]$, $a \neq e$ and hence $v \notin \{u_m : m \in \omega\}$ so in particular supp $v \neq \emptyset$. If $k = \min$ supp v, then $k \leq n$. We have $x(v) \in X(s_{k-1}^k)$ and $X(u_n) \subseteq X(u_k)$. Now $X(s_{k-1}^k) \cap X(u_k) = \emptyset$ and so $a = x(v) \notin X(u_n)$. Thus $X(u_n) \subseteq x[u_n F]$. $\qquad\square$

From this point until the statement of Zelenuk's Theorem, G will denote a discrete group with identity e and C will denote a finite subsemigroup of G^*.

Definition 7.5. (a) $C^\sim = \{x \in \beta G : xC \subseteq C\}$.
   (b) φ is the filter of subsets U of G for which $C \subseteq \overline{U}$.
   (c) $\varphi^\sim$ is the filter of subsets U of G for which $C^\sim \subseteq \overline{U}$.

Observe that $C^\sim$ is a semigroup and $e \in C^\sim$. Note also that $\varphi = \bigcap C$ and $\varphi^\sim = \bigcap C^\sim$.

Lemma 7.6. *We can define a left invariant topology on G for which $\varphi^\sim$ is the filter of neighborhoods of e. If we also have $xC = C$ for every $x \in C^\sim$, then this topology has a basis of clopen sets.*

Proof. Given $U \in \varphi$, we put $U^\frown = \{a \in G : aC \subseteq \overline{U}\}$ and observe that $e \in U^\frown$.
   We begin by showing that

(i) for each $U \in \varphi$, $U^\frown \in \varphi^\sim$,
(ii) $\{U^\frown : U \in \varphi\}$ is a base for the filter $\varphi^\sim$, and
(iii) for all $U \in \varphi$ and all $a \in U^\frown$, $a^{-1}U^\frown \in \varphi^\sim$.

To verify (i), suppose that $x \in C^\sim \setminus \overline{U^\frown}$. Since $U^\frown \notin x$,

$$G \setminus U^\frown = \{a \in G : ay \notin \overline{U} \text{ for some } y \in C\} \in x.$$

Since C is finite, there exists $y \in C$ such that $\{a \in G : ay \notin \overline{U}\} \in x$ and hence $xy \notin \overline{U}$. This contradicts the assumption that $x \in C^\sim$.

Since (i) holds, to verify (ii) it will be sufficient to show that $\bigcap_{U \in \varphi} \overline{U^\frown} = C^\sim$. (For then if one had some $V \in \varphi^\sim$ such that for all $U \in \varphi$, $U^\frown \setminus V \neq \emptyset$ one would have that $\{\overline{U^\frown \setminus V} : U \in \varphi\}$ is a collection of closed subsets of βG with the finite intersection property, and hence there would be some $x \in \bigcap_{U \in \varphi} \overline{U^\frown \setminus V}$.)

Suppose, on the contrary, that there is an element $x \in \bigcap_{U \in \varphi} \overline{U^\frown} \setminus C^\sim$. Then $xy \notin C$ for some $y \in C$. We can choose $U \in \varphi$ such that $xy \notin \overline{U}$. This implies that $x \notin \overline{U^\frown}$, a contradiction.

To verify (iii), let $U \in \varphi$ and $a \in U^{\widehat{}}$ be given and suppose that $a^{-1}U^{\widehat{}} \notin \varphi^{\sim}$. Then there is an element $y \in C^{\sim} \backslash \overline{a^{-1}U^{\widehat{}}}$. Then

$$G \backslash a^{-1}U^{\widehat{}} = \{b \in G : abz \notin \overline{U} \text{ for some } z \in C\} \in y.$$

Since C is finite, we can choose $z \in C$ such that $\{b \in G : abz \notin \overline{U}\} \in y$. This implies that $ayz \notin \overline{U}$ contradicting the assumptions that $yz \in C$ and $a \in U^{\widehat{}}$.

Now that (i), (ii), and (iii) have been verified, let $\mathcal{B} = \{aU^{\widehat{}} : U \in \varphi\}$. We claim that $\mathcal{B}$ is a basis for a left invariant topology on G with $\varphi^{\sim}$ as the set of neighborhoods of e.

To see that $\mathcal{B}$ is a basis for a topology, let $U, V \in \varphi$, let $a, b, c \in G$, and assume that $c \in aU^{\widehat{}} \cap bV^{\widehat{}}$. Now $a^{-1}c \in U^{\widehat{}}$ so by (iii), $c^{-1}aU^{\widehat{}} \in \varphi^{\sim}$ so by (ii) we may pick $W_1 \in \varphi$ such that $W_1^{\widehat{}} \subseteq c^{-1}aU^{\widehat{}}$. Similarly, we may pick $W_2 \in \varphi$ such that $cW_2^{\widehat{}} \subseteq bV^{\widehat{}}$ and hence $c(W_1 \cap W_2)^{\widehat{}} \subseteq aU^{\widehat{}} \cap bV^{\widehat{}}$.

That the topology generated by $\mathcal{B}$ is left invariant is trivial.

To see that $\varphi^{\sim}$ is the set of neighborhoods of e, notice that by (ii), each member of $\varphi^{\sim}$ is a neighborhood of e. So let W be a neighborhood of e and pick $a \in G$ and $U \in \varphi$ such that $e \in aU^{\widehat{}} \subseteq W$. Then $a^{-1} \in U^{\widehat{}}$ so by (iii), $aU^{\widehat{}} \in \varphi^{\sim}$ and hence $W \in \varphi^{\sim}$.

Finally we suppose that $xC = C$ for every $x \in C^{\sim}$. We shall show that in this case, for each $U \in \varphi$, $U^{\widehat{}}$ is closed.

To this end let $U \in \varphi$ and let $a \in G \backslash U^{\widehat{}}$. We show that $V = G \backslash a^{-1}U^{\widehat{}} \in \varphi^{\sim}$ so that aV is a neighborhood of a missing $U^{\widehat{}}$. So let $y \in C^{\sim}$. We show that $y \in \overline{V}$. Now $ayC = aC \not\subseteq \overline{U}$ since $a \notin U^{\widehat{}}$ so pick $z \in C$ such that $ayz \notin U^{\widehat{}}$. Then $\{b \in G : abz \notin \overline{U}\} \in y$ and $\{b \in G : abz \notin \overline{U}\} \subseteq \{b \in G : abC \not\subseteq \overline{U}\} = V$.

Since each λ_a is a homeomorphism we have shown that $\mathcal{B}$ consists of clopen sets.
$\square$

Lemma 7.7. *Assume that $xC = C$ for every $x \in C^{\sim}$. The following statements are equivalent.*

(a) *The topology defined in Lemma 7.6 is Hausdorff.*

(b) $\{a \in G : aC \subseteq C\} = \{e\}$.

(c) $\bigcap \varphi^{\sim} = \{e\}$.

Proof. (a) implies (b). Let $a \in G \backslash \{e\}$ and pick $U \in \varphi^{\sim}$ such that $a \notin U$. Since $C^{\sim} \subseteq \overline{U}$, one has $a \notin C^{\sim}$.

(b) implies (c). Let $a \in G \backslash \{e\}$. Then by assumption $a \notin C^{\sim}$. Also, $C^{\sim} = \bigcap_{x \in C} \rho_x^{-1}[C]$ so $C^{\sim}$ is closed in βG. Pick $U \subseteq G$ such that $C^{\sim} \subseteq \overline{U}$ and $a \notin U$. Then $U \in \varphi^{\sim}$ so $a \notin \bigcap \varphi^{\sim}$.

(c) implies (a). Let $a \in G \backslash \{e\}$. By Lemma 7.6 there is some clopen $U \in \varphi^{\sim}$ such that $a \notin U$. Then U and $G \backslash U$ are disjoint neighborhoods of e and a.
$\square$

We shall now assume, until the statement of Zelenuk's Theorem, that C is a finite subgroup of G^*. We shall denote the identity of C by u. We note that, since C is a group, $xC = C$ for every $x \in C^{\sim}$.

Lemma 7.8. *If G has no nontrivial finite subgroups, then $\bigcap \varphi^{\sim} = \{e\}$.*

Proof. Observe that $\bigcap \varphi^{\sim} = C^{\sim} \cap G = \{a \in G : aC = C\}$. (For if $a \in C^{\sim} \cap G$, then for each $U \in \varphi^{\sim}$, $a \in U$. And if $a \in G \backslash C^{\sim}$, then for each $p \in C^{\sim}$, $G \backslash \{a\} \in p$ so $G \backslash \{a\} \in \varphi^{\sim}$.) Therefore $\bigcap \varphi^{\sim}$ is a subgroup of G so is either $\{e\}$ or infinite.

So suppose that $\bigcap \varphi^{\sim}$ is infinite. By the pigeonhole principle, there exists $a \neq b$ in G such that $au = bu$, contradicting Lemma 6.28. $\qquad\square$

Lemma 7.9. *There is a left invariant topology on G with a basis of clopen sets such that $\varphi^{\sim}$ is the filter of neighborhoods of e. If G has no nontrivial finite subgroups, then the topology is Hausdorff.*

Proof. This is immediate from Lemmas 7.6, 7.7, and 7.8. $\qquad\square$

Definition 7.10. (a) Fix a family $\langle U_y \rangle_{y \in C}$ of pairwise disjoint subsets of G such that $U_y \in y$ for every $y \in C$.

(b) For each $y \in C$, we put $A_y = \{a \in G : az \in \overline{U_{yz}}$ for every $z \in C\}$.

Observe that $A_y = \bigcap_{z \in C} \{a \in G : a^{-1}U_{yz} \in z\}$ and that $e \in A_u$.

Lemma 7.11. *For each $y \in C$, $A_y \in y$, and if y and w are distinct members of C, then $A_y \cap A_w = \emptyset$.*

Proof. Let $y \in C$. For each $z \in C$, $U_{yz} \in yz$ so $\{a \in G : a^{-1}U_{yz} \in z\} \in y$. Thus $\bigcap_{z \in C} \{a \in G : a^{-1}U_{yz} \in z\} \in y$.

Now let y and w be distinct members of C and suppose that $a \in A_y \cap A_w$. Then $U_y = U_{yu} \in au$ and $U_w = U_{wu} \in au$ so $U_y \cap U_w \neq \emptyset$, a contradiction. $\qquad\square$

Definition 7.12. (a) $X = \bigcup_{y \in C} A_y$.

(b) We define $f : X \to C$ by stating that $f(a) = y$ if $a \in A_y$.

(c) For each $z \in C$ and $a \in X$, $V_z(a) = \{b \in A_z : ab \in A_{f(a)z}\}$.

(d) For each $a \in X$, $V(a) = \bigcup_{z \in C} V_z(a)$.

Notice that for each $a \in X$, $V(a) \subseteq X$ and $aV(a) \subseteq X$. Notice also that $f(e) = u$ so that for each $z \in C$, $V_z(e) = A_z$ and consequently $V(e) = X$.

Lemma 7.13. *For each $a \in X$ and each $b \in V(a)$, $f(a)f(b) = f(ab)$.*

Proof. Let $a \in X$ and let $b \in V(a)$. Pick $z \in C$ such that $b \in V_z(a)$. Since $V_z(a) \subseteq A_z$, we have that $f(b) = z$. Then $ab \in A_{f(a)z} = A_{f(a)f(b)}$ so $f(ab) = f(a)f(b)$. $\qquad\square$

Lemma 7.14. *For each $a \in X$, $V(a) \in \varphi^{\sim}$.*

Proof. We begin by showing that

(i) for each $y \in C$ and each $a \in G$, $a \in A_y$ if and only if $au \in \overline{A_y}$ and

(ii) for all $y, z \in C$ and each $a \in A_y$, $az \in \overline{A_{yz}}$.

To verify (i), suppose first that $a \in A_y$. If $au \notin \overline{A_y}$, then $\{b \in G : ab \notin A_y\} \in u$. Now $ab \notin A_y$ implies that $abz \notin \overline{U_{yz}}$ for some $z \in C$. Since C is finite, we may suppose that there exists $z \in C$ such that $\{b \in G : abz \notin \overline{U_{yz}}\} \in u$. This implies that $auz = az \notin \overline{U_{yz}}$, a contradiction.

Now suppose that $au \in \overline{A_y}$. Then $\{b \in G : ab \in A_y\} \in u$ so for each $z \in C$, $\{b \in G : abz \in \overline{U_{yz}}\} \in u$. Thus, for each $z \in C$, $az = auz \in \overline{U_{yz}}$ so that $a \in A_y$.

To verify (ii), suppose instead that $a^{-1}A_{yz} \notin z$ so that $G \backslash a^{-1}A_{yz} \in z$. Now

$$G \backslash a^{-1}A_{yz} = \bigcup_{w \in C}\{b \in G : abw \notin \overline{U_{yzw}}\}$$

so pick $w \in C$ such that $\{b \in G : abw \notin \overline{U_{yzw}}\} \in z$. Then $azw \notin \overline{U_{yzw}}$ so $a \notin A_y$, a contradiction.

Now, having established (i) and (ii), let $a \in X$. We show that $V(a) \in \varphi^{\sim}$. So suppose instead that $V(a) \notin \varphi^{\sim}$ and pick $x \in C^{\sim} \backslash V(a)$. Let $y = f(a)$ and let $z = xu$. Since $x \in C^{\sim}$, $z \in C$, so $x \notin V_z(a)$ and hence

$$G \backslash V_z(a) = \{b \in G : b \notin A_z \text{ or } ab \notin A_{yz}\} \in x.$$

By (i), $\{b \in G : b \notin A_z \text{ or } ab \notin A_{yz}\} = \{b \in G : bu \notin \overline{A_z} \text{ or } abu \notin \overline{A_{yz}}\}$ so either $\{b \in G : bu \notin \overline{A_z}\} \in x$ or $\{b \in G : abu \notin \overline{A_{yz}}\} \in x$. That is, $xu \notin \overline{A_z}$ or $axu \notin \overline{A_{yz}}$. Since $xu = z$ this says $z \notin \overline{A_z}$, which is impossible, or $az \notin \overline{A_{yz}}$ which contradicts (ii). $\qquad \square$

Corollary 7.15. *X is open in the left invariant topology defined on G by taking $\varphi^{\sim}$ as the base of neighborhoods of e.*

Proof. For every $a \in X$, we have $aV(a) \subseteq X$ by the definition of $V(a)$. $\qquad \square$

Lemma 7.16. *For every non-empty $Y \subseteq X$ which is open in the topology defined by $\varphi^{\sim}$, we have $f[Y] = C$.*

Proof. If $U \in \varphi^{\sim}$, then, for every $y \in C$, $U \cap A_y \neq \emptyset$ because $U \in y$ and $A_y \in y$. So $f[U] = C$. If $a \in Y$, then $aU \subseteq Y$ for some $U \in \varphi^{\sim}$ satisfying $U \subseteq V(a)$. By Lemma 7.13 $f[aU] = f(a)f[U]$ so $C = f(a)C = f(a)f[U] = f[aU] \subseteq f[Y]$. $\qquad \square$

Theorem 7.17 (Zelenuk's Theorem). *If G is a countable discrete group with no nontrivial finite subgroups, then G^* contains no nontrivial finite groups.*

Proof. We assume that $C \subseteq G^*$ is a finite group satisfying $C \simeq \mathbb{Z}_p$ for some integer $p > 1$ and derive a contradiction.

Let γ be an isomorphism from C onto $\mathbb{Z}_p$ and for each $i \in \mathbb{Z}_p$, let $y_i = \gamma^{-1}(i)$. Let $h = \gamma \circ f$. Then $h : X \to \mathbb{Z}_p$ and $h(a) = i$ if and only if $f(a) = y_i$.

We assume that G has the topology produced in Lemma 7.9. Then by Lemmas 7.9, 7.13, 7.14, and 7.16 and Corollary 7.15 (and some other observations made after the definitions) the hypotheses of Lemma 7.4 are satisfied. So we presume we have chosen $x(t)$ and $X(t)$ for each $t \in F$ as guaranteed by Lemma 7.4.

Now $A_{y_1} \in y_1$ and if $a \in A_{y_1}$, then $f(a) = y_1$ so $h(a) = 1$. If $a = x(t)$, then $h(a) = g_p(t)$ by condition (8) of Lemma 7.4 and so $A_{y_1} \subseteq \{x(t) : t \in F \text{ and } g_p(t) = 1\}$ and hence

$$\{x(t) : t \in F \text{ and } c(t) \equiv 1 \pmod{p}\} = \{x(t) : t \in F \text{ and } g_p(t) = 1\} \in y_1.$$

For $r \in \{0, 1, \ldots, p - 1\}$, let $B_r = \{x(t) : t \in F \text{ and } c(t) \equiv rp + 1 \pmod{p^2}\}$. Now $A_{y_1} \subseteq \bigcup_{r=0}^{p-1} B_r$ and by condition (9) of Lemma 7.4 $B_r \cap B_k = \emptyset$ when $r \neq k$ so we may pick the unique $r \in \{0, 1, \ldots, p - 1\}$ such that $B_r \in y_1$.

Now for each $n \in \mathbb{N}$, $X(u_n)$ is open and $e = x(u_n) \in X(u_n)$ so $X(u_n)$ is a neighborhood of e and hence $X(u_n) \in \varphi^{\sim} = \bigcap C^{\sim} \subseteq \bigcap C \subseteq y_1$. Given $t \in F$ and $n \in \mathbb{N}$ one has $\min \operatorname{supp} t > n$ if and only if $t \in u_n F \setminus \{u_m : m \in \mathbb{N}\}$. Also, for each $n \in \mathbb{N}$, $x[u_n F] = X(u_n)$ by condition (12) of Lemma 7.4 so $\{x(t) : t \in F \text{ and } \min \operatorname{supp} t > n\} = X(u_n) \setminus \{e\} \in y_1$.

For each $n \in \mathbb{N}$, let

$$\begin{aligned} A_n = \{x(s_1 + s_2 + \cdots + s_n) : \text{ for each } i \in \{1, 2, \ldots, n\}, \ s_i \in F, \\ c(s_i) \equiv rp + 1 \pmod{p^2}, \\ \text{and if } i < n, \ s_i << s_{i+1}\}. \end{aligned}$$

We show by induction on n that $A_n \in y_1{}^n$. Since $A_1 = B_r \in y_1$, the assertion is true for $n = 1$.

So let $n \in \mathbb{N}$ and assume that $A_n \in y_1{}^n$. We claim that

$$A_n \subseteq \{a \in G : a^{-1} A_{n+1} \in y_1\}$$

so that $A_{n+1} \in y_1{}^n y_1 = y_1{}^{n+1}$. So let $a \in A_n$ and pick $s_1 << s_2 << \cdots << s_n$ in F such that each $c(s_i) \equiv rp + 1 \pmod{p^2}$ and $a = x(s_1 + s_2 + \cdots + s_n)$. Let

$$D = \{x(t) : t \in F \text{ and } \min \operatorname{supp} t > \max \operatorname{supp} s_n + 1\}.$$

Then $B_r \cap D \in y_1$ so it suffices to show that $B_r \cap D \subseteq a^{-1} A_{n+1}$. To this end, let $t \in F$ such that $\min \operatorname{supp} t > \max \operatorname{supp} s_n + 1$ and $c(t) \equiv rp + 1 \pmod{p^2}$. By condition (10) of Lemma 7.4,

$$x(s_1 + s_2 + \cdots + s_n) x(t) = x(s_1 + s_2 + \cdots + s_n + t)$$

so $x(s_1 + s_2 + \cdots + s_n) x(t) \in A_{n+1}$ as required.

Now $y_1{}^{p+1} = y_1$ so $A_{p+1} \in y_1$ so pick some $a \in A_{p+1} \cap B_r$. Then

$$a = x(s_1 + s_2 + \cdots + s_{p+1})$$

where each $c(s_i) \equiv rp + 1 \pmod{p^2}$ and $s_1 << s_2 << \cdots << s_{p+1}$. Then

$$\begin{aligned} c(s_1 + s_2 + \cdots + s_{p+1}) &\equiv (rp + 1) \cdot (p + 1) \pmod{p^2} \\ &\equiv (r + 1)p + 1 \pmod{p^2} \end{aligned}$$

and hence $a \in B_{r+1}$ so $B_r \cap B_{r+1} \neq \emptyset$, a contradiction. $\qquad\square$

Corollary 7.18. $\mathbb{N}^*$ *contains no nontrivial finite subgroups.*

Proof. Since $\mathbb{Z}^*$ contains no nontrivial finite subgroups, neither does $\mathbb{N}^*$. □

Recall that a *partial multiplication* on a set X is a function mapping some subset Z of $X \times X$ to X. Given a partial multiplication on X and points a and b of X, we say that ab *is defined* if and only if $(a, b) \in Z$. We shall sometimes express the same fact by saying "$ab \in X$". (The latter terminology is convenient when the partial multiplication on X is induced by a multiplication on a larger structure.)

Definition 7.19. Let X be a topological space with a distinguished element e and a partial multiplication. We shall say that X is a *local left group* if there is a left topological group G in which X can be topologically embedded so that the following conditions hold:

  (i) e is the identity of G,
 (ii) the partial multiplication defined on X is that induced by the multiplication of G,
(iii) for every $a \in X$, there is a neighborhood $V(a)$ of e in X for which $aV(a) \subseteq X$, and
 (iv) for every $a \in X$, $aX \cap X$ is a neighborhood of a in X.

We shall say that X is a *regular local left group* if X can be embedded in a Hausdorff zero dimensional left topological group G so that these four conditions hold.

Notice that if G is a left topological group with identity e, then every open neighborhood of e in G is a local left group. Notice also that in any local left group, one may presume that $V(e) = X$.

Definition 7.20. Let X and Y be local left groups. We shall say that a mapping $k : X \to Y$ is a *local homomorphism* if, for each $a \in X$, there is a neighborhood $V(a)$ of e in X such that $b \in V(a)$ implies that $ab \in X$, $k(a)k(b) \in Y$ and $k(ab) = k(a)k(b)$. We shall say that k is a *local isomorphism* if it is a bijective local homomorphism and k^{-1} is a local homomorphism.

Lemma 7.21. *Let X and Y be local left groups with distinguished elements e and f respectively and let $k : X \to Y$ be a local homomorphism. Then $k(e) = f$. If k is continuous at e, then k is continuous on all of X.*

Proof. Pick groups G and H containing X and Y respectively as guaranteed by the definition of local left group. For each $a \in X$ pick $V(a)$ as guaranteed by the definition of local homomorphism. Since $e \in X$ and $e \in V(e)$, one has $k(e) = k(ee) = k(e)k(e)$ so $k(e)$ is an idempotent in H and thus $k(e) = f$.

Now assume that k is continuous at e and let $a \in X$. Let W be a neighborhood of $k(a)$ in Y and pick open $U \subseteq H$ such that $k(a) \in U \cap Y \subseteq W$. Then $Y \cap k(a)^{-1}U$ is a neighborhood of f in Y so pick a neighborhood T of e in X such that $k[T] \subseteq Y \cap k(a)^{-1}U$ and pick open $R \subseteq G$ such that $e \in R \cap X \subseteq V(a) \cap T$. Then $aR \cap X$

is a neighborhood of a in X and, by the definition of local left group, $aX \cap X$ is a neighborhood of a in X. We claim that $k[aR \cap aX \cap X] \subseteq W$. To see this let $c \in aR \cap aX \cap X$ and pick $b \in R \cap X$ such that $c = ab$. Then $b \in V(a) \cap T$ so $k(c) = k(ab) = k(a)k(b)$ and $k(b) \in k(a)^{-1}U$. Thus $k(c) \in U \cap Y \subseteq W$. $\square$

Theorem 7.22. *Let X and Y be countable regular local left groups without isolated points. Then there is a local isomorphism $k : X \to Y$. If Y is first countable, then k can be chosen to be continuous. If X and Y are both first countable, then k can be chosen to be a homeomorphism.*

Proof. We shall apply Lemma 7.4 with $p = 1$, so that the functions h and g_p are trivial.

Let e and f denote the distinguished elements of X and Y respectively. We can define $x(t) \in X$ and $X(t) \subseteq X$ for every $t \in F$, so that the conditions in the statement of Lemma 7.4 are satisfied. We can also define $y(t) \in Y$ and $Y(t) \subseteq Y$ so that these conditions are satisfied with x replaced by y and X replaced by Y.

Define $k : X \to Y$ by $k\big(x(t)\big) = y(t)$ for each $t \in F$. By condition (9) of Lemma 7.4, $x(t) = x(s)$ if and only if $y(t) = y(s)$ so k is well defined and one-to-one. By conclusion (12) of Lemma 7.4, k is defined on all of X and $k[X] = Y$.

Now let $a \in X$, pick $t \in F$ such that $a = x(t)$, and let $n = l(t) + 1$. Pick (since X is a local left group) a neighborhood $V_1(a)$ such that $ab \in X$ for all $b \in V_1(a)$ and let $V(a) = V_1(a) \cap X(u_n)$. By conditions (1) and (2) of Lemma 7.4 and the fact that $x(u_n) = e$, $V(a)$ is a neighborhood of e in X. Let $b \in V(a)$. Since $b \in V_1(a)$, $ab \in X$. By condition (12) of Lemma 7.4 pick $v \in u_n F$ such that $b = x(v)$. Then by condition (10) of Lemma 7.4, $x(t + v) = x(t)x(v) = ab$ and $y(t + v) = y(t)y(v) = k(a)k(b)$ and consequently $k(ab) = k(a)k(b)$ as required.

Thus k is a bijective local homomorphism. Since $k^{-1} : Y \to X$ is characterized by $k\big(y(t)\big) = x(t)$ for each $t \in F$, an identical argument establishes that k^{-1} is a local homomorphism.

Now assume that Y is first countable, and let $\{W_n : n \in \mathbb{N}\}$ be a neighborhood base at f. Then we may assume that for each $n \in \mathbb{N}$, $Y(u_{n+1}) \subseteq W_n$ by condition (13) of Lemma 7.4. Given any $n \in \mathbb{N}$, by condition (12) of Lemma 7.4, $Y(u_{n+1}) = y[u_{n+1}F] = k\big[x[u_{n+1}F]\big] = k[X(u_{n+1})]$ so $X(u_{n+1})$ is a neighborhood of e contained in $k^{-1}[W_n]$. Thus k is continuous at e so, by Lemma 7.21, k is continuous on X.

Similarly, if X is first countable, we deduce that k^{-1} is continuous. $\square$

Exercise 7.1.1. Let G be a countable group with no nontrivial finite groups. Show that G^* contains no nontrivial compact groups. (Hint: An infinite compact subset of βG cannot be homogeneous by Theorem 6.38.)

Exercise 7.1.2. Let G be an infinite commutative group which does contain a nontrivial finite subgroup. Show that G^* also contains a nontrivial finite subgroup.

Exercise 7.1.3. Let $p \in \mathbb{N}^*$. Let $p_1 = p$ and for $n \in \mathbb{N}$, let $p_{n+1} = p_n + p$. (We cannot use np for the sum of p with itself n times because np is the product of n with p in $(\beta\mathbb{N}, \cdot)$. As we shall see in Corollary 17.22, np is never equal to the sum of p with itself n times, if $n > 1$ and $p \in \mathbb{N}^*$.) Show that, if $p_n = p$ for some $n > 1$ in $\mathbb{N}$, p must be idempotent.

7.2 Semigroups Isomorphic to $\mathbb{H}$

We remind the reader that $\mathbb{H}$ denotes $\bigcap_{n=1}^{\infty} c\ell_{\beta\mathbb{N}}(2^n\mathbb{N})$. Recall from Chapter 6 that a good deal is known of the structure of $\mathbb{H}$. (More information will be found in Section 7.3.)

We show in this section that copies of $\mathbb{H}$ arise in many contexts. In particular, G^* contains copies of $\mathbb{H}$ whenever G is a countable abelian group (Corollary 7.30) or a countable free group (Corollary 7.31).

Definition 7.23. Let X be a subset of a semigroup. A function $\psi : \omega \to X$ will be called an $\mathbb{H}$-*map* if it is bijective and if $\psi(m + n) = \psi(m)\psi(n)$ whenever $m, n \in \mathbb{N}$ satisfy max supp$(m) + 1 <$ min supp(n).

(We remind the reader that, if $n \in \omega$, supp$(n) \in \mathcal{P}_f(\omega)$ is defined by the equation $n = \Sigma\{2^i : i \in \text{supp}(n)\}$.)

In the following theorem, and again in Theorem 7.28, we shall be dealing with two topologies at the same time. The spaces βG and βX are constructed by taking G and X to be discrete, while the "neighborhoods" refer to the left invariant topology on G and the topology it induces on X.

Theorem 7.24. *Let G be a group with a left invariant zero dimensional Hausdorff topology, and let X be a countable subspace of G which contains the identity e of G and has no isolated points. Suppose also that, for each $a \in X$, $aX \cap X$ is a neighborhood of a in X and that, for each $a \in X$, there is a neighborhood $V(a)$ of e in X, with $V(e) = X$, for which $aV(a) \subseteq X$.*

Then there is a countable set $\{V_n : n \in \mathbb{N}\}$ of neighborhoods of e in X for which $Y = \bigcap_{n=1}^{\infty} c\ell_{\beta X} V_n\backslash\{e\}$ is a subsemigroup of βG. Furthermore, there is an $\mathbb{H}$-map $\psi : \omega \to X$ such that $\widetilde{\psi}$ defines an isomorphism from $\mathbb{H}$ onto Y.

In the case in which the filter of neighborhoods of e in X has a countable base, Y can be taken to be $\bigcap\{c\ell_{\beta X} W : W$ is a neighborhood of e in $X\}\backslash\{e\}$.

Proof. We apply Lemma 7.4 with $p = 1$ so that $|\mathbb{Z}_p| = 1$. We define $h : X \to \mathbb{Z}_1$ to be the constant map. We then observe that the hypotheses of Lemma 7.4 are satisfied, and hence $x(t)$ and $X(t)$ can be defined for every $t \in F$ so that the conditions stated in this lemma will hold.

We put $V_n = X(u_n)$ and $Y = \bigcap_{n=1}^{\infty} c\ell_{\beta X} X(u_n)\backslash\{e\}$. We show that Y is a semigroup by applying Theorem 4.20 with $\mathcal{A} = \{X(u_n)\backslash\{e\} : n \in \mathbb{N}\}$. Now for each $n \in \mathbb{N}$, $X(u_n)\backslash\{e\} = x[u_n F]\backslash\{e\} = \{x(t) : t \in F$ and min supp $(t) > n\}$ by condition (12) of Lemma 7.4. Given any $m \in \mathbb{N}$ and any $v \in u_m F\backslash\{e\}$, let $n = \max$ supp $(v) + 2$. Then if $w \in u_n F\backslash\{e\}$ we have $v + w \in u_m F\backslash\{e\}$ and $x(v + w) = x(v)x(w)$ by condition (10) of Lemma 7.4 so $x(v)(x[u_n F]\backslash\{e\}) \subseteq x[u_m F]\backslash\{e\}$ as required by Theorem 4.20.

We define $\theta : F \to \omega$ by stating that $\theta(t) = \Sigma\{2^i : i \in \operatorname{supp} t\}$. By condition (9) of Lemma 7.4, for every $t_1, t_2 \in F$, we have $\theta(t_1) = \theta(t_2)$ if and only if $x(t_1) = x(t_2)$. Thus we can define a bijective mapping $\psi : \omega \to X$ for which $\psi \circ \theta = x$. The mapping $\widetilde{\psi} : \beta\omega \to \beta X$ is then also bijective (by Exercise 3.4.1). By condition (10) of Lemma 7.4, $\psi(m + n) = \psi(m)\psi(n)$ whenever $m, n \in \omega$ satisfy $\max \operatorname{supp}(m) + 1 < \min \operatorname{supp}(n)$, for we then have $m = \theta(s)$ and $n = \theta(t)$ for some $s, t \in F$ with $s << t$. So ψ is an $\mathbb{H}$-map. By Lemma 6.3 $\widetilde{\psi}(p + q) = \widetilde{\psi}(p)\widetilde{\psi}(q)$ for every $p, q \in \mathbb{H}$.

Thus $\widetilde{\psi}_{|\mathbb{H}}$ is a homomorphism. Furthermore,

$$\widetilde{\psi}[\mathbb{H}] = \bigcap_{n=1}^{\infty} c\ell_{\beta X}\, \psi[2^n\mathbb{N}] = \bigcap_{n=1}^{\infty} c\ell_{\beta X}\, x[u_n F]\backslash\{e\} = Y.$$

This establishes that $\widetilde{\psi}$ defines a continuous isomorphism from $\mathbb{H}$ onto Y.

Finally, if there is a countable base $\{W_n : n \in \mathbb{N}\}$ for the neighborhoods of e in X, the sets $X(u_n)$ can be chosen so that $X(u_{n+1}) \subseteq W_n$ for every $n \in \mathbb{N}$ (by condition (13) of Lemma 7.4). So $Y = \bigcap\{c\ell_{\beta X}\, W : W \text{ is a neighborhood of } e \text{ in } X\}\backslash\{e\}$. $\square$

Lemma 7.25. *Let $\mathcal{T}$ be a left invariant zero dimensional Hausdorff topology on $(\mathbb{Z}, +)$ with a countable base and assume that ω has no isolated points in this topology. Then the hypotheses of Theorem 7.24 hold for $G = \mathbb{Z}$ and $X = \omega$.*

Proof. Given any $a \in X$, $a + X$ is a cofinite subset of the Hausdorff space X and hence is open in X. For each $a \in X$ let $V(a) = X$. $\square$

Theorem 7.24 has several applications to subsemigroups of $\beta\mathbb{N}$. The following is one example and others are given in the exercises.

Corollary 7.26. *The set $\bigcap_{n=1}^{\infty} c\ell_{\beta\mathbb{N}}(n\mathbb{N})$ is algebraically and topologically isomorphic to $\mathbb{H}$.*

Proof. Let $\mathcal{B} = \{a + n\mathbb{Z} : a \in \mathbb{Z} \text{ and } n \in \mathbb{N}\}$. Given $a, b, c \in \mathbb{Z}$ and $n, m \in \mathbb{N}$, if $c \in (a + n\mathbb{Z}) \cap (b + m\mathbb{Z})$, then $c \in c + nm\mathbb{Z} \subseteq (a + n\mathbb{Z}) \cap (b + m\mathbb{Z})$ so $\mathcal{B}$ is a basis for a left invariant topology $\mathcal{T}$ on $\mathbb{Z}$. To see that $\mathcal{T}$ is zero dimensional, notice that if $c \in \mathbb{Z}\backslash(a + n\mathbb{Z})$, then $c + n\mathbb{Z}$ is a neighborhood of c missing $(a + n\mathbb{Z})$. To see that $\mathcal{T}$ is Hausdorff, let a and b be distinct members of $\mathbb{Z}$ and pick $n \in \mathbb{N}$ with $n > |a - b|$. Then $(a + n\mathbb{Z}) \cap (b + n\mathbb{Z}) = \emptyset$.

Thus by Lemma 7.25, Theorem 7.24 applies. So

$$\bigcap\{c\ell_{\beta\omega}(W\backslash\{0\}) : W \text{ is a neighborhood of } 0 \text{ in } \omega\} = \bigcap_{n \in \mathbb{N}}(c\ell_{\beta\mathbb{N}}\, n\mathbb{N})$$

is algebraically and topologically isomorphic to $\mathbb{H}$. $\square$

Lemma 7.27. *Let G be a countably infinite subgroup of a compact metric topological group C. Then with the relative topology, G is a Hausdorff zero dimensional first countable topological group without isolated points.*

Proof. Let d be the metric of C. That G is a Hausdorff first countable topological group is immediate. To see that G is zero dimensional, let $x \in G$ and let U be a

neighborhood of x in G. Since G is countable, for only countably many $r \in \mathbb{R}$ is
$\{y \in C : d(x, y) = r\} \cap G \neq \emptyset$. Pick $r \in \mathbb{R}$ such that $\{y \in C : d(x, y) = r\} \cap G = \emptyset$
and $\{y \in G : d(x, y) < r\} \subseteq U$. Then $\{y \in G : d(x, y) < r\}$ is a clopen neighborhood
of x in G which is contained in U.

To see that G has no isolated points, it suffices to show that the identity e of G is
not isolated. Let U be an open neighborhood of e in C. Then $\{xU : x \in G\}$ is an open
cover of $c\ell\, G$. (Given $y \in c\ell\, G$, $yU^{-1} \cap G \neq \emptyset$. If $x \in yU^{-1} \cap G$, then $y \in xU$.)
Pick finite $F \subseteq G$ such that $G \subseteq \bigcup_{x \in F} xU$. Pick $x \in F$ such that $xU \cap G$ is infinite.
Then $U \cap G = U \cap x^{-1}G$ is infinite. $\qquad\square$

In the following theorem, the condition that C be metrizable is not really necessary,
because any countable topological group which can be mapped injectively into a compact
topological group can also be mapped injectively into a compact metrizable topological
group. However, the proof of this fact would take us rather far from our subject; and it
is not needed in any of our applications.

Theorem 7.28. *Let G be a countably infinite discrete group which can be mapped
into a compact metrizable topological group C by an injective homomorphism h. Let
$V = G^* \cap \widetilde{h}^{-1}[\{1\}]$, where 1 denotes the identity of C and $\widetilde{h} : \beta G \to C$ denotes the
continuous extension of h. Then V is a compact G_δ subsemigroup of G^* which contains
all the idempotents of G^*. Furthermore, there is an $\mathbb{H}$-map $\psi : \omega \to G$ such that
$\widetilde{\psi} : \beta\omega \to \beta G$ defines an isomorphism from $\mathbb{H}$ onto V. In addition, for every pair of
distinct elements $a, b \in G$, $aV \cap Vb = \emptyset$.*

Proof. Let e denote the identity of G. By Lemma 7.27, with the relative topology, $h[G]$
is a Hausdorff zero dimensional first countable left topological group without isolated
points. Thus, giving G the topology $\{h^{-1}[U] : U \text{ is open in } C\}$, h becomes a topological
embedding and G enjoys all of these same properties.

For each $a \in G$, let $V(a) = G$. Then G satisfies the hypotheses of Theorem 7.24
with $X = G$. Let $V = \bigcap\{c\ell_{\beta G}\, W : W \text{ is a neighborhood of } e \text{ in } G\}\backslash\{e\}$. (Notice
that one is again considering two topologies on G. The space βG is constructed taking
G to be discrete, while the "neighborhoods" of e are with respect to the topology
just introduced.) By Theorem 7.24, V is a semigroup algebraically and topologically
isomorphic to $\mathbb{H}$. Further, since G is first countable, V is a G_δ in βG.

To see that $V = G^* \cap \widetilde{h}^{-1}[\{1\}]$, note that, if $p \in G^*$, then $p \in V$ if and only
if $h^{-1}[U] \in p$ for every neighborhood U of 1 in C, and $h^{-1}[U] \in p$ if and only if
$\widetilde{h}(p) \in \overline{U}$. Then, since $\widetilde{h}$ is a homomorphism by Corollary 4.22, V contains all of the
idempotents of G^*.

Now let $a, b \in G$ and assume that $aV \cap Vb \neq \emptyset$. Pick x and y in V such that
$ax = yb$. Since $\widetilde{h} : \beta G \to C$ is a homomorphism and $\widetilde{h}(x) = \widetilde{h}(y) = 1$, we have
$h(a) = h(b)$. So $a = b$, because h is injective. $\qquad\square$

Lemma 7.29. *Let G be an abelian group with identity e. For every $a \neq e$ in G, there
is a homomorphism h from G to the circle group $\mathbb{T} = \{z \in \mathbb{C} : |z| = 1\}$ for which
$h(a) \neq 1$.*

Proof. Let $C = \{a^n : n \in \mathbb{Z}\}$ be the cyclic group generated by a. We define f on C by stating that $f(a^n) = \exp(in)$ if C is infinite, and $f(a^n) = \exp\left(\dfrac{2n\pi i}{k}\right)$ if C has order k. We shall show that f can be extended to G. Let $\mathcal{A} = \{(h, H) : H$ is a subgroup of $G, C \subseteq H, h$ is a homomorphism from H to $\mathbb{T}$ and $f \subseteq h\}$. Ordering $\mathcal{A}$ by inclusion on both coordinates, we obtain a maximal member (h, H) of $\mathcal{A}$ by Zorn's Lemma.

We claim that $H = G$. Suppose instead that $H \neq G$ and pick $x \in G\backslash H$. Let $H' = \{x^n y : n \in \mathbb{Z}$ and $y \in H\}$. Then H' is a subgroup of G properly containing H. We show that h can be extended to a homomorphism h' from H' to G, contradicting the maximality of (h, H).

Assume first that there is no $n \in \mathbb{Z}\backslash\{0\}$ such that $x^n \in H$. Since members of H' have a unique expression of the form $x^n y$ with $n \in \mathbb{Z}$ and $y \in H$, one may define a homomorphism h' on H' by $h'(x^n y) = h(y)$.

Otherwise, we choose m to be the first positive integer for which $x^m \in H$ and choose $\theta \in \mathbb{R}$ satisfying $\exp(i\theta) = h(x^m)$. We observe that, for any $n \in \mathbb{Z}$, we have $x^n \in H$ if and only if n is a multiple of m. We can now extend h to H' by stating that $h'(x^n y) = \exp\left(\dfrac{ni\theta}{m}\right)h(y)$. To see that this is well defined, suppose that $x^n y = x^k z$ where $n, k \in \mathbb{Z}$ and $y, z \in H$. Then $k - n = qm$ for some $q \in \mathbb{Z}$ and so $y = (x^m)^q z$. Thus $h(y) = \left(h(x^m)\right)^q h(z) = \exp(qi\theta)h(z) = \exp\left(\dfrac{(k-n)i\theta}{m}\right)h(z)$. Thus $\exp\left(\dfrac{ni\theta}{m}\right)h(y) = \exp\left(\dfrac{ki\theta}{m}\right)h(z)$. $\qquad\square$

Corollary 7.30. *Let G be a commutative countable discrete group. Then there is a compact G_δ subsemigroup V of G^* which contains all the idempotents of G^*. Furthermore, there is an $\mathbb{H}$-map $\psi : \omega \to G$ such that $\widetilde{\psi} : \beta\omega \to \beta G$ defines an isomorphism from $\mathbb{H}$ onto V. In addition, V has the property that $aV \cap V = \emptyset$ for every $a \in G$ other than the identity.*

Proof. Let e denote the identity of G. By Lemma 7.29, for every $a \neq e$ in G, there is a homomorphism $h_a : G \to \mathbb{T}$ for which $h_a(a) \neq 1$. We put $C = {}^G\mathbb{T}$, and observe that C is a compact topological group and, being the countable product of metric spaces, is metrizable. We define $h : G \to C$ by $\left(h(x)\right)_a = h_a(x)$. Then the hypotheses of Theorem 7.28 are satisfied and so the conclusion follows. $\qquad\square$

Corollary 7.31. *Let G denote the free group on a countable set of generators. There is a compact G_δ subsemigroup V of G^* which contains all the idempotents of G^*. Furthermore, there is an $\mathbb{H}$-map $\psi : \omega \to G$ such that $\widetilde{\psi} : \beta\omega \to \beta G$ defines an isomorphism from $\mathbb{H}$ onto V. In addition, $aV \cap Vb = \emptyset$ for every pair of distinct elements $a, b \in G$.*

Proof. We shall use $\emptyset$ to denote the identity of G. By Theorem 1.23, for every $a \in G\backslash\{\emptyset\}$ there is a finite group F_a and a homomorphism $h_a : G \to F_a$ for which $h_a(a) \neq \emptyset$. We put $C = \bigtimes_{a \in G} F_a$, where each F_a has the discrete topology. Then C is a compact topological group and, being the countable product of metric spaces, is metrizable. We define $h : G \to C$ by $(h(x))_a = h_a(x)$. Then the hypotheses of Theorem 7.28 are satisfied and so the conclusion follows. $\qquad\square$

Theorem 7.32. *Let G be an infinite countable discrete group which can be mapped by an injective homomorphism into a compact metrizable topological group. Then the minimal left ideals of βG are homeomorphic to those of $\beta\omega$. Furthermore, if L is any minimal left ideal of $\beta\omega$, there is a homeomorphism from L onto a minimal left ideal M of βG which maps $L \cap \mathbb{H}$ isomorphically onto a subsemigroup of M containing all the idempotents of M.*

Proof. By Theorem 7.28, there exist a compact subsemigroup V of G^* which contains all the idempotents of G^* and an $\mathbb{H}$-map $\psi : \omega \to G$ for which $\widetilde{\psi} : \beta\omega \to \beta G$ defines a continuous isomorphism from $\mathbb{H}$ onto V.

By Theorem 2.11, all the minimal left ideals of $\beta\omega$ are homeomorphic to each other, and so are all those of βG. Thus it suffices to establish the final statement of the theorem.

Let L be a minimal left ideal of $\beta\omega$ and pick an idempotent $p \in L$. By Theorem 1.38, p is a minimal idempotent in $\beta\omega$. Since all idempotents of $\beta\omega$ are in $\mathbb{H}$ by Lemma 6.8, all idempotents of βG are in V, and $\widetilde{\psi}_{|\mathbb{H}}$ is an isomorphism onto V, $\widetilde{\psi}(p)$ is a minimal idempotent in βG. Thus $M = (\beta G)\widetilde{\psi}(p)$ is a minimal left ideal of βG.

We claim that for each $m \in \omega$, $\widetilde{\psi}(m+p) = \psi(m)\widetilde{\psi}(p)$. Indeed, $\widetilde{\psi} \circ \lambda_m$ and $\lambda_{\psi(m)} \circ \widetilde{\psi}$ are continuous functions agreeing on $\{n \in \mathbb{N} : \min \operatorname{supp}(n) > \max \operatorname{supp}(m)+1\}$, which is a member of p. So $\widetilde{\psi}[\omega + p] = G\widetilde{\psi}(p)$. Thus

$$\widetilde{\psi}[L] = \widetilde{\psi}[\beta\omega + p] = c\ell_{\beta G}\ \widetilde{\psi}[\omega + p] = c\ell_{\beta G}\big(G\widetilde{\psi}(p)\big) = (\beta G)\widetilde{\psi}(p) = M.$$

Since $\widetilde{\psi}$ is a bijection (by Exercise 3.4.1), L is homeomorphic to M. Since $\widetilde{\psi}_{|\mathbb{H}}$ is an isomorphism and V contains all of the idempotents of βG, one has $\widetilde{\psi}[L \cap \mathbb{H}]$ is a subsemigroup of M containing all of the idempotents of M. $\qquad\square$

Exercise 7.2.1. Show that $\bigcap_{n \in \mathbb{Z}} c\ell_{\beta\mathbb{Z}}(2^n\mathbb{Z})$ is algebraically and topologically isomorphic to $\mathbb{H}$.

Exercise 7.2.2. Show that $\bigcap_{n \in \mathbb{N}} c\ell_{\beta\mathbb{N}}(3^n\mathbb{N})$ is algebraically and topologically isomorphic to $\mathbb{H}$.

Exercise 7.2.3. Show that $\bigcap_{n \in \mathbb{N}} c\ell_{\beta\mathbb{Z}}(3^n\mathbb{Z})$ is algebraically and topologically isomorphic to $\mathbb{H}$.

Exercise 7.2.4. Show that $\mathbb{N}^*$ is not algebraically and topologically isomorphic to $\mathbb{H}$. (Hint: Consider Exercises 6.1.1 and 6.1.2.)

Exercise 7.2.5. Suppose that $G = \bigoplus_{i \in \mathbb{N}} G_i$, where each G_i is a nontrivial countable discrete group. (The *direct sum* of the groups G_i, is the subgroup of $\times_{i \in \mathbb{N}} G_i$ containing the elements equal to the identity on all but finitely many coordinates.) Let e_i denote the identity of G_i and let $\pi_i : G \to G_i$ denote the natural projection map. If $U_i = \{a \in G : \pi_j(a) = e_j \text{ for all } j \leq i\}$, show that $\bigcap_{i \in \mathbb{N}} c\ell_{\beta G}\ U_i \setminus \{e\}$ is algebraically and topologically isomorphic to $\mathbb{H}$.

7.3 Free Semigroups and Free Groups in βS

We shall show that there is a high degree of algebraic freedom in βS. If S is any cancellative discrete semigroup with cardinality κ, then S^* contains algebraic copies of the free group on 2^{2^κ} generators.

Recall that a sequence $\langle x_n \rangle_{n=1}^{\infty}$ in a semigroup S has *distinct finite products* provided that, whenever $F, G \in \mathcal{P}_f(\mathbb{N})$ and $\prod_{n \in F} x_n = \prod_{n \in G} x_n$, one has $F = G$.

Theorem 7.33. *Let S be a discrete semigroup and let $\langle x_n \rangle_{n=1}^{\infty}$ be a sequence in S which has distinct finite products. If $A = \{x_n : n \in \mathbb{N}\}$, then $|A^*| = 2^c$ and the elements of A^* generate a free subsemigroup in S^*.*

Proof. By Corollary 3.57 $|A^*| = 2^c$.

We define a mapping $c : \mathrm{FP}(\langle x_n \rangle_{n=1}^{\infty}) \to \mathbb{N}$ and mappings $f_i : \mathrm{FP}(\langle x_n \rangle_{n=1}^{\infty}) \to A$ for each $i \in \mathbb{N}$ as follows. Given $y \in \mathrm{FP}(\langle x_n \rangle_{n=1}^{\infty})$, there is a unique $F \in \mathcal{P}_f(\mathbb{N})$ such that $y = \prod_{m \in F} x_m$. Let $c(y) = |F|$ and write $F = \{m_1, m_2, \ldots, m_{c(y)}\}$, where $m_1 < m_2 < \cdots < m_{c(y)}$. If $i \in \{1, 2, \ldots, c(y)\}$, let $f_i(y) = x_{m_i}$ and otherwise let $f_i(y) = x_1$.

Now let $k \in \mathbb{N}$ and let $p_1, p_2, \ldots, p_k \in A^*$. If $i \in \{1, 2, \ldots, k\}$, then

$$\widetilde{f_i}(p_1 p_2 \ldots p_k) = \widetilde{f_i}\Big(\lim_{x_{n_1} \to p_1} \lim_{x_{n_2} \to p_2} \ldots \lim_{x_{n_k} \to p_k} x_{n_1} x_{n_2} \ldots x_{n_k} \Big)$$
$$= \lim_{x_{n_1} \to p_1} \lim_{x_{n_2} \to p_2} \ldots \lim_{x_{n_k} \to p_k} f_i(x_{n_1} x_{n_2} \ldots x_{n_k})$$
$$= \lim_{x_{n_1} \to p_1} \lim_{x_{n_2} \to p_2} \ldots \lim_{x_{n_k} \to p_k} x_{n_i}$$
$$= p_i .$$

Similarly,

$$\widetilde{c}(p_1 p_2 \ldots p_k) = \widetilde{c}\Big(\lim_{x_{n_1} \to p_1} \lim_{x_{n_2} \to p_2} \ldots \lim_{x_{n_k} \to p_k} x_{n_1} x_{n_2} \ldots x_{n_k} \Big)$$
$$= k .$$

Now assume that $k, m \in \mathbb{N}$, $p_1, p_2, \ldots, p_k, q_1, q_2, \ldots, q_m \in A^*$, and further $p_1 p_2 \ldots p_k = q_1 q_2 \ldots q_m$. Then $k = \widetilde{c}(p_1 p_2 \ldots p_k) = \widetilde{c}(q_1 q_2 \ldots q_m) = m$ and given $i \in \{1, 2, \ldots, k\}$, $p_i = \widetilde{f_i}(p_1 p_2 \ldots p_k) = \widetilde{f_i}(q_1 q_2 \ldots q_m) = q_i$.

Thus A^* generates a free subsemigroup of βS. $\qquad \square$

As a consequence of Theorem 7.33 (with $S = \mathbb{N}$) we have the following example.

Example 7.34. *If $A = \{2^n : n \in \mathbb{N}\}$, the elements of A^* generate a free subsemigroup of $(\beta \mathbb{N}, +)$.*

Theorem 7.35. *Let κ be an infinite cardinal and let $\langle a_\lambda \rangle_{\lambda < \kappa}$ be a κ-sequence in a discrete semigroup S which has distinct finite products. Let*

$$T = \bigcap_{\lambda < \kappa} c\ell\big(\mathrm{FP}(\langle a_\mu \rangle_{\lambda < \mu < \kappa}) \big).$$

Then T is a compact subsemigroup of βS and every maximal group in the smallest ideal of T contains an algebraic copy of the free group on 2^{2^κ} generators.

Proof. To see that T is a subsemigroup of βS, we apply Theorem 4.20. So let $\lambda < \kappa$ and let $x \in \mathrm{FP}(\langle a_\mu \rangle_{\lambda < \mu < \kappa})$ be given. Pick a finite subset L of $\{\mu : \lambda < \mu < \kappa\}$ such that $x = \prod_{\mu \in L} a_\mu$. Let $\gamma = \max L$. Then $x \cdot \mathrm{FP}(\langle a_\mu \rangle_{\gamma < \mu < \kappa}) \subseteq \mathrm{FP}(\langle a_\mu \rangle_{\lambda < \mu < \kappa})$.

Let p be a minimal idempotent in T. We recall that this implies that pTp is a group, by Theorem 1.59.

Let $A = \{a_\mu : \mu < \kappa\}$ and note that $U_\kappa(A) \subseteq T$. (To see this, let $q \in U_\kappa(A)$ and let $\lambda < \kappa$ be given. Then $\{a_\mu : \lambda < \mu < \kappa\} \in q$ and $\{a_\mu : \lambda < \mu < \kappa\} \subseteq \mathrm{FP}(\langle a_\mu \rangle_{\lambda < \mu < \kappa})$ so $q \in c\ell\big(\mathrm{FP}(\langle a_\mu \rangle_{\lambda < \mu < \kappa})\big)$.) Let G denote the subgroup of T generated by $p \cdot U_\kappa(A) \cdot p$. We shall show that G is isomorphic to the free group with 2^{2^κ} generators.

We choose any set in one-one correspondence with $U_\kappa(A)$, denoting by σ_q the element in this set corresponding to the element q in $U_\kappa(A)$. (Of course we can choose $U_\kappa(A)$ itself, but it may be helpful to think of the set as distinct from $U_\kappa(A)$.) Let F be the free group generated by $\{\sigma_q : q \in U_\kappa(A)\}$. By the universal property of free groups (Lemma 1.22), there is a homomorphism $g : F \to G$ for which $g(\sigma_q) = pqp$. We shall show that g is an isomorphism. It will be sufficient to prove that g is one-to-one on every subgroup of F which is generated by finitely many elements of $\{\sigma_q : q \in U_\kappa(A)\}$, because any two elements of F belong to such a subgroup of F.

Let $q_1, q_2, \ldots, q_n$ be distinct elements of $U_\kappa(A)$ and let D be the subgroup of F generated by $\{\sigma_{q_1}, \sigma_{q_2}, \ldots, \sigma_{q_n}\}$. For each $i \in \{1, 2, \ldots, n\}$ choose $B_i \in q_i$ such that $B_i \cap B_j = \emptyset$ if $i \neq j$. We may assume $\bigcup_{i=1}^n B_i = A$ because we can replace B_1 by $A \setminus \bigcup_{i=2}^n B_i$.

We define a mapping $h : \mathrm{FP}(\langle a_\mu \rangle_{\mu < \kappa}) \to F$ as follows. First, we define h on A. If $a_\mu \in B_i$, then $h(a_\mu) = \sigma_{q_i}$. Now given $r \in \mathrm{FP}(\langle a_\mu \rangle_{\mu < \kappa})$ there is a unique $L \in \mathcal{P}_f(\kappa)$ such that $r = \prod_{\mu \in L} a_\mu$. Define $h(r) = \prod_{\mu \in L} h(a_\mu)$. (Recall that this product is taken in increasing order of indices.) Now extend h arbitrarily so that $h : S \to F$.

By Theorem 2.24 F can be embedded in a compact topological group $\widetilde{F}$, and we shall assume that $F \subseteq \widetilde{F}$. Let $\widetilde{h} : \beta S \to \widetilde{F}$ be the continuous extension of h. Note that $h[B_i] = \{\sigma_{q_i}\}$ so that $\widetilde{h}(q_i) = \sigma_{q_i}$.

We shall show that $\widetilde{h}_{|T}$ is a homomorphism. By Theorem 4.21 it suffices to show that for each $x \in \mathrm{FP}(\langle a_\mu \rangle_{\mu < \kappa})$ there is some $\lambda < \kappa$ so that for all $y \in \mathrm{FP}(\langle a_\mu \rangle_{\lambda < \mu < \kappa})$, $h(x \cdot y) = h(x) \cdot h(y)$. Given $x \in \mathrm{FP}(\langle a_\mu \rangle_{\mu < \kappa})$ with $x = \prod_{\mu \in L} a_\mu$, let $\lambda = \max L$. Then given $y \in \mathrm{FP}(\langle a_\mu \rangle_{\lambda < \mu < \kappa})$, if $y = \prod_{t \in M} a_\mu$, one has $\min M > \max L$ so

$$
\begin{aligned}
h(x \cdot y) &= h(\textstyle\prod_{\mu \in L \cup M} a_\mu) \\
&= \textstyle\prod_{\mu \in L \cup M} h(a_\mu) \\
&= \textstyle\prod_{\mu \in L} h(a_\mu) \cdot \prod_{\mu \in M} h(a_\mu) \\
&= h(x) \cdot h(y).
\end{aligned}
$$

Thus $\widetilde{h} \circ g$ is a homomorphism. Further, given $i \in \{1, 2, \ldots, n\}$, one has $\widetilde{h}\big(g(\sigma_{q_i})\big) = \widetilde{h}(p \cdot q_i \cdot p) = \widetilde{h}(p) \cdot \widetilde{h}(q_i) \cdot \widetilde{h}(p) = \widetilde{h}(q_i) = \sigma_{q_i}$. (Since p is an idempotent in T, $\widetilde{h}(p)$ is the identity of F.) Since $\widetilde{h} \circ g$ agrees with the identity on the generators of D, it agrees with the identity on D and consequently g is injective on D as required. $\square$

Corollary 7.36. *Every maximal group in the smallest ideal of* $\mathbb{H}$ *contains a free group on* 2^c *generators.*

Proof. The sequence $\langle 2^t \rangle_{t < \omega}$ has distinct finite sums and

$$\mathbb{H} = \bigcap_{n < \omega} c\ell\big(\mathrm{FS}(\langle 2^t \rangle_{n < t < \omega})\big)$$

so Theorem 7.35 applies.   □

Corollary 7.37. *Every maximal group in the smallest ideal of* $(\beta\mathbb{N}, +)$ *contains a free group on* 2^c *generators.*

Proof. By Lemma 6.8, $\mathbb{H}$ contains all of the idempotents of $(\beta\mathbb{N}, +)$ and consequently, $K(\beta\mathbb{N}) \cap \mathbb{H} \neq \emptyset$. Thus by Theorem 1.65, $K(\mathbb{H}) = K(\beta\mathbb{N}) \cap \mathbb{H}$.   □

Corollary 7.38. *Let S be an infinite right cancellative and weakly left cancellative semigroup. Every neighborhood of every idempotent in S^* contains an algebraic copy of the free group on* 2^c *generators.*

Proof. This follows from Theorem 6.32 and Corollary 7.36.   □

Note that we do not claim that every idempotent in S^* is the identity of a free group on 2^c generators. In fact, according to Theorem 12.42 it is consistent with ZFC that there are idempotents $p \in \mathbb{N}^*$ for which the maximal group $H(p)$ is just a copy of $\mathbb{Z}$.

Corollary 7.39. *Let S be an infinite discrete semigroup with cardinality κ which is right cancellative and weakly left cancellative. Then βS contains an algebraic copy of the free group on* 2^{2^κ} *generators.*

Proof. By Lemma 6.31 there is a sequence $\langle a_\lambda \rangle_{\lambda < \kappa}$ which has distinct finite products so Theorem 7.35 applies.   □

Corollary 7.40. *Let G be a countably infinite group which can be mapped into a compact metrizable topological group by an injective homomorphism. Then every maximal group in the smallest ideal of βG contains a free group on* 2^c *generators.*

Proof. By Theorem 7.28, there is a subsemigroup $\widetilde{G}$ of βG which is isomorphic to $\mathbb{H}$ and contains all of the idempotents of G^*. Thus by Theorem 1.65, $K(\widetilde{G}) = \widetilde{G} \cap K(\beta G)$. Thus, if p is any idempotent in $K(\beta G)$, p is minimal in a copy of $\mathbb{H}$ so $p\widetilde{G}p$ contains a free group on 2^c generators by Corollary 7.36.   □

We remind the reader that a subspace Y of a topological space X is said to be C^*-*embedded* in X if every continuous function from Y into $[0, 1]$ can be extended to a continuous function defined on X.

Lemma 7.41. *Let S be a discrete semigroup and let p be an idempotent in βS. Then Sp is extremally disconnected and is C^*-embedded in $(\beta S)p$. In fact, any continuous* *function from Sp to a compact Hausdorff space extends to a continuous function defined on $(\beta S)p$.*

Proof. Let U and V be disjoint open subsets of Sp. Define $f : S \to \{-1, 0, 1\}$ by stating that $f(s) = -1$ if $sp \in U$, $f(s) = 1$ if $sp \in V$, and $f(s) = 0$ otherwise. Let $x \in U$. Then $xp = x \in U$ so $\{s \in S : sp \in U\} \in x$ so $\widetilde{f}(x) = -1$, where $\widetilde{f} : \beta S \to \{-1, 0, 1\}$ denotes the continuous extension of f. Similarly, $\widetilde{f}(y) = 1$ for every $y \in V$. Hence $\bar{U} \cap \bar{V} = \emptyset$, and so Sp is extremally disconnected.

To show that Sp is C^*-embedded, let $\varphi : Sp \to Y$ be a continuous mapping into a compact Hausdorff space. The map $\tau : S \to Y$ defined by $\tau(s) = \varphi(sp)$ extends to a continuous map $\widetilde{\tau} : \beta S \to Y$. Let $t \in S$. Then $\lim_{s \to p} tsp = tpp = tp$ so

$$
\begin{aligned}
\widetilde{\tau}(tp) &= \lim_{s \to p} \tau(ts) \\
&= \lim_{s \to p} \varphi(tsp) \\
&= \varphi(\lim_{s \to p} tsp) \\
&= \varphi(tp)
\end{aligned}
$$

so the restriction of $\widetilde{\tau}$ to $(\beta S)p$ is an extension of φ. $\square$

Let S be an infinite discrete semigroup. We know that, for every idempotent p in βS, the subsemigroup $p(\beta S)p$ of βS is a group if and only if p is in the smallest ideal of βS by Theorem 1.59. In this case, it is the maximal group containing p.

We have also seen that the maximal groups in the smallest ideal of βS are disjoint and that they are all algebraically isomorphic (by Theorem 1.64). We shall show that, in spite of being algebraically isomorphic, they lie in at least $2^{\mathfrak{c}}$ different homeomorphism classes, provided that they are infinite.

If S is weakly left cancellative and has cardinality κ, then there are $2^{2^{\kappa}}$ maximal groups in the smallest ideal of βS (by Corollary 6.43).

Theorem 7.42. *Let S be a discrete commutative semigroup and let L be a minimal left ideal in βS. If L is infinite, the maximal groups in L lie in at least $2^{\mathfrak{c}}$ homeomorphism classes.*

Proof. Notice that if e is an idempotent in L, then the group $e(\beta S)e = eL$. Suppose that e and f are idempotents in L and that there is a homeomorphism $\varphi : eL \to fL$. We shall show that, given any $x \in eL$ and any $y \in fL$, there is a homeomorphism of L to itself which maps x to y. Notice that by Theorem 4.23 $Se = eS$ and $Sf = fS$. In particular, $Se = See = eSe \subseteq eL$.

We know, by Lemma 7.41, that $\varphi_{|Se}$ has a continuous extension $\widetilde{\varphi}$ defined on L. Now eS is dense in eL, because $c\ell_{\beta S}(eS) = c\ell_{\beta S}(Se) = (\beta S)e = L$ and $eL \subseteq L$. So $\widetilde{\varphi}$ coincides with φ on eL.

Now $\tau = \varphi^{-1}{}_{|Sf}$ also has a continuous extension $\widetilde{\tau}$ defined on L. Since $\widetilde{\tau} \circ \widetilde{\varphi}$ and $\widetilde{\varphi} \circ \widetilde{\tau}$ are the identity maps on eL and fL respectively and since these sets are dense in L, it follows that they are also the identity maps on L. So $\widetilde{\varphi}$ is a homeomorphism.

There is an element z of fL for which $\widetilde{\varphi}(x)z = y$ because fL is a group. Now the map ρ_z defines a homeomorphism on L (by Theorem 2.11), and so $\rho_z \circ \widetilde{\varphi}$ is a homeomorphism of L to itself which maps x to y.

The conclusion now follows from the fact that the points of L lie in at least 2^c different homeomorphism classes by Theorem 6.38. $\square$

Exercise 7.3.1. Let G be any given finite group. Show that there is an infinite discrete semigroup S for which the maximal groups in the smallest ideal of βS are isomorphic to G and are equal to the minimal left ideals of βS. (Hint: Choose T to be an infinite right zero semigroup and put $S = T \times G$.)

Exercise 7.3.2. Let S denote an infinite discrete commutative group. Prove that none of the maximal groups in the smallest ideal of βS can be closed. (Hint: Use Theorem 6.38 and the fact that each maximal group in the smallest ideal of βS is infinite.)

Notes

The material in Section 7.1 as well as Theorem 7.24 are due to E. Zelenuk [246].

I. Protasov has generalised Zelenuk's Theorem by characterizing the finite subgroups of βG, where G denotes a countable group. Every finite subgroup of βG has the form Hp for some finite subgroup H of G and some idempotent p in βG which commutes with all the elements of H. [201]

Our proof of Lemma 7.29 is based on the proof in [114, p.441]

Corollary 7.36 is from [152], a result of collaboration with J. Pym. Corollary 7.38 extends a result of A. Lisan in [178].

Chapter 8

Cancellation

It may be the case that βS does not have any interesting algebraic properties. For example, if S is a right zero or a left zero semigroup, then βS is as well and there is nothing very interesting to be said about the algebra of βS.

However, the presence of cancellation properties in S produces a rich algebraic structure in βS. To cite one example: we saw in Chapter 6 that, if S is an infinite cancellative semigroup with cardinality κ, then βS contains $2^{2^{\kappa}}$ disjoint left ideals. We also saw in Chapter 7 that βS contains copies of the free group on $2^{2^{\kappa}}$ generators.

In the first three sections of this chapter, we shall describe elements of βS which are right or left cancelable. If S is a cancellative semigroup with cardinality κ, we shall show that βS contains $2^{2^{\kappa}}$ right cancelable elements. If we also suppose that S is countable, the set of right cancelable elements of βS contains a dense open subset of S^*. In the important case in which S is $\mathbb{N}$ or a countable group, we can strengthen this statement and assert that the set of elements of βS that are cancelable on both sides contains a dense open subset of S^*.

In the final section, we shall suppose that S is $\mathbb{N}$ or a countable group and shall discuss the smallest compact subsemigroup of S^* containing a given right cancelable element. We shall show that these semigroups have a rich algebraic structure. Their properties lead to new insights about βS, allowing us to prove that $\overline{K(\beta S)}$ contains infinite chains of idempotents, as well as $\leq_R$-maximal idempotents.

We remark that, if S is an infinite discrete semigroup, βS may well contain a rich set of cancelable elements, but cannot be a cancellative semigroup. If it were, S would be cancellative and S^* would contain two disjoint left ideals, L_1 and L_2, by Theorem 6.42. We could choose an idempotent p in L_1. Then for any $x \in \beta S$, the equation $xp = xpp$ would imply that $x = xp$ and hence that $\beta S = L_1$. This would contradict our assumption that $\emptyset \neq L_2 \subseteq \beta S \backslash L_1$.

8.1 Cancellation Involving Elements of S

Lemma 8.1. *Suppose that S is a discrete semigroup. If s is a left cancelable element of S, it is also a left cancelable element of βS. The analogous statement holds for right cancelable elements as well.*

Proof. Let $s \in S$ be left cancelable. Since the map λ_s is injective on S, its extension to βS is injective as well (by Exercise 3.4.1). In the same way, if $s \in S$ is right cancelable, ρ_s is injective on βS. $\square$

Recall that we showed in Lemma 6.28 that if s is a cancelable element of the semigroup S, $t \in S$, and $p \in \beta S$ one has:

(i) if S is right cancellative and $sp = tp$, then $s = t$, and
(ii) if S is left cancellative and $ps = pt$, then $s = t$.

As a consequence the following holds.

Corollary 8.2. *Suppose that S is a discrete cancellative semigroup. If s and t are distinct elements of S, then, for every $p \in \beta S$, $sp \neq tp$ and $ps \neq pt$.*

Proof. This is an immediate consequence of Lemma 6.28. $\square$

We now consider some of the properties of left cancellative semigroups. In the following example, we show that the assumption that S is left cancellative is not enough to imply that, for every pair of distinct elements $s, t \in S$ and every $p \in \beta S$, $ps \neq pt$.

Example 8.3. *There is a countable left cancellative semigroup S containing distinct elements g and h such that $pg = ph$ for some $p \in \beta S$.*

Proof. We take S to be the set of all injective functions $f : \mathbb{N} \to \mathbb{N}$ for which there exist $m, r \in \mathbb{N}$ such that $f(n) = 2^r n$ whenever $n \geq m$. We observe that S is a semigroup under composition and that S is countable and left cancellative.

We define $g, h \in S$ as follows: if $n \geq 2$, $g(n) = h(n) = 2n$, while $g(1) = 1$ and $h(1) = 2$.

We shall show that, for every finite partition $\mathcal{F}$ of S, there exists $A \in \mathcal{F}$ such that $Ag \cap Ah \neq \emptyset$. Let

$$E = \{f \in S : f(1) \text{ and } f(2) \text{ are odd}, f(1) < f(2), \text{ and } f(n) = 2n \text{ for all } n \geq 3\}.$$

For each $A \in \mathcal{F}$, let $\widehat{A} = \{(f(1), f(2)) : f \in E \cap A\}$. Given odd integers $a < b$, there is a unique member f of E with $(f(1), f(2)) = (a, b)$. So

$$\{(a, b) : a, b \in 2\mathbb{N} - 1 \text{ and } a < b\} = \bigcup_{A \in \mathcal{F}} \widehat{A}.$$

By Ramsey's Theorem (Theorem 5.6), we can choose $A \in \mathcal{F}$ and odd integers a, b, and c with $a < b < c$ and $\{(a, b), (b, c)\} \subseteq \widehat{A}$. There exist $f, k \in E \cap A$ such that $(a, b) = (f(1), f(2))$ and $(b, c) = ((k(1), k(2))$. Then $k(g(1)) = k(1) = b = f(2) = f(h(1))$ and, for every $n \geq 2$, $k(g(n)) = k(2n) = 4n = f(2n) = f(h(n))$. So $kg = fh$ and hence $Ag \cap Ah \neq \emptyset$.

By Theorem 5.7, there is an ultrafilter $p \in \beta S$ such that $Bg \cap Bh \neq \emptyset$ for every $B \in p$. Since $\{Bg : B \in p\}$ is a base for pg and $\{Bh : B \in p\}$ is a base for ph, it follows that $pg = ph$. $\square$

We saw in Theorem 3.35 that if $f : S \rightarrow S$, $p \in \beta S$, and $\tilde{f}(p) = p$, then $\{x \in S : f(x) = x\} \in p$. It might seem reasonable to conjecture that if $f, k : S \rightarrow S$, $p \in \beta S$, and $\tilde{f}(p) = \tilde{k}(p)$, then $\{x \in S : f(x) = k(x)\} \in p$. However, the functions $\rho_g, \rho_h : S \rightarrow S$ and the point $p \in \beta S$ of Example 8.3 provide a counterexample.

In the next lemma we show that an example of this kind cannot occur with two commuting elements of S.

Lemma 8.4. *Let S be a left cancellative discrete semigroup and let s and t be distinct elements of S such that $st = ts$. Then, for every $p \in \beta S$, $ps \neq pt$.*

Proof. We can define an equivalence relation $\equiv$ on S by stating that $u \equiv v$ if and only if $us^n = vs^n$ for some $n \in \mathbb{N}$. (To verify transitivity, note that if $us^n = vs^n$ and $vs^m = ws^m$, then $us^{n+m} = ws^{n+m}$.) Let $\theta : S \rightarrow S/\equiv$ denote the canonical projection. We shall define a mapping $f : \theta[S] \rightarrow \theta[S]$ which has no fixed points.

For each $u \in S$, we put $f(\theta(us)) = \theta(ut)$. To show that this is well defined, suppose that $\theta(us) = \theta(vs)$. Then $us^n = vs^n$ for some $n \in \mathbb{N}$, and hence $us^n t = vs^n t$ so that $uts^n = vts^n$. So $\theta(ut) = \theta(vt)$. Thus f is well defined on $\theta[Ss]$. To complete the definition of f, we put $f(\theta(w)) = \theta(ss)$ for every $w \in S\backslash\theta^{-1}[\theta[Ss]]$. We observe that $\tilde{f}(\tilde{\theta}(ps)) = \tilde{\theta}(pt)$ for every $p \in \beta S$, where $\tilde{\theta} : \beta S \rightarrow \beta(S/\equiv)$. (One has that $\tilde{f} \circ \tilde{\theta} \circ \rho_s$ and $\tilde{\theta} \circ \rho_t$ are continuous functions agreeing on S, hence on βS.)

We claim that f has no fixed points. Certainly if $w \in S\backslash\theta^{-1}[\theta[Ss]]$, then $f(\theta(w)) \neq \theta(w)$. Now suppose that $\theta(us) = \theta(ut)$ for some $u \in S$. Then $uss^n = uts^n$ for some n so that $us^n s = us^n t$ and hence $s = t$, contradicting our assumption that s and t are distinct.

It now follows from Lemma 3.33, that $\theta[S]$ can be partitioned into three sets A_0, A_1, A_2, such that $f[A_i] \cap A_i = \emptyset$ for every $i \in \{0, 1, 2\}$. Suppose that $p \in \beta S$ satisfies $ps = pt$. We can choose i such that $\theta^{-1}[A_i] \in ps$. Then $A_i \in \tilde{\theta}(ps)$ and $f[A_i] \in \tilde{f}(\tilde{\theta}(ps)) = \tilde{\theta}(pt) = \tilde{\theta}(ps)$ so that $A_i \cap f[A_i] \neq \emptyset$, a contradiction. $\square$

Lemma 8.5. *Let S be a discrete left cancellative semigroup and let s and t be distinct elements of S. If $p \in \beta S$, then $sp = tp$ if and only if $\{u \in S : su = tu\} \in p$.*

Proof. Let $E = \{u \in S : su = tu\}$. We suppose that $sp = tp$ and that $E \notin p$. We shall define a mapping $f : S \rightarrow S$ which has no fixed points.

For every $u \in S\backslash E$, we put $f(su) = tu$. We observe that this is well defined, because every element in sS has a unique expression of the form su. Furthermore, $f(su) \neq su$. To complete the definition of f, for each $v \in S\backslash s \cdot (S\backslash E)$, we choose $f(v)$ to be an arbitrary element of $s(S\backslash E)$. This is possible, because our assumption that $E \notin p$ implies that $S\backslash E \neq \emptyset$.

By Theorem 3.34, $\tilde{f} : \beta S \rightarrow \beta S$ has no fixed point. However, $f(su) = tu$ if $u \in S\backslash E$ and hence $\tilde{f}(sp) = tp = sp$, a contradiction.

For the converse implication observe that λ_s and λ_t are continuous functions agreeing on a member of p. $\square$

Comment 8.6. Let $(S \cdot)$ be a semigroup and define an operation $*$ on S by $x * y = y \cdot x$. Then for all $x \in S$ and all $p \in \beta S$ one has $x * p = p\text{-}\lim_{y \in S}(x * y) = p\text{-}\lim_{y \in S}(y \cdot x) = p \cdot x$ and, similarly, $p * x = x \cdot p$. Consequently, the left-right switches of Example 8.3 and Lemmas 8.4 and 8.5 remain valid.

Exercise 8.1.1. Let S be any semigroup. Show that the set of right cancelable elements of S is either empty or a subsemigroup of S, and that the same is true of the set of left cancelable elements. Show that the complement of the first is either empty or a right ideal in S, and that the complement of the second is either empty or a left ideal of S.

8.2 Right Cancelable Elements in βS

In this section, we shall show that cancellation assumptions about S imply the existence of a rich set of right cancelable elements in S^*. We shall also show that right cancelability is closely related to topological properties and to the Rudin–Keisler order.

The following theorem gives an intrinsic characterization of ultrafilters which are right cancelable in βS for any infinite semigroup S. It is intriguing that this is an intrinsic property, characterizing a single ultrafilter without reference to any others, since right cancelability is defined in terms of the set of all ultrafilters in βS. (Another intrinsic characterization will be given in Theorem 8.19 in the case that S is either $(\mathbb{N}, +)$ or a countable group.)

Theorem 8.7. *Let S be an infinite semigroup, let $p \in \beta S$, and let T be an infinite subset of S. Then $sp \neq rp$ whenever s and r are distinct members of $\overline{T}$ if and only if for every $A \subseteq T$ there is some $B \subseteq S$ such that $A = \{x \in T : x^{-1}B \in p\}$. In particular, p is right cancelable in βS if and only if for every $A \subseteq S$, there is some $B \subseteq S$ such that $A = \{x \in S : x^{-1}B \in p\}$.*

Proof. The sufficiency is easy. Given $r \neq s$ in $\overline{T}$, pick $A \in r \setminus s$. Pick $B \subseteq S$ such that $A \cap T = \{x \in T : x^{-1}B \in p\}$. Then $B \in r \cdot p \setminus s \cdot p$.

To prove the necessity, assume that ρ_p is injective on $\overline{T}$ and let $A \subseteq T$. Since $\rho_{p|\overline{T}}$ is a homeomorphism, $\overline{A}p$ is a clopen subset of $\overline{T}p$. So by Theorem 3.23 there exists a set $B \subseteq T$ for which $\overline{A}p = \overline{B} \cap \overline{T}p$. If $x \in T$, we have

$$x \in A \Leftrightarrow xp \in \overline{B}$$
$$\Leftrightarrow x^{-1}B \in p$$

so $A = \{x \in T : x^{-1}B \in p\}$. $\square$

Now right cancellation in βS can be made to fail in trivial ways. For example, if S has two distinct left identities a and b, then for all $p \in \beta S$, $ap = bp$. Consequently, one may also wish to know for which points $p \in S^*$ is p right cancelable in S^* (meaning of course that whenever q and r are distinct elements of S^*, $qp \neq rp$). The proof of the following theorem is a routine modification of the proof of Theorem 8.7, and we leave it as an exercise.

Theorem 8.8. *Let S be an infinite semigroup and let $p \in S^*$. Then p is right cancelable in S^* if and only if for every $A \subseteq S$, there is some $B \subseteq S$ such that $|A \triangle \{x \in S : x^{-1}B \in p\}| < \omega$.*

Proof. This is Exercise 8.2.1. $\square$

We remind the reader that a subset D of a topological space X is *strongly discrete* if and only if there is a family $\langle U_a \rangle_{a \in D}$ of pairwise disjoint open subsets of X such that $a \in U_a$ for every $a \in D$. It is easy to see that if D is countable and X is regular, then D is strongly discrete if and only if it is discrete.

Lemma 8.9. *Let S be a discrete semigroup, let T be an infinite subset of S, and let $p \in \beta S$ have the property that $ap \neq bp$ whenever a and b are distinct elements of T.*

    (i) *If $qp \neq rp$ whenever q and r are distinct members of $\overline{T}$, then Tp is discrete in βS.*

    (ii) *If Tp is strongly discrete in βS, then $qp \neq rp$ whenever q and r are distinct members of $\overline{T}$.*

Proof. (i) If Tp is not discrete in βS, there is an element $a \in T$ such that $ap \in c\ell((T \setminus \{a\})p)$. Since ρ_p is continuous, $c\ell((T \setminus \{a\})p) = (c\ell(T \setminus \{a\}))p$ and so $ap = xp$ for some $x \in \overline{(T \setminus \{a\})} = \overline{T} \setminus \{a\}$.

    (ii) Suppose that Tp is strongly discrete in βS and pick a family $\langle B_a \rangle_{a \in T}$ of subsets of S such that $B_a \cap B_b = \emptyset$ whenever a and b are distinct members of T and $ap \in \overline{B_a}$ for each $a \in T$. Let $A \subseteq T$ and let $C = \bigcup_{a \in A} B_a$. Then $A = \{s \in T : s^{-1}C \in p\}$. (If $a \in A$, then $a^{-1}B_a \subseteq a^{-1}C$ and if $a \in T \setminus A$, then $a^{-1}B_a \cap a^{-1}C = \emptyset$.) Thus by Theorem 8.7, $qp \neq rp$ whenever q and r are distinct members of $\overline{T}$. $\square$

Theorem 8.10. *Let S be an infinite discrete right cancellative and weakly left cancellative semigroup with cardinality κ. Then the set of right cancelable elements of βS contains a dense open subset of $U_\kappa(S)$. In particular, S^* contains 2^{2^κ} elements which are right cancelable in βS.*

Proof. We apply Theorem 6.30 with $R = S$. By this theorem, every subset T of S with $|T| = \kappa$ contains a subset V with $|V| = \kappa$ such that for every uniform ultrafilter p on V, $ap \neq bp$ when a and b are distinct members of S and Sp is strongly discrete in βS. By Lemma 8.9, this implies that every uniform ultrafilter on V is right cancelable in βS. Since the sets of the form $\overline{T} \cap U_\kappa(S)$ provide a base for the open subsets of $U_\kappa(S)$, it follows that every non-empty open subset of $U_\kappa(S)$ contains a non-empty open subset of $U_\kappa(S)$ whose elements are all right cancelable in βS. Thus the set of right cancelable elements of βS contains a dense open subset of $U_\kappa(S)$.

    Finally, the fact that there are 2^{2^κ} right cancelable elements of βS follows from the fact that $|U_\kappa(V)| = 2^{2^\kappa}$ if $|V| = \kappa$ (by Theorem 3.58.) $\square$

We now show that, if S is countable, right cancelability in βS is equivalent to several other properties. (If $S = T$ in Theorem 8.11, then statement (3) says that p is right cancelable in βS.)

Theorem 8.11. *Let S be a semigroup, let $p \in \beta S$, and let T be an infinite subset of S. Consider the following statements. Statements (1) and (2) are equivalent and imply each of statements (3), (4), (5), and (6) which are equivalent. Statements (7), (8), and (9) are equivalent and are implied by each of statements (3), (4), (5), and (6). If T is countable, all nine statements are equivalent.*

(1) *$ap \neq bp$ whenever a and b are distinct members of T and Tp is strongly discrete.*

(2) *There is a function $f : S \to S$ such that for every $q \in \overline{T}$, $\widetilde{f}(q \cdot p) = q$.*

(3) *$qp \neq rp$ whenever q and r are distinct members of $\overline{T}$.*

(4) *For every $A \subseteq T$ there exists $B \subseteq S$ such that $A = \{s \in T : s^{-1}B \in p\}$.*

(5) *For every pair of disjoint subsets A and B of T, $\overline{Ap} \cap \overline{Bp} = \emptyset$.*

(6) *$\rho_{p|\overline{T}} : \overline{T} \to \overline{T}p$ is a homeomorphism.*

(7) *$\rho_{p|T} : T \to Tp$ is a homeomorphism.*

(8) *$ap \neq bp$ whenever a and b are distinct members of T and Tp is discrete.*

(9) *For every $a \in T$ and every $q \in \overline{T}\backslash\{a\}$, $ap \neq qp$.*

Proof. To see that (1) implies (2) assume that (1) holds. Then there is a family $\langle B_a \rangle_{a \in T}$ of pairwise disjoint subsets of T such that $ap \in \overline{B_a}$ for every $a \in T$. We define $f : S \to S$ by stating that $f(s) = a$ if $s \in B_a$, defining f arbitrarily on $S\backslash \bigcup_{a \in S} B_a$. Since, given $a \in T$, f is identically equal to a on B_a and since $ap \in \overline{B_a}$, it follows that $\widetilde{f}(ap) = a$ for every $a \in T$. Hence, for every $q \in \overline{T}$, we have $\widetilde{f} \circ \rho_p$ agrees with the identity on a member of q and thus $\widetilde{f}(qp) = q$.

To see that (2) implies (1) pick a function f as guaranteed by (2). For each $a \in T$ one has that $\widetilde{f} \circ \lambda_a(p) = a$ and $\{a\}$ is open in S so pick a member B_a of p such that $\widetilde{f} \circ \lambda_a[\overline{B_a}] = \{a\}$. If $a \neq b$, then $f[aB_a] = \{a\} \neq \{b\} = f[bB_b]$ so $aB_a \cap bB_b = \emptyset$.

That (1) implies (3) is Lemma 8.9 (ii). That (3) and (4) are equivalent is Theorem 8.7.

To see that (4) implies (5), let A and B be disjoint subsets of T and pick $C \subseteq S$ such that $A = \{s \in T : s^{-1}C \in p\}$. Then $Ap \subseteq \overline{C}$ and $Bp \subseteq \overline{S\backslash C}$.

To see that (5) implies (3), let q and r be distinct members of $\overline{T}$ and pick $A \in q\backslash r$. Then $qp \in \overline{(A \cap T)p}$ and $rp \in \overline{(T\backslash A)p}$ so $qp \neq rp$.

Statements (3) and (6) are equivalent because a continuous function with compact domain (onto a Hausdorff space, which all of our hypothesized spaces are) is a homeomorphism if and only if it is one to one.

That statement (6) implies statement (7) is trivial as is the equivalence of statements (7) and (8).

To see that statement (8) implies statement (9), let $a \in T$ and let $q \in \overline{T}\backslash\{a\}$. Then $qp \in \overline{(T\backslash\{a\})p} = \overline{(T\backslash\{a\})}p = \overline{(T\backslash\{a\})p} = \overline{Tp\backslash\{ap\}}$ and $ap \notin \overline{Tp\backslash\{ap\}}$.

To see that statement (9) implies statement (8), notice that trivially $ap \neq bp$ whenever $a \neq b$ in T. If Tp is not discrete then for some $a \in T$ one has $ap \in \overline{Tp\backslash\{ap\}} = \overline{(T\backslash\{a\})p}$, a contradiction.

Finally, if S is countable one has that statement (8) implies statement (1).   □

We do not know whether the requirement that S is countable is needed for the equivalence of all nine statements in Theorem 8.11.

Question 8.12. *Do there exist a semigroup S (necessarily uncountable) and a point p of βS such that p is right cancelable in βS but Sp is not strongly discrete?*

Question 8.13. *Do there exist a semigroup S (necessarily uncountable) and a point p of βS such that $ap \neq bp$ whenever a and b are distinct members of S and Sp is discrete, but p is not right cancelable in βS?*

Recall that a point x of a topological space X is a *weak P-point* of X provided that whenever D is a countable subset of $X \backslash \{x\}$, $x \notin c\ell\, D$.

Corollary 8.14. *Assume that S is a countable cancellative semigroup. Then every weak P-point in S^* is right cancelable in βS.*

Proof. Suppose that p is a weak P-point in S^*. We shall show that for every $a \in S$ and every $q \in \beta S \backslash \{a\}$ one has $ap \neq qp$. The required result will then follow from Theorem 8.11.

Suppose instead that we have $a \in S$ and $q \in \beta S \backslash \{a\}$ such that $ap = qp$. By Corollary 8.2, $q \in S^*$. By Lemma 8.1, λ_a is one to one, and hence $\lambda_{a|S^*}$ is a homeomorphism from S^* onto aS^* so that ap is a weak P-point of aS^*.

Let $B = \{s \in S \backslash \{a\} : s^{-1}aS \in p\}$. Since $aS \in ap = qp$, one has $B \in q$ so that $ap = qp \in c\ell(Bp)$. But now, given $s \in B$ one has $sp \in \overline{aS} = a\beta S$ so $Bp \subseteq a\beta S$. By Corollary 4.33 $Bp \subseteq S^*$ so $Bp \subseteq aS^*$. Since $ap \notin Bp$ we have a contradiction to the fact that ap is a weak P-point of aS^*.   □

In certain cases we shall see that the following necessary condition for right cancelability is also sufficient.

Lemma 8.15. *Let S be a weakly left cancellative semigroup and let $p \in \beta S$. If p is right cancelable in βS, then $p \notin S^*p$.*

Proof. Suppose that $p = xp$ for some $x \in S^*$ and pick $a \in S$. Then $ax \in S^*$ by Theorem 4.31, so $ax \neq a$ while $ap = axp$, a contradiction.   □

We pause to introduce a notion which will be used in our next characterization.

Definition 8.16. Let S be a discrete space. The *Rudin–Keisler order* $\leq_{RK}$ on βS is defined by stating that for any $p, q \in \beta S$, $p \leq_{RK} q$ if and only if there is a function $f : S \to S$ for which $\widetilde{f}(q) = p$.

If $p, q \in \beta S$ we shall write $p \approx_{RK} q$ if $p \leq_{RK} q$ and $q \leq_{RK} p$ and we shall write $p <_{RK} q$ if $p \leq_{RK} q$ and $q \not\leq_{RK} p$.

Theorem 8.17. *Let S be a discrete space. The following statements are equivalent.*

(a) $p \approx_{RK} q$.

(b) $p \leq_{RK} q$ *and whenever* $f : S \to S$ *and* $\widetilde{f}(q) = p$, *there exists some* $Q \in q$ *such that* $f_{|Q}$ *is injective.*

(c) *There exist* $f : S \to S$ *and* $Q \in q$ *such that* $\widetilde{f}(q) = p$ *and* $f_{|Q}$ *is injective.*

(d) *There is a bijection* $g : S \to S$ *such that* $\widetilde{g}(q) = p$.

Proof. (a) implies (b). Let $f : S \to S$ such that $\widetilde{f}(q) = p$. Since $q \leq_{RK} p$, pick $g : S \to S$ such that $\widetilde{g}(p) = q$. Then $\widetilde{g \circ f}(q) = \widetilde{g}(\widetilde{f}(q)) = q$ so by Theorem 3.35, $Q = \{a \in S : g(f(a)) = a\} \in q$.

That (b) implies (c) is trivial.

(c) implies (d). By passing to a subset if necessary we may presume that $|S \backslash Q| = |S \backslash f[Q]|$. Thus we may choose a bijection $g : S \to S$ such that $g_{|Q} = f_{|Q}$.

(d) implies (a). Since $\widetilde{g}(q) = p$, $p \leq_{RK} q$. Since $\widetilde{g^{-1}}(p) = q$, $q \leq_{RK} p$. $\square$

In the case S is a countable group (or $(\mathbb{N}, +)$) we see that we can add to the equivalent conditions of Theorem 8.11.

Theorem 8.18. *Suppose that S is either $(\mathbb{N}, +)$ or a countable group. For every $p \in S^*$ the following statements are equivalent.*

(1) *p is right cancelable in βS.*

(2) *$p \notin S^* p$.*

(3) *There is no idempotent $e \in S^*$ for which $p = ep$.*

(4) *For every $x \in S^*$, $x <_{RK} xp$ and $p <_{RK} xp$.*

(5) *For every $x \in S^*$, $p <_{RK} xp$.*

Proof. That (1) implies (2) is Lemma 8.15. To see that (2) implies (1), it suffices by Theorem 8.11 to show that for each $a \in S$ and each $q \in \beta S \backslash \{a\}$, $ap \neq qp$ so let $a \in S$ and $q \in \beta S \backslash \{a\}$ and suppose that $ap = qp$. In case S is a countable group, we have $p = a^{-1}qp$ and by Corollary 4.33 $a^{-1}q \in S^*$. In case S is $(\mathbb{N}, +)$, the equation becomes $a + p = q + p$ so, in $\beta \mathbb{Z}$, $p = -a + q + p$. By Exercise 4.3.5, $\mathbb{N}^*$ is a left ideal of $\beta \mathbb{Z}$ so $-a + q \in \mathbb{N}^*$.

The equivalence of (2) and (3) follows from the fact that $\{x \in S^* : p = xp\}$ is a compact subsemigroup of S^* if and only if it is non-empty. Thus, by Theorem 2.5, this set is non-empty if and only if it contains an idempotent.

To show that (1) implies (4), suppose that p is right cancelable in βS. By Theorem 8.11, there is a family $\langle U_a \rangle_{a \in S}$ of pairwise disjoint open subsets of βS such that $ap \in U_a$ for every $a \in S$. We put $P_a = \{b \in S : ab \in U_a\}$ and observe that $P_a \in p$ for every $a \in S$. Every element $s \in \bigcup_{a \in S} a P_a$ can be expressed uniquely in the form $s = ab$ with $a \in S$ and $b \in P_a$. We define functions $f, g : S \to S$ by stating that $f(s) = a$

and $g(s) = b$ if s is expressed in this way, defining f and g arbitrarily on $S \backslash \bigcup_{a \in S} aP_a$. Let $x \in S^*$. Since $f(ab) = a$ and $g(ab) = b$ if $a \in S$ and $b \in P_a$, we have

$$\widetilde{f}(xp) = f(x\text{-}\lim_{a \in S} p\text{-}\lim_{b \in P_a} ab) = x\text{-}\lim_{a \in S} p\text{-}\lim_{b \in P_a} f(ab) = x\text{-}\lim_{a \in S} p\text{-}\lim_{b \in P_a} a = x$$

and

$$\widetilde{g}(xp) = g(x\text{-}\lim_{a \in S} p\text{-}\lim_{b \in P_a} ab) = x\text{-}\lim_{a \in S} p\text{-}\lim_{b \in P_a} g(ab) = x\text{-}\lim_{a \in S} p\text{-}\lim_{b \in P_a} b = p.$$

So $x \leq_{RK} xp$ and $p \leq_{RK} xp$.

If $x \approx_{RK} xp$, then by Theorem 8.17 there is a set $B \in xp$ such that $f_{|B}$ is injective. Let $X = \{a \in S : a^{-1}B \in p\}$. Then $X \in x$ so is nonempty. Pick $a \in X$ and, since $p \in S^*$, pick distinct elements b and c of $a^{-1}B \cap P_a$. Then ab and ac are distinct elements of B so $f(ab) \neq f(ac)$. But $ab, ac \in aP_a$ so $f(ab) = a = f(ac)$, a contradiction.

Similarly, if $p \approx_{RK} xp$, then there is a set $C \in xp$ such that $f_{|C}$ is injective. Let $Y = \{a \in S : a^{-1}C \in p\}$. Then $Y \in x$ and $x \in S^*$, so pick distinct a and b in Y. Pick $c \in a^{-1}C \cap b^{-1}C \cap P_a \cap P_b$. Then ac and bc are distinct members of C so $g(ac) \neq g(bc)$. But $ac \in aP_a$ and $bc \in bP_b$, so $g(ac) = c = g(bc)$, a contradiction.

Trivially (4) implies (5) and (5) implies (2). $\square$

The equivalence of statements (1), (4), and (5) of Theorem 8.18 will be established under weaker assumptions in Theorem 11.8.

We saw in Theorem 8.7 that for any semigroup S and any $p \in \beta S$, p is right cancelable in βS if and only if for every $A \subseteq S$ there is some $B \subseteq S$ such that $A = \{x \in S : x^{-1}B \in p\}$. We see now that, in the case S is a countable group, it is sufficient to know this for $A = \{e\}$.

Theorem 8.19. *Suppose that S is either $(\mathbb{N}, +)$ or a countable group. An element $p \in \beta S$ is right cancelable in βS if and only if there exists $B \in p$ such that, for every $a \in S$ distinct from the identity, $a^{-1}B \notin p$.*

Proof. Necessity. In case S is a group with identity e, let $A = \{e\}$ and pick B as guaranteed by Theorem 8.7. In case S is $(\mathbb{N}, +)$, apply 8.7 with $A = \{1\}$, and if C is such that $\{1\} = \{x \in \mathbb{N} : -x + C \in p\}$, let $B = -1 + C$.

Sufficiency. Pick such B and suppose that p is not right cancelable. Then by Theorem 8.18 pick $x \in S^*$ such that $p = xp$. Since $B \in p$, one has $\{a \in S : a^{-1}B \in p\} \in x$ so there is some a distinct from the identity such that $a^{-1}B \in p$, a contradiction. $\square$

In the next theorem, we show that every ultrafilter in $\beta \mathbb{N}$ is right cancelable on a large open subset of $\beta \mathbb{N}$.

Theorem 8.20. *Suppose that S is either $(\mathbb{N}, +)$ or a countable group. Let $p \in \beta S$. There is a dense open subset U of βS such that $U \cap S^*$ is dense in S^* and $xp \neq yp$ whenever x and y are distinct elements of U.*

Proof. For each $a \in S$, let $C_a = \{x \in S^* : ap = xp\}$. Then C_a is a closed subset of S^*. We claim that it is nowhere dense in S^*. To see this, suppose we have some infinite subset A of S such that $\overline{A} \cap S^* \subseteq C_a$. Then by Corollaries 8.2 and 4.33, Ap is an infinite subset of S^* so we can choose an infinite subset T of $A \backslash \{a\}$ for which Tp is discrete in βS. Now by Theorem 8.11, $qp \neq rp$ whenever q and r are distinct members of $\overline{T}$, so for at most one $q \in \overline{T}$ is $qp = ap$ and thus there exists $r \in \overline{T} \cap S^*$ such that $rp \neq ap$ so that $r \notin C_a$, a contradiction.

Now $\bigcup_{a \in S} C_a$ is nowhere dense in S^* by Corollary 3.37. Let $U = \beta S \backslash \overline{\bigcup_{a \in S} C_a}$. Then U is a dense open subset of βS for which $U \cap S^*$ is dense in S^*.

It is immediate that, for every $a \in S$ and every $x \in U$, $ap \neq xp$. Suppose that x and y are distinct elements of U. We can choose disjoint subsets X and Y of S satisfying $X \in x$, $Y \in y$, $\overline{X} \subseteq U$, and $\overline{Y} \subseteq U$. We observe that $xp \in \overline{X}p$ and $yp \in \overline{Y}p$. So the equation $xp = yp$ would imply, by Theorem 3.40, that $ap = y'p$ for some $a \in X$ and some $y' \in \overline{Y}$ or that $x'p = bp$ for some $x' \in \overline{X}$ and some $b \in Y$. Since this is not possible, $xp \neq yp$. $\square$

We now show that the semigroup generated by a right cancelable element of S^* provides, under certain conditions on S, an example of a semigroup whose closure fails dramatically to be a semigroup.

Theorem 8.21. *Let S be a discrete countable cancellative semigroup and let $p \in S^*$ be a right cancelable element of βS. Let $T = \{p^n : n \in \mathbb{N}\}$ and $X = cl_{\beta G} T \backslash T$. Then the elements of X generate a free subsemigroup of βS.*

Proof. We shall use the fact that S^* is an ideal of βS (by Theorem 4.36).

We first note that, if $a \in S$, $n \in \mathbb{N}$, and $y \in X$, then $ap^n \notin (\beta S)y$. This follows from the observation that $y \in cl\{p^r : r > n\} \subseteq S^* p^n$ and that, since p^n is right cancelable in βS, $ap^n \notin S^* p^n$.

We now show that the equation $ax = uy$, where $x, y \in X$, $a \in S$ and $u \in \beta S$, implies that $a = u$ and $x = y$. We may assume that $a \neq u$, because $a = u$ implies that $x = y$ by Lemma 8.1.

We note that $ax \in cl(aT)$ and $uy \in cl((S \backslash \{a\})y)$. It follows from Theorem 3.40 that $ap^n = vy$ for some $n \in \mathbb{N}$ and some $v \in \beta S$, or else $az = by$ for some $z \in X$ and some $b \in S \backslash \{a\}$. The first possibility cannot hold because $ap^n \notin (\beta S)y$, and so we may assume the second.

Now $az \in cl(aT)$ and $by \in cl(bT)$. So another application of Theorem 3.40 shows that $az' = by'$, where $z', y' \in cl\,T$ and either z' or y' is in T. We have seen that $z' \in T$ implies that $y' \in T$ and vice-versa. So we have $ap^m = bp^n$ for some $m, n \in \mathbb{N}$. However, this cannot hold if $m = n$ (by Lemma 6.28). It cannot hold if $m < n$, because this would imply that $a = bp^{n-m} \in S^*$. Similarly, it cannot hold if $m > n$. So $ax = uy$ does imply that $a = u$ and $x = y$.

We can therefore assert that the elements of X are right cancelable in βS, by condition (9) of Theorem 8.11. Furthermore, for any $x, y \in X$, $x \neq y$ implies that $(\beta S)x \cap (\beta S)y = \emptyset$, by Theorem 6.19.

To see that X generates a free subsemigroup of βS, suppose that $m, n \in \mathbb{N}$ and that $x_1 x_2 \ldots x_m = y_1 y_2 \ldots y_n$ where each x_i and each y_j is in X and the sequences $(x_1, x_2, \ldots, x_m)$ and $(y_1, y_2, \ldots, y_n)$ are different. We also suppose that we have chosen an equation of this kind for which $m+n$ is as small as possible. Since $(\beta S)x_m \cap (\beta S)y_n = \emptyset$ if $x_m \neq y_n$, we have that $x_m = y_n$. We claim that $m, n > 1$. To see this, we note that $m = 1$ implies that $n > 1$ and that $ax_1 = ay_1 y_2 \ldots y_n$ for any $a \in S$. Since x_1 is right cancelable in βS, this results in the contradiction that $a \in S^*$. Using the fact that x_m is right cancelable once again, we have $x_1 x_2 \ldots x_{m-1} = y_1 y_2 \ldots y_{n-1}$. This contradicts our assumption that $m + n$ is as small as possible.   □

We shall show in Corollary 8.26 that right cancelable elements can occur in the closure of the smallest ideal of βS.

Theorem 8.22. *Let S be a discrete countably infinite cancellative semigroup and let T be an infinite subsemigroup of S. Then there are elements p in $\overline{E(K(\beta T))}$ for which $Sp \cap S^* S^* = \emptyset$.*

Proof. We recall that S^* is an ideal of βS (by Theorem 4.36). Let $f : S \to \omega$ denote the function guaranteed by Lemma 6.47. Since $f[T] = \omega$, $\widetilde{f}[\beta T] = \beta \omega$.

We claim that

(∗) for all $p \in S^*$ and all $x \in \beta S$, $\widetilde{f}(xp) \in \{-1 + \widetilde{f}(p), \widetilde{f}(p), 1 + \widetilde{f}(p)\}$.

To establish (∗), it suffices to establish the same statement with $x \in S$. So let $x \in S$ and suppose that $\widetilde{f}(xp) \notin \{-1 + \widetilde{f}(p), \widetilde{f}(p), 1 + \widetilde{f}(p)\}$. Pick $B \in \widetilde{f}(xp)$ such that $\widetilde{f}(p) \notin \bigcup_{i=-1}^{1}(i + \overline{B})$. Pick $C \in xp$ such that $\widetilde{f}[\overline{C}] \subseteq \overline{B}$ and pick $D \in p$ such that $\widetilde{f}[\overline{D}] \cap \bigcup_{i=-1}^{1}(i + \overline{B}) = \emptyset$. Pick $s \in D \cap x^{-1}C$ with $f(s) > f(x) + 1$, which one can do by Lemma 6.47(2). Then by Lemma 6.47(3), one has some $i \in \{-1, 0, 1\}$ such that $f(xs) = i + f(s)$. Now $xs \in C$ so $f(xs) \in B$ and thus $f(s) \in -i + B$, contradicting the fact that $s \in D$.

We observe that, for any minimal left ideal L of $\beta \mathbb{N}$, we have $\mathbb{Z} + L \subseteq L$. This follows from the fact that any $x \in L$ has a left identity $e \in L$ (by Theorem 1.64). So, for every $n \in \mathbb{Z}$, $n + x = (n + e) + x \in \beta \mathbb{N} + L \subseteq L$. Thus, for any minimal left ideal L of $\mathbb{N}^*$, $f^{-1}[L]$ is a left ideal of S^*.

We can choose a sequence $\langle L_n \rangle_{n=1}^{\infty}$ of disjoint minimal left ideals of $\mathbb{N}^*$ (by Theorem 6.9). For each $n \in \mathbb{N}$, we can choose an idempotent $e_n \in \widetilde{f}^{-1}[L_n] \cap K(\beta T)$ (by Corollary 2.6), because $f^{-1}[L_n] \cap \beta T$ is a left ideal of βT. (We know that $f^{-1}[L_n] \cap \beta T \neq \emptyset$ because $\widetilde{f}[\beta T] = \beta \omega$.) We may suppose that the set $\{\widetilde{f}(e_n) : n \in \mathbb{N}\}$ is discrete, because this could be ensured by replacing $\langle e_n \rangle_{n=1}^{\infty}$ by a subsequence. Let $E = \{e_n : n \in \mathbb{N}\}$. We claim that $\widetilde{f}$ is one-to-one on $c\ell E$. To see this, let p and q be distinct members of $c\ell E$ and suppose that $\widetilde{f}(p) = \widetilde{f}(q)$. Pick disjoint $A \in p$ and $B \in q$. Then $\widetilde{f}(p) \in \widetilde{f}[\overline{A} \cap c\ell E] = \widetilde{f}[c\ell(\overline{A} \cap E)] = c\ell \, \widetilde{f}[\overline{A} \cap E]$ and $\widetilde{f}(q) \in c\ell \, \widetilde{f}[\overline{B} \cap E]$ so by Theorem 3.40 we may assume without loss of generality that for some $e_m \in \overline{A}$, $\widetilde{f}(e_m) \in c\ell \, \widetilde{f}[\overline{B} \cap E]$, contradicting the fact that $\{\widetilde{f}(e_n) : n \in \mathbb{N}\}$ is discrete.

Let p be any point of accumulation of E. To show that $Sp \cap S^* S^* = \emptyset$, suppose on the contrary that $ap = xy$ for some $a \in S$ and some $x, y \in S^*$.

Let $C = \{z \in c\ell E : az \in (\beta S)y\}$. Then by $(*)$ $z \in C$ implies that $i + \widetilde{f}(z) = j + \widetilde{f}(y)$ for some $i, j \in \{-1, 0, 1\}$. So C is finite, because $\widetilde{f}$ is injective on $c\ell E$. We can choose a clopen neighborhood U of p such that $U \cap (C\backslash\{p\}) = \emptyset$. If $M = \{n \in \mathbb{N} : e_n \in U$ and $e_n \neq p\}$, then $p \in c\ell\{e_n : n \in M\}$.

There is at most one element $b \in S$ for which $ap = by$ (by Lemma 6.28). Let $S' = S\backslash\{b\}$ if such an element exists and let $S' = S$ otherwise.

Now $ap \in c\ell\{ae_n : n \in M\}$ and $xy \in c\ell(S'y)$. It follows from Theorem 3.40 that $ae_n \in (\beta S)y$ for some $n \in M$, or $az = cy$ for some $z \in c\ell\{e_n : n \in M\}$ and some $c \in S'$. The first possibility is ruled out since if $n \in M$, then $e_n \notin C$. So we assume the second. Now the equation $az = cy$ implies that $z \in C$. Since $z \in c\ell\{e_n : n \in M\} \subseteq U$, one has $z = p$. However, the equation $ap = cy$ is ruled out by our choice of S'. $\square$

Corollary 8.23. *Let S be a discrete countably infinite cancellative semigroup and let T be an infinite subsemigroup of S. There are elements p in $\overline{E(K(\beta T))}$ which are not in $S^* S^*$.*

Proof. Pick $a \in S$. If $p \in S^* S^*$, then $ap \in Sp \cap S^* S^*$ by Theorem 4.31. $\square$

Corollary 8.24. *If S is a discrete infinite right cancellative and weakly left cancellative semigroup, then the set of idempotents in S^* is not closed.*

Proof. By Corollary 8.23, the set of idempotents in $\beta\mathbb{N}$ is not closed and therefore the set of idempotents in $\mathbb{H}$ is not closed. Now S^* is a compact semigroup (by Theorem 4.31) and therefore contains an idempotent. It follows from Theorem 6.32 that S^* contains a copy of $\mathbb{H}$. $\square$

Corollary 8.25. *Let S be a discrete countably infinite cancellative semigroup. Then $K(\beta S)$ is not closed in βS.*

Proof. This follows from Corollary 8.23 (with $S = T$) and the observation that $K(\beta S) \subseteq S^* S^*$, because $S^* S^*$ is an ideal of βS (by Theorem 4.36). $\square$

Corollary 8.26. *Let S be a discrete countably infinite cancellative semigroup and let T be an infinite subsemigroup of S. There are right cancelable elements of βS in $\overline{E(K(\beta T))}$.*

Proof. This is immediate from Theorem 8.22, Corollary 8.2, and condition (9) of Theorem 8.11. $\square$

Exercise 8.2.1. Prove Theorem 8.8 by modifying the proof of Theorem 8.7.

Exercise 8.2.2. Let S be any discrete semigroup and let $x, p \in \beta S$. Show that:
  (i) If p is right cancelable, $xp \in K(\beta S)$ if and only if $x \in K(\beta S)$.
  (ii) If p is left cancelable, $px \in K(\beta S)$ if and only if $x \in K(\beta S)$. (Recall that every element in $K(\beta S)$ belongs to a group contained in $K(\beta S)$ by Theorem 1.64).

Exercise 8.2.3. Prove that the topological center of $\mathbb{N}^*$ is empty. (Hint: Suppose that x is in the topological center of $\mathbb{N}^*$. Show that x is in the center of $\beta\mathbb{N}$ by choosing an element $y \in \mathbb{N}^*$ which is right cancelable in $\beta\mathbb{N}$. For every $z \in \beta\mathbb{N}$ we have $x + z + y = \lambda_x \big(\lim_{n \to z} (n + y) \big) \lim_{n \to z} (x + n + y)$ and $z + x + y = \lim_{n \to z} (n + x + y)$, where n denotes a positive integer.)

8.3 Right Cancellation in $\beta\mathbb{N}$ and $\beta\mathbb{Z}$

In the case of the semigroups $(\mathbb{N}, +)$ and $(\mathbb{Z}, +)$, there are additional characterizations of right cancelability.

Theorem 8.27. *Let $p \in \beta\mathbb{N}$. Then p is right cancelable in $\beta\mathbb{N}$ if and only if there is an increasing sequence $\langle x_n \rangle_{n=1}^{\infty}$ in $\mathbb{N}$ such that for every $k \in \mathbb{N}$,*

$$\{x_n : x_n + k < x_{n+1}\} \in p.$$

Proof. Necessity. Choose by Theorem 8.19 some $B \in p$ such that $-k + B \notin p$ for each $k \in \mathbb{N}$ and let $\langle x_n \rangle_{n=1}^{\infty}$ enumerate B in increasing order. Then given $k \in \mathbb{N}$, $B \setminus \bigcup_{i=1}^{k} (-i + B) \in p$ and $B \setminus \bigcup_{i=1}^{k} (-i + B) \subseteq \{x_n : x_n + k < x_{n+1}\}$.

  Sufficiency. Let $B = \{x_n : n \in \mathbb{N}\}$. Then given $k \in \mathbb{N}$,

$$(-k + B) \cap \{x_n : x_n + k < x_{n+1}\} = \emptyset$$

so $-k + B \notin p$. $\qquad\square$

The proof of the following theorem is nearly identical, and we leave it as an exercise.

Theorem 8.28. *Let $p \in \beta\mathbb{N}$. Then p is right cancelable in $\beta\mathbb{Z}$ if and only if there is an increasing sequence $\langle x_n \rangle_{n=1}^{\infty}$ in $\mathbb{N}$ such that for every $k \in \mathbb{N}$, $\{x_n : x_{n-1} + k < x_n < x_{n+1} - k\} \in p$.*

Proof. This is Exercise 8.3.1 $\qquad\square$

In the next example, we show that the condition of Theorem 8.28 is really stronger than the condition of Theorem 8.27.

Example 8.29. *There is an element $p \in \beta\mathbb{N}$ which is right cancelable in $\beta\mathbb{N}$ but not in $\beta\mathbb{Z}$.*

Proof. We choose an idempotent $q \in \beta\mathbb{N}$ for which $\mathrm{FS}(\langle 3^t \rangle_{t=k}^{\infty}) \in q$ for each $k \in \mathbb{N}$. (This is possible by Lemma 5.11.) We put $p = -q + q$ and shall show that p is right cancelable in $\beta\mathbb{N}$ but p is not right cancelable in $\beta\mathbb{Z}$. (We recall that $-q \in \beta\mathbb{Z}$ is the ultrafilter generated by $\{-A : A \in q\}$). By Exercise 4.3.5, $\mathbb{N}^*$ is a left ideal of $\beta\mathbb{Z}$ so $p \in \mathbb{N}^*$.

It is easy to check that $-q \in E(\beta\mathbb{Z})$ and thus $-q + p = p$ so that $p \in \mathbb{Z}^* + p$ and hence by Theorem 8.18, p is not right cancelable in $\beta\mathbb{Z}$.

Suppose that p is not right cancelable in $\beta\mathbb{N}$. Then $p = z + p$ for some $z \in \mathbb{N}^*$, by Theorem 8.18. That is $-q + q = z + -q + q$. Let

$$A = \{ \ \Sigma_{t \in F} 3^t - \Sigma_{t \in H} 3^t + k : $$
$$k \in \mathbb{N}, \ F, H \in \mathcal{P}_f(\mathbb{N}), \ \min F > \max H + 1, \text{ and } k < 3^{\min H - 1} \}$$

and let

$$B = \{ \Sigma_{t \in F} 3^t - \Sigma_{t \in H} 3^t : F, H \in \mathcal{P}_f(\mathbb{N}) \text{ and } \min F > \max H + 1 \}.$$

Using Theorem 4.15, one easily shows that $A \in z + -q + q$ and $B \in -q + q$.

Now, elements of B are easy to recognize by their ternary representations. That is, if $F, H \in \mathcal{P}_f(\mathbb{N})$, $j = \min F > \max H + 1$, $i = \min H$, and $\max F = \ell$, then $\Sigma_{t \in F} 3^t - \Sigma_{t \in H} 3^t = \Sigma_{t=i}^{\ell} a_t 3^t$, where $a_i = a_{j-1} = 2$, $a_t \in \{1, 2\}$ for all $t \in \{i, i+1, \ldots, j-1\}$, and $a_t \in \{0, 1\}$ if $t \geq j$. That is each such an element has a highest order 2 and a lowest order 2, which is its lowest nonzero ternary digit, and no 0's between these positions. No element of A can fit that description, so $A \cap B = \emptyset$, a contradiction. $\qquad\square$

Corollary 8.26 yields some surprising information about $\mathbb{H}$. We know from Example 2.16 that in a compact right topological semigroup S, $c\ell K(S)$ need not be a left ideal. On the other hand, if S is a discrete semigroup, we know from Theorems 2.15 and 2.17 that $c\ell K(\beta S)$ is always a two sided ideal of βS.

Theorem 8.30. *The closure of $K(\mathbb{H})$ is not a left ideal of $\mathbb{H}$.*

Proof. By Corollary 8.26 (with $T = \mathbb{N}$ and $S = \mathbb{Z}$), there is an element $x \in \overline{E(K(\beta\mathbb{N}))}$ which is right cancelable in $\beta\mathbb{Z}$. By Theorem 8.27, there is an increasing sequence $\langle b_n \rangle_{n=1}^{\infty}$ in $\mathbb{N}$ such that for each $k \in \mathbb{N}$, $\{b_n : b_{n+1} > b_n + k\} \in x$.

Now $K(\mathbb{H}) = \mathbb{H} \cap K(\beta\mathbb{N})$ by Theorem 1.65. So, by Lemma 6.8, $E(K(\mathbb{H})) = E(K(\beta\mathbb{N}))$ and thus $x \in c\ell K(\mathbb{H})$. We shall show that there is some $w \in \mathbb{H}$ such that $w + x \notin c\ell K(\mathbb{H})$.

Recall that for $a \in \mathbb{N}$, $\text{supp}(a)$ denotes the binary support of a. For $a, b \in \mathbb{N}$, write $a << b$ if and only if $\max \text{supp}(a) < \min \text{supp}(b)$. For $a \in \mathbb{N}$, let $f(a) = \max \text{supp}(a)$. Let

$$X = \{a + b_n : a \in \mathbb{N}, \ b_{n+1} > b_n + 2^{f(a)+1}, \text{ and } a << b_n\}.$$

We claim that $\beta\mathbb{N} + x \subseteq \overline{X}$ for which it suffices to show that $\mathbb{N} + x \subseteq \overline{X}$. So let $a \in \mathbb{N}$. Then $\{b_n : b_{n+1} > b_n + 2^{f(a)+1}\} \in x$ and, since $x \in \mathbb{H}$, $\{w \in \mathbb{N} : a << w\} \in x$. Since $\{b_n : b_{n+1} > b_n + 2^{f(a)+1}\} \cap \{w \in \mathbb{N} : a << w\} \subseteq -a + X$, $a + x \in \overline{X}$ as required.

Now each $y \in X$ has a unique representation in the form $a + b_n$ with $b_{n+1} > b_n + 2^{f(a)+1}$ so we may define $g : X \to \mathbb{N}$ by $g(a + b_n) = a$ and extend g arbitrarily to $\mathbb{N}$. Let $B = \{b_n : n \in \mathbb{N}\}$. Then given any $y \in \beta\mathbb{N}$, one has

$$\widetilde{g}(y + x) = \widetilde{g}\big(y\text{-}\lim_{a \in \mathbb{N}} x\text{-}\lim_{b \in B}(a + b)\big) = y\text{-}\lim_{a \in \mathbb{N}} x\text{-}\lim_{b \in B} g(a + b) = y\text{-}\lim_{a \in \mathbb{N}} a = y.$$

We now show that

(∗) if $z \in \mathbb{H} \cap K(\beta\mathbb{N}) \cap \overline{X}$, then $\widetilde{g}(z) \in K(\beta\mathbb{N})$.

To see this, let $z \in \mathbb{H} \cap K(\beta\mathbb{N}) \cap \overline{X}$ and pick an idempotent $e \in K(\beta\mathbb{N})$ such that $z = e + z$. Now if $u, a \in \mathbb{N}$ and $u \ll a$, then $f(u + a) = f(a)$ so if $b_{n+1} > b_n + 2^{f(a)+1}$ and $a \ll b_n$, then $u + a + b_n \in X$ and $a + b_n \in X$ so $g(u + a + b_n) = u + a = u + g(a + b_n)$.

We now show that $\widetilde{g}(z) = e + \widetilde{g}(z)$ so that $\widetilde{g}(z) \in K(\beta\mathbb{N})$ as required. Suppose instead that $\widetilde{g}(z) \neq e + \widetilde{g}(z)$ and pick disjoint open neighborhoods U of $\widetilde{g}(z)$ and V of $e + \widetilde{g}(z)$. Pick an open neighborhood R of z such that $\widetilde{g}[R] \subseteq U$. Then $e + z \in R$ so pick $C_1 \in e$ such that $\overline{C_1} + z \subseteq R$ and pick $C_2 \in e$ such that $\overline{C_2} + \widetilde{g}(z) \subseteq V$. Pick $u \in C_1 \cap C_2$ and let $m = f(u) + 1$. Pick $A_1 \in z$ such that $u + \overline{A_1} \subseteq R$ and pick a neighborhood W of $\widetilde{g}(z)$ such that $u + W \subseteq V$. Pick $A_2 \in z$ such that $\widetilde{g}[\overline{A_2}] \subseteq W$. Then $A_1 \cap A_2 \cap \mathbb{N}2^m \cap X \in z$ so pick $y \in A_1 \cap A_2 \cap \mathbb{N}2^m \cap X$. Since $y \in X$, pick a and n such that $y = a + b_n$, $b_{n+1} > b_n + 2^{f(a)+1}$, and $a \ll b_n$. Since $a \ll b_n$, $m \leq \min \operatorname{supp}(y) = \min \operatorname{supp}(a + b_n) = \min \operatorname{supp}(a)$ so $u \ll a$. Now $y \in A_1$ so $u + a + b_n \in R$ so $g(u + a + b_n) \in U$. Also, $y \in A_2$ so $g(a + b_n) \in W$ so $u + g(a + b_n) \in V$. This is a contradiction since $g(a + b_n) = u + g(a + b_n)$ so (∗) is established.

Now pick any $w \in \mathbb{H} \setminus c\ell\, K(\beta\mathbb{N})$ (such as $w \in \{2^n : n \in \mathbb{N}\}^*$). We claim that $w + x \notin c\ell\, K(\mathbb{H})$ so suppose instead that $w + x \in c\ell\, K(\mathbb{H}) = c\ell\big(\mathbb{H} \cap K(\beta\mathbb{N})\big)$. Then since $\beta\mathbb{N} + x \subseteq \overline{X}$ we have $w + x \in c\ell\big(\mathbb{H} \cap K(\beta\mathbb{N})\big) \cap \overline{X}$. Since $\overline{X}$ is clopen, $c\ell\big(\mathbb{H} \cap K(\beta\mathbb{N})\big) \cap \overline{X} = c\ell\big(\mathbb{H} \cap K(\beta\mathbb{N}) \cap \overline{X}\big)$. Thus

$$w = \widetilde{g}(w + x) \in \widetilde{g}[c\ell(\mathbb{H} \cap K(\beta\mathbb{N}) \cap \overline{X})] \subseteq c\ell\, K(\beta\mathbb{N}),$$

by (∗). This contradiction completes the proof.   □

Corollary 8.31. $c\ell\, K(\mathbb{H}) \nsubseteq \mathbb{H} \cap c\ell\, K(\beta\mathbb{N})$.

Proof. We know that $\mathbb{H} \cap c\ell\, K(\beta\mathbb{N})$ is an ideal of $\mathbb{H}$ by Theorems 2.15 and 2.17.   □

We do not know whether it is possible for the sum of two elements in $\beta\mathbb{N} \setminus \big(K(\beta\mathbb{N})\big)$ to be in $K(\beta\mathbb{N})$; nor do we know the corresponding possibility for the closure of $K(\beta\mathbb{N})$. However, something can be said in answer to these questions if one of the elements is right cancelable. The answer for the case of $K(\beta\mathbb{N})$ was in fact given in Exercise 8.2.2.

Theorem 8.32. *Let p be a right cancelable element in $\beta\mathbb{N} \setminus \overline{K(\beta\mathbb{N})}$. For every $q \in \beta\mathbb{N}$, $q + p \in \overline{K(\beta\mathbb{N})}$ if and only if $q \in \overline{K(\beta\mathbb{N})}$.*

Proof. It follows from Theorem 2.15 that $q + p \in \overline{K(\beta\mathbb{N})}$ if $q \in \overline{K(\beta\mathbb{N})}$. So we shall suppose that $q + p \in \overline{K(\beta\mathbb{N})}$.

By Theorem 8.27, there is a set $P \in p$ which can be arranged as an increasing sequence $\langle x_n \rangle_{n=1}^{\infty}$ with the property that, for every $k \in \mathbb{N}$, $\{x_n : x_n + k < x_{n+1}\} \in p$. It is convenient to suppose that $x_1 = 1$. Since $p \notin \overline{K(\beta\mathbb{N})}$, we may suppose that $\overline{P} \cap K(\beta\mathbb{N}) = \emptyset$.

For each $a \in \mathbb{N}$, let $C_a = \{x_n : 2a + x_n < x_{n+1}\}$ and let $B = \bigcup_{a \in \mathbb{N}}(a + C_a)$. By Theorem 4.15, $B \in q + p$. We can define functions $f, g : \mathbb{N} \to \omega$ by stating that $f(r) = x_n$ if and only if $x_n \leq r < x_{n+1}$ and $g(r) = r - f(r)$. We observe that, if $b \in B$ is expressed as $b = a + x_n$, where $a \in \mathbb{N}$ and $2a + x_n < x_{n+1}$, then $f(b) = x_n$ and $g(b) = a$.

This implies that

$$\widetilde{g}(q + p) = q\text{-}\lim_{a \in \mathbb{N}} p\text{-}\lim_{x_n \in C_a} g(a + x_n) = q\text{-}\lim_{a \in \mathbb{N}} a = q.$$

We shall show that, for any $z \in \overline{B} \cap K(\beta\mathbb{N})$, $\widetilde{g}(z) \in K(\beta\mathbb{N})$. We first show that $\widetilde{g}(z) \in \mathbb{N}^*$. To see this, suppose that $\widetilde{g}(z) = m \in \mathbb{N}$. For every $b \in B \cap g^{-1}[\{m\}]$, we have $b = g(b) + f(b) = m + f(b)$. Since $B \cap g^{-1}[\{m\}] \in z$, it follows that $z = m + \widetilde{f}(z)$. So $m + \widetilde{f}(z) \in K(\beta\mathbb{N})$ and hence $\widetilde{f}(z) = -m + (m + \widetilde{f}(z)) \in K(\beta\mathbb{N})$. Now $\widetilde{f}(z) \in \overline{P}$, and so this contradicts our assumption that $K(\beta\mathbb{N}) \cap \overline{P} = \emptyset$.

Since $z \in K(\beta\mathbb{N})$, there is an idempotent $e \in K(\beta\mathbb{N})$ for which $z = e + z$. We claim that for each $a \in \mathbb{N}$, $D_a = \{b \in \mathbb{N} : g(a + b) = a + g(b)\} \in z$. To see this, let $a \in \mathbb{N}$. Since $\widetilde{g}(z) \in \mathbb{N}^*$ and $B \in z$ one has $\{b \in B : g(b) > a\} \in z$. We claim that $\{b \in B : g(b) > a\} \subseteq D_a$. To this end, let $b \in B$ with $g(b) > a$ and let $f(b) = x_n$. Then $b = g(b) + x_n$ and since $b \in B$, $2g(b) + x_n < x_{n+1}$ and thus $x_n < a + b < x_{n+1}$ so $g(a + b) = a + b - x_n = a + g(b)$. Thus

$$\widetilde{g}(z) = \widetilde{g}(e + z)$$
$$= e\text{-}\lim_{a \in \mathbb{N}} z\text{-}\lim_{b \in D_a} g(a + b)$$
$$= e\text{-}\lim_{a \in \mathbb{N}} z\text{-}\lim_{b \in D_a} (a + g(b))$$
$$= e\text{-}\lim_{a \in \mathbb{N}} z\text{-}\lim_{b \in D_a} (\lambda_a \circ g)(b)$$
$$= e\text{-}\lim_{a \in \mathbb{N}} (a + \widetilde{g}(z))$$
$$= e + \widetilde{g}(z).$$

This shows that $\widetilde{g}(z) \in K(\beta\mathbb{N})$, as claimed.

Since $q + p \in \overline{K(\beta\mathbb{N}) \cap \overline{B}}$, it follows that

$$q = \widetilde{g}(q + p) \in \widetilde{g}[\overline{K(\beta\mathbb{N}) \cap \overline{B}}] \subseteq c\ell\,\widetilde{g}[K(\beta\mathbb{N}) \cap \overline{B}] \subseteq \overline{K(\beta\mathbb{N})}.$$

$\square$

Exercise 8.3.1. Prove Theorem 8.28.

Exercise 8.3.2. Show that there is an element of $\beta\mathbb{N}$ which is right cancelable, but not left cancelable, in $\beta\mathbb{N}$. (Hint: Show that the element p produced in Example 8.29 is not left cancelable.)

Exercise 8.3.3. Show that every left cancelable element of $\beta\mathbb{N}$ is also left cancelable in $\beta\mathbb{Z}$. (Hint: Use the fact that there are elements of $\beta\mathbb{N}$ that are right cancelable in $\beta\mathbb{Z}$.)

8.4 Left Cancelable Elements in βS

Left cancelable elements in βS are harder to characterize than right cancelable elements, because the maps λ_p are discontinuous and are therefore harder to work with than the maps ρ_p. However, there are many semigroups S for which βS does contain a rich set of left cancelable elements. These include the important case in which $S = \mathbb{N}$.

We shall derive several results for countable semigroups which can be algebraically embedded in the topological center of a compact cancellative right topological semigroup, in particular for those which can be algebraically embedded in compact topological groups. We recall that this is a property of all discrete countable semigroups which are commutative and cancellative, and of many non-commutative semigroups as well, including the free group on any countable set of generators. (See the proofs of Corollaries 7.30 and 7.31.)

Lemma 8.33. *Suppose that S is a discrete semigroup, that C is a compact cancellative right topological semigroup and that there is an injective homomorphism $h : S \to \Lambda(C)$. If D is a countable subset of S for which $h[D]$ is discrete, then every element in $c\ell_{\beta S} D$ is left cancelable in βS.*

Proof. We note that our assumptions imply that S is cancellative. We also note that $\widetilde{h} : \beta S \to C$ is a homomorphism (by Corollary 4.22).

Suppose that $p \in c\ell_{\beta S} D$ and that $px = py$, where x and y are distinct elements of βS. Then $\widetilde{h}(p)\widetilde{h}(x) = \widetilde{h}(p)\widetilde{h}(y)$ so $\widetilde{h}(x) = \widetilde{h}(y)$.

Now $px \in \overline{Dx}$ and $py \in \overline{Dy}$. By Theorem 3.40, we may suppose without loss of generality that $ax = qy$ for some $a \in D$ and some $q \in \overline{D}$. Now $q \neq a$, by Lemma 8.1, and so $D\backslash\{a\} \in q$. Now

$$\widetilde{h}(q) \in \widetilde{h}[\overline{D\backslash\{a\}}] = c\ell(\widetilde{h}[D\backslash\{a\}]) = c\ell(h[D]\backslash\{h(a)\})$$

and consequently $h(a)\widetilde{h}(x) = \widetilde{h}(q)\widetilde{h}(y) \in c\ell(h[D]\backslash\{h(a)\})\widetilde{h}(y)$ and so by right cancellation $h(a) \in \overline{h[D]\backslash\{h(a)\}}$ contradicting our assumption that $h[D]$ is discrete. $\square$

Theorem 8.34. *Let S be a countable discrete semigroup which can be algebraically embedded in the topological center of a compact cancellative right topological semigroup. Then the set of cancelable elements of βS contains a dense open subset of S^*.*

Proof. Suppose that C is a compact cancellative right topological semigroup and that $h : S \to \Lambda(C)$ is an injective homomorphism. Every infinite subset T of S contains an infinite subset D for which $h[D]$ is discrete. By Lemma 8.33, every element of $c\ell_{\beta S} D$ is left cancelable in βS. So the set of left cancelable elements of βS contains a dense open subset of S^*. The same statement holds for the set of right cancelable elements of βS, by Theorem 8.10. $\square$

Theorem 8.35. *Suppose that S is a countable discrete semigroup, that C is a compact cancellative right topological semigroup and that there is an injective homomorphism*

$h : S \to \Lambda(C)$. *Let D be a countable subset of S with the property that $\tilde{h}$ is constant on $\overline{D} \cap S^*$. Then every element of $\overline{D}$ is cancelable in βS.*

Proof. We note that $\tilde{h} : \beta S \to C$ is a homomorphism (by Corollary 4.22) and that S is cancellative.

Every element of D is cancelable in βS, by Lemma 8.1. So it is sufficient to prove that every element of $\overline{D} \cap S^*$ is cancelable in βS.

Suppose that $p \in \overline{D} \cap S^*$ and that $px = py$, where x and y are distinct elements of βS. This implies that $\tilde{h}(x) = \tilde{h}(y)$. There is at most one element $s \in D$ for which $h(s) = \tilde{h}(p)$. Let B denote D with this element deleted, if it exists. Now $px \in \overline{Bx}$ and $py \in \overline{By}$. By Theorem 3.40, we may suppose that $ax = qy$ for some $a \in B$ and some $q \in \overline{B}$. Now $a \neq q$, by Lemma 8.1. Furthermore, $\tilde{h}(ax) = \tilde{h}(qy)$ and so $h(a)\tilde{h}(x) = \tilde{h}(q)\tilde{h}(y)$ so by right cancellation $h(a) = \tilde{h}(q)$. This implies that $q \notin D$ and hence that $\tilde{h}(q) = \tilde{h}(p)$, contradicting our assumption that $h(a) \neq \tilde{h}(p)$. So p is left cancelable in βS.

We now show that, if $a \in S$, $p \in \overline{D} \cap S^*$, and $z \in \beta S \backslash \{a\}$, then $ap \neq zp$. It will then follow from Theorem 8.11 that p is also right cancelable in βS. By Corollary 8.2, $ap \neq zp$ for $z \in S \backslash \{a\}$.

Suppose then that $ap = zp$ for some $z \in S^*$. Since $ap \in \overline{aD}$ and $zp \in \overline{(S \backslash \{a\})p}$, it follows from Theorem 3.40 that $ac = rp$ for some $c \in D$ and some $r \in \overline{S \backslash \{a\}}$ or $aq = bp$ for some $q \in \overline{D} \cap S^*$ and some $b \in S \backslash \{a\}$. But the first possibility cannot occur because S^* is a left ideal of βS by Corollary 4.33 so the second possibility must hold. Then $\tilde{h}(q) = \tilde{h}(p)$. So $aq = bp$ implies that $h(a) = h(b)$ and thus that $a = b$, a contradiction. $\square$

Given $n \in \mathbb{N}$, let $h_n : \mathbb{Z} \to \mathbb{Z}_n$ denote the canonical projection so that $\tilde{h}_n : \beta \mathbb{Z} \to \mathbb{Z}_n$ denotes its continuous extension.

Corollary 8.36. *Suppose that $p \in \mathbb{Z}^*$ and that there is a set $A \in p$ and an infinite subset M of $\mathbb{N}$ such that $\tilde{h}_n(x) = \tilde{h}_n(p)$ for every $x \in \overline{A} \cap \mathbb{Z}^*$ and every $n \in M$. Then p is cancelable in $(\beta \mathbb{Z}, +)$.*

Proof. Let C denote the compact topological group $\times_{n \in M} \mathbb{Z}_n$. We can define an embedding $h : \mathbb{Z} \to C$ by stating that $\pi_n(h(r)) = h_n(r)$ for every $n \in M$. It then follows from Theorem 8.35 that p is cancelable in $\beta \mathbb{Z}$. $\square$

Recall that a point x of a topological space X is a *P-point* if and only if whenever $\langle U_n \rangle_{n=1}^{\infty}$ is a sequence of neighborhoods of x one has that $\bigcap_{n=1}^{\infty} U_n$ is a neighborhood of x.

Corollary 8.37. *Let p be a P-point of $\mathbb{Z}^*$. Then there is a set $A \in p$ such that every $q \in \overline{A} \cap \mathbb{Z}^*$ is cancelable in $\beta \mathbb{Z}$.*

Proof. For each $n \in \mathbb{N}$ let $B_n = \{m \in \mathbb{N} : h_n(m) = \tilde{h}_n(p)\}$. Then each $B_n \in p$ so $\bigcap_{n=1}^{\infty} \overline{B_n}$ is a neighborhood of p so pick $A \in p$ such that $\overline{A} \cap \mathbb{Z}^* \subseteq \bigcap_{n=1}^{\infty} \overline{B_n}$. Then Corollary 8.36 applies to A and any $q \in \overline{A} \cap \mathbb{Z}^*$. $\square$

Corollary 8.38. *Let $\langle x_n \rangle_{n=1}^{\infty}$ be an infinite sequence in $\mathbb{N}$ with the property that x_{n+1} is a multiple of x_n for every $n \in \mathbb{N}$. Then every ultrafilter in $\mathbb{N}^* \cap \overline{\{x_n : n \in \mathbb{N}\}}$ is cancelable in $(\beta\mathbb{Z}, +)$.*

Proof. This is Exercise 8.4.1.   □

By Ramsey's Theorem (Theorem 5.6), every infinite sequence in $\mathbb{N}$ contains an infinite subsequence which satisfies the hypothesis of Corollary 8.38 or the hypothesis of the following theorem. So these two results taken together provide a rich set of left cancelable elements of $\beta\mathbb{N}$.

Theorem 8.39. *Let $\langle x_n \rangle_{n=1}^{\infty}$ be an infinite sequence in $\mathbb{N}$ with the property that, for every $m \neq n$ in $\mathbb{N}$, x_n is not a multiple of x_m. Then every ultrafilter in $\overline{\{x_n : n \in \mathbb{N}\}}$ is left cancelable in $(\beta\mathbb{N}, +)$.*

Proof. Let $P = \{x_n : n \in \mathbb{N}\}$, let $p \in \overline{P}$, and suppose that u and v are distinct elements of $\beta\mathbb{N}$ for which $p + u = p + v$.

We now observe that we may suppose that $u, v \in \bigcap_{n \in \mathbb{N}} \overline{n\mathbb{N}}$. To see this, note that there is a cancelable element r of $\beta\mathbb{N}$ for which $u + r \in \bigcap_{n \in \mathbb{N}} \overline{n\mathbb{N}}$ by Exercise 8.4.3. The equation $p + u = p + v$ implies that $\tilde{h}_n(u) = \tilde{h}_n(v)$ for every $n \in \mathbb{N}$, and hence we also have $v + r \in \bigcap_{n \in \mathbb{N}} \overline{n\mathbb{N}}$. Since $u + r \neq v + r$, we can replace u and v by $u + r$ and $v + r$ respectively.

We define functions $f : \mathbb{N} \to \mathbb{N}$ and $g : \mathbb{N} \to \omega$ as follows: If some x_k divides n, then $f(n) = x_m$ where m is the first index such that x_m divides n. If no x_k divides n, then $f(n) = 1$. Let $g(n) = n - f(n)$.

If $s \in \bigcap_{m=1}^{n} x_m \mathbb{N}$, then $f(x_n + s) = x_n$ and $g(x_n + s) = s$. For each $n \in \mathbb{N}$, let $B_n = \bigcap_{m=1}^{n} x_m \mathbb{N}$. Then each $B_n \in u$ so

$$\tilde{g}(p + u) = p\text{-}\lim_{x_n \in P}\ u\text{-}\lim_{s \in B_n} g(x_n + s) = p\text{-}\lim_{x_n \in P}\ u\text{-}\lim_{s \in B_n} s = u.$$

Similarly, $\tilde{g}(p + v) = v$. This contradicts our assumption that $p + u = p + v$.   □

Example 8.40. *There is an element of $\beta\mathbb{N}$ which is left cancelable, but not right cancelable, in $\beta\mathbb{N}$.*

Proof. As in Example 8.29, we choose q to be an idempotent in $\beta\mathbb{N}$ for which $FS(\langle 3^n \rangle_{n=k}^{\infty}) \in q$ for every $k \in \mathbb{N}$. We choose $y \in \mathbb{N}^*$ with $\{3^n : n \in \mathbb{N}\} \in y$, and we put $x = q + -q + y$. We shall show that x is left cancelable in $\beta\mathbb{N}$.

Suppose that u and v are distinct elements of $\beta\mathbb{N}$ for which $x + u = x + v$. We may suppose that $u, v \in \bigcap_{n \in \mathbb{N}} \overline{n\mathbb{N}}$. To see this, we observe that by Exercise 8.4.3, there is a cancelable element r of $\beta\mathbb{N}$ such that $u + r \in \bigcap_{n \in \mathbb{N}} \overline{n\mathbb{N}}$. Since r is cancelable, $u + r \neq v + r$. Since $x + u = x + v$ implies that $\tilde{h}_n(u)\tilde{h}_n(v)$ for every $n \in \mathbb{N}$, we also have $v + r \in \bigcap_{n \in \mathbb{N}} \overline{n\mathbb{N}}$. So we can replace u and v by $u + r$ and $v + r$ respectively.

Pick U and V disjoint members of u and v respectively. Let

$$A = \{n + 3^\ell - \Sigma_{t \in H} 3^t + \Sigma_{t \in F} 3^t : \ell \in \mathbb{N},\ n \in \mathbb{N}3^{\ell+1} \cap U,$$
$$H, F \in \mathcal{P}_f(\mathbb{N}),\ \ell > \max H, \text{ and } \min H > \max F\}$$

and

$$B = \{n + 3^\ell - \Sigma_{t \in H} 3^t + \Sigma_{t \in F} 3^t : \ell \in \mathbb{N},\ n \in \mathbb{N}3^{\ell+1} \cap V,$$
$$H, F \in \mathcal{P}_f(\mathbb{N}),\ \ell > \max H, \text{ and } \min H > \max F\}.$$

Using Theorem 4.15, one easily shows that $A \in q + -q + y + u = x + u$ and $B \in q + -q + y + v = x + v$.

Now consider the ternary expansion of any number s of the form $n + 3^\ell - \Sigma_{t \in H} 3^t + \Sigma_{t \in F} 3^t$ where $\ell \in \mathbb{N}$, $n \in \mathbb{N}3^{\ell+1}$, $H, F \in \mathcal{P}_f(\mathbb{N})$, $\ell > \max H$, and $\min H = j > \max F$. Pick a number $k \in \mathbb{N}$ such that $n < 3^{k+1}$. Then $s = \Sigma_{t=0}^k a_t 3^t$ where $a_t \in \{0, 1\}$ if $t \in \{0, 1, \ldots, j-1\}$, $a_j = 2$, $a_t \in \{1, 2\}$ if $t \in \{j+1, j+2, \ldots, \ell-1\}$, $a_\ell = 0$, and $a_t \in \{0, 1, 2\}$ if $t \in \{\ell, \ell+1, \ldots, k\}$. That is, knowing that s can be written in this form, one can read ℓ just by looking at the ternary expansion, because ℓ is the position of the lowest order 0 occurring above the lowest order 2. Since ℓ is determined from the form of the ternary expansion, so is n. That is, $n = \Sigma_{t=\ell+1}^k a_t 3^t$. Since $U \cap V = \emptyset$, one concludes that $A \cap B = \emptyset$, a contradiction.

So x is left cancelable in $\beta\mathbb{N}$. It is not right cancelable, because $q + x = x$. $\square$

Theorem 8.41. *Let S be a discrete countable cancellative semigroup and let p be a P-point in S^*. Then p is cancelable in βS.*

Proof. Suppose that $px = py$, where x and y are distinct elements of βS.

Let $D = \{a \in S : ax \in Sy\}$. If $a \in D$, there is a unique element $b \in S$ for which $ax = by$ (by Corollary 8.2). Thus we can define a function $f : D \to S$ by stating that $ax = f(a)y$ for every $a \in D$. We observe that f has no fixed points, by Lemma 8.1. We can extend f to a function $g : S \to S$ which also has no fixed points. By Lemma 3.33, there is a set $A \in p$ such that $A \cap g[A] = \emptyset$. Similarly, if $E = \{b \in S : by \in Sx\}$, we can choose a function $h : S \to S$ such that $by = h(b)x$ for every $b \in E$ and $B \cap h[B] = \emptyset$ for some $B \in p$.

We note that the equation $sx = py$ has at most one solution with $s \in S$, and that the same is true of the equation $px = sy$ (by Corollary 8.2). Hence we can choose $P \in p$ such that $P \subseteq A \cap B$ and, for every $a \in P$, $ax \neq py$ and $px \neq ay$. For each $a \in P$, let $C_a = \{u \in S^* : ax = uy \text{ or } ux = ay\}$. Since each C_a is closed ($C_a = \rho_y^{-1}[\{ax\}] \cup \rho_x^{-1}[\{ay\}]$), $p \notin C_a$, and p is a P-point of S^*, it follows that $p \notin \bigcup_{a \in P} C_a$. So we can choose $Q \in p$ with $Q \subseteq P$, such that $\overline{Q} \cap C_a = \emptyset$ for every $a \in P$.

Now $px \in \overline{Q}x$ and $py \in \overline{Q}y$. It follows without loss of generality from Theorem 3.40, that $ax = uy$ for some $a \in Q$ and some $u \in \overline{Q}$. This equation implies that $u \notin S^*$, since otherwise we should have $u \in \overline{Q} \cap C_a$. So $u \in S$ and $u = g(a)$, and hence $u \in A \cap g[A]$. This contradicts the fact that $A \cap g[A] = \emptyset$.

We have thus shown that p is left cancelable in βS. By Corollary 8.14, p is also right cancelable in βS. $\square$

The reader may have noticed that all the criteria for left cancelability given in this section, apart from that of being a P-point in an arbitrary countable cancellative semigroup, describe a clopen subset of S^* all of whose elements are left cancelable. Thus these criteria cannot be used to find left cancelable elements in $\overline{K(\beta S)}$. There are several questions about left cancelable elements of $\beta \mathbb{N}$ which remain open, although the corresponding questions for right cancelable elements are easily answered.

Question 8.42. *Is every element in* $\mathbb{N}^* \backslash (\mathbb{N}^* + \mathbb{N}^*)$ *left cancelable?*

Question 8.43. *Are weak P-points in* $\mathbb{N}^*$ *left cancelable?*

Question 8.44. *Are there left cancelable elements in* $\overline{K(\beta S)}$?

Exercise 8.4.1. Prove Corollary 8.38.

Exercise 8.4.2. Let $\mathbb{H}^\circ$ denote the interior of $\mathbb{H}$ in $\mathbb{N}^*$. Prove that $\mathbb{H}^\circ$ is dense in $\mathbb{H}$ and that every element of $\mathbb{H}^\circ$ is cancelable in $\beta \mathbb{Z}$.

Exercise 8.4.3. Let $p \in \beta \mathbb{N}$. Show that there is a cancelable element $r \in \beta \mathbb{N}$ for which $\widetilde{h}_n(r + p) = \widetilde{h}_n(p + r) = 0$ for every $n \in \mathbb{N}$. (Hint: For every $k \in \mathbb{N}$, $\{m \in \mathbb{Z} : h_k(m) = \widetilde{h}_k(-p)\} \in -p$. Thus for each $n \in \mathbb{N}$, one can choose $m_n \in \bigcap_{k=1}^n \{m \in \mathbb{Z} : h_k(m) = \widetilde{h}_k(-p)\}$. Show that one can assume that this $m_n \in \mathbb{N}$ and that $h_k(m_n + p) = 0$ for every $k \in \{1, 2, \ldots, n\}$. The set $A = \{m_n : n \in \mathbb{N}\}$ then satisfies the hypotheses of Corollary 8.36.)

8.5 Compact Semigroups Determined by Right Cancelable Elements in Countable Groups

In this section we assume that G is a countably infinite discrete group and we discuss the smallest compact subsemigroup of βG containing a given right cancelable element of βG. We show that it has a rich structure. Furthermore, all the subsemigroups of βG arising in this way are algebraically and topologically isomorphic to ones which arise in $\beta \mathbb{N}$.

We shall use some special notation in this section. Throughout this section, we shall suppose that G is a countable group and shall denote its identity by e. We shall also suppose that we have chosen an element $p \in G^*$ which is right cancelable in βG.

We assume that we have arranged the elements of G as a sequence, $\langle s_n \rangle_{n=1}^\infty$, with $e = s_1$. We shall write $a < b$ if $a, b \in G$ and a precedes b in this sequence.

Given $F, H \subseteq G$, let $F^{-1} H = \bigcup_{t \in F} t^{-1} H$. We can choose $P \in p$ with the property that $e \notin P$ and $F^{-1} P \notin p$ whenever $F \in \mathcal{P}_f(G \backslash \{e\})$ (by Theorem 8.19).

If $A \subseteq G$, we use FP(A) to mean the set of finite products of distinct terms of A written in increasing order. That is, if $A = \{s_t : t \in B\}$, then FP(A) = FP($\langle s_t \rangle_{t \in B}$).

We shall assume that P has been arranged as an increasing sequence $\langle b_n \rangle_{n=1}^\infty$. (That is, if $b_n = s_t$ and $b_{n+1} = s_k$, then $t < k$.)

Definition 8.45. Let $n \in \mathbb{N}$.

(a) $F_n = \{u^{-1}v : u, v \in \mathrm{FP}(\{a \in G : a \leq b_n\})$ and $u \neq v\}$.

(b) $P_n = P \backslash (F_n \cup F_n^{-1} P)$.

Observe that for each $n \in \mathbb{N}$, $P_n \in p$ and $\{a \in G : a \leq b_n\} \backslash \{e\} \subseteq F_n$.

Definition 8.46. (a) $T = \{b_{n_1} b_{n_2} \ldots b_{n_k} : $ for each $i \in \{2, 3, \ldots, k\}$, $b_{n_i} \in P_{n_{i-1}}\}$.

(b) For each $n \in \mathbb{N}$,

$$T_n = \{b_{n_1} b_{n_2} \ldots b_{n_k} : \text{ for each } i \in \{2, 3, \ldots, k\}, \ b_{n_i} \in P_{n_{i-1}} \text{ and } b_{n_1} \in P_n\}$$

(c) $T_\infty = \bigcap_{n=1}^{\infty} \overline{T_n}$.

We shall call a product of the form $b_{n_1} b_{n_2} \ldots b_{n_k}$ as given in the definition of T a *P*-product. Notice that $P \subseteq T$. (If $i = 0$, we define the empty product $b_{n_1} b_{n_2} \ldots b_{n_i}$ to be e.)

Definition 8.47. Let S be a discrete semigroup and let $q \in \beta S$.

$$C_q = \bigcap \{D \subseteq \beta S : D \text{ is a compact subsemigroup of } \beta S \text{ and } q \in D\}.$$

Notice that C_q is the smallest compact subsemigroup of βS which contains q. We may denote this by $C_q(\beta S)$ in a context in which more than one semigroup S is being considered.

Lemma 8.48. *Suppose that $k, l \in \mathbb{N}$, that $b_{m_1} b_{m_2} \ldots b_{m_k}$ and $b_{n_1} b_{n_2} \ldots b_{n_l}$ are P-products and that $a b_{m_1} b_{m_2} \ldots b_{m_k} = b_{n_1} b_{n_2} \ldots b_{n_l}$ for some $a \in G$ satisfying $a < b_{m_1}$ and $a^{-1} < b_{m_1}$. Then $l \geq k$ and, if $i = l - k$, we have $a = b_{n_1} b_{n_2} \ldots b_{n_i}$ and $m_j = n_{i+j}$ for every $j \in \{1, 2, \ldots, k\}$.*

Proof. We shall first show that $b_{m_k} = b_{n_l}$. Suppose that $n_{l-1} \geq m_{k-1}$. We then have $b_{n_l} = u^{-1} v b_{m_k}$, where $u = b_{n_1} b_{n_2} \ldots b_{n_{l-1}}$ and $v = a b_{m_1} b_{m_2} \ldots b_{m_{k-1}}$. This implies that $u = v$, because otherwise we should have $b_{n_l} \in F_{n_{l-1}}^{-1} P$, contradicting our assumption that $b_{n_l} \in P_{n_{l-1}}$. However, $u = v$ implies that $b_{m_k} = b_{n_l}$.

Similarly, if $m_{k-1} \geq n_{l-1}$, the equation $b_{m_k} = v^{-1} u b_{n_l}$ implies that $b_{m_k} = b_{n_l}$.

So $b_{m_k} = b_{n_l}$ and these terms can be cancelled from the equation and the argument repeated until we have $a = b_{n_1} b_{n_2} \ldots b_{n_i}$ if $i = l - k \geq 0$, or else $a b_{m_1} b_{m_2} \ldots b_{m_j} = e$ if $j = k - l > 0$. The first possibility is what we wish to prove, and so it will be sufficient to rule out the second. This is done by noting that, if $j > 1$, since $b_{m_j} \neq e$, the equation $a b_{m_1} b_{m_2} \ldots b_{m_j} = e$ implies that $b_{m_j} \in F_{m_{j-1}}$, which is a contradiction. If $j = 1$, it implies that $a^{-1} = b_{m_1}$, contradicting our assumption that $a^{-1} < b_{m_1}$. $\square$

Lemma 8.49. *The expression for an element of T as a P-product is unique.*

Proof. We apply Lemma 8.48 with $a = e$. We observe that we cannot express e as a *P*-product $b_{n_1} b_{n_2} \ldots b_{n_i}$ with $i > 0$. To see this, note that otherwise, if $i = 1$, we should have $b_{n_1} = e$, contradicting our assumption that $e \notin P$. If $i > 1$, we should have $b_{n_i} \in F_{n_{i-1}}$, contradicting our definition of a P-product.

Thus, if $b_{m_1} b_{m_2} \ldots b_{m_k} = b_{n_1} b_{n_2} \ldots b_{n_l}$, where these are P-products, Lemma 8.48 implies that $k = l$ and that $m_i = n_i$ for every $i \in \{1, 2, \ldots, k\}$.   □

Lemma 8.49 justifies the following definition.

Definition 8.50. Define $\psi : T \to \mathbb{N}$ by stating that $\psi(x) = k$ if $x = b_{n_1} b_{n_2} \ldots b_{n_k}$, where this is a P-product.

As usual $\widetilde{\psi} : \overline{T} :\to \beta\mathbb{N}$ is the continuous extension of ψ.

Theorem 8.51. *Let G be a countably infinite discrete group and let $p \in G^*$ be right cancelable in βG. Then T_∞ is a compact subsemigroup of G^* which contains C_p. Furthermore, $\widetilde{\psi}$ is a homomorphism on T_∞ satisfying $\widetilde{\psi}(p) = 1$ and $\widetilde{\psi}[C_p] = \beta\mathbb{N}$.*

Proof. Let $x \in T_m$ and express x as a P-product: $x = b_{m_1} b_{m_2} \ldots b_{m_k}$. If $n > m_k$ and if $y \in T_n$, then $xy \in T_m$ and $\psi(xy) = \psi(x) + \psi(y)$. It follows from Theorems 4.20 and 4.21 that T_∞ is a subsemigroup of G^* and that $\widetilde{\psi}$ is a homomorphism on T_∞.

For every $n \in \mathbb{N}$ and every $A \in p$, we can choose $b \in P_n \cap A$. So $A \cap T_n \cap \widetilde{\psi}^{-1}[\{1\}] \neq \emptyset$. It follows that

$$\emptyset \neq \bigcap_{A \in p} \bigcap_{n \in \mathbb{N}} c\ell(A \cap T_n \cap \widetilde{\psi}^{-1}[\{1\}]) \subseteq \{p\} \cap T_\infty \cap \widetilde{\psi}^{-1}[\{1\}].$$

This shows that $p \in T_\infty$ and hence that $C_p \subseteq T_\infty$. It also shows that $\widetilde{\psi}(p) = 1$ and hence that $\widetilde{\psi}[C_p] = \beta\mathbb{N}$, because $\widetilde{\psi}[C_p]$ is a compact subsemigroup of $\beta\mathbb{N}$ which contains 1.   □

Theorem 8.51 allows us to see that C_p has a rich algebraic structure.

Corollary 8.52. *Let G be a countably infinite discrete group and let $p \in G^*$ be right cancelable in βG. The semigroup C_p has 2^c minimal left ideals and 2^c minimal right ideals. Each of these contains 2^c idempotents.*

Proof. We apply Theorem 6.44. The result then follows because, if L is a left ideal of $\beta\mathbb{N}$, $\widetilde{\psi}^{-1}[L] \cap C_p$ is a left ideal of C_p; and the corresponding remark holds for right ideals. We recall that every left ideal of C_p contains a minimal left ideal and every right ideal of C_p contains a minimal right ideal, and the intersection of any left ideal and any right ideal contains an idempotent (by Corollary 2.6 and Theorem 2.7).   □

We remind the reader that, if e and f are idempotents in a semigroup, we write $f \leq e$ if $ef = fe = f$.

Lemma 8.53. *Let e_1 and e_2 be idempotents in $\beta\mathbb{N}$ with $e_2 < e_1$, and let f_1 be an idempotent in C_p for which $\widetilde{\psi}(f_1) = e_1$. There is an idempotent f_2 in C_p for which $f_2 < f_1$ and $\widetilde{\psi}(f_2) = e_2$.*

Proof. First notice that $\widetilde{\psi}^{-1}[\{e_2\}] \cap C_p f_1 \neq \emptyset$, because $(\widetilde{\psi}^{-1}[\{e_2\}] \cap C_p) f_1$ is contained in this set. So $\widetilde{\psi}^{-1}[\{e_2\}] \cap C_p f_1$ is a compact semigroup and thus contains an idempotent f (by Theorem 2.5). We put $f_2 = f_1 f$. It is easy to check that f_2 is an idempotent satisfying $\widetilde{\psi}(f_2) = e_2$ and $f_2 < f_1$. $\qquad\square$

Corollary 8.54. *Let G be a countably infinite discrete group and let $p \in G^*$ be right cancelable in βG.*

The semigroup C_p contains infinite decreasing chains of idempotents.

Proof. There is a decreasing sequence of idempotents $\langle e_n \rangle_{n=1}^{\infty}$ in $\beta\mathbb{N}$ by Theorem 6.12. By Lemma 8.53 we can inductively choose a decreasing sequence $\langle f_n \rangle_{n=1}^{\infty}$ of idempotents in C_p for which $\widetilde{\psi}(f_n) = e_n$ for every $n \in \mathbb{N}$. $\qquad\square$

Corollary 8.55. *Let G be a countably infinite discrete group and let $p \in G^*$ be right cancelable in βG. Every maximal group in the smallest ideal of C_p contains a copy of the free group on 2^c generators.*

Proof. Let q be a minimal idempotent in C_p. Then $\widetilde{\psi}(q)$ is a minimal idempotent in $\beta\mathbb{N}$ by Lemma 8.53 and so $(\widetilde{\psi}(q))\beta\mathbb{N}(\widetilde{\psi}(q))$ contains a group F which is a copy of the free group on 2^c generators (by Corollary 7.36). For each generator x in F, we can choose $y \in qC_pq$ such that $\widetilde{\psi}(y) = x$. These elements generate a group in qC_pq which can be mapped homomorphically onto F and which is therefore free. $\qquad\square$

Comment 8.56. If we do not specify that p is right cancelable, the preceding results may no longer hold. For example, if p is an idempotent, C_p is just a singleton. Less trivially, if $p = 1 + e$, where e is a minimal idempotent in $\beta\mathbb{Z}$, C_p is contained in the minimal left ideal $\beta\mathbb{Z} + e$ of $\beta\mathbb{Z}$ and contains no chains of idempotents of length greater than 1.

Theorem 8.57. *Let G be a countably infinite discrete group and let $p \in G^*$ be right cancelable in βG. The semigroup C_p does not meet $K(\beta G)$.*

Proof. We can choose $x \in \overline{P} \cap G^*$ with $x \neq p$. We can then choose $Q \subseteq P$ such that $Q \in x$ and $Q \notin p$. We put $P' = P \backslash Q$. Let $\langle b_n' \rangle_{n=1}^{\infty}$ enumerate P' in increasing order. Then define F_n', P_n', T', T_n', and T_∞' in terms of P' analogously to Definitions 8.45 and 8.46. Notice that for each n, $F_n \subseteq F_n'$, and hence $P_n' \subseteq P_n$, $T_n' \subseteq T_n$, and $T_\infty' \subseteq T_\infty$.

We claim that $T_\infty' \cap K(\beta G) = \emptyset$. Suppose instead that we have some $y \in T_\infty' \cap K(\beta G)$. Then by Theorem 1.67, $y \in (\beta G)xy$. So pick some $u \in \beta G$ such that $y = uxy$. For each $a \in G$, the set $X_a = \{b_n : b_n \in Q, b_n > a, \text{ and } b_n > a^{-1}\} \in x$. For each $b_n \in X_a$, we have $T_n \in y$. So $\{ab_n v : a \in G, b_n \in X_a, \text{ and } v \in T_n\} \in uxy$, by Theorem 4.15. Since $T' \in y$, there must be elements $a \in G$, $b_n \in X_a$, and $v \in T_n$ for which $ab_n v \in T'$. We note that $b_n v$ is a P-product and so, by Lemma 8.48, it follows that $b_n \in P'$. This contradicts our assumption that $b_n \in Q$ and establishes that $T_\infty' \cap K(\beta G) = \emptyset$.

Now $C_p \subseteq T'_\infty$ (by Theorem 8.51 with P' in place of P) and hence $C_p \cap K(\beta G) = \emptyset$.
$\square$

Corollary 8.58. *If G is a countably infinite discrete group, there is a right cancelable element p in βG for which $C_p \subseteq c\ell(K(\beta G)) \backslash K(\beta G)$.*

Proof. We can choose a right cancelable element p of βG in $\overline{K(\beta G)}$ (by Corollary 8.26). Since $\overline{K(\beta G)}$ is a compact subsemigroup of βG (by Theorem 4.44), $C_p \subseteq \overline{K(\beta G)}$. By Theorem 8.57, $C_p \cap K(\beta G) = \emptyset$. $\square$

Lemma 8.59. *Let G be a countably infinite discrete group, let $p \in G^*$ be right cancelable in βG, and let q be an idempotent in C_p. Then there is some idempotent $r \in (\beta G)p$ which is $\leq_R$-maximal in G^* such that $q \leq_R r$.*

Proof. We can choose a $\leq_R$-maximal idempotent r in G^* for which $q \leq_R r$ by Theorem 2.12.

Suppose that $r \notin (\beta G)p$. There is then a set $E \in r$ such that $(\beta G)p \cap \overline{E} = \emptyset$. For every $a \in G$, we have $E \notin ap$ and hence $B_a = a^{-1}(G \backslash E) \in p$.

We consider the set A of all P-products $b_{n_1} b_{n_2} \ldots b_{n_k}$ with the property that, for each $i \in \{1, 2, \ldots, k\}$, $b_{n_i} \notin E$ and, if $i > 1$, $b_{n_i} \in B_u$ for every $u \in F_{n_{i-1}}$. We observe that, for a P-product of this form, we have $\prod_{j=1}^{i-1} b_{n_j} \in A$ for every $i \in \{2, 3, \ldots, k\}$. We also note that $A \cap E = \emptyset$.

For each $n \in \mathbb{N}$, we define A_n to be the set of all P-products $b_{n_1} b_{n_2} \ldots b_{n_k}$ in $A \cap T_n$ for which $b_{n_1} \in B_u$ whenever $u \in F_n$. We shall show that $\bigcap_{n \in \mathbb{N}} \overline{A_n}$ is a compact semigroup which contains p. That it is a compact semigroup follows from Theorem 4.20 and the observation that, if $u \in A_m$ is expressed as a P-product $b_{n_1} b_{n_2} \ldots b_{n_k}$, then $uA_n \subseteq A_m$ whenever $n > n_k$. Now, for each $n \in \mathbb{N}$, we have $A_n \in p$, because $\bigcap_{u \in F_n} \{b_r : b_r \in P_n \text{ and } b_r \in B_u\} \in p$. So $p \in \bigcap_{n \in \mathbb{N}} \overline{A_n}$.

Since $\bigcap_{n \in \mathbb{N}} \overline{A_n}$ is a compact semigroup which contains p, $C_p \subseteq \bigcap_{n \in \mathbb{N}} \overline{A_n}$ and so $q \in \bigcap_{n \in \mathbb{N}} \overline{A_n}$. For each $a \in G$, let Q_a denote the set of P-products $b_{n_1} b_{n_2} \ldots b_{n_k}$ with $b_{n_1} > a$ and $b_{n_1} > a^{-1}$. Then $Q_a \in q$ and so $\{au : a \in E \text{ and } u \in Q_a\} \in rq$ (by Theorem 4.15). Since $A \in q = rq$, there exists $a \in E$ and $u \in Q_a$ such that $au \in A$. By Lemma 8.48, it follows that $a \in A$. This contradicts the assumption that $a \in E$. $\square$

The following theorem may be surprising because all idempotents in $K(\beta \mathbb{N})$ are $\leq_R$-minimal (by Theorems 1.36 and 1.38).

Theorem 8.60. *Let G be a countably infinite discrete group. For every $q \in G^* \backslash K(\beta G)$, there is a $\leq_R$-maximal idempotent of G^* in $\overline{K(\beta G)} \cap (\beta G)q$.*

Proof. By Theorem 6.56 we can choose an element $x \in G^*$ for which xq is right cancelable in βG. By Corollary 8.26, there is an element $p \in \overline{K(\beta G)}$ which is right cancelable in βG. Since pxq is also right cancelable, it follows from Lemma 8.59 that there is a $\leq_R$-maximal idempotent of G^* in $(\beta G)pxq$. Now $\overline{K(\beta G)}$ is an ideal of βG (by Theorem 4.44), and so $(\beta G)pxq \subseteq \overline{K(\beta G)}$. $\square$

Theorem 8.61. *Let G be a countably infinite discrete group and let $p \in G^*$ be right cancelable in βG. There is an injective mapping $\phi : T \to \mathbb{N}$ with the following properties:*

(1) $\widetilde{\phi}[T_\infty] \subseteq \mathbb{H}$.
(2) $\widetilde{\phi}$ *defines an isomorphism from T_∞ onto $\widetilde{\phi}[T_\infty]$.*
(3) $\widetilde{\phi}(p) \in \overline{\{2^n : n \in \mathbb{N}\}}$.

Proof. We define ϕ by stating that $\phi(b_{n_1} b_{n_2} \ldots b_{n_k}) = \Sigma_{i=1}^k 2^{n_i}$ whenever $b_{n_1} b_{n_2} \ldots b_{n_k}$ is a P-product. Then ϕ is well defined by Lemma 8.49, and is trivially injective. We note that, if $b_{n_1} b_{n_2} \ldots b_{n_k} \in T_n$, then $b_{n_1} \in P_n$ and so $b_{n_1} \notin \{a \in G : a \leq b_n\}$ and hence $n_1 > n$. It follows that $\phi[T_n] \subseteq 2^n \mathbb{N}$ and that $\widetilde{\phi}[T_\infty] \subseteq \mathbb{H}$.

Suppose that $x \in T$ is expressed as a P-product $b_{n_1} b_{n_2} \ldots b_{n_k}$. Then, if $n > n_k$ and $y \in T_n$, we have $\phi(xy) = \phi(x) + \phi(y)$. It follows from Theorem 4.21 that $\widetilde{\phi}$ is a homomorphism on T_∞. Since $\widetilde{\phi}$ is injective (by Exercise 3.4.1), it follows that $\widetilde{\phi}$ defines an isomorphism from T_∞ onto $\widetilde{\phi}[T_\infty]$.

Now $\phi[P] \subseteq \{2^n : n \in \mathbb{N}\}$. Since $P \in p$, $\{2^n : n \in \mathbb{N}\} \in \widetilde{\phi}(p)$. $\square$

Corollary 8.62. *Let G be a countably infinite discrete group and let $p \in G^*$ be right cancelable in βG. There is an element $q \in \mathbb{N}^*$, for which $\{2^n : n \in \mathbb{N}\} \in q$ and $C_p(\beta G)$ is algebraically and topologically isomorphic to $C_q(\beta \mathbb{N})$.*

Proof. We put $q = \widetilde{\phi}(p)$, where ϕ is the mapping defined in Theorem 8.61. Then $\widetilde{\phi}[C_p(\beta G)]$ is a compact subsemigroup of $\beta \mathbb{N}$ which contains q, and hence $\widetilde{\phi}[C_p(\beta G)] \supseteq C_q(\beta \mathbb{N})$. Similarly, $\widetilde{\phi}^{-1}[C_q(\beta \mathbb{N})] \cap T_\infty$ is a compact subsemigroup of $\beta \mathbb{N}$ containing p and so $C_p(\beta G) \subseteq \widetilde{\phi}^{-1}[C_q(\beta \mathbb{N})]$. So $\widetilde{\phi}[C_p(\beta G)] = C_q(\beta \mathbb{N})$ and $\widetilde{\phi}$ defines an isomorphism from $C_p(\beta G)$ onto $C_q(\beta \mathbb{N})$. $\square$

Theorem 8.63. *Let G be a countably infinite discrete group and let $p \in G^*$ be right cancelable in βG. Then T_∞ is algebraically and topologically isomorphic to $\mathbb{H}$.*

Proof. By Theorem 7.24 (taking $X = G$ and $V(a) = G$ for all $a \in G$), it is sufficient to show that we can define a left invariant zero dimensional topology on G which has the sets $\{e\} \cup T_n$ as a basis of neighborhoods of e.

As we observed in the proof of Theorem 8.51, if $x \in T_m$ is expressed as a P-product $b_{m_1} b_{m_2} \ldots b_{m_k}$, then $x T_n \subseteq T_m$ if $n > m_k$.

Let $\mathcal{B} = \{\{x\} \cup x T_n : x \in G \text{ and } n \in \mathbb{N}\}$. We claim that $\mathcal{B}$ is a basis for a topology on G such that for each $x \in G$, $\{\{x\} \cup x T_n : n \in \mathbb{N}\}$ is a neighborhood basis at x. To see this, let $x, y \in G$, let $n, k \in \mathbb{N}$, and let $a \in (\{x\} \cup x T_n) \cap (\{y\} \cup y T_k)$.

(i) If $a = x = y$, let $r = \max\{n, k\}$.
(ii) If $a = x = y b_{m_1} b_{m_2} \ldots b_{m_i}$ where $b_{m_1} b_{m_2} \ldots b_{m_i}$ is a P-product and $b_{m_1} \in P_k$, let $r = \max\{n, m_i + 1\}$.
(iii) If $a = x b_{l_1} b_{l_2} \ldots b_{l_j} = y$ where $b_{l_1} b_{l_2} \ldots b_{l_j}$ is a P-products and $b_{l_1} \in P_n$ let $r = \max\{l_j + 1, k\}$.

(iv) If $a = xb_{l_1}b_{l_2}\ldots b_{l_j} = yb_{m_1}b_{m_2}\ldots b_{m_i}$ where $b_{l_1}b_{l_2}\ldots b_{l_j}$ and $b_{m_1}b_{m_2}\ldots b_{m_i}$ are P-products, $b_{l_1} \in P_n$, and $b_{m_1} \in P_k$, let $r = \max\{l_j + 1, m_i + 1\}$.

Then $\{a\} \cup aT_r \subseteq (\{x\} \cup xT_n) \cap (\{y\} \cup yT_k)$ as required. The topology generated by $\mathcal{B}$ is clearly left invariant.

Suppose that $a \in G\backslash T_m$. If we choose n such that $a < b_n$ and $a < b_n^{-1}$, it follows from Lemma 8.48 that $aT_n \cap T_m = \emptyset$. Thus the sets $\{e\} \cup T_n$ are clopen in our topology.

We now claim that $\bigcap_{n\in\mathbb{N}} T_n = \emptyset$, because $a < b_n$ implies that $a \notin T_n$. To see this, suppose that $a = b_{n_1}b_{n_2}\ldots b_{n_l}$, where this is a P-product and $b_{n_1} \in P_n$. We note that $b_{n_1} > b_n$, since otherwise we should have $b_{n_1} \in F_n$. So $b_{n_1} > a$ and therefore $l > 1$. We then have $b_{n_l} \in F_{n_{l-1}}$, contradicting our definition of P-product.

This shows that our topology is Hausdorff, and completes the proof. □

Lemma 8.64. *Let G be a countably infinite discrete group and let p be a right cancelable element of G^*. Suppose that $x \in \beta G$ and $y \in T_\infty$. Then $xy \in \overline{T}$ implies that $x \in \overline{T}$.*

Proof. Let $X \in x$ and let Z denote the set of all products of the form $ab_{n_1}b_{n_2}\ldots b_{n_k}$, where $b_{n_1}b_{n_2}\ldots b_{n_k}$ is a P-product, $a \in X$, $a < b_{n_1}$ and $a^{-1} < b_{n_1}$. By Theorem 4.15, $Z \in xy$. Hence, if $xy \in \overline{T}$, $T \cap Z \neq \emptyset$. It then follows from Lemma 8.48, that $T \cap X \neq \emptyset$ and hence that $T \in x$. □

So far, in this section, we have restricted our attention to βG, where G denotes a countably infinite discrete group. However, some of the results obtained have applications to $(\beta\mathbb{N}, +)$ and $(\beta\mathbb{N}, \cdot)$, the following theorem being an example.

Theorem 8.65. *Let S be a countably infinite discrete semigroup which can be embedded in a countable discrete group G. Then $\overline{K(\beta S)}$ contains $2^{\mathfrak{c}}$ nonminimal idempotents, infinite chains of idempotents, and idempotents which are $\leq_R$-maximal in G^*.*

Proof. By Corollary 8.26 (with T and S replaced by S and G respectively), there is an element $p \in \overline{K(\beta S)}$ which is right cancelable in βG. Taking this p to be the element held fixed at the start of this section, we can choose the set P so that $P \subseteq S$. Consequently $T \subseteq S$ so that $\overline{T} \subseteq \beta S$.

Now $\overline{K(\beta S)}$ is a compact subsemigroup of βS, by Theorem 4.44, and is therefore a compact subsemigroup of βG. So $C_p(\beta G) \subseteq \overline{K(\beta S)}$. By Corollary 8.54, $C_p(\beta G)$ contains infinite chains of idempotents, and so the same statement is true of $\overline{K(\beta S)}$.

Let e_1 be any non-minimal idempotent in $\beta\mathbb{N}$. Since e_1 is not minimal, we can choose an idempotent e_2 in $\beta\mathbb{N}$ for which $e_2 < e_1$. By Lemma 8.53, we can then choose idempotents f_1 and f_2 in $C_p(\beta G)$ satisfying $f_2 < f_1$, $\widetilde{\psi}(f_1) = e_1$, and $\widetilde{\psi}(f_2) = e_2$. Since $f_1, f_2 \in \overline{K(\beta S)}$, f_1 is a nonminimal idempotent in $\overline{K(\beta S)}$. There are $2^{\mathfrak{c}}$ possible choices of e_1 (by Corollary 6.33) and therefore $2^{\mathfrak{c}}$ possible choices of f_1. So $\overline{K(\beta S)}$ contains $2^{\mathfrak{c}}$ nonminimal idempotents.

By Lemma 8.59, if q is an idempotent in $C_p(\beta G)$, there is an idempotent $r \in (\beta G)p$ which is $\leq_R$-maximal in G^* and satisfies $q \leq_R r$. Since $q = rq$ and $q \in T_\infty$, it follows from Lemma 8.64 that $r \in \overline{T} \subseteq \beta S$.

We know that $r = xp$ for some $x \in \beta G$. By Lemma 8.64, $x \in \overline{T}$ and so $r \in (\beta S)p$. Since $\overline{K(\beta S)}$ is an ideal of βS, it follows that $r \in \overline{K(\beta S)}$.   □

Most of the results in this section also hold for the semigroups $C_p(\beta \mathbb{N})$, where p denotes a right cancelable element of $\mathbb{N}^*$. The proofs have to be modified, since $\mathbb{N}$ is not a group, but they are essentially similar to the proofs given above. We leave the details to the reader in the following exercise.

Exercise 8.5.1. Let p be a right cancelable element in $\beta \mathbb{N}$. By Theorem 8.27, there is a set $P \in p$ which can be arranged as an increasing sequence $\langle b_n \rangle_{n=1}^{\infty}$ with the property that, for each $k \in \mathbb{N}$, $P_k = \{b_n : b_n + k < b_{n+1}\} \in p$. We define T to be the set of all sums of the form $b_{n_1} + b_{n_2} + \cdots + b_{n_k}$, where, for each $i \in \{2, 3, \ldots, k\}$, $b_{n_i+1} - b_{n_i} > 1 + 2 + 3 + \cdots + b_{n_{i-1}}$. We shall refer to a sum of this kind as a P-sum. We define T_n to be the set of sums of all P-sums for which $b_{n_1+1} - b_{n_1} > 1 + 2 + \cdots + n$ and we put $T_{\infty} = \bigcap_{n \in \mathbb{N}} \overline{T_n}$. Prove the following statements:

(1) The expression of an integer in T as a P-sum is unique.

(2) T_{∞} is a compact subsemigroup of $\beta \mathbb{N}$ which contains C_p.

(3) There is a homomorphism mapping C_p onto $\beta \mathbb{N}$.

(4) C_p contains $2^{\mathfrak{c}}$ minimal left ideals and $2^{\mathfrak{c}}$ minimal right ideals, and each of these contains $2^{\mathfrak{c}}$ idempotents.

(5) C_p contains infinite chains of idempotents.

(6) There is an element $q \in \overline{\{2^n : n \in \mathbb{N}\}}$ for which C_q is algebraically and topologically isomorphic to C_p.

(7) T_{∞} is algebraically and topologically isomorphic to $\mathbb{H}$.

Notes

Most of the results of Sections 8.1, 8.2, 8.3, and 8.4 are from [46] (a result of collaboration with A. Blass), [125], [154], and [228]. Theorem 8.22 extends a result of H. Umoh in [235]. Theorem 8.30 is from [129], where it had a much longer proof, which did however provide an explicit description of the elements of $c\ell K(\mathbb{H})$.

Theorem 8.63 is due to I. Protasov. Most of the other theorems in Section 8.5 were proved for $\beta \mathbb{N}$ in [88], a result of collaboration with A. El-Mabhouh and J. Pym, [157], and [228], and were proved for countable groups by I. Protasov.

Chapter 9

Idempotents

We remind the reader that, if p and q are idempotents in a semigroup $(S, \cdot)$, we write:

$$p \leq_L q \quad \text{if} \quad p \cdot q = p;$$
$$p \leq_R q \quad \text{if} \quad q \cdot p = p;$$
$$p \leq q \quad \text{if} \quad p \cdot q = q \cdot p = p.$$

These relations are reflexive and transitive, and the third is anti-symmetric as well.

We may write $p < q$ if $p \leq q$ and $p \neq q$. We may also write $p <_L q$ if $p \leq_L q$ and $q \nleq_L p$, and $p <_R q$ if $p \leq_R q$ and $q \nleq_R p$.

Let p be an idempotent in a semigroup S. We shall say that p is *maximal* if, for every $q \in E(S)$, $p \leq q$ implies that $p = q$. We shall say that p is *right maximal* if, for every $q \in E(S)$, $p \leq_R q$ implies that $q \leq_R p$ and that p is *left maximal* if, for every $q \in E(S)$, $p \leq_L q$ implies that $q \leq_L p$.

We do not need to introduce separate terms for the minimal elements of the three relations defined above, because we saw in Theorem 1.36 that, for any idempotent in S, being minimal for one of these relations is equivalent to being minimal for each of the others. This is also equivalent to being in the smallest ideal of S, if S has a minimal left ideal which contains an idempotent (by Theorem 1.59). In this case, for every idempotent $q \in S$, there exists an idempotent $p \in K(S)$ satisfying $p \leq q$ (by Theorem 1.60).

We have also seen that the study of the relation $\leq$ on $E(\beta\mathbb{N})$ is equivalent to studying commutativity of idempotents in $\beta\mathbb{N}$. Two idempotents in $\beta\mathbb{N}$ are comparable for this relation if and only if they commute (by Corollary 6.24). In fact, this statement holds in βS, if S is any countable discrete semigroup which can be embedded in a group.

There are seemingly simple questions about the order of idempotents in βS which are very hard to answer. At the end of Section 9.3, we shall list some simple and obvious questions which have remained open for several years.

9.1 Right Maximal Idempotents

Recall from Theorem 2.12 that right maximal idempotents exist in every compact right topological semigroup. We shall see now that there are 2^c right maximal idempotents in $\mathbb{N}^*$. This contrasts with the fact that we do not know of any ZFC proof that there are any left maximal idempotents in $\beta\mathbb{N}$.

Theorem 9.1. *There are 2^c right maximal idempotents in $\mathbb{N}^*$.*

Proof. We remind the reader that, for any $n \in \mathbb{N}$, $\operatorname{supp}(n) \in \mathcal{P}_f(\omega)$ is defined by the equation $n = \Sigma_{i\in\operatorname{supp}(n)} 2^i$. We define $\gamma : \mathbb{N} \to \omega$ by stating that $\gamma(n) = 2^m$ where $m = \min\operatorname{supp}(n)$. If $k \in \mathbb{N}$ and $r > \min\operatorname{supp}(k)$, then for all $n \in 2^r\mathbb{N}$ one has $\gamma(k + n) = \gamma(k)$. Thus, given $p \in \beta\mathbb{N}$ and $q \in \mathbb{H}$,

$$\begin{aligned}
\widetilde{\gamma}(p + q) &= p\text{-}\lim_{k\in\mathbb{N}} q\text{-}\lim_{n\in\mathbb{N}} \gamma(k + n) \\
&= p\text{-}\lim_{k\in\mathbb{N}} \gamma(k) \\
&= \widetilde{\gamma}(p).
\end{aligned}$$

We note that γ is the identity map on $\{2^n : n \in \omega\}$ and hence that $\widetilde{\gamma}$ is the identity map on $c\ell_{\beta\mathbb{N}}\{2^n : n \in \mathbb{N}\}$. Let $X = c\ell_{\beta\mathbb{N}}\{2^n : n \in \omega\} \cap \mathbb{N}^*$. Then $|X| = 2^c$ by Theorem 3.59.

For each $x \in X$, let $D_x = \widetilde{\gamma}^{-1}[\{x\}]\cap\mathbb{H}$. Then D_x is a compact subsemigroup of $\mathbb{H}$ and the sets D_x are pairwise disjoint. By Theorem 2.12, we can choose an idempotent q_x in D_x which is a right maximal idempotent in the semigroup D_x.

We shall show that q_x is a right maximal idempotent in $\mathbb{N}^*$. To see this, let p be an idempotent in $\mathbb{N}^*$ for which $p + q_x = q_x$. Then $\widetilde{\gamma}(p) = \widetilde{\gamma}(p + q_x) = \widetilde{\gamma}(q_x) = x$ and so $p \in D_x$. This implies that $q_x + p = p$ and thus that q_x is a right maximal idempotent in $\mathbb{N}^*$ $\square$

We now show that right maximal idempotents in $\mathbb{Z}^*$ have remarkable properties.

Lemma 9.2. *Let G be a countable discrete group and let q be a right maximal idempotent in G^*. Then, if p is any element of βG which is not right cancelable in βG, $pq = q$ implies that $qp = p$.*

Proof. By Theorem 8.18, there is an idempotent $e \in G^*$ for which $ep = p$. Assume that $pq = q$. Then $eq = epq = pq = q$ so, since q is right maximal, $qe = e$ so $qp = qep = ep = p$. $\square$

Lemma 9.3. *Let S be a countable right cancellative discrete semigroup. Then every compact right zero subsemigroup of βS is finite.*

Proof. We first note that, if $t \in S$ has a left identity $s \in S$, then s has to be a right identity for S. To see this, let $u \in S$. Then $ust = ut$ so by right cancellation $us = u$.

Let D denote the set of right identities of S. If $q \in \overline{D}$, then $pq = p$ for every $p \in \beta S$, because $pq = p\text{-}\lim_{u\in S} q\text{-}\lim_{v\in D} uv = p\text{-}\lim_{u\in S} u = p$.

Suppose that C is an infinite compact right zero subsemigroup of βS. Then $q \in C$ implies that $D \notin q$, because otherwise we should have $xq = x$ for every $x \in C$. This implies that $x = q$, because $xq = q$, and hence that $|C| = 1$.

Since C is infinite, pick an infinite sequence $\langle p_n \rangle_{n=1}^\infty$ of distinct elements of C and let p be any accumulation point of the sequence. There is at most one element p_n in the sequence for which $p_n = p$, and so we can delete this element (if it exists) and assume that $p_n \neq p$ for every $n \in \mathbb{N}$. Since $p \in C$, $p = pp \in c\ell_{\beta S}\big((S \backslash D)p\big) \cap c\ell_{\beta S}\{p_n : n \in \mathbb{N}\}$. It follows from Theorem 3.40, that $sp = q$ for some $q \in c\ell_{\beta \mathbb{N}}\{p_n : n \in \mathbb{N}\}$ and some $s \in S \backslash D$, or else $p_n = xp$ for some $n \in \mathbb{N}$ and some $x \in \beta S$.

The first possibility implies that $sq = spq = qq = q$. It follows from Theorem 3.35 that $\{t \in S : st = t\} \in q$ and hence that $\{t \in S : st = t\} \neq \emptyset$ and that $s \in D$, a contradiction.

The second possibility implies that $p = p_n p = xpp = xp = p_n$, again a contradiction. □

Theorem 9.4. *Suppose that G is a countable discrete group. Let q be a right maximal idempotent in G^* and let $C = \{p \in G^* : pq = q\}$. Then C is a finite right zero subsemigroup of βG and every member of C is a right maximal idempotent in G^*.*

Proof. C is a compact right topological semigroup and so $K(C) \neq \emptyset$. We note that $q \in K(C)$ because $\{q\} = K(C)q \subseteq K(C)$. Since $\{q\} = Cq$, $\{q\}$ is a minimal left ideal of C. Thus all the minimal left ideals of C are singletons, by Theorem 1.64, and so the elements of $K(C)$ are all idempotent. Now every minimal right ideal of C intersects every minimal left ideal of C. It follows by Theorem 1.61 that C has a unique minimal right ideal R and therefore that $K(C) = R = E(R)$ (by Theorem 1.64). By Lemma 1.30(b), $E(R)$ is a right zero semigroup and so $K(C)$ is a right zero semigroup.

We now claim that $E(C) = K(C)$. To see this, let $p \in E(C)$. Since $pq = q$ and q is right maximal, we have $qp = p$ and so $p \in K(C)$.

It follows that C contains no $\leq$-chains of idempotents with more than one element, because all elements of $K(C)$ are $\leq$-minimal in C. Now any compact subsemigroup of G^* which contains a right cancelable element, must contain infinite $\leq$-chains of idempotents (by Corollary 8.54). So C cannot contain any right cancelable elements of βG.

We shall show that $C = K(C)$. To see this, let $p \in C$. Since p is not right cancelable in βG and $pq = q$, it follows from Lemma 9.2, that $qp = p$ and hence that $p \in K(C)$. Thus, by Lemma 9.3, C is finite.

Finally, let $p \in C$ and assume that $r \in G^*$ and $p \leq_R r$. Then $rq = r(pq) = (rp)q = pq = q$ and thus $r \in C$ so $pr = r$. □

Lemma 9.5. *Suppose that G is a countable discrete group. Let q be a right maximal idempotent in G^* and let $C = \{p \in \beta G : pq = q\}$. Then, for every $x, y \in \beta G$, the equation $xq = yq$ implies that $x \in yC$ or $y \in xC$.*

Proof. Suppose that $x \notin yC$ and $y \notin xC$. Then, for each $p \in C$, $x \neq yp$ and so there are disjoint clopen subsets U_p and V_p of βG for which $x \in U_p$ and $yp \in V_p$. If $X_p = G \cap U_p$ and $Y_p = G \cap \rho_p^{-1}[V_p]$, then $X_p \in x$, $Y_p \in y$ and $\overline{X_p} \cap (Y_p p) = \emptyset$, because $\overline{X_p} \subseteq U_p$ and $Y_p p \subseteq V_p$. Similarly, there exist $X_p' \in x$ and $Y_p' \in y$ for which

$(X'_p p) \cap \overline{Y'_p} = \emptyset$. Let $X = \bigcap_{p \in C}(X_p \cap X'_p)$ and $Y = \bigcap_{p \in C}(Y_p \cap Y'_p)$. By Lemma 6.28, $C = \{e\} \cup \{p \in G^* : pq = q\}$, so by Theorem 9.4, C is finite and so $X \in x$ and $Y \in y$.

Assume that $xq = yq$. Since $xq \in \overline{Xq}$ and $yq \in \overline{Yq}$, it follows from Theorem 3.40 that $bq = uq$ for some $b \in Y$ and some $u \in \overline{X}$, or else $vq = aq$ for some $v \in \overline{Y}$ and some $a \in X$. Assume without loss of generality that $bq = uq$ for some $b \in Y$ and some $u \in \overline{X}$. Then $b^{-1}uq = q$ and hence $b^{-1}u \in C$. If $p = b^{-1}u$, then $bp = u$, contradicting our assumption that $b \in Y_p$ and $u \in \overline{X_p}$, while $\overline{X_p} \cap (Y_p p) = \emptyset$. □

Theorem 9.6. *Suppose that G is a countable discrete group. Let q be a right maximal idempotent in G^*, let $C = \{p \in \beta G : pq = q\}$, and let $n = |C|$. Suppose that x is a given element of βG and that $Y = \{y \in \beta G : yq = xq\}$. Then Y has either n or $n - 1$ elements.*

Proof. Let $C' = C \setminus \{e\}$, where e denotes the identity of G. We note that by Lemma 6.28, $C' = \{p \in G^* : pq = q\}$ so C' is a finite right zero semigroup, by Theorem 9.4, and in particular all elements of C' are idempotents. For every $p \in C'$, xp is the unique element in $Y \cap (\beta G)p$. To see this, we note that xp is in this set because $xpq = xq$. On the other hand, if y is in this set, then $y = yp = yqp = xqp = xp$.

We shall show that the sets $(\beta G)p$, where $p \in C'$, are pairwise disjoint. If $p, p' \in C'$, then $(\beta G)p \cap (\beta G)p' \neq \emptyset$ implies that $p \in (\beta G)p'$ or $p' \in (\beta G)p$ by Corollary 6.20. Now $p \in (\beta G)p'$ implies that $pp' = p$ and hence that $p = p'$, because $pp' = p'$. Thus $|Y \cap (\beta G)C'| = |C'| = n - 1$.

We claim that there is at most one element of Y in $\beta G \setminus ((\beta G)C')$. Suppose that y and z are distinct elements of Y which are in this set. Since $yq = zq$, $y \in zC = zC' \cup \{z\}$ or $z \in yC = yC' \cup \{y\}$ (by Lemma 9.5). However, neither of these possibilities can hold, because $y \notin zC'$, $z \notin yC'$ and $y \neq z$.

Thus $|Y|$ is either n or $n - 1$, as claimed. □

We now show that right maximal idempotents occur in every neighborhood of every idempotent in S^*, if S is a countable cancellative semigroup.

Theorem 9.7. *Let S be a countable cancellative semigroup and let U be a G_δ subset of S^*. If U contains an idempotent, then U contains a right maximal idempotent of S^*.*

Proof. We may suppose that S has an identity e, since an identity can be adjoined to any semigroup. We assume that S has been given a sequential ordering (i.e., a well ordering of order type ω), with e as the first element. If $a, b \in S$, we shall write $a \prec b$ if a precedes b in this ordering. For each $a \in S$, we put $L(a) = \{b \in S : b \preceq a\}$. For each $F \in \mathcal{P}_f(S)$, $\prod F$ will denote the product of the elements of F with these arranged in increasing order. We define $\prod \emptyset$ to be e. If $B \subseteq S$, FP(B) will, as usual, denote $\{\prod F : F \in \mathcal{P}_f(B)\}$.

Let $p \in U$ be an idempotent. For each $A \in p$, recall that $A^* = \{s \in A : A \in sp\}$. We remind the reader that $A^* \in p$ and that, for every $s \in A^*$, $s^{-1}A^* \in p$ (by Lemma 4.14).

We can choose a decreasing sequence $\langle A_n \rangle_{n=1}^{\infty}$ of infinite subsets of S satisfying $A_n \in p$ and $\overline{A_n} \cap S^* \subseteq U$ for every n.

We shall inductively define a sequence $\langle a_n \rangle_{n=1}^{\infty}$ in S. We choose any $a_1 \in A_1^{\star}$. We then suppose that $a_1, a_2, \ldots, a_k$ have been chosen so that the following conditions are satisfied for every $i, j \in \{1, 2, \ldots, k\}$:

  (i) If $i < j$, then $a_i \prec a_j$,
 (ii) If $i \le j$, then $\mathrm{FP}(\langle a_n \rangle_{n=i}^{j}) \subseteq A_i^{\star}$, and
(iii) If $i > 1$, then $\mathrm{FP}\big(L(a_{i-1})\big) \cap \mathrm{FP}\big(L(a_{i-1})\big)a_i = \emptyset$.

Now, if $b, c \in \mathrm{FP}\big(L(a_k)\big)$, since S is left cancellative there is at most one element $s \in S$ for which $b = cs$ and hence there are only a finite number of solutions of all the equations of this form. Thus we can choose

$$a_{k+1} \in A_{k+1}^{\star} \cap \bigcap_{i=1}^{k} \bigcap \{b^{-1} A_i^{\star} : b \in \mathrm{FP}(\langle a_n \rangle_{n=i}^{k})\},$$

with the property that $a_k \prec a_{k+1}$ and $b \ne ca_{k+1}$ whenever $b, c \in \mathrm{FP}\big(L(a_k)\big)$. The induction hypotheses are then satisfied.

The sequence $\langle a_n \rangle_{n=1}^{\infty}$ having been chosen, we note that, for every $m \in \mathbb{N}$, we have $\mathrm{FP}(\langle a_n \rangle_{n=m}^{\infty}) \subseteq A_m$.

We now show that if $b \in S \backslash \{e\}$, $F, G \in \mathcal{P}_f(\{a_n : n \in \mathbb{N}\})$ with $\min F = a_t$, $b \preceq a_{t-1}$, and $b \cdot \prod F = \prod G$, then $F \not\subseteq G$ and $b = \prod(G \backslash F)$.

We first establish that with b, F, and G as above, one has $\max F = \max G$. Let $a_i = \max F$ and let $a_j = \max G$. If $i < j$, then $b \cdot \prod F \in \mathrm{FP}\big(L(a_{j-1})\big)$ and $\prod(G \backslash \{a_j\}) \in \mathrm{FP}\big(L(a_{j-1})\big)$ (even if $G \backslash \{a_j\} = \emptyset$) so $\prod G \in \mathrm{FP}\big(L(a_{j-1})\big)a_j$, contradicting hypothesis (iii). If $j < i$, then $\prod G \in \mathrm{FP}\big(L(a_{i-1})\big)$ and $b \cdot \prod(F \backslash \{a_i\}) \in \mathrm{FP}\big(L(a_{i-1})\big)$. (If $F = \{a_i\}$, this is because $b \preceq a_{t-1} = a_{i-1}$.) Thus, if $j < i$, then $b \cdot \prod F \in \mathrm{FP}\big(L(a_{i-1})\big)a_i$, again contradicting hypothesis (iii).

Now we establish that $F \not\subseteq G$ and $b = \prod(G \backslash F)$ by induction on $|F|$. If $F = \{a_t\}$, then $ba_t = \prod(G \backslash \{a_t\})a_t$ so, since S is right cancellative, $b = \prod(G \backslash \{a_t\})$. Since $b \ne e$, $G \backslash \{a_t\} \ne \emptyset$ so $F \not\subseteq G$.

Now assume that $|F| > 1$ and the statement is true for smaller sets. We observe that $|G| > 1$. (Indeed, if $G = \{a_j\}$, then we have $b \cdot \prod(F \backslash \{a_j\}) = e$, so if $a_i = \max(F \backslash \{a_j\})$, we have $b \cdot \prod(F \backslash \{a_j\}) \in \mathrm{FP}\big(L(a_{i-1})\big)a_i \cap \mathrm{FP}\big(L(a_{i-1})\big)$, contradicting hypothesis (iii).) Let $F' = F \backslash \{a_j\}$ and $G' = G \backslash \{a_j\}$ where $a_j = \max F = \max G$. Then by right cancellation, $b \cdot \prod F' = \prod G'$ so $F' \not\subseteq G'$ and $b = \prod(G' \backslash F') = \prod(G \backslash F)$.

Now there is an idempotent $q \in \bigcap_{m=1}^{\infty} \overline{\mathrm{FP}(\langle a_n \rangle_{n=m}^{\infty})}$ by Lemma 5.11. By Theorem 2.12, there is a right maximal idempotent $r \in S^*$ for which $rq = q$. We shall show that $r \in U$.

Let $R \in r$ and let $m \in \mathbb{N}$. We show that $R \cap \mathrm{FP}(\langle a_n \rangle_{n=m}^{\infty}) \ne \emptyset$. For each $b \in R \backslash \{e\}$, pick $k(b) \in \mathbb{N}$ such that $b \prec a_{k(b)-1}$. Then by Theorem 4.15,

$$\bigcup_{b \in R \backslash \{e\}} b \cdot \mathrm{FP}(\langle a_n \rangle_{n=k(b)}^{\infty}) \in rq = q.$$

Since also $\mathrm{FP}(\langle a_n \rangle_{n=m}^{\infty}) \in q$, pick $b \in R \backslash \{e\}$, $F \in \mathcal{P}_f(\{a_n : n \ge k(b)\})$, and $G \in \mathcal{P}_f(\{a_n : n \ge m\})$ such that $b \cdot \prod F = \prod G$. Then $F \not\subseteq G$ and $b = \prod(G \backslash F) \in \mathrm{FP}(\langle a_n \rangle_{n=m}^{\infty})$.

Since for each $R \in r$ and $m \in \mathbb{N}$, $R \cap \mathrm{FP}(\langle a_n \rangle_{n=m}^{\infty}) \neq \emptyset$, we have that

$$r \in S^* \cap \bigcap_{m=1}^{\infty} \overline{\mathrm{FP}(\langle a_n \rangle_{n=m}^{\infty})} \subseteq S^* \cap \bigcap_{m=1}^{\infty} \overline{A_m} \subseteq U. \qquad \square$$

We do not know whether the sum of two elements in $\mathbb{N}^* \backslash K(\beta\mathbb{N})$ can be in $K(\beta\mathbb{N})$, or whether the sum of two elements in $\mathbb{N}^* \backslash \overline{K(\beta\mathbb{N})}$ can be in $\overline{K(\beta\mathbb{N})}$. We do know that, if p is a right cancelable element of $\beta\mathbb{N}$, then, for any $q \in \beta\mathbb{N}$, $q + p \in K(\beta\mathbb{N})$ implies that $q \in K(\beta\mathbb{N})$ by Exercise 8.2.2, and that, if p is a right cancelable element of $\beta\mathbb{N} \backslash \overline{K(\beta\mathbb{N})}$, then $q + p \in K(\beta\mathbb{N})$ implies that $q \in \overline{K(\beta\mathbb{N})}$ by Theorem 8.32. There are idempotents p in $\beta\mathbb{N}$ which also have this property. As a consequence of the next theorem, one sees that if p is a right maximal idempotent in $\mathbb{N}^*$ and if $q \in \beta\mathbb{N} \backslash K(\beta\mathbb{N})$, then $q + p \notin K(\beta\mathbb{N})$. The corresponding statement for $\overline{K(\beta\mathbb{N})}$ is also true: if p is a right maximal idempotent in $\beta\mathbb{N} \backslash \overline{K(\beta\mathbb{N})}$ and if $q \in \beta\mathbb{N} \backslash \overline{K(\beta\mathbb{N})}$, then $q + p \notin \overline{K(\beta\mathbb{N})}$. We shall not prove this here. However, a proof can be found in [159].

Theorem 9.8. *Let G be a countably infinite discrete group and let p be a right maximal idempotent in G^*. Then, if $q \in \beta G \backslash K(\beta G)$, $qp \notin K(\beta G)$.*

Proof. Suppose that $qp \in K(\beta G)$. Then $qp = eqp$ for some minimal idempotent $e \in \beta G$ by Theorem 1.64. Let $C = \{x \in G^* : xp = p\}$. By Theorem 9.4, C is a finite right zero semigroup.

Now $q \notin (\beta G)C$. To see this we observe that, if $q = ur$ for some $u \in \beta G$ and some $r \in C$, then $q = qr = q(pr) = (qp)r \in K(\beta G)$, contradicting our assumption that $q \notin K(\beta G)$. Since $(\beta G)C$ is compact, we can choose $Q \in q$ satisfying $\overline{Q} \cap (\beta G)C = \emptyset$. Since $q \notin K(\beta G)$, $q \neq eq$ and so we may also suppose that $Q \notin eq$.

We also claim that $eq \notin (\beta G)C$. If we assume the contrary, then $eq = ur$ for some $u \in \beta G$ and some $r \in C$. By Corollary 6.20, this implies that $q \in (\beta G)r$ or $r \in (\beta G)q$. We have ruled out the first possibility, and the second implies that $p = rp \in (\beta G)qp \subseteq K(\beta G)$. However, by Exercise 9.1.4, $p \notin K(\beta G)$. So we can choose $Y \in eq$ satisfying $\overline{Y} \cap (\beta G)C = \emptyset$ and $Y \cap Q = \emptyset$.

Now $qp \in c\ell(Qp)$ and $eqp \in c\ell(Yp)$. It follows from Theorem 3.40 that $ap = yp$ for some $a \in \overline{Q}$ and some $y \in \overline{Y}$, or else $xp = bp$ for some $x \in \overline{Q}$ and some $b \in Y$.

If we assume the first possibility, the fact that $a \neq y$ implies that $y \in G^*$, by Lemma 6.28 and thus $a^{-1}y \in G^*$ by Theorem 4.31. The equation $p = a^{-1}yp$ then implies that $a^{-1}y \in C$ and hence that $y \in (\beta G)C$, contradicting our choice of Y. The second possibility results in a contradiction in a similar way. $\qquad \square$

Definition 9.9. Let S be a semigroup and let p be an idempotent in S. Then p is a *strongly right maximal idempotent* of S if and only if the equation $xp = p$ has the unique solution $x = p$ in S.

Observe that trivially a strongly right maximal idempotent is right maximal.

We now show that strongly right maximal idempotents exist in $\mathbb{N}^*$, by giving a proof which illustrates the power of the methods introduced by E. Zelenuk. As we shall see in Theorem 12.39, strongly summable ultrafilters are strongly right maximal. However, by Corollary 12.38, the existence of strongly summable ultrafilters cannot be deduced in ZFC. It was an open question for several years whether the existence of strongly right maximal idempotents in $\mathbb{N}^*$ could be deduced in ZFC. This question has now been answered by I. Protasov.

We do not know whether it can be shown in ZFC that there are any right maximal idempotents in $\mathbb{N}^*$ which are *not* strongly right maximal. However, E. Zelenuk has shown that Martin's Axiom does imply the existence of idempotents of this kind [247].

Theorem 9.10. *There is a strongly right maximal idempotent in $\mathbb{N}^*$. Furthermore, for every right maximal idempotent e in $\mathbb{Z}^*$, there is an $\mathbb{H}$-map ψ from ω onto a subset of $\mathbb{Z}$ for which $\widetilde{\psi}^{-1}(e)$ is defined and is a strongly right maximal idempotent in $\mathbb{N}^*$.*

Proof. We know that there are right maximal idempotents in $\mathbb{Z}^*$ by Theorem 2.12. Let e be an idempotent of this kind, let $C = \{x \in \mathbb{Z}^* : x + e = e\}$, and let $C^\sim = \{x \in \beta\mathbb{Z} : x + C \subseteq C\}$. By Theorem 9.4 C is a finite right zero subsemigroup of $\beta\mathbb{Z}$ so in particular, $x + C = C$ for every $x \in C$. Then $C^\sim = \{0\} \cup C$. (Certainly $\{0\} \cup C \subseteq C^\sim$. Assume $x \in C^\sim \setminus \{0\}$. Then by Lemma 6.28, $x \notin \mathbb{Z}$. Let $y = x + e$. Then $y \in C$ so $x + e = x + e + e = y + e = e$.) Let $\varphi^\sim = \bigcap C^\sim = \{U \subseteq \mathbb{Z} : C^\sim \subseteq \overline{U}\}$. By Lemmas 7.6 and 7.7, there is a left invariant zero dimensional topology τ on $\mathbb{Z}$ for which $\varphi^\sim$ is the filter of neighborhoods of 0.

We choose a set $U \subseteq \mathbb{Z}$ such that 0 and e are in $\overline{U}$ but $f \notin \overline{U}$ if $f \in C^\sim \setminus \{0, e\}$. We note that for every $f \in C^\sim \setminus \{0, e\}$, $\{n \in \mathbb{Z} : n + f \in \overline{\mathbb{Z} \setminus U}\} \in e$ because $e + f = f \in \overline{\mathbb{Z} \setminus U}$. We put

$$V = U \cap \bigcap_{f \in C^\sim \setminus \{0, e\}} \{n \in \mathbb{Z} : n + f \in \overline{\mathbb{Z} \setminus U}\}$$

and observe that $0 \in V$ and $V \in e$. Let $X = V^\star = \{n \in V : -n + V \in e\}$ and note that $0 \in X$ and $X \in e$. Let $n \in X$. We shall show that

(i) there exists $W(n) \in \varphi^\sim$ for which $n + (W(n) \cap X) \subseteq X$,

(ii) $(n + X) \cap X$ is a neighborhood of n in X for the relative topology induced by τ, and

(iii) n is not isolated in the relative topology on X induced by τ.

Before verifying (i), (ii), and (iii), note that if $x_W \in W$ for each $W \in \varphi^\sim$, then the net $\langle x_W \rangle_{W \in \varphi^\sim}$ (directed by reverse inclusion) has a cluster point f in $C^\sim$. To see this, suppose instead that for each $f \in C^\sim$ one has some $U(f) \in f$ and some $W(f) \in \varphi^\sim$ such that $x_{W'} \notin U(f)$ for all $W' \in \varphi^\sim$ with $W' \subseteq W(f)$. Let $W' = \left(\bigcup_{f \in C^\sim} U(f)\right) \cap \left(\bigcap_{f \in C^\sim} W(f)\right)$. Then $W' \in \varphi^\sim$. Pick $f \in C^\sim$ such that $x_{W'} \in U(f)$. Then since $W' \subseteq W(f)$, we have a contradiction.

To verify (i), suppose that for every $W \in \varphi^\sim$, there exists $r_W \in W \cap X$ such that $n + r_W \notin X$. Pick a cluster point $f \in C^\sim$ of the net $\langle r_W \rangle_{W \in \varphi^\sim}$. Since each $r_W \in X$ one has $f \in \overline{X}$, and thus $f = 0$ or $f = e$. Since each $r_W \in \mathbb{Z} \setminus (-n + X)$ one has $f \in \overline{\mathbb{Z} \setminus (-n + X)}$. But $0 \in -n + X$ and by Lemma 4.14, $-n + X = -n + V^\star \in e$ so $e \in \overline{-n + X}$, a contradiction.

To verify (ii) we show that there is some $W \in \varphi^\sim$ such that $(n + W) \cap X \subseteq (n + X) \cap X$. Suppose instead that for each $W \in \varphi^\sim$ there is some $s_W \in W$ such that $n + s_W \in X$

but $s_W \notin X$. Pick a cluster point $f \in C^{\sim}$ of the net $\langle s_W \rangle_{W \in \varphi^{\sim}}$. Since each $s_W \in \mathbb{Z} \setminus X$, $f \notin \overline{X}$ and since each $s_W \in -n + X, n + f \in \overline{X}$. But since $f \notin \overline{X}$, one has $f \notin \{0, e\}$ so, since $n \in V, n + f \notin \overline{U} \supseteq \overline{X}$, a contradiction.

To verify (iii), let $W \in \varphi^{\sim}$. We show that there exists $m \in W \setminus \{0\}$ such that $n + m \in X$ (so that $(n + W) \cap X \neq \{n\}$). Since $W \in \varphi^{\sim}, e \in C^{\sim} \subseteq \overline{W}$ so $W \setminus \{0\} \in e$. Also, $n \in V^{\star}$ so by Lemma 4.14, $-n + V^{\star} \in e$. Pick $m \in (W \setminus \{0\}) \cap (-n + V^{\star})$.

Having established (i), (ii), and (iii), for each $n \in X$ pick $W(n) \in \varphi^{\sim}$ as guaranteed by (i), choosing $W(0) = X$. Now we have shown that the hypotheses of Theorem 7.24 are satisfied with $G = \mathbb{Z}$ and $V(n) = W(n) \cap X$. Thus there is a countable set $\{V_n : n \in \mathbb{N}\}$ of neighborhoods of 0 in X for which $Y = \bigcap_{n=1}^{\infty} c\ell_{\beta X} V_n \setminus \{0\}$ is a subsemigroup of $\beta \mathbb{Z}$ and there is an $\mathbb{H}$-map $\psi : \omega \to X$ such that $\psi_{|\mathbb{H}}$ is an isomorphism from $\mathbb{H}$ onto Y. For each n, there is some $W_n \in \varphi^{\sim}$ such that $W_n \cap X \subseteq V_n$ and thus each $V_n \in e$ so $e \in Y$.

Now the equation $x + e = e$ has the unique solution $x = e$ in $\overline{X} \setminus \{0\}$. It follows that the equation $x + \widetilde{\psi}^{-1}(e) = \widetilde{\psi}^{-1}(e)$ has the unique solution $x = \widetilde{\psi}^{-1}(e)$ in $\mathbb{H}$ and hence that $\widetilde{\psi}^{-1}(e)$ is a strongly right maximal idempotent in $\mathbb{H}$. Since by Lemma 6.8 all idempotents of $\mathbb{N}^*$ are in $\mathbb{H}$, $\widetilde{\psi}^{-1}(e)$ is a strongly right maximal idempotent in $\mathbb{N}^*$. □

Theorem 9.11. *There are $2^{\mathfrak{c}}$ strongly right maximal idempotents in $\mathbb{H}$. Consequently there are $2^{\mathfrak{c}}$ strongly right maximal idempotents in $\mathbb{N}^*$.*

Proof. We know that there are $2^{\mathfrak{c}}$ right maximal idempotents in $\mathbb{Z}^*$ by Theorem 9.1. For each idempotent e of this kind, there is an $\mathbb{H}$ map ψ_e from ω onto a subset of $\mathbb{Z}$ for which $\widetilde{\psi_e}^{-1}(e)$ is defined and is a strongly right maximal idempotent in $\mathbb{H}$ (by Theorem 9.10). For any given e, there are at most $\mathfrak{c}$ right maximal idempotents f in $\mathbb{Z}^*$ for which $\widetilde{\psi_e}^{-1}(e) = \widetilde{\psi_f}^{-1}(f)$, because this equation implies that $\widetilde{\psi_f}(\widetilde{\psi_e}^{-1}(e)) = f$, so that $\widetilde{\psi_f}$ is a $\mathbb{H}$ map taking $\widetilde{\psi_e}^{-1}(e)$ to f, and there are at most $\mathfrak{c}$ $\mathbb{H}$-maps from ω to subsets of $\mathbb{Z}$. Thus there are $2^{\mathfrak{c}}$ distinct elements of the form $\widetilde{\psi_e}^{-1}(e)$. □

Corollary 9.12. *Let G be a countably infinite discrete group which can be algebraically embedded in a metrizable compact topological group. Then there are $2^{\mathfrak{c}}$ strongly right maximal idempotents in G^*.*

Proof. This follows immediately from Theorem 9.11 and Theorem 7.28. □

Exercise 9.1.1. Which idempotent in $(\beta \mathbb{Z}, +)$ is the unique $\leq_R$-maximal idempotent and the unique $\leq_L$-maximal idempotent?

Exercise 9.1.2. Let S be a semigroup and let $p, q \in E(S)$. Show that $p \leq_R q$ if and only if $pS \subseteq qS$ and that $p \leq_L q$ if and only if $Sp \subseteq Sq$.

Exercise 9.1.3. Let q be an element of $\beta \mathbb{N}$ which is not right cancelable in $\beta \mathbb{N}$. Show that there is a right maximal idempotent $p \in \beta \mathbb{N}$ for which $p + q = q$. (Hint: Use Theorems 8.18 and 2.12.)

The following exercise contrasts with the fact that we do not know whether idempotents in $K(\beta\mathbb{Z})$ can be left maximal.

Exercise 9.1.4. Let G be a countably infinite discrete group. Show that no idempotent in $K(\beta G)$ can be right maximal in G^*. (Hint: Use Theorems 6.44 and 9.4.)

Exercise 9.1.5. Let p and q be any two given elements of $\beta\mathbb{Z}$. Show that the equation $x + p = q$ either has $2^{\mathfrak{c}}$ solutions in $\beta\mathbb{Z}$ or else has only a finite number. (Hint: You may wish to use Theorem 3.59.)

9.2 Topologies Defined by Idempotents

In this section we investigate left invariant topologies induced on a group G in each of two natural ways by idempotents in G^*. The first of these is the topology induced on G by the map $a \mapsto ap$ from G to Gp.

Theorem 9.13. *Let G be an infinite discrete group and let p be an idempotent in G^*. There is a left invariant zero dimensional Hausdorff topology on G such that the filter $\varphi^{\sim}$ of neighborhoods of the identity e consists of the subsets U of G for which $\{x \in \beta G : xp = p\} \subseteq c\ell_{\beta G}\, U$. This topology is extremally disconnected and is the same as the topology on G induced by the mapping $(\rho_p)_{|G} : G \to G^*$. Furthermore, if $(\rho_p)_{|G}$ and $(\rho_q)_{|G}$ induce the same topology on G, then $p\beta G = q\beta G$. Consequently there are at least $2^{\mathfrak{c}}$ distinct topologies which arise in this way.*

Proof. Let $\theta_p = (\rho_p)_{|G}$. Notice that by Lemma 6.28, θ_p is injective. In particular $\{a \in G : ap = p\} = \{e\}$ so by Lemmas 7.6 and 7.7, with $C = \{p\}$ (so that $C^{\sim} = \{x \in \beta G : xp = p\}$), we can define a zero dimensional Hausdorff left invariant topology on G, for which $\varphi^{\sim}$ is the filter of neighborhoods of e.

We show first that the topology we have defined on G is the one induced by θ_p. Let $\mathcal{T}$ be the topology defined by $\varphi^{\sim}$ and let $\mathcal{V}$ be the topology induced by θ_p. In the proof of Lemma 7.6 we showed in statement (ii) that the sets of the form $U^{\frown} = \{a \in G : ap \in c\ell_{\beta G}\, U\}$ where $U \in p$ form a basis for the neighborhoods of e in $\mathcal{T}$ and in statement (iii) that each $U^{\frown} \in \mathcal{T}$. Hence $\{bU^{\frown} : b \in G \text{ and } U \in p\}$ forms a basis for $\mathcal{T}$. Given $b \in G$ and $U \in p$, $bU^{\frown} = \theta_p^{-1}\big[\lambda_{b^{-1}}^{-1}[c\ell_{\beta G}\, U]\big]$ so $\mathcal{T} \subseteq \mathcal{V}$. To see that $\mathcal{V} \subseteq \mathcal{T}$, let $U \in \mathcal{V}$ and let $b \in U$. Pick $V \subseteq G$ such that $b \in \theta_p^{-1}[c\ell_{\beta G}\, V] \subseteq U$. Then $b^{-1}V \in p$ and $b \in b(b^{-1}V)^{\frown} \subseteq U$.

Since Gp is extremally disconnected as a subspace of βG (by Lemma 7.41), it follows that G is extremally disconnected.

Now suppose that p and q are idempotents in G^* which define the same topology on G. Then the map $\psi = \theta_q \circ \theta_p^{-1}$ is a homeomorphism from Gp to Gq for which $\psi(ap) = aq$ for all $a \in G$. Now $q = \psi(p) = \psi(pp) = \lim_{a \to p} \psi(ap) = \lim_{a \to p} aq = pq$ so q belongs to the principal right ideal of βG defined by p. Similarly, since

$\theta_p \circ \theta_q^{-1}$ is a homeomorphism, p belongs to the principal right ideal of βG defined by q. Thus $p\beta G = q\beta G$. Now βG contains at least $2^{\mathfrak{c}}$ disjoint minimal right ideals (by Corollary 6.41). Choosing an idempotent in each will produce at least $2^{\mathfrak{c}}$ different topologies on G. $\qquad\square$

The second method of inducing a topology on G by an idempotent p is more direct, taking $\{A \cup \{e\} : a \in p\}$ as a neighborhood base for e.

Theorem 9.14. *Let G be an infinite discrete group with identity e and let $p \in G^*$. Let $\mathcal{N} = \{A \cup \{e\} : A \in p\}$ and let $\mathcal{T} = \{U \subseteq G : \text{for all } a \in U, a^{-1}U \in p\}$. Then $\mathcal{T}$ is the finest left invariant topology on G such that each neighborhood of e is a member of $\mathcal{N}$. The filter of neighborhoods of e is equal to $\mathcal{N}$ if and only if p is idempotent. In this case, the topology $\mathcal{T}$ is Hausdorff.*

Proof. It is immediate that $\mathcal{T}$ is a left invariant topology on G. (For this fact, one only needs p to be a filter on G.) Let V be a neighborhood of e with respect to $\mathcal{T}$. Pick $U \in \mathcal{T}$ such that $e \in U \subseteq V$. Then $U \in p$ and thus $V \in p$. Also $V = V \cup \{e\}$ so $V \in \mathcal{N}$.

Now let $\mathcal{V}$ be a left invariant topology on G such that every neighborhood of e with respect to $\mathcal{V}$ is a member of $\mathcal{N}$. To see that $\mathcal{V} \subseteq \mathcal{T}$, let $U \in \mathcal{V}$ and let $a \in U$. Then $a^{-1}U$ is a neighborhood of e with respect to $\mathcal{V}$ so $a^{-1}U = A \cup \{e\}$ for some $A \in p$ and thus $a^{-1}U \in p$.

Let $\mathcal{M}$ be the set of neighborhoods of e with respect to $\mathcal{T}$. Assume first that $\mathcal{M} = \mathcal{N}$. To see that $pp = p$, let $A \in p$. Then $A \cup \{e\}$ is a neighborhood of e so pick $U \in \mathcal{T}$ such that $e \in U \subseteq A \cup \{e\}$. Then $U \in p$. We claim that $U \subseteq \{a \in G : a^{-1}A \in p\}$. To this end, let $a \in U$. Then $a^{-1}U \in p$ and $a^{-1}U \subseteq a^{-1}A \cup \{a^{-1}\}$ so $a^{-1}A \cup \{a^{-1}\} \in p$ so $a^{-1}A \in p$.

Now assume that $pp = p$. To see that $\mathcal{M} = \mathcal{N}$, let $V \in \mathcal{N}$. Then $V \in p$. Let $B = V^* = \{x \in V : x^{-1}V \in p\}$. Then $B \in p$. We show that $B \cup \{e\} \in \mathcal{T}$. (Then $e \in B \cup \{e\} \subseteq V$ so $V \in \mathcal{M}$.) To see this, let $x \in B \cup \{e\}$. Now $e^{-1}B = B \in p$ and, if $x \neq e$, then by Lemma 4.14, $x^{-1}B \in p$.

Finally assume that $pp = p$ and let $a \neq e$. Then $ap \neq ep$ by Lemma 6.28, so pick $B \in p \backslash ap$. Let $A = B \backslash (a^{-1}B \cup \{a, a^{-1}\})$. Then $aA \cup \{a\}$ and $A \cup \{e\}$ are disjoint neighborhoods of a and e respectively. $\qquad\square$

The following theorem tells us, among other things, that the topologies determined by p in Theorems 9.13 and 9.14 agree if and only if p is strongly right maximal.

We saw in Corollary 9.12 that, if G is a countable group which can be embedded in a compact metrizable topological group, there are $2^{\mathfrak{c}}$ strongly right maximal idempotents in G^*.

Theorem 9.15. *Let G be a group with identity e and let p be an idempotent in G^*. Let $\mathcal{T}$ be the left invariant topology on G such that $\mathcal{N} = \{A \cup \{e\} : A \in p\}$ is the filter of neighborhoods of e. Then $\mathcal{T}$ is Hausdorff and the following statements are equivalent.*

(a) *The topology $\mathcal{T}$ is regular.*

(b) *The idempotent p is strongly right maximal in G^*.*

(c) *The map $(\rho_p)_{|G}$ is a homeomorphism from $(G, \mathcal{T})$ onto Gp.*

(d) *The topology $\mathcal{T}$ is regular and extremally disconnected.*

(e) *The topology $\mathcal{T}$ is zero dimensional.*

Proof. The topology $\mathcal{T}$ is Hausdorff by Theorem 9.14.

(a) implies (b). Let $q \in G^*$ such that $qp = p$ and suppose that $q \neq p$. Pick $B \in q \backslash p$ with $e \notin B$. Now $G \backslash B = (G \backslash B) \cup \{e\}$ is a neighborhood of e so pick a neighborhood U of e which is closed with respect to $\mathcal{T}$ such that $U \subseteq G \backslash B$. Then $U \in p = qp$ so pick $b \in B$ such that $b^{-1}U \in p$. Since U is closed, $G \backslash U$ is a neighborhood of b so $b^{-1}(G \backslash U) \in p$, a contradiction.

(b) implies (c). By Theorem 9.13, the topology induced on G by the map $(\rho_p)_{|G}$ has $\varphi^\sim = \{U \subseteq G : \{x \in \beta G : xp = p\} \subseteq c\ell_{\beta G}\ U\}$ as the filter of neighborhoods of e. Now, by assumption $\{x \in \beta G : xp = p\} = \{p, e\}$. (By Lemma 6.28, $ap \neq p$ if $a \in G \backslash \{e\}$.) Thus, given $U \subseteq G$, $\{x \in \beta G : xp = p\} \subseteq c\ell_{\beta G}\ U$ if and only if $U \in p$ and $e \in U$. Consequently $\varphi^\sim = \mathcal{N}$ so this topology is $\mathcal{T}$.

(c) implies (d). By Theorem 9.13, the topology induced on G by the map $(\rho_p)_{|G}$ is regular and extremally disconnected.

Each of the implications (d) implies (e) and (e) implies (a) is trivial.   □

Theorem 9.16. *Let G be a discrete group with identity e, and let p be an idempotent in G^*. Let $\mathcal{T}$ be the left invariant topology defined on G by taking $\{A \cup \{e\} : A \in p\}$ as a base for the neighborhoods of e. Let $\mathcal{V}$ be any topology on G for which G has no isolated points. Then $\mathcal{V}$ cannot be strictly finer than $\mathcal{T}$.*

Proof. By Theorem 9.14, $\mathcal{T} = \{U \subseteq G : \text{for all } a \in U, a^{-1}U \in p\}$. Suppose that $\mathcal{T} \not\subseteq \mathcal{V}$ and pick $V \in \mathcal{V} \backslash \mathcal{T}$. Since $V \notin \mathcal{T}$ pick $a \in V$ such that $a^{-1}V \notin p$. Then $G \backslash a^{-1}V \in p$ so $(G \backslash a^{-1}V) \cup \{e\}$ is a neighborhood of e with respect to the topology $\mathcal{T}$ and thus $a\big((G \backslash a^{-1}V) \cup \{e\}\big) = (G \backslash V) \cup \{a\}$ is a neighborhood of a with respect to $\mathcal{T}$, and thus also with respect to $\mathcal{V}$. But then $\big((G \backslash V) \cup \{a\}\big) \cap V = \{a\}$ is a neighborhood of a with respect to $\mathcal{V}$, a contradiction.   □

Corollary 9.17. *Let G be an infinite group. If p is a strongly right maximal idempotent in G^*, then the left invariant topology $\mathcal{T}$ on G defined by taking $\{A \cup \{e\} : A \in p\}$ as the filter of neighborhoods of the identity e is homogeneous, zero dimensional, Hausdorff, extremally disconnected, and maximal among all topologies without isolated points. Distinct strongly right maximal idempotents give rise to distinct topologies.*

Proof. Since the topology is left invariant, it is trivially homogeneous. By Theorem 9.15, $\mathcal{T}$ is Hausdorff, zero dimensional, and extremally disconnected. By Theorem 9.16, $\mathcal{T}$ is maximal among all topologies without isolated points. By Theorem 9.15, $\mathcal{T}$ is the topology induced on G by the function $(\rho_p)_{|G}$. Given distinct right maximal idempotents p and q, one has $p \neq qp$ so $p \notin q\beta G$. Thus, by Theorem 9.13, $(\rho_p)_{|G}$ and $(\rho_q)_{|G}$ induce distinct topologies on G.   □

Combined with Corollary 9.12, the following Corollary shows that there are 2^c distinct homeomorphism classes of topologies on $\mathbb{Z}$ that are homogeneous, zero dimensional, Hausdorff, extremally disconnected, and maximal among all topologies without isolated points.

Corollary 9.18. *Let G be an infinite group. Let*

$$\kappa = |\{p \in G^* : p \text{ is a strongly right maximal idempotent in } G^*\}|.$$

If $2^{|G|} < \kappa$, then there are at least κ distinct homeomorphism classes of topologies on G that are homogeneous, zero dimensional, Hausdorff, extremally disconnected, and maximal among all topologies without isolated points.

Proof. By Corollary 9.17 distinct strongly right maximal idempotents give rise to distinct topologies on G, each of which is homogeneous, zero dimensional, Hausdorff, extremally disconnected, and maximal among all topologies without isolated points. Since any homeomorphism class has at most $|G|^{|G|} = 2^{|G|}$ members, the conclusion follows. $\square$

The use of ultrafilters leads to partition theorems for left topological groups. The following theorem and Exercises 9.2.5 and 9.2.6 illustrate results of this kind.

Let X be a topological space. Recall that an ultrafilter p on X is said to *converge* to a point x of X if and only if p contains the filter of neighborhoods of x.

Theorem 9.19. *Let G be a countably infinite discrete group with identity e. Suppose that there is a left invariant topology τ on G for which there is a right cancelable ultrafilter $p \in G^*$ converging to e. Then G can be partitioned into ω disjoint subsets which are all τ-dense in G.*

Proof. Let $\phi : G \times G \to \mathbb{N}$ be a bijection. For each $b \in G$ let $X_b = \{ap^{\phi(a,b)} : a \in G\}$. Notice that if $a, b, c, d \in G$ and $ap^{\phi(a,b)} = cp^{\phi(c,d)}$, then $\phi(a,b) = \phi(c,d)$ so that $a = c$ and $b = d$. (If $ap^n = cp^m$ where, say, $n > m$, one has by the right cancelability of p that $ap^{n-m} = c \in G$, while G^* is a right ideal of βG by Theorem 4.31.)

We claim that for each $b \in G$, $c\ell_{\beta G} X_b \cap c\ell_{\beta G} \left(\bigcup_{d \in G \setminus \{b\}} X_d \right) = \emptyset$. Suppose instead that for some $b \in G$, $c\ell_{\beta G} X_b \cap c\ell_{\beta G} \left(\bigcup_{d \in G \setminus \{b\}} X_d \right) \neq \emptyset$. Then by Theorem 3.40, either $X_b \cap c\ell_{\beta G} \left(\bigcup_{d \in G \setminus \{b\}} X_d \right) \neq \emptyset$ or $c\ell_{\beta G} X_b \cap \left(\bigcup_{d \in G \setminus \{b\}} X_d \right) \neq \emptyset$. Since the latter implies a version of the former, we may assume that we have some $x \in X_b \cap c\ell_{\beta G} \left(\bigcup_{d \in G \setminus \{b\}} X_d \right)$. Then for some $a \in G$, $x = ap^{\phi(a,b)}$ and $x \in c\ell_{\beta G}\{cp^{\phi(c,d)} : c \in G \text{ and } d \in G \setminus \{b\}\}$. As we have already observed, $ap^{\phi(a,b)} \neq cp^{\phi(c,d)}$ if $d \neq b$, so $x \in c\ell_{\beta G}\{cp^m : c \in G \text{ and } m > \phi(a,b)\}$. Let $n = \phi(a,b)$. Then $ap^n \in c\ell_{\beta G}\{cp^m : c \in G \text{ and } m > n\} \subseteq G^*p^n$ so $p^n \in a^{-1}G^*p^n \subseteq G^*p^n$. Thus, by Theorem 8.18 p^n is not right cancelable, a contradiction.

Since $c\ell_{\beta G} X_b \cap c\ell_{\beta G} \left(\bigcup_{d \in G \setminus \{b\}} X_d \right) = \emptyset$ for every $b \in G$, we can choose a family $\langle A_b \rangle_{b \in G}$ of pairwise disjoint subsets of G such that $X_b \subseteq c\ell_{\beta G} A_b$ for each b. We may replace A_e by $G \setminus \bigcup_{b \in G \setminus \{e\}} A_b$ so that $\{A_b : b \in G\}$ is a partition of G.

Now let $b \in G$. We claim that A_b is τ-dense in G. Notice that for each $n \in \mathbb{N}$, p^n converges to e because (by Exercise 9.2.3) the ultrafilters in G^* that converge to e form a subsemigroup of βG. Let V be a nonempty τ-open subset of G and pick $a \in V$. Then $a^{-1}V$ is a τ-neighborhood of e so $a^{-1}V \in p^{\phi(a,b)}$. Since also $A_b \in ap^{\phi(a,b)}$, we have $V \cap A_b \neq \emptyset$.   □

Corollary 9.20. *Let G be a group with a left invariant topology τ with respect to which G has no isolated points. If there is a countable basis for τ then G can be partitioned into ω disjoint subsets which are all τ-dense in G.*

Proof. Let $\{V_n : n \in \mathbb{N}\}$ be a basis for the neighborhoods of the identity e of G. By Theorem 3.36 the interior in G^* of $\bigcap_{n=1}^{\infty} V_n^*$ is nonempty and thus by Theorem 8.10 there is some $p \in \bigcap_{n=1}^{\infty} V_n^*$ which is right cancelable in βG (where βG is the Stone–Čech compactification of the discrete space G). Thus Theorem 9.19 applies.   □

Exercise 9.2.1. Let S be any discrete semigroup and let p be any idempotent in S^*. Prove that the left ideal $(\beta S)p$ is extremally disconnected. (Hint: Consider Lemma 7.41.)

Exercise 9.2.2. Let G be an infinite discrete group and let p and q be idempotents in G^*. Show that the following statements are equivalent.

(a) $q \leq_R p$.
(b) The function $\psi : Gp \to Gq$ defined by $\psi(ap) = aq$ is continuous.
(c) The topology induced on G by $(\rho_p)_{|G}$ is finer than or equal to the one induced by $(\rho_q)_{|G}$.

(Hint: Consider the proof of Theorem 9.13.)

Exercise 9.2.3. Let G be a group with identity e and let τ be a non-discrete left invariant topology on G. Let $X_\tau = \bigcap\{cl_{\beta G}(U\backslash\{e\}) : U$ is a τ-neighborhood of e in $G\}$. (So X_τ is the set of nonprincipal ultrafilters on G which converge to e.) Show that X_τ is a compact subsemigroup of G^*. (Hint: Use Theorem 4.20.) Show that if τ' is also a non-discrete left invariant topology on G, then $\tau = \tau'$ if and only if $X_\tau = X_{\tau'}$.

The following exercise shows that topologies defined by idempotents can be characterized by a simple topological property.

Exercise 9.2.4. Let G be a group with identity e and let p be an idempotent in G^*. Let $\mathcal{T}$ be the left invariant topology defined on G by choosing the sets of the form $U \cup \{e\}$, where $U \in p$, as the neighborhoods of e. Show that for any $A \subseteq G$ and any $x \in G$, $x \in cl_{\mathcal{T}} A$ if and only if either $x \in A$ or $x^{-1}A \in p$. Then show that, for any two disjoint subsets A and B of G, we have $cl_{\mathcal{T}} A \cap cl_{\mathcal{T}} B = (A \cap cl_{\mathcal{T}} B) \cup (cl_{\mathcal{T}} A \cap B)$.
    Conversely, suppose that $\mathcal{T}$ is a left invariant Hausdorff topology on G without isolated points such that for any two disjoint subsets A and B of G, one has $cl_{\mathcal{T}} A \cap cl_{\mathcal{T}} B = (A \cap cl_{\mathcal{T}} B) \cup (cl_{\mathcal{T}} A \cap B)$. Let $p = \{A \subseteq G : e \in cl_{\mathcal{T}}(A\backslash\{e\})\}$. Show that

p is an idempotent in G^* such that the $\mathcal{T}$-neighborhoods of e are the sets of the form $U \cup \{e\}$, where $U \in p$.

Exercise 9.2.5. Let G be a countably infinite discrete group with identity e and let p be an idempotent in G^*. Let τ be the left invariant topology defined on G by choosing the sets of the form $U \cup \{e\}$, where $U \in p$, as the neighborhoods of e. Show that G cannot be partitioned into two disjoint sets which are both τ-dense in G.

Exercise 9.2.6. Let G be a countably infinite discrete group with identity e and let p be a right maximal idempotent in G^*. Let $D = \{q \in G^* : qp = p\}$. Then D is a finite right zero subsemigroup of G^* by Theorem 9.4. Let τ be the left invariant topology defined on G by choosing the subsets U of G for which $D \cup \{e\} \subseteq c\ell_{\beta G} U$ as the neighborhoods of the identity. Suppose that $|D| = n$. Show that G can be partitioned into n disjoint τ-dense subsets, but cannot be partitioned into $n + 1$ disjoint τ-dense subsets. (Hint: Use Corollary 6.20 to show that $(\beta G)q \cap (\beta G)q' = \emptyset$ if q and q' are distinct elements of C.)

The following exercise provides a contrast to Ellis' Theorem (Corollary 2.39).

Exercise 9.2.7. Let p be an idempotent in $\mathbb{N}^*$ and let $\mathcal{T} = \{U \subseteq \mathbb{Z} :$ for all $a \in U$, $-a + U \in p\}$. Then by Theorem 9.14, $(\mathbb{Z}, +, \mathcal{T})$ is a semitopological semigroup which is algebraically a group. Prove that it is not a topological semigroup. (Hint: Either $\{2^{2n}(2k + 1) : n, k \in \omega\} \in p$ or $\{2^{2n+1}(2k + 1) : n, k \in \omega\} \in p$.)

9.3 Chains of Idempotents

We saw in Corollary 6.34 that non-minimal idempotents exist in S^* whenever S is right cancellative and weakly left cancellative. In this section, we shall show that every non-minimal idempotent p in $\mathbb{Z}^*$ lies immediately above $2^{\mathfrak{c}}$ non-minimal idempotents, in the sense that there are $2^{\mathfrak{c}}$ non-minimal idempotents $q \in \mathbb{Z}^*$ satisfying $q < p$, which are maximal subject to this condition. This will allow us to construct ω_1-sequences of idempotents in $\mathbb{Z}^*$, with the property that each idempotent in the sequence corresponding to a non-limit ordinal is maximal subject to being less than its predecessor. Whether any infinite increasing chains of idempotents exist in $\mathbb{Z}^*$ is a difficult open question.

Theorem 9.21. *Let G be a countably infinite discrete group, let p be any non-minimal idempotent in G^*, and let $A \subseteq G$ with $\overline{A} \cap K(\beta G) \neq \emptyset$. Then there is a set $Q \subseteq (E(G^*) \cap \overline{A}) \backslash K(\beta G)$ such that*

(1) $|Q| = 2^{\mathfrak{c}}$,

(2) *each $q \in Q$ satisfies $q < p$, and*

(3) *each $q \in Q$ is maximal subject to the condition that $q < p$.*

Proof. By Theorems 6.56 and 6.58, there is an infinite subset B of A such that the following statements hold for every $x \in B^*$:

(i) $p \notin (\beta G)xp$,

(ii) xp is right cancelable in βG,

(iii) $(\beta G)xp$ is maximal subject to being a principal left ideal of βG strictly contained in $(\beta G)p$, and

(iv) for all distinct x and x' in B^*, $(\beta G)xp$ and $(\beta G)x'p$ are disjoint.

Let x be a given element of B^* and let C denote the smallest compact subsemigroup of βG which contains xp. Pick an idempotent $u \in C$. By Theorem 8.57, u is not minimal in G^*. We can choose an idempotent $q \in (\beta G)xp$ which is $\leq_R$-maximal in G^* and satisfies $qu = u$ (by Lemma 8.59). Let $v = pq$. Using the fact that $qp = q$, it is easy to check that v is idempotent and that $v \leq p$. Now $v \neq p$, because $v \in (\beta G)xp$ and $p \notin (\beta G)xp$, and so $v < p$.

Let $D = \{w \in E(G^*) : v < w < p\}$. Then $D \subseteq \{w \in \beta G : wq = pq\}$ since, if $v < w < p$, one has $pq = v = wv = wpq = wq$. Thus, by Theorem 9.6, D is finite. If $D \neq \emptyset$, choose w maximal in D. If $D = \emptyset$, let $w = v$. We then have an idempotent $w \in G^*$ which satisfies $w < p$ and is maximal subject to this condition.

We shall show that $w \in (\beta G)xp$. Since $v \in (\beta G)w \cap (\beta G)xp$, it follows from Corollary 6.20, that $w \in (\beta G)xp$ or $xp \in (\beta G)w$. The first possibility is what we wish to prove, and so we may assume the second. We also have $w \in (\beta G)p$ and so $(\beta G)xp \subseteq (\beta G)w \subseteq (\beta G)p$. Now $(\beta G)xp$ is maximal subject to being a principal left ideal of βG strictly contained in $(\beta G)p$. Thus $(\beta G)w = (\beta G)xp$ or $(\beta G)w = (\beta G)p$. The first possibility implies that $w \in (\beta G)xp$, as claimed. The second can be ruled out because it implies that $p \in (\beta G)w$ and hence that $pw = p$, contradicting our assumption that $pw = w$.

We now claim that $w \notin K(\beta G)$. To see this, since $v \leq w$ it suffices to show that $v \notin K(\beta G)$. We note that $xvu = xpqu = xpu \in C$ and that $C \cap K(\beta G) = \emptyset$ (by Theorem 8.57). So $xvu \notin K(\beta G)$ and therefore $v \notin K(\beta G)$.

Thus we have found a non-minimal idempotent $w \in (\beta G)xp$ which satisfies $w < p$, and is maximal subject to this condition. Our theorem now follows from the fact that there are 2^c possible choices of x, because $|B^*| = 2^c$ (by Theorem 3.59), and that $(\beta G)xp$ and $(\beta G)x'p$ are disjoint if x and x' are distinct elements in B^*. □

Lemma 9.22. *Let G be a countably infinite discrete group. Let $\langle q_n \rangle_{n=1}^{\infty}$ be a sequence of idempotents in G^* such that, for every $n \in \mathbb{N}$, $q_{n+1} <_L q_n$. If q is any limit point of the sequence $\langle q_n \rangle_{n=1}^{\infty}$, then $q \in (\beta G)q_n$ for every $n \in \mathbb{N}$ and q is right cancelable in βG.*

Proof. For every $n \in \mathbb{N}$, we have $q_r \in (\beta G)q_n$ whenever $r > n$. Since $(\beta G)q_n$ is closed in βG, it follows that $q \in (\beta G)q_n$.

Suppose that q is not right cancelable in βG. Then $q = uq$ for some idempotent $u \in G^*$ (by Theorem 8.18). So $q \in c\ell\big((G \backslash \{e\})q\big)$ and $q \in c\ell\{q_n : n \in \mathbb{N}\}$, where e denotes the identity of G. It follows from Theorem 3.40 that $aq = q'$ for some

$a \in G\backslash\{e\}$ and some $q' \in c\ell\{q_n : n \in \mathbb{N}\}$, or else $q_n = xq$ for some $n \in \mathbb{N}$ and some $x \in \beta G$.

Assume first that $q_n = xq$ for some $n \in \mathbb{N}$ and some $x \in \beta G$. Then $q \in \beta G q_{n+1}$ so $q_n \in \beta G q_{n+1}$ and thus, $q_n \leq_L q_{n+1}$, a contradiction.

Thus we have $aq = q'$ for some $a \in G\backslash\{e\}$ and some $q' \in c\ell\{q_n : n \in \mathbb{N}\}$. Since $aq \in c\ell\{aq_n : n \in \mathbb{N}\}$ and $q' \in c\ell\{q_n : n \in \mathbb{N}\}$, another application of Theorem 3.40, allows us to deduce that $aq_n = q''$ or else $aq'' = q_n$ for some $n \in \mathbb{N}$ and some $q'' \in c\ell\{q_n : n \in \mathbb{N}\}$. Since the equation $aq'' = q_n$ implies that $a^{-1}q_n = q''$, we need only refute the first of these equations. Assume first that $q'' = q_m$ for some $m \in \mathbb{N}$. Then $q_m = aq_n = aq_nq_n = q_mq_n$ so $q_m \leq_L q_n$. Also $aq_nq_m = q_mq_m = q_m = aq_n$ so by Lemma 8.1, $q_nq_m = q_n$ so that $q_n \leq_L q_m$. Thus $m = n$ and consequently $aq_n = eq_n$ so by Lemma 6.28, $a = e$, a contradiction.

Thus q'' is a limit point of the sequence $\langle q_n \rangle_{n=1}^{\infty}$. So, as already established, $q'' \in (\beta G)q_{n+1}$ and thus $q_n = a^{-1}q'' \in (\beta G)q_{n+1}$ so that $q_n \leq_L q_{n+1}$, a contradiction.   □

Theorem 9.23. *Let G be a countably infinite discrete group and let p be any non-minimal idempotent in G^*. There is an ω_1- sequence $\langle p_\alpha \rangle_{\alpha < \omega_1}$ of distinct non-minimal idempotents in G^* with the following properties:*

(1) *$p_0 = p$,*

(2) *for every $\alpha, \beta \in \omega_1$, $\alpha < \beta$ implies that $p_\beta <_L p_\alpha$,*

(3) *for each non-limit ordinal α in ω_1, $p_\alpha < p_{\alpha-1}$ and p_α is maximal subject to satisfying this relation, and*

(4) *for each limit ordinal $\alpha \neq 0$ in ω_1, p_α is a right maximal idempotent in G^*.*

Proof. We construct the sequence inductively, first putting $p_0 = p$. We then assume that $\beta \in \omega_1$ and that p_α has been defined for every $\alpha < \beta$ so that $\langle p_\alpha \rangle_{\alpha < \beta}$ has the required properties.

If β is not a limit ordinal, we can choose p_β to be a non-minimal idempotent which satisfies $p_\beta < p_{\beta-1}$ and is maximal subject to this condition by Theorem 9.21. Then $\langle p_\alpha \rangle_{\alpha \leq \beta}$ has the required properties.

Suppose then that β is a limit ordinal. We can choose a cofinal sequence $\langle \alpha_n \rangle_{n=1}^{\infty}$ in β. Let q be a limit point of the sequence $\langle p_{\alpha_n} \rangle_{n=1}^{\infty}$. By Lemma 9.22, $q \in (\beta G)p_{\alpha_n}$ for every $n \in \mathbb{N}$ and so $q \in (\beta G)p_\alpha$ for every $\alpha < \beta$. Furthermore, q is right cancelable in βG. So we can choose an idempotent $p_\beta \in (\beta G)q$ which is $\leq_R$-maximal in G^* (by Lemma 8.59). The fact that $p_\beta \in (\beta G)p_\alpha$ for every $\alpha < \beta$, implies that $p_\beta \leq_L p_\alpha$. Furthermore, if $\alpha < \beta$, we can choose $\gamma \in \omega_1$ satisfying $\alpha < \gamma < \beta$. We then have $p_\beta \leq_L p_\gamma <_L p_\alpha$ and this implies that $p_\beta <_L p_\alpha$. Thus the sequence $\langle p_\alpha \rangle_{\alpha \leq \beta}$ has the required properties.   □

Remark 9.24. *Theorems 9.21 and 9.23 are valid with $(\mathbb{N}, +)$ in place of G. This is an immediate consequence of the fact that $\mathbb{N}^*$ is a left ideal of $(\beta\mathbb{Z}, +)$ by Exercise 4.3.5.*

We conclude this section by listing some open questions which are tantalizingly easy to formulate, but very hard to answer.

We shall see, as a consequence of Theorems 12.29 and 12.45, that it is consistent that there are idempotents p in $\beta\mathbb{N}$ such that, if $p = q + r$, then both q and r are in $\mathbb{Z} + p$. In particular, such idempotents are both $\leq_R$-maximal and $\leq_L$-maximal.

Question 9.25. *Can it be shown in ZFC that $\mathbb{Z}^*$ contains left maximal idempotents?*

As a consequence of Exercise 9.1.4 we know that no minimal idempotent in $\mathbb{Z}^*$ can be right maximal.

Question 9.26. *Are there any idempotents in $\mathbb{Z}^*$ which are both minimal and maximal?*

If $\langle p_n \rangle_{n=1}^\infty$ is a sequence of idempotents in $\beta\mathbb{Z}$ such that $p_n <_L p_{n+1}$ for each n, then $\langle \beta\mathbb{Z} + p_n \rangle_{n=1}^\infty$ is a strictly increasing chain of principal left ideals. We saw in Chapter 6 that the question of whether such a chain of left ideals exists is itself a difficult open problem.

Question 9.27. *Is there an infinite increasing $\leq_L$-chain of idempotents in $\beta\mathbb{Z}$?*

Exercise 9.3.1. Suppose that κ is a cardinal for which there exists a sequence $\langle p_\alpha \rangle_{\alpha<\kappa}$ in $\beta\mathbb{Z}$ with the properties listed in the statement of Theorem 9.23. Show that $\kappa \leq \mathfrak{c}$. (Hint: First show that, for each $\alpha \in \kappa$, $p_\alpha \notin c\ell\{p_\beta : \beta > \alpha\}$. This implies that there is a clopen subset U_α of $\beta\mathbb{Z}$ such that $p_\alpha \in U_\alpha$ and $p_\beta \notin U_\alpha$ if $\beta > \alpha$.)

9.4 Identities in βS

If S is any discrete semigroup, then an identity e of S is also an identity of βS. This follows from the fact that the equations $se = s$ and $es = s$, which are valid for every $s \in S$, extend to βS because the maps ρ_e and λ_e are continuous on βS. In this section, we shall see that this is the only way in which an identity can arise in βS. We shall show, in fact, that βS cannot have a unique right identity which is a member of S^*.

If S is a weakly left cancellative semigroup, S^* is a left ideal in βS (by Theorem 4.31) and βS can have no right identities in S^*. However, a simple example of a semigroup S in which S^* has right identities, although S has none, is provided by putting $S = (\mathbb{N}, \wedge)$. In this case, every element of S^* is a right identity for βS. (See Exercise 4.1.11.)

In contrast, we shall see that, if S is commutative, then no element of S^* can be a left identity for βS.

Theorem 9.28. *Let S be a discrete semigroup. If βS has a right identity $q \in S^*$, then βS has at least 2^{2^κ} right identities in S^*, where $\kappa = \|q\|$.*

Proof. For each $s \in S$, let $I_s = \{t \in S : st = s\}$. Since $sq = s$, we have $q \in \lambda_s^{-1}[\{s\}]$. This is an open subset of βS and so $I_s = S \cap \lambda_s^{-1}[\{s\}] \in q$. Choose $A \in q$ with $|A| = \kappa$. Then $\{A \cap I_s : s \in A\}$ is a collection of κ sets with the κ-uniform finite intersection property. It follows from Theorem 3.62 that $\bigcap_{s \in A} \overline{A \cap I_s}$ contains 2^{2^κ} κ-uniform ultrafilters. We shall show that every ultrafilter p in this set is a right identity for βS.

To see this, let $s \in S$. We can choose $t \in A \cap I_s$, because A and I_s are both in q. Then $I_t \in p$. Now $I_t \subseteq I_s$, because $u \in I_t$ implies that $su = (st)u = s(tu) = st = s$. So $I_s \in p$. Hence $sp \in c\ell\{su : u \in I_s\} = \{s\}$ and therefore $sp = s$. It follows from the continuity of ρ_p that $xp = x$ for every $x \in \beta S$. $\qquad\square$

Corollary 9.29. *Let S be any discrete semigroup. If $q \in \beta S$ is a unique right identity for βS, then $q \in S$.*

Corollary 9.30. *Let S be any discrete semigroup. If $q \in \beta S$ is an identity for βS, then $q \in S$.*

Theorem 9.31. *Let S be a left cancellative discrete semigroup. If p is a right identity for βS, then $p \in S$ and p is an identity for S and hence for βS.*

Proof. Let $s \in S$. Since $sp = s$, we must have $p \in S$, because S^* is a left ideal of βS (by Theorem 4.31). This implies that p is a left identity for S because, for every $t \in S$, we have $st = spt$ and so $t = pt$. So p is an identity for S, and we have noted that this implies that p is an identity for βS. $\qquad\square$

Theorem 9.32. *Let S be a commutative discrete semigroup and let $q \in S^*$. Then q cannot be a left identity for βS.*

Proof. If q is a left identity for βS, then $qs = s$ for every $s \in S$. Since $qs = \lim_{t \to q} ts = \lim_{t \to q} st = sq$, we have $sq = s$ for every $s \in S$. By the continuity of ρ_q, this implies that $xq = x$ for every $x \in \beta S$. So q is an identity for βS, contradicting Corollary 9.30. $\qquad\square$

We obtain a partial analogue to Theorem 9.28.

Theorem 9.33. *Let S be a discrete semigroup. If βS has a left identity $q \in S^*$ with $\|q\| = \kappa$, then S^* has at least 2^{2^κ} elements p with the property that $ps = s$ for every $s \in S$.*

Proof. The proof of Theorem 9.28 may be copied, with left-right switches, except for the last sentence. $\qquad\square$

We now see that, if S is countable, any right identity on S^* is close to being a right identity on βS.

Theorem 9.34. *Let S be any discrete countable semigroup. If S^* has a right identity q, then S has a left ideal V such that q is a right identity on V and $S \backslash V$ is finite.*

Proof. Let $U = S^* \backslash c\ell(Sq \cap S^*)$. For each $s \in S$, $\{sq\} \cap S^*$ is either a singleton or empty. Thus, by Corollary 3.37, $Sq \cap S^*$ is nowhere dense so that U is a dense open subset of S^*.

Let $T = \{t \in S : tq \in S\}$. We claim that $S \backslash T$ is finite. To see this, suppose the contrary. Then there is an element $x \in U \cap (S \backslash T)^*$, and we can choose $X \in x$ satisfying $X^* \subseteq U$ and $X \subseteq S \backslash T$. We have $x \in \overline{X}$ and $xq \in \overline{Xq}$. Since $x = xq$, it follows from Theorem 3.40 that $s = yq$ for some $s \in X$ and some $y \in \overline{X}$, or $y = sq$ for some $s \in X$ and some $y \in X^*$. The second possibility cannot hold because $X^* \subseteq U$. So we may assume the first. Then $s \in \overline{Xq}$ and so, since s is isolated in βS, we must have $s = tq$ for some $t \in X$. This contradicts the assumption that $X \subseteq S \backslash T$. So $S \backslash T$ is finite and therefore $T^* = S^*$.

Since q is idempotent, $Tq \subseteq T$. So ρ_q maps T to T. It follows from Theorem 3.35 that, if $V = \{t \in T : tq = t\}$, then $V \in x$ for every $x \in T^*$. So $T \backslash V$ is finite and therefore $S \backslash V$ is finite. Furthermore, it is clear that V is a left ideal of S.   □

Corollary 9.35. *Let S be any discrete countable semigroup. If S^* has a right identity, then S^* has at least 2^c right identities.*

Proof. Let $q \in S^*$ be a right identity for S^* and let V be the left ideal of S guaranteed by Theorem 9.34. By Theorem 9.28, with V in place of S, there are 2^c elements of S^* which are right identities on $\overline{V}$. If p is one of these elements and if $x \in S^*$, then the fact that $vp = v$ for every $v \in V$ implies that $xp = x$, because $x \in \overline{V}$.   □

Exercise 9.4.1. Let S be a right cancellative discrete semigroup and let p be a left identity for βS. Prove that $p \in S$ and that p is an identity for S and hence for βS. (This is Theorem 9.31 with the words "left" and "right" interchanged.)

Notes

Theorem 9.10 was proved by I. Protasov (in a personal communication).

Topologies of the kind discussed in Theorem 9.13 have been studied by T. Papazyan [191].

Theorem 9.15, Theorem 9.16, Corollary 9.17, and Theorem 9.19 are due to I. Protasov (in personal communications). Corollary 9.17 answers an old unpublished question of E. van Douwen, who wanted to know if there were any homogeneous regular topologies that are maximal among topologies with no isolated points.

The semigroups X_τ, defined in Exercise 9.2.3, were introduced by E. Zelenuk in [246].

Theorem 9.28 is due to J. Baker, A. Lau, and J. Pym, in [9].

Chapter 10

Homomorphisms

In this chapter, we shall consider continuous homomorphisms defined on subsemigroups of βS, where S denotes a countable discrete semigroup which is commutative and cancellative.

Given two semigroups which are also topological spaces, we shall say that they are algebraically and topologically equivalent or that one is a copy of the other, if there is a mapping from one to the other which is both a homeomorphism and an algebraic isomorphism.

We have seen that copies of $\mathbb{H}$ occur everywhere, because of the remarkable facility with which homomorphisms can be defined on $\mathbb{H}$. (See Theorem 6.32, for example).

In contrast, if S and T are countably infinite discrete semigroups which are commutative and cancellative, there are surprisingly few continuous injective homomorphisms mapping βT into βS or T^* into S^*, and none at all mapping βT into S^*. In particular, there are no topological and algebraic copies of $\beta \mathbb{N}$ in $\mathbb{N}^*$. Of course, any injective homomorphism from T into S does determine a continuous injective homomorphism from βT into βS, as well as a continuous injective homomorphism from T^* into S^* (by Exercise 3.4.1 and Theorem 4.8). We shall see in Theorems 10.30 and 10.31 that it is true that every continuous injective homomorphism from βT into βS, and almost true that every continuous injective homomorphism from T^* into S^*, must arise in this way as the extension of an injective homomorphism from T into S.

If S and T are groups and if L and M are principal left ideals in S^* and T^* respectively, defined by nonminimal idempotents, we shall see that any mapping between L and M which is both an isomorphism and a homeomorphism must arise from an isomorphism between S and T. We shall also see that any two distinct principal left ideals of $\beta \mathbb{N}$ defined by nonminimal idempotents cannot be copies of each other. Whether this statement holds for minimal left ideals is an open question.

We remind the reader that, if S is a commutative discrete semigroup, then S is contained in the center of βS (by Theorem 4.23). We shall frequently use this fact without giving any further reference.

We shall use additive notation throughout this chapter, except on one occasion where we present two results about $(\mathbb{N}, \cdot)$, because $(\mathbb{N}, +)$ is the most important semigroup to which our discussion applies and because this notation in convenient for some of the proofs.

10.1 Homomorphisms to the Circle Group

In this section we develop some algebraic information related to natural homomorphisms from a subsemigroup of $(\mathbb{R}, +)$, the primary example being $\mathbb{N}$, to the circle group $\mathbb{T}$. This information is of interest for its own sake and will be useful in Section 10.2 and again in Section 16.4 where we produce a large class of examples of sets with special combinatorial properties.

We take the circle group $\mathbb{T}$ to be $\mathbb{R}/\mathbb{Z}$ and shall use $\oplus$ for the group operation in $\mathbb{T}$. We define π to be the natural projection of $\mathbb{R}$ onto $\mathbb{T}$. If $t_1, t_2 \in \mathbb{T}$ we shall write $t_1 \prec t_2$ if there exists $u_1, u_2 \in \mathbb{R}$ such that $\pi(u_1) = t_1$, $\pi(u_2) = t_2$ and $u_1 < u_2 < u_1 + \frac{1}{2}$.

Lemma 10.1. *Let S be an infinite discrete semigroup and let $h : S \to \mathbb{T}$ be an injective homomorphism, with $\widetilde{h}$ denoting its continuous extension. Let*

$$U = \{x \in S^* : \{s \in S : \widetilde{h}(x) \prec h(s)\} \in x\},$$
$$D = \{x \in S^* : \{s \in S : h(s) \prec \widetilde{h}(x)\} \in x\}, \text{ and}$$
$$Z = \{x \in S^* : \widetilde{h}(x) = 0\}.$$

Then $U \cap D = \emptyset$, $U \cup D = S^*$, and $U \cap Z$ and $D \cap Z$ are non-empty. Furthermore, U and D are both right ideals of βS.

Proof. Let $x \in S^*$, let $T_1 = \{s \in S : \widetilde{h}(x) \prec h(s)\}$ and $T_2 = \{s \in S : h(s) \prec \widetilde{h}(x)\}$. Then $T_1 \cap T_2 = \emptyset$. Since $|S \backslash (T_1 \cup T_2)| \leq 1$, $T_1 \in x$ or $T_2 \in x$. Thus $x \notin U \cap D$ and $x \in U \cup D$.

We now show that $U \cap Z$ and $D \cap Z$ are non-empty. Let p be an idempotent in S^*. Since $\widetilde{h}$ is a homomorphism (by Corollary 4.22), $p \in Z$. Suppose that $p \in U$. Then we can choose a sequence $\langle t_n \rangle_{n=1}^{\infty}$ in $h[S]$ for which $\pi(0) \prec t_n$ for every n and $t_n \to \pi(0)$. Now $\widetilde{h}[\beta S]$ is a compact subsemigroup of $\mathbb{T}$ and is therefore a group (by Exercise 2.2.3). Hence, for each $n \in \mathbb{N}$, we can choose $y_n \in \beta S$ for which $\widetilde{h}(y_n) = -t_n$. Let $y \in S^*$ be a limit point of the sequence $\langle y_n \rangle_{n=1}^{\infty}$. Then $y \in Z$ and $\widetilde{h}(y_n) = -t_n \prec \pi(0)$ for every n. For each $n \in \mathbb{N}$, $y_n \in c\ell(\{s \in S : h(s) \prec \pi(0)\})$ and so $y \in c\ell(\{s \in S : h(s) \prec \pi(0)\})$. Thus $\{s \in S : h(s) \prec \pi(0)\} \in y$ and $y \in D$. This shows that $p \in U$ implies that $D \cap Z \neq \emptyset$. Similarly, $p \in D$ implies that $U \cap Z \neq \emptyset$. So $U \cap Z$ and $D \cap Z$ are non-empty.

We now show that U is a right ideal of βS. Suppose that $x \in U$ and that $y \in \beta S$. Choose u and v in $\mathbb{R}$ such that $\pi(u) = \widetilde{h}(x)$ and $\pi(v) = \widetilde{h}(y)$. Let $X = \{s \in S : \widetilde{h}(x) \prec h(s)\}$. Then $X \in x$. Given $a \in X$, let $\overline{a} \in (u, u + \frac{1}{2})$ satisfy $\pi(\overline{a}) = h(a)$. Put $Y_a = \{s \in S : h(s) \in \pi[(v+u-\overline{a}, v+u+\frac{1}{2}-\overline{a})]\}$. Now $\pi[(v+u-\overline{a}, v+u+\frac{1}{2}-\overline{a})]$ is a neighborhood of $\pi(v) = \widetilde{h}(y)$ in $\mathbb{T}$, and so $Y_a \in y$. By Theorem 4.15, $\{a + b : a \in X$ and $b \in Y_a\} \in x + y$. We claim that $\{a + b : a \in X$ and $b \in Y_a\} \subseteq \{s \in S : \widetilde{h}(x + y) \prec h(s)\}$ so that $x + y \in U$ as required. To this end let $a \in X$ and $b \in Y_a$. Pick $w \in (v + u - \overline{a}, v + u + \frac{1}{2} - \overline{a})$ such that $\pi(w) = h(b)$. Since $\widetilde{h}$ and π are homomorphisms we have $\widetilde{h}(x + y) = \pi(u + v)$ and $h(a + b) = \pi(\overline{a} + w)$. Since $u + v < \overline{a} + w < u + v + \frac{1}{2}$, we have $\widetilde{h}(x + y) \prec h(a + b)$ as required.

This establishes that U is a right ideal of βS. The proof that D is a right ideal of βS is essentially the same. $\square$

The following notation does not reflect its dependence on the choice of the semi-group S.

Definition 10.2. Let S be a subsemigroup of $(\mathbb{R}, +)$ and let α be a positive real number. We define functions $g_\alpha : S \to \mathbb{Z}$, $f_\alpha : S \to [-\frac{1}{2}, \frac{1}{2})$, and $h_\alpha : S \to \mathbb{T}$ as follows for $x \in S$:

$$g_\alpha(x) = \lfloor x\alpha + \tfrac{1}{2} \rfloor;$$
$$f_\alpha(x) = x\alpha - g_\alpha(x);$$
$$h_\alpha(x) = \pi\big(f_\alpha(x)\big).$$

We shall use $\widetilde{g_\alpha}$, $\widetilde{f_\alpha}$, and $\widetilde{h_\alpha}$ to denote the continuous extensions of these functions which map βS to $\beta \mathbb{Z}$, $[-\frac{1}{2}, \frac{1}{2}]$, and $\mathbb{T}$ respectively, where S has the discrete topology. These exist because the spaces $\beta \mathbb{Z}$, $[-\frac{1}{2}, \frac{1}{2}]$, and $\mathbb{T}$ are compact.

Note that $g_\alpha(x)$ is the nearest integer to $x\alpha$ if $x\alpha \notin \mathbb{Z} + \frac{1}{2}$. We shall use the fact that, for any $p \in \beta S$ and any neighborhood U of $\widetilde{f_\alpha}(p)$, the fact that $\widetilde{f_\alpha}$ is continuous implies that $\widetilde{f_\alpha}^{-1}[U]$ is a neighborhood of p and hence that $f_\alpha^{-1}[U] \in p$ (by Theorem 3.22).

Lemma 10.3. *Let S be a subsemigroup of $(\mathbb{R}, +)$ and let $\alpha > 0$. The map h_α is a homomorphism from S to $\mathbb{T}$. Consequently $\widetilde{h_\alpha} : \beta S \to \mathbb{T}$ is also a homomorphism.*

Proof. We note that $h_\alpha(x) = \pi(\alpha x)$ and so h_α is a homomorphism. The second conclusion then follows by Corollary 4.22. $\square$

Definition 10.4. Let S be a subsemigroup of $(\mathbb{R}, +)$ and let $\alpha > 0$.

(a) $U_\alpha = \{p \in \beta S : \{x \in S : \widetilde{f_\alpha}(p) < f_\alpha(x)\} \in p\}$.
(b) $D_\alpha = \{p \in \beta S : \{x \in S : \widetilde{f_\alpha}(p) > f_\alpha(x)\} \in p\}$.
(c) $Z_\alpha = \{p \in \beta S : \widetilde{f_\alpha}(p) = 0\}$.
(d) $X_\alpha = U_\alpha \cap Z_\alpha$.
(e) $Y_\alpha = D_\alpha \cap Z_\alpha$.

Observe that U_α and D_α are the set of points of βS for which $\widetilde{f_\alpha}$ approaches its value from above and below respectively.

Definition 10.5. Let S be a subsemigroup of $(\mathbb{R}, +)$ and let $\alpha \in \mathbb{R}$. Then α is *irrational with respect to S* if and only if

$$\alpha \notin \left\{\frac{n}{a-b} : n \in \mathbb{Z},\ a, b \in S,\ \text{and}\ a \neq b\right\} \cup \left\{\frac{n}{a} : n \in \mathbb{Z}\ \text{and}\ a \in S \backslash \{0\}\right\}.$$

Saying that α is irrational with respect to S means that h_α is one-to-one on S and that $0 \notin h_\alpha[S \backslash \{0\}]$.

Notice that "irrational with respect to $\mathbb{N}$" is simply "irrational". Notice also that if $|S| < \mathfrak{c}$, then the set of numbers irrational with respect to S is dense in $\mathbb{R}$.

Remark 10.6. *Suppose that S is a subsemigroup of $(\mathbb{R}, +)$ and that $\alpha > 0$ is irrational with respect to S. Let U, D and Z be the sets defined in Lemma 10.1 with $h = h_\alpha$. Then $U = U_\alpha$, $D = D_\alpha$ and $Z = Z_\alpha \backslash \{0\}$.*

Lemma 10.7. *Let S be a subsemigroup of $(\mathbb{R}, +)$ and let $\alpha > 0$ be irrational with respect to S. Then $U_\alpha \cup D_\alpha = S^*$.*

Proof. This is immediate from Lemma 10.1 and Remark 10.6. $\qquad\qquad\qquad\square$

The following result is of interest because, given the continuity of ρ_p, it is usually easier to describe left ideals of βS. (See for example Theorem 5.20.)

Theorem 10.8. *Let S be a subsemigroup of $(\mathbb{R}, +)$ and let $\alpha > 0$ be irrational with respect to S. Then U_α and D_α are right ideals of βS.*

Proof. This is immediate from Lemma 10.1 and Remark 10.6. $\qquad\qquad\qquad\square$

In the following lemma, all we care about is that $c\ell_{\beta S}(\mathbb{N}a) \cap U_\alpha \neq \emptyset$ and that $c\ell_{\beta S}(\mathbb{N}a) \cap D_\alpha \neq \emptyset$. We get the stronger conclusion for free, however.

Lemma 10.9. *Let S be a subsemigroup of $(\mathbb{R}, +)$ and let $\alpha > 0$ be irrational with respect to S. Then for each $a \in S \backslash \{0\}$, $c\ell_{\beta S}(\mathbb{N}a) \cap X_\alpha \neq \emptyset$ and $c\ell_{\beta S}(\mathbb{N}a) \cap Y_\alpha \neq \emptyset$.*

Proof. We apply Lemma 10.1 with h replaced by h_α and S replaced by $\mathbb{N}a$. Then $c\ell_{\beta S}(\mathbb{N}a) \cap X_\alpha = U \cap Z \neq \emptyset$ and $c\ell_{\beta S}(\mathbb{N}a) \cap Y_\alpha = D \cap Z \neq \emptyset$. $\qquad\square$

For the remainder of this section we restrict our attention to $\mathbb{N}$.

Notice that, while h_α is a homomorphism on $(\beta\mathbb{N}, +)$, it is not a homomorphism on $(\beta\mathbb{N}, \cdot)$. However, close to 0 it is better behaved.

Theorem 10.10. *Let $S = \mathbb{N}$ and let α be a positive irrational number. Then X_α and Y_α are left ideals of $(\beta\mathbb{N}, \cdot)$.*

Proof. We establish the statement for X_α, the proof for Y_α being nearly identical.

Let $p \in X_\alpha$ and let $q \in \beta\mathbb{N}$. We first observe that, for any $m, n \in \mathbb{N}$, if $|f_\alpha(n)| < \frac{1}{2m}$, then $f_\alpha(mn) = mf_\alpha(n)$.

To see that $qp \in X_\alpha$, let $\epsilon > 0$ be given (with $\epsilon \leq \frac{1}{2}$). For each $m \in \mathbb{N}$, let $B_m = \{n \in \mathbb{N} : 0 < f_\alpha(n) < \frac{\epsilon}{m}\}$ and note that $B_m \in p$. Thus by Theorem 4.15, $\{mn : m \in \mathbb{N}$ and $n \in B_m\} \in qp$. since $\{mn : m \in \mathbb{N}$ and $n \in B_m\} \subseteq \{k \in \mathbb{N} : 0 < f_\alpha(k) < \epsilon\}$, it follows that $qp \in X_\alpha$. $\qquad\qquad\qquad\square$

Of course, as the kernel of a homomorphism, Z_α is a subsemigroup of $(\beta\mathbb{N}, +)$ for any $\alpha > 0$. Consequently, by Theorem 10.8, if α is irrational, then X_α and Y_α are subsemigroups of $(\beta\mathbb{N}, +)$.

Corollary 10.11. *Let $S = \mathbb{N}$ and let α be a positive irrational number. Then Z_α is a subsemigroup of $(\beta\mathbb{N}, \cdot)$.*

Proof. Since α is irrational, $\{n \in \mathbb{N} : f_\alpha(n) = 0\} = \emptyset$ so by Lemma 10.7, $Z_\alpha = X_\alpha \cup Y_\alpha$ so the result follows from Theorem 10.10. $\qquad\square$

The following result is our main tool to be used in Section 16.4 in deriving nontrivial explicit examples of IP* sets and central* sets.

Theorem 10.12. *Let $S = \mathbb{N}$ and let $\alpha > 0$. Then $\widetilde{g_\alpha}$ is an isomorphism and a homeomorphism from Z_α onto $Z_{1/\alpha}$ with inverse $\widetilde{g_{1/\alpha}}$. If α is irrational, then $\widetilde{g_\alpha}$ takes X_α onto $Y_{1/\alpha}$ and takes Y_α onto $X_{1/\alpha}$.*

Proof. We show that

(1) if $p, q \in Z_\alpha$, then $\widetilde{g_\alpha}(p + q) = \widetilde{g_\alpha}(p) + \widetilde{g_\alpha}(q)$,

(2) if $p \in Z_\alpha$, then $\widetilde{g_\alpha}(p) \in Z_{1/\alpha}$,

(3) if $p \in Z_\alpha$, then $\widetilde{g_{1/\alpha}}\big(\widetilde{g_\alpha}(p)\big) = p$,

(4) if $p \in X_\alpha$, then $\widetilde{g_\alpha}(p) \in Y_{1/\alpha}$, and

(5) if $p \in Y_\alpha$, then $\widetilde{g_\alpha}(p) \in X_{1/\alpha}$.

Then using the fact that the same assertions are valid with $1/\alpha$ replacing α proves the theorem. Notice that the hypotheses of statements (4) and (5) are nonvacuous only when α is irrational.

To verify (1), let $p, q \in Z_\alpha$ and let $0 < \epsilon < \frac{1}{4}$. Let $A = \{m \in \mathbb{N} : |g_\alpha(m) - \alpha m| < \epsilon\}$. If n and m are in A, then $|g_\alpha(m) + g_\alpha(n) - \alpha(m + n)| < 2\epsilon < \frac{1}{2}$ so $g_\alpha(m) + g_\alpha(n) = g_\alpha(m + n)$. Also, $A \in p$ and $A \in q$ so

$$\widetilde{g_\alpha}(p+q) = p\text{-}\lim_{m\in A} q\text{-}\lim_{n\in A} g_\alpha(m+n) = p\text{-}\lim_{m\in A} q\text{-}\lim_{n\in A}\big(g_\alpha(m)+g_\alpha(n)\big) = \widetilde{g_\alpha}(p)+\widetilde{g_\alpha}(q).$$

To verify (2) and (3), let $p \in Z_\alpha$ and let $q = \widetilde{g_\alpha}(p)$. Let $\epsilon \in (0, \frac{\alpha}{2})$ and let $B = \{n \in \mathbb{N} : |f_\alpha(n)| < \epsilon\}$. For any $n \in \mathbb{N}$, we have $|f_\alpha(n)| < \epsilon$ if and only if $|m - n\alpha| < \epsilon$ where $m = g_\alpha(n)$. Now $|m - n\alpha| < \epsilon$ if and only if $|m\frac{1}{\alpha} - n| < \frac{\epsilon}{\alpha}$. The second inequality establishes that n is the integer closest to $m\frac{1}{\alpha}$ and hence that $n = g_{\frac{1}{\alpha}}(m)$. Thus we have seen that $n \in B$ implies both $|f_{\frac{1}{\alpha}}\big(g_\alpha(n)\big)| < \frac{\epsilon}{\alpha}$ and $n = g_{\frac{1}{\alpha}}\big(g_\alpha(n)\big)$. So $|\widetilde{f_{\frac{1}{\alpha}}}(q)| = p\text{-}\lim_{n\in B} |f_{\frac{1}{\alpha}}\big(g_\alpha(n)\big)| \le \frac{\epsilon}{\alpha}$ and $p = p\text{-}\lim_{n\in B} n = p\text{-}\lim_{n\in B} g_{\frac{1}{\alpha}}\big(g_\alpha(n)\big) = \widetilde{g_{\frac{1}{\alpha}}}(q)$. This establishes (2) and (3).

To verify (4), let $p \in X_\alpha$, let $\epsilon > 0$ be given with $\epsilon < 1/2$, and let $B = \{n \in \mathbb{N} : -\epsilon < f_{1/\alpha}(n) < 0\}$. We need to show $B \in g_\alpha(p)$. Pick $\delta > 0$ with $\delta < \epsilon \cdot \alpha$. Let $C = \{n \in \mathbb{N} : 0 < f_\alpha(n) < \delta\}$. We show $g_\alpha[C] \subseteq B$. Let $n \in C$ and let $m = g_\alpha(n)$ so that $m < \alpha n < m + \delta$. Then $m/\alpha < n < m/\alpha + \delta/\alpha < m/\alpha + \epsilon$ so $n - \epsilon < m/\alpha < n$ and hence $m \in B$ as required. This establishes (4), and (5) is shown in just the same way. $\qquad\square$

Exercise 10.1.1. It is a fact that if F and G are disjoint finite nonempty subsets of $\mathbb{R}$ and $F \cup G$ is linearly independent over $\mathbb{Q}$, then $\bigcap_{\alpha \in F} U_\alpha \cap \bigcap_{\alpha \in G} D_\alpha \neq \emptyset$. Use this fact, together with Theorem 10.8 to show that $(\beta \mathbb{N}, +)$ has a collection of $2^\mathfrak{c}$ pairwise disjoint right ideals. (This fact also follows from Theorem 6.44.)

Exercise 10.1.2. Let $\alpha > 0$ and let $p, q \in \beta \mathbb{N}$. Prove that $\widetilde{g_\alpha}(p + q) \in \{\widetilde{g_\alpha}(p) + \widetilde{g_\alpha}(q), \widetilde{g_\alpha}(p) + \widetilde{g_\alpha}(q) - 1, \widetilde{g_\alpha}(p) + \widetilde{g_\alpha}(q) + 1\}$.

Exercise 10.1.3. Let $\alpha > 0$ be irrational, let $k \in \mathbb{N}$ with $k > \alpha$, and let $\delta = k - \alpha$. Prove that

   (i) for all $p \in \beta \mathbb{N}$, $\widetilde{h_\alpha}(p) = -\widetilde{h_\delta}(p)$,

   (ii) $Z_\alpha = Z_\delta$,

   (iii) $X_\alpha = Y_\delta$, and

   (iv) $Y_\alpha = X_\delta$.

10.2 Homomorphisms from βT into S^*

We know that topological copies of $\beta \mathbb{N}$ are abundant in $\mathbb{N}^*$. Indeed, every infinite compact subset of $\mathbb{N}^*$ contains a copy of this kind, by Theorem 3.59. However, as we shall show in this section, there are no copies of the right topological semigroup $\beta \mathbb{N}$ in $\mathbb{N}^*$. In fact, if $\phi : \beta \mathbb{N} \to \mathbb{N}^*$ is a continuous homomorphism, then $\phi[\beta \mathbb{N}]$ is finite and $|\phi[\mathbb{N}^*]| = 1$.

We shall show that, if S and T are countable, commutative, cancellative semigroups and if S can be embedded in the unit circle, then we can often assert that $\phi[T^*]$ is a finite group whenever $\phi : \beta T \to S^*$ is a continuous homomorphism.

Lemma 10.13. *Let S be an infinite discrete semigroup which can be algebraically embedded in $\mathbb{T}$ and let V be an infinite subsemigroup of S. Then, for every x and p in S^*, there exists $y \in V^*$ for which $x + y + p \neq y + p + x$.*

Proof. Let $h : S \to \mathbb{T}$ be an injective homomorphism, with $\widetilde{h} : \beta S \to \mathbb{T}$ denoting its continuous extension. Let U and D be defined as in Lemma 10.1. By this lemma, applied to V instead of S, if $U' = \{y \in V^* : \{s \in V : \widetilde{h}(y) \prec h(s)\} \in y\}$ and $D' = \{y \in V^* : \{s \in V : h(s) \prec \widetilde{h}(y)\} \in y\}$, then U' and D' are non-empty.

Let $x, p \in S^*$. If $x \in U$, we can choose $y \in D'$. Since $D' \subseteq D$ and U and D are right ideals of βS, $x + y + p \in U$ and $y + p + x \in D$ and so $x + y + p \neq y + p + x$. Similarly, if $x \in D$, we can choose $y \in U'$ and deduce that $x + y + p \neq y + p + x$. $\square$

Lemma 10.14. *Suppose that S is a countably infinite discrete semigroup which can be algebraically embedded in $\mathbb{T}$ and that T is a countably infinite commutative weakly left cancellative semigroup. Let $\phi : \beta T \to S^*$ be a continuous homomorphism. Then $K(\phi[T^*])$ is a finite group. Furthermore, for every $c \in T$, $\{b \in T : \phi(b + c) \in K(\phi[T^*])\}$ is infinite.*

Proof. Observe that by Theorem 4.23, if $a \in S$, then a commutes with every member of βS. Also, if $t \in T$, then t commutes with every member of βT and hence $\phi(t)$ commutes with every member of $\phi[\beta T]$.

We shall show that if p is any minimal idempotent in $\phi[T^*]$ and $q \in T^*$ with $\phi(q) = p$, then

(1) there exists $t \in T$ such that $\phi(t) \in \phi[T^*] + p$ and

(2) given any $c \in T$ and any $B \in q$, there exists $t \in T$ such that $\phi(b) + \phi(c) \in K(\phi[T^*])$.

To see this, let such p and q be given, let $c \in T$, and let $B \in q$. Let $V = \{a \in S : a + \phi(q) \in \phi[\beta T]\}$. We claim that V is finite. Suppose instead that V is infinite, and notice that V is a subsemigroup of S. (Given $a, b \in V, a + b + \phi(q) = a + b + \phi(q) + \phi(q) = a + \phi(q) + b + \phi(q)$.) For every $y \in V^*$, we have $y + \phi(q) \in \phi[\beta T]$, by the continuity of $\rho_{\phi(q)}$. Then for every $y \in V^*$, we have $\phi(c) + y + \phi(q) = y + \phi(q) + \phi(c)$ which contradicts Lemma 10.13.

Choose $A \in \phi(c)$ for which $A \cap V = \emptyset$. Now $\phi(c) + \phi(q) = \phi(q) + \phi(c)$, $\phi(c) + \phi(q) \in c\ell(A + \phi(q))$, and $\phi(q) + \phi(c) \in c\ell(\phi[B] + \phi(b))$. It follows from Theorem 3.40 that one of the two following possibilities must hold:

(i) $a + \phi(q) = \phi(v) + \phi(c)$ for some $a \in A$ and some $v \in \beta T$;

(ii) $u + \phi(q) = \phi(b) + \phi(c)$ for some $b \in B$ and some $u \in \beta S$.

The first possibility is ruled out by the assumption that $A \cap V = \emptyset$. So we may suppose that (ii) holds. Let $t = b + c$. Then $\phi(t) \in \beta S + p$ and so $\phi(t) + p = \phi(t)$. Since $\phi(t) = \phi(t + q) + \phi(q)$ and $t + q \in T^*$ (by Theorem 4.31), $\phi(t) \in \phi[T^*] + p \subseteq K(\phi[T^*])$. Thus (1) and (2) hold.

Now let p be a minimal idempotent in $\phi[T^*]$ and pick by (1) $t \in T$ such that $\phi(t) \in \phi[T^*] + p$. Let $F = \phi[T^*] + \phi(t)$. Then, since $t \in T$, $F = \phi(t) + \phi[T^*]$ so F is both a minimal left ideal and a minimal right ideal of $\phi[T^*]$, and is therefore a group (by Theorem 1.61). Since F is compact, it must be finite, because an infinite compact subset of βS cannot be homogeneous by 6.38.

If p' is any other minimal idempotent in $\phi[T^*]$, there exists $t' \in T$ such that $\phi(t') \in \phi[T^*] + p'$. Now $\phi(t + t') = \phi(t' + t)$, and so the minimal left ideals $\phi[T^*] + p' = \phi[T^*] + \phi(t')$ and $\phi[T^*] + p = \phi[T^*] + \phi(t)$ of $\phi[\beta T]$ intersect. They are therefore equal. This shows that F is the unique minimal left ideal of $\phi[T^*]$ and so $F = K(\phi[T^*])$ (by Theorem 1.64).

The fact that $\{b \in T : \phi(b + c) \in K(\phi[T^*])\}$ is infinite for each $c \in T$ follows immediately from (2). $\square$

Theorem 10.15. *Suppose that S is a countably infinite discrete semigroup which can be algebraically embedded in the unit circle. Let T be a countably infinite commutative weakly left cancellative semigroup with the property that any ideal of T has finite complement and let $\phi : \beta T \to S^*$ be a continuous homomorphism. Then $\phi[\beta T]$ is finite and $\phi[T^*]$ is a finite group.*

Proof. By Lemma 10.14, if $F = K(\phi[T^*])$, then F is a finite group. Let $I = \phi^{-1}[F]$. We claim that $I \cap T$ is an ideal of T. It follows from Lemma 10.14 that $I \cap T \neq \emptyset$. If $c \in I \cap T$, then $\phi(c) = \phi(c) + p$ for some minimal idempotent p of $\phi[T^*]$. There exists $q \in T^*$ for which $\phi(q) = p$. For every $b \in T$, we have $\phi(b + c) = \phi(b) + \phi(c) = \phi(b) + \phi(c) + \phi(q) = \phi(b + q + c) = \phi(b + q) + \phi(c) \in K(\phi[T^*])$. So $b + c \in I \cap T$ and we have shown that $I \cap T$ is an ideal of T.

Thus $T \backslash I$ is finite. Given $q \in T^*$, $I \in q$ and so $\phi[T^*] \subseteq F$. Since $\phi[\beta T] \subseteq \phi[T \backslash I] \cup F$, $\phi[\beta T]$ is finite. Since $\phi[T^*]$ is a finite cancellative semigroup, it is a group by Exercise 1.3.1. □

Notice that the hypotheses of Theorem 10.15 are satisfied in each of the following three cases.

  (i) T is a countably infinite commutative group;
  (ii) $T = (\mathbb{N}, +)$; or
  (iii) $T = (\mathbb{N}, \vee)$.

The following lemma is simple and well known.

Lemma 10.16. *Let G be a commutative group with identity 0 which contains no elements of finite order apart from 0. If $|G| \leq \mathfrak{c}$, then G can be embedded in $(\mathbb{R}, +)$.*

Proof. Let $|G| = \kappa$. Suppose that $G \backslash \{0\}$ has been arranged as a κ-sequence $\langle a_\alpha \rangle_{\alpha < \kappa}$ and that, for each $\beta < \kappa$, G_β denotes the subgroup of G generated by $\{a_\alpha : \alpha \leq \beta\}$.

We define $f_0 : G_0 \to \mathbb{R}$ by putting $f_0(na_0) = n$ for each $n \in \mathbb{Z}$. We then make the inductive assumption that $0 < \beta < \kappa$ and that we have defined an injective homomorphism $f_\alpha : G_\alpha \to \mathbb{R}$ for every $\alpha < \beta$. We also assume that $f_\alpha \subseteq f_{\alpha'}$ whenever $\alpha < \alpha' < \beta$. We shall show that we can define f_β so that these properties hold with β replaced by $\beta + 1$.

Let $H = \bigcup_{\alpha < \beta} G_\alpha$ and $f = \bigcup_{\alpha < \beta} f_\alpha$. Then $f : H \to \mathbb{R}$ is an injective homomorphism. Notice that $G_\beta = H + \mathbb{Z}a_\beta$.

Assume first that $\mathbb{Z}a_\beta \cap H = \{0\}$. Then we can choose $s \in \mathbb{R} \backslash (\mathbb{Q}f[H])$, because $|\mathbb{Q}f[H]| < \mathfrak{c}$. We define $f_\beta : H + \mathbb{Z}a_\beta \to \mathbb{R}$ by stating that $f_\beta(b + ma) = f(b) + ms$ for every $b \in H$ and every $m \in \mathbb{Z}$. Then f_β is well defined because $\mathbb{Z}a_\beta \cap H = \{0\}$ and is easily seen to be an injective homomorphism.

Thus we may assume that $ka_\beta = c$ for some $k \in \mathbb{Z} \backslash \{0\}$ and some $c \in H$. We define $f_\beta : H + \mathbb{Z}a_\beta \to \mathbb{R}$ by stating that $f_\beta(b + ma_\beta) = f(b) + \frac{m}{k}f(c)$ for every $b \in H$ and every $m \in \mathbb{Z}$.

To see that f_β is well defined, suppose that $b + ma_\beta = b' + m'a_\beta$, where $b, b' \in H$ and $m, m' \in \mathbb{Z}$. Then $k(b' - b) = k(m - m')a_\beta = (m - m')c$ and so $k(f(b') - f(b)) = (m - m')f(c)$ and $f(b) + \frac{m}{k}f(c) = f(b') + \frac{m'}{k}f(c)$.

It is easy to see that f_β is a homomorphism. To see that it is injective, suppose that $f_\beta(b + ma_\beta) = 0$ for some $b \in H$ and some $m \in \mathbb{Z}$. Then $f(b) + \frac{m}{k}f(c) = 0$ and so $f(kb + mc) = 0$. Since f is injective, $kb + mc = 0$. That is to say, $k(b + ma_\beta) = 0$ and so $b + ma_\beta$ is of finite order in G and hence $b + ma_\beta = 0$.

Thus we can define an injective homomorphism $f_\beta : G_\beta \to \mathbb{R}$ for every $\beta < \kappa$, so that $f_\beta \subseteq f_{\beta'}$ whenever $\beta < \beta' < \kappa$. Then $\bigcup_{\beta < \kappa} f_\beta : G \to \mathbb{R}$ is an injective homomorphism. $\square$

Theorem 10.17. *Let G be a countably infinite commutative group which contains no non-trivial finite subgroups, and let T be a countably infinite commutative weakly left cancellative semigroup with the property that any ideal of T has finite complement. If $\phi : \beta T \to G^*$ is a continuous homomorphism, then $\phi[\beta T]$ is finite and $|\phi[T^*]| = 1$.*

Proof. By Lemma 10.16, G can be embedded in $\mathbb{R}$ and therefore in $\mathbb{T}$. (If S is a countable subsemigroup of $\mathbb{R}$, we can choose $\alpha \in \mathbb{R}$ such that α is irrational with respect to S. The mapping h_α then embeds S in $\mathbb{T}$). Thus it follows from Theorem 10.15 that $\phi[\beta T]$ is finite and $\phi[T^*]$ is a group. By Theorem 7.17, G^* contains no non-trivial finite subgroups and so $|\phi[T^*]| = 1$. $\square$

Theorem 10.18. *If $\phi : \beta\mathbb{N} \to \mathbb{N}^*$ is a continuous homomorphism, $\phi[\beta\mathbb{N}]$ is finite and $|\phi[\mathbb{N}^*]| = 1$.*

Proof. This follows from Theorem 10.17 with $T = \mathbb{N}$ and $G = \mathbb{Z}$. $\square$

Question 10.19. *Is there any continuous homomorphism $\phi : \beta\mathbb{N} \to \mathbb{N}^*$ for which $|\phi[\beta\mathbb{N}]| \neq 1$?*

We note that, in view of Theorem 10.18, there are two simple equivalent versions of Question 10.19

Corollary 10.20. *The following statements are equivalent.*

(a) *There is a nontrivial continuous homomorphism from $\beta\mathbb{N}$ to $\mathbb{N}^*$.*
(b) *There exist $p \neq q$ in $\mathbb{N}^*$ such that $p + p = q + p = p + q = q + q = q$.*
(c) *There is a finite subsemigroup of $\mathbb{N}^*$ whose elements are not all idempotent.*

Proof. (a) implies (b). Let $\phi : \beta\mathbb{N} \to \mathbb{N}^*$ be a nontrivial continuous homomorphism. Then by Theorem 10.18, $\phi[\beta\mathbb{N}]$ is finite and there is some $q \in \mathbb{N}^*$ such that $\phi[\mathbb{N}^*] = \{q\}$. There is a largest $x \in \mathbb{N}$ such that $\phi(x) \neq q$. (Since ϕ is nontrivial, there is some such x, and if infinitely many x have $\phi(x) \neq q$ and $r \in c\ell\{x \in \mathbb{N} : \phi(x) \neq q\} \cap \mathbb{N}^*$, then $\phi(r) \neq q$.) Let $p = \phi(x)$.

(b) implies (c). The set $\{p, q\}$ is such a subsemigroup.

(c) implies (a). Let S be a finite subsemigroup of $\mathbb{N}^*$ whose elements are not all idempotent and pick $p \in S$ which is not an idempotent. Define a homomorphism f from $\mathbb{N}$ to S by stating that $f(n)$ is the sum of p with itself n times and let ϕ be the continuous extension of f to $\beta\mathbb{N}$. Then ϕ is a homomorphism by Corollary 4.22 and ϕ is nontrivial because $\{p\}$ is not a semigroup. $\square$

Question 10.21. *Which discrete semigroups S have the property that any continuous homomorphism from βS to S^* must have a finite image?*

10.3 Homomorphisms from T^* into S^*

A good deal of specialized notation will apply throughout this section. S and T will denote countably infinite discrete semigroups which are commutative and cancellative. G will denote the group generated by S and 0 will denote the identity of G. (So G is a countably infinite commutative group in which every element can be expressed as the difference of two elements of S). We shall regard S as embedded in G and βS as embedded in βG.

We shall assume that T has an identity. (This is no real restriction, because an identity can be adjoined to any semigroup).

We shall assume that we have chosen sequential orderings for G and T. For each $m \in \mathbb{N}$, G_m will denote the set whose elements are the first m elements of G, and T_m will denote the set whose elements are the first m elements of T.

We shall suppose that $\phi : T^* \to S^*$ is a continuous injective homomorphism. We choose q to be a nonminimal idempotent in T^* and we put $p = \phi(q)$. (The existence of a nonminimal idempotent in T^* follows from Corollary 6.34.)

Lemma 10.22. $\phi[K(T^*)] = K(\phi[T^*])$.

Proof. The mapping ϕ defines an isomorphism from T^* onto $\phi[T^*]$. $\square$

Lemma 10.23. *For every* $s \in S$ *and every* $x \in \beta S$, $s + x \in K(\beta S)$ *implies that* $x \in K(\beta S)$.

Proof. Suppose that $s + x \in K(\beta S)$. Since $K(\beta S)$ is a union of groups (by Theorem 1.64), there is an idempotent $u \in K(\beta S)$ for which $s + x + u = s + x$. By Lemma 8.1 this implies that $x + u = x$ and hence that $x \in K(\beta S)$. $\square$

Lemma 10.24. *There is a subset* $Y = \{t_n : n \in \mathbb{N}\}$ *of* T *with the following properties:*
  (i) *For every* $y \in Y^*$ *and every* $b, c \in T$, $\phi(b + q) \notin \beta G + \phi(c + y) + p$.
  (ii) *For every* m *and* n *in* $\mathbb{N}$, *every* $a \in G_m$, *and every* $b, c \in T_m$, *if* $m < n$, *then* $\phi(b + t_m + q) \neq a + \phi(c + t_n + q)$.

Proof. We can choose a minimal idempotent $u \in T^*$ satisfying $u = q + u = u + q$ (by Theorem 1.60). We note that $\phi(u)$ is in the smallest ideal of $\phi[T^*]$ by Lemma 10.22.

We claim that, for every $b \in T$, $\phi(b + q) \notin \beta G + \phi(u)$. To see this, we note that $\phi(b + q) \in \beta G + \phi(u)$ implies that $\phi(b + q) = \phi(b + q) + \phi(u) \in K(\phi[T^*])$. By Lemma 10.22, this would imply that $b + q \in K(T^*)$ and hence that $q \in K(T^*)$ (by Lemma 10.23). This contradicts our choice of q.

So for each $b \in T$, $\phi(b + q)$ has a clopen neighborhood W_b in βG disjoint from the compact subset $\beta G + \phi(u)$ of βG. For each $a \in G$ and each $b, c \in T$, let

$$E_{a,b,c} = \{x \in T^* : a + \phi(c + x) + p \in W_b\}.$$

By the continuity of the map $x \mapsto a + \phi(c + x) + p$, $E_{a,b,c}$ is a clopen subset of T^*. We note that $u \notin E_{a,b,c}$, because $a + \phi(c + u) + p = a + \phi(c + u + u) + \phi(q) =$

$a + \phi(c + u) + \phi(u + q)$. Since $u + q = u$, $a + \phi(c + u) + p \in \beta G + \phi(u) \subseteq \beta G \backslash W_b$. Thus $\bigcup_{a \in G} \bigcup_{b \in T} \bigcup_{c \in T} E_{a,b,c} \neq T^*$, because u does not belong to this union. It follows from Exercise 3.4.5 that there is a non-empty open subset U of T^* disjoint from $\bigcup_{a \in G} \bigcup_{b \in T} \bigcup_{c \in T} E_{a,b,c}$. We can choose an infinite subset V of T for which $V^* \subseteq U$.

We claim that, if $y \in V^*$ and $b, c \in T$, then $\phi(b + q) \notin \beta G + \phi(c + y) + p$. Suppose, on the contrary, that $\phi(b + q) \in \beta G + \phi(c + y) + p$. Since W_b is a neighborhood of $\phi(b + q)$ in βG and we are assuming that $\phi(b + q) \in c\ell(G + \phi(c + y) + p)$, we must have $a + \phi(c + y) + p \in W_b$ for some $a \in G$. This implies that $y \in E_{a,b,c}$ and contradicts our assumptions that $y \in U$ and $U \cap E_{a,b,c} = \emptyset$.

To obtain property (ii), we shall enumerate V as $\{v_n : n \in \mathbb{N}\}$ and then replace $\langle v_n \rangle_{n=1}^\infty$ by a subsequence $\langle t_n \rangle_{n=1}^\infty$ defined inductively.

We put $t_1 = v_1$. We then assume that we have chosen elements $t_1, t_2, \ldots, t_k$ in $\{v_n : n \in \mathbb{N}\}$ with the following property: if $i, j \in \{1, 2, \ldots, k\}$, $a \in G_i$, $b, c \in T_i$, and $i < j$, then $\phi(b + t_i + q) \neq a + \phi(c + t_j + q)$.

Suppose that $i \in \{1, 2, \ldots, k\}$. If $a \in G$ and $b, c \in T$, we claim there are at most a finite number of values of n for which $\phi(b + t_i + q) = a + \phi(c + v_n + q)$. To see this, suppose that this set of values of n is infinite. Then the corresponding elements v_n have a limit point y in V^* satisfying $\phi(b + t_i + q) = a + \phi(c + y + q) = a + \phi(c + y) + p$. We have seen that this cannot occur.

It follows that we can choose v_n with the property that $\phi(b + t_i + q) \neq a + \phi(c + v_n + q)$ whenever $i \in \{1, 2, \ldots, k\}$, $a \in G_k$ and $b, c \in T_k$. We put $t_{k+1} = v_n$. This shows that we can define a sequence $\langle t_n \rangle_{n=1}^\infty$ with the property specified in (ii).

We then put $Y = \{t_n : n \in \mathbb{N}\}$. The fact that $Y^* \subseteq V^*$ implies that property (i) holds as well. $\square$

Lemma 10.25. *Let Y denote the set guaranteed by Lemma 10.24, let $y \in Y^*$ and let $t \in T$. Then $\phi(t + y) + p$ is right cancelable in βG.*

Proof. If we suppose the contrary, then $\phi(t + y) + p = x + \phi(t + y) + p$ for some $x \in G^*$ (by Theorem 8.18). Now $\phi(t + y) + p \in c\ell(\{\phi(t + t_n + q) : n \in \mathbb{N}\})$, because $\phi(t + y) + p = \phi(t + y + q)$ and $t + y + q \in c\ell(\{t + t_n + q : n \in \mathbb{N}\})$. We also have $x + \phi(t + y) + p \in c\ell((G \backslash \{0\}) + \phi(t + y) + p)$. It follows from Theorem 3.40 that $\phi(t + t_n + q) = x' + \phi(t + y) + p$ for some $n \in \mathbb{N}$ and some $x' \in \beta G$, or else $\phi(t + y' + q) = a + \phi(t + y) + p$ for some $y' \in Y^*$ and some $a \neq 0$ in G.

The first possibility contradicts condition (i) of Lemma 10.24, and so we shall assume the second. Let $m \in \mathbb{N}$ satisfy $a, -a \in G_m$ and $t \in T_m$. Since $\phi(t + y' + q) \in c\ell(\{\phi(t + t_n + q) : n > m\})$ and $a + \phi(t + y) + p \in c\ell(\{a + \phi(t + t_n + q) : n > m\})$, a second application of Theorem 3.40 allows us to deduce that $\phi(t + t_n + q) = a + \phi(t + y'' + q)$ or $\phi(t + t_n + q) = -a + \phi(t + y'' + q)$ for some $n > m$ in $\mathbb{N}$ and some $y'' \in c\ell(\{t_n : n > m\})$. Now, by condition (i) of Lemma 10.24, neither of these equations can hold if $y'' \in Y^*$. So $y'' = t_r$ for some $r > m$. By condition (ii) of Lemma 10.24, neither of these equations can hold if $r \neq n$. However, they cannot hold if $r = n$, by Lemma 6.28.

Thus $\phi(t + y) + p$ is right cancelable in βG. $\square$

Lemma 10.26. *Let $Y = \{t_n : n \in \mathbb{N}\}$ be the set guaranteed by Lemma 10.24. Suppose that $\phi(b + y + q) \in \beta G + \phi(c + y + q)$ for some $b, c \in T$ and some $y \in Y^*$. Then $\phi(c + q) \in G + \phi(b + q)$.*

Proof. Since $\phi(b+y+q) \in c\ell(\phi[b + Y + q])$ and $\phi(b+y+q) \in c\ell(G+\phi(c+y+q))$, it follows from Theorem 3.40 that $\phi(b + t_n + q) \in \beta G + \phi(c + y + q)$ for some $n \in \mathbb{N}$, or else $\phi(b + z + q) = a + \phi(c + y + q)$ for some $z \in Y^*$ and some $a \in G$. The first possibility is ruled out by condition (i) of Lemma 10.24, and so we assume the second.

We choose $m \in \mathbb{N}$ such that $a, -a \in G_m$ and $b, c \in T_m$. Now $\phi(b + z + q) \in c\ell(\{\phi(b + t_n + q) : n > m\})$ and $a + \phi(c+y+q) \in c\ell(a + \{\phi(c + t_n + q) : n > m\})$. So a second application of Theorem 3.40 shows that there exists $n > m$ in $\mathbb{N}$ and $v \in c\ell(\{t_n : n > m\})$ satisfying $\phi(b + t_n + q) = a + \phi(c + v + q)$ or $\phi(b + v + q) = a + \phi(c + t_n + q)$. By condition (i) of Lemma 10.24, neither of these equations can hold if $v \in Y^*$. So $v = t_r$ for some $r > m$. However, by condition (ii) of Lemma 10.24, neither of these equations can hold if $r \neq n$. Thus $r = n$.

So $\phi(b+t_n+q) = a'+\phi(c+t_n+q)$, where $a' \in \{a, -a\}$. By adding $\phi(y+q)$ on the right of this equation, we see that $\phi(b+q)+\phi(t_n+y+q) = a'+\phi(c+q)+\phi(t_n+y+q)$. By Lemma 10.25, $\phi(t_n + y + q)$ is right cancelable in βG. So $\phi(b+q) = a'+\phi(c+q)$ and $\phi(c + q) \in G + \phi(b + q)$. $\qquad\square$

Lemma 10.27. *For every $c \in T$, $\phi(c + q) \in G + p$.*

Proof. Let Y be the set guaranteed by Lemma 10.24 and let $y \in Y^*$. For every $b \in T$, we have $\phi(c+q)+\phi(b+y+q) = \phi(b+q)+\phi(c+y+q)$. It follows from Corollary 6.20 that $\phi(b + y + q) \in \beta G + \phi(c + y + q)$ or $\phi(c + y + q) \in \beta G + \phi(b + y + q)$. In either case, Lemma 10.26 implies that $\phi(c + q) \in G + \phi(b + q)$. By choosing b to be the identity of T, we deduce that $\phi(c + q) \in G + p$. $\qquad\square$

In the statement of the following theorem, we remind the reader of the standing hypotheses that have been applying throughout this section.

Theorem 10.28. *Let S and T be countably infinite, commutative, and cancellative semigroups and assume that T has an identity. Let G be the group generated by S and assume that $\phi : T^* \to S^*$ is a continuous injective homomorphism. Then there is an injective homomorphism $f : T \to G$ such that $\widetilde{f}(x) = \phi(x)$ for every $x \in T^*$.*

Proof. By Lemma 10.27, for each $t \in T$, there exists $a \in G$ for which $\phi(t+q) = a+p$. This element of G is unique, by Lemma 6.28. We define $f : T \to G$ by stating that $\phi(t + q) = f(t) + p$. It is easy to check that f is an injective homomorphism, and so $\widetilde{f}$ is also an injective homomorphism (by Exercise 3.4.1 and Corollary 4.22).

Since $\phi(t+q) = f(t)+p$ for every $t \in T$, it follows by continuity that $\phi(x+q) = \widetilde{f}(x) + p$ for every $x \in \beta T$.

Let Y be the set guaranteed by Lemma 10.24 and let $y \in Y^*$. For every $x \in T^*$, we have $\phi(x) + \phi(y + q) = \phi(x + y + q) = \widetilde{f}(x + y) + p = \widetilde{f}(x) + \widetilde{f}(y) + p = \widetilde{f}(x) + \phi(y + q)$. Now $\phi(y + q)$ is right cancelable in βG, by Lemma 10.25. So $\widetilde{f}(x) = \phi(x)$. $\qquad\square$

Corollary 10.29. *Let f denote the mapping defined in Theorem* 10.28 *and let* $T' = f^{-1}[S]$. *Then T' is a subsemigroup of T for which $T \backslash T'$ is finite.*

Proof. It is immediate that T' is a subsemigroup of T. By Theorem 10.28, $\widetilde{f}(x) = \phi(x) \in S^*$ for every $x \in T^*$. Thus, if $x \in T^*$, then $S \in \widetilde{f}(x)$ so $T' = f^{-1}[S] \in x$ by Lemma 3.30. Consequently $T \backslash T'$ is finite. □

Theorem 10.30. *Let S and T be countably infinite, commutative, and cancellative semigroups and assume that T has an identity and that $\phi : T^* \to S^*$ is a continuous injective homomorphism. There is a subsemigroup T' of T for which $T \backslash T'$ is finite, and an injective homomorphism $h : T' \to S$ for which $\widetilde{h}(x) = \phi(x)$ for every $x \in T^*$.*

Proof. Let f denote the mapping defined in Theorem 10.28. We put $T' = f^{-1}[S]$ and let $h = f_{|T'}$. Our claim then follows from Corollary 10.29 and Theorem 10.28. □

The necessity of introducing T' in Theorem 10.30, is illustrated by choosing $T = \omega$ and $S = \mathbb{N}$. Then T^* and S^* are isomorphic, but there are no injective homomorphisms from T to S. In this example, $T' = T \backslash \{0\}$.

In the case in which S is a group we have $S = G$ and hence $T' = f^{-1}[S] = f^{-1}[G] = T$.

Theorem 10.31. *Let S and T be countably infinite, commutative, and cancellative semigroups and assume that T has an identity. Suppose that $\theta : \beta T \to \beta S$ is a continuous injective homomorphism. Then there is an injective homomorphism $f : T \to S$ for which $\theta = \widetilde{f}$.*

Proof. Let $\phi = \theta_{|T^*}$. We observe that $\phi[T^*] \subseteq S^*$, because $x \in T^*$ implies that $\phi(x)$ is not isolated in βS and hence that $\phi(x) \notin S$. By Theorem 10.28 there is an injective homomorphism $f : T \to G$ such that $\phi(x) = \widetilde{f}(x)$ for every $x \in T^*$.

Let Y denote the set guaranteed by Lemma 10.24 and let $y \in Y^*$. For each $t \in T$, we have $\theta(t + y + q) = \theta(t) + \theta(y + q) = \theta(t) + \phi(y + q)$ and also $\theta(t + y + q) = \widetilde{f}(t + y + q) = f(t) + \widetilde{f}(y + q) = f(t) + \phi(y + q)$. By Lemma 10.25, $\phi(y + q)$ is right cancelable in βG. So $f(t) = \theta(t)$. Now $f(t) \in G$ and $\theta(t) \in \beta S$ so $f(t) \in S$. □

Theorem 10.32. *Let S and T be countably infinite, commutative, and cancellative semigroups and assume that T has an identity. Then S^* does not contain any topological and algebraic copies of βT.*

Proof. This is an immediate consequence of Theorem 10.31. □

We do not know the answer to the following question. (Although, of course, algebraic copies of $\mathbb{Z}$ are plentiful in $\mathbb{N}^*$, as are topological ones.)

Question 10.33. *Does $\mathbb{N}^*$ contain an algebraic and topological copy of $\mathbb{Z}$?*

Exercise 10.3.1. Show that the only topological and algebraic copies of $\mathbb{N}^*$ in $\mathbb{N}^*$ are the sets of the form $(k\mathbb{N})^*$, where $k \in \mathbb{N}$.

Exercise 10.3.2. Show that the only topological and algebraic copies of $\mathbb{Z}^*$ in $\mathbb{Z}^*$ are the sets of the form $(k\mathbb{Z})^*$, where $k \in \mathbb{Z}\backslash\{0\}$.

Exercise 10.3.3. Show that the only topological and algebraic copies of $(\mathbb{N}^*, +)$ in $(\mathbb{N}^*, \cdot)$ are those induced by mappings of the form $n \mapsto k^n$ from $\mathbb{N}$ to itself, where $k \in \mathbb{N}\backslash\{1\}$.

Exercise 10.3.4. Let S be an infinite discrete cancellative semigroup. Show that S^* contains an algebraic and topological copy of $\mathbb{N}$. (Hint: By Theorem 8.10, there is an element $p \in S^*$ which is right cancelable in βS. Define p_n for each $n \in \mathbb{N}$ by stating that $p_1 = p$ and $p_{n+1} = p_n + p$ for every $n \in \mathbb{N}$. Then $\{p_n : n \in \mathbb{N}\}$ is an algebraic and topological copy of $\mathbb{N}$.)

Exercise 10.3.5. Show that $\mathbb{N}^*$ contains an algebraic and topological copy of ω. (Hint: Let p be an idempotent in $\mathbb{N}^*$ for which $FS(\langle 3^n \rangle_{n=1}^{\infty}) \in p$ and let $q \in c\ell(\{2 \cdot 3^n : n \in \mathbb{N}\}) \cap \mathbb{N}^*$. Define x_n for every $n \in \omega$ by stating that $x_0 = p$, $x_1 = p + q + p$ and $x_{n+1} = x_n + x_1$ for every $n \in \mathbb{N}$. Then $\{x_n : n \in \omega\}$ is an algebraic and topological copy of ω. To see that $\{x_n : n \in \omega\}$ is discrete, consider the alterations of 1's and 2's in the ternary expansion of integers. For example, $\{\sum_{t \in F_1} 3^t + \sum_{t \in F_2} 2 \cdot 3^t + \sum_{t \in F_3} 3^t : \max F_1 < \min F_2 \text{ and } \max F_2 < \min F_3\} \in x_1$.)

10.4 Isomorphisms Defined on Principal Left and Right Ideals

In this section, we shall discuss continuous isomorphisms between principal left ideals or principal right ideals defined by idempotents.

Throughout section, S and T will denote countably infinite discrete semigroups, which are commutative and cancellative, G will denote the group generated by S and H will denote the group generated by T.

Lemma 10.34. *Suppose that p and q are nonminimal idempotents in S^* and T^* respectively, and that $\psi : q + \beta T + q \to p + \beta S + p$ is a continuous isomorphism. Let $S' = \{a \in G : a + p \in \beta S\}$ and $T' = \{b \in H : b + q \in \beta T\}$. Then $\psi(q) = p$ and there is an isomorphism $f : T' \to S'$ such that $\psi(t + q) = f(t) + p$ for every $t \in T'$.*

Proof. Since q is in the center of $q + \beta T + q$, $\psi(q)$ is in the center of $p + \beta S + p$. So $\psi(q) = a + p$ for some $a \in G$ (by Theorem 6.63). Since $\psi(q)$ is idempotent, it follows from Lemma 6.28 that a is the identity of G and hence that $p = \psi(q)$.

For every $t \in T'$, $t + q = q + (t + q) + q$ is in $q + \beta T + q$ and is in the center of this semigroup. So $\psi(t + q)$ is in the center of $p + \beta S + p$. It follows from Theorem 6.63 that $\psi(t + q) = s + p$ for some $s \in G$. We note that $s \in S'$ and that the element s is unique (by Lemma 6.28). So we can define a mapping $f : T' \to S'$ such that $\psi(t + q) = f(t) + p$ for every $t \in T'$.

It is easy to see that f is injective and a homomorphism. To see that f is surjective, let $s \in S'$. Since $s + p$ is in the center of $p + \beta S + p$, $s + p = \psi(x)$ for some x in the center of $q + \beta T + q$. We must have $x = t + q$ for some $t \in T'$ by Theorem 6.63, and this implies that $f(t) = s$. $\qquad\square$

Lemma 10.35. *Let p and q be nonminimal idempotents in S^* and T^* respectively. Let $S' = \{a \in G : a+p \in \beta S\}$ and let $T' = \{b \in H : b+q \in \beta T\}$. If $\psi : \beta T+q \to \beta S+p$ is a continuous isomorphism, then $\psi(q)$ is a nonminimal idempotent, $\beta S + p = \beta S + \psi(q)$, and there is an isomorphism $f : T' \to S'$ for which $\psi(t + q) = f(t) + \psi(q)$ for every $t \in T'$.*

Proof. We first show that $\beta S + p = \beta S + \psi(q)$. On the one hand, $\psi(q) \in \beta S + p$ and so $\beta S + \psi(q) \subseteq \beta S + p$. On the other hand, $\psi^{-1}(p) = \psi^{-1}(p) + q$ and so $p = \psi(\psi^{-1}(p) + q) = p + \psi(q)$ and therefore $\beta S + p \subseteq \beta S + \psi(q)$. It follows that $\psi(q)$ is a nonminimal idempotent (by Theorem 1.59).

We also observe that $\{a \in G : a+p \in \beta S\} = \{a \in G : a+\psi(q) \in \beta S\}$. To see, for example, that $\{a \in G : a+p \in \beta S\} \subseteq \{a \in G : a+\psi(q) \in \beta S\}$, let $a \in G$ and assume that $a + p \in \beta S$. Then $a + \psi(q) = a + \psi(q) + p = \psi(q) + a + p \in \beta S + \beta S \subseteq \beta S$.

The result now follows from Lemma 10.34 and the observation that ψ defines a continuous isomorphism from $q + \beta T + q$ onto $\psi(q) + \beta S + \psi(q)$. (To see that $\psi(q)+\beta S+\psi(q) \subseteq \psi(q+\beta T+q)$, let $r \in \psi(q)+\beta S+\psi(q)$. Then $r = r+\psi(q)$ so $r = \psi(v)$ for some $v \in \beta T+q$. Then $q+v \in q+\beta T+q$ and $\psi(q+v) = \psi(q)+r = r$.) $\qquad\square$

We omit the proof of the following lemma, since it is essentially similar to the preceding proof.

Lemma 10.36. *Let p and q be nonminimal idempotents in S^* and T^* respectively. Let $S' = \{a \in G : a+p \in \beta S\}$ and let $T' = \{b \in H : b+q \in \beta T\}$. If $\psi : q+\beta T \to p+\beta S$ is a continuous isomorphism, then $\psi(q)$ is a nonminimal idempotent, $p+\beta S = \psi(q)+ \beta S$, and there is an isomorphism $f : T' \to S'$ such that $\psi(t + q) = f(t) + \psi(q)$ for every $t \in T'$.*

Lemma 10.37. *Suppose that $y \in G^*\backslash K(\beta G)$. Let $f : H \to G$ be an isomorphism. Then there is an element $x \in T^*$ for which $\widetilde{f}(x) + y$ is right cancelable in βG.*

Proof. We note that $\widetilde{f} : \beta H \to \beta G$ is an isomorphism and so $\widetilde{f}[K(\beta H)] = K(\beta G)$. By Lemma 6.65, $\overline{T} \cap K(\beta H) \neq \emptyset$, and so $\overline{f[T]} \cap K(\beta G) = \widetilde{f}[\overline{T} \cap K(\beta H)] \neq \emptyset$. It follows from Theorem 6.56 (with $S = G$) that there is an element $v \in G^*$ such that $f[T] \in v$ and $v + y$ is right cancelable in βG. Since $v \in \widetilde{f}[\overline{T}]$, $v = \widetilde{f}(x)$ for some $x \in T^*$. $\qquad\square$

Theorem 10.38. *Suppose that G and H are countable discrete commutative groups and that L and M are principal left ideals in βG and βH respectively, defined by nonminimal idempotents. If $\psi : M \to L$ is a continuous isomorphism, there is an isomorphism $f : H \to G$ for which $\psi = \widetilde{f}_{|M}$.*

Proof. Let $L = \beta G + p$ and $M = \beta H + q$, where p and q are nonminimal idempotents in G^* and H^* respectively. By applying Lemma 10.35 with $G = S = S'$ and $H = T = T'$, we see that $\psi(q)$ is nonminimal and there is an isomorphism $f : H \to G$ such that $\psi(t + q) = f(t) + \psi(q)$ for every $t \in H$. By continuity, this implies that $\psi(x + q) = \tilde{f}(x) + \psi(q)$ for every $x \in \beta H$.

For $x \in \beta H$, we have $\psi(q + x + q) = \tilde{f}(q + x) + \psi(q) = \tilde{f}(q) + \tilde{f}(x) + \psi(q)$ and also $\psi(q + x + q) = \psi(q) + \psi(x + q) = \psi(q) + \tilde{f}(x) + \psi(q)$. So $\tilde{f}(q) + \tilde{f}(x) + \psi(q) = \psi(q) + \tilde{f}(x) + \psi(q)$ for every $x \in \beta H$. By Lemma 10.37, we can choose $x \in \beta H$ such that $\tilde{f}(x) + \psi(q)$ is right cancelable in βG. Thus $\psi(q) = \tilde{f}(q)$. This implies that $\psi(x + q) = \tilde{f}(x + q)$ for every $x \in \beta H$. $\qquad\square$

Theorem 10.39. *Suppose that L and M are principal left ideals in $\beta \mathbb{N}$ defined by nonminimal idempotents. If $\psi : M \to L$ is a continuous isomorphism then $M = L$ and ψ is the identity map.*

Proof. Let p and q be nonminimal idempotents in $\beta \mathbb{N}$ for which $L = \beta \mathbb{N} + p$ and $M = \beta \mathbb{N} + q$. Then $L = \beta \mathbb{Z} + p$ and $M = \beta \mathbb{Z} + q$, since, given $r \in \beta \mathbb{Z}, r + p = r + p + p \in \beta \mathbb{N} + p$ because $\mathbb{N}^*$ is a left ideal in $\beta \mathbb{Z}$ (by Exercise 4.3.5). So L and M are principal left ideals in $\beta \mathbb{Z}$ defined by nonminimal idempotents in $\mathbb{Z}^*$. It follows from Theorem 10.38 that there is an isomorphism $f : \mathbb{Z} \to \mathbb{Z}$ for which $\psi = \tilde{f}_{|M}$. Now there are precisely two isomorphisms from $\mathbb{Z}$ to itself: the identity map $t \mapsto t$ and the map $t \mapsto -t$. We can rule out the possibility that f is the second of these maps, because this would imply that $\tilde{f}(q) \notin \beta \mathbb{N}$. Hence f is the identity map and so is ψ. $\qquad\square$

Theorem 10.40. *Let G and H be countable commutative groups and let L and M be principal right ideals in G^* and H^* respectively, defined by nonminimal idempotents. Suppose that there is a continuous isomorphism $\psi : M \to L$. Then there is an isomorphism $f : H \to G$ for which $\tilde{f}[M] = L$.*

Proof. Let p and q be nonminimal idempotents in G^* and H^* respectively for which $L = p + \beta G$ and $M = q + \beta H$. By applying Lemma 10.36 with $G = S = S'$ and $H = T = T'$, we see that $\psi(q)$ is nonminimal and there is an isomorphism $f : H \to G$ such that $\psi(t + q) = f(t) + \psi(q)$. It follows, by continuity, that $\psi(x + q) = \tilde{f}(x) + \psi(q)$ for every $x \in M$. Putting $x = q$ shows that $\psi(q) = \tilde{f}(q) + \psi(q)$.

On the other hand, we also have, for every $x \in \beta H$, that $\psi(q + x + q) = \tilde{f}(q) + \tilde{f}(x) + \psi(q)$ and $\psi(q + x + q) = \psi(q) + \psi(q + x + q) = \psi(q) + \tilde{f}(q) + \tilde{f}(x) + \psi(q)$. So $\tilde{f}(q) + \tilde{f}(x) + \psi(q) = \psi(q) + \tilde{f}(q) + \tilde{f}(x) + \psi(q)$ for every $x \in \beta H$. We can choose $x \in H^*$ for which $\tilde{f}(x) + \psi(q)$ is right cancelable in βG (by Lemma 10.37) and so $\tilde{f}(q) = \psi(q) + \tilde{f}(q)$.

It follows easily that $\psi(q) + \beta G = \tilde{f}(q) + \beta G$. We also know that $\psi(q) + \beta G = L$ (by Lemma 10.36). So $\tilde{f}[M] = \tilde{f}(q) + \beta G = \psi(q) + \beta G = L$. $\qquad\square$

The preceding results in this section tell us nothing about principal left or right ideals defined by minimal idempotents.

If S is any discrete semigroup, we know that any two minimal left ideals in βS are both homeomorphic and isomorphic by Theorems 1.64 and 2.11. However, the maps which usually define homeomorphisms between minimal left ideals are different from those which define isomorphisms. It may be the case that one minimal left ideal cannot be mapped onto another by a continuous isomorphism. It is tantalizing that we do not know the answer to either of the following questions.

Question 10.41. *Are there two minimal left ideals in $\beta\mathbb{N}$ with the property that one cannot be mapped onto the other by a continuous homomorphism?*

Question 10.42. *Are there two distinct minimal left ideals in $\beta\mathbb{N}$ with the property that one can be mapped onto the other by a continuous homomorphism?*

Notes

Most of the result of Section 10.1 are from [33], results of collaboration with V. Bergelson and B. Kra. Theorem 10.8 is due to J. Baker and P. Milnes in [10].

Theorem 10.18 is from [228] and answers a question of E. van Douwen in [80].

Most of the results in Section 10.3 were proved in collaboration with A. Maleki in [180].

Chapter 11

The Rudin–Keisler Order

Recall that if S is a discrete space, then the *Rudin–Keisler order* $\leq_{RK}$ on βS is defined by stating that, for any $p, q \in \beta S$, $p \leq_{RK} q$ if and only if there is a function $f : S \to S$ for which $\widetilde{f}(q) = p$. Also recall that if $p, q \in \beta S$, we write $p <_{RK} q$ if $p \leq_{RK} q$ and $q \nleq_{RK} p$, and we write $p \approx_{RK} q$ if $p \leq_{RK} q$ and $q \leq_{RK} p$.

As a consequence of Theorem 8.17 one notes that while the relation $\leq_{RK}$ is clearly reflexive and transitive, it is not anti-symmetric. The relation $\leq_{RK}$ does induce an antisymmetric partial order on the set of equivalence classes of the equivalence relation $\approx_{RK}$.

The Rudin–Keisler order has proved to be an important tool in analyzing spaces of ultrafilters. It shows how one ultrafilter can be essentially different from another.

There are deep and difficult theorems about the Rudin–Keisler order which we shall not attempt to prove in this chapter. Our aim is to indicate some of the connections between the Rudin–Keisler order and the algebraic structure of βS. If S is a discrete semigroup, one would expect that the Rudin–Keisler order on βS would have little relation to the algebra of βS, because any bijective mapping from one subset of S to another induces a Rudin–Keisler equivalence between ultrafilters, and a bijection may have no respect for algebraic structure. Nevertheless, there are algebraic properties which are related to the Rudin–Keisler order. For example, if S is countable and cancellative, an element $q \in S^*$ is right cancelable in βS if and only if $p <_{RK} pq$ and $q <_{RK} pq$ for every $p \in S^*$ as we shall see in Theorem 11.8.

Exercise 11.0.1. Let S be a discrete space and let $p \in \beta S$. Show that $p \leq q$ for every $q \in \beta S$ if and only if $p \in S$.

Exercise 11.0.2. Let S be a countable discrete space and let $p \in \beta S$. Show that $\{q \in \beta S : q \leq_{RK} p\}$ has cardinality at most $\mathfrak{c}$.

Exercise 11.0.3. Let S be a discrete semigroup and let $a \in S$ and $p \in \beta S$. Show that $ap \leq_{RK} p$ and $pa \leq_{RK} p$. Show that $ap \approx_{RK} p$ if S is left cancellative and that $pa \approx_{RK} p$ if S is right cancellative.

Exercise 11.0.4. Let S be a countable discrete space and let p be a P-point in S^*. If $q \in S^*$ satisfies $q \leq_{RK} p$, show that q is also a P-point in S^*.

11.1 Connections with Right Cancelability

In this section we shall explore the connections between the relation $\leq_{RK}$ and right cancelability in βS. These arise from the fact that both concepts are related to the topological property of strong discreteness.

Definition 11.1. Let S be a discrete space and let $p, q \in \beta S$. The *tensor product* $p \otimes q$ of p and q is defined by

$$p \otimes q = \{A \subseteq S \times S : \{s : \{t : (s, t) \in A\} \in q\} \in p\}.$$

The reader is asked to show in Exercise 11.1.1 that $p \otimes q$ is an ultrafilter on $S \times S$.

If $A \subseteq S \times S$, then $A \in p \otimes q$ if and only if A contains a set of the form $\{(s, t) : s \in P$ and $t \in Q_s\}$, where $P \in p$ and $Q_s \in q$ for each $s \in P$. Note that, if $s, t \in S$, we have $s \otimes t = (s, t)$ where, as usual, we identify the point (s, t) with the principal ultrafilter it generates.

If we choose any bijection $\phi : S \times S \rightarrow S$, we can regard $p \otimes q$ as being an ultrafilter on S by identifying it with $\widetilde{\phi}(p \otimes q)$, where $\widetilde{\phi} : \beta(S \times S) \rightarrow \beta S$ is the continuous extension of ϕ. In particular, if $s, t \in S$, then $s \otimes t$ can be identified with the element $\phi(s, t)$ of S.

In this section, we shall sometimes think of $\otimes$ as being a binary operation on S, assuming that a bijection ϕ has been chosen. Of course, different bijections will result in $p \otimes q$ being identified with different elements of βS. These will, however, all be Rudin–Keisler equivalent. So it does not matter which one we choose for studying properties of the relation $\leq_{RK}$.

Lemma 11.2. *Let S be a discrete space and let $p, q \in S^*$. Then $p <_{RK} p \otimes q$ and $q <_{RK} p \otimes q$.*

Proof. If π_1 and π_2 are the projection maps from $S \times S$ onto S, it is easy to see that $\widetilde{\pi}_1(p \otimes q) = p$ and $\widetilde{\pi}_2(p \otimes q) = q$. So $p \leq_{RK} p \otimes q$ and $q \leq_{RK} p \otimes q$. Now there is no member of $p \otimes q$ on which π_1 or π_2 is injective, and so $p \not\approx_{RK} p \otimes q$ and $q \not\approx_{RK} p \otimes q$ by Theorem 8.17. $\square$

Lemma 11.3. *Let S be a discrete space and let $p, q \in \beta S$. Then*

$$p \otimes q = \lim_{s \to p} \lim_{t \to q} (s, t)$$

where s and t denote elements of S and the limits are taken to be in $\beta(S \times S)$.

Proof. This is Exercise 11.1.2. $\square$

Recall that in Section 4.1, we extended an arbitrary binary operation on S to βS.

Lemma 11.4. *Let S be a discrete space and let $*$ be a binary operation defined on S. Let $p, q, x, y \in \beta S$. If $x \leq_{RK} p$ and $y \leq_{RK} q$, then $x * y \leq_{RK} p \otimes q$.*

Proof. Let $f : S \to S$ and $g : S \to S$ be functions for which $\widetilde{f}(p) = x$ and $\widetilde{g}(q) = y$. We define $h : S \times S \to S$ by $h(s, t) = f(s) * g(t)$. Then

$$\widetilde{h}(p \otimes q) = \lim_{s \to p} \lim_{t \to q} h(s, t) = \lim_{s \to p} \lim_{t \to q} f(s) * g(t) = \widetilde{f}(p) * \widetilde{g}(q) = x * y. \qquad \square$$

Corollary 11.5. *Let S be a discrete space and let $*$ be a binary operation defined on S. For every $p, q \in \beta S$, we have $p * q \leq_{RK} p \otimes q$. Furthermore, if $h : S \times S \to S$ is defined by $h(s, t) = s * t$, then $\widetilde{h}(p \otimes q) = p * q$.*

Proof. The proof is the same as the proof of Lemma 11.4, with f and g taken to be the identity maps. $\qquad \square$

Corollary 11.5 shows that, for any $p, q \in \beta S$, $p \otimes q$ is a $\leq_{RK}$ upper bound of the set of all elements of the form $p * q$, where $*$ denotes a binary operation on S. We now investigate the possibility that $p * q \approx_{RK} p \otimes q$.

When we say that an arbitrary binary operation $*$ on S is left cancellative, we mean exactly the same thing as when the operation is assumed to be associative. That is for each $s \in S$, the map $t \mapsto s * t$ is injective.

Theorem 11.6. *Let S be a discrete space and let $*$ be a left cancellative binary operation defined on S. Then, for every $p, q \in \beta S$, the following are equivalent:*

(1) *$p * q \approx_{RK} p \otimes q$.*
(2) *There is a set $D \in p$ for which $D * q$ is strongly discrete and $s * q \neq t * q$ whenever s and t are distinct members of D.*

Proof. (1) implies (2). If $p * q \approx_{RK} p \otimes q$, there is a set $A \in p \otimes q$ on which the mapping $(s, t) \mapsto s * t$ from $S \times S$ to S is injective (by Corollary 11.5 and Theorem 8.17.) We may suppose that $A = \bigcup_{s \in D}(\{s\} \times E_s)$, where $D \in p$ and $E_s \in q$ for each $s \in D$. Then for each $s \in S$, $c\ell_{\beta S}(s * E_s)$ is a neighborhood of $s * q$ in βS. If s and s' are distinct elements of D, $s * E_s$ and $s' * E_{s'}$ are disjoint, and so $c\ell_{\beta S}(s * E_s)$ and $c\ell_{\beta S}(s' * E_{s'})$ are disjoint.

(2) implies (1). Suppose that for some $D \in p$, $D * q$ is strongly discrete and $s * q \neq t * q$ whenever s and t are distinct members of D. Then, for each $s \in D$, there is a neighborhood U_s of $s * q$ in βS such that $U_s \cap U_{s'} = \emptyset$ whenever s and s' are distinct elements of D. For each $s \in D$, we can choose $E_s \in q$ for which $s * E_s \subseteq U_s$. If $A = \bigcup_{s \in D}(\{s\} \times E_s)$, then $A \in p \otimes q$ and, since $s * E_s \cap s' * E_{s'} = \emptyset$ when $s \neq s'$ and $s * t \neq s * t'$ when $t \neq t'$, the mapping $(s, t) \mapsto s * t$ is injective on A. So $p * q \approx_{RK} p \otimes q$. $\qquad \square$

Corollary 11.7. *Let S be a countable left cancellative discrete semigroup and let q be a right cancelable element of βS. Then $pq \approx_{RK} p \otimes q$ for every $p \in \beta S$.*

Proof. By Theorem 8.11, Sq is strongly discrete in βS and $sq \neq tq$ whenever $s \neq t$ in S, and so our claim follows from Theorem 11.6. $\qquad \square$

We are now able to establish the equivalence of statements (1), (4) and (5) of Theorem 8.18 under weaker hypotheses.

Theorem 11.8. *Suppose that S is a countable cancellative discrete semigroup and that $q \in S^*$. Then the following statements are equivalent.*

(1) *q is right cancelable in βS.*

(2) *$p <_{RK} pq$ and $q <_{RK} pq$ for every $p \in S^*$.*

(3) *$q <_{RK} pq$ for every $p \in S^*$.*

(4) *$q \not\approx_{RK} pq$ for every $p \in S^*$.*

Proof. (1) implies (2). If q is right cancelable in βS, it follows from Lemma 11.2 and Corollary 11.7 that $p <_{RK} pq$ and $q <_{RK} pq$ for every $p \in S^*$.

That (2) implies (3) and (3) implies (4) is trivial.

(4) implies (1). Suppose that q is not right cancelable in βS. By Theorem 8.11, there exist $a \in S$ and $p \in \beta S \setminus \{a\}$ such that $aq = pq$. By Lemma 6.28, $p \notin S$. We have $aq \approx_{RK} q$ by Exercise 11.0.3. So $q \approx_{RK} pq$, a contradiction. $\square$

Theorem 11.9. *If S is an infinite discrete space, there are no elements of S^* which are maximal in the Rudin–Keisler order. In fact, if $|S| = \kappa$, every element of S^* has 2^{2^κ} distinct Rudin–Keisler successors.*

Proof. We may suppose that S is a group, because there are groups of cardinality κ. (The direct sum of κ copies of $\mathbb{Z}_2$ provides an example.)

It then follows from Theorem 6.30 that there is a subset P of S^* such that $|P| = 2^{2^\kappa}$, $ap \neq bp$ when $a \neq b$ in S and Sp is strongly discrete in βS for every $p \in P$, and $(\beta S)p \cap (\beta S)q = \emptyset$ whenever p and q are distinct elements of P.

By Lemma 11.2 and Theorem 11.6, $x <_{RK} xp$ for every $x \in S^*$ and every $p \in P$. Since $(\beta S)p \cap (\beta S)q = \emptyset$, $xp \neq xq$ if p and q are distinct elements of P. $\square$

Theorem 11.10. *Let S be an infinite discrete cancellative semigroup. There is no element of S^* whose Rudin–Keisler predecessors form a subsemigroup of βS.*

Proof. Let $p \in S^*$ and let $L = \{x \in \beta S : x \leq_{RK} p\}$. Let $\kappa = \|p\|$ and let $R \in p$ have cardinality κ. We can find a subset V of S with cardinality κ such that $aq \neq bq$ when $a \neq b$ in R and Rq is strongly discrete in βS for every uniform ultrafilter q on V by Theorem 6.30. Since R can be mapped bijectively onto V, we can choose a uniform ultrafilter q on V for which $p \approx_{RK} q$. By Theorem 11.6 and Lemma 11.2, we have $p <_{RK} pq$. Since $p, q \in L$ and $pq \notin L$, L is not a subsemigroup of βS. $\square$

Theorem 11.11. *Let S be an infinite discrete left cancellative semigroup. Let $D \subseteq S$ and let $q \in S^*$. Suppose that Dq is strongly discrete in βS and that $sq \neq tq$ whenever s and t are distinct elements of D. Suppose also that $x, y \in S^*$, $p \in \overline{D} \cap S^*$, $x \leq_{RK} p$, and $y \leq_{RK} q$. Then $xy \leq_{RK} pq$.*

Proof. By Theorem 11.6, $pq \approx_{RK} p \otimes q$. Now $xy \leq_{RK} p \otimes q$ by Lemma 11.4. $\square$

Corollary 11.12. *Let S be a discrete countable left cancellative semigroup and let q be a right cancelable element of βS. Then, for every $x, p \in S^*$, $x \leq_{RK} p$ implies that $xq \leq_{RK} pq$.*

Proof. By Theorem 8.11(1), Theorem 11.11 applies with y replaced by q. $\qquad\square$

We now prove a converse to Corollary 11.12.

Theorem 11.13. *Let S be a countable cancellative semigroup. Let q be a right cancelable element of βS and let $p \in \beta S$ satisfy $p \leq_{RK} q$. If $x, y \in S^*$ and $xq \leq_{RK} yp$, then $x \leq_{RK} y$.*

Proof. Let $f : S \to S$ be a function for which $\widetilde{f}(yp) = xq$. If $q \in S$, then $p \in S$ and so $x \approx_{RK} xq \leq_{RK} yp \approx_{RK} y$ by Exercise 11.0.3, so we may assume that $q \in S^*$.

Let $B = \{b \in S : \widetilde{f}(bp) \in Sq\}$. For each $b \in B$, there is a unique $c \in S$ for which $\widetilde{f}(bp) = cq$, by Lemma 6.28. So we can define a function $g : S \to S$ by putting $g(b) = c$ if $b \in B$, and extending g arbitrarily to $S \backslash B$. We may suppose that $\widetilde{g}(y) \neq x$, since otherwise $x \leq_{RK} y$ as we wish to prove. So there is a set $V \in y$ for which $g[V] \notin x$. We choose $V \subseteq B$ in the case in which $B \in y$ and we choose $V \subseteq S \backslash B$ if $B \notin y$.

We now claim that

(*) for each $X \in x$ and each $Y \in y$, if $X \subseteq S \backslash g[V]$ and $Y \subseteq V$, then $aq = \widetilde{f}(zp)$ for some $a \in X$ and some $z \in \overline{Y}$.

To see this, suppose instead that (*) fails and pick $X \in x$ and $Y \in y$ witnessing this failure. Now $xq \in c\ell(Xq) \cap c\ell \, \widetilde{f}[Yp]$ so by Theorem 3.40 and the assumption that (*) fails, we must have $wq = \widetilde{f}(bp)$ for some $w \in \overline{X}$ and some $b \in Y$. Then $wq \leq_{RK} bp$, $bp \approx_{RK} p$ by Exercise 11.0.3, and $p \leq_{RK} q$ by assumption. If $w \in S^*$, then by Theorem 11.8, $q <_{RK} wq$, a contradiction. Thus $w \in S$ and so $b \in B$ and $w = g(b)$. Also $b \in Y \subseteq V$ so that $w \in g[V] \cap X$, contradicting the fact that $g[V] \cap X = \emptyset$. Thus (*) is established.

Let $A = \{a \in S : aq \in \widetilde{f}[(\beta S)p]\}$. If $A \notin x$, then letting $X = S \backslash (A \cup g[V])$ and $Y = V$ one obtains a contradiction to (*), and so $A \in x$. For each $a \in A$, let $C_a = \{z \in \beta S : aq = \widetilde{f}(zp)\}$. Then $C_a = (\widetilde{f} \circ \rho_p)^{-1}[\{aq\}]$ so C_a is closed. We claim that for each $a \in A$, $C_a \cap c\ell \left(\bigcup_{b \in A \backslash \{a\}} C_b \right) = \emptyset$. Indeed, by Theorem 8.11, $aq \notin c\ell\{bq : b \in A \backslash \{a\}\}$ so pick a neighborhood W of aq which misses $\{bq : b \in A \backslash \{a\}\}$. Given $z \in C_a$, pick $D \in z$ such that $\widetilde{f}[\overline{D}q] \subseteq W$. Then $\overline{D} \cap \bigcup_{b \in A \backslash \{a\}} C_b = \emptyset$.

Thus one can pick for each $a \in A$ a clopen set V_a such that $C_a \subseteq V_a$ and $V_a \cap \left(\bigcup_{b \in A \backslash \{a\}} C_b \right) = \emptyset$. Let A be sequentially ordered by $<$. For the first element a_0 of A, let $U_{a_0} = V_{a_0}$. For later elements, let $U_a = V_a \backslash \bigcup_{b < a} V_b$. Then for each $a \in A$, U_a is clopen, $C_a \subseteq U_a$, and $U_a \cap U_b = \emptyset$ if $a \neq b$.

Define $h : S \to S$ by $h(s) = a$ if $s \in U_a$, defining h arbitrarily on $S \backslash \bigcup_{a \in A} U_a$. We claim that $\widetilde{h}(y) = x$. Suppose instead that $\widetilde{h}(y) \neq x$ and pick $X \in x$ such that $\widetilde{h}(y) \notin \overline{X}$ and pick $Y \in y$ such that $\widetilde{h}[\overline{Y}] \cap X = \emptyset$. We may assume that $X \subseteq S \backslash g[V]$

and $Y \subseteq V$. Then by (*), $aq = \widetilde{f}(zp)$ for some $a \in X$ and some $z \in \overline{Y}$. So $z \in U_a$ and thus $\widetilde{h}(z) = a$, contradicting the fact that $\widetilde{h}[\overline{Y}] \cap X = \emptyset$. $\square$

Corollary 11.14. *Let S be a countable cancellative semigroup and let $p \in \beta S$. The following statements are equivalent.*

(1) *p is right cancelable in βS.*

(2) *For every $x, y \in \beta S$, $xp \leq_{RK} yp$ implies that $x \leq_{RK} y$.*

(3) *For every $x, y \in \beta S$, $xp \approx_{RK} yp$ implies that $x \approx_{RK} y$.*

(4) *For every $x \in S^*$, $p <_{RK} xp$.*

Proof. If $p \in S$, then (1) is true by Lemma 8.1, (2) and (3) are true by Exercise 11.0.3, and (4) is true by Exercises 11.0.1 and 11.0.3. Thus we may assume that $p \in S^*$.

The equivalence of (1) and (4) follows from Theorem 11.8. To see that (1) implies (2), let $x, y \in \beta S$ with $xp \leq_{RK} yp$. If $x, y \in S^*$, then $x \leq_{RK} y$ by Theorem 11.13 and if $x \in S$, them $x \leq_{RK} y$ by Exercise 11.0.1. Thus it suffices to show that one cannot have $x \in S^*$ and $y \in S$. Indeed in this case $p <_{RK} xp$ by Theorem 11.8 and $yp \approx_{RK} p$ by Exercise 11.0.3 so one would conclude that $p <_{RK} p$.

It is trivial that (2) implies (3). To see that (3) implies (1), assume that (3) holds and suppose that p is not right cancelable in βS. Pick $x \neq y$ in βS such that $xp = yp$ and pick $X \in x$ and $Y \in y$ such that $X \cap Y = \emptyset$. Now $xp \in \overline{Xp}$ and $yp \in \overline{Yp}$ so it follows from Theorem 3.40 that $sp = zp$, where $s \in X$ and $z \in \overline{Y}$ or $s \in Y$ and $z \in \overline{X}$. By (3), this implies that $s \approx_{RK} z$ and hence, by Exercise 11.0.1, that $z \in S$. This contradicts Lemma 6.28, because $s \neq z$. $\square$

Theorem 11.15. *Let S be a countable discrete space and let $*$ be a left cancellative binary operation defined on S. Suppose that $p, q \in S$ and that $A \in p$ has the property that $A * q$ is discrete in βS and that $a * q \neq b * q$ whenever a and b are distinct elements of A. Then $p * q \leq_{RK} q * p$ implies that p and q are Rudin–Keisler comparable.*

Proof. We suppose that $p * q \leq_{RK} q * p$, and note that we then have $p \otimes q \leq_{RK} q * p$ by Theorem 11.6 and the fact that "discrete" and "strongly discrete" are the same in this case.

Let π_1 and π_2 denote the projection maps of $S \times S$ onto S. Then $\widetilde{\pi}_1(x \otimes q) = x$ and $\widetilde{\pi}_2(x \otimes q) = q$ for every $x \in \beta S$.

Let $f : S \to S \times S$ be a function for which $\widetilde{f}(q * p) = p \otimes q$. If $P \in p$ and $Q \in q$, $p \otimes q$ belongs to each of the sets $c\ell(P \otimes q)$ and $c\ell(\widetilde{f}[Q * p])$. It follows from Theorem 3.40 that either

(i) $\widetilde{f}(b * p) = x \otimes q$ for some $b \in Q$ and some $x \in \overline{P}$ or

(ii) $a \otimes q = \widetilde{f}(z * p)$ for some $a \in P$ and some $z \in \overline{Q}$.

Now if (i) holds for some P and Q, then $\widetilde{\pi}_2 \circ \widetilde{f} \circ \lambda_b$ is a continuous extension of the function $s \mapsto \pi_2\big(f(b * s)\big)$ taking p to q so that $q \leq_{RK} p$. We may therefore suppose that (ii) holds for every $P \in p$ and every $Q \in q$.

Let $B = \{b \in S : \tilde{\pi}_1(\tilde{f}(b * p)) \in S\}$ and define $g : S \rightarrow S$ by stating that $g(b) = \tilde{\pi}_1(\tilde{f}(b * p))$ if $b \in B$ and defining g arbitrarily on $S \backslash B$. We claim that $\tilde{g}(q) = p$, so suppose instead that $\tilde{g}(q) \neq p$. Pick $P \in p$ and $Q \in q$ such that $\overline{\tilde{g}[Q]} \cap \overline{P} = \emptyset$. By (ii) pick some $a \in P$ and some $z \in \overline{Q}$ such that $a \otimes q = \tilde{f}(z * p)$. Thus $a = \tilde{\pi}_1(a \otimes q) = \tilde{\pi}_1(\tilde{f}(z * p))$ and a is isolated so there is some $R \in z$ with $\tilde{\pi}_1[\tilde{f}[\overline{R} * p]] = \{a\}$. Pick $b \in R \cap Q$. Then $\tilde{\pi}_1(\tilde{f}(b * p)) = a$ so $b \in B$ and thus $g(b) = a$, a contradiction.   □

Corollary 11.16. *Let S be a countable discrete space and let $p, q \in S^*$. If $p \otimes q \leq_{RK} q \otimes p$, then p and q are Rudin–Keisler comparable.*

Proof. The hypotheses of Theorem 11.15 are clearly satisfied if $\otimes$ is regarded as a binary operation on S via a fixed bijection between $S \times S$ and S.   □

Corollary 11.17. *Let S be a discrete countable cancellative semigroup and let q be a right cancelable element of S^*. If $p \in S^*$ and $pq \leq_{RK} qp$, then p and q are Rudin–Keisler comparable.*

Proof. Theorem 11.15 applies by Theorem 8.11(1).   □

If S is an infinite discrete semigroup, we know that βS is rarely commutative. In the following corollary, we show something stronger than non-commutativity in the case in which S is countable and cancellative. In this case, there are elements p and q of S^* for which pq and qp are not Rudin–Keisler comparable.

Corollary 11.18. *Let S be a countable infinite discrete cancellative semigroup. There are elements $p, q \in S^*$ for which pq and qp are not Rudin–Keisler comparable.*

Proof. By Theorem 6.36, there are elements p and q of S^* which are not Rudin–Keisler comparable. We may suppose that p and q are right cancelable in βS, because there is an infinite subset T of S with the property that every element of T^* is right cancelable in βS (by Theorem 8.10). Since S can be mapped bijectively onto T, p and q are Rudin–Keisler equivalent to elements in T^*. Our claim then follows from Corollary 11.17.
□

Exercise 11.1.1. Let S be a discrete space and let $p, q \in \beta S$. Show that $p \otimes q \in \beta(S \times S)$.

Exercise 11.1.2. Prove Lemma 11.3.

Exercise 11.1.3. We cannot normally use the most obvious function to establish a relation of the form $p \otimes q \leq_{RK} q \otimes p$. Let $f : \mathbb{N} \times \mathbb{N} \rightarrow \mathbb{N} \times \mathbb{N}$ be defined by $f(s, t) = (t, s)$. Show that, for any $p, q \in \mathbb{N}^*$, $\tilde{f}(q \otimes p) \neq p \otimes q$.

Exercise 11.1.4. Let S be an infinite cancellative semigroup. Show that there is no element p in S^* for which $\{q \in \beta S : q \approx_{RK} p\}$ is a subsemigroup of βS. (Hint: Consider the proof of Theorem 11.10.)

11.2 Connections with Left Cancelability in $\mathbb{N}^*$

Because βS is a right topological semigroup, left cancelability in βS is more difficult to handle than right cancelability. As we have seen, it is easy to derive connections between the Rudin–Keisler order and right cancelability. However, whether the analogous results hold for left cancelability remains a difficult open question, even in $\beta \mathbb{N}$. For example, we do not know whether $q <_{RK} p+q$ for every $q \in \mathbb{N}^*$, whenever p is a left cancelable element of $\mathbb{N}^*$. However, we shall see that there is a rich set of left cancelable elements p in $\mathbb{N}^*$ for which this statement is true.

Recall that, for each $k \in \mathbb{N}$, the natural homomorphism $h_k : \mathbb{Z} \to \mathbb{Z}_k$ has a continuous extension $\widetilde{h}_k : \beta \mathbb{Z} \to \mathbb{Z}_k$ which is also a homomorphism (by Theorem 4.8). If $x, y \in \beta \mathbb{N}$ and $k \in \mathbb{N}$, we may write $x \equiv y \pmod{k}$ if $\widetilde{h}_k(x) = \widetilde{h}_k(y)$.

Theorem 11.19. *Suppose that $p \in \mathbb{N}^*$ and that there exists $A \in p$ with the property that $x \equiv p \pmod{k}$ for every $x \in A^*$ and every $k \in \mathbb{N}$. Then $p <_{RK} p + q$ and $q <_{RK} p + q$ for every $q \in \mathbb{N}^*$.*

Proof. Let $q \in \mathbb{N}^*$. We observe that there is at most one $a \in A$ for which $a + q \in A^* + q$. To see this, suppose that $a, b \in A$ satisfy $a + q \in A^* + q$ and $b + q \in A^* + q$. Then $h_k(a) = h_k(b)$ for every $k \in \mathbb{N}$ and so $a = b$. We may replace A by $A \backslash \{a\}$, if such an element a exists. So we may suppose that $A + q$ is disjoint from $A^* + q$.

We claim that $a + q \neq b + q$ whenever $a \neq b$ in A and $A + q$ is strongly discrete in $\beta \mathbb{N}$. Otherwise we should have $b + q = x + q$ for some $b \in A$ and some $x \in \overline{A \backslash \{b\}}$, by Theorem 8.11. Now this cannot hold if $x \in A^*$; neither can it hold if $x \in A$ (by Lemma 6.28). Thus our theorem follows from Theorem 11.6 and Lemma 11.2. □

Comment 11.20. The set of ultrafilters p in $\mathbb{N}^*$ which satisfy the hypotheses of Theorem 11.19 contains a dense open subset of $\mathbb{N}^*$. To see this, let U be any non-empty clopen subset of $\mathbb{N}^*$ and let $q \in U$. For each $k \in \mathbb{N}$, let $V_k = \{x \in \beta \mathbb{N} : x \equiv q \pmod{k}\}$. Then V_k is a clopen subset of $\beta \mathbb{N}$. Since $q \in U \cap \bigcap_{k=1}^{\infty} V_k$, this set is a non-empty G_δ-subset of $\mathbb{N}^*$ and therefore has a non-empty interior in $\mathbb{N}^*$ (by Theorem 3.36). Any ultrafilter p in the interior of $U \cap \bigcap_{k=1}^{\infty} V_k$ satisfies the hypotheses of Theorem 11.19.

Theorem 11.21. *Let $\langle a_n \rangle_{n=1}^{\infty}$ be an infinite increasing sequence in $\mathbb{N}$. Suppose that either of the two following conditions is satisfied:*

  (i) *For every $n \in \mathbb{N}$, a_{n+1} is a multiple of a_n.*

  (ii) *For every $m, n \in \mathbb{N}$ with $m \neq n$, a_n is not a multiple of a_m.*

If $A = \{a_n : n \in \mathbb{N}\}$ and $p \in A^*$, then $p <_{RK} p + q$ and $q <_{RK} p + q$ for every $q \in \mathbb{N}^*$.

Proof. We first consider the case in which $q \in \bigcap_{n=1}^{\infty} \overline{n\mathbb{N}}$. For each $n \in \mathbb{N}$, let $B_n = \bigcap_{k=1}^{n+1} a_k \mathbb{N}$ so that $B_n \in q$.

If condition (i) is satisfied, we define $f : \mathbb{N} \to \mathbb{N}$ by stating that

$$f(n) = \max\{a_m : a_m | n\} \quad \text{if} \quad a_m | n \text{ for some } m,$$

defining f arbitrarily otherwise. For every $n \in \mathbb{N}$ and every $b \in B_n$, $f(a_n + b) = a_n$. So $a_m + B_m$ and $a_n + B_n$ are disjoint if $m \neq n$. Since $a_m + B_m \in a_m + q$, the set $A + q$ is strongly discrete in $\beta\mathbb{N}$ and our claim follows from Theorem 11.6 and Lemma 11.2.

If condition (ii) is satisfied, we define $g : \mathbb{N} \to \mathbb{N}$ by stating that

$$g(n) = \min\{a_m : a_m \mid n\} \quad \text{if} \quad a_m \mid n \text{ for some } m,$$

defining g arbitrarily otherwise. For every $n \in \mathbb{N}$ and every $b \in B_n$, $g(a_n + b) = a_n$. So we can again deduce that $A + q$ is discrete and that Theorem 11.6 and Lemma 11.2 apply.

Now let q be an arbitrary element of $\mathbb{N}^*$. For each $n \in \mathbb{N}$, we can choose $b_n \in \mathbb{N}$ satisfying $b_n + q \equiv 0 \pmod{k}$ for every $k \in \{1, 2, \ldots, n\}$. To see this, let $n \in \mathbb{N}$ be given. Then $\{c \in \mathbb{N} : c \equiv q \pmod{k} \text{ for every } k \in \{1, 2, \ldots, n\}\} \in q$ so pick c in this set. Pick $m \in \mathbb{N}$ such that $n!m > c$ and let $b_n = n!m - c$. Let $B = \{b_n : n \in \mathbb{N}\}$ and let $r \in B^*$. Then $q + r \in \bigcap_{n=1}^{\infty} \overline{n\mathbb{N}}$ since for each k, $\{b : b + q \equiv 0 \pmod{k}\} \in r$.

By what we have already proved, with $q + r$ in place of q, we know that $a + q + r \neq b + q + r$ when a and b are distinct members of A (by Lemma 6.28) and $A + q + r$ is discrete in $\beta\mathbb{N}$. This is equivalent to stating that $a + q + r \neq x + q + r$ if $a \in A$ and $x \in \overline{A}\backslash\{a\}$, by Theorem 8.11. This implies that $a + q \neq x + q$ for every $a \in A$ and every $x \in \overline{A}\backslash\{a\}$ and hence that $A + q$ is strongly discrete. So Theorem 11.6 and Lemma 11.2 apply once again.   □

Comment 11.22. By Ramsey's Theorem (Theorem 5.6), every infinite sequence in $\mathbb{N}$ contains an infinite subsequence satisfying the hypotheses of Theorem 11.21. So the set of ultrafilters $p \in \mathbb{N}^*$ to which this theorem applies also contains a dense open subset of $\mathbb{N}^*$.

Theorem 11.23. *Let S be a discrete countable cancellative semigroup and let p be a P-point in S^*. Then $p <_{RK} pq$ and $q <_{RK} pq$ for every $q \in S^*$.*

Proof. Let $q \in S^*$. We note that there is at most one element a of S for which $aq = pq$, by Lemma 6.28. Let $S' = \{a \in S : aq \neq pq\}$. So $S' = S$ or $|S\backslash S'| = 1$.

For each $b \in S'$, let $C_b = \{x \in S^* : xq = bq\}$. Then C_b is a compact subset of S^*. Since $p \notin C_b$ and since p is a P-point in S^*, $p \notin \bigcup_{b \in S'} C_b$. It follows that there is a member V of p such that $V \subseteq S'$ and $V \cap \bigcup_{b \in S'} C_b = \emptyset$. We claim that for any $b \in V$ and any $x \in \overline{V}\backslash\{b\}$, we have $bq \neq xq$. To see this, note that the equation $bq = xq$ cannot hold if $x \in S$ by Lemma 6.28. Also if $x \in S^*$, then $x \notin C_b$ so $bq \neq xq$.

It follows from Theorem 8.11 that Vq is strongly discrete in βS. So Theorem 11.6 and Lemma 11.2 apply.   □

Comment 11.24. We observe that the ultrafilters $p \in \mathbb{N}^*$ to which Theorems 11.19, 11.21, and 11.23 apply are left cancelable in $\beta\mathbb{Z}$. This is a consequence of Corollaries 8.36, 8.37, and 8.38 and Theorem 8.39.

Question 11.25. *If p is a left cancelable ultrafilter in $\mathbb{N}^*$, must it be true that $p <_{RK} p + q$ and $q <_{RK} p + q$ for every $q \in \mathbb{N}^*$?*

11.3 Further Connections with the Algebra of βS

We remind the reader that a family $\mathcal{F}$ of subsets of a given set is said to be almost disjoint if the intersection of any two distinct members of $\mathcal{F}$ is finite.

Theorem 11.26. *Let A be a subset of $\mathbb{N}^*$. If $|A| \leq \mathfrak{c}$, the elements of A have a common $\leq_{RK}$-successor in $\mathbb{H}$.*

Proof. We index A as $\langle p_x \rangle_{x \in \mathbb{R}}$, allowing repetitions if $|A| < \mathfrak{c}$. Let $\langle E_x \rangle_{x \in \mathbb{R}}$ be an almost disjoint family of infinite subsets of $\{2^n : n \in \mathbb{N}\}$. (Such a family exists by Theorem 3.56.) For each $x \in \mathbb{R}$, we choose $q_x \in \overline{E_x} \cap \mathbb{N}^*$ such that $q_x \approx_{RK} p_x$. For each $F \in \mathcal{P}_f(\mathbb{R})$, where $\mathcal{P}_f(\mathbb{R})$ denotes the set of non-empty finite subsets of $\mathbb{R}$, we put $s_F = \Sigma_{x \in F} q_x$, with the terms in the sum occurring in the order of increasing indices. We order $\mathcal{P}_f(\mathbb{R})$ by set inclusion and choose q to any limit point of the net $\langle s_F \rangle_{F \in \mathcal{P}_f(\mathbb{R})}$ in $\beta \mathbb{N}$, noting that $q \in \mathbb{H}$.

Recall that $\mathrm{supp}(n)$ is the binary support of n. For each $x \in \mathbb{R}$, we define $f_x : \mathbb{N} \to \mathbb{N}$ by stating that

$$f_x(n) = \min(E_x \cap \{2^m : m \in \mathrm{supp}(n)\}) \quad \text{if} \quad E_x \cap \{2^m : m \in \mathrm{supp}(n)\} \neq \emptyset$$

and defining $f_x(n)$ arbitrarily otherwise.

We shall show that $\widetilde{f_x}(q) = q_x$ for each $x \in \mathbb{R}$. We suppose that this equation does not hold. Then we can choose $B \in q_x$ such that $f_x^{-1}[B] \notin q$, and we can choose $Q \in q$ such that $f_x^{-1}[B] \cap Q = \emptyset$.

Since q is a limit point of the net $\langle s_F \rangle_{F \in \mathcal{P}_f(\mathbb{R})}$, there exists $F \in \mathcal{P}_f(\mathbb{R})$ such that $x \in F$ and $Q \in s_F$. We can choose a disjoint family $\langle B_y \rangle_{y \in F}$ of subsets of $\mathbb{N}$ satisfying $B_x \subseteq E_x \cap B$ and, for every $y \in F$, $B_y \subseteq E_y$, $B_y \in q_y$, and $B_y \cap E_x = \emptyset$ if $y \neq x$. Let M denote the set of all integers m of the form $m = \Sigma_{y \in F} n_y$, where $n_y \in B_y$ for each $y \in F$. We observe that this expression for m is unique, since each n_y is a power of 2, and that $f_x(m) = n_x \in B$.

Suppose that F is arranged in increasing order. By Exercise 4.1.6, $M \in \Sigma_{y \in F} q_y = s_F$. Thus we can choose $n \in M \cap Q$. Since $f_x(n) \in B$, it follows that $n \in f_x^{-1}[B] \cap Q$, contradicting our assumption that this set is empty. So $\widetilde{f_x}(q) = q_x$ and we have $p_x \approx_{RK} q_x \leq_{RK} q$ for every $x \in \mathbb{R}$. $\qquad\square$

Corollary 11.27. *Let A be a subset of $\mathbb{N}^*$ with $|A| \leq \mathfrak{c}$. The elements of A have a common $\leq_{RK}$-successor in any given minimal left ideal of $\beta \mathbb{N}$, and they also have a common $\leq_{RK}$-successor in any given minimal right ideal of $\beta \mathbb{N}$. Furthermore, there is a left ideal L of $\beta \mathbb{N}$ and a right ideal R of $\beta \mathbb{N}$ such that $x <_{RK} y$ for every $x \in A$ and every $y \in L \cup R$.*

Proof. By Theorem 11.26, the elements of A have a common $\leq_{RK}$-successor q in $\beta \mathbb{N}$. We may suppose that $x <_{RK} q$ for every $x \in A$, by Theorem 11.9. Choose $p \in \overline{\{n! : n \in \mathbb{N}\}} \cap \mathbb{N}^*$ such that $p \approx_{RK} q$.

Now p is right cancelable in $\beta \mathbb{N}$, by Corollary 8.38. So $p <_{RK} u + p$ for every $u \in \mathbb{N}^*$, by Theorem 11.8. We also have $p <_{RK} p + u$ for every $u \in \mathbb{N}^*$, by

Theorem 11.21. We can choose u to lie in any given left ideal or in any given right ideal of $\beta\mathbb{N}$.

Let $L = \beta\mathbb{N} + p$ and $R = p + \beta\mathbb{N}$. Then $x <_{RK} q \approx_{RK} p \leq_{RK} y$ for every $x \in A$ and every $y \in L \cup R$.   □

Corollary 11.28. *Every minimal left ideal of $\beta\mathbb{N}$ contains an increasing $\leq_{RK}$-chain of order type $\mathfrak{c}^+$, and so does every minimal right ideal of $\beta\mathbb{N}$.*

Proof. Chains of this kind can be constructed by an obvious transfinite induction, using Corollary 11.27 and Theorem 11.9.   □

Theorem 11.29. *There are at most $\mathfrak{c}$ elements of $\mathbb{N}^*$ whose $\leq_{RK}$-successors form a subsemigroup of $\beta\mathbb{N}$.*

Proof. Suppose that $p \in \mathbb{N}^*$ has the property that its $\leq_{RK}$-successors from a subsemigroup of $\beta\mathbb{N}$. We shall show that $q \in K(\beta\mathbb{N})$ implies that $p \leq_{RK} q$.

To see this, let L be the minimal left ideal of $\beta\mathbb{N}$ containing q. By Corollary 11.27, there is an element $u \in L$ for which $p \leq_{RK} u$. There is also a minimal left ideal M of $\beta\mathbb{N}$ such that $p \leq_{RK} v$ for every $v \in M$. So $p \leq_{RK} y$ for every $y \in M + u$ since the successors of p form a semigroup. However, $M + u$ is a left ideal contained in L and thus $M + u = L$. Consequently $p \leq_{RK} q$.

The result now follows from the fact that any given element of $\mathbb{N}^*$ can have at most $\mathfrak{c} \leq_{RK}$-predecessors (by Exercise 11.0.2).   □

11.4 The Rudin–Frolík Order

There are other useful topological order relations on spaces of ultrafilters, one of these being the Rudin–Frolík order.

Definition 11.30. Let S be a countable discrete space. The *Rudin–Frolík order* $\sqsubseteq$ on βS is defined by stating that, if $p, q \in \beta S$, $p \sqsubseteq q$ if there is an injective function $f : S \to \beta S$ for which $f[S]$ is discrete in βS and $\widetilde{f}(p) = q$.

If $p, q \in \beta S$, we may write $p \sqsubset q$ if $p \sqsubseteq q$ and $q \not\sqsubseteq p$.

It is immediate that the relation $\sqsubseteq$ is reflexive, and we shall see in the next lemma that it is transitive. It is not, however, anti-symmetric.

Lemma 11.31. *Let S be a countable discrete space. The relation $\sqsubseteq$ on βS is transitive.*

Proof. Let $g : S \to \beta S$ be an injective function for which $g[S]$ is discrete in βS, and let X be a countable discrete subset of βS. Then $\widetilde{g}[X]$ is also discrete, because it follows from Exercise 11.4.1 that $\widetilde{g}$ defines a homeomorphism from βS onto $c\ell_{\beta S}\, g[S]$.

Suppose now that $p, q, r \in \beta S$ and that $p \sqsubseteq q$ and $q \sqsubseteq r$. So there are injective functions f and g from S to βS for which $f[S]$ and $g[S]$ are discrete, $\widetilde{f}(p) = q$, and $\widetilde{g}(q) = r$. Then $\widetilde{g} \circ f$ is injective by Exercise 11.4.1, $\widetilde{g}[f[S]]$ is discrete, and $(\widetilde{g \circ f})(p) = \widetilde{g}(\widetilde{f}(p)) = r$. So $p \sqsubseteq r$.   □

We now observe that the assertion that $p \sqsubseteq q$ is stronger than the assertion that $p \leq_{RK} q$.

Lemma 11.32. *Let S be a countable discrete space and suppose that $p, q \in \beta S$ satisfy $p \sqsubseteq q$. Then $p \leq_{RK} q$.*

Proof. This is an immediate consequence of Lemma 6.37 and the fact that this is a purely set theoretic statement.   □

We now see that there are elements $q \in S^*$ which are not right cancelable, but have the property that $p \sqsubseteq pq$ for every $p \in S^*$. We remind the reader that an idempotent $q \in S^*$ is strongly right maximal if the equation $xq = q$ has the unique solution $x = q$ in S^*. (And strongly right maximal idempotents exist by Theorem 9.10.)

Lemma 11.33. *Let G be a discrete countable group and let q be a right maximal idempotent in G^*. Let $C = \{x \in G^* : xq = q\}$. Then, for every $p \in \beta G$, either $p \in (\beta G)C$ or there exists $P \in p$ such that Pq is discrete in βG.*

Proof. Suppose that $p \notin \overline{(\beta G)C}$. Since C is finite (by Theorem 9.4), $(\beta G)C$ is compact. So we can choose $P \in p$ for which $\overline{P} \cap (\beta G)C = \emptyset$.

We claim that Pq is discrete. If we assume the contrary, then $aq \in c\ell((P\backslash\{a\})q)$ for some $a \in P$. This implies that $aq = xq$ for some $x \in \overline{P}\backslash\{a\}$. Then $a^{-1}x \in C$ so $x \in aC$, contradicting our choice of P.   □

Theorem 11.34. *Let G be a discrete countable group and let q be a strongly right maximal idempotent in G^*. Then $p \sqsubseteq pq$ for every $p \in G^*$. In fact, for every $p \in G^*$, either $p \sqsubset pq$ and $q <_{RK} pq$ or $p = pq$.*

Proof. Suppose that $p \in G^*$ and that $p \neq pq$. Let $C = \{x \in G^* : xq = q\}$. Then $C = \{q\}$ and so we are assuming that $p \notin (\beta G)C$, because $p \in (\beta G)q$ implies that $p = pq$. It follows from Lemma 11.33 that we can choose $P \in p$ such that Pq is discrete (and hence, being countable, strongly discrete) in βS. Further $aq \neq bq$ when a and b are distinct elements of P by Lemma 6.28.

So $p \sqsubset pq$ and $q <_{RK} pq$, by Exercise 11.4.5, Lemma 11.2 and Theorem 11.6.   □

Exercise 11.4.1. Let S be a countable discrete space and let $f : S \rightarrow \beta S$ be an injective function for which $f[S]$ is discrete in βS. Show that $\widetilde{f}$ is also injective, where $\widetilde{f} : \beta S \rightarrow \beta S$ denotes the continuous extension of f. (Hint: let x and y be distinct elements of βS and let X and Y be disjoint subsets of S for which $X \in x$ and $Y \in y$. Observe that $f[X] \cap \overline{f[Y]} = \overline{f[X]} \cap f[Y] = \emptyset$ and apply Theorem 3.40.)

Exercise 11.4.2. Let S be a countable discrete space and let $p \in \beta S$. Show that $|\{q \in \beta S : q \sqsubseteq p\}| \leq \mathfrak{c}$.

Exercise 11.4.3. Let $p, q \in \beta\mathbb{N}$. If $p \sqsubseteq q$ and $q \sqsubseteq p$, we shall say that p and q are Rudin–Frolík equivalent. Show that p and q are Rudin–Frolík equivalent if and only if they are Rudin–Keisler equivalent. (Hint: Use Theorem 8.17 and Lemma 11.32)

Exercise 11.4.4. Let p be a weak P-point in $\mathbb{N}^*$. Show that, for any $q \in \mathbb{N}^*$, $q \sqsubseteq p$ implies that $p \leq_{RK} q$. Deduce that p is minimal in $\mathbb{N}^*$ for the Rudin–Frolík order.

Exercise 11.4.5. Let S be a discrete countable semigroup and let $p, q \in S^*$. Suppose that there is a set $P \in p$ such that Pq is discrete in βS and $aq \neq bq$ whenever a and b are distinct elements of P. Show that $p \sqsubset pq$. (Hint: Once you have shown that $p \sqsubseteq pq$, apply Lemma 11.2, Theorem 11.6, and Lemma 11.32 to deduce that one cannot have $pq \sqsubseteq p$.)

Exercise 11.4.6. Let S be a discrete countable semigroup and let $q \in S^*$ be right cancelable in βS. Show that $p \sqsubset pq$ for every $p \in S^*$. (Hint: Use Exercise 11.4.5 and Theorem 8.11.)

Exercise 11.4.7. Show that every element of $\mathbb{N}^*$ has $2^{\mathfrak{c}}$ distinct Rudin–Frolík successors. (Hint: use Exercise 11.4.6, Theorem 6.30, and Theorem 8.11.)

Exercise 11.4.8. Let S be a discrete countable cancellative semigroup and let $p \in S^*$. Show that p is right cancelable in βS if and only if there is an injective function $f : S \to \beta S$ for which $f[S]$ is discrete in βS and $\widetilde{f}(q) = qp$ for every $q \in S^*$. (Hint: Use Theorem 8.11. For the sufficiency consider the proof of Lemma 6.37. For the necessity consider ρ_p.)

Notes

Some of the results of Sections 11.1 and 11.2 are from [103], obtained in collaboration with S. García-Ferreira. Theorem 11.10 contrasts strongly with the situation that holds for the *Comfort order* of ultrafilters. Given an ultrafilter p on a set S, a space X is said to be *p-compact* if and only if whenever $f : S \to X$, $p\text{-}\lim_{s \in S} f(s)$ exists in X. One then defines $p \leq_C q$ for p and q in βS if and only if every q-compact space is p-compact. It is easy to see that $p \leq_{RK} q$ implies that $p \leq_C q$. It is shown in [103] that given any semigroup S and any $p \in \beta S$, $\{q \in \beta S : q \leq_C p\}$ is a subsemigroup of βS.

Corollary 11.27 and Exercise 11.4.8 are based on results in [46], a result of collaboration with A. Blass.

Exercise 11.4.4 shows that the existence of elements of $\mathbb{N}^*$ which are Rudin–Frolík minimal can be proved in ZFC, because the existence of weak P-points in $\mathbb{N}^*$ can be proved in ZFC. (See [170].) This contrasts with the Rudin–Keisler order. The existence of elements of $\mathbb{N}^*$ which are minimal in the Rudin–Keisler order does follow from the Continuum Hypothesis, but cannot be demonstrated in ZFC. In fact, any ultrafilter in $\mathbb{N}^*$ which is minimal in the Rudin–Keisler order has to be a P-point in $\mathbb{N}^*$. (See [69].)

The Rudin–Keisler order was introduced and studied by M. Rudin in [211] and [212] and independently by H. Keisler in lectures at UCLA in 1967.

The Rudin–Frolík order was implicitly introduced by Z. Frolík in [95] and some of its basic properties were established by M. Rudin in [212].

There are several very interesting properties of the Rudin–Keisler and Rudin–Frolík orders that we have not presented because they do not relate to the algebra of βS. Included among these is the equivalence of the following three statements for $p \in \beta S$:

(a) p is minimal in the Rudin–Keisler order.

(b) p is a *selective* ultrafilter. (That is, given any $f : S \to S$, there is some $A \in p$ such that the restriction of f to A is either one-to-one or constant.)

(c) p is a *Ramsey* ultrafilter. (That is, for every $n \in \mathbb{N}$ and every partition $\{P_0, P_1\}$ of $[S]^n$, there exist some $A \in p$ and some $i \in \{0, 1\}$ such that $[A]^n \subseteq P_i$.)

Another interesting result is that the Rudin–Keisler equivalence classes (which are the same as the Rudin–Frolík equivalence classes by Exercise 11.4.3) of Rudin–Frolík predecessors of a given element of $\beta \mathbb{N}$ are linearly ordered by $\leq_{RK}$. A systematic presentation of these and other results can be found in Sections 9 and 16 of [69].

Chapter 12

Ultrafilters Generated by Finite Sums

We have seen in Theorem 5.8 that if p is an idempotent in $(\beta\mathbb{N}, +)$ and $A \in p$, then A contains $FS(\langle x_n \rangle_{n=1}^{\infty})$ for some sequence $\langle x_n \rangle_{n=1}^{\infty}$ in $\mathbb{N}$. However, it is not necessarily true that $FS(\langle x_n \rangle_{n=1}^{\infty}) \in p$. We investigate in this chapter the existence and properties of ultrafilters satisfying the requirement that each $A \in p$ contains some $FS(\langle x_n \rangle_{n=1}^{\infty})$ with $FS(\langle x_n \rangle_{n=1}^{\infty}) \in p$.

12.1 Martin's Axiom

We shall have several proofs in this chapter for which Martin's Axiom is a hypothesis. Since we expect that Martin's Axiom may be unfamiliar to many of our readers, we include in this section an elementary introduction.

We use the terminology which is standard when dealing with Martin's Axiom, although some of it may seem strange at first glance.

Definition 12.1. A *partially ordered set* is a pair $(P, \leq)$ where P is a nonempty set and $\leq$ is a transitive and reflexive relation, that is a *partial order*.

Notice that a partial order is not required here to be anti-symmetric. (However, the partial orders that we will use are anti-symmetric.)

Definition 12.2. Let $(P, \leq)$ be a partially ordered set. A *filter in P* is a subset G of P such that

(1) $G \neq \emptyset$,

(2) for every $a, b \in G$, there is some $c \in G$ such that $c \leq a$ and $c \leq b$, and

(3) for every $a \in G$ and $b \in P$, if $a \leq b$, then $b \in G$.

Thus, if S is a set, a filter *on S* is a filter *in* the partially ordered set $(\mathcal{P}(S)\setminus\{\emptyset\}, \subseteq)$.

Definition 12.3. Let $(P, \leq)$ be a partially ordered set. A subset D of P is *dense* in P if and only if for every $a \in P$ there is some $b \in D$ such that $b \leq a$.

Thus, in more usual terminology, a subset of P is dense if and only if it is cofinal downward.

Definition 12.4. Let $(P, \leq)$ be a partially ordered set.

(a) Elements a and b of P are *compatible* if and only if there is some $c \in P$ such that $c \leq a$ and $c \leq b$. (Otherwise they are *incompatible*.)

(b) $(P, \leq)$ is a *c.c.c. partially ordered set* if and only if whenever D is a subset of P consisting of pairwise incompatible elements, one has that D is countable.

In the above definition, "c.c.c." stands for "countable chain condition". However, it actually asserts that all *antichains* are countable.

Definition 12.5. (a) Let κ be an infinite cardinal. Then $\text{MA}(\kappa)$ is the assertion that whenever $(P, \leq)$ is a c.c.c. partially ordered set and $\mathcal{D}$ is a collection of at most κ dense subsets of P, there is some filter G in P such that for all $D \in \mathcal{D}$, $G \cap D \neq \emptyset$.

(b) *Martin's Axiom* is the assertion that for all cardinals κ, if $\omega \leq \kappa < \mathfrak{c}$, then $\text{MA}(\kappa)$.

We first note that Martin's Axiom follows from the continuum hypothesis.

Theorem 12.6. $\text{MA}(\omega)$ *is true.*

Proof. Let $(P, \leq)$ be a partially ordered set and let $\{A_n : n \in \mathbb{N}\}$ be a set of dense subsets of P. Choose $a_1 \in A_1$. Inductively, having chosen $a_n \in A_n$, pick $a_{n+1} \in A_{n+1}$ such that $a_{n+1} \leq a_n$. Let $G = \{b \in P : \text{there is some } n \in \mathbb{N} \text{ with } a_n \leq b\}$. $\qquad\square$

Notice that, for the truth of $\text{MA}(\omega)$ one does not need that $(P, \leq)$ is a c.c.c. partial order. However, the next theorem (together with Exercise 12.1.1) shows that the c.c.c. requirement cannot be deleted.

Theorem 12.7. *There exist a partially ordered set $(P, \leq)$ and a collection $\mathcal{D}$ of ω_1 dense subsets of P such that no filter in P meets every member of $\mathcal{D}$.*

Proof. Let $P = \{f : \text{there exists } A \in \mathcal{P}_f(\mathbb{N}) \text{ such that } f : A \to \omega_1\}$. For $f, g \in P$ agree that $f \leq g$ if and only if $g \subseteq f$. For each $\alpha < \omega_1$, let $D_\alpha = \{f \in P : \alpha \text{ is in the range of } f\}$ and let $\mathcal{D} = \{D_\alpha : \alpha < \omega_1\}$. Trivially each D_α is dense in P.

Suppose now that we have a filter G such that $G \cap D_\alpha \neq \emptyset$ for each $\alpha < \omega_1$. Let $g = \bigcup G$. We claim that g is a function. To this end, let a, b, c be given with $(a, b) \in g$ and $(a, c) \in g$. Pick $f, h \in G$ such that $(a, b) \in f$ and $(a, c) \in h$ and pick $k \in G$ such that $k \leq f$ and $k \leq h$. Then $(a, b) \in k$ and $(a, c) \in k$ so $b = c$. But now g is a function whose domain is contained in $\mathbb{N}$ and whose range is all of ω_1, which is impossible. $\qquad\square$

We now illustrate a typical use of Martin's Axiom. The partially ordered set used is similar to those used in succeeding sections.

Recall that a set $\mathcal{A}$ of subsets of a set S is an *almost disjoint family* if and only if whenever A and B are distinct members of $\mathcal{A}$, $|A \cap B| < \omega$.

Definition 12.8. Let S be a set. A set $\mathcal{A}$ of subsets of S is a *maximal almost disjoint family* if and only if it is an almost disjoint family which is not properly contained in any other almost disjoint family.

Theorem 12.9. *Let $\mathcal{A}$ be a maximal almost disjoint family of infinite subsets of a set S. Then $\mathcal{A}$ is not countably infinite.*

Proof. This is Exercise 12.1.2. $\square$

As a consequence of Theorem 12.9, one sees immediately that the continuum hypothesis implies that any infinite maximal almost disjoint family of subsets of $\mathbb{N}$ must have cardinality $\mathfrak{c}$. We shall see that the same result follows from Martin's Axiom.

Lemma 12.10. *Let $\mathcal{A}$ be a family of infinite subsets of $\mathbb{N}$ with the property that, for every $\mathcal{F} \in \mathcal{P}_f(\mathcal{A})$, $\mathbb{N} \setminus \bigcup \mathcal{F}$ is infinite. Let $|\mathcal{A}| = \kappa$. If $\mathrm{MA}(\kappa)$, then there is an infinite subset B of $\mathbb{N}$ such that $A \cap B$ is finite for every $A \in \mathcal{A}$.*

Proof. Let $Q = \{(H, \mathcal{F}) : H \in \mathcal{P}_f(\mathbb{N}) \text{ and } \mathcal{F} \in \mathcal{P}_f(\mathcal{A})\}$. Order Q by agreeing that $(H', \mathcal{F}') \leq (H, \mathcal{F})$ if and only if

(1) $H \subseteq H'$,

(2) $\mathcal{F} \subseteq \mathcal{F}'$, and

(3) $(H' \setminus H) \cap \bigcup \mathcal{F} = \emptyset$.

It is routine to verify that $\leq$ is a partial order on Q (which is in fact antisymmetric).

Next notice that if $(H, \mathcal{F})$ and $(H', \mathcal{F}')$ are incompatible elements of Q, then $H \neq H'$ (since otherwise $(H, \mathcal{F} \cup \mathcal{F}') \leq (H, \mathcal{F})$ and $(H, \mathcal{F} \cup \mathcal{F}') \leq (H', \mathcal{F}')$). Thus, since $|\mathcal{P}_f(\mathbb{N})| = \omega$, it follows that $(Q, \leq)$ is a c.c.c. partially ordered set.

Now, for each $A \in \mathcal{A}$, let $D(A) = \{(H, \mathcal{F}) \in Q : A \in \mathcal{F}\}$. Then given any $(H, \mathcal{F}) \in Q$, one has $(H, \mathcal{F} \cup \{A\}) \in D(A)$ and $(H, \mathcal{F} \cup \{A\}) \leq (H, \mathcal{F})$. So $D(A)$ is dense.

For each $n \in \mathbb{N}$, let $E(n) = \{(H, \mathcal{F}) \in Q : H \setminus \{1, 2, \ldots, n\} \neq \emptyset\}$. Given any $(H, \mathcal{F}) \in Q$, we can choose $m > n$ such that $m \notin \bigcup \mathcal{F}$. Then $(H \cup \{m\}, \mathcal{F}) \in E(n)$ and $(H \cup \{m\}, \mathcal{F}) \leq (H, \mathcal{F})$. So $E(n)$ is dense in Q

By $\mathrm{MA}(\kappa)$, there is a filter G in Q such that $G \cap D(A) \neq \emptyset$ for each $A \in \mathcal{A}$ and $G \cap E(n) \neq \emptyset$ for each $n \in \mathbb{N}$. Let

$$B = \bigcup \{H : \text{there is some } \mathcal{F} \text{ with } (H, \mathcal{F}) \in G\}.$$

Since $G \cap E(n) \neq \emptyset$ for each $n \in \mathbb{N}$, it follows that B is infinite. We shall show that $A \cap B$ is finite for every $A \in \mathcal{A}$. To this end, let $A \in \mathcal{A}$ and pick $(H, \mathcal{F}) \in G \cap D(A)$. We claim that $A \cap B \subseteq H$. Let $x \in A \cap B$ and pick some $(H', \mathcal{F}') \in G$ such that $x \in H'$. Since $(H, \mathcal{F}) \in G$ and $(H', \mathcal{F}') \in G$, pick $(H'', \mathcal{F}'') \in G$ such that $(H'', \mathcal{F}'') \leq (H, \mathcal{F})$ and $(H'', \mathcal{F}'') \leq (H', \mathcal{F}')$. Then $x \in H' \subseteq H''$, $(H'' \setminus H) \cap \bigcup \mathcal{F} = \emptyset$, and $A \in \mathcal{F}$. So $x \in H$ as required. $\square$

Corollary 12.11. *Let* $\kappa < \mathfrak{c}$ *and let* $\langle F_\alpha \rangle_{\alpha < \kappa}$ *be a family of closed subsets of* $\mathbb{N}^*$. *Assuming* $\mathrm{MA}(\kappa)$, *if* $\bigcup_{\alpha < \kappa} F_\alpha \neq \mathbb{N}^*$, *then* $c\ell_{\mathbb{N}^*}\left(\bigcup_{\alpha < \kappa} F_\alpha\right) \neq \mathbb{N}^*$.

Proof. Let $x \in \mathbb{N}^* \setminus \bigcup_{\alpha < \kappa} F_\alpha$. For each $\alpha < \kappa$, we can choose a clopen subset D_α of $\mathbb{N}^*$ such that $F_\alpha \subseteq D_\alpha$ and $x \notin D_\alpha$. We have $D_\alpha = A_\alpha^*$ for some $A_\alpha \subseteq \mathbb{N}$ by Theorem 3.23. Now $\bigcup_{\alpha < \kappa} D_\alpha \neq \mathbb{N}^*$. Hence, if $\mathcal{A} = \{A_\alpha : \alpha < \kappa\}$, then $\mathcal{A}$ satisfies the hypotheses of Lemma 12.10. So there is an infinite subset B of $\mathbb{N}$ such that $A_\alpha \cap B$ is finite for every $\alpha < \kappa$. Our claim then follows from the fact that B^* is a non-empty clopen subset of $\mathbb{N}^*$ disjoint from each D_α and therefore disjoint from each F_α. □

Corollary 12.12. *Let* $\kappa < \mathfrak{c}$ *and let* $\langle G_\alpha \rangle_{\alpha < \kappa}$ *be a family of open subsets of* $\mathbb{N}^*$. *Assuming* $\mathrm{MA}(\kappa)$, *if* $\bigcap_{\alpha < \kappa} G_\alpha \neq \emptyset$, *then* $\mathrm{int}_{\mathbb{N}^*}\left(\bigcap_{\alpha < \kappa} G_\alpha\right) \neq \emptyset$.

Proof. This is the dual of Corollary 12.11. □

Theorem 12.13. *Let* $\mathcal{A}$ *be an infinite almost disjoint family of infinite subsets of* $\mathbb{N}$ *with* $|\mathcal{A}| = \kappa$. *If* $\mathrm{MA}(\kappa)$, *then* $\mathcal{A}$ *is not a maximal almost disjoint family.*

Proof. This follows immediately from Lemma 12.10. □

Corollary 12.14. *Let* $\mathcal{A}$ *be an infinite maximal almost disjoint family of subsets of* $\mathbb{N}$ *and assume Martin's Axiom. Then* $|\mathcal{A}| = \mathfrak{c}$.

Proof. By Theorem 12.13 one cannot have $|\mathcal{A}| < \mathfrak{c}$ and, since $\mathcal{A} \subseteq \mathcal{P}(\mathbb{N})$, $|\mathcal{A}| \leq \mathfrak{c}$. □

We also see as a consequence of the next corollary that Martin's axiom is the strongest assertion of its kind which is not simply false.

Corollary 12.15. $\mathrm{MA}(\mathfrak{c})$ *is false.*

Proof. Let $\mathcal{B}$ be an infinite set of pairwise disjoint infinite subsets of $\mathbb{N}$. A routine application of Zorn's Lemma yields a maximal almost disjoint family $\mathcal{A}$ with $\mathcal{B} \subseteq \mathcal{A}$. Necessarily $|\mathcal{A}| \leq \mathfrak{c}$ so Theorem 12.13 together with the assumption of $\mathrm{MA}(\mathfrak{c})$ yield a contradiction. □

Exercise 12.1.1. Prove that the partially ordered set in Theorem 12.7 is not a c.c.c. partially ordered set.

Exercise 12.1.2. Prove Theorem 12.9.

Exercise 12.1.3. Recall that a P-point in a topological space is a point x with the property that any countable intersection of neighborhoods of x is again a neighborhood of x. It is a consequence of Theorems 12.29, 12.35, and 12.36, that Martin's Axiom implies that there are P-points in $\mathbb{N}^*$. Provide a direct proof of this fact. (Hint: Let $\langle \mathcal{F}_\alpha \rangle_{\alpha < \mathfrak{c}}$ be an enumeration of all countable families of clopen subsets of $\mathbb{N}^*$. Use Corollary 12.11 to construct an increasing family $\langle D_\alpha \rangle_{\alpha < \mathfrak{c}}$ of proper clopen subsets of $\mathbb{N}^*$ with the property that, for every $\alpha < \mathfrak{c}$, either $\bigcap \mathcal{F}_\alpha \subseteq D_\alpha$ or $\mathbb{N}^* \setminus \bigcap \mathcal{F}_\alpha \subseteq D_\alpha$.)

Exercise 12.1.4. Show that Martin's Axiom implies that $\mathbb{N}^*$ satisfies the following strong form of the Baire Category Theorem: If $\mathcal{F}$ is a family of nowhere dense subsets of $\mathbb{N}^*$ with $|\mathcal{F}| < \mathfrak{c}$, then $\bigcup \mathcal{F}$ is nowhere dense in $\mathbb{N}^*$.

(Hint: Let $\kappa < \mathfrak{c}$ and let $\langle F_\alpha \rangle_{\alpha < \kappa}$ be a family of nowhere dense subsets of $\mathbb{N}^*$. Use Corollary 12.11 to construct an increasing family $\langle D_\alpha \rangle_{\alpha < \kappa}$ of proper clopen subsets of $\mathbb{N}^*$ for which $F_\alpha \subseteq D_\alpha$ for every $\alpha < \kappa$.)

12.2 Strongly Summable Ultrafilters — Existence

We deal in this section with the relationships among three classes of ultrafilters on $\mathbb{N}$. One of these classes, namely the idempotents, we have already investigated extensively.

Definition 12.16. Let $p \in \beta\mathbb{N}$.

(a) The ultrafilter p is *strongly summable* if and only if for every $A \in p$, there is a sequence $\langle x_n \rangle_{n=1}^\infty$ in $\mathbb{N}$ such that $\mathrm{FS}(\langle x_n \rangle_{n=1}^\infty) \subseteq A$ and $\mathrm{FS}(\langle x_n \rangle_{n=1}^\infty) \in p$.

(b) The ultrafilter p is *weakly summable* if and only if for every $A \in p$, there is a sequence $\langle x_n \rangle_{n=1}^\infty$ in $\mathbb{N}$ such that $\mathrm{FS}(\langle x_n \rangle_{n=1}^\infty) \subseteq A$.

Recall that $\Gamma = c\ell\{p \in \beta\mathbb{N} : p + p = p\}$.

Theorem 12.17. *The set* $\Gamma = \{p \in \beta\mathbb{N} : p \text{ is weakly summable}\}$. *In particular, every idempotent is weakly summable.*

Proof. This is an immediate consequence of Theorem 5.12. $\qquad\square$

We set out to show that every strongly summable ultrafilter is an idempotent.

Lemma 12.18. *Let p be a strongly summable ultrafilter. For each $A \in p$ there is an increasing sequence $\langle x_n \rangle_{n=1}^\infty$ in $\mathbb{N}$ such that $\mathrm{FS}(\langle x_n \rangle_{n=1}^\infty) \subseteq A$ and for each $m \in \mathbb{N}$, $\mathrm{FS}(\langle x_n \rangle_{n=m}^\infty) \in p$.*

Proof. For $i \in \{1, 2\}$, let $C_i = \{3^n(3k+i) : n, k \in \omega\}$. (That is, C_i is the set of positive integers whose rightmost nonzero ternary digit is i.) Pick $i \in \{1, 2\}$ such that $C_i \in p$.

Define $f : \mathbb{N} \to \omega$ by $f(3^n(3k+j)) = n$ for all $n, k \in \omega$ and $j \in \{1, 2\}$. Notice that if $x, y \in C_i$ and $f(x) = f(y)$, then $x + y \notin C_i$. Consequently, if $\{x, y, x + y\} \subseteq C_i$, then $f(x + y) = \min\{f(x), f(y)\}$.

Let $A \in p$ and pick a sequence $\langle x_n \rangle_{n=1}^\infty$ such that $\mathrm{FS}(\langle x_n \rangle_{n=1}^\infty) \in p$ and

$$\mathrm{FS}(\langle x_n \rangle_{n=1}^\infty) \subseteq A \cap C_i.$$

By the observation above, the sequence $\langle x_n \rangle_{n=1}^\infty$ is one to one, so we may presume that it is increasing. Let $m \in \mathbb{N}$ and suppose that $\mathrm{FS}(\langle x_n \rangle_{n=m}^\infty) \notin p$. Then $m > 1$. Now

$$\mathrm{FS}(\langle x_n \rangle_{n=1}^\infty) = \mathrm{FS}(\langle x_n \rangle_{n=1}^{m-1}) \cup \bigcup\nolimits_{F \subseteq \{1,2,\ldots,m-1\}} \left(\Sigma_{n \in F} x_n + \mathrm{FS}(\langle x_n \rangle_{n=m}^\infty) \right)$$

where $\sum_{n\in\emptyset} x_n = 0$. Since $FS(\langle x_n\rangle_{n=1}^{m-1})$ is finite and, by assumption, $\sum_{n\in\emptyset} x_n + FS(\langle x_n\rangle_{n=m}^{\infty}) \notin p$, pick F with $\emptyset \neq F \subseteq \{1, 2, \ldots, m-1\}$ such that $\sum_{n\in F} x_n + FS(\langle x_n\rangle_{n=m}^{\infty}) \in p$. Given any $y \in \sum_{n\in F} x_n + FS(\langle x_n\rangle_{n=m}^{\infty})$, one has that for some $G \in \mathcal{P}_f(\{m, m+1, m+2, \ldots\})$, $y = \sum_{n\in F} x_n + \sum_{n\in G} x_n$, where

$$\{\textstyle\sum_{n\in F} x_n, \sum_{n\in G} x_n, \sum_{n\in F} x_n + \sum_{n\in G} x_n\} \subseteq C_i$$

so, as observed above, $f(y) \leq f(\sum_{n\in F} x_n)$.

Pick a sequence $\langle y_n\rangle_{n=1}^{\infty}$ with $FS(\langle y_n\rangle_{n=1}^{\infty}) \subseteq \sum_{n\in F} x_n + FS(\langle x_n\rangle_{n=m}^{\infty})$ and $FS(\langle y_n\rangle_{n=1}^{\infty}) \in p$. Then for each $k \in \mathbb{N}$, $f(y_k) \leq f(\sum_{n\in F} x_n)$ so pick some $k \neq \ell$ such that $f(y_k) = f(y_\ell)$. Then $y_k + y_\ell \notin C_i$, a contradiction. $\square$

Theorem 12.19. *Let p be a strongly summable ultrafilter. Then $p = p + p$.*

Proof. Let $A \in p$ and pick by Lemma 12.18 a sequence $\langle x_n\rangle_{n=1}^{\infty}$ in $\mathbb{N}$ such that $FS(\langle x_n\rangle_{n=1}^{\infty}) \subseteq A$ and for each $m \in \mathbb{N}$, $FS(\langle x_n\rangle_{n=m}^{\infty}) \in p$. It suffices to show that $FS(\langle x_n\rangle_{n=1}^{\infty}) \subseteq \{y \in \mathbb{N} : -y + A \in p\}$, so let $y \in FS(\langle x_n\rangle_{n=1}^{\infty})$ and pick $F \in \mathcal{P}_f(\mathbb{N})$ such that $y = \sum_{n\in F} x_n$. Let $m = \max F + 1$. Then $FS(\langle x_n\rangle_{n=m}^{\infty}) \in p$ and $FS(\langle x_n\rangle_{n=m}^{\infty}) \subseteq -y + A$. $\square$

Lemma 12.20. *Let p be a strongly summable ultrafilter. For every $A \in p$ there is a sequence $\langle x_n\rangle_{n=1}^{\infty}$ in $\mathbb{N}$ such that*

(1) $FS(\langle x_n\rangle_{n=1}^{\infty}) \subseteq A$,

(2) *for each $m \in \mathbb{N}$, $FS(\langle x_n\rangle_{n=m}^{\infty}) \in p$, and*

(3) *for each $n \in \mathbb{N}$, $x_{n+1} > 4\sum_{t=1}^{n} x_t$.*

Proof. Let $A \in p$. For $i \in \{0, 1, 2\}$ let

$$B_i = \bigcup_{n=0}^{\infty} \{2^{3n+i}, 2^{3n+i} + 1, 2^{3n+i} + 2, \ldots, 2^{3n+i+1} - 1\}$$

and pick $i \in \{0, 1, 2\}$ such that $B_i \in p$. Pick by Lemma 12.18 an increasing sequence $\langle x_n\rangle_{n=1}^{\infty}$ in $\mathbb{N}$ such that $FS(\langle x_n\rangle_{n=1}^{\infty}) \subseteq A \cap B_i$ and for each $m \in \mathbb{N}$, $FS(\langle x_n\rangle_{n=m}^{\infty}) \in p$. We claim that $\langle x_n\rangle_{n=1}^{\infty}$ is as required, so suppose not and pick the first n such that $x_{n+1} \leq 4\sum_{t=1}^{n} x_t$. Pick $k \in \omega$ such that $2^{3k+i} \leq x_n < 2^{3k+i+1}$. Then in fact $\sum_{t=1}^{n} x_t < 2^{3k+i+1}$. (If $n = 1$ this is immediate. Otherwise, $x_n > 4\sum_{t=1}^{n-1} x_t$ so $\sum_{t=1}^{n-1} x_t < 2^{3k+i-1}$ so $\sum_{t=1}^{n} x_t < 2^{3k+i+1} + 2^{3k+i-1} < 2^{3k+i+2}$ and, since $\sum_{t=1}^{n} x_t \in B_i$, $\sum_{t=1}^{n} x_t < 2^{3k+i+1}$.) Now $x_{n+1} \leq 4\sum_{t=1}^{n} x_t < 2^{3k+i+3}$ and $x_{n+1} \in B_i$ so $x_{n+1} < 2^{3k+i+1}$. Since also $x_n < x_{n+1}$ we have that $2^{3k+i+1} < x_n + x_{n+1} < 2^{3k+i+2}$, so $x_n + x_{n+1} \notin B_i$, a contradiction. $\square$

The following easy theorem shows that not nearly all idempotents are strongly summable.

Theorem 12.21. *Let $p \in c\ell\, K(\beta\mathbb{N}, +)$. Then p is not strongly summable.*

Proof. Suppose that p is strongly summable and pick by Lemma 12.20 a sequence $\langle x_n \rangle_{n=1}^{\infty}$ such that $FS(\langle x_n \rangle_{n=1}^{\infty}) \in p$, and for each $n \in \mathbb{N}$, $x_{n+1} > 4 \sum_{t=1}^{n} x_t$. But then $FS(\langle x_n \rangle_{n=1}^{\infty})$ is not piecewise syndetic, contradicting Corollary 4.41.   □

Theorem 12.22. *There is a weakly summable ultrafilter which is not an idempotent.*

Proof. This was shown in Corollary 8.24.   □

Theorem 12.23. *The set Γ is not a semigroup. In fact, there exist idempotents p and q in $\beta\mathbb{N}$ such that $p + q \notin \Gamma$.*

Proof. This was shown in Exercise 6.1.4.   □

We now introduce a special kind of strongly summable ultrafilter. Its definition is somewhat complicated, but we shall prove in Section 12.4 that these idempotents can only be written in a trivial way as sums of other ultrafilters.

Recall that for $y \in \mathbb{N}$ we define the binary support of y by $y = \sum_{n \in \text{supp}(y)} 2^n$. In a similar fashion, for elements y of $FS(\langle t! \rangle_{t=1}^{\infty})$ we define the factorial support of y.

Definition 12.24. (a) For $y \in FS(\langle t! \rangle_{t=1}^{\infty})$, fsupp$(y)$ is defined by $y = \sum_{t \in \text{fsupp}(y)} t!$.

(b) The ultrafilter $p \in \beta\mathbb{N}$ is a *special strongly summable ultrafilter* if and only if

(1) for each $m \in \mathbb{N}$, $FS(\langle n! \rangle_{n=m}^{\infty}) \in p$,

(2) for each $A \in p$, there is an increasing sequence $\langle x_n \rangle_{n=1}^{\infty}$ such that $FS(\langle x_n \rangle_{n=1}^{\infty}) \in p$, $FS(\langle x_n \rangle_{n=1}^{\infty}) \subseteq A$, and for each $n \in \mathbb{N}$, x_n divides x_{n+1}, and

(3) for each infinite $L \subseteq \mathbb{N}$, there is a sequence $\langle x_n \rangle_{n=1}^{\infty}$ such that $FS(\langle x_n \rangle_{n=1}^{\infty}) \subseteq FS(\langle t! \rangle_{t=1}^{\infty})$, $FS(\langle x_n \rangle_{n=1}^{\infty}) \in p$, and $L \setminus \bigcup \{\text{fsupp}(y) : y \in FS(\langle x_n \rangle_{n=1}^{\infty})\}$ is infinite.

We shall show that Martin's Axiom implies that special strongly summable ultrafilters exist. Indeed, we shall show that Martin's Axiom implies that any family of subsets of $FS(\langle n! \rangle_{n=1}^{\infty})$ which is contained in an idempotent and has cardinality less than $\mathfrak{c}$, is contained in a special strongly summable idempotent.

Definition 12.25. (a) $\mathcal{S}$ denotes the set of infinite subsets S of $\mathbb{N}$ with the property that m divides n whenever $m, n \in S$ and $m < n$. $\mathcal{S}_f$ denotes the set of non-empty finite subsets of $\mathbb{N}$ which have the same property.

(b) If X is a subset of $\mathbb{N}$ which can be arranged as an increasing sequence $\langle x_n \rangle_{n=1}^{\infty}$, we put $FS_m(X) = FS(\langle x_n \rangle_{n=m}^{\infty})$ for each $m \in \mathbb{N}$. We put $FS_{\infty}(X) = \bigcap_{m \in \mathbb{N}} c\ell_{\beta\mathbb{N}}(FS_m(X))$.

Lemma 12.26. *Let $\mathcal{F}$ denote a family of subsets of $\mathbb{N}$ with the finite intersection property, such that $m\mathbb{N} \in \mathcal{F}$ for every $m \in \mathbb{N}$. Suppose that $B \in \mathcal{F}$ and that $\overline{B}$ contains an idempotent p. Suppose also that $B^{\star} = \{b \in B : -b + B \in p\} \in \mathcal{F}$ and that $-b + B^{\star} \in \mathcal{F}$ for every $b \in B^{\star}$. Let $\omega \leq \kappa < \mathfrak{c}$ and assume that $|\mathcal{F}| \leq \kappa$. Then it* *follows from* $\mathrm{MA}(\kappa)$ *that there exists* $X \in \mathcal{S}$ *such that* $\mathrm{FS}(X) \subseteq B$ *and* $X \cap A \neq \emptyset$ *for every* $A \in \mathcal{F}$.

Proof. We may suppose that $\mathcal{F}$ is closed under finite intersections.

Let $Q = \{F \in \mathcal{S}_f : \mathrm{FS}(F) \subseteq B^\star\}$. We define a partial order on Q by stating that $F' \leq F$ if $F \subseteq F'$.

Since Q is countable, it is trivial that it satisfies the c.c.c. condition.

For each $A \in \mathcal{F}$, let $D(A) = \{F \in Q : F \cap A \neq \emptyset\}$. To see that $D(A)$ is dense in Q, let $F \in Q$ and let $m = \prod F$. We can choose $a \in A \cap B^\star \cap \bigcap_{b \in \mathrm{FS}(F)}(-b + B^\star) \cap m\mathbb{N}$. Then $F \cup \{a\} \leq F$ and $F \in D(A)$.

Thus it follows from Martin's Axiom that there is a filter Φ in Q such that $\Phi \cap D(A) \neq \emptyset$ for every $A \in \mathcal{F}$. Let $X = \bigcup \Phi$.

We shall show that X has the properties required. To see that X is infinite, let $m \in \mathbb{N}$ and pick $F \in \Phi \cap D(m\mathbb{N})$. Then $X \cap m\mathbb{N} \neq \emptyset$. Since Φ is a filter, any two elements x and y of X belong to a set $F \in \Phi$. So, if $x < y$, x divides y. Thus $X \in \mathcal{S}$. Furthermore, if H is any finite subset of X, $H \subseteq F$ for some $F \in \Phi$ and so $\mathrm{FS}(H) \subseteq B$. Thus $\mathrm{FS}(X) \subseteq B$. Finally, for any $A \in \mathcal{F}$, there exists $F \in \Phi \cap D(A)$ and so $X \cap A \neq \emptyset$. $\square$

Lemma 12.27. *Let p be an idempotent in $\mathbb{N}^*$, let $\mathcal{F} \subseteq p$, and let $\omega \leq \kappa < \mathfrak{c}$. If $|\mathcal{F}| \leq \kappa$, then $\mathrm{MA}(\kappa)$ implies that there exists $X \in \mathcal{S}$ such that $\mathrm{FS}_\infty(X) \subseteq \bigcap_{A \in \mathcal{F}} \overline{A}$.*

Proof. Let $\mathcal{G} = \mathcal{F} \cup \{m\mathbb{N} : m \in \mathbb{N}\} \cup \{B^\star : B \in \mathcal{F}\} \cup \{-b + B^\star : B \in \mathcal{F} \text{ and } b \in B^\star\}$ and note that by Lemmas 4.14 and 6.6, $\mathcal{G} \subseteq p$. Let $\mathcal{H}$ be the set of finite intersections of members of $\mathcal{G}$ and note that $\mathcal{H} \subseteq p$ and $|\mathcal{H}| \leq \kappa$.

Well order $\mathcal{F}$ as $\langle A_\lambda \rangle_{\lambda \leq \kappa}$ (with repetition if $|\mathcal{F}| < \kappa$). Pick by Lemma 12.26 (with $\mathcal{H}$ in place of $\mathcal{F}$ and A_0 in place of B) some $X_0 \in \mathcal{S}$ for which $\mathrm{FS}(X_0) \subseteq A_0$, and $X_0 \cap A \neq \emptyset$ for every $A \in \mathcal{H}$.

We then make the inductive assumption that $0 < \beta \leq \kappa$ and that we have defined $X_\alpha \in \mathcal{S}$ for every $\alpha < \beta$ so that the following conditions are satisfied:

(a) $\mathrm{FS}(X_\alpha) \subseteq A_\alpha$ and $X_\alpha \cap A \neq \emptyset$ for every $A \in \mathcal{H}$ and

(b) if $\delta < \alpha$, then $|X_\alpha \setminus X_\delta| < \omega$.

Using condition (b) and the facts that $\mathcal{H}$ is closed under finite intersections and that $m\mathbb{N} \in \mathcal{H}$ for every $m \in \mathbb{N}$ one sees that

(*) for any finite nonempty $\mathcal{D} \subseteq \mathcal{H} \cup \{X_\alpha : \alpha < \beta\}$, there exist $A \in \mathcal{H}$ and $\alpha < \beta$ with $A \cap X_\alpha \subseteq \bigcap \mathcal{D}$.

In particular, by condition (a), $\{A \cap X_\alpha : A \in \mathcal{H} \text{ and } \alpha < \beta\}$ has the finite intersection property.

We apply Lemma 12.26, with $\{A \cap X_\alpha : A \in \mathcal{H} \text{ and } \alpha < \beta\}$ in place of $\mathcal{F}$ and A_β in place of B. Pick $W_\beta \in \mathcal{S}$ such that $\mathrm{FS}(W_\beta) \subseteq A_\beta$ and $W_\beta \cap A \cap X_\alpha \neq \emptyset$ for every $A \in \mathcal{H}$ and every $\alpha < \beta$. By the observation (*), $\bigcap\{\overline{W_\beta \cap A \cap X_\alpha} : A \in \mathcal{H} \text{ and } \alpha < \beta\} \neq \emptyset$ so by Corollary 12.12, there exists an infinite subset X_β of $\mathbb{N}$ such that $X_\beta{}^\star \subseteq \overline{W_\beta \cap A \cap X_\alpha}$ for every $A \in \mathcal{H}$ and every $\alpha < \beta$. We may suppose that

$X_\beta \subseteq W_\beta$. We may also suppose that $X_\beta \in \mathcal{S}$, because $X_\beta \cap m\mathbb{N} \neq \emptyset$ for every $m \in \mathbb{N}$, and so X_β contains a set in $\mathcal{S}$.

It is clear that conditions (a) and (b) are satisfied with β in place of α.

We put $X = X_\kappa$. If $\alpha \leq \kappa$, $X \backslash X_\alpha$ is finite and so there is some $n \in \mathbb{N}$ such that $FS_n(X) \subseteq FS(X_\alpha)$ and thus $FS_\infty(X) \subseteq \overline{FS(X_\alpha)} \subseteq \overline{A_\alpha}$. $\qquad\square$

Corollary 12.28. *Let $\mathcal{F}$ be a family of subsets of $\mathbb{N}$ contained in an idempotent p for which $FS(\langle n! \rangle_{n=1}^\infty) \in p$. Let L be any infinite subset of $\mathbb{N}$ and let $\omega \leq \kappa < \mathfrak{c}$. If $MA(\kappa)$ and $|\mathcal{F}| \leq \kappa$, then there exists $X \in \mathcal{S}$ with $FS(X) \subseteq FS(\langle n! \rangle_{n=1}^\infty)$ such that $FS_\infty(X) \subseteq \overline{A}$ for every $A \in \mathcal{F}$ and $L \backslash \bigcup_{x \in FS(X)} fsupp(x)$ is infinite.*

Proof. We may suppose that $\mathcal{F}$ is closed under finite intersections. By Exercise 12.2.1, we may suppose that $FS(\langle n! \rangle_{n=m}^\infty) \in \mathcal{F}$ for every $m \in \mathbb{N}$.

By Lemma 12.27, there exists $Y \in \mathcal{S}$ such that $FS_\infty(Y) \subseteq \overline{A}$ for every $A \in \mathcal{F}$. We claim that $Y \cap FS(\langle n! \rangle_{n=m}^\infty) \neq \emptyset$ for every $m \in \mathbb{N}$. To see this, let $m \in \mathbb{N}$. Since $FS_\infty(Y) = \bigcap_{k \in \mathbb{N}} \overline{FS_k(Y)} \subseteq \overline{(FS\langle n! \rangle_{n=m}^\infty)}$, there exists $k \in \mathbb{N}$ for which $FS_k(Y) \subseteq FS(\langle n! \rangle_{n=m}^\infty)$.

Thus we can inductively choose a sequence $\langle x_n \rangle_{n=1}^\infty$ in $Y \cap FS(\langle n! \rangle_{n=1}^\infty)$ with the property that, for each $n \in \mathbb{N}$, $\max fsupp(x_n) < \ell_n < \min fsupp(x_{n+1})$ for some $\ell_n \in L$. If $X = \{x_n : n \in \mathbb{N}\}$, then X has the properties required. $\qquad\square$

Notice that, while the previous results only use $MA(\kappa)$ for a fixed $\kappa < \mathfrak{c}$, the following theorem uses the full strength of Martin's Axiom.

Theorem 12.29. *Let $\mathcal{F}$ be a family of subsets of $\mathbb{N}$ contained in some idempotent $q \in \mathbb{N}^*$ for which $FS(\langle n! \rangle_{n=1}^\infty) \in q$. If $|\mathcal{F}| < \mathfrak{c}$, Martin's Axiom implies that there is a special strongly summable ultrafilter p for which $\mathcal{F} \subseteq p$.*

Proof. We assume Martin's Axiom.

Let $\langle Y_\alpha \rangle_{\alpha < \mathfrak{c}}$ be an enumeration of $\mathcal{P}(\mathbb{N})$ and let $\langle L_\alpha \rangle_{\alpha < \mathfrak{c}}$ be an enumeration of the infinite subsets of $\mathbb{N}$.

We can choose $Z_0 \in \{Y_0, \mathbb{N} \backslash Y_0\}$ such that $Z_0 \in q$. By Corollary 12.28 applied to $\mathcal{F} \cup \{Z_0\}$, we can choose $X_0 \in \mathcal{S}$ with $FS(X_0) \subseteq FS(\langle n! \rangle_{n=1}^\infty)$ such that $FS_\infty(X_0) \subseteq \overline{A \cap Z_0}$ for every $A \in \mathcal{F}$ and $L_0 \backslash \bigcup_{x \in FS(X_0)} fsupp(x)$ is infinite. We then make the inductive assumption that $0 < \beta < \mathfrak{c}$ and that $X_\alpha \in \mathcal{S}$ with $FS(X_\alpha) \subseteq FS(\langle n! \rangle_{n=1}^\infty)$ has been defined for every $\alpha < \beta$ so that the following conditions hold:

(a) $FS_\infty(X_\alpha) \subseteq \overline{Y_\alpha}$ or $FS_\infty(X_\alpha) \subseteq \overline{\mathbb{N} \backslash Y_\alpha}$;

(b) if $\delta < \alpha$, then $FS_\infty(X_\alpha) \subseteq FS_\infty(X_\delta)$; and

(c) $L_\alpha \backslash \bigcup_{x \in FS(X_\alpha)} fsupp(x)$ is infinite.

By Lemma 5.11, for each $\alpha < \beta$, $FS_\infty(X_\alpha)$ is a compact subsemigroup of $\mathbb{N}^*$ and hence so is $\bigcap_{\alpha < \beta} FS_\infty(X_\alpha)$ which therefore contains an idempotent r. Since $FS(X_0) \subseteq FS(\langle n! \rangle_{n=1}^\infty)$, $FS(\langle n! \rangle_{n=1}^\infty) \in r$. We can choose $Z_\beta \in \{Y_\beta, \mathbb{N} \backslash Y_\beta\}$ satisfying $Z_\beta \in r$. By Corollary 12.28 (applied to $\{FS_m(X_\alpha) : \alpha < \beta \text{ and } m \in \mathbb{N}\} \cup \{Z_\beta\}$ in place of $\mathcal{F}$) we can choose $X_\beta \in \mathcal{S}$ with $\mathrm{FS}(X_\alpha) \subseteq \mathrm{FS}(\langle n! \rangle_{n=1}^\infty)$ such that $L_\beta \setminus \bigcup_{x \in \mathrm{FS}(X_\beta)} \mathrm{fsupp}(x)$ is infinite, $\mathrm{FS}_\infty(X_\beta) \subseteq \overline{Z_\beta}$, and $\mathrm{FS}_\infty(X_\beta) \subseteq \overline{\mathrm{FS}_m(X_\alpha)}$ for every $\alpha < \beta$ and every $m \in \mathbb{N}$. Thus $\mathrm{FS}_\infty(X_\beta) \subseteq \mathrm{FS}_\infty(X_\alpha)$ for every $\alpha < \beta$. So conditions (a)–(c) are satisfied with β in place of α.

This shows that we can define $X_\alpha \in \mathcal{S}$ for every $\alpha < \mathfrak{c}$ so that conditions (a)-(c) are satisfied. We put $p = \{B \subseteq \mathbb{N} : \mathrm{FS}_\infty(X_\alpha) \subseteq \overline{B}$ for some $\alpha < \mathfrak{c}\}$. It is clear that p is a filter. For every $Y \subseteq \mathbb{N}$, $Y \in p$ or $\mathbb{N} \setminus Y \in p$, and so p is an ultrafilter. Since $\mathrm{FS}_\infty(X_0) \subseteq \bigcap_{A \in \mathcal{F}} \overline{A}$, we have $\mathcal{F} \subseteq p$.

Since $\mathrm{FS}(X_0) \subseteq \mathrm{FS}(\langle n! \rangle_{n=1}^\infty)$, we have by Exercise 12.2.1 that for each $m \in \mathbb{N}$, $\mathrm{FS}(\langle n! \rangle_{n=m}^\infty) \in p$. Given $A \in p$, $A = Y_\alpha$ for some α and so $\mathrm{FS}_\infty(X_\alpha) \subseteq \overline{A}$ and thus for some m, $\mathrm{FS}_m(X_\alpha) \subseteq A$. Condition (3) of the definition holds directly, and so we have that p is a special strongly summable ultrafilter. □

Exercise 12.2.1. Show that if p is an idempotent in $\mathbb{N}^*$ and $\mathrm{FS}(\langle n! \rangle_{n=1}^\infty) \in p$, then for all $m \in \mathbb{N}$, $\mathrm{FS}(\langle n! \rangle_{n=m}^\infty) \in p$. (Hint: Use Lemma 6.6 and consider the proof of Lemma 12.18.)

12.3 Strongly Summable Ultrafilters — Independence

In the last section we saw that Martin's Axiom implies that strongly summable ultrafilters exist. In this section we show that their existence cannot be deduced in ZFC.

Definition 12.30. $\mathcal{U}$ is a *union ultrafilter* if and only if $\mathcal{U}$ is an ultrafilter on $\mathcal{P}_f(\mathbb{N})$ and for every $\mathcal{A} \in \mathcal{U}$ there is a sequence $\langle F_n \rangle_{n=1}^\infty$ of pairwise disjoint members of $\mathcal{P}_f(\mathbb{N})$ such that $\mathrm{FU}(\langle F_n \rangle_{n=1}^\infty) \in \mathcal{U}$ and $\mathrm{FU}(\langle F_n \rangle_{n=1}^\infty) \subseteq \mathcal{A}$.

It would appear that union ultrafilters and strongly summable ultrafilters are essentially equivalent notions. This is indeed true, as we shall shortly see. The correspondence in one direction is in fact immediate.

Theorem 12.31. *Let $\mathcal{U}$ be a union ultrafilter. Let*

$$p = \left\{ \left\{ \textstyle\sum_{t \in F} 2^{t-1} : F \in \mathcal{A} \right\} : \mathcal{A} \in \mathcal{U} \right\}.$$

Then p is a strongly summable ultrafilter.

Proof. The function $f : \mathcal{P}_f(\mathbb{N}) \to \mathbb{N}$ defined by $f(F) = \sum_{t \in F} 2^{t-1}$ is a bijection, so p is trivially an ultrafilter. To see that p is strongly summable, let $A \in p$. Let $\mathcal{A} = \{F \in \mathcal{P}_f(\mathbb{N}) : \sum_{t \in F} 2^{t-1} \in A\}$. Then $\mathcal{A} \in \mathcal{U}$ so pick a sequence $\langle F_n \rangle_{n=1}^\infty$ of pairwise disjoint members of $\mathcal{P}_f(\mathbb{N})$ such that $\mathrm{FU}(\langle F_n \rangle_{n=1}^\infty) \in \mathcal{U}$ and $\mathrm{FU}(\langle F_n \rangle_{n=1}^\infty) \subseteq \mathcal{A}$. For each n, let $x_n = \sum_{t \in F_n} 2^{t-1}$. Then $\mathrm{FS}(\langle x_n \rangle_{n=1}^\infty) \in p$ and $\mathrm{FS}(\langle x_n \rangle_{n=1}^\infty) \subseteq A$. □

To establish the correspondence in the reverse direction, we need some purely arithmetic lemmas. They are somewhat stronger than needed here, but they will be used in their full strength in the next section.

Lemma 12.32. *Let* $\langle x_n \rangle_{n=1}^{\infty}$ *be a sequence in* $\mathbb{N}$ *such that, for each* $n \in \mathbb{N}$, $x_{n+1} > 4 \sum_{t=1}^{n} x_t$. *Then each member of* $\mathbb{N}$ *can be written in at most one way as a linear combination of* x_n's *with coefficients from* $\{1, 2, 3, 4\}$.

Proof. Suppose not and let z be the smallest counterexample. Then (taking as usual $\sum_{t \in \emptyset} x_t = 0$) one has

$$z = \sum_{t \in F_1} x_t + \sum_{t \in F_2} 2x_t + \sum_{t \in F_3} 3x_t + \sum_{t \in F_4} 4x_t$$
$$= \sum_{t \in G_1} x_t + \sum_{t \in G_2} 2x_t + \sum_{t \in G_3} 3x_t + \sum_{t \in G_4} 4x_t$$

where $\{F_1, F_2, F_3, F_4\}$ and $\{G_1, G_2, G_3, G_4\}$ are each pairwise disjoint and not each $F_i = G_i$. Let $n = \max \bigcup_{i=1}^{4} (F_i \cup G_i)$ and assume without loss of generality that $n \in \bigcup_{i=1}^{4} F_i$. Then $n \notin \bigcup_{i=1}^{4} G_i$, since if it were we would have $z - x_n$ as a smaller counterexample. Thus

$$z = \sum_{t \in F_1} x_t + \sum_{t \in F_2} 2x_t + \sum_{t \in F_3} 3x_t + \sum_{t \in F_4} 4x_t$$
$$\geq x_n$$
$$> 4 \sum_{t=1}^{n-1} x_t$$
$$\geq \sum_{t \in G_1} x_t + \sum_{t \in G_2} 2x_t + \sum_{t \in G_3} 3x_t + \sum_{t \in G_4} 4x_t$$
$$= z,$$

a contradiction. $\square$

Lemma 12.33. *Let* $\langle x_n \rangle_{n=1}^{\infty}$ *be a sequence in* $\mathbb{N}$ *such that, for each* $n \in \mathbb{N}$, $x_{n+1} > 4 \sum_{t=1}^{n} x_t$. *Then each member of* $\mathbb{Z}$ *can be written in at most one way as a linear combination of* x_n's *with coefficients from* $\{-2, -1, 1, 2\}$.

Proof. Assume that we have

$$\sum_{t \in F_1} 2x_t + \sum_{t \in F_2} x_t - \sum_{t \in F_3} x_t - \sum_{t \in F_4} 2x_t$$
$$= \sum_{t \in G_1} 2x_t + \sum_{t \in G_2} x_t - \sum_{t \in G_3} x_t - \sum_{t \in G_4} 2x_t$$

where $\{F_1, F_2, F_3, F_4\}$ and $\{G_1, G_2, G_3, G_4\}$ are each pairwise disjoint. Then

$$\sum_{t \in F_1 \cap G_4} 4x_t + \sum_{t \in (F_1 \cap G_3) \cup (F_2 \cap G_4)} 3x_t$$
$$+ \sum_{t \in (F_1 \setminus (G_3 \cup G_4)) \cup (G_4 \setminus (F_1 \cup F_2)) \cup (F_2 \cap G_3)} 2x_t + \sum_{t \in (F_2 \setminus (G_3 \cup G_4)) \cup (G_3 \setminus (F_1 \cup F_2))} x_t$$
$$= \sum_{t \in G_1 \cap F_4} 4x_t + \sum_{t \in (G_1 \cap F_3) \cup (G_2 \cap F_4)} 3x_t$$
$$+ \sum_{t \in (G_1 \setminus (F_3 \cup F_4)) \cup (F_4 \setminus (G_1 \cup G_2)) \cup (G_2 \cap F_3)} 2x_t + \sum_{t \in (G_2 \setminus (F_3 \cup F_4)) \cup (F_3 \setminus (G_1 \cup G_2))} x_t.$$

Thus, by Lemma 12.32, one has

$$F_1 \cap G_4 = G_1 \cap F_4,$$
$$(F_1 \cap G_3) \cup (F_2 \cap G_4) = (G_1 \cap F_3) \cup (G_2 \cap F_4),$$
$$\big(F_1 \setminus (G_3 \cup G_4)\big) \cup \big(G_4 \setminus (F_1 \cup F_2)\big) \cup (F_2 \cap G_3) = \big(G_1 \setminus (F_3 \cup F_4)\big) \cup \big(F_4 \setminus (G_1 \cup G_2)\big)$$
$$\cup (G_2 \cap F_3),$$
$$\big(F_2 \setminus (G_3 \cup G_4)\big) \cup \big(G_3 \setminus (F_1 \cup F_2)\big) = \big(G_2 \setminus (F_3 \cup F_4)\big) \cup \big(F_3 \setminus (G_1 \cup G_2)\big).$$

It is routine to establish from these equations that $F_i = G_i$ for each $i \in \{1, 2, 3, 4\}$.□

Lemma 12.34. *Let $\langle x_n \rangle_{n=1}^{\infty}$ be a sequence in $\mathbb{N}$ such that, for each $n \in \mathbb{N}$, $x_{n+1} > 4 \sum_{t=1}^{n} x_t$. If $\{a, b, a + b\} \subseteq \mathrm{FS}(\langle x_n \rangle_{n=1}^{\infty})$, $a = \sum_{t \in F} x_t$, and $b = \sum_{t \in G} x_t$, then $F \cap G = \emptyset$.*

Proof. Pick $H \in \mathcal{P}_f(\mathbb{N})$ such that $a + b = \sum_{t \in H} x_t$. Then

$$\sum_{t \in H} x_t = \sum_{t \in F \triangle G} x_t + \sum_{t \in F \cap G} 2 x_t$$

so by Lemma 12.32, $F \cap G = \emptyset$. □

Theorem 12.35. *Let p be a strongly summable ultrafilter and pick by Lemma 12.20 a sequence $\langle x_n \rangle_{n=1}^{\infty}$ in $\mathbb{N}$ such that $\mathrm{FS}(\langle x_n \rangle_{n=1}^{\infty}) \in p$ and for each $n \in \mathbb{N}$, $x_{n+1} > 4 \sum_{t=1}^{n} x_t$. Define $\varphi : \mathrm{FS}(\langle x_n \rangle_{n=1}^{\infty}) \to \mathcal{P}_f(\mathbb{N})$ by $\varphi(\sum_{t \in F} x_t) = \{x_t : t \in F\}$. Let $\mathcal{U} = \{\mathcal{A} \subseteq \mathcal{P}_f(\mathbb{N}) : \varphi^{-1}[\mathcal{A}] \in p\}$. Then $\mathcal{U}$ is a union ultrafilter.*

Proof. Notice first that by Lemma 12.32, the function φ is well defined. One has then immediately that $\mathcal{U}$ is an ultrafilter. To see that $\mathcal{U}$ is a union ultrafilter, let $\mathcal{A} \in \mathcal{U}$. Let $A = \varphi^{-1}[\mathcal{A}]$. Then $A \in p$ so pick a sequence $\langle y_n \rangle_{n=1}^{\infty}$ such that $\mathrm{FS}(\langle y_n \rangle_{n=1}^{\infty}) \subseteq A$ and $\mathrm{FS}(\langle y_n \rangle_{n=1}^{\infty}) \in p$. For $n \in \mathbb{N}$, let $F_n = \varphi(y_n)$. By Lemma 12.34, one has for each $H \in \mathcal{P}_f(\mathbb{N})$ that $\varphi(\sum_{n \in H} y_n) = \bigcup_{n \in H} F_n$. Consequently, one has that $\mathrm{FU}(\langle F_n \rangle_{n=1}^{\infty}) \subseteq \mathcal{A}$ and $\mathrm{FU}(\langle F_n \rangle_{n=1}^{\infty}) \in \mathcal{U}$. □

Theorem 12.36. *Let $\mathcal{U}$ be a union ultrafilter and let $q = \widetilde{\max}(\mathcal{U})$, where $\widetilde{\max} : \beta(\mathcal{P}_f(\mathbb{N})) \to \beta\mathbb{N}$ is the continuous extension of $\max : \mathcal{P}_f(\mathbb{N}) \to \mathbb{N}$. Then q is a P-point of $\mathbb{N}^*$.*

Proof. Notice that q is a nonprincipal ultrafilter because $\mathcal{U}$ is and the function $\max$ is finite to one on $\mathcal{P}_f(\mathbb{N})$. To see that q is a P-point, let $\langle A_n \rangle_{n=1}^{\infty}$ be a sequence of members of q. We need to show that there is some $B \in q$ such that $B^* \subseteq \bigcap_{n=1}^{\infty} A_n^*$. We may presume that $A_1 = \mathbb{N}$ and that $A_{n+1} \subseteq A_n$ and $n \notin A_{n+1}$ for all n. (In particular $\bigcap_{n=1}^{\infty} A_n = \emptyset$.) Define $f : \mathbb{N} \to \mathbb{N}$ by $f(x) = \max\{n \in \mathbb{N} : x \in A_n\}$. Let

$$\mathcal{B}_0 = \{F \in \mathcal{P}_f(\mathbb{N}) : f(\max F) \leq \min F\}$$

and

$$\mathcal{B}_1 = \{F \in \mathcal{P}_f(\mathbb{N}) : f(\max F) > \min F\}$$

and pick $i \in \{0, 1\}$ such that $\mathcal{B}_i \in \mathcal{U}$.

Pick a pairwise disjoint sequence $\langle F_n \rangle_{n=1}^{\infty}$ in $\mathcal{P}_f(\mathbb{N})$ such that $\mathrm{FU}(\langle F_n \rangle_{n=1}^{\infty}) \subseteq \mathcal{B}_i$ and $\mathrm{FU}(\langle F_n \rangle_{n=1}^{\infty}) \in \mathcal{U}$. We may presume that $\min F_n < \min F_{n+1}$ for all n. Since $\mathrm{FU}(\langle F_n \rangle_{n=1}^{\infty}) \in \mathcal{U}$, we have that $\{\max G : G \in \mathrm{FU}(\langle F_n \rangle_{n=1}^{\infty})\} \in q$. Since

$$\{\max F_n : n \in \mathbb{N}\} = \{\max G : G \in \mathrm{FU}(\langle F_n \rangle_{n=1}^{\infty})\},$$

we have $\{\max F_n : n \in \mathbb{N}\} \in q$.

We show first that $i \neq 0$, so suppose instead that $i = 0$. Pick $k \in \mathbb{N}$ such that for all $n \geq k$, $\max F_n > \max F_1$. Since q is nonprincipal $\{\max F_n : n \in \mathbb{N} \text{ and } n \geq k\} \in q$. Let $\ell = \min F_1$. Now $A_{\ell+1} \in q$ so pick $n \geq k$ such that $\max F_n \in A_{\ell+1}$. Then

$$\ell + 1 \leq f(\max F_n) = f\big(\max(F_n \cup F_1)\big) \leq \min(F_n \cup F_1) = \min F_1 = \ell,$$

a contradiction.

Thus $i = 1$. Let $B = \{\max F_n : n \in \mathbb{N}\}$. Now, if $n, k \in \mathbb{N}$ and $\max F_n \in B \backslash A_k$, then $n \leq \min F_n < f(\max F_n) < k$ so $|B \backslash A_k| < k$. Thus $B^* \subseteq \bigcap_{n=1}^{\infty} A_n{}^*$ as required. □

We have need of the following fact which it would take us too far afield to prove.

Theorem 12.37. *The existence of P-points in $\mathbb{N}^*$ cannot be established in ZFC.*

Proof. [222, VI, §4]. □

Corollary 12.38. *The existence of strongly summable ultrafilters can not be established in ZFC.*

Proof. By Theorem 12.37 it is consistent relative to ZFC that there are no P-points in $\mathbb{N}^*$. Theorem 12.35 shows that if strongly summable ultrafilters exist, so do union ultrafilters, while Theorem 12.36 shows that if union ultrafilters exist so do P-points in $\mathbb{N}^*$. □

Exercise 12.3.1. As in Theorem 12.31, let $\mathcal{U}$ be a union ultrafilter and let $p = \big\{\{\Sigma_{t \in F} 2^{t-1} : F \in \mathcal{A}\} : \mathcal{A} \in \mathcal{U}\big\}$. Define $f : \mathcal{P}_f(\mathbb{N}) \to \mathbb{N}$ by $f(F) = \Sigma_{t \in F} 2^{t-1}$. Prove that $\widetilde{f}(\mathcal{U}) = p$.

Exercise 12.3.2. As in Theorem 12.35, let p be a strongly summable ultrafilter and pick a sequence $\langle x_n \rangle_{n=1}^{\infty}$ in $\mathbb{N}$ such that $\mathrm{FS}(\langle x_n \rangle_{n=1}^{\infty}) \in p$ and for each $n \in \mathbb{N}$, $x_{n+1} > 4 \Sigma_{t=1}^{n} x_t$. Define $\varphi : \mathrm{FS}(\langle x_n \rangle_{n=1}^{\infty}) \to \mathcal{P}_f(\mathbb{N})$ by $\varphi(\Sigma_{t \in F} x_t) = \{x_t : t \in F\}$ and extend φ arbitrarily to the rest of $\mathbb{N}$. Let $\mathcal{U} = \{\mathcal{A} \subseteq \mathcal{P}_f(\mathbb{N}) : \varphi^{-1}[\mathcal{A}] \in p\}$. Prove that $\widetilde{\varphi}(p) = \mathcal{U}$.

12.4 Algebraic Properties of Strongly Summable Ultrafilters

Recall that for any discrete semigroup $(S, +)$, an idempotent p in βS is *strongly right maximal* in βS if and only if $\{q \in \beta S : q + p = p\} = \{p\}$. We saw in Theorem 9.10 that strongly right maximal idempotents in $\beta \mathbb{N}$ exist.

Theorem 12.39. *Let $p \in \beta \mathbb{N}$ be a strongly summable ultrafilter. Then p is a strongly right maximal idempotent.*

Proof. By Theorem 12.19, p is an idempotent. Let $q \neq p$ be given. We show that $q + p \neq p$. Pick $A \in p \setminus q$ and pick by Lemma 12.20 a sequence $\langle x_n \rangle_{n=1}^{\infty}$ in $\mathbb{N}$ such that $\mathrm{FS}(\langle x_n \rangle_{n=1}^{\infty}) \subseteq A$, for each $m \in \mathbb{N}$, $\mathrm{FS}(\langle x_n \rangle_{n=m}^{\infty}) \in p$, and $x_{n+1} > 4 \sum_{t=1}^{n} x_t$ for all n. It suffices to show that

$$\{y \in \mathbb{N} : -y + \mathrm{FS}(\langle x_n \rangle_{n=1}^{\infty}) \in p\} \subseteq \mathrm{FS}(\langle x_n \rangle_{n=1}^{\infty}).$$

(In fact equality holds, but we do not care about that.) Indeed, one has then that $\{y \in \mathbb{N} : -y + \mathrm{FS}(\langle x_n \rangle_{n=1}^{\infty}) \in p\} \notin q$ so $\mathrm{FS}(\langle x_n \rangle_{n=1}^{\infty}) \notin q + p$ and thus $p \neq q + p$.

To this end, let $a \in \mathbb{N}$ such that $-a + \mathrm{FS}(\langle x_n \rangle_{n=1}^{\infty}) \in p$. Then

$$\left(-a + \mathrm{FS}(\langle x_n \rangle_{n=1}^{\infty})\right) \cap \mathrm{FS}(\langle x_n \rangle_{n=1}^{\infty}) \in p.$$

So pick a sequence $\langle y_n \rangle_{n=1}^{\infty}$ such that

$$\mathrm{FS}(\langle y_n \rangle_{n=1}^{\infty}) \subseteq \left(-a + \mathrm{FS}(\langle x_n \rangle_{n=1}^{\infty})\right) \cap \mathrm{FS}(\langle x_n \rangle_{n=1}^{\infty}).$$

Pick F and G in $\mathcal{P}_f(\mathbb{N})$ such that $y_1 = \sum_{t \in F} x_t$ and $y_2 = \sum_{t \in G} x_t$. By Lemma 12.34, $F \cap G = \emptyset$. Also, $a + y_1$ and $a + y_2$ are in $\mathrm{FS}(\langle x_n \rangle_{n=1}^{\infty})$. So pick H and K in $\mathcal{P}_f(\mathbb{N})$ such that $a + y_1 = \sum_{t \in H} x_t$ and $a + y_2 = \sum_{t \in K} x_t$. Then

$$a = \sum_{t \in H} x_t - \sum_{t \in F} x_t = \sum_{t \in K} x_t - \sum_{t \in G} x_t.$$

So $\sum_{t \in G \triangle H} x_t + \sum_{t \in G \cap H} 2x_t = \sum_{t \in F \triangle K} x_t + \sum_{t \in F \cap K} 2x_t$. Thus, by Lemma 12.32 $G \triangle H = F \triangle K$ and $G \cap H = F \cap K$. Since $F \cap G = \emptyset$, one concludes from these equations that $F \subseteq H$. So $a = \sum_{t \in H \setminus F} x_t$ and hence $a \in \mathrm{FS}(\langle x_n \rangle_{n=1}^{\infty})$ as required. $\square$

We saw in Corollary 7.36 that maximal groups in $K(\beta\mathbb{N})$ are as large as possible (and highly noncommutative). That is, they all contain a copy of the free group on $2^{\mathfrak{c}}$ generators. We set out to show now that if p is strongly summable, then $H(p)$ is as small (and commutative) as possible, namely a copy of $\mathbb{Z}$.

Lemma 12.40. *Let $\langle x_n \rangle_{n=1}^{\infty}$ be a sequence in $\mathbb{N}$ such that for each n, $x_{n+1} > 4 \sum_{t=1}^{n} x_t$. Let F, G_1, and G_2 be finite subsets of $\mathbb{N}$ such that $G_1 \cap G_2 = \emptyset$ and $F \cup G_1 \cup G_2 \neq \emptyset$. Let $a = \min(F \cup G_1 \cup G_2)$ and let*

$$t = \sum_{n \in F} x_n - \sum_{n \in G_1} x_n - \sum_{n \in G_2} 2x_n.$$

Then $t = 0$ or $|t| > \dfrac{x_a}{2}$.

Proof. Assume that $t \neq 0$. Let $F' = F \setminus (G_1 \cup G_2)$, let $G_1' = (G_1 \setminus F) \cup (G_2 \cap F)$, and let $G_2' = G_2 \setminus F$. Then F', G_1', and G_2' are pairwise disjoint and

$$t = \sum_{n \in F'} x_n - \sum_{n \in G_1'} x_n - \sum_{n \in G_2'} 2x_n.$$

Since $t \neq 0$, $F' \cup G_1' \cup G_2' \neq \emptyset$. Let $b = \max(F' \cup G_1' \cup G_2')$ and notice that $b \geq a$. Then if $b \in F'$, we have

$$
\begin{aligned}
t &\geq x_b - \textstyle\sum_{n \in G_1'} x_n - \sum_{n \in G_2'} 2x_n \\
&\geq x_b - \textstyle\sum_{n=1}^{b-1} 2x_n \\
&> x_b - \frac{x_b}{2} = \frac{x_b}{2} \geq \frac{x_a}{2}.
\end{aligned}
$$

Otherwise $b \in (G_1' \cup G_2')$ so that

$$
\begin{aligned}
t &\leq \textstyle\sum_{n \in F'} x_n - x_b \\
&\leq \textstyle\sum_{n=1}^{b-1} x_n - x_b \\
&< -\frac{3x_b}{4} \leq -\frac{3x_a}{4}.
\end{aligned}
$$

$\square$

The following lemma is exceedingly technical and we apologize in advance for that fact. It will be invoked twice in the proof of the main result of this section.

Lemma 12.41. *Let $\langle x_n \rangle_{n=1}^{\infty}$, $\langle y_n \rangle_{n=1}^{\infty}$, and $\langle z_n \rangle_{n=1}^{\infty}$ be sequences in $\mathbb{N}$ and assume that for each n, $x_{n+1} > 4 \sum_{t=1}^{n} x_t$. Let $\ell \in \mathbb{N}$ and assume that whenever $k, j \geq \ell$, one has some $H_{k,j} \in \mathcal{P}_f(\mathbb{N})$ such that $z_k + y_j = \sum_{n \in H_{k,j}} x_n$. Let $b = \max H_{\ell,\ell}$ and assume that whenever $\min\{k, j\} > \ell$ one has $\min H_{k,j} > b$. Let $t = (\sum_{n \in H_{\ell,\ell} \setminus H_{\ell,\ell+1}} x_n) - y_\ell$ and for $k > \ell$, let $F_k = H_{k,\ell} \cap H_{k,\ell+1}$ and $G_k = H_{\ell,k} \cap H_{\ell+1,k}$. Then $|t| < \frac{x_{b+1}}{4}$ and for each $k > \ell$, $z_k = (\sum_{n \in F_k} x_n) + t$ and $y_k = (\sum_{n \in G_k} x_n) - t$.*

Proof. We first show that for any $k > \ell$, $H_{k,\ell} \setminus H_{k,\ell+1} = H_{\ell,\ell} \setminus H_{\ell,\ell+1}$ and $H_{\ell,k} \setminus H_{\ell+1,k} = H_{\ell,\ell} \setminus H_{\ell+1,\ell}$. Indeed, consider the equations

$$
\begin{aligned}
z_\ell + y_\ell &= \textstyle\sum_{n \in H_{\ell,\ell}} x_n \\
z_\ell + y_{\ell+1} &= \textstyle\sum_{n \in H_{\ell,\ell+1}} x_n \\
z_k + y_\ell &= \textstyle\sum_{n \in H_{k,\ell}} x_n
\end{aligned}
$$

and

$$
z_k + y_{\ell+1} = \textstyle\sum_{n \in H_{k,\ell+1}} x_n.
$$

From the first two equations we have

$$
y_{\ell+1} - y_\ell = \textstyle\sum_{n \in H_{\ell,\ell+1} \setminus H_{\ell,\ell}} x_n - \sum_{n \in H_{\ell,\ell} \setminus H_{\ell,\ell+1}} x_n
$$

while the last two equations yield

$$
y_{\ell+1} - y_\ell = \textstyle\sum_{n \in H_{k,\ell+1} \setminus H_{k,\ell}} x_n - \sum_{n \in H_{k,\ell} \setminus H_{k,\ell+1}} x_n.
$$

Thus by Lemma 12.33, one has that $H_{k,\ell} \setminus H_{k,\ell+1} = H_{\ell,\ell} \setminus H_{\ell,\ell+1}$. Likewise, working with $z_{\ell+1} - z_\ell$, one gets that $H_{\ell,k} \setminus H_{\ell+1,k} = H_{\ell,\ell} \setminus H_{\ell+1,\ell}$.

Now let $s = (\sum_{n \in H_{\ell,\ell} \setminus H_{\ell+1,\ell}} x_n) - z_\ell$. Then

$$-\frac{x_{b+1}}{4} < -\sum_{n=1}^{b} x_n$$
$$\leq -\sum_{n \in H_{\ell,\ell}} x_n$$
$$= -(z_\ell + y_\ell) < -y_\ell$$
$$\leq \sum_{n \in H_{\ell,\ell} \setminus H_{\ell,\ell+1}} x_n - y_\ell = t$$
$$< \sum_{n \in H_{\ell,\ell}} x_n$$
$$\leq \sum_{n=1}^{b} x_n < \frac{x_{b+1}}{4}$$

and so $|t| < \dfrac{x_{b+1}}{4}$. Similarly, $|s| < \dfrac{x_{b+1}}{4}$ and thus $|s + t| < \dfrac{x_{b+1}}{2}$.

Now let $k > \ell$ be given. Then $z_k + y_\ell = \sum_{n \in H_{k,\ell}} x_n$ so

$$z_k = \sum_{n \in H_{k,\ell} \cap H_{k,\ell+1}} x_n + \sum_{n \in H_{k,\ell} \setminus H_{k,\ell+1}} x_n - y_\ell$$
$$= \sum_{n \in F_k} x_n + \sum_{n \in H_{\ell,\ell} \setminus H_{\ell,\ell+1}} x_n - y_\ell$$
$$= \sum_{n \in F_k} x_n + t.$$

Also $z_\ell + y_k = \sum_{n \in H_{\ell,k}} x_n$ so

$$y_k = \sum_{n \in H_{\ell,k} \cap H_{\ell+1,k}} x_n + \sum_{n \in H_{\ell,k} \setminus H_{\ell+1,k}} x_n - z_\ell$$
$$= \sum_{n \in G_k} x_n + \sum_{n \in H_{\ell,\ell} \setminus H_{\ell+1,\ell}} x_n - z_\ell$$
$$= \sum_{n \in G_k} x_n + s.$$

To complete the proof, it therefore suffices to show that $s = -t$. From the derivations above we have that $z_k + y_k = \sum_{n \in F_k} x_n + \sum_{n \in G_k} x_n + t + s$ and we also know that $z_k + y_k = \sum_{n \in H_{k,k}} x_n$. Thus $\sum_{n \in H_{k,k}} x_n - \sum_{n \in F_k} x_n - \sum_{n \in G_k} x_n = s + t$. Let $K_1 = (F_k \setminus G_k) \cup (G_k \setminus F_k)$ and let $K_2 = F_k \cap G_k$. Then

$$s + t = \sum_{n \in H_{k,k}} x_n - \sum_{n \in K_1} x_n - \sum_{n \in K_2} 2x_n$$

and $\min(H_{k,k} \cup K_1 \cup K_2) = \min(H_{k,k} \cup F_k \cup G_k) \geq b + 1$ because $F_k \subseteq H_{k,\ell+1}$ and $G_k \subseteq H_{\ell+1,k}$. Since $|s + t| < \dfrac{x_{b+1}}{2}$, Lemma 12.40 tells us that $s + t = 0$ as required. □

As a consequence of the next theorem (and Theorem 12.29) it is consistent with ZFC that there are maximal groups of $(\beta \mathbb{N}, +)$ that are just copies of $\mathbb{Z}$. We do not know whether the existence of such small maximal groups can be established in ZFC.

Theorem 12.42. *Let p be a strongly summable ultrafilter. If $q \in \mathbb{N}^*$ and there exists $r \in \mathbb{N}^*$ such that $q + r = r + q = p$, then $q \in \mathbb{Z} + p$. In particular, $H(p) = \mathbb{Z} + p$.*

Proof. Let us notice first that since, by Theorem 4.23, $\mathbb{Z}$ is contained in the center of $(\beta \mathbb{Z}, +)$, $\mathbb{Z} + p$ is a subgroup of $(\beta \mathbb{Z}, +)$. And, by Exercise 4.3.5, $\mathbb{Z} + p \subseteq \beta \mathbb{N}$. Consequently $\mathbb{Z} + p \subseteq H(p)$. The "in particular" conclusion then follows from the first assertion immediately, because any member of $H(p)$ is in $\mathbb{N}^*$ and is invertible with respect to p.

So assume we have q and r in $\mathbb{N}^*$ such that $q + r = r + q = p$ and suppose that $q \notin \mathbb{Z} + p$. Then likewise, $p \notin \mathbb{Z} + q$. Since $p \notin \{1 + q, q, -1 + q\}$, pick $A_1 \in p$ such that $-1 + A_1 \notin q$, $A_1 \notin q$, and $1 + A_1 \notin q$. By Lemma 12.20 we may presume (by passing to a subset) that $A_1 = \text{FS}(\langle x_{1,n} \rangle_{n=1}^\infty)$ where for each n, $x_{1,n+1} > 4 \sum_{j=1}^n x_{1,j}$, and for each m, $\text{FS}(\langle x_{1,n} \rangle_{n=m}^\infty) \in p$. Let $B_1 = \{x \in \mathbb{N} : -x + A_1 \in q\}$. Since $p = r + q$, $B_1 \in r$. Pick $y_1 \in B_1$. Let

$$C_1 = (-y_1 + A_1) \cap \{x \in \mathbb{N} : -x + A_1 \in r\} \cap \bigcap_{i=-1}^1 (\mathbb{N} \setminus (-i + A_1)).$$

Since $y_1 \in B_1$ and $p = q + r$, $C_1 \in q$. Pick $z_1 > y_1$ with $z_1 \in C_1$. Since $y_1 + z_1 \in A_1$, pick $J_{1,1,1} \in \mathcal{P}_f(\mathbb{N})$ such that $y_1 + z_1 = \sum_{n \in J_{1,1,1}} x_{1,n}$. Let $b_1 = \max J_{1,1,1}$ and let $a_1 = b_1 + 1$. Note that a_1 is the smallest member of $\mathbb{N}$ with $x_{1,a_1} > \sum_{n \in J_{1,1,1}} x_{1,n}$.

Assume now that $k > 1$ and we have chosen $A_{k-1} = \text{FS}(\langle x_{k-1,n} \rangle_{n=1}^\infty) \in p$, $B_{k-1} \in r$, $C_{k-1} \in q$, $y_{k-1} \in B_{k-1}$, $z_{k-1} \in C_{k-1} \subseteq \{x \in \mathbb{N} : -x + A_{k-1} \in r\}$, b_{k-1}, and a_{k-1} such that for each n, $x_{k-1,n+1} > 4 \sum_{j=1}^n x_{k-1,j}$, and for each m, $\text{FS}(\langle x_{k-1,n} \rangle_{n=m}^\infty) \in p$. Then $\text{FS}(\langle x_{k-1,n} \rangle_{n=b_{k-1}+1}^\infty) \in p$ and $\text{FS}(\langle x_{1,n} \rangle_{n=a_{k-1}+2}^\infty) \in p$. Also, $p \notin \{t + q : t \in \mathbb{Z}$ and $|t| \le k\}$. Thus we may choose by Lemma 12.20, $A_k = \text{FS}(\langle x_{k,n} \rangle_{n=1}^\infty) \in p$ where for each n, $x_{k,n+1} > 4 \sum_{j=1}^n x_{k,j}$, for each m, $\text{FS}(\langle x_{k,n} \rangle_{n=m}^\infty) \in p$, $A_k \notin t + q$ for any $t \in \mathbb{Z}$ with $|t| \le k$, and

$$A_k \subseteq \text{FS}(\langle x_{k-1,n} \rangle_{n=b_{k-1}+1}^\infty) \cap \text{FS}(\langle x_{1,n} \rangle_{n=a_{k-1}+2}^\infty).$$

Note that $A_k \subseteq A_{k-1}$.

Now $A_k \in p = r + q$. Let

$$B_k = (-z_{k-1} + A_{k-1}) \cap \{x \in \mathbb{N} : -x + A_k \in q\} \cap B_{k-1}.$$

Then $B_k \in r$. Pick $y_k \in B_k$ with $y_k > z_{k-1}$. Let

$$C_k = C_{k-1} \cap (-y_k + A_k) \cap \{x \in \mathbb{N} : -x + A_k \in r\} \cap \bigcap_{i=-k}^k (\mathbb{N} \setminus (-i + A_k)).$$

Then $C_k \in q$. Pick $z_k \in C_k$ with $z_k > y_k$. Let a_k be the smallest member of $\mathbb{N}$ with $x_{1,a_k} > y_k + z_k$. Pick $J_{k,k,k} \in \mathcal{P}_f(\mathbb{N})$ with $y_k + z_k = \sum_{n \in J_{k,k,k}} x_{k,n}$ and let $b_k = \max J_{k,k,k}$.

The inductive construction being complete, let $\ell, k, j \in \mathbb{N}$ be given with $\ell \le \min\{j, k\}$.

> If $k \ge j$ we have $z_k \in C_k \subseteq C_j \subseteq -y_j + A_j$ so $y_j + z_k \in A_j$.
> If $k < j$ we have $y_j \in B_j \subseteq B_{k+1} \subseteq -z_k + A_k$ so $y_j + z_k \in A_k$.

Consequently $y_j + z_k \in A_\ell$. Further, if $i > \ell$, then $A_i \subseteq A_{\ell+1} \subseteq \text{FS}(\langle x_{\ell,n} \rangle_{n=b_\ell+1}^\infty)$. Consequently, we may choose $J_{\ell,k,j} \in \mathcal{P}_f(\mathbb{N})$ such that $y_j + z_k = \sum_{n \in J_{\ell,k,j}} x_{\ell,n}$ and, if $\ell < \min\{j, k\}$, then $\min J_{\ell,k,j} > b_\ell$. Note that, if $\ell = k = j$, then the definition of $J_{\ell,k,j}$ agrees with the one given earlier.

Now let $t = (\sum_{n \in J_{1,1,1} \setminus J_{1,1,2}} x_{1,n}) - y_1$. For $k > 1$, let $F_k = J_{1,k,1} \cap J_{1,k,2}$ and $G_k = J_{1,1,k} \cap J_{1,2,k}$. By Lemma 12.41 we have that $|t| < \frac{1}{4} x_{1,b_1+1}$ and for each $k > 1$, $z_k = \sum_{n \in F_k} x_{1,n} + t$ and $y_k = \sum_{n \in G_k} x_{1,n} - t$.

Next let $\ell = |t|$ (or $\ell = 2$ if $|t| \leq 1$). Let $t' = (\sum_{n \in J_{\ell,\ell,\ell} \setminus J_{\ell,\ell,\ell+1}} x_{\ell,n}) - y_\ell$. For $k > \ell$, let $F_k' = J_{\ell,k,\ell} \cap J_{\ell,k,\ell+1}$ and $G_k' = J_{\ell,\ell,k} \cap J_{\ell,\ell+1,k}$. Then again by Lemma 12.41 we have that for each $k > \ell$, $z_k = \sum_{n \in F_k'} x_{\ell,n} + t'$ and $y_k = \sum_{n \in G_k'} x_{\ell,n} - t'$. To complete the proof we show that $t = t'$. Indeed, once we have done this we will have $z_{\ell+1} = \sum_{n \in F_{\ell+1}'} x_{\ell,n} + t \in t + A_\ell$ while $z_{\ell+1} \in C_{\ell+1} \subseteq C_\ell \subseteq (\mathbb{N} \setminus (t + A_\ell))$, a contradiction.

Now, given n, $x_{\ell,n} \in A_\ell \subseteq \text{FS}(\langle x_{1,m} \rangle_{m=1}^\infty)$ so pick T_n such that $x_{\ell,n} = \sum_{m \in T_n} x_{1,m}$. Also, if $n \neq k$, then $x_{\ell,n} + x_{\ell,k} \in \text{FS}(\langle x_{1,m} \rangle_{m=1}^\infty)$ so by Lemma 12.34, $T_n \cap T_k = \emptyset$.

Let $k = \ell + 1$. We show that if $n \in J_{\ell,k,k}$, then $\min T_n \geq a_\ell + 2$. Indeed, $z_k + y_k \in A_k \subseteq \text{FS}(\langle x_{1,m} \rangle_{m=a_\ell+2}^\infty)$ so pick H with $z_k + y_k = \sum_{m \in H} x_{1,m}$ and $\min H \geq a_\ell + 2$. On the other hand, using at the second equality the fact that the T_n's are pairwise disjoint, we have $z_k + y_k = \sum_{j \in J_{\ell,k,k}} x_{\ell,n} = \sum_{m \in L} x_{1,m}$ where $L = \bigcup_{n \in J_{\ell,k,k}} T_n$. Thus $L = H$ and in particular each $\min T_n \geq \min H \geq a_\ell + 2$.

Now let $F_k'' = \bigcup_{n \in F_k'} T_n$ and let $G_k'' = \bigcup_{n \in G_k'} T_n$. Since $F_k' \subseteq J_{\ell,k,\ell+1} = J_{\ell,k,k}$ and $G_k' \subseteq J_{\ell,\ell+1,k} = J_{\ell,k,k}$, we have $\min F_k'' \geq a_\ell + 2$ and $\min G_k'' \geq a_\ell + 2$. Note also that $z_k = \sum_{n \in F_k''} x_{1,n} + t'$ and $y_k = \sum_{n \in G_k''} x_{1,n} - t'$. Now let

$$L_1 = \{n \in F_k : n \geq a_\ell + 2\},$$
$$L_2 = \{n \in F_k : n < a_\ell + 2\},$$
$$M_1 = \{n \in G_k : n \geq a_\ell + 2\},$$

and

$$M_2 = \{n \in G_k : n < a_\ell + 2\}.$$

Then $z_k = \sum_{n \in L_1} x_{1,n} + \sum_{n \in L_2} x_{1,n} + t$ and $y_k = \sum_{n \in M_1} x_{1,n} + \sum_{n \in M_2} x_{1,n} - t$. So, equating both expressions for z_k, we have

$$\sum_{n \in F_k''} x_{1,n} - \sum_{n \in L_1} x_{1,n} = \sum_{n \in L_2} x_{1,n} + t - t'.$$

Likewise, from the expressions for y_k we have

$$\sum_{n \in G_k''} x_{1,n} - \sum_{n \in M_1} x_{1,n} = \sum_{n \in M_2} x_{1,n} + t' - t.$$

Since $\min(F_k'' \cup L_1) \geq a_\ell + 2$ and $\min(G_k'' \cup M_1) \geq a_\ell + 2$, we have by Lemma 12.40 that

(1) $\sum_{n \in L_2} x_{1,n} + t - t' = 0$ or $|\sum_{n \in L_2} x_{1,n} + t - t'| \geq \frac{1}{2} x_{1,a_\ell+2}$ and

(2) $\sum_{n \in M_2} x_{1,n} + t' - t = 0$ or $|\sum_{n \in M_2} x_{1,n} + t' - t| \geq \frac{1}{2} x_{1,a_\ell+2}$.

Now $0 \leq \sum_{n \in L_2} x_{1,n} \leq \sum_{n=1}^{a_\ell+1} x_{1,n} < \frac{1}{4} x_{1,a_\ell+2}$. We have already seen that

$$|t| < \frac{1}{4}x_{1,b_1+1} = \frac{1}{4}x_{1,a_1} \leq \frac{1}{4}x_{1,a_\ell} < \frac{1}{16}x_{1,a_\ell+1} < \frac{1}{64}x_{1,a_\ell+2},$$

$$t' = \left(\sum_{n\in J_{\ell,\ell,\ell}\setminus J_{\ell,\ell,k}} x_{\ell,n}\right) - y_\ell < \sum_{n\in J_{\ell,\ell,\ell}} x_{\ell,n} = y_\ell + z_\ell < x_{1,a_\ell}, \text{ and}$$

$$t' = \left(\sum_{n\in J_{\ell,\ell,\ell}\setminus J_{\ell,\ell,k}} x_{\ell,n}\right) - y_\ell \geq -y_\ell > -(z_\ell + y_\ell) > -x_{1,a_\ell}$$

so $|t'| < x_{1,a_\ell} < \frac{1}{16}x_{1,a_\ell+2}$. Thus $|\sum_{n\in L_2} x_{1,n} + t - t'| < \frac{21}{64}x_{1,a_\ell+2} < \frac{1}{2}x_{1,a_\ell+2}$
so $\sum_{n\in L_2} x_{1,n} + t - t' = 0$. Likewise, $\sum_{n\in M_2} x_{1,n} + t' - t = 0$. Thus $\sum_{n\in L_2} x_{1,n} + \sum_{n\in M_2} x_{1,n} = 0$ so that $L_2 = M_2 = \emptyset$ and hence $t = t'$ as required.   □

We show in the remainder of this section that it is consistent with the usual axioms of set theory that there exist idempotents in $(\beta\mathbb{N}, +)$ which can only be written as a sum of elements of $\beta\mathbb{N}$ in a trivial fashion.

Lemma 12.43. *Let $\langle x_n\rangle_{n=1}^\infty$ be an increasing sequence in $\mathbb{N}$ with the property that for each $n \in \mathbb{N}$, x_n divides x_{n+1}. If $a, b, m \in \mathbb{N}$, $a \leq x_m$, x_{m+1} divides b, and $a + b \in \mathrm{FS}(\langle x_n\rangle_{n=1}^\infty)$, then $a \in \mathrm{FS}(\langle x_n\rangle_{n=1}^\infty)$ and $b \in \mathrm{FS}(\langle x_n\rangle_{n=1}^\infty)$.*

Proof. Pick $F \in \mathcal{P}_f(\mathbb{N})$ such that $a + b = \sum_{t\in F} x_t$. Let $G = F \cap \{1, 2, \ldots, m\}$ and let $H = F\setminus\{1, 2, \ldots, m\}$. Then $a - \sum_{t\in G} x_t = \sum_{t\in H} x_t - b$ (where $\sum_{t\in\emptyset} x_t = 0$). Then x_{m+1} divides $\sum_{t\in H} x_t - b$ while $|a - \sum_{t\in G} x_t| < x_{m+1}$, so $a - \sum_{t\in G} x_t = \sum_{t\in H} x_t - b = 0$.   □

Lemma 12.44. *Let p be a special strongly summable ultrafilter, let $q \in \beta\mathbb{N}$, and let $r \in \bigcap_{n=1}^\infty c\ell(\mathbb{N}n)$. If $p = q + r$, then $q = r = p$.*

Proof. By Theorem 12.39 it suffices to show that $r = p$. So suppose not and pick $A \in p\setminus r$. Pick an increasing sequence $\langle x_n\rangle_{n=1}^\infty$ such that $\mathrm{FS}(\langle x_n\rangle_{n=1}^\infty) \subseteq A$, $\mathrm{FS}(\langle x_n\rangle_{n=1}^\infty) \in p$, and for each $n \in \mathbb{N}$, x_n divides x_{n+1}. Then $\mathrm{FS}(\langle x_n\rangle_{n=1}^\infty) \in q + r$ so pick $a \in \mathbb{N}$ such that $-a + \mathrm{FS}(\langle x_n\rangle_{n=1}^\infty) \in r$. Pick $m \in \mathbb{N}$ such that $x_m \geq a$. Pick

$$b \in \left(-a + \mathrm{FS}(\langle x_n\rangle_{n=1}^\infty)\right) \cap \mathbb{N}x_{m+1} \cap (\mathbb{N}\setminus A).$$

Then $a + b \in \mathrm{FS}(\langle x_n\rangle_{n=1}^\infty)$, $a \leq x_m$, and x_{m+1} divides b so by Lemma 12.43, $b \in \mathrm{FS}(\langle x_n\rangle_{n=1}^\infty) \subseteq A$, a contradiction.   □

Of course, any idempotent p in $\beta\mathbb{N}$ can be written as $(p - n) + (p + n)$ for any $n \in \mathbb{Z}$. We see now that if p is a special strongly summable ultrafilter, then this is the only way that p can be written as a sum.

Theorem 12.45. *Let p be a special strongly summable ultrafilter and let $q, r \in \beta\mathbb{N}$. If $p = q + r$, then q and r are in $\mathbb{Z} + p$.*

Proof. We claim that it suffices to show that $r \in \mathbb{Z} + \bigcap_{n=1}^\infty c\ell(\mathbb{N}n)$. For assume we have $k \in \mathbb{Z}$ such that $-k + r \in \bigcap_{n=1}^\infty c\ell(\mathbb{N}n)$. Then $(k + q) + (-k + r) = p$ so by Lemma 12.44, $k + q = -k + r = p$.

Thus we suppose that $r \notin \mathbb{Z} + \bigcap_{n=1}^{\infty} c\ell(\mathbb{N}n)$, so that also $q \notin \mathbb{Z} + \bigcap_{n=1}^{\infty} c\ell(\mathbb{N}n)$. Define

$$\alpha : \mathbb{N} \to \underset{t=1}{\overset{\infty}{\times}} \{0, 1, \ldots, t\}$$

by the equation $x = \Sigma_{t=1}^{\infty} \alpha(x)(t) \cdot t!$ (It is easy to see that any positive integer has a unique such *factorial* expansion.) Let $\widetilde{\alpha}$ be the continuous extension of α to $\beta\mathbb{N}$.

Let $M = \{t \in \mathbb{N} : \widetilde{\alpha}(q)(t) \neq t\}$ and let $L = \{t \in \mathbb{N} : \widetilde{\alpha}(q)(t) \neq 0\}$. We claim that both M and L are infinite. To see that M is infinite, suppose instead we have some $k \in \mathbb{N}$ such that for all $t > k$, $\widetilde{\alpha}(q)(t) = t$. Let $z = (k+1)! - \Sigma_{t=1}^{k} \widetilde{\alpha}(q)(t) \cdot t!$. We claim that $q + z \in \bigcap_{n=1}^{\infty} c\ell(\mathbb{N}n)$, a contradiction. To see this, let $n \in \mathbb{N}$ be given with $n > k$. We show that $-z + \mathbb{N}(n+1)! \in q$. Now, by the continuity of $\widetilde{\alpha}$,

$$\mathbb{N}(n+1)! + \Sigma_{t=1}^{n} \widetilde{\alpha}(q)(t) \cdot t! = \{x \in \mathbb{N} : \text{for all } t \in \{1, 2, \ldots, n\},$$
$$\alpha(x)(t) = \widetilde{\alpha}(q)(t)\} \in q.$$

We claim that

$$\mathbb{N}(n+1)! + \Sigma_{t=1}^{n} \widetilde{\alpha}(q)(t) \cdot t! \subseteq -z + \mathbb{N}(n+1)!.$$

To see this, let $w \in \mathbb{N}(n+1)! + \Sigma_{t=1}^{n} \widetilde{\alpha}(q)(t) \cdot t!$. Then

$$w = a \cdot (n+1)! + \Sigma_{t=k+1}^{n} t \cdot t! + \Sigma_{t=1}^{k} \widetilde{\alpha}(q)(t) \cdot t!$$

so

$$w + z = a \cdot (n+1)! + \Sigma_{t=k+1}^{n} t \cdot t! + (k+1)! = (a+1) \cdot (n+1)!$$

as required.

Similarly, if L is finite so that for some k, one has $\widetilde{\alpha}(q)(t) = 0$ for all $t > k$, and $z = \Sigma_{t=1}^{k} \widetilde{\alpha}(q)(t) \cdot t!$, then $q - z \in \bigcap_{n=1}^{\infty} c\ell(\mathbb{N}n)$.

Since p is a special strongly summable ultrafilter, pick a sequence $\langle z_n \rangle_{n=1}^{\infty}$ such that $\mathrm{FS}(\langle z_n \rangle_{n=1}^{\infty}) \subseteq \mathrm{FS}(\langle t! \rangle_{t=1}^{\infty})$, $\mathrm{FS}(\langle z_n \rangle_{n=1}^{\infty}) \in p$ and $M \setminus \bigcup \{\mathrm{fsupp}(y) : y \in \mathrm{FS}(\langle z_n \rangle_{n=1}^{\infty})\}$ is infinite. Let $K = \bigcup \{\mathrm{fsupp}(y) : y \in \mathrm{FS}(\langle z_n \rangle_{n=1}^{\infty})\}$.

Now $\mathrm{FS}(\langle z_n \rangle_{n=1}^{\infty}) \in q + r$ so pick $a \in \mathbb{N}$ such that $-a + \mathrm{FS}(\langle z_n \rangle_{n=1}^{\infty}) \in r$. Pick $m \in \mathbb{N}$ such that $a < m!$ and pick $\ell > s > m$ such that $\ell \in M \setminus K$ and $s \in L$. Now $\mathrm{FS}(\langle t! \rangle_{t=\ell+1}^{\infty}) \in p$ so

$$\{x \in \mathbb{N} : -x + \mathrm{FS}(\langle t! \rangle_{t=\ell+1}^{\infty}) \in r\} \cap (\mathbb{N}(\ell+1)! + \Sigma_{t=1}^{\ell} \widetilde{\alpha}(q)(t) \cdot t!) \in q$$

so pick $b \in \mathbb{N}(\ell+1)! + \Sigma_{t=1}^{\ell} \widetilde{\alpha}(q)(t) \cdot t!$ such that $-b + \mathrm{FS}(\langle t! \rangle_{t=\ell+1}^{\infty}) \in r$. Pick

$$x \in \left(-b + \mathrm{FS}(\langle t! \rangle_{t=\ell+1}^{\infty})\right) \cap \left(-a + \mathrm{FS}(\langle z_n \rangle_{n=1}^{\infty})\right).$$

Then $b + x = \Sigma_{t \in G} t!$ where $\min G \geq \ell + 1$ and for some $d \in \mathbb{N}$,

$$b = d \cdot (\ell+1)! + \Sigma_{t=1}^{\ell} \widetilde{\alpha}(q)(t) \cdot t!.$$

Since $s \in L$, $\widetilde{\alpha}(q)(s) \neq 0$ so

$$\Sigma_{t=1}^{s} \widetilde{\alpha}(q)(t) \cdot t! - a \geq s! - a > 0$$

and

$$\Sigma_{t=1}^{s} \widetilde{\alpha}(q)(t) \cdot t! - a < \Sigma_{t=1}^{s} \widetilde{\alpha}(q)(t) \cdot t! < (s+1)!$$

so $\Sigma_{t=1}^{s} \widetilde{\alpha}(q)(t) \cdot t! - a = \Sigma_{t=1}^{s} c_t \cdot t!$ where each $c_t \in \{0, 1, \ldots, t\}$ and not all c_t's are 0.

Now $b + x \in \mathrm{FS}(\langle z_n \rangle_{n=1}^{\infty}) + (b - a)$ so for some $H \in \mathcal{P}_f(\mathbb{N})$,

$$
\begin{aligned}
\Sigma_{t \in G}\, t! &= b + x \\
&= \Sigma_{n \in H} z_n + (b - a) \\
&= \Sigma_{n \in H} z_n + d \cdot (\ell + 1)! + \Sigma_{t=1}^{\ell} \widetilde{\alpha}(q)(t) \cdot t! - a \\
&= \Sigma_{n \in H} z_n + d \cdot (\ell + 1)! + \Sigma_{t=s+1}^{\ell} \widetilde{\alpha}(q)(t) \cdot t! + \Sigma_{t=1}^{s} \widetilde{\alpha}(q)(t) \cdot t! - a \\
&= \Sigma_{n \in H} z_n + d \cdot (\ell + 1)! + \Sigma_{t=s+1}^{\ell} \widetilde{\alpha}(q)(t) \cdot t! + \Sigma_{t=1}^{s} c_t \cdot t!.
\end{aligned}
$$

Pick $I \in \mathcal{P}_f(\mathbb{N})$ such that $\Sigma_{n \in H} z_n = \Sigma_{t \in I}\, t!$ and let $I_1 = I \cap \{1, 2, \ldots, \ell\}$ and $I_2 = I \backslash \{1, 2, \ldots, \ell\}$. Then

$$\Sigma_{t \in G}\, t! - d \cdot (\ell + 1)! - \Sigma_{t \in I_2}\, t! = \Sigma_{t \in I_1}\, t! + \Sigma_{t=s+1}^{\ell} \widetilde{\alpha}(q)(t) \cdot t! + \Sigma_{t=1}^{s} c_t \cdot t!.$$

Now $(\ell + 1)!$ divides the left hand side of this equation. On the other hand, $\ell \notin K$ so $\ell \notin I_1$ and $\ell \in M$ so $\widetilde{\alpha}(q)(\ell) \leq \ell - 1$. Thus

$$
\begin{aligned}
0 &< \Sigma_{t \in I_1}\, t! + \Sigma_{t=s+1}^{\ell} \widetilde{\alpha}(q)(t) \cdot t! + \Sigma_{t=1}^{s} c_t \cdot t! \\
&\leq \Sigma_{t=1}^{\ell-1} t! + \Sigma_{t=1}^{\ell-1} t \cdot t! + (\ell - 1) \cdot \ell! \\
&= \Sigma_{t=2}^{\ell-1} t! + \ell \cdot \ell! \\
&< (\ell + 1)!,
\end{aligned}
$$

a contradiction.　　$\square$

Notes

The presentation of Martin's Axiom is based on that by K. Kunen in [171]. See the book by D. Fremlim [94] for substantial information about Martin's Axiom.

It was shown in 1972 in [117] that the (then unproved) Finite Sums Theorem together with the continuum hypothesis implied the existence of an ultrafilter p on $\mathbb{N}$ such that, for every $A \in p$, $\{x \in \mathbb{N} : -x + A \in p\} \in p$. And A. Taylor (unpublished) showed that the continuum hypothesis could be weakened to Martin's Axiom. Of course, we now know that such an ultrafilter is simply an idempotent in $(\beta \mathbb{N}, +)$. (See the notes to Chapter 5 for the history of the discovery of this fact.) With this knowledge it appeared that nothing interesting remained in [117]. Then, in 1985, E. van Douwen noted in conversation that in fact the ultrafilter produced there had the stronger property that it was generated by sets of the form $FS(\langle x_n \rangle_{n=1}^{\infty})$, and wondered if the existence of such an ultrafilter could be proved in ZFC. This question led to the investigation of strongly summable ultrafilters, beginning with [131].

Union ultrafilters were introduced by A. Blass in [44]. Theorem 12.37, one of three nonelementary results used in this book that we do not prove, is due to S. Shelah. Corollary 12.38 is due to P. Matet [181]. The proof of Corollary 12.38 presented here is from [47], a result of collaboration with A. Blass.

Theorem 12.39 is from [46], a result of collaboration with A. Blass.

Theorem 12.42 is from [136].

Some of the results of this chapter have been generalised in [151] (a result of collaboration with I. Protasov) to countably infinite abelian groups. If G is a group of this kind, the existence of strongly summable nonprincipal ultrafilters on G follows from Martin's Axiom, but cannot be demonstrated in ZFC. If G can be embedded in the unit circle, then any strongly summable ultrafilter $p \in G^*$ has the property that the equation $p + x = p$ has the unique solution $x = p$ in G^*, and so does the equation $x + p = p$. It also follows from Martin's Axiom that there are nonprincipal strongly summable ultrafilters $p \in G^*$ with the remarkable algebraic property that, for any $x, y \in G^*$, $x + y = p$ implies that x and y are both in $G + p$.

If G is a countably infinite Boolean group, the existence of a nonprincipal strongly summable ultrafilter p on G has an interesting consequence. The ultrafilter p can be used to define a non-discrete extremally disconnected topology on G for which G is a topological group. It is not known whether the existence of a topological group of this kind can be demonstrated in ZFC.

Chapter 13

Multiple Structures in βS

We deal in this chapter with the relationships among different operations on the same set S, as for example the semigroups $(\mathbb{N}, +)$ and $(\mathbb{N}, \cdot)$. We also include a section considering the relationships between the left continuous and right continuous extensions of an operation $\cdot$ on S.

13.1 Sums Equal to Products in $\beta\mathbb{Z}$

The earliest applications of the theory of compact right topological semigroups to Ramsey Theory involved the semigroups $(\mathbb{N}, +)$ and $(\mathbb{N}, \cdot)$. (See Chapters 5 and 17 for several examples.) In investigating these applications, the question arose whether the equation $q + p = s \cdot r$ has any solutions in $\mathbb{N}^*$, or indeed in $\mathbb{Z}^*$. These questions remain open, but we present several partial results in this section.

We begin with the only nontrivial instance of the distributive law known to hold in $\beta\mathbb{Z}$. (For more information about the distributive law see Section 13.2.)

Since we are dealing with two operations, the notations λ_p and ρ_p are now ambiguous. We adopt the convention in Sections 13.1 and 13.2 that $\rho_p(q) = q + p$ and $\lambda_p(q) = p + q$ and introduce the notations $r_p(q) = q \cdot p$ and $\ell_p(q) = p \cdot q$.

If $p \in \beta\mathbb{Z}$, $-p$ denotes $(-1) \cdot p = \{-A : A \in p\}$.

Lemma 13.1. *Let $n \in \mathbb{Z}$ and let $p, q \in \beta\mathbb{Z}$. Then $n \cdot (p + q) = n \cdot p + n \cdot q$ and $(p + q) \cdot n = p \cdot n + q \cdot n$.*

Proof. Since, by Theorem 4.23, $\mathbb{Z}$ is contained in the center of $(\beta\mathbb{Z}, \cdot)$, the second assertion follows from the first.

Now $n \cdot (p + q) = (\ell_n \circ \rho_q)(p)$ and $n \cdot p + n \cdot q = (\rho_{n \cdot q} \circ \ell_n)(p)$ so it suffices to show that the functions $\ell_n \circ \rho_q$ and $\rho_{n \cdot q} \circ \ell_n$ agree on $\mathbb{Z}$. So let $m \in \mathbb{Z}$ be given. Then

$$(\ell_n \circ \rho_q)(m) = n \cdot (m + q) = (\ell_n \circ \lambda_m)(q)$$

and

$$(\rho_{n \cdot q} \circ \ell_n)(m) = n \cdot m + n \cdot q = (\lambda_{n \cdot m} \circ \ell_n)(q).$$

Since the distributive law holds in $\mathbb{Z}$, the functions $\ell_n \circ \lambda_m$ and $\lambda_{n \cdot m} \circ \ell_n$ agree on $\mathbb{Z}$. $\square$

We have occasionally used the canonical mapping from $\mathbb{Z}$ to $\mathbb{Z}_n$ before. In this chapter it will be used in several proofs, so we fix some notation for it.

Definition 13.2. Let $n \in \mathbb{N}$. Then $h_n : \mathbb{Z} \to \mathbb{Z}_n$ denotes the canonical mapping.

Notice that the continuous extension $\widetilde{h}_n : \beta\mathbb{Z} \to \mathbb{Z}_n$ of h_n is a homomorphism both from $(\beta\mathbb{Z}, +)$ to $(\mathbb{Z}_n, +)$ and from $(\beta\mathbb{Z}, \cdot)$ to $(\mathbb{Z}_n, \cdot)$ (by Lemma 2.14).

Definition 13.3. Let b be a real number satisfying $b > 1$. We define $\phi_b : \mathbb{N} \to \omega$ by putting $\phi_b(n) = \lfloor \log_b(n) \rfloor$. As usual, $\widetilde{\phi}_b : \beta\mathbb{N} \to \beta\omega$ denotes the continuous extension of ϕ_b.

Lemma 13.4. *Let $p \in \mathbb{N}^*$ and let $b > 1$.*
(i) $\widetilde{\phi}_b(q + p) \in \widetilde{\phi}_b(p) + \{-1, 0, 1\}$ for every $q \in \beta\mathbb{Z}$;
(ii) $\widetilde{\phi}_b(qp) \in \widetilde{\phi}_b(q) + \widetilde{\phi}_b(p) + \{0, 1\}$ for every $q \in \beta\mathbb{N}$.

Proof. (i) Choose $\epsilon > 0$ with the property that $|\log_b(1 + t)| < 1$ if $t \in (-\epsilon, \epsilon)$. If $m \in \mathbb{Z}$, $n \in \mathbb{N}$, and $|m| < \epsilon n$, we have $\log_b(m + n) = \log_b n + \log_b(1 + \frac{m}{n})$ and hence $\phi_b(m + n) \in \phi_b(n) + \{-1, 0, 1\}$.

For each $i \in \{-1, 0, 1\}$, we define $M_i \subseteq \mathbb{Z}$ by putting $m \in M_i$ if $\{n \in \mathbb{N} : \phi_b(m+n) = \phi_b(n)+i\} \in p$. Since the sets M_i partition $\mathbb{Z}$, we can choose $j \in \{-1, 0, 1\}$ such that $M_j \in q$. For each $m \in M_j$, let $N_m = \{n \in \mathbb{N} : \phi_b(m + n) = \phi_b(n) + j\} \in p$. We have:

$$\widetilde{\phi}_b(q + p) = q\text{-}\lim_{m \in M_j} p\text{-}\lim_{n \in N_m} \phi_b(m + n)$$
$$= q\text{-}\lim_{m \in M_j} p\text{-}\lim_{n \in N_m} (\phi_b(n) + j)$$
$$= \widetilde{\phi}_b(p) + j.$$

(ii) This follows in the same way from the observation that, for every $m, n \in \mathbb{N}$, $\phi_b(mn) \in \phi_b(m) + \phi_b(n) + \{0, 1\}$. $\square$

Theorem 13.5. *Let $p \in \mathbb{Z}^*$, let $q \in \beta\mathbb{Z}$, let $n \in \mathbb{Z}$ and assume that $n \cdot p = q + p$. Then $n = 1$ and thus $p = q + p$. If $q \in \mathbb{Z}$, then $q = 0$.*

Proof. We may assume that $p \in \mathbb{N}^*$, since otherwise, by Lemma 13.1, we could replace p by $-p$ and q by $-q$. Now by Exercise 4.3.5, $q + p \in \mathbb{N}^*$. And trivially $0 \cdot p = 0$, while $n \in -\mathbb{N}$ implies that $n \cdot p \in -\mathbb{N}^*$. Thus $n \in \mathbb{N}$.

Suppose that $n > 1$. We can then choose $b \in \mathbb{R}$ satisfying $1 < b^2 < n$. So $\phi_b(n) \geq 2$. It follows from Lemma 13.4 that $\widetilde{\phi}_b(n \cdot p) = \widetilde{\phi}_b(p) + k$ for some $k \geq 2$ in $\mathbb{N}$. On the other hand, $\widetilde{\phi}_b(q + p) \in \widetilde{\phi}_b(p) + \{-1, 0, 1\}$. By Lemma 6.28, this is a contradiction. $\square$

It is easy to find solutions to the equation $q + p = n \cdot r$ where $n \in \mathbb{Z}$. The next two results are concerned with multiple solutions to such an equation for fixed p and r. Such multiple solutions are used in Theorem 13.9 to characterize the existence of q and s in $\mathbb{Z}^*$ with $q + p = s \cdot r$.

Lemma 13.6. *Let* $p, r \in \mathbb{Z}^*$, *let* $s, t \in \beta\mathbb{Z}$, *and let* $m, n \in \mathbb{Z}$. *If* $s + p = m \cdot r$ *and* $t + p = n \cdot r$, *then* $m = n$.

Proof. We may suppose that $p \in \mathbb{N}^*$, since otherwise, by Lemma 13.1, we could replace p, s, t, and r by $-p, -s, -t$, and $-r$ respectively. This implies that $s + p$ and $t + p$ are in $\mathbb{N}^*$, by Exercise 4.3.5, and hence that $m \neq 0$ and $n \neq 0$. Therefore r, m, and n are all in $\beta\mathbb{N}$ or all in $-\beta\mathbb{N}$. We may suppose that $r, m, n \in \beta\mathbb{N}$, since otherwise we could replace r, m, and n by $-r, -m$, and $-n$ respectively.

If $n \neq m$, we may suppose without loss of generality that $n > m$. We can choose $b \in \mathbb{R}$ satisfying $b > 1$ and $mb^4 < n$. So $\phi_b(n) \geq \phi_b(m) + 4$.

By Lemma 13.4, $\widetilde{\phi}_b(n \cdot r) = \widetilde{\phi}_b(m \cdot r) + k$ for some $k \geq 3$ in $\mathbb{N}$. We also know that $\widetilde{\phi}_b(s + p)$ and $\widetilde{\phi}_b(t + p)$ are both in $\widetilde{\phi}_b(p) + \{-1, 0, 1\}$. So $\widetilde{\phi}_b(t + p) = \widetilde{\phi}_b(s + p) + \ell$ for some $\ell \in \{-2, -1, 0, 1, 2\}$. Since $\ell \neq k$, this contradicts Lemma 6.28.

This establishes that $n = m$. $\square$

Lemma 13.7. *Let* $p, r \in \mathbb{Z}^*$ *and let* $m, n, m', n' \in \mathbb{Z}$. *If* $m + p = n \cdot r$ *and* $m' + p = n' \cdot r$, *then* $(m, n) = (m', n')$.

Proof. By Lemma 13.6, $n = n'$. It then follows from Lemma 6.28 that $m = m'$. $\square$

As we remarked at the beginning of this section, we are primarily concerned with two questions, namely whether there exist solutions to the equation $q + p = s \cdot r$ in $\mathbb{N}^*$ and whether there exist such solutions in $\mathbb{Z}^*$. Now, of course, an affirmative answer to the first question implies an affirmative answer to the second. Further, the reader is asked to show in Exercise 13.1.1 that if $p, r, s \in \mathbb{Z}^*$ and there exists $q \in \mathbb{Z}^*$ such that $q + p = s \cdot r$, then there exists $q' \in \mathbb{Z}^*$ such that $q' + |p| = |s| \cdot |r|$, where $|p|$ has its obvious meaning. Thus the questions are nearly the same. However, it is conceivable that one could have $p, r, s \in \mathbb{N}^*$ and $q \in -\mathbb{N}^*$ such that $q + p = s \cdot r$ but not have any $q' \in \mathbb{N}^*$ such that $q' + p = s \cdot r$. Accordingly, we address both questions.

Lemma 13.8. *Let* $p, q, r \in \mathbb{Z}^*$ *and let*

$$H = \{m \in \mathbb{Z} : \text{there exists } t \in \beta\mathbb{Z} \text{ such that } m + p = t \cdot r\}.$$

There exists $s \in \mathbb{Z}^*$ *such that* $q + p = s \cdot r$ *if and only if* $H \in q$.

Proof. Necessity. Pick $s \in \mathbb{Z}^*$ such that $q + p = s \cdot r$. We note that, by Lemma 13.6, there is at most one $a \in \mathbb{Z}$ for which $a \cdot r \in \beta\mathbb{Z} + p$. Thus, if $S = \{n \in \mathbb{Z} : n \cdot r \notin \beta\mathbb{Z} + p\}$, we have $S \in s$. Suppose that $H \notin q$ and let $B = \mathbb{Z} \backslash H \in q$. Since $q + p \in c\ell(B + p)$ and $s \cdot r \in c\ell(S \cdot r)$, it follows from Theorem 3.40, that $m + p \in \beta\mathbb{Z} \cdot r$ for some $m \in B$, or else $n \cdot r \in \beta\mathbb{Z} + p$ for some $n \in S$. The first possibility can be ruled out because it implies that $m \in B \cap H$. The second can be ruled out by our choice of S.

Sufficiency. We note that there is at most one $m \in H$ for which $m + p \in \mathbb{Z} \cdot r$ (by Lemma 13.7). Let $H' = \{m \in \mathbb{Z} : m + p \in \mathbb{Z}^* \cdot r\}$. Then $H' \in q$ and hence $q + p \in c\ell_{\beta\mathbb{Z}}(H' + p) \subseteq c\ell_{\beta\mathbb{Z}}(\mathbb{Z}^* \cdot r) = \mathbb{Z}^* \cdot r$. $\square$

Theorem 13.9. *Let $p, r \in \mathbb{Z}^*$, let*

$$G = \{q \in \mathbb{Z}^* : \text{there exists } s \in \mathbb{Z}^* \text{such that } q + p = s \cdot r\}$$

and let

$$H = \{m \in \mathbb{Z} : \text{there exists } t \in \beta\mathbb{Z} \text{ such that } m + p = t \cdot r\}.$$

Then the following statements are equivalent.

(a) $G \neq \emptyset$.

(b) $|H| = \omega$.

(c) $|G| = 2^c$.

Proof. (a) implies (b). Pick $q \in G$. By Lemma 13.8, $H \in q$ so, since $q \in \mathbb{Z}^*$, $|H| = \omega$.

(b) implies (c). By Theorem 3.58, $|\{q \in \mathbb{Z}^* : H \in q\}| = 2^c$ so by Lemma 13.8, $|G| = 2^c$.

That (c) implies (a) is trivial. $\qquad\square$

Theorem 13.10. *Let $p, r \in \mathbb{N}^*$, let*

$$G = \{q \in \mathbb{N}^* : \text{there exists } s \in \mathbb{N}^* \text{ such that } q + p = s \cdot r\}$$

and let

$$H = \{m \in \mathbb{N} : \text{there exists } t \in \beta\mathbb{N} \text{ such that } m + p = t \cdot r\}.$$

Then the following statements are equivalent.

(a) $G \neq \emptyset$.

(b) $|H| = \omega$.

(c) $|G| = 2^c$.

Proof. (a) implies (b). Pick $q \in G$. By Lemma 13.8,

$$\{m \in \mathbb{Z} : \text{there exists } t \in \beta\mathbb{Z} \text{ such that } m + p = t \cdot r\} \in q.$$

Since also $\mathbb{N} \in q$, one has

$$\{m \in \mathbb{N} : \text{there exists } t \in \beta\mathbb{Z} \text{ such that } m + p = t \cdot r\} \in q.$$

Now, given $m \in \mathbb{N}$ and $t \in \beta\mathbb{Z}$ such that $m + p = t \cdot r$, one has $m + p \in \mathbb{N}^*$, so one also must have $t \in \beta\mathbb{N}$. (Otherwise $t \cdot r \in -\mathbb{N}^* \cup \{0\}$.) Thus $H \in q$ and therefore $|H| = \omega$.

(b) implies (c). By Theorem 3.58, $|\{q \in \mathbb{N}^* : H \in q\}| = 2^c$. Further,

$$H \subseteq \{m \in \mathbb{Z} : \text{there exists } t \in \beta\mathbb{Z} \text{ such that } m + p = t \cdot r\}.$$

So by Lemma 13.8, if $q \in \mathbb{N}^*$ and $H \in q$, then there is some $s \in \mathbb{Z}^*$ such that $q + p = s \cdot r$. Since r and $q + p$ are in $\mathbb{N}^*$, $s \in \mathbb{N}^*$, so that $q \in G$. $\qquad\square$

To put Theorems 13.9 and 13.10 in context, notice that we do not know whether the set H ever has two members, let alone infinitely many members. Notice also that if H, p, and r are as in Theorem 13.10, then $H + p \subseteq \beta\mathbb{N} \cdot r$. Theorem 13.12 characterizes the existence of such H.

Definition 13.11. Let $\emptyset \neq H \subseteq \mathbb{N}$. Then

$$\mathcal{T}_H = \{A \subseteq \mathbb{N} : \text{ there exists } F \in [H]^{<\omega} \text{ such that } \mathbb{N} = \bigcup_{n \in F}(-n + A)\}.$$

Theorem 13.12. *Let $H \subseteq \mathbb{N}$ with $|H| > 1$. Then there exist $p, r \in \mathbb{N}^*$ such that $H + p \subseteq \beta\mathbb{N} \cdot r$ if and only if there is a choice of $x_A \in \mathbb{N}$ for each $A \in \mathcal{T}_H$ such that $\{x_A^{-1}A : A \in \mathcal{T}_H\}$ has the infinite finite intersection property.*

Proof. Sufficiency. Pick by Corollary 3.14 $r \in \mathbb{N}^*$ such that $\{x_A^{-1}A : A \in \mathcal{T}_H\} \subseteq r$. Recall that $C(r) = \{A \subseteq \mathbb{N} : \text{ for all } x \in \mathbb{N}, \ x^{-1}A \in r\}$ and that, by Theorem 6.18, $\beta\mathbb{N} \cdot r = \{p \in \beta\mathbb{N} : C(r) \subseteq p\}$. We claim that $\{-n + B : B \in C(r) \text{ and } n \in H\}$ has the finite intersection property. Since $C(r)$ is a filter, it suffices to let $B \in C(r)$ and $F \in \mathcal{P}_f(H)$ and show that $\bigcap_{n \in F}(-n + B) \neq \emptyset$. So suppose instead that $\bigcap_{n \in F}(-n + B) = \emptyset$. Let $A = \mathbb{N} \backslash B$. Then $A \in \mathcal{T}_H$ so $x_A^{-1}A \in r$ so $B \notin C(r)$, a contradiction.

Pick $p \in \beta\mathbb{N}$ such that $\{-n + B : B \in C(r) \text{ and } n \in H\} \subseteq p$. Then for each $n \in H, C(r) \subseteq n + p$ so by Theorem 6.18, $n + p \in \beta\mathbb{N} \cdot r$. Since, by Corollary 4.33, $\beta\mathbb{N} \cdot r \subseteq \mathbb{N}^*$, it follows that $p \in \mathbb{N}^*$.

Necessity. Pick p and r in $\mathbb{N}^*$ such that $H + p \subseteq \beta\mathbb{N} \cdot r$. Given $A \in \mathcal{T}_H$, it suffices to show that there is some $x_A \in \mathbb{N}$ such that $x_A^{-1}A \in r$. So let $A \in \mathcal{T}_H$ and pick $F \in [H]^{<\omega}$ such that $\mathbb{N} = \bigcup_{n \in F}(-n + A)$. Pick $n \in F$ such that $-n + A \in p$ and pick $s \in \beta\mathbb{N}$ such that $n + p = s \cdot r$. Then $A \in n + p = s \cdot r$ so $\{x \in \mathbb{N} : x^{-1}A \in r\} \in s$ so pick $x_A \in \mathbb{N}$ such that $x_A^{-1}A \in r$. $\qquad\square$

By Lemma 13.8 and Theorem 13.10, if there exist p, q, r, and s in $\mathbb{N}^*$ with $q + p = s \cdot r$, then $\{q \in \mathbb{N}^* : q + p = s \cdot r \text{ for some } p, s, r \in \mathbb{N}^*\}$ has nonempty interior in $\mathbb{N}^*$. By way of contrast, the set of points occupying any of the other positions of such an equation must be topologically small.

Theorem 13.13. *Let*

$$P = \{p \in \mathbb{N}^* : q + p = s \cdot r \text{ for some } q, s, r \in \mathbb{N}^*\},$$
$$S = \{s \in \mathbb{N}^* : q + p = s \cdot r \text{ for some } q, p, r \in \mathbb{N}^*\}, \text{ and}$$
$$R = \{r \in \mathbb{N}^* : q + p = s \cdot r \text{ for some } q, p, s \in \mathbb{N}^*\}.$$

Then each of P, S, and R is nowhere dense in $\mathbb{N}^*$.

Proof. To see that P is nowhere dense in $\mathbb{N}^*$, we show that $P \subseteq \mathbb{Z} + \mathbb{N}^* \cdot \mathbb{N}^*$. (We have that $\mathbb{N}^* \cdot \mathbb{N}^*$ is nowhere dense in $\mathbb{N}^*$ by Theorem 6.35 and so $\mathbb{Z} + \mathbb{N}^* \cdot \mathbb{N}^*$ is nowhere dense in $\mathbb{N}^*$ by Corollary 3.37.) Let $p \in P$ and pick q, s, and r in $\mathbb{N}^*$ such that $q + p = s \cdot r$. By Lemma 13.8, $\{m \in \mathbb{N} : m + p \in \beta\mathbb{N} \cdot r\} \in q$ and by Lemma 13.7 at most one $m \in \mathbb{N}$ has $m + p \in \mathbb{N} \cdot r$. Thus $p \in \mathbb{Z} + \mathbb{N}^* \cdot \mathbb{N}^*$ as claimed.

To see that S is nowhere dense in $\mathbb{N}^*$, suppose instead that there is an infinite subset B of $\mathbb{N}$ such that $B^* \subseteq c\ell_{\beta\mathbb{N}} S$ and choose $x \in B^*$. Choose a sequence $\langle b_n \rangle_{n=1}^{\infty}$ in B with the property that $h_m(b_n) = \widetilde{h_m}(x)$ for every $n \in \mathbb{N}$ and every $m \in \{1, 2, \ldots, n\}$. Let $A = \{b_n : n \in \mathbb{N}\}$. Then for every $s \in A^*$ and every $m \in \mathbb{N}, \widetilde{h_m}(s) = \widetilde{h_m}(x)$.

Pick $s \in A^* \cap S$ and pick q, p, and r in $\mathbb{N}^*$ such that $q + p = s \cdot r$. By Lemma 13.6, there is at most one $n \in \mathbb{N}$ for which $n \cdot r \in \beta\mathbb{N} + p$. If such an n exists, let $C = A\backslash\{n\}$, and if not let $C = A$. Let $D = \{m \in \mathbb{N} : m + p \in C^* \cdot r\}$. We claim that $D \in q$ so suppose instead that $\mathbb{N}\backslash D \in q$. Then $q + p \in \overline{(\mathbb{N}\backslash D) + p}$ and $s \cdot r \in \overline{C \cdot r}$ so by Theorem 3.40 either $n \cdot r \in \overline{(\mathbb{N}\backslash D) + p}$ for some $n \in C$ or $m + p = s' \cdot r$ for some $m \in \mathbb{N}\backslash D$ and some $s' \in C^*$. Both possibilities are excluded and so $D \in q$ as claimed. Choose $m, k \in D$ with $k < m$. Now $\widetilde{h_m}$ is constant on $C^* \cdot r$, so $\widetilde{h_m}(m+p) = \widetilde{h_m}(k+p)$ and so $h_m(m) = h_m(k)$, a contradiction.

To see that R is nowhere dense in $\mathbb{N}^*$ we show first that $R \subseteq \bigcup_{n\in\mathbb{N}} \ell_n^{-1}[\mathbb{Z}+\mathbb{N}^*\cdot\mathbb{N}^*]$. Let $r \in R$ and let $q + p = s \cdot r$, where q, p, $s \in \mathbb{N}^*$. By Lemma 13.8, we can choose $m < m'$ in $\mathbb{N}$ such that $m + p = x \cdot r$ and $m' + p = y \cdot r$ for some x, $y \in \beta\mathbb{N}$. Putting $a = m' - m$, we have $a + x \cdot r = y \cdot r$. Now $a + x \cdot r \in \overline{a + \mathbb{N} \cdot r}$ and $y \cdot r \in \overline{\mathbb{N} \cdot r}$. So it follows from Theorem 3.40 that we may suppose that the equation $a + x \cdot r = y \cdot r$ holds with $x \in \mathbb{N}$ or $y \in \mathbb{N}$. By Lemma 13.7, this equation cannot hold with x and y both in $\mathbb{N}$. (Let $p = x \cdot r$. If $a + x \cdot r = y \cdot r$, then $a + p = y \cdot r$ and $0 + p = x \cdot r$.) So there exists $n \in \mathbb{N}$ for which $n \cdot r \in \mathbb{Z} + \mathbb{N}^* \cdot \mathbb{N}^*$. Thus $R \subseteq \bigcup_{n\in\mathbb{N}} \ell_n^{-1}[\mathbb{Z} + \mathbb{N}^* \cdot \mathbb{N}^*]$ as claimed.

As we have observed, $\mathbb{Z} + \mathbb{N}^* \cdot \mathbb{N}^*$ is nowhere dense in $\mathbb{N}^*$. Now if M is a nowhere dense subset of $\mathbb{N}^*$ and $n \in \mathbb{N}$, then $\ell_n^{-1}[M]$ is also nowhere dense in $\mathbb{N}^*$. Otherwise, if $\ell_n^{-1}[M]$ were dense in A^* for some infinite subset A of $\mathbb{N}$, M would be dense in $\ell_n[A^*] = (n \cdot A)^*$. So $\bigcup_{n\in\mathbb{N}} \ell_n^{-1}[\mathbb{Z}+\mathbb{N}^*\cdot\mathbb{N}^*]$ is nowhere dense in $\mathbb{N}^*$ by Corollary 3.37. $\square$

The following result establishes that one cannot find p, q, r, and s in $\mathbb{Z}^*$ with $q + p = s \cdot r$ in the most familiar places.

Theorem 13.14. *Let $p, q, r, s \in \mathbb{Z}^*$. If $|\{a \in \mathbb{N} : a\mathbb{Z} \in r\}| = \omega$, then $q + p \neq s \cdot r$.*

Proof. Suppose instead that $q + p = s \cdot r$. By Theorem 13.9, pick $m \neq n$ in $\mathbb{Z}$ such that $m + p \in \beta\mathbb{Z} \cdot r$ and $n + p = \beta\mathbb{Z} \cdot r$. Pick $a > \max\{|m|, |n|\}$ such that $a\mathbb{Z} \in r$. Then $\widetilde{h_a}(r) = 0$ and so $\widetilde{h_a}(m + p) = \widetilde{h_a}(n + p) = 0$. This implies that $h_a(m) = h_a(n)$, a contradiction. $\square$

Corollary 13.15. $K(\beta\mathbb{N}, \cdot) \cap K(\beta\mathbb{N}, +) = \emptyset$.

Proof. Suppose one has some $r \in K(\beta\mathbb{N}, \cdot) \cap K(\beta\mathbb{N}, +)$. Pick a minimal left ideal L of $(\beta\mathbb{N}, \cdot)$ such that $r \in L$. Then by Lemma 1.52, $L = Lr$ so $r = s \cdot r$ for some $s \in L \subseteq \mathbb{N}^*$. Similarly $r = q + r$ for some $q \in \mathbb{N}^*$. But $\bigcap_{n=1}^{\infty} \overline{n\mathbb{N}}$ is an ideal of $(\beta\mathbb{N}, \cdot)$. Thus $K(\beta\mathbb{N}, \cdot) \subseteq \bigcap_{n=1}^{\infty} \overline{n\mathbb{N}}$ and consequently $r \in \bigcap_{n=1}^{\infty} \overline{n\mathbb{N}}$. But then by Theorem 13.14, $q + r \neq s \cdot r$, a contradiction. $\square$

By way of contrast with Corollary 13.15, we shall see in Corollary 16.25 that $K(\beta\mathbb{N}, \cdot) \cap c\ell\, K(\beta\mathbb{N}, +) \neq \emptyset$.

We conclude this section with a result in a somewhat different vein, showing that a certain linear equation cannot be solved.

Lemma 13.16. *Let* $u \neq v$ *in* $\mathbb{Z}\backslash\{0\}$. *There do not exist* $p \in \bigcap_{n\in\mathbb{N}} \overline{n\mathbb{N}}$ *and* $q \in \mathbb{N}^*$ *such that*

$$uq + p = vq + p.$$

Proof. Suppose we have such elements p and q and pick a prime $r > |u| + |v|$. For $i \in \{1, 2, \ldots, r-1\}$, let $C_i = \{r^n(rk + i) : n, k \in \omega\}$. Then C_i is the set of $x \in \mathbb{N}$ whose rightmost nonzero digit in the base r expansion is i. Pick i such that $C_i \in q$.

For each $m \in \mathbb{N}$, let $f(m)$ denote the largest integer in ω for which $r^{f(m)}$ is a factor of m. Since $r^{f(m)+1}\mathbb{N} \in p$, it follows from Theorem 4.15 that $\{um + s : m \in C_i$ and $s \in r^{f(m)+1}\mathbb{N}\} \in uq + p$ and $\{vn + t : n \in C_i$ and $t \in r^{f(n)+1}\mathbb{N}\} \in vq + p$. Thus there exist $m, n \in C_i$, $s \in r^{f(m)+1}\mathbb{N}$, and $t \in r^{f(n)+1}\mathbb{N}$ such that $um + s = vn + t$. We observe that $f(um + s) = f(m)$, because $r^{f(m)}$ is a factor of $um + s$, but $r^{f(m)+1}$ is not. Similarly, $f(vn + t) = f(n)$. So $f(m) = f(n) = k$, say.

We have $m \equiv ir^k \pmod{r^{k+1}}$ and $n \equiv ir^k \pmod{r^{k+1}}$. So $um + s \equiv uir^k \pmod{r^{k+1}}$ and $vn + t \equiv vir^k \pmod{r^{k+1}}$. This implies that $u \equiv v \pmod{r}$, a contradiction. $\square$

Theorem 13.17. *Let* $u \neq v$ *in* $\mathbb{Z}\backslash\{0\}$. *There do not exist* $p, q \in \mathbb{N}^*$ *such that*

$$uq + p = vq + p.$$

Proof. Suppose we have such elements p and q. Pick $r \in K(\beta\mathbb{N}, +)$. Then $p + r \in K(\beta\mathbb{N}, +)$ so by Theorem 1.64 there is some maximal group G in $K(\beta\mathbb{N}, +)$ with $p + r \in G$. Let e be the identity of G and let s be the (additive) inverse of $p + r$ in G. Then adding $r + s$ on the right in the equation $uq + p = vq + p$, one obtains $uq + e = vq + e$. This contradicts Lemma 13.16, because $h_n(e) = 0$ for every $n \in \mathbb{N}$, and so $e \in \bigcap_{n\in\mathbb{N}} \overline{n\mathbb{N}}$. $\square$

Exercise 13.1.1. Given $p \in \beta\mathbb{Z}$, define

$$|p| = \begin{cases} p & \text{if } p \in \beta\mathbb{N} \\ -p & \text{if } p \in \beta\mathbb{Z}\backslash\beta\mathbb{N}. \end{cases}$$

Prove that if $p, r, s \in \mathbb{Z}^*$ and there exists $q \in \mathbb{Z}^*$ such that $q + p = s \cdot r$, then there exists $q' \in \mathbb{Z}^*$ such that $q' + |p| = |s| \cdot |r|$.

13.2 The Distributive Laws in $\beta\mathbb{Z}$

We saw in Lemma 13.1 that if $n \in \mathbb{Z}$ and $p, q \in \beta\mathbb{Z}$, then $n \cdot (p+q) = n \cdot p + n \cdot q$. This is the only nontrivial instance of the distributive law known to hold involving members of $\mathbb{Z}^*$. We establish in this section that any other instances, if they exist, are rare indeed.

Theorem 13.18. *Let* $n \in \mathbb{Z}\backslash\{0\}$ *and let* $k, m \in \mathbb{Z}$. *The following statements are equivalent.*

(a) *There exists $p \in \mathbb{Z}^*$ such that $m \cdot p + n \cdot p = k \cdot p$.*

(b) $k = n$ *and either $m = 0$ or $m = n$.*

Proof. (a) implies (b). Assume that we have $p \in \mathbb{Z}^*$ such that $m \cdot p + n \cdot p = k \cdot p$. As in the proof of Theorem 13.5 and Lemma 13.6, we can assume that $p \in \mathbb{N}^*$ and $n, k \in \mathbb{N}$. We show first that $k = n$ so suppose instead that $k \neq n$. Choose a real number $b > 1$ for which $nb^3 < k$ or $kb^3 < n$. So $|\widetilde{\phi}_b(n) - \widetilde{\phi}_b(k)| \geq 3$. By Lemma 13.4, $\widetilde{\phi}_b(m \cdot p + n \cdot p) \in \phi_b(n) + \widetilde{\phi}_b(p) + \{-1, 0, 1, 2\}$ and $\widetilde{\phi}_b(k \cdot p) \in \phi_b(k) + \widetilde{\phi}_b(p) + \{0, 1\}$. It follows from Lemma 6.28, that $|\phi_b(m) - \phi_b(n)| \leq 2$. This contradiction shows that $k = n$.

Suppose now that $m \neq 0$. Let $r \in \mathbb{N}$ be a prime satisfying $r > |m| + n$. For any $a \in \mathbb{N}$, we have $h_{r^a}(m)\widetilde{h}_{r^a}(p) + h_{r^a}(n)\widetilde{h}_{r^a}(p) = h_{r^a}(n)\widetilde{h}_{r^a}(p)$. Since r is not a factor of $|m|$, it follows that $\widetilde{h}_{r^a}(p) = 0$. That is, $r^a\mathbb{N} \in p$ for each $a \in \mathbb{N}$.

For $i \in \{1, 2, \ldots, r - 1\}$, let

$$A_i = \{r^l(tr + i) : l, t \in \omega\}.$$

(Thus A_i is the set of positive integers whose rightmost nonzero digit in the base r expansion is i.) Pick $i \in \{1, 2, \ldots, r - 1\}$ for which $A_i \in p$.

For each $x \in \mathbb{N}$, let $f(x)$ denote the largest integer in ω for which $r^{f(x)}$ is a factor of x. We have $\{mx + s : x \in A_i$ and $s \in r^{f(x)+1}\mathbb{N}\} \in m \cdot p + n \cdot p$ (by Theorem 4.15) and $nA_i \in n \cdot p$. Thus there exist $x, y \in A_i$ and $s \in r^{f(x)+1}\mathbb{N}$ for which $mx + s = ny$. We note that $f(mx + s) = f(x)$ and $f(ny) = f(y)$. So $f(x) = f(y) = \ell$, say. Now $x \equiv ir^\ell \pmod{r^{\ell+1}}$ and $y \equiv ir^l \pmod{r^{\ell+1}}$. So $mx + s \equiv mir^\ell \pmod{r^{\ell+1}}$ and $ny \equiv nir^l \pmod{r^{\ell+1}}$. It follows that $m \equiv n \pmod{r}$ and hence that $m = n$.

(b) implies (a). Assume that $k = n$. If $m = 0$, then $m \cdot p + n \cdot p = k \cdot p$ for all $p \in \beta\mathbb{Z}$. If $m = n$ and $p + p = p$, then by Lemma 13.1, $n \cdot p + n \cdot p = n \cdot (p + p) = n \cdot p$. $\square$

Corollary 13.19. *Let $p \in \mathbb{N}^*$. Then $p \neq (-p) + p$.*

Proof. This is immediate from Theorem 13.18. $\square$

Corollary 13.20. *Let $p \in \mathbb{Z}^*$ and let $n, m \in \mathbb{Z}\backslash\{0\}$. Then $(n + m) \cdot p \neq n \cdot p + m \cdot p$ and $p \cdot (n + m) \neq p \cdot n + p \cdot m$.*

Proof. Since, by Theorem 4.23, $\mathbb{Z}$ is contained in the center of $(\beta\mathbb{Z}, \cdot)$, it suffices to show that $(n + m) \cdot p \neq n \cdot p + m \cdot p$. Suppose instead that $(n + m) \cdot p = n \cdot p + m \cdot p$. Then by Theorem 13.18, $n + m = m$, so $n = 0$, a contradiction. $\square$

Theorem 13.18 has special consequences for the semigroup $\mathbb{H}$.

Theorem 13.21. *Let $q, r \in \mathbb{H}$ and let $p \in \mathbb{Z}^*$. Then $p \cdot (q + r) \neq p \cdot q + p \cdot r$ and $(p + q) \cdot r \neq p \cdot r + q \cdot r$. In particular there are no instances of the validity of either distributive law in $\mathbb{H}$.*

Proof. By Lemma 6.8, $q + r \in \mathbb{H}$, so by Theorem 13.14, $p \cdot (q + r) \notin \mathbb{Z}^* + \mathbb{Z}^*$. Since $(\mathbb{Z}\backslash\{0\}, \cdot)$ is cancellative, we have by Corollary 4.33 that $\mathbb{Z}^*$ is a subsemigroup of $(\beta\mathbb{Z}, \cdot)$ and so $p \cdot q + p \cdot r \in \mathbb{Z}^* + \mathbb{Z}^*$.

Similarly $(p + q) \cdot r \notin \mathbb{Z}^* + \mathbb{Z}^*$ and $p \cdot r + q \cdot r \in \mathbb{Z}^* + \mathbb{Z}^*$. $\qquad\square$

We now set out to establish the topological rarity of instances of a distributive law in $\mathbb{N}^*$, if indeed any such instances exist at all.

Definition 13.22. Let $A \subseteq \mathbb{N}$. A is a *doubly thin* set if and only if A is infinite and whenever (m, n) and (k, l) are distinct elements of $\mathbb{N} \times \omega$, one has $|(n+mA) \cap (l+kA)| < \omega$.

Lemma 13.23. *Let B be an infinite subset of $\mathbb{N}$. There is a doubly thin set $A \subseteq B$.*

Proof. Enumerate $\mathbb{N} \times \omega$ as $\langle(m(k), n(k))\rangle_{k=1}^{\infty}$ and pick $s_1 \in B$. Inductively let $k \in \mathbb{N}$ and assume that $s_1, s_2, \ldots, s_k$ have been chosen. Pick $s_{k+1} > s_k$ such that

$$s_{k+1} \in B \setminus \left\{ \frac{s_t \cdot m(l) + n(l) - n(i)}{m(i)} : i, l, t \in \{1, 2, \ldots, k\} \right\}.$$

Let $A = \{s_k : k \in \mathbb{N}\}$.

Then A is infinite. Suppose that one has some $l \neq i$ such that

$$|(n(l) + m(l)A) \cap (n(i) + m(i)A)| = \omega.$$

Now, since $(m(l), n(l)) \neq (m(i), n(i))$, there is at most one $x \in A$ such that $n(l) + m(l)x = n(i) + m(i)x$. If there is such an $x = s_v$, let $b = \max\{v, i, l\}$, and otherwise let $b = \max\{i, l\}$.

Pick $z \in (n(l) + m(l)A) \cap (n(i) + m(i)A)$ with $z > n(l) + m(l)s_b$ and $z > n(i) + m(i)s_b$ and pick $t, k \in \mathbb{N}$ such that $z = n(l) + m(l)s_t = n(i) + m(i)s_k$ and notice that $t > b$ and $k > b$. Then $s_t \neq s_k$ so assume without loss of generality that $t < k$. But then

$$s_k = \frac{n(l) + m(l)s_t - n(i)}{m(i)},$$

a contradiction. $\qquad\square$

Lemma 13.24. *Let A be a doubly thin subset of $\mathbb{N}$, let $p \in A^*$, and let $B \subseteq \mathbb{N} \times \omega$. Then $c\ell\{n + m \cdot p : (m, n) \in B\} \cap c\ell\{n + m \cdot p : (m, n) \in (\mathbb{N} \times \omega)\backslash B\} = \emptyset$.*

Proof. Suppose that $q \in c\ell\{n+m \cdot p : (m, n) \in B\} \cap c\ell\{n+m \cdot p : (m, n) \in (\mathbb{N} \times \omega)\backslash B\}$. Order $\mathbb{N} \times \omega$ by $\prec$ in order type ω. (That is, $\prec$ linearly orders $\mathbb{N} \times \omega$ so that each element has only finitely many predecessors.) Let

$$C = \bigcup_{(m,n) \in B} \left((n + mA) \setminus \bigcup_{(m',n') \prec (m,n)} (n' + m'A) \right)$$

and let

$$D = \bigcup_{(m,n) \in (\mathbb{N} \times \omega)\backslash B} \left((n + mA) \setminus \bigcup_{(m',n') \prec (m,n)} (n' + m'A) \right).$$

Notice that $C \cap D = \emptyset$. Thus without loss of generality we have $\mathbb{N}\backslash C \in q$. Pick $(m, n) \in B$ such that $n + m \cdot p \in \overline{\mathbb{N}\backslash C}$. Now $n + mA \in n + m \cdot p$, $n + m \cdot p \in \mathbb{N}^*$, and for every $(n', m') \prec (n, m)$, $|(n' + m'A) \cap (n + mA)| < \omega$ since A is doubly thin. Thus

$$(n + mA) \backslash \bigcup_{(m', n') \prec (m, n)} (n' + m'A) \in n + m \cdot p$$

so $C \in n + m \cdot p$, a contradiction. $\qquad\square$

We have one final preliminary.

Lemma 13.25. *Let A be a doubly thin subset of $\mathbb{N}$ and let $p \in A^*$. Then $\mathbb{N}^* \cdot p$, $\mathbb{N}^* + \mathbb{N}^* \cdot p$, and $\mathbb{N}^* \cdot (\mathbb{N}^* + p)$ have pairwise disjoint closures.*

Proof. By Lemma 13.24, it suffices to observe that

$$\mathbb{N}^* \cdot p \subseteq c\ell\{0 + m \cdot p : m \in \mathbb{N}\},$$
$$\mathbb{N}^* + \mathbb{N}^* \cdot p \subseteq c\ell\{n + m \cdot p : n, m \in \mathbb{N} \text{ and } n < m\}, \text{ and}$$
$$\mathbb{N}^* \cdot (\mathbb{N}^* + p) \subseteq c\ell\{n + m \cdot p : n, m \in \mathbb{N} \text{ and } n \geq m\}.$$

The first observation is immediate. For the second, just notice that if $q, r \in \mathbb{N}^*$, then

$$q + r \cdot p = q\text{-}\lim_{n \in \mathbb{N}}(r\text{-}\lim_{m \in \mathbb{N}}(n + m \cdot p)).$$

For the last observation, let $q, r \in \mathbb{N}^*$, let $B \in q \cdot (r + p)$, and pick $m \in \mathbb{N}$ such that $m^{-1}B \in r + p$. Then $\{n \in \mathbb{N} : -n + m^{-1}B \in p\} \in r$ so pick n such that $-n + m^{-1}B \in p$. Then $m \cdot n + m \cdot p = m \cdot (n + p) \in \overline{B}$. $\qquad\square$

Theorem 13.26. *The set $\{p \in \mathbb{N}^* : \mathbb{N}^* \cdot p, \mathbb{N}^* + \mathbb{N}^* \cdot p, \text{ and } \mathbb{N}^* \cdot (\mathbb{N}^* + p) \text{ have pairwise disjoint closures}\}$ has dense interior in $\mathbb{N}^*$.*

Proof. Let B be an infinite subset of $\mathbb{N}$. Pick by Lemma 13.23 a doubly thin subset A of B. Then apply Lemma 13.25. $\qquad\square$

Corollary 13.27. *The set $\{p \in \mathbb{N}^* : \text{for all } q, r \in \mathbb{N}^*, (q + r) \cdot p \neq q \cdot p + r \cdot p \text{ and } r \cdot (q + p) \neq r \cdot q + r \cdot p\}$ has dense interior in $\mathbb{N}^*$.*

Proof. Given $p, q, r \in \mathbb{N}^*$,

$$(q + r) \cdot p \in \mathbb{N}^* \cdot p,$$
$$q \cdot p + r \cdot p \in \mathbb{N}^* + \mathbb{N}^* \cdot p,$$
$$r \cdot (q + p) \in \mathbb{N}^* \cdot (\mathbb{N}^* + p), \text{ and}$$
$$r \cdot q + r \cdot p \in \mathbb{N}^* + \mathbb{N}^* \cdot p,$$

so Theorem 13.26 applies. $\qquad\square$

13.3 Ultrafilters on $\mathbb{R}$ Near 0

We have dealt throughout this book with the Stone–Čech compactification of a discrete space, especially a discrete semigroup. One may naturally wonder why we restrict our attention to discrete spaces, especially given that the Stone–Čech compactification exists for a much larger class of spaces. We shall see in Theorem 21.47 that for many nondiscrete semigroups S, including $(\mathbb{R}, +)$, the Stone–Čech compactification of S cannot be made into a semigroup compactification of S.

Of course, if S is a topological semigroup, one can give the set S the discrete topology and proceed as we always have. It would seem that by doing that, one would lose any information about the original topological semigroup. We shall see however in Section 17.5 that we can get information about partitions of the real interval $(0, \infty)$, every member of which is measurable, or every member of which is a Baire set, by first giving $(0, \infty)$ the discrete topology and passing to its Stone–Čech compactification. Because of these interesting applications, we shall investigate the algebra of $(\beta \mathbb{R}_d, +)$ and $(\beta \mathbb{R}_d, \cdot)$, where $\mathbb{R}_d$ is the set $\mathbb{R}$ with the discrete topology. We shall be primarily concerned with $(\beta \mathbb{R}_d, +)$.

Definition 13.28. (a) Given a topological space X, X_d is the set X endowed with the discrete topology.

(b) Let $\alpha : \beta \mathbb{R}_d \to [-\infty, \infty]$ be the continuous extension of the identity function.

(c) $B(\mathbb{R}) = \{ p \in \beta \mathbb{R}_d : \alpha(p) \notin \{-\infty, \infty\} \}$.

(c) Given $x \in \mathbb{R}$,
$$x^+ = \{ p \in B(\mathbb{R}) : \alpha(p) = x \text{ and } (x, \infty) \in p \} \text{ and}$$
$$x^- = \{ p \in B(\mathbb{R}) : \alpha(p) = x \text{ and } (-\infty, x) \in p \}.$$

(d) $U = \bigcup_{x \in \mathbb{R}} x^+$ and $D = \bigcup_{x \in \mathbb{R}} x^-$.

The set $B(\mathbb{R})$ is the set of "bounded" ultrafilters on $\mathbb{R}$. That is, an ultrafilter $p \in \beta \mathbb{R}_d$ is in $B(\mathbb{R})$ if and only if there is some $n \in \mathbb{N}$ with $[-n, n] \in p$. We collect some routine information about the notions defined above.

Lemma 13.29. (a) *Let $x \in \mathbb{R}$ and let $p \in \beta \mathbb{R}_d$. Then $p \in x^+$ if and only if for every $\epsilon > 0$, $(x, x + \epsilon) \in p$. Also, $p \in x^-$ if and only if for every $\epsilon > 0$, $(x - \epsilon, x) \in p$.*

(b) *Let $p, q \in B(\mathbb{R})$, let $x = \alpha(p)$, and let $y = \alpha(q)$. If $p \in x^+$, then $p + q \in (x + y)^+$. If $p \in x^-$, then $p + q \in (x + y)^-$.*

(c) *$B(\mathbb{R}) \backslash \mathbb{R} = U \cup D$ and U and D are disjoint right ideals of $(B(\mathbb{R}), +)$. In particular, $B(\mathbb{R})$ is not commutative.*

(d) *The set 0^+ is a compact subsemigroup of $(B(\mathbb{R}), +)$.*

(e) *If $x \in \mathbb{R}$ and $p \in \beta \mathbb{R}_d$, then $x + p = p + x$.*

(f) *The set 0^+ is a two sided ideal of the semigroup $(\beta(0, 1)_d, \cdot)$.*

Proof. The proof of (a) is a routine exercise, (e) is a consequence of Theorem 4.23, and (c) and (d) follow from (b). We establish (b) and (f).

To verify (b) assume first that $p \in x^+$. To see that $p + q \in (x + y)^+$, let $\epsilon > 0$ be given and let $A = (x + y, x + y + \epsilon)$. To see that $A \in p + q$, we show that

$(x, x+\epsilon) \subseteq \{z \in \mathbb{R} : -z + A \in q\}$. So let $z \in (x, x+\epsilon)$. Let $\delta = \min\{z-x, x+\epsilon-z\}$. Since $\alpha(q) = y$, we have $(y-\delta, y+\delta) \in q$ and $(y-\delta, y+\delta) \subseteq -z+A$ so $-z+A \in q$ as required. The proof that $p + q \in (x + y)^-$ if $p \in x^-$ is nearly identical.

To verify (f), let $p \in 0^+$ and let $q \in \beta(0, 1)_d$. To see that $p \cdot q \in 0^+$ and $q \cdot p \in 0^+$, let $\epsilon > 0$ be given. Since $(0, \epsilon) \cdot (0, 1) = (0, \epsilon)$, we have $(0, \epsilon) \subseteq \{x \in (0, 1) : x^{-1}(0, \epsilon) \in q\}$ (so that $(0, \epsilon) \in p \cdot q$) and $(0, 1) \subseteq \{x \in (0, 1) : x^{-1}(0, \epsilon) \in p\}$ (so that $(0, \epsilon) \in q \cdot p$). □

We shall restrict our attention to 0^+, and it is natural to ask why. On the one hand, it is a subsemigroup of $(\beta\mathbb{R}_d, +)$ and for any other $x \in \mathbb{R}$, x^+ and x^- are not semigroups. On the other hand we see in the following theorem that 0^+ holds all of the algebraic structure of $(B(\mathbb{R}), +)$ not already revealed by $\mathbb{R}$.

Theorem 13.30. (a) $(0^+, +)$ *and* $(0^-, +)$ *are isomorphic.*

(b) *The function* $\varphi : \mathbb{R}_d \times (\{0\} \cup 0^+ \cup 0^-) \to B(\mathbb{R})$ *defined by* $\varphi(x, p) = x + p$ *is a continuous isomorphism onto* $(B(\mathbb{R}), +)$.

Proof. (a) Define $\tau : 0^+ \to 0^-$ by $\tau(p) = -p$, where $-p = \{-A : A \in p\}$. It is routine to verify that τ takes 0^+ one-to-one onto 0^-. Let $p, q \in 0^+$. To see that $\tau(p+q) = \tau(p)+\tau(q)$ it suffices, since $\tau(p+q)$ and $\tau(p)+\tau(q)$ are both ultrafilters, to show that $\tau(p + q) \subseteq \tau(p) + \tau(q)$. So let $A \in \tau(p + q)$. Then $-A \in p + q$ so

$$B = \{x \in \mathbb{R} : -x + -A \in q\} \in p$$

and hence $-B \in \tau(p)$. Then $-B \subseteq \{x \in \mathbb{R} : -x + A \in \tau(q)\}$ so $A \in \tau(p) + \tau(q)$ as required.

(b) To see that φ is a homomorphism, let (x, p) and (y, q) be in $\mathbb{R}_d \times (\{0\} \cup 0^+ \cup 0^-)$. Since $p + y = y + p$, we have $\varphi(x, p) + \varphi(y, q) = \varphi(x + y, p + q)$.

To see that φ is one-to-one, assume we have $\varphi(x, p) = \varphi(y, q)$. By Lemma 13.29(b) we have $x = \alpha(x + p) = \alpha(y + q) = y$. Then $x + p = x + q$ so $p = -x + x + q = q$.

To see that φ is onto $B(\mathbb{R})$, let $q \in B(\mathbb{R})$ and let $x = \alpha(q)$. Let $p = -x + q$. Then $q = x + p = \varphi(x, p)$.

To see that φ is continuous, let $(x, p) \in \mathbb{R}_d \times (\{0\} \cup 0^+ \cup 0^-)$ and let $A \in x + p$. Then $-x + A \in p$ so $\{x\} \times \left(c\ell_{\beta\mathbb{R}_d}(-x + A)\right)$ is a neighborhood of (x, p) contained in $\varphi^{-1}[c\ell\,\beta\mathbb{R}_d A]$. □

As we have remarked, 0^+ has an interesting and useful multiplicative structure. But much is known of this structure because, by Lemma 13.29(f), 0^+ is a two sided ideal of $(\beta(0, 1)_d, \cdot)$, so $K(\beta(0, 1)_d, \cdot) \subseteq 0^+$ and hence by Theorem 1.65, $K(0^+, \cdot) = K(\beta(0, 1)_d, \cdot)$.

On the other hand, 0^+ is far from being an ideal of $(B(S), +)$, so no general results apply to $(0^+, +)$ beyond those that apply to any compact right topological semigroup. We first give an easy characterization of idempotents in $(0^+, +)$, whose proof we leave as an exercise.

Theorem 13.31. *There exists* $p = p + p$ *in* 0^+ *with* $A \in p$ *if and only if there is some sequence* $\langle x_n \rangle_{n=1}^\infty$ *in* $(0, 1)$ *such that* $\sum_{n=1}^\infty x_n$ *converges and* $FS(\langle x_n \rangle_{n=1}^\infty) \subseteq A$.

Proof. This is Exercise 13.3.1.   □

As a compact right topological semigroup, 0^+ has a smallest two sided ideal by Theorem 2.8. We now turn our attention to characterizing the smallest ideal of 0^+ and its closure.

If $(S, +)$ is a discrete semigroup we know from Theorem 4.39 that a point $p \in \beta S$ is in the smallest ideal of βS if and only if, for each $A \in p$, $\{x \in S : -x + A \in p\}$ is syndetic (and also if and only if for all $q \in \beta S$, $p \in \beta S + q + p$). We obtain in Theorem 13.33 a nearly identical characterization of $K(0^+, +)$.

Definition 13.32. A subset B of $(0, 1)$ is *syndetic near* 0 if and only if for every $\epsilon > 0$ there exist some $F \in \mathcal{P}_f((0, \epsilon))$ and some $\delta > 0$ such that $(0, \delta) \subseteq \bigcup_{t \in F}(-t + B)$.

Theorem 13.33. *Let $p \in 0^+$. The following statements are equivalent.*

(a) $p \in K(0^+, +)$.

(b) *For all $A \in p$, $\{x \in (0, 1) : -x + A \in p\}$ is syndetic near 0.*

(c) *For all $r \in 0^+$, $p \in 0^+ + r + p$.*

Proof. (a) implies (b). Let $A \in p$, let $B = \{x \in (0, 1) : -x + A \in p\}$, and suppose that B is not syndetic near 0. Pick $\epsilon > 0$ such that for all $F \in \mathcal{P}_f((0, \epsilon))$ and all $\delta > 0$, $(0, \delta) \setminus \bigcup_{t \in F}(-t + B) \neq \emptyset$.

Let $\mathcal{G} = \{(0, \delta) \setminus \bigcup_{t \in F}(-t + B) : F \in \mathcal{P}_f((0, \epsilon))$ and $\delta > 0\}$. Then $\mathcal{G}$ has the finite intersection property so pick $r \in \beta(0, 1)_d$ with $\mathcal{G} \subseteq r$. Since $\{(0, \delta) : \delta > 0\} \subseteq r$ we have $r \in 0^+$.

Pick a minimal left ideal L of 0^+ with $L \subseteq 0^+ + r$, by Corollary 2.6. Since $K(0^+)$ is the union of all of the minimal right ideals of 0^+ by Theorem 2.8, pick a minimal right ideal R of 0^+ with $p \in R$. Then $L \cap R$ is a group by Theorem 2.7, so let q be the identity of $L \cap R$. Then $R = q + 0^+$, so $p \in q + 0^+$ so $p = q + p$ so $B \in q$. Also $q \in 0^+ + r$ so pick $w \in 0^+$ such that $q = w + r$. Then $(0, \epsilon) \in w$ and $\{t \in (0, 1) : -t + B \in r\} \in w$ so pick $t \in (0, \epsilon)$ such that $-t + B \in r$. But $(0, 1) \setminus (-t + B) \in \mathcal{G} \subseteq r$, a contradiction.

(b) implies (c). Let $r \in 0^+$. For each $A \in p$, let $B(A) = \{x \in (0, 1) : -x + A \in p\}$ and let $C(A) = \{t \in (0, 1) : -t + B(A) \in r\}$. Observe that for any $A_1, A_2 \in p$, one has $B(A_1 \cap A_2) = B(A_1) \cap B(A_2)$ and $C(A_1 \cap A_2) = C(A_1) \cap C(A_2)$.

We claim that for every $A \in p$ and every $\epsilon > 0$, $C(A) \cap (0, \epsilon) \neq \emptyset$. To see this, let $A \in p$ and $\epsilon > 0$ be given and pick $F \in \mathcal{P}_f((0, \epsilon))$ and $\delta > 0$ such that $(0, \delta) \subseteq \bigcup_{t \in F}(-t + B(A))$. Since $(0, \delta) \in r$ we have $\bigcup_{t \in F}(-t + B(A)) \in r$ and hence there is some $t \in F$ with $-t + B(A) \in r$. Then $t \in C(A) \cap (0, \epsilon)$.

Thus $\{(0, \epsilon) \cap C(A) : \epsilon > 0$ and $A \in p\}$ has the finite intersection property so pick $q \in \beta(0, 1)_d$ with $\{(0, \epsilon) \cap C(A) : \epsilon > 0$ and $A \in p\} \subseteq q$. Then $q \in 0^+$. We claim that $p = q + r + p$ for which it suffices (since both are ultrafilters) to show that $p \subseteq q + r + p$. Let $A \in p$ be given. Then $\{t \in (0, 1) : -t + B(A) \in r\} = C(A) \in q$ so $B(A) \in q + r$, so $A \in q + r + p$ as required.

(c) implies (a). Pick $r \in K(0^+)$.   □

If $(S, +)$ is any discrete semigroup, we know from Corollary 4.41 that, given any discrete semigroup S and any $p \in \beta S$, p is in the closure of the smallest ideal of βS if and only if each $A \in p$ is piecewise syndetic. We obtain a similar result in 0^+. Modifying the definition of "piecewise syndetic" to apply to 0^+ is somewhat less straightforward than was the case with "syndetic".

Definition 13.34. A subset A of $(0, 1)$ is *piecewise syndetic near* 0 if and only if there exist sequences $\langle F_n \rangle_{n=1}^{\infty}$ and $\langle \delta_n \rangle_{n=1}^{\infty}$ such that

(1) for each $n \in \mathbb{N}$, $F_n \in \mathcal{P}_f\big((0, 1/n)\big)$ and $\delta_n \in (0, 1/n)$, and

(2) for all $G \in \mathcal{P}_f\big((0, 1)\big)$ and all $\mu > 0$ there is some $x \in (0, \mu)$ such that for all $n \in \mathbb{N}$, $\big(G \cap (0, \delta_n)\big) + x \subseteq \bigcup_{t \in F_n}(-t + A)$.

Theorem 13.35. *Let $A \subseteq (0, 1)$. Then $K(0^+) \cap c\ell_{\beta(0,1)_d} A \neq \emptyset$ if and only if A is piecewise syndetic near* 0.

Proof. Necessity. Pick $p \in K(0^+) \cap c\ell_{\beta(0,1)_d} A$ and let $B = \{x \in (0, 1) : -x + A \in p\}$. By Theorem 13.33, B is syndetic near 0. Inductively for $n \in \mathbb{N}$ pick $F_n \in \mathcal{P}_f\big((0, 1/n)\big)$ and $\delta_n \in (0, 1/n)$ (with $\delta_n \le \delta_{n-1}$ if $n > 1$) such that $(0, \delta_n) \subseteq \bigcup_{t \in F_n}(-t + B)$.

Let $G \in \mathcal{P}_f\big((0, 1)\big)$ be given. If $G \cap (0, \delta_1) = \emptyset$, the conclusion is trivial, so assume $G \cap (0, \delta_1) \neq \emptyset$ and let $H = G \cap (0, \delta_1)$. For each $y \in H$, let

$$m(y) = \max\{n \in \mathbb{N} : y < \delta_n\}.$$

For each $y \in H$ and each $n \in \{1, 2, \ldots, m(y)\}$, we have $y \in \bigcup_{t \in F_n}(-t + B)$ so pick $t(y, n) \in F_n$ such that $y \in -t(y, n) + B$. Then given $y \in H$ and $n \in \{1, 2, \ldots, m(y)\}$, one has $-(t(y, n) + y) + A \in p$.

Now let $\mu > 0$ be given. Then $(0, \mu) \in p$ so pick

$$x \in (0, \mu) \cap \bigcap_{y \in H} \bigcap_{n=1}^{m(y)} \big(-(t(y, n) + y) + A\big).$$

Then given $n \in \mathbb{N}$ and $y \in G \cap (0, \delta_n)$, one has $y \in H$ and $n \le m(y)$ so $t(y, n) + y + x \in A$ so

$$y + x \in -t(y, n) + A \subseteq \bigcup_{t \in F_n}(-t + A).$$

Sufficiency. Pick $\langle F_n \rangle_{n=1}^{\infty}$ and $\langle \delta_n \rangle_{n=1}^{\infty}$ satisfying (1) and (2) of Definition 13.34. Given $G \in \mathcal{P}_f\big((0, 1)\big)$ and $\mu > 0$, let

$$C(G, \mu) = \{x \in (0, \mu) : \text{for all } n \in \mathbb{N}, \big(G \cap (0, \delta_n)\big) + x \subseteq \bigcup_{t \in F_n}(-t + A)\}.$$

By assumption each $C(G, \mu) \neq \emptyset$. Further, given G_1 and G_2 in $\mathcal{P}_f\big((0, 1)\big)$ and $\mu_1, \mu_2 > 0$, one has $C(G_1 \cup G_2, \min\{\mu_1, \mu_2\}) \subseteq C(G_1, \mu_1) \cap C(G_2, \mu_2)$ so

$$\{C(G, \mu) : G \in \mathcal{P}_f\big((0, 1)\big) \text{ and } \mu > 0\}$$

has the finite intersection property so pick $p \in \beta(0, 1)_d$ with $\{C(G, \mu) : G \in \mathcal{P}_f\big((0, 1)\big)$ and $\mu > 0\} \subseteq p$. Note that since each $C(G, \mu) \subseteq (0, \mu)$, one has $p \in 0^+$.

Now we claim that for each $n \in \mathbb{N}$, $0^+ + p \subseteq c\ell_{\beta(0,1)_d}\left(\bigcup_{t \in F_n}(-t + A)\right)$, so let $n \in \mathbb{N}$ and let $q \in 0^+$. To show that $\bigcup_{t \in F_n}(-t + A) \in q + p$, we show that

$$(0, \delta_n) \subseteq \{y \in (0, 1) : -y + \bigcup_{t \in F_n}(-t + A) \in p\}.$$

So let $y \in (0, \delta_n)$. Then $C(\{y\}, \delta_n) \in p$ and $C(\{y\}, \delta_n) \subseteq -y + \bigcup_{t \in F_n}(-t + A)$.

Now pick $r \in (0^+ + p) \cap K(0^+)$ (since $0^+ + p$ is a left ideal of 0^+). Given $n \in \mathbb{N}$, $\bigcup_{t \in F_n}(-t + A) \in r$ so pick $t_n \in F_n$ such that $-t_n + A \in r$. Now each $t_n \in F_n \subseteq (0, 1/n)$ so $\lim_{n \to \infty} t_n = 0$ so pick $q \in 0^+ \cap c\ell_{\beta(0,1)_d}\{t_n : n \in \mathbb{N}\}$. Then $q + r \in K(0^+)$ and $\{t_n : n \in \mathbb{N}\} \subseteq \{t \in (0, 1) : -t + A \in r\}$ so $A \in q + r$. $\square$

Since $(0^+, +)$ is a compact right topological semigroup, the closure of any right ideal is again a right ideal. Consequently $c\ell_{0^+} K(0^+) = c\ell_{\beta(0,1)_d} K(0^+)$ is a right ideal of 0^+. On the other hand, if S is any discrete semigroup, we know from Theorem 4.44 that the closure of $K(\beta S)$ is a two sided ideal of βS. We do not know whether $c\ell_{0^+} K(0^+)$ is a left ideal of 0^+, but would conjecture that it is not.

Exercise 13.3.1. Prove Theorem 13.31. (Hint: Consider Lemma 5.11 and either proof of Theorem 5.8.)

13.4 The Left and Right Continuous Extensions of One Operation

Throughout this book, we have taken the operation $\cdot$ on βS to be the operation making $(\beta S, \cdot)$ a right topological semigroup with S contained in its topological center.

Of course, the extension can also be accomplished in the reverse order to that used in Section 4.1. That is, given $x \in S$ and $p \in \beta S$ we can define $p \diamond x = p\text{-}\lim_{y \in S} y \cdot x$ (so that if $p \in S$, then $p \diamond x = p \cdot x$) and given $p, q \in \beta S$,

$$p \diamond q = q\text{-}\lim_{x \in S} p \diamond x = q\text{-}\lim_{x \in S}(p\text{-}\lim_{y \in S} y \cdot x).$$

Then $(\beta S, \diamond)$ is a compact left topological semigroup and for each $x \in S$, the function $r_x : \beta S \to \beta S$ defined by $r_x(p) = p \diamond x$ is continuous. Further, Theorems 2.7 through 2.11 apply to $(\beta S, \diamond)$ with "left" and "right" interchanged.

Notice that, given $p, q \in \beta S$ and $A \subseteq S$,

$A \in p \diamond q$ if and only if $\{x \in S : Ax^{-1} \in p\} \in q$, where $Ax^{-1} = \{y \in S : yx \in A\}$.

Definition 13.36. Let $(S, \cdot)$ be a semigroup. Then $\diamond$ is the extension of $\cdot$ to βS with the property that for each $p \in \beta S$, the function ℓ_p is continuous, and for each $x \in S$, the function r_x is continuous, where for $q \in \beta S$, $\ell_p(q) = p \diamond q$ and $r_x(q) = q \diamond x$.

We see that if S is commutative, then there is no substantive difference between $(\beta S, \cdot)$ and $(\beta S, \diamond)$.

Theorem 13.37. *Let $(S, \cdot)$ be a commutative semigroup. Then for all $p, q \in \beta S$, $p \diamond q = q \cdot p$. In particular, $K(\beta S, \cdot) = K(\beta S, \diamond)$ and, if H is a subset of βS, then H is a subsemigroup of $(\beta S, \cdot)$ if and only if H is a subsemigroup of $(\beta S, \diamond)$.*

Proof. This is Exercise 13.4.1. □

We show now that if S is not commutative, then both of the "in particular" conclusions may fail.

Theorem 13.38. *Let S be the free semigroup on countably many generators. Then there is a subset H of βS such that*

(1) *H is a subsemigroup of $(\beta S, \cdot)$ and*

(2) *for every $p, q \in \beta S$, $p \diamond q \notin H$.*

Proof. Let the generators of S be $\{y_n : n \in \mathbb{N}\}$. For each $k \in \mathbb{N}$, let

$$M_k = \{y_{n(1)} y_{n(2)} \cdots y_{n(l)} : l \in \mathbb{N} \text{ and } k \leq n(1) < n(2) < \cdots < n(l)\},$$

and let $H = \bigcap_{k=1}^{\infty} c\ell\, M_k$.

To see that H is a subsemigroup of $(\beta S, \cdot)$, let $p, q \in H$ and let $k \in \mathbb{N}$. We need to show that $M_k \in p \cdot q$. To see this, we show that $M_k \subseteq \{x \in S : x^{-1} M_k \in q\}$, so let $x \in M_k$ and pick $l \in \mathbb{N}$ and $n(1), n(2), \ldots, n(l) \in \mathbb{N}$ such that $k \leq n(1) < n(2) < \cdots < n(l)$ and $x = y_{n(1)} y_{n(2)} \cdots y_{n(l)}$. Then $M_{n(l)+1} \in q$ and $M_{n(l)+1} \subseteq x^{-1} M_k$.

To verify conclusion (2), let $p, q \in \beta S$. For each $k \in \mathbb{N}$, let

$$R_k = \{y_k\} \cup S y_k \cup y_k S \cup S y_k S = \{x \in S : y_k \text{ occurs in } x\}.$$

Then R_k is an ideal of S so by (the left-right switch of) Corollary 4.18, $\overline{R_k}$ is an ideal of $(\beta S, \diamond)$. Since $R_k \cap M_{k+1} = \emptyset$ for each k, one has that $p \diamond q \notin H$ if $p \in \bigcup_{k=1}^{\infty} \overline{R_k}$.

So assume that $p \notin \bigcup_{k=1}^{\infty} \overline{R_k}$. We claim that $M_1 \notin p \diamond q$, so suppose instead that $M_1 \in p \diamond q$ and let $B = \{x \in S : M_1 x^{-1} \in p\}$. Then $B \in q$ so $B \neq \emptyset$ so pick $x \in B$ and pick $k \in \mathbb{N}$ such that y_k occurs in x. Then $M_1 x^{-1} \setminus \bigcup_{t=1}^{k} R_t \in p$ so pick $z \in M_1 x^{-1} \setminus \bigcup_{t=1}^{k} R_t$ and pick $j \in \mathbb{N}$ such that y_j occurs in z. Then $j > k$ so $zx \notin M_1$, a contradiction. □

For the left-right switch of Theorem 13.38, see Exercise 13.4.2.

We see now that we can have $K(\beta S, \diamond) \neq K(\beta S, \cdot)$. Recall that after Definition 4.38 we noted that there are really two notions of "piecewise syndetic", namely the notion of *right piecewise syndetic* which was defined in Definition 4.38, and the notion of *left piecewise syndetic*. A subset A of the semigroup S is *left piecewise syndetic* if and only if there is some $G \in \mathcal{P}_f(S)$ such that for every $F \in \mathcal{P}_f(S)$ there is some $x \in S$ with $x \cdot F \subseteq \bigcup_{t \in G} A t^{-1}$.

Lemma 13.39. *Let S be the free semigroup on two generators. There is a subset A of S which is left piecewise syndetic but is not right piecewise syndetic.*

Proof. Let the generators of S be a and b. For each $n \in \mathbb{N}$, let $W_n = \{x \in S : l(x) \leq n\}$, where $l(x)$ is the length of x. Let $A = \bigcup_{n=1}^{\infty} b^n W_n a$.

Suppose that A is right piecewise syndetic and pick $G \in \mathcal{P}_f(S)$ such that for every $F \in \mathcal{P}_f(S)$ there is some $x \in S$ with $F \cdot x \subseteq \bigcup_{t \in G} t^{-1} A$. Let $m = \max\{l(t) : t \in G\} + 1$ and let $F = \{a^m\}$. Pick $x \in S$ such that $a^m x \in \bigcup_{t \in G} t^{-1} A$ and pick $t \in G$ such that $t a^m x \in A$. Pick $n \in \mathbb{N}$ such that $t a^m x \in b^n W_n a$. Then $t = b^n v$ for some $v \in S \cup \{\emptyset\}$ so $n \leq l(t)$. But now, the length of $t a^m x$ is at least $n + m + 1$ while the length of any element of $b^n W_n a$ is at most $2n + 1$. Since $m \geq l(t) + 1 \geq n + 1$, this is a contradiction.

To see that A is left piecewise syndetic, let $G = \{a\}$ and let $F \in \mathcal{P}_f(S)$ be given. Pick $n \in \mathbb{N}$ such that $F \subseteq W_n$ and let $x = b^n$. Then $x \cdot F \cdot a \subseteq b^n W_n a \subseteq A$ so $x \cdot F \subseteq A a^{-1}$ as required. $\qquad \square$

For a contrast to Lemma 13.39, see Exercise 13.4.3

Theorem 13.40. *Let S be the free semigroup on two generators. Then*

$$K(\beta S, \diamond) \setminus c\ell\, K(\beta S, \cdot) \neq \emptyset \quad and \quad K(\beta S, \cdot) \setminus c\ell\, K(\beta S, \diamond) \neq \emptyset.$$

Proof. We establish the first statement, leaving the second as an exercise. Pick by Lemma 13.39 a set $A \subseteq S$ which is left piecewise syndetic but not right piecewise syndetic. Then by Theorem 4.40, $\overline{A} \cap K(\beta S, \cdot) = \emptyset$ so $\overline{A} \cap c\ell\, K(\beta S, \cdot) = \emptyset$. On the other hand, by the left-right switched version of Theorem 4.40, $\overline{A} \cap K(\beta S, \diamond) \neq \emptyset$. $\qquad \square$

We do not know whether $K(\beta S, \diamond) \cap K(\beta S, \cdot) = \emptyset$ for some semigroup S or even whether this is true for the free semigroup on two generators. The following theorem tells us that in any case the two smallest ideals are not too far apart.

Theorem 13.41. *Let S be any semigroup. Then $K(\beta S, \diamond) \cap c\ell\, K(\beta S, \cdot) \neq \emptyset$ and $c\ell\, K(\beta S, \diamond) \cap K(\beta S, \cdot) \neq \emptyset$.*

Proof. We establish the first statement only, the other proof being nearly identical. Let $p \in K(\beta S, \cdot)$ and let $q \in K(\beta S, \diamond)$. Then $p \diamond q \in K(\beta S, \diamond)$. We show that $p \diamond q \in c\ell\, K(\beta S, \cdot)$. Given $s \in S$, the continuous functions ρ_s and r_s agree on S, hence on βS, so $p \diamond s = r_s(p) = \rho_s(p) = p \cdot s$. Thus

$$p \diamond q = \ell_p(q\text{-}\lim_{s \in S} s)$$

$$= q\text{-}\lim_{s \in S} p \diamond s$$

$$= q\text{-}\lim_{s \in S} p \cdot s.$$

For each $s \in S$, $p \cdot s \in K(\beta S, \cdot) \subseteq c\ell\, K(\beta S, \cdot)$, so $p \diamond q \in c\ell\, K(\beta S, \cdot)$. $\qquad \square$

Exercise 13.4.1. Prove Theorem 13.37.

Exercise 13.4.2. Let S be the free semigroup on countably many generators. Prove that there is a subset H of βS such that

(1) H is a subsemigroup of $(\beta S, \diamond)$ and

(2) for every $p, q \in \beta S$, $p \cdot q \notin H$.

Exercise 13.4.3. Let S be any semigroup and assume that $r \in \mathbb{N}$ and $S = \bigcup_{i=1}^{r} A_i$. Prove that there is some $i \in \{1, 2, \ldots, r\}$ such that A_i is both left piecewise syndetic and right piecewise syndetic. (Hint: Use Theorems 4.40 and 13.41.)

Notes

Most of the material in Section 13.1 is based on results from [124], except for Theorem 13.17 which is based on a result of B. Balcar and P. Kalāšek in [13] and on a personal communication from A. Maleki. In [124] it was also proved that the equation $q + p = q \cdot p$ has no solutions in $\mathbb{N}^*$. That proof is quite complicated so is not included here.

Theorem 13.18, Corollary 13.20, and Theorem 13.21 are from [138] and Corollary 13.19 is from [117]. Theorem 13.26 and Corollary 13.27 are due to E. van Douwen in [80].

Most of the material in Section 13.3 is from [143], results obtained in collaboration with I. Leader.

Theorem 13.38 is from [87], a result of collaboration with A. El-Mabhouh and J. Pym. Theorems 13.40 and 13.41 are due to P. Anthony in [2].

Part III

Combinatorial Applications

Chapter 14

The Central Sets Theorem

In this chapter we derive the powerful Central Sets Theorem and some of its consequences. Other consequences will be obtained in later chapters. (Recall that a subset A of a semigroup S is *central* if and only if A is a member of some minimal idempotent in βS.) The notion of "central" originated in topological dynamics, and its connection with concepts in this field will be discussed in Chapter 19. We shall introduce the ideas of the proof gently by proving progressively stronger theorems.

14.1 Van der Waerden's Theorem

Our first introduction to the idea of the proof of the Central Sets Theorem is via van der Waerden's Theorem.

Recall from Theorem 4.40 that given a subset A of a discrete semigroup S, $\overline{A} \cap K(\beta S) \neq \emptyset$ if and only if A is piecewise syndetic. Thus, in particular, any central set is piecewise syndetic.

Theorem 14.1. *Let A be a piecewise syndetic set in the semigroup $(\mathbb{N}, +)$ and let $\ell \in \mathbb{N}$. There exist $a, d \in \mathbb{N}$ such that $\{a, a + d, a + 2d, \ldots, a + \ell d\} \subseteq A$.*

Proof. Let $Y = \times_{t=0}^{\ell} \beta \mathbb{N}$. Then by Theorem 2.22, $(Y, +)$ is a compact right topological semigroup and if $\vec{x} \in \times_{t=0}^{\ell} \mathbb{N}$, then $\lambda_{\vec{x}}$ is continuous. Consequently Y is a semigroup compactification of $\times_{t=0}^{\ell} \mathbb{N}$.

Let

$$E^{\triangleleft} = \big\{(a, a + d, a + 2d, \ldots, a + \ell d) : a \in \mathbb{N} \text{ and } d \in \mathbb{N} \cup \{0\}\big\}$$

and let

$$I^{\triangleleft} = \{(a, a + d, a + 2d, \ldots, a + \ell d) : a, d \in \mathbb{N}\}.$$

Let $E = c\ell_Y E^{\triangleleft}$ and let $I = c\ell_Y I^{\triangleleft}$.

Observe that $E^{\triangleleft}$ is a subsemigroup of $\times_{t=0}^{\ell} \mathbb{N}$ and that $I^{\triangleleft}$ is an ideal of $E^{\triangleleft}$. Thus by Theorem 4.17 E is a subsemigroup of Y and I is an ideal of E.

By Theorem 2.23, $K(Y) = \times_{t=0}^{\ell} K(\beta\mathbb{N})$. We now claim that $E \cap K(Y) \neq \emptyset$ so that by Theorem 1.65, $K(E) = E \cap K(Y)$. In fact we show that

(*) if $p \in K(\beta\mathbb{N})$ and $\overline{p} = (p, p, \ldots, p)$, then $\overline{p} \in E$.

To establish (*), let U be a neighborhood of $\overline{p}$ and for $t \in \{0, 1, \ldots, \ell\}$ pick $A_t \in p$ such that $\times_{t=0}^{\ell} \overline{A_t} \subseteq U$. Then $\bigcap_{t=0}^{\ell} A_t \in p$ so $\bigcap_{t=0}^{\ell} A_t \neq \emptyset$ so pick $a \in \bigcap_{t=0}^{\ell} A_t$. Then $(a, a, \ldots, a) \in U \cap E$.

Since $K(E) = E \cap K(Y) = E \cap \times_{t=0}^{\ell} K(\beta\mathbb{N})$ and I is an ideal of E, we have $E \cap \times_{t=0}^{\ell} K(\beta\mathbb{N}) \subseteq I$. Now since A is piecewise syndetic, pick by Theorem 4.40 some $p \in \overline{A} \cap K(\beta\mathbb{N})$ and let $\overline{p} = (p, p, \ldots, p)$. Then by (*) $\overline{p} \in E \cap \times_{t=0}^{\ell} K(\beta\mathbb{N}) \subseteq I$. Let $U = \times_{t=0}^{\ell} \overline{A}$. Then U is a neighborhood of $\overline{p}$ so $U \cap I^{\triangleleft} \neq \emptyset$ so pick $a, d \in \mathbb{N}$ such that $(a, a + d, a + 2d, \ldots, a + \ell d) \in U$. □

The above proof already illustrates the most startling fact about the Central Sets Theorem proof, namely how much one gets for how little. That is, one establishes the statement (*) based on the trivial fact that $E^{\triangleleft}$ contains the diagonal of $\times_{t=0}^{\ell} \mathbb{N}$. Then one concludes immediately that $\overline{p} \in I$ and hence all members of p contain interesting configurations. It is enough to make someone raised on the work ethic feel guilty.

Note that if $r \in \mathbb{N}$ and $\mathbb{N} = \bigcup_{i=1}^{r} A_i$, then some A_i is piecewise syndetic. This can be seen in at least two ways. One can verify combinatorially that the union of two sets which are not piecewise syndetic is not piecewise syndetic. Somewhat easier from our point of view is to note that $\beta\mathbb{N} = \bigcup_{i=1}^{r} \overline{A_i}$ so some $\overline{A_i} \cap K(\beta\mathbb{N}) \neq \emptyset$ so Theorem 4.40 applies.

Corollary 14.2 (van der Waerden [238]). *Let $r \in \mathbb{N}$ and let $\mathbb{N} = \bigcup_{i=1}^{r} A_i$. For each $\ell \in \mathbb{N}$ there exist $i \in \{1, 2, \ldots, r\}$ and $a, d \in \mathbb{N}$ such that $\{a, a + d, \ldots, a + \ell d\} \subseteq A_i$.*

Proof. Some A_i is piecewise syndetic so Theorem 14.1 applies.   □

In Exercise 5.1.1 the reader was asked to show that the above version of van der Waerden's Theorem implies the apparently stronger version wherein one i is chosen which "works" for all ℓ. In fact the current method of proof yields the stronger version directly.

Corollary 14.3 (van der Waerden [238]). *Let $r \in \mathbb{N}$ and let $\mathbb{N} = \bigcup_{i=1}^{r} A_i$. There exists $i \in \{1, 2, \ldots, r\}$ such that for each $\ell \in \mathbb{N}$ there exists $a, d \in \mathbb{N}$ such that $\{a, a + d, \ldots, a + \ell d\} \subseteq A_i$.*

Proof. Some A_i is piecewise syndetic so Theorem 14.1 applies.   □

One may make the aesthetic objection to the proof of Theorem 14.1 that one requires a different space Y for each length of arithmetic progression. In fact, one may do the proof once for all values of ℓ by starting with an infinite product. (See Exercise 14.1.2.) We shall use such an approach in our proof of the Central Sets Theorem.

As a consequence of van der Waerden's Theorem, we establish the existence of a combinatorially interesting ideal of $(\beta\mathbb{N}, +)$ and of $(\beta\mathbb{N}, \cdot)$, namely the set of ultrafilters every member of which contains arbitrarily long arithmetic progressions.

Definition 14.4. $\mathcal{AP} = \{p \in \beta\mathbb{N} :$ for every $A \in p$ and every $\ell \in \mathbb{N}$, there exist $a, d \in \mathbb{N}$ such that $\{a, a + d, a + 2d, \ldots, a + \ell d\} \subseteq A\}$.

Theorem 14.5. $\mathcal{AP}$ is an ideal of $(\beta\mathbb{N}, +)$ and of $(\beta\mathbb{N}, \cdot)$.

Proof. By Theorems 4.40 and 14.1 $K(\beta\mathbb{N}, +) \subseteq \mathcal{AP}$ so $\mathcal{AP} \neq \emptyset$. To complete the proof, let $p \in \mathcal{AP}$ and let $q \in \beta\mathbb{N}$. Given $A \in q + p$ and $\ell \in \mathbb{N}$, pick $x \in \mathbb{N}$ such that $-x + A \in p$. Pick $a, d \in \mathbb{N}$ such that $\{a, a + d, a + 2d, \ldots, a + \ell d\} \subseteq -x + A$. Then $\{x + a, x + a + d, x + a + 2d, \ldots, x + a + \ell d\} \subseteq A$. Given $A \in p + q$ and $\ell \in \mathbb{N}$, pick $a, d \in \mathbb{N}$ such that $\{a, a + d, a + 2d, \ldots, a + \ell d\} \subseteq \{x \in \mathbb{N} : -x + A \in q\}$. Then $\bigcap_{t=0}^{\ell}(-(a + td) + A) \in q$ so pick $x \in \bigcap_{t=0}^{\ell}(-(a + td) + A)$. Then $\{x + a, x + a + d, x + a + 2d, \ldots, x + a + \ell d\} \subseteq A$.

Given $A \in qp$ and $\ell \in \mathbb{N}$, pick $x \in \mathbb{N}$ such that $x^{-1}A \in p$. Pick $a, d \in \mathbb{N}$ such that $\{a, a+d, a+2d, \ldots, a+\ell d\} \subseteq x^{-1}A$. Then $\{xa, xa+xd, xa+2xd, \ldots, xa+\ell xd\} \subseteq A$. Given $A \in pq$ and $\ell \in \mathbb{N}$, pick $a, d \in \mathbb{N}$ such that $\{a, a + d, a + 2d, \ldots, a + \ell d\} \subseteq \{x \in \mathbb{N} : x^{-1}A \in q\}$. Then $\bigcap_{t=0}^{\ell}(a + td)^{-1}A \in q$ so pick $x \in \bigcap_{t=0}^{\ell}(a + td)^{-1}A$. Then $\{xa, xa + xd, xa + 2xd, \ldots, xa + \ell xd\} \subseteq A$. $\square$

Exercise 14.1.1. Using compactness (see Section 5.5) and Corollary 14.2 prove that given any $k, \ell \in \mathbb{N}$ there exists $n \in \mathbb{N}$ such that whenever $\{1, 2, \ldots, n\}$ is r-colored, there must be a length ℓ monochrome arithmetic progression.

Exercise 14.1.2. Prove Theorem 14.1 as follows. Let $Y = \times_{t=0}^{\infty}\beta\mathbb{N}$. Let $E^{\triangleleft} = \{(a, a+d, a+2d, \ldots) : a \in \mathbb{N}$ and $d \in \mathbb{N}\cup\{0\}\}$ and let $I^{\triangleleft} = \{(a, a+d, a+2d, \ldots) : a, d \in \mathbb{N}\}$. Let $E = c\ell_Y E^{\triangleleft}$ and let $I = c\ell_Y I^{\triangleleft}$. Prove that E is a subsemigroup of Y and that I is an ideal of E. Then prove

(∗) if $p \in K(\beta\mathbb{N})$ and $\overline{p} = (p, p, p, \ldots)$, then $\overline{p} \in E$.

Pick $p \in K(\beta\mathbb{N}) \cap \overline{A}$, let $\overline{p} = (p, p, p, \ldots)$, and show that $\overline{p} \in I$. Given $\ell \in \mathbb{N}$ let $U = \{\vec{q} \in Y :$ for all $i \in \{0, 1, \ldots, \ell\}, q_i \in \overline{A}\}$ and show that $U \cap I^{\triangleleft} \neq \emptyset$.

14.2 The Hales–Jewett Theorem

In this section we prove the powerful generalization of van der Waerden's theorem which is due to A. Hales and R. Jewett [113]. In this case it is worth noting that even though we work with a very noncommutative semigroup, namely the free semigroup on a finite alphabet, the proof is nearly identical to the proof of van der Waerden's Theorem. (There is a much more significant contrast between the commutative and noncommutative versions of the Central Sets Theorem.)

Definition 14.6. Let S be the free semigroup on the alphabet A and let v be a variable which is not a member of A.

(a) A *variable word* $w(v)$ is a member of the free semigroup on the alphabet $A \cup \{v\}$ in which v occurs.

(b) $S(v) = \{w(v) : w(v)$ is a variable word$\}$.

(c) Given a variable word $w(v)$ and a member a of A, $w(a)$ is the word in which each occurrence of v is replaced by an occurrence of a.

Recall that the free semigroup on the alphabet A was formally defined (Definition 1.3) as the set of functions with domain $\{0, 1, \ldots, n\}$ for some $n \in \omega$ and range contained in A. Given a member w of the free semigroup on an alphabet A and a "letter" $a \in A$, when we say that a *occurs* in w we mean formally that a is in the range of (the function) w. Given $w(v) \in S(v)$, and $a \in A$, one then has formally that $w(a)$ is the function with the same domain as $w(v)$ such that for i in this domain,

$$w(a)_i = \begin{cases} w(v)_i & \text{if } w(v)_i \neq v \\ a & \text{if } w(v)_i = v. \end{cases}$$

Informally, if $A = \{1, 2, 3, 4, 5\}$ and $w(v) = 13vv3251v4$, then $w(3) = 1333325134$ and $w(2) = 1322325124$.

Theorem 14.7. *Let A be a finite nonempty alphabet, let S be the free semigroup over A, and let B be a piecewise syndetic subset of S. Then there is a variable word $w(v)$ such that $\{w(a) : a \in A\} \subseteq B$.*

Proof. Let $\ell = |A|$ and write $A = \{a_1, a_2, \ldots, a_\ell\}$. Let $Y = \bigtimes_{t=1}^{\ell} \beta S$. Denote the operation of Y (as well as that of S) by juxtaposition. Then by Theorem 2.22, Y is a compact right topological semigroup and if $\vec{x} \in \bigtimes_{t=1}^{\ell} S$, then $\lambda_{\vec{x}}$ is continuous. Thus Y is a semigroup compactification of $\bigtimes_{t=1}^{\ell} S$.

Let $I^{\triangleleft} = \{(w(a_1), w(a_2), \ldots, w(a_\ell)) : w(v) \in S(v)\}$ and let

$$E^{\triangleleft} = I^{\triangleleft} \cup \{(w, w, \ldots, w) : w \in S\}.$$

Let $E = c\ell_Y E^{\triangleleft}$ and let $I = c\ell_Y I^{\triangleleft}$.

Observe that $E^{\triangleleft}$ is a subsemigroup of $\bigtimes_{t=1}^{\ell} S$ and that $I^{\triangleleft}$ is an ideal of $E^{\triangleleft}$. Thus by Theorem 4.17 E is a subsemigroup of Y and I is an ideal of E.

By Theorem 2.23 we have $K(Y) = \bigtimes_{t=1}^{\ell} K(\beta S)$. We now claim that $E \cap K(Y) \neq \emptyset$ so that by Theorem 1.65, $K(E) = E \cap K(Y)$. In fact we show that

(∗) if $p \in K(\beta S)$ and $\overline{p} = (p, p, \ldots, p)$, then $\overline{p} \in E$.

To establish (∗), let U be a neighborhood of $\overline{p}$ and for $t \in \{1, 2, \ldots, \ell\}$ pick $B_t \in p$ such that $\bigtimes_{t=1}^{\ell} \overline{B_t} \subseteq U$. Then $\bigcap_{t=1}^{\ell} B_t \in p$ so $\bigcap_{t=1}^{\ell} B_t \neq \emptyset$ so pick $w \in \bigcap_{t=1}^{\ell} B_t$. Then $(w, w, \ldots, w) \in U \cap E$.

Since $K(E) = E \cap K(Y) = E \cap \bigtimes_{t=1}^{\ell} K(\beta S)$ and I is an ideal of E, we have $E \cap \bigtimes_{t=1}^{\ell} K(\beta S) \subseteq I$. Now since B is piecewise syndetic, pick by Theorem 4.40 some

$p \in \overline{B} \cap K(\beta S)$ and let $\overline{p} = (p, p, \ldots, p)$. Then by $(*)$ $\overline{p} \in E \cap \times_{t=1}^{\ell} K(\beta S) \subseteq I$. Let $U = \times_{t=1}^{\ell} \overline{B}$. Then U is a neighborhood of $\overline{p}$ so $U \cap I^{\lhd} \neq \emptyset$ so pick $w(v) \in S(v)$ such that $(w(a_1), w(a_2), \ldots, w(a_\ell)) \in U$.   □

Corollary 14.8 (Hales–Jewett [113]). *Let A be a finite nonempty alphabet, let S be the free semigroup over A, let $r \in \mathbb{N}$ and let $S = \bigcup_{i=1}^{r} B_i$. Then there exist $i \in \{1, 2, \ldots, r\}$ and a variable word $w(v)$ such that $\{w(a) : a \in A\} \subseteq B_i$.*

Proof. Some B_i is piecewise syndetic so Theorem 14.7 applies.   □

Exercise 14.2.1. Using compactness (see Section 5.5) and Corollary 14.8 prove that given any finite nonempty alphabet A and any $r \in \mathbb{N}$ there is some $n \in \mathbb{N}$ such that whenever the length n words over A are r-colored there must be a variable word $w(v)$ such that $\{w(a) : a \in A\}$ is monochrome.

Exercise 14.2.2. Derive van der Waerden's Theorem (Corollary 14.2) from the Hales–Jewett Theorem (Corollary 14.8). (Hint: Consider for example the alphabet $\{0, 1, 2, 3, 4\}$ and the following length 5 arithmetic progression written in base 5: 20100010134, 20111011134, 20122012134, 20133013134, 20144014134.)

14.3 The Commutative Central Sets Theorem

In this section we derive the Central Sets Theorem for commutative semigroups, as well as some of its immediate corollaries. (One is Corollary 14.13 and two more are exercises.)

We first encode some of the important ideas that underlie the proof of the Central Sets Theorem, some of which we have illustrated in the proofs of van der Waerden's Theorem and the Hales–Jewett Theorem. The lemma is more general than needed here because we shall use it again in later chapters. (We remind the reader that we have been using the notation $\overline{x}$ for a function constantly equal to x.)

Lemma 14.9. *Let J be a set, let $(D, \leq)$ be a directed set, and let $(S, \cdot)$ be a semigroup. Let $\langle T_i \rangle_{i \in D}$ be a decreasing family of nonempty subsets of S such that for each $i \in D$ and each $x \in T_i$ there is some $j \in D$ such that $x \cdot T_j \subseteq T_i$. Let $Q = \bigcap_{i \in D} c\ell_{\beta S} T_i$. Then Q is a compact subsemigroup of βS. Let $\langle E_i \rangle_{i \in D}$ and $\langle I_i \rangle_{i \in D}$ be decreasing families of nonempty subsets of $\times_{t \in J} S$ with the following properties:*

(a) *for each $i \in D$, $I_i \subseteq E_i \subseteq \times_{t \in J} T_i$,*
(b) *for each $i \in D$ and each $\vec{x} \in I_i$ there exists $j \in D$ such that $\vec{x} \cdot E_j \subseteq I_i$, and*
(c) *for each $i \in D$ and each $\vec{x} \in E_i \setminus I_i$ there exists $j \in D$ such that $\vec{x} \cdot E_j \subseteq E_i$ and $\vec{x} \cdot I_j \subseteq I_i$.*

Let $Y = \times_{t \in J} \beta S$, let $E = \bigcap_{i \in D} c\ell_Y E_i$, and let $I = \bigcap_{i \in D} c\ell_Y I_i$. Then E is a subsemigroup of $\times_{t \in J} Q$ and I is an ideal of E. If, in addition, either

(d) *for each $i \in D$, $T_i = S$ and $\{a \in S : \overline{a} \notin E_i\}$ is not piecewise syndetic, or*

(e) *for each $i \in D$ and each $a \in T_i$, $\overline{a} \in E_i$,*

then given any $p \in K(Q)$, one has $\overline{p} \in E \cap K(\times_{t \in J} Q) = K(E) \subseteq I$.

Proof. It follows from Theorem 4.20 that Q is a subsemigroup of βS.

By condition (a) and the fact that $\langle I_i \rangle_{i \in D}$ is decreasing we have $\emptyset \neq I \subseteq E \subseteq \times_{t \in J} Q$. To complete the proof that E is a subsemigroup of $\times_{t \in J} Q$ and I is an ideal of E, we let $\vec{p}, \vec{q} \in E$ and show that $\vec{p} \cdot \vec{q} \in E$ and if either $\vec{p} \in I$ or $\vec{q} \in I$, then $\vec{p} \cdot \vec{q} \in I$. To this end, let U be an open neighborhood of $\vec{p} \cdot \vec{q}$ and let $i \in D$ be given. We show that $U \cap E_i \neq \emptyset$ and if $\vec{p} \in I$ or $\vec{q} \in I$, then $U \cap I_i \neq \emptyset$. Pick a neighborhood V of $\vec{p}$ such that $V \cdot \vec{q} \subseteq U$ and pick $\vec{x} \in E_i \cap V$ with $\vec{x} \in I_i$ if $\vec{p} \in I$. If $\vec{x} \in I_i$ pick $j \in D$ such that $\vec{x} \cdot E_j \subseteq I_i$. If $\vec{x} \in E_i \setminus I_i$, pick $j \in D$ such that $\vec{x} \cdot E_j \subseteq E_i$ and $\vec{x} \cdot I_j \subseteq I_i$. Now $\vec{x} \cdot \vec{q} \in U$ so pick a neighborhood W of $\vec{q}$ such that $\vec{x} \cdot W \subseteq U$ and pick $\vec{y} \in W \cap E_j$ with $\vec{y} \in I_j$ if $\vec{q} \in I$. Then $\vec{x} \cdot \vec{y} \in U \cap E_i$ and if either $\vec{p} \in I$ or $\vec{q} \in I$, then $\vec{x} \cdot \vec{y} \in U \cap I_i$.

To complete the proof, assume that (d) or (e) holds. It suffices to establish

(∗) if $p \in K(Q)$, then $\overline{p} \in E$.

Indeed, assume we have established (∗). Then $\overline{p} \in E \cap \times_{t \in J} K(Q)$ and $\times_{t \in J} K(Q) = K(\times_{t \in J} Q)$ by Theorem 2.23. Then by Theorem 1.65, $K(E) = E \cap K(\times_{t \in J} Q)$ and, since I is an ideal of E, $K(E) \subseteq I$.

To establish (∗), let $p \in K(Q)$ be given. To see that $\overline{p} \in E$, let $i \in D$ be given and let U be a neighborhood of $\overline{p}$. Pick $F \in \mathcal{P}_f(J)$ and for each $t \in F$ pick some $A_t \in p$ such that $\bigcap_{t \in F} \pi_t^{-1}[c\ell_{\beta S} A_t] \subseteq U$.

Assume now that (d) holds. Since $p \in K(\beta S)$ and $\{a \in S : \overline{a} \notin E_i\}$ is not piecewise syndetic, we have by Theorem 4.40 that $\{a \in S : \overline{a} \notin E_i\} \notin p$ and hence $\{a \in S : \overline{a} \in E_i\} \in p$. Pick $a \in (\bigcap_{t \in F} A_t) \cap \{a \in S : \overline{a} \in E_i\}$. Then $\overline{a} \in U \cap E_i$.

Finally assume that (e) holds. Since $p \in c\ell \, T_i$, pick $a \in (\bigcap_{t \in F} A_t) \cap T_i$. Then $\overline{a} \in U \cap E_i$.   □

Definition 14.10. Φ is the set of all functions $f : \mathbb{N} \to \mathbb{N}$ for which $f(n) \leq n$ for all $n \in \mathbb{N}$.

We are now ready to prove the Central Sets Theorem (for commutative semigroups).

Theorem 14.11 (Central Sets Theorem). *Let S be a commutative semigroup, let A be a central subset of S, and for each $\ell \in \mathbb{N}$, let $\langle y_{\ell,n} \rangle_{n=1}^{\infty}$ be a sequence in S. There exist a sequence $\langle a_n \rangle_{n=1}^{\infty}$ in S and a sequence $\langle H_n \rangle_{n=1}^{\infty}$ in $\mathcal{P}_f(\mathbb{N})$ such that $\max H_n < \min H_{n+1}$ for each $n \in \mathbb{N}$ and such that for each $f \in \Phi$, $\mathrm{FP}(\langle a_n \cdot \prod_{t \in H_n} y_{f(n),t} \rangle_{n=1}^{\infty}) \subseteq A$.*

Proof. Let $Y = \times_{\ell=1}^{\infty} \beta S$. By Theorem 2.22, Y is a compact right topological semigroup and $\times_{\ell=1}^{\infty} S \subseteq \Lambda(Y)$. Given $a \in S$ and $H \in \mathcal{P}_f(\mathbb{N})$ let

$$\vec{x}(a, H) = (a \cdot \textstyle\prod_{t \in H} y_{1,t} \, , \, a \cdot \prod_{t \in H} y_{2,t} \, , \, a \cdot \prod_{t \in H} y_{3,t} \, , \, \dots)$$

(so that $\vec{x}(a, H) \in \times_{\ell=1}^{\infty} S \subseteq Y$).

For each $i \in \mathbb{N}$, let $T_i = S$, let $I_i = \{\vec{x}(a, H) : a \in S, H \in \mathcal{P}_f(\mathbb{N})$, and min $H \geq i\}$, and let $E_i = I_i \cup \{\overline{a} : a \in S\}$. We claim that T_i, E_i, and I_i satisfy hypotheses (a), (b), (c), and (d) of Lemma 14.9. Indeed, (a) and (d) hold trivially. To verify (b), let $i \in \mathbb{N}$ and $\vec{x} \in I_i$ be given. Pick $a \in S$ and $H \in \mathcal{P}_f(\mathbb{N})$ such that min $H \geq i$ and $\vec{x} = \vec{x}(a, H)$. Let $j = \max H + 1$. Then, since S is commutative, one sees easily that $\vec{x} \cdot E_j \subseteq I_i$. To verify (c), let $i \in \mathbb{N}$ and $\vec{x} \in E_i \backslash I_i$ be given. Pick $a \in S$ such that $\vec{x} = \overline{a}$. Then $\vec{x} \cdot E_i \subseteq E_i$ and $\vec{x} \cdot I_i \subseteq I_i$.

Since A is central in S, pick an idempotent $p \in K(\beta S)$ such that $A \in p$. By Lemma 14.9 $\overline{p} \in I = \bigcap_{i \in \mathbb{N}} c\ell_Y I_i$.

We now construct the sequences $\langle a_n \rangle_{n=1}^{\infty}$ and $\langle H_n \rangle_{n=1}^{\infty}$ inductively. The construction is reminiscent of the proof of Theorem 5.8.

Let $A^{\star} = A^{\star}(p) = \{x \in A : x^{-1}A \in p\}$ and let $U = \{\vec{q} \in Y : q_1 \in \overline{A^{\star}}\}$. By Theorem 4.12 $A^{\star} \in p$ so U is a neighborhood of $\overline{p}$. Since also $\overline{p} \in I$, pick $\vec{z} \in U \cap I_1$. Pick $a_1 \in S$ and $H_1 \in \mathcal{P}_f(\mathbb{N})$ such that $\vec{z} = \vec{x}(a_1, H_1)$. Since $\vec{x}(a_1, H_1) \in U$ we have that $a_1 \cdot \prod_{t \in H_1} y_{1,t} \in A^{\star}$.

Inductively, let $n > 1$ and assume that we have chosen $\langle a_m \rangle_{m=1}^{n-1}$ and $\langle H_m \rangle_{m=1}^{n-1}$ so that, if $m \in \{1, 2, \ldots, n - 2\}$, then $\max H_m < \min H_{m+1}$, and so that for any $f \in \Phi$, $\mathrm{FP}(\langle a_m \cdot \prod_{t \in H_m} y_{f(m),t} \rangle_{m=1}^{n-1}) \subseteq A^{\star}$. Let $k = \max H_{n-1} + 1$. Let

$$L = \bigcup_{f \in \Phi} \mathrm{FP}(\langle a_m \cdot \prod_{t \in H_m} y_{f(m),t} \rangle_{m=1}^{n-1}).$$

By Lemma 4.14 we have for each $b \in L$ that $b^{-1}A^{\star} \in p$. Let $B = A^{\star} \cap \bigcap_{b \in L} b^{-1}A^{\star}$. Since L is finite (although Φ isn't) we have that $B \in p$. Let $U = \{\vec{q} \in Y :$ for all $\ell \in \{1, 2, \ldots, n\}, q_\ell \in \overline{B}\}$. Then U is a neighborhood of $\overline{p}$ so pick $\vec{z} \in U \cap I_k$. Pick $a_n \in S$ and $H_n \in \mathcal{P}_f(\mathbb{N})$ with min $H_n \geq k$ such that $\vec{z} = \vec{x}(a_n, H_n)$.

To complete the proof we need to show that for any $f \in \Phi$,

$$\mathrm{FP}(\langle a_m \cdot \prod_{t \in H_m} y_{f(m),t} \rangle_{m=1}^{n}) \subseteq A^{\star}.$$

So let $f \in \Phi$ and let $\emptyset \neq F \subseteq \{1, 2, \ldots, n\}$. If $n \notin F$, then

$$\prod_{m \in F} (a_m \cdot \prod_{t \in H_m} y_{f(m),t}) \in \mathrm{FP}(\langle a_m \cdot \prod_{t \in H_m} y_{f(m),t} \rangle_{m=1}^{n-1}) \subseteq A^{\star},$$

so assume that $n \in F$. Now $a_n \cdot \prod_{t \in H_n} y_{f(n),t} \in B$ since $\vec{x}(a_n, H_n) = \vec{z} \in U$. If $F = \{n\}$, then we have $\prod_{m \in F}(a_m \cdot \prod_{t \in H_m} y_{f(m),t}) = a_n \cdot \prod_{t \in H_n} y_{f(n),t} \in B \subseteq A^{\star}$. Otherwise, let $G = F \backslash \{n\}$ and let $b = \prod_{m \in G}(a_m \cdot \prod_{t \in H_m} y_{f(m),t})$. Then $b \in L$ so $a_n \cdot \prod_{t \in H_n} y_{f(n),t} \in B \subseteq b^{-1}A^{\star}$ so $\prod_{m \in F}(a_m \cdot \prod_{t \in H_m} y_{f(m),t}) = b \cdot a_n \cdot \prod_{t \in H_n} y_{f(n),t} \in A^{\star}$. $\square$

Of course we have the corresponding partition result.

Corollary 14.12. *Let S be a commutative semigroup and for each $\ell \in \mathbb{N}$, let $\langle y_{\ell,n} \rangle_{n=1}^{\infty}$ be a sequence in S. Let $r \in \mathbb{N}$ and let $S = \bigcup_{i=1}^{r} A_i$. There exist $i \in \{1, 2, \ldots, r\}$, a sequence $\langle a_n \rangle_{n=1}^{\infty}$ in S and a sequence $\langle H_n \rangle_{n=1}^{\infty}$ in $\mathcal{P}_f(\mathbb{N})$ such that $\max H_n < \min H_{n+1}$ for each $n \in \mathbb{N}$ and for each $f \in \Phi$, $\mathrm{FP}(\langle a_n \cdot \prod_{t \in H_n} y_{f(n),t} \rangle_{n=1}^{\infty}) \subseteq A_i$.*

Proof. Some A_i is central.   □

A special case of the Central Sets Theorem is adequate for many applications. See Exercise 14.3.2.

It is not surprising, since the proof uses an idempotent, that one gets the Finite Products Theorem (Corollary 5.9) as a corollary to the Central Sets Theorem (in the case S is commutative). For a still simple, but more interesting, corollary consider the following extension of van der Waerden's Theorem, in which the increment can be chosen from the finite sums of any prespecified sequence.

Corollary 14.13. *Let $\langle x_n \rangle_{n=1}^{\infty}$ be a sequence in $\mathbb{N}$, let $r \in \mathbb{N}$, and let $\mathbb{N} = \bigcup_{i=1}^{r} A_i$. Then there is some $i \in \{1, 2, \ldots, r\}$ such that for all $\ell \in \mathbb{N}$ there exist $a \in \mathbb{N}$ and $d \in \mathrm{FS}(\langle x_n \rangle_{n=1}^{\infty})$ with $\{a, a+d, a+2d, \ldots, a+\ell d\} \subseteq A_i$.*

Proof. Pick $i \in \{1, 2, \ldots, r\}$ such that A_i is central in $(\omega, +)$. For each $k, n \in \mathbb{N}$, let $y_{k,n} = (k-1) \cdot x_n$. Pick sequences $\langle a_n \rangle_{n=1}^{\infty}$ and $\langle H_n \rangle_{n=1}^{\infty}$ as guaranteed by Theorem 14.11. Given $\ell \in \mathbb{N}$, pick any $m > \ell$ and let $a = a_m$ and let $d = \Sigma_{t \in H_m} x_t$. Now, given $k \in \{0, 1, \ldots, \ell\}$, pick any $f \in \Phi$ such that $f(m) = k+1$. Then

$$a + kd = a_m + \Sigma_{t \in H_m} y_{f(m),t} \in \mathrm{FS}(\langle a_n \cdot \Sigma_{t \in H_n} y_{f(n),t} \rangle_{n=1}^{\infty}) \subseteq A_i.$$
                                                                              □

Exercise 14.3.1. Let $\langle x_n \rangle_{n=1}^{\infty}$ be a sequence in $\mathbb{N}$, let $r \in \mathbb{N}$, and let $\mathbb{N} = \bigcup_{i=1}^{r} A_i$. Prove that there is some $i \in \{1, 2, \ldots, r\}$ such that for all $\ell \in \mathbb{N}$ there exist $a \in \mathbb{N}$ and $d \in \mathrm{FS}(\langle x_n \rangle_{n=1}^{\infty})$ and $c \in \mathrm{FS}(\langle x_n{}^2 \rangle_{n=1}^{\infty})$ with $\{a, a+d, a+2d, \ldots, a+\ell d\} \cup \{a+c, a+2c, \ldots, a+\ell c\} \subseteq A_i$.

Exercise 14.3.2. Let S be a commutative semigroup, let A be central in S, let $k \in \mathbb{N}$ and for each $\ell \in \{1, 2, \ldots, k\}$, let $\langle y_{\ell,n} \rangle_{n=1}^{\infty}$ be a sequence in S. Prove (as a corollary to Theorem 14.11) that there exist a sequence $\langle a_n \rangle_{n=1}^{\infty}$ and for each $\ell \in \{1, 2, \ldots, k\}$ a product subsystem $\langle z_{\ell,n} \rangle_{n=1}^{\infty}$ of $\langle y_{\ell,n} \rangle_{n=1}^{\infty}$ such that for each $\ell \in \{1, 2, \ldots, k\}$, $\mathrm{FP}(\langle a_n \cdot z_{\ell,n} \rangle_{n=1}^{\infty}) \subseteq A$. Show in fact that the product subsystems can be chosen in a uniform fashion. That is, there is a sequence $\langle H_n \rangle_{n=1}^{\infty}$ in $\mathcal{P}_f(\mathbb{N})$ with $\max H_n < \min H_{n+1}$ for each n such that for each $\ell \in \{1, 2, \ldots, k\}$ and each $n \in \mathbb{N}$, $z_{\ell,n} = \prod_{t \in H_n} y_{\ell,t}$.

14.4 The Noncommutative Central Sets Theorem

In this section we generalize the Central Sets Theorem (Theorem 14.11) to arbitrary semigroups. The statement of the Central Sets Theorem in this generality is considerably more complicated, because the "a" in the conclusion must be split up into many parts. (In the event that the semigroup S is commutative, the conclusion of Theorem 14.15 reduces to that of Theorem 14.11.)

Definition 14.14. Let $m \in \mathbb{N}$. Then

$$\mathcal{I}_m = \{(H_1, H_2, \ldots, H_m) \in \mathcal{P}_f(\mathbb{N})^m : \text{if } m > 1 \text{ and}$$
$$t \in \{1, 2, \ldots, m-1\}, \text{ then } \max H_t < \min H_{t+1}\}.$$

Theorem 14.15. *Let S be a semigroup, let A be a central subset of S, and for each $\ell \in \mathbb{N}$, let $\langle y_{\ell,n} \rangle_{n=1}^{\infty}$ be a sequence in S. Given $\ell, m \in \mathbb{N}$, $\vec{a} \in S^{m+1}$, and $\vec{H} \in \mathcal{I}_m$, let*

$$w(\vec{a}, \vec{H}, \ell) = \left(\prod_{i=1}^{m} (a_i \cdot \prod_{t \in H_i} y_{\ell,t}) \right) \cdot a_{m+1}.$$

There exist sequences $\langle m(n) \rangle_{n=1}^{\infty}$, $\langle \vec{a}_n \rangle_{n=1}^{\infty}$, and $\langle \vec{H}_n \rangle_{n=1}^{\infty}$ such that

(1) for each $n \in \mathbb{N}$, $m(n) \in \mathbb{N}$, $\vec{a}_n \in S^{m(n)+1}$, $\vec{H}_n \in \mathcal{I}_{m(n)}$, and $\max H_{n,m(n)} < \min H_{n+1,1}$, and

(2) for each $f \in \Phi$, $\mathrm{FP}(\langle w(\vec{a}_n, \vec{H}_n, f(n)) \rangle_{n=1}^{\infty}) \subseteq A$.

Proof. Let $Y = \times_{\ell=1}^{\infty} \beta S$. Given $m \in \mathbb{N}$, $\vec{a} \in S^{m+1}$, and $\vec{H} \in \mathcal{I}_m$ let

$$\vec{z}(\vec{a}, \vec{H}) = (w(\vec{a}, \vec{H}, 1), w(\vec{a}, \vec{H}, 2), w(\vec{a}, \vec{H}, 3), \ldots).$$

(Observe that if S is commutative, $a = \prod_{i=1}^{m+1} a_i$, $H = \bigcup_{i=1}^{m} H_i$, and $\vec{x}(a, H)$ is as defined in the proof of Theorem 14.11, then $\vec{x}(a, H) = \vec{z}(\vec{a}, \vec{H})$.)

For each $i \in \mathbb{N}$, let $T_i = S$, let $I_i = \{\vec{z}(\vec{a}, \vec{H}) : \text{there exists } m \in \mathbb{N} \text{ such that } \vec{a} \in S^{m+1}, \vec{H} \in \mathcal{I}_m, \text{ and } \min H_1 \geq i\}$ and let $E_i = I_i \cup \{\vec{a} : a \in S\}$. We claim that T_i, E_i, and I_i satisfy hypotheses (a), (b), (c), and (d) of Lemma 14.9. Indeed, (a) and (d) hold trivially.

To verify (b), let $i \in \mathbb{N}$ and $\vec{x} \in I_i$ be given. Pick $m \in \mathbb{N}$, $\vec{a} \in S^{m+1}$, and $\vec{H} \in \mathcal{I}_m$ such that $\min H_1 \geq i$ and $\vec{x} = \vec{z}(\vec{a}, \vec{H})$. Let $j = \max H_m + 1$. To see that $\vec{x} \cdot E_j \subseteq I_i$, let $\vec{y} \in E_j$. If $\vec{y} = \vec{s}$ for some $s \in S$, let $\vec{b} = (a_1, a_2, \ldots, a_m, a_{m+1} \cdot s)$. Then $\vec{x} \cdot \vec{y} = \vec{z}(\vec{b}, \vec{H}) \in I_i$. So assume $\vec{y} \in I_j$, pick $k \in \mathbb{N}$, $\vec{b} \in S^{k+1}$, and $\vec{G} \in \mathcal{I}_k$ such that $\min G_1 \geq j$ and $\vec{y} = \vec{z}(\vec{b}, \vec{G})$. Let $\vec{c} = (a_1, a_2, \ldots, a_m, a_{m+1} \cdot b_1, b_2, \ldots, b_{k+1})$ and let $\vec{K} = (H_1, H_2, \ldots, H_m, G_1, G_2, \ldots, G_k)$. Then $\vec{c} \in S^{k+m+1}$, $\vec{K} \in \mathcal{I}_{m+k}$ and $\vec{x} \cdot \vec{y} = \vec{z}(\vec{c}, \vec{K}) \in I_i$.

To verify (c), let $i \in \mathbb{N}$ and $\vec{x} \in E_i \setminus I_i$ be given. Pick $a \in S$ such that $\vec{x} = \vec{a}$. Then as above one easily sees that $\vec{x} \cdot E_i \subseteq E_i$ and $\vec{x} \cdot I_i \subseteq I_i$.

Since A is central in S, pick an idempotent $p \in K(\beta S)$ such that $A \in p$. By Lemma 14.9 $\overline{p} \in I = \bigcap_{i \in \mathbb{N}} c\ell_Y I_i$.

Now we construct the sequences $\langle m(n) \rangle_{n=1}^{\infty}$, $\langle \vec{a}_n \rangle_{n=1}^{\infty}$, and $\langle \vec{H}_n \rangle_{n=1}^{\infty}$ inductively. Write $A^\star = A^\star(p)$. Let $U = \{\vec{q} \in Y : q_1 \in \overline{A^\star}\}$. By Theorem 4.12 $A^\star \in p$ so U is a neighborhood of $\overline{p}$. Since also $\overline{p} \in I$, pick $\vec{y} \in U \cap I_1$. Pick $m(1) \in \mathbb{N}$, $\vec{a}_1 \in S^{m(1)+1}$, and $\vec{H}_1 \in \mathcal{I}_{m(1)}$ such that $\vec{y} = \vec{z}(\vec{a}_1, \vec{H}_1)$. Since $\vec{z}(\vec{a}_1, \vec{H}_1) \in U$ we have that $w(\vec{a}_1, \vec{H}_1, 1) \in A^\star$.

Inductively, let $n > 1$ and assume that we have chosen $\langle m(j) \rangle_{j=1}^{n-1}$, $\langle \vec{a}_j \rangle_{j=1}^{n-1}$, and $\langle \vec{H}_j \rangle_{j=1}^{n-1}$ so that, if $j \in \{1, 2, \ldots, n-2\}$, then $\max H_{j,m(j)} < \min H_{j+1,1}$, and so that for any $f \in \Phi$, $\mathrm{FP}(\langle w(\vec{a}_j, \vec{H}_j, f(j)) \rangle_{j=1}^{n-1}) \subseteq A^\star$.

Let $k = \max H_{n-1,m(n-1)} + 1$. Let $E = \bigcup_{f \in \Phi} \mathrm{FP}(\langle w(\vec{a}_j, \vec{H}_j, f(j))\rangle_{j=1}^{n-1})$. By Lemma 4.14 we have for each $b \in E$ that $b^{-1}A^{\star} \in p$. Let $B = A^{\star} \cap \bigcap_{b \in E} b^{-1}A^{\star}$. Since E is finite we have that $B \in p$. Let $U = \{\vec{q} \in Y : \text{for all } \ell \in \{1, 2, \ldots, n\},$ $q_\ell \in \overline{B}\}$. Then U is a neighborhood of $\overline{p}$ so pick $\vec{y} \in U \cap I_k$. Pick $m(n) \in \mathbb{N}$, $\vec{a}_n \in S^{m(n)+1}$ and $\vec{H}_n \in \mathcal{I}_{m(n)}$ with $\min H_{n,1} \geq k$ such that $\vec{y} = \vec{z}(\vec{a}_n, \vec{H}_n)$.

To complete the proof we need to show that for any $f \in \Phi$,

$$\mathrm{FP}(\langle w(\vec{a}_j, \vec{H}_j, f(j))\rangle_{j=1}^{n}) \subseteq A^{\star}.$$

So let $f \in \Phi$ and let $\emptyset \neq F \subseteq \{1, 2, \ldots, n\}$. If $n \notin F$, then

$$\textstyle\prod_{j \in F} w(\vec{a}_j, \vec{H}_j, f(j)) \in \mathrm{FP}(\langle w(\vec{a}_j, \vec{H}_j, f(j))\rangle_{j=1}^{n-1}) \subseteq A^{\star},$$

so assume that $n \in F$. Now $w(\vec{a}_n, \vec{H}_n, f(n)) \in B$ since $\vec{z}(\vec{a}_n, \vec{H}_n) = \vec{y} \in U$. If $F = \{n\}$, then we have $\prod_{j \in F} w(\vec{a}_j, \vec{H}_j, f(j)) = w(\vec{a}_n, \vec{H}_n, f(n)) \in B \subseteq A^{\star}$. Otherwise, let $G = F \setminus \{n\}$ and let $b = \prod_{j \in G} w(\vec{a}_j, \vec{H}_j, f(j))$. Then $b \in E$ so $w(\vec{a}_n, \vec{H}_n, f(n)) \in B \subseteq b^{-1}A^{\star}$ so $\prod_{j \in F} w(\vec{a}_j, \vec{H}_j, f(j)) = b \cdot w(\vec{a}_n, \vec{H}_n, f(n)) \in A^{\star}$. $\square$

Exercise 14.4.1. Derive the Hales–Jewett Theorem (Theorem 14.7) as a corollary to the noncommutative Central Sets Theorem (Theorem 14.15).

14.5 A Combinatorial Characterization of Central Sets

Recall from Theorem 5.12 that members of idempotents in βS are completely characterized by the fact that they contain $\mathrm{FP}(\langle x_n\rangle_{n=1}^{\infty})$ for some sequence $\langle x_n\rangle_{n=1}^{\infty}$. Accordingly, it is natural to ask whether the Central Sets Theorem characterizes members of minimal idempotents, that is, central sets. We see now that it does not.

Lemma 14.16. *Let $n, m, k \in \mathbb{N}$ and for each $i \in \{1, 2, \ldots, n\}$, let $\langle y_{i,t}\rangle_{t=1}^{\infty}$ be a sequence in $\mathbb{N}$. Then there exists $H \in \mathcal{P}_f(\mathbb{N})$ with $\min H > m$ such that for each $i \in \{1, 2, \ldots, n\}$, $\Sigma_{t \in H} y_{i,t} \in \mathbb{N}2^k$.*

Proof. Choose an infinite set $G_1 \subseteq \mathbb{N}$ such that for all $t, s \in G_1$, $y_{1,t} \equiv y_{1,s} \pmod{2^k}$. Inductively, given $i \in \{1, 2, \ldots, n - 1\}$ and G_i, choose an infinite subset G_{i+1} of G_i such that for all $t, s \in G_{i+1}$, $y_{i+1,t} \equiv y_{i+1,s} \pmod{2^k}$. Then for all $i \in \{1, 2, \ldots, n\}$ and all $t, s \in G_n$ one has $y_{i,t} \equiv y_{i,s} \pmod{2^k}$. Now pick $H \subseteq G_n$ with $\min H > m$ and $|H| = 2^k$. $\square$

Recall from Definition 6.2 that given $n \in \mathbb{N}$, $\mathrm{supp}(n) \in \mathcal{P}_f(\omega)$ is defined by $n = \Sigma_{i \in \mathrm{supp}(n)} 2^i$.

Lemma 14.17. *There is a set $A \subseteq \mathbb{N}$ such that*
  (a) *A is not piecewise syndetic in $(\mathbb{N}, +)$.*

(b) *For all $x \in A$ there exists $n \in \mathbb{N}$ such that $\emptyset \neq A \cap \mathbb{N}2^n \subseteq -x + A$.*

(c) *For each $n \in \mathbb{N}$ and any n sequences $\langle y_{1,t} \rangle_{t=1}^{\infty}, \langle y_{2,t} \rangle_{t=1}^{\infty}, \ldots, \langle y_{n,t} \rangle_{t=1}^{\infty}$ in $\mathbb{N}$ there exist $a \in \mathbb{N}$ and $H \in \mathcal{P}_f(\mathbb{N})$ such that for all $i \in \{1, 2, \ldots, n\}$, $a + \Sigma_{t \in H} \, y_{i,t} \in A \cap \mathbb{N}2^n$.*

Proof. For each $k \in \mathbb{N}$ let $B_k = \{2^k, 2^k + 1, 2^k + 2, \ldots, 2^{k+1} - 1\}$ and let $A = \{n \in \mathbb{N} :$ for each $k \in \mathbb{N}$, $B_k \setminus \mathrm{supp}(n) \neq \emptyset\}$. Then one recognizes that $n \in A$ by looking at the binary expansion of n and noting that there is at least one 0 between positions 2^k and 2^{k+1} for each $k \in \mathbb{N}$.

To show that A is not piecewise syndetic we need to show that for each $g \in \mathbb{N}$ there is some $b \in \mathbb{N}$ such that for any $x \in \mathbb{N}$ there is some $y \in \{x + 1, x + 2, \ldots, x + b\}$ with $\{y + 1, y + 2, \ldots, y + g\} \cap A = \emptyset$. To this end let $g \in \mathbb{N}$ be given and pick $k \in \mathbb{N}$ such that $2^{2^k} > g$. Let $b = 2^{2^{k+1}}$. Let $x \in \mathbb{N}$ be given and pick the least $a \in \mathbb{N}$ such that $a \cdot 2^{2^{k+1}} - 2^{2^k} > x$ and let $y = a \cdot 2^{2^{k+1}} - 2^{2^k}$. Then $x < y \leq x + b$ and for each $t \in \{1, 2, \ldots, 2^{2^k} - 1\}$ one has $\{2^k, 2^k + 1, 2^k + 2, \ldots, 2^{k+1} - 1\} \subseteq \mathrm{supp}(y + t)$ so $y + t \notin A$.

To verify conclusion (b), let $x \in A$ and pick $k \in \mathbb{N}$ such that $2^{2^k - 1} > x$. Let $n = 2^k$. Then $\emptyset \neq A \cap \mathbb{N}2^n \subseteq -x + A$.

Finally, to verify (c) let $n \in \mathbb{N}$ and let sequences $\langle y_{1,t} \rangle_{t=1}^{\infty}, \langle y_{2,t} \rangle_{t=1}^{\infty}, \ldots, \langle y_{n,t} \rangle_{t=1}^{\infty}$ be given. We first observe that by Lemma 14.16 we can choose $H \in \mathcal{P}_f(\mathbb{N})$ such that for each $i \in \{1, 2, \ldots, \}$, $\Sigma_{t \in H} \, y_{i,t} \in \mathbb{N}2^{n+1}$.

Next we observe that given any $z_1, z_2, \ldots, z_n$ in $\mathbb{N}$ and any k with $2^k > n$, there exists $r \in B_k$ such that $B_k \setminus \mathrm{supp}(2^r + z_i) \neq \emptyset$ for each $i \in \{1, 2, \ldots, n\}$. Indeed, if $r \in B_k$ and $B_k \subseteq \mathrm{supp}(2^r + z)$ then $\mathrm{supp}(z) \cap B_k = B_k \setminus \{r\}$. Consequently $|\{r \in B_k :$ there is some $i \in \{1, 2, \ldots, n\}$ with $B_k \subseteq \mathrm{supp}(2^r + z_i)\}| \leq n$.

For $i \in \{1, 2, \ldots, n\}$, let $z_{0,i} = \Sigma_{t \in H} \, y_{i,t}$. Pick the least ℓ such that $2^\ell > n$. Now given i we have $2^{n+1} | z_{0,i}$ and $2^{\ell - 1} < n + 1$ so $2^{\ell - 1} \in B_{\ell - 1} \setminus \mathrm{supp}(z_{0,i})$. Pick $r_0 \in B_\ell$ such that $B_\ell \setminus \mathrm{supp}(2^{r_0} + z_{0,i}) \neq \emptyset$ for each $i \in \{1, 2, \ldots, n\}$ and let $z_{1,i} = z_{0,i} + 2^{r_0}$. Inductively choose $r_j \in B_{\ell + j}$ such that $B_{\ell + j} \setminus \mathrm{supp}(2^{r_j} + z_{j,i}) \neq \emptyset$ for each $i \in \{1, 2, \ldots, n\}$ and let $z_{j+1,i} = z_{j,i} + 2^{r_j}$. Continue the induction until $\ell + j = k$ where $2^{2^k} > \Sigma_{t \in H} \, y_{i,t}$ for each $i \in \{1, 2, \ldots, n\}$ and let $a = 2^{r_0} + 2^{r_1} + \cdots + 2^{r_{k-\ell}}$. $\square$

Theorem 14.18. *There is a set $A \subseteq \mathbb{N}$ such that A satisfies the conclusion of the Central Sets Theorem (Theorem 14.11) for $(\mathbb{N}, +)$ but $K(\beta\mathbb{N}, +) \cap c\ell \, A = \emptyset$.*

Proof. Let A be as in Lemma 14.17. By conclusion (a) of Lemma 14.17 and Theorem 4.40 we have that $K(\beta\mathbb{N}, +) \cap c\ell \, A = \emptyset$.

For each $\ell \in \mathbb{N}$, let $\langle y_{\ell,n} \rangle_{n=1}^{\infty}$ be a sequence in $\mathbb{N}$. Choose by condition (c) of Lemma 14.17 some a_1 and H_1 with $a_1 + \Sigma_{t \in H_1} \, y_{1,t} \in A$. Inductively, given a_n and H_n, let $\ell = \max H_n$ and pick $m > n$ in $\mathbb{N}$ such that for all $i \in \{1, 2, \ldots, n\}$, $a_n + \Sigma_{t \in H_n} \, y_{i,t} < 2^{2^m}$. Pick by condition (c) of Lemma 14.17 (applied to the sequences $\langle y_{1,\ell+t} \rangle_{t=1}^{\infty}, \langle y_{2,\ell+t} \rangle_{t=1}^{\infty}, \ldots, \langle y_{n+1,\ell+t} \rangle_{t=1}^{\infty}$, so that $\min H_{n+1} > \ell = \max H_n$) some a_{n+1} and H_{n+1} such that for all $i \in \{1, 2, \ldots, n+1\}$, $a_{n+1} + \Sigma_{t \in H_{n+1}} \, y_{i,t} \in A \cap \mathbb{N}2^{2^m}$.

Observe that if $x, y \in A$ and for some k, $x < 2^{2^k}$ and $2^{2^k} \mid y$, then $x + y \in A$. Consequently for $f \in \Phi$ one has $\mathrm{FS}(\langle a_n + \Sigma_{t \in H_n}\, y_{f(n),t}\rangle_{n=1}^{\infty}) \subseteq A$. $\qquad\square$

We now proceed to derive a combinatorial characterization of central sets. This characterization involves the following generalization of the notion of a piecewise syndetic set.

Definition 14.19. Let $(S, \cdot)$ be a semigroup and let $\mathcal{A} \subseteq \mathcal{P}(S)$. Then $\mathcal{A}$ is *collectionwise piecewise syndetic* if and only if there exist functions $G : \mathcal{P}_f(\mathcal{A}) \to \mathcal{P}_f(S)$ and $x : \mathcal{P}_f(\mathcal{A}) \times \mathcal{P}_f(S) \to S$ such that for all $F \in \mathcal{P}_f(S)$ and all $\mathcal{F}$ and $\mathcal{H}$ in $\mathcal{P}_f(\mathcal{A})$ with $\mathcal{F} \subseteq \mathcal{H}$ one has $F \cdot x(\mathcal{H}, F) \subseteq \bigcup_{t \in G(\mathcal{F})} t^{-1}(\bigcap \mathcal{F})$.

Note that a subset A is piecewise syndetic if and only if $\{A\}$ is collectionwise piecewise syndetic. An alternate characterization is: $\mathcal{A}$ is collectionwise piecewise syndetic if and only if there exists a function $G : \mathcal{P}_f(\mathcal{A}) \to \mathcal{P}_f(S)$ such that

$$\left\{y^{-1}\big(G(\mathcal{F})\big)^{-1}\big(\bigcap \mathcal{F}\big) : y \in S \text{ and } \mathcal{F} \in \mathcal{P}_f(\mathcal{A})\right\}$$

has the finite intersection property.

(Here $y^{-1}\big(G(\mathcal{F})\big)^{-1}\big(\bigcap \mathcal{F}\big) = \bigcup_{t \in G(\mathcal{F})} y^{-1}t^{-1}\big(\bigcap \mathcal{F}\big)$, see Exercise 14.5.1.)

In the event that the family and the semigroup are both countable, we have a considerably simpler characterization of collectionwise piecewise syndetic.

Lemma 14.20. *Assume the semigroup* $S = \{a_n : n \in \mathbb{N}\}$ *and* $\{A_n : n \in \mathbb{N}\} \subseteq \mathcal{P}(S)$. *The family* $\{A_n : n \in \mathbb{N}\}$ *is collectionwise piecewise syndetic if and only if there exist a sequence* $\langle g(n)\rangle_{n=1}^{\infty}$ *in* $\mathbb{N}$ *and a sequence* $\langle y_n\rangle_{n=1}^{\infty}$ *in* S *such that for all* $n, m \in \mathbb{N}$ *with* $m \geq n$,

$$\{a_1 y_m, a_2 y_m, \ldots, a_m y_m\} \subseteq \bigcup_{j=1}^{g(n)} a_j^{-1}\big(\bigcap_{i=1}^{n} A_i\big).$$

Proof. For the necessity, pick functions G and x as guaranteed by Definition 14.19. Given $n \in \mathbb{N}$, let

$$g(n) = \min\{k \in \mathbb{N} : G(\{A_1, A_2, \ldots, A_n\}) \subseteq \{a_1, a_2, \ldots, a_k\}\}$$

and $y_n = x(\{A_1, A_2, \ldots, A_n\}, \{a_1, a_2, \ldots, a_n\})$. Then if $m \geq n$, letting

$$\mathcal{F} = \{A_1, A_2, \ldots, A_n\}, \quad \mathcal{H} = \{A_1, A_2, \ldots, A_m\}, \quad \text{and} \quad F = \{a_1, a_2, \ldots, a_m\},$$

one has $\mathcal{F} \subseteq \mathcal{H}$ and

$$\begin{aligned}
\{a_1 y_m, a_2 y_m, \ldots, a_m y_m\} &= F \cdot x(\mathcal{H}, F) \\
&\subseteq \bigcup_{t \in G(\mathcal{F})} t^{-1}\big(\bigcap \mathcal{F}\big) \\
&\subseteq \bigcup_{j=1}^{g(n)} a_j^{-1}\big(\bigcap_{i=1}^{n} A_i\big).
\end{aligned}$$

For the sufficiency, let $\langle g(n)\rangle_{n=1}^{\infty}$ and $\langle y_n\rangle_{n=1}^{\infty}$ be as in the statement of the lemma. Given $\mathcal{F} \in \mathcal{P}_f(\{A_n : n \in \mathbb{N}\})$, let $m = \max\{k \in \mathbb{N} : A_k \in \mathcal{F}\}$ and let $G(\mathcal{F}) =$

$\{a_1, a_2, \ldots, a_{g(m)}\}$. Given $\mathcal{H} \in \mathcal{P}_f(\{A_n : n \in \mathbb{N}\})$ and $F \in \mathcal{P}_f(S)$, let $m = \max\{k \in \mathbb{N} : A_k \in \mathcal{H}$ or $a_k \in F\}$ and let $x(\mathcal{H}, F) = y_m$. Then if $\mathcal{F}, \mathcal{H} \in \mathcal{P}_f(\{A_n : n \in \mathbb{N}\})$, $F \in \mathcal{P}_f(S)$, $\mathcal{F} \subseteq \mathcal{H}$, $m = \max\{k \in \mathbb{N} : A_k \in \mathcal{H}$ or $a_k \in F\}$, and $n = \max\{k \in \mathbb{N} : A_k \in \mathcal{F}\}$, then $n \leq m$ so

$$F \cdot x(\mathcal{H}, F) \subseteq \{a_1 y_m, a_2 y_m, \ldots, a_m y_m\}$$
$$\subseteq \bigcup_{j=1}^{g(n)} a_j^{-1} \left(\bigcap_{i=1}^{n} A_i\right)$$
$$\subseteq \bigcup_{t \in G(\mathcal{F})} t^{-1}\left(\bigcap \mathcal{F}\right). \qquad \square$$

Theorem 14.21. *Let $(S, \cdot)$ be an infinite semigroup and let $\mathcal{A} \subseteq \mathcal{P}(S)$. There exists $p \in K(\beta S)$ with $\mathcal{A} \subseteq p$ if and only if $\mathcal{A}$ is collectionwise piecewise syndetic.*

Proof. Necessity. Pick $p \in K(\beta S)$ such that $\mathcal{A} \subseteq p$. For each $\mathcal{F} \in \mathcal{P}_f(\mathcal{A})$, $\bigcap \mathcal{F} \in p$, so by Theorem 4.39, $B(\mathcal{F}) = \{t \in S : t^{-1}(\bigcap \mathcal{F}) \in p\}$ is syndetic. Pick $G(\mathcal{F}) \in \mathcal{P}_f(S)$ such that $S = \bigcup_{t \in G(\mathcal{F})} t^{-1} B(\mathcal{F})$. For each $y \in S$ and each $\mathcal{F} \in \mathcal{P}_f(\mathcal{A})$, pick $t(y, \mathcal{F}) \in G(\mathcal{F})$ such that $y \in (t(y, \mathcal{F}))^{-1} B(\mathcal{F})$ (so that $(t(y, \mathcal{F}) \cdot y)^{-1}(\bigcap \mathcal{F}) \in p$). Given $\mathcal{H} \in \mathcal{P}_f(\mathcal{A})$ and $F \in \mathcal{P}_f(S)$, $\{(t(y, \mathcal{F}) \cdot y)^{-1}(\bigcap \mathcal{F}) : y \in F$ and $\emptyset \neq \mathcal{F} \subseteq \mathcal{H}\}$ is a finite subset of p so pick

$$x(\mathcal{H}, F) \in \bigcap\{(t(y, \mathcal{F}) \cdot y)^{-1}\left(\bigcap \mathcal{F}\right) : y \in F \text{ and } \emptyset \neq \mathcal{F} \subseteq \mathcal{H}\}.$$

To see that the functions G and x are as required, let $F \in \mathcal{P}_f(S)$ and let $\mathcal{F}, \mathcal{H} \in \mathcal{P}_f(\mathcal{A})$ with $\mathcal{F} \subseteq \mathcal{H}$. To see that $F \cdot x(\mathcal{H}, F) \subseteq \bigcup_{t \in G(\mathcal{F})} t^{-1}(\bigcap \mathcal{F})$, let $y \in F$. Then $x(\mathcal{H}, F) \in (t(y, \mathcal{F}) \cdot y)^{-1}(\bigcap \mathcal{F})$, so

$$y \cdot x(\mathcal{H}, F) \in (t(y, \mathcal{F}))^{-1}\left(\bigcap \mathcal{F}\right) \subseteq \bigcup_{t \in G(\mathcal{F})} t^{-1}\left(\bigcap \mathcal{F}\right).$$

Sufficiency. Pick $G : \mathcal{P}_f(\mathcal{A}) \to \mathcal{P}_f(S)$ and $x : \mathcal{P}_f(\mathcal{A}) \times \mathcal{P}_f(S) \to S$ as guaranteed by the definition. For each $\mathcal{F} \in \mathcal{P}_f(\mathcal{A})$ and each $y \in S$, let

$$D(\mathcal{F}, y) = \{x(\mathcal{H}, F) : \mathcal{H} \in \mathcal{P}_f(\mathcal{A}), F \in \mathcal{P}_f(S), \mathcal{F} \subseteq \mathcal{H}, \text{ and } y \in F\}.$$

Then $\{D(\mathcal{F}, y) : \mathcal{F} \in \mathcal{P}_f(\mathcal{A})$ and $y \in S\}$ has the finite intersection property so pick $u \in \beta S$ such that $\{D(\mathcal{F}, y) : \mathcal{F} \in \mathcal{P}_f(\mathcal{A})$ and $y \in S\} \subseteq u$. We claim that

(∗) for each $\mathcal{F} \in \mathcal{P}_f(\mathcal{A})$, $S \cdot u \subseteq \overline{\bigcup_{t \in G(\mathcal{F})} t^{-1}\left(\bigcap \mathcal{F}\right)}$.

To this end, let $\mathcal{F} \in \mathcal{P}_f(\mathcal{A})$ and let $y \in S$. Then $D(\mathcal{F}, y) \subseteq \{x : y \cdot x \in \bigcup_{t \in G(\mathcal{F})} t^{-1}(\bigcap \mathcal{F})\}$ so $\bigcup_{t \in G(\mathcal{F})} t^{-1}(\bigcap \mathcal{F}) \in y \cdot u$ as required.

By (∗) we have that for each $\mathcal{F} \in \mathcal{P}_f(\mathcal{A})$, $\beta S \cdot u \subseteq \overline{\bigcup_{t \in G(\mathcal{F})} t^{-1}(\bigcap \mathcal{F})}$. Since $\beta S \cdot u$ is a left ideal of βS, pick a minimal left ideal L of βS such that $L \subseteq \beta S \cdot u$ and pick $q \in L$. Then for each $\mathcal{H} \in \mathcal{P}_f(\mathcal{A})$ we have $\bigcup_{t \in G(\mathcal{H})} t^{-1}(\bigcap \mathcal{H}) \in q$ so pick $t(\mathcal{H}) \in G(\mathcal{H})$ such that $(t(\mathcal{H}))^{-1}(\bigcap \mathcal{H}) \in q$.

For each $\mathcal{F} \in \mathcal{P}_f(\mathcal{A})$, let $E(\mathcal{F}) = \{t(\mathcal{H}) : \mathcal{H} \in \mathcal{P}_f(\mathcal{A})$ and $\mathcal{F} \subseteq \mathcal{H}\}$. Then $\{E(\mathcal{F}) : \mathcal{F} \in \mathcal{P}_f(\mathcal{A})\}$ has the finite intersection property. So pick $w \in \beta S$ such that $\{E(\mathcal{F}) : \mathcal{F} \in \mathcal{P}_f(\mathcal{A})\} \subseteq w$. Let $p = w \cdot q$. Then $p \in L \subseteq K(\beta S)$. To see that $\mathcal{A} \subseteq p$, let $A \in \mathcal{A}$. Since $E(\{A\}) \in w$, it suffices to show that $E(\{A\}) \subseteq \{t \in S : t^{-1}A \in q\}$, so let $\mathcal{H} \in \mathcal{P}_f(\mathcal{A})$ with $\{A\} \subseteq \mathcal{H}$. Then $\left(t(\mathcal{H})\right)^{-1}\left(\bigcap \mathcal{H}\right) \in q$ and $\left(t(\mathcal{H})\right)^{-1}\left(\bigcap \mathcal{H}\right) \subseteq \left(t(\mathcal{H})\right)^{-1}A$, so $\left(t(\mathcal{H})\right)^{-1}A \in q$. $\qquad\square$

Our combinatorial characterization of *central* is based on an analysis of the proofs of Theorem 5.8. The important thing to notice about these proofs is that when one chooses x_n one in fact has a large number of choices. That is, one can draw a tree, branching infinitely often at each node, so that any path through that tree yields a sequence $\langle x_n \rangle_{n=1}^{\infty}$ with $\mathrm{FP}(\langle x_n \rangle_{n=1}^{\infty}) \subseteq A$. (Recall that in $\mathrm{FP}(\langle x_n \rangle_{n=1}^{\infty})$, the products are taken in increasing order of indices.)

We formalize the notion of "tree" below. We recall that each ordinal is the set of its predecessors. (So $3 = \{0, 1, 2\}$ and $0 = \emptyset$ and, if f is the function $\{(0, 3), (1, 5), (2, 9), (3, 7), (4, 5)\}$, then $f_{|3} = \{(0, 3), (1, 5), (2, 9)\}$.)

Definition 14.22. T is a *tree in* A if and only if T is a set of functions and for each $f \in T$, domain$(f) \in \omega$ and range$(f) \subseteq A$ and if domain$(f) = n > 0$, then $f_{|n-1} \in T$. T is a *tree* if and only if for some A, T is a tree in A.

The last requirement in the definition is not essential. Any set of functions with domains in ω can be converted to a tree by adding in all restrictions to initial segments. We include the requirement in the definition for aesthetic reasons — it is not nice for branches at some late level to appear from nowhere.

Definition 14.23. (a) Let f be a function with domain$(f) = n \in \omega$ and let x be given. Then $f^\frown x = f \cup \{(n, x)\}$.

(b) Given a tree T and $f \in T$, $B_f = B_f(T) = \{x : f^\frown x \in T\}$.

(c) Let $(S, \cdot)$ be a semigroup and let $A \subseteq S$. Then T is a *$*$-tree in* A if and only if T is a tree in A and for all $f \in T$ and all $x \in B_f$, $B_{f^\frown x} \subseteq x^{-1}B_f$.

(d) Let $(S, \cdot)$ be a semigroup and let $A \subseteq S$. Then T is an *FP-tree in* A if and only if T is a tree in A and for all $f \in T$, $B_f = \{\prod_{t \in F} g(t) : g \in T$ and $f \not\subseteq g$ and $\emptyset \neq F \subseteq \mathrm{dom}(g) \setminus \mathrm{dom}(f)\}$.

The idea of the terminology is that an FP-tree is a tree of finite products. It is this notion which provides the most fundamental combinatorial characterization of the notion of "central". A $*$-tree arises more directly from the proof outlined above.

Lemma 14.24. *Let* $(S, \cdot)$ *be an infinite semigroup and let* $A \subseteq S$. *Let* p *be an idempotent in* βS *with* $A \in p$. *There is an FP-tree* T *in* A *such that for each* $f \in T$, $B_f \in p$.

Proof. We shall define the initial segments $T_n = \{f \in T : \mathrm{dom}(f) = n\}$ inductively. Let $T_0 = \{\emptyset\}$ (of course) and let $C_\emptyset = A^\star$. Then $C_\emptyset \in p$ by Theorem 4.12. Let $T_1 = \big\{\{(0, x)\} : x \in C_\emptyset\big\}$.

Inductively assume that we have $n \in \mathbb{N}$ and have defined T_n so that for each $f \in T_n$ one has $\mathrm{FP}(\langle f(t)\rangle_{t=0}^{n-1}) \subseteq A^\star$. Given $f \in T_n$, write $P_f = \mathrm{FP}(\langle f(t)\rangle_{t=0}^{n-1})$, let $C_f = A^\star \cap \bigcap_{x \in P_f} x^{-1} A^\star$, and note that by Lemma 4.14, $C_f \in p$. Let $T_{n+1} = \{f^\frown y : f \in T_n$ and $y \in C_f\}$. Then given $g \in T_{n+1}$, one has $\mathrm{FP}(\langle g(t)\rangle_{t=0}^{n}) \subseteq A^\star$.

The induction being complete, let $T = \bigcup_{n=0}^\infty T_n$. Then T is a tree in A. One sees immediately from the construction that for each $f \in T$, $B_f = C_f$. We need to show that for each $f \in T$ one has $B_f = \{\prod_{t \in F} g(t) : g \in T$ and $f \not\subseteq g$ and $\emptyset \neq F \subseteq \mathrm{dom}(g)\backslash \mathrm{dom}(f)\}$. Given $f \in T$ and $x \in B_f$, let $g = f^\frown x$ and let $F = \mathrm{dom}(g)\backslash \mathrm{dom}(f)$ (which is a singleton). Then $x = \prod_{t \in F} g(t)$.

For the other inclusion we first observe that if $f, h \in T$ with $f \subseteq h$ then $P_f \subseteq P_h$ so $B_h \subseteq B_f$. Let $f \in T_n$ and let $x \in \{\prod_{t \in F} g(t) : g \in T$ and $f \not\subseteq g$ and $\emptyset \neq F \subseteq \mathrm{dom}(g)\backslash \mathrm{dom}(f)\}$. Pick $g \in T$ with $f \not\subseteq g$ and pick F with $\emptyset \neq F \subseteq \mathrm{dom}(g)\backslash \mathrm{dom}(f)$ such that $x = \prod_{t \in F} g(t)$. First assume $F = \{m\}$. Then $m \geq n$. Let $h = g_{|m}$. Then $f \subseteq h$ and $h^\frown x = g_{|m+1} \in T$. Hence $x \in B_h \subseteq B_f$ as required. Now assume $|F| > 1$, let $m = \max F$, and let $G = F\backslash\{m\}$. Let $h = g_{|m}$, let $w = \prod_{t \in G} g(t)$, and let $y = g(m)$. Then $y \in B_h$. Let $P_f = \mathrm{FP}(\langle f(t)\rangle_{t=0}^{n-1})$ and $P_h = \mathrm{FP}(\langle h(t)\rangle_{t=0}^{m-1})$. We need to show that $w \cdot y \in B_f$. That is, we need $w \cdot y \in A^\star$ and for all $z \in P_f$, $w \cdot y \in z^{-1} A^\star$. Now $w \in P_h$ and $y \in B_h$ so $y \in w^{-1} A^\star$ so $w \cdot y \in A^\star$. Let $z \in P_f$. Then $z \cdot w \in P_h$ and $y \in B_h$ so $y \in (z \cdot w)^{-1} A^\star$ and so $w \cdot y \in z^{-1} A^\star$. $\qquad\square$

Theorem 14.25. *Let $(S, \cdot)$ be an infinite semigroup and let $A \subseteq S$. Statements* (a), (b), (c), *and* (d) *are equivalent and are implied by statement* (e). *If S is countable, then all five statements are equivalent.*

(a) *A is central.*

(b) *There is a FP-tree T in A such that $\{B_f : f \in T\}$ is collectionwise piecewise syndetic.*

(c) *There is a $*$-tree T in A such that $\{B_f : f \in T\}$ is collectionwise piecewise syndetic.*

(d) *There is a downward directed family $\langle C_F\rangle_{F \in I}$ of subsets of A such that*

   (i) *for each $F \in I$ and each $x \in C_F$ there exists $G \in I$ with $C_G \subseteq x^{-1} C_F$ and*

   (ii) *$\{C_F : F \in I\}$ is collectionwise piecewise syndetic.*

(e) *There is a decreasing sequence $\langle C_n\rangle_{n=1}^\infty$ of subsets of A such that*

   (i) *for each $n \in \mathbb{N}$ and each $x \in C_n$, there exists $m \in \mathbb{N}$ with $C_m \subseteq x^{-1} C_n$ and*

   (ii) *$\{C_n : n \in \mathbb{N}\}$ is collectionwise piecewise syndetic.*

Proof. (a) implies (b). Pick an idempotent $p \in K(\beta S)$ with $A \in p$. By Lemma 14.24 pick an FP-tree with $\{B_f : f \in T\} \subseteq p$. By Theorem 14.21 $\{B_f : f \in T\}$ is collectionwise piecewise syndetic.

(b) implies (c). Let T be an FP-tree. Then given $f \in T$ and $x \in B_f$, we claim that $B_{f^\frown x} \subseteq x^{-1} B_f$. To this end let $y \in B_{f^\frown x}$ and pick $g \in T$ and $F \subseteq \mathrm{dom}(g)\backslash \mathrm{dom}(f^\frown x)$ such that $f^\frown x \subseteq g$ and $y = \prod_{t \in F} g(t)$. Let $n = \mathrm{dom}(f)$ and let $G = F \cup \{n\}$. Then $x \cdot y = \prod_{t \in G} g(t)$ and $G \subseteq \mathrm{dom}(g)\backslash \mathrm{dom}(f)$, so $x \cdot y \in B_f$ as required.

(c) implies (d). Let T be the given $*$-tree. Since $\{B_f : f \in T\}$ is collectionwise piecewise syndetic, so is $\{\bigcap_{f \in F} B_f : F \in \mathcal{P}_f(T)\}$. (This can be seen directly or by invoking Theorem 14.21.) Let $I = \mathcal{P}_f(T)$ and for each $F \in I$, let $C_F = \bigcap_{f \in F} B_f$. Then $\{C_F : F \in I\}$ is collectionwise piecewise syndetic, so (ii) holds. Let $F \in I$ and let $x \in C_F$. Let $G = \{f^\frown x : f \in F\}$. Then $G \in I$. Now for each $f \in F$ we have $B_{f^\frown x} \subseteq x^{-1} B_f$ so $C_G \subseteq x^{-1} C_F$.

(d) implies (a). Let $M = \bigcap_{F \in I} c\ell\, C_F$. By Theorem 4.20 M is a subsemigroup of βS. Since $\{C_F : F \in I\}$ is collectionwise piecewise syndetic, we have by Theorem 14.21 that $M \cap K(\beta S) \neq \emptyset$ so we may pick a minimal left ideal L of βS with $L \cap M \neq \emptyset$. Then $L \cap M$ is a compact subsemigroup of βS which thus contains an idempotent, and this idempotent is necessarily minimal.

That (e) implies (d) is trivial.

Finally assume that S is countable. We show that (c) implies (e). So let T be the given $*$-tree in A. Then T is countable so enumerate T as $\langle f_n \rangle_{n=1}^\infty$. For each $n \in \mathbb{N}$, let $C_n = \bigcap_{k=1}^n B_{f_k}$. Then $\{C_n : n \in \mathbb{N}\}$ is collectionwise piecewise syndetic. Let $n \in \mathbb{N}$ and let $x \in C_n$. Pick $m \in \mathbb{N}$ such that $\{f_k^\frown x : k \in \{1, 2, \ldots, n\}\} \subseteq \{f_t : t \in \{1, 2, \ldots, m\}\}$. Then $C_m = \bigcap_{t=1}^m B_{f_t} \subseteq \bigcap_{k=1}^n B_{f_k^\frown x} \subseteq \bigcap_{k=1}^n x^{-1} B_{f_k} = x^{-1} C_n$. □

We close this section by pointing out a Ramsey-theoretic consequence of the characterization.

Corollary 14.26. *Let S be an infinite semigroup, let $r \in \mathbb{N}$, and let $S = \bigcup_{i=1}^r A_i$. There exist $i \in \{1, 2, \ldots, r\}$ and an FP-tree T in A_i such that $\{B_f : f \in T\}$ is collectionwise piecewise syndetic.*

Proof. Pick an idempotent $p \in K(\beta S)$ and pick $i \in \{1, 2, \ldots, r\}$ such that $A_i \in p$. Apply Theorem 14.25. □

Exercise 14.5.1. Let S be a semigroup. Prove that a family $\mathcal{A} \subseteq \mathcal{P}(S)$ is collectionwise piecewise syndetic if and only if there exists a function $G : \mathcal{P}_f(\mathcal{A}) \to \mathcal{P}_f(S)$ such that $\{y^{-1}\big(G(\mathcal{F})\big)^{-1}\big(\bigcap \mathcal{F}\big) : y \in S \text{ and } \mathcal{F} \in \mathcal{P}_f(\mathcal{A})\}$ has the finite intersection property. (Where $y^{-1}\big(G(\mathcal{F})\big)^{-1}\big(\bigcap \mathcal{F}\big) = \bigcup_{t \in G(\mathcal{F})} y^{-1} t^{-1}\big(\bigcap \mathcal{F}\big)$.)

Notes

The original Central Sets Theorem is due to Furstenberg [98, Proposition 8.21] and applied to the semigroup $(\mathbb{N}, +)$. It used a different but equivalent definition of *central*. See Theorem 19.27 for a proof of the equivalence of the notions of central. This original version also allowed the (finitely many) sequences $\langle y_{\ell,n} \rangle_{n=1}^\infty$ to take values in $\mathbb{Z}$. Since any idempotent minimal in $(\beta\mathbb{N}, +)$ is also minimal in $(\beta\mathbb{Z}, +)$, and hence any central set in $(\mathbb{N}, +)$ is central in $(\mathbb{Z}, +)$, the original version follows from Theorem 14.11.

The idea for the proof of the Central Sets Theorem (as well as the proofs in this chapter of van der Waerden's Theorem and the Hales–Jewett Theorem) is due to Furstenberg and Katznelson in [99], where it was developed in the context of enveloping semigroups. The idea to convert this proof into a proof in βS is due to V. Bergelson, and the construction in this context first appeared in [27].

Corollary 14.13 can in fact be derived from the Hales–Jewett Theorem (Corollary 14.8). See for example [28, p. 434].

The material in Section 14.5 is from [148], a result of collaboration with A. Maleki, except for Theorem 14.21 which is from [147], a result of collaboration with A. Lisan.

Chapter 15

Partition Regularity of Matrices

In this chapter we present several applications of the Central Sets Theorem (Theorem 14.11) and of its proof.

15.1 Image Partition Regular Matrices

Many of the classical results of Ramsey Theory are naturally stated as instances of the following problem. Given $u, v \in \mathbb{N}$ and a $u \times v$ matrix A with non-negative integer entries, is it true that whenever $\mathbb{N}$ is finitely colored there must exist some $\vec{x} \in \mathbb{N}^v$ such that the entries of $A\vec{x}$ are monochrome?

Consider for example van der Waerden's Theorem (Corollary 14.2). The arithmetic progression $\{a, a + d, a + 2d, a + 3d\}$ is precisely the set of entries of

$$\begin{pmatrix} 1 & 0 \\ 1 & 1 \\ 1 & 2 \\ 1 & 3 \end{pmatrix} \cdot \begin{pmatrix} a \\ d \end{pmatrix}.$$

Also Schur's Theorem (Theorem 5.3) and the case $m = 3$ of Hilbert's Theorem (Theorem 5.2) guarantee an affirmative answer in the case of the following two matrices:

$$\begin{pmatrix} 1 & 0 \\ 0 & 1 \\ 1 & 1 \end{pmatrix} \qquad \begin{pmatrix} 1 & 1 & 0 & 0 \\ 1 & 0 & 1 & 0 \\ 1 & 1 & 1 & 0 \\ 1 & 0 & 0 & 1 \\ 1 & 1 & 0 & 1 \\ 1 & 0 & 1 & 1 \\ 1 & 1 & 1 & 1 \end{pmatrix}$$

This suggests the following natural definition. We remind the reader that for $n \in \mathbb{N}$ and x in a semigroup $(S, +)$, nx means the sum of x with itself n times, i.e. the additive version of x^n. We use additive notation here because it is most convenient for the matrix manipulations. Note that the requirement that S have an identity is not substantive since one may be added to any semigroup. We add the requirement so that $0x$ will make sense. (See also Exercise 15.1.1, where the reader is asked to show that the central sets in S are not affected by the adjoining of a 0.)

Definition 15.1. Let $(S, +)$ be a semigroup with identity 0, let $u, v \in \mathbb{N}$, and let A be a $u \times v$ matrix with entries from ω. Then A is *image partition regular over S* if and only if whenever $r \in \mathbb{N}$ and $S = \bigcup_{i=1}^{r} E_i$, there exist $i \in \{1, 2, \ldots, r\}$ and $\vec{x} \in (S \setminus \{0\})^v$ such that $A\vec{x} \in E_i{}^u$.

It is obvious that one must require in Definition 15.1 that the vector $\vec{x}$ not be constantly 0. We make the stronger requirement because in the classical applications one wants all of the entries to be non-zero. (Consider van der Waerden's Theorem with increment 0.)

We have restricted our matrix A to have nonnegative entries, since in an arbitrary semigroup $-x$ may not mean anything. In Section 15.4 where we shall deal with image partition regularity over $\mathbb{N}$ we shall extend the definition of image partition regularity to allow entries from $\mathbb{Q}$.

Definition 15.2. Let $u, v \in \mathbb{N}$ and let A be a $u \times v$ matrix with entries from $\mathbb{Q}$. Then A satisfies the *first entries condition* if and only if no row of A is $\vec{0}$ and whenever $i, j \in \{1, 2, \ldots, u\}$ and $k = \min\{t \in \{1, 2, \ldots, v\} : a_{i,t} \neq 0\} = \min\{t \in \{1, 2, \ldots, v\} : a_{j,t} \neq 0\}$, then $a_{i,k} = a_{j,k} > 0$. An element b of $\mathbb{Q}$ is a *first entry* of A if and only if there is some row i of A such that $b = a_{i,k}$ where $k = \min\{t \in \{1, 2, \ldots, v\} : a_{i,t} \neq 0\}$.

Given any family $\mathcal{R}$ of subsets of a set S, one can define the set $\mathcal{R}^*$ of all sets that meet every member of $\mathcal{R}$. We shall investigate some of these in Chapter 16. For now we need the notion of central* sets.

Definition 15.3. Let S be a semigroup and let $A \subseteq S$. Then A is a *central* set of S if and only if $A \cap C \neq \emptyset$ for every central set C of S.

Lemma 15.4. *Let S be a semigroup and let $A \subseteq S$. Then the following statements are equivalent.*

(a) *A is a central* set.*

(b) *A is a member of every minimal idempotent in βS.*

(c) *$A \cap C$ is a central set for every central set C of S.*

Proof. (a) implies (b). Let p be a minimal idempotent in βS. If $A \notin p$ then $S \setminus A$ is a central set of S which misses A.

(b) implies (c). Let C be a central set of S and pick a minimal idempotent p in βS such that $C \in p$. Then also $A \in p$ so $A \cap C \in p$.

That (c) implies (a) is trivial. □

In the following theorem we get a conclusion far stronger than the assertion that matrices satisfying the first entries condition are image partition regular. The stronger conclusion is of some interest in its own right. More importantly, the stronger conclusion is needed as an induction hypothesis in the proof.

Theorem 15.5. *Let $(S, +)$ be an infinite commutative semigroup with identity 0, let $u, v \in \mathbb{N}$, and let A be a $u \times v$ matrix with entries from ω which satisfies the first entries condition. Let C be central in S. If, for every first entry c of A, cS is a central* set, then there exist sequences $\langle x_{1,n} \rangle_{n=1}^{\infty}$, $\langle x_{2,n} \rangle_{n=1}^{\infty}$, $\ldots$, $\langle x_{v,n} \rangle_{n=1}^{\infty}$ in S such that for every $F \in \mathcal{P}_f(\mathbb{N})$, $\vec{x}_F \in (S \backslash \{0\})^v$ and $A \vec{x}_F \in C^u$, where*

$$
\vec{x}_F = \begin{pmatrix} \Sigma_{n \in F} \, x_{1,n} \\ \Sigma_{n \in F} \, x_{2,n} \\ \vdots \\ \Sigma_{n \in F} \, x_{v,n} \end{pmatrix}.
$$

Proof. $S \backslash \{0\}$ is an ideal of S so by Corollary 4.18, 0 is not a minimal idempotent, and thus we may presume that $0 \notin C$. We proceed by induction on v. Assume first $v = 1$. We can assume A has no repeated rows, so in this case we have $A = (c)$ for some $c \in \mathbb{N}$ such that cS is central*. Then $C \cap cS$ is a central set so pick by Theorem 14.11 (with the sequence $y_{1,n} = 0$ for each n and the function f constantly equal to 1) some sequence $\langle k_n \rangle_{n=1}^{\infty}$ with $\text{FS}(\langle k_n \rangle_{n=1}^{\infty}) \subseteq C \cap cS$. (In fact here we could get by with an appeal to Theorem 5.8.) For each $n \in \mathbb{N}$ pick some $x_{1,n} \in S$ such that $k_n = c x_{1,n}$. The sequence $\langle x_{1,n} \rangle_{n=1}^{\infty}$ is as required. (Given $F \in \mathcal{P}_f(\mathbb{N})$, $\Sigma_{n \in F} \, k_n \neq 0$ so $\Sigma_{n \in F} \, x_{1,n} \neq 0$.)

Now let $v \in \mathbb{N}$ and assume the theorem is true for v. Let A be a $u \times (v+1)$ matrix with entries from ω which satisfies the first entries condition, and assume that whenever c is a first entry of A, cS is a central* set. By rearranging the rows of A and adding additional rows to A if need be, we may assume that we have some $r \in \{1, 2, \ldots, u-1\}$ and some $d \in \mathbb{N}$ such that

$$
a_{i,1} = \begin{cases} 0 & \text{if } i \in \{1, 2, \ldots, r\} \\ d & \text{if } i \in \{r+1, r+2, \ldots, u\}. \end{cases}
$$

Let B be the $r \times v$ matrix with entries $b_{i,j} = a_{i,j+1}$. Pick sequences $\langle z_{1,n} \rangle_{n=1}^{\infty}$, $\langle z_{2,n} \rangle_{n=1}^{\infty}$, $\ldots$, $\langle z_{v,n} \rangle_{n=1}^{\infty}$ in S as guaranteed by the induction hypothesis for the matrix B. For each $i \in \{r+1, r+2, \ldots, u\}$ and each $n \in \mathbb{N}$, let

$$
y_{i,n} = \Sigma_{j=2}^{v+1} a_{i,j} \cdot z_{j-1,n}
$$

and let $y_{r,n} = 0$ for all $n \in \mathbb{N}$. (For other values of i, one may let $y_{i,n}$ take on any value in S at all.)

Now $C \cap dS$ is central, so pick by Theorem 14.11 sequences $\langle k_n \rangle_{n=1}^{\infty}$ in S and $\langle H_n \rangle_{n=1}^{\infty}$ in $\mathcal{P}_f(\mathbb{N})$ such that $\max H_n < \min H_{n+1}$ for each n and for each $i \in \{r, r+1, \ldots, u\}$,

$$
\text{FS}(\langle k_n + \Sigma_{t \in H_n} \, y_{i,t} \rangle_{n=1}^{\infty}) \subseteq C \cap dS.
$$

(If $\langle k'_n \rangle_{n=1}^{\infty}$ and $\langle H'_n \rangle_{n=1}^{\infty}$ are as given by Theorem 14.11, let $k_n = k'_{n+u}$ and $H_n = H'_{n+u}$ for every n, and for $i \in \{r, r+1, \ldots, u\}$, let $f_i(n) = n$ if $n < i$ and let $f_i(n) = i$ if $n \geq i$.)

Note in particular that each $k_n = k_n + \Sigma_{t \in H_n} y_{r,t} \in C \cap dS$, so pick $x_{1,n} \in S$ such that $k_n = dx_{1,n}$. For $j \in \{2, 3, \ldots, v + 1\}$, let $x_{j,n} = \Sigma_{t \in H_n} z_{j-1,t}$. We claim that the sequences $\langle x_{j,n} \rangle_{n=1}^{\infty}$ are as required. To see this, let $F \in \mathcal{P}_f(\mathbb{N})$ be given. We need to show that for each $j \in \{1, 2, \ldots, v + 1\}$, $\Sigma_{n \in F} x_{j,n} \neq 0$ and for each $i \in \{1, 2, \ldots, u\}$,

$$\Sigma_{j=1}^{v+1} a_{i,j} \Sigma_{n \in F} x_{j,n} \in C.$$

For the first assertion note that if $j > 1$, then $\Sigma_{n \in F} x_{j,n} = \Sigma_{t \in G} z_{j-1,t}$ where $G = \bigcup_{n \in F} H_n$. If $j = 1$, then $d \Sigma_{n \in F} x_{1,n} = \Sigma_{n \in F}(k_n + \Sigma_{t \in H_n} y_{r,t}) \in C$.

To establish the second assertion, let $i \in \{1, 2, \ldots, u\}$ be given.

Case 1. $i \leq r$. Then

$$\begin{aligned}
\Sigma_{j=1}^{v+1} a_{i,j} \Sigma_{n \in F} x_{j,n} &= \Sigma_{j=2}^{v+1} a_{i,j} \Sigma_{n \in F} \Sigma_{t \in H_n} z_{j-1,t} \\
&= \Sigma_{j=1}^{v} b_{i,j} \Sigma_{t \in G} z_{j,t} \in C
\end{aligned}$$

where $G = \bigcup_{n \in F} H_n$.

Case 2. $i > r$. Then

$$\begin{aligned}
\Sigma_{j=1}^{v+1} a_{i,j} \Sigma_{n \in F} x_{j,n} &= d \Sigma_{n \in F} x_{1,n} + \Sigma_{j=2}^{v+1} a_{i,j} \Sigma_{n \in F} x_{j,n} \\
&= \Sigma_{n \in F} dx_{1,n} + \Sigma_{n \in F} \Sigma_{j=2}^{v+1} a_{i,j} \Sigma_{t \in H_n} z_{j-1,t} \\
&= \Sigma_{n \in F} dx_{1,n} + \Sigma_{n \in F} \Sigma_{t \in H_n} \Sigma_{j=2}^{v+1} a_{i,j} z_{j-1,t} \\
&= \Sigma_{n \in F}(k_n + \Sigma_{t \in H_n} y_{i,t}) \in C.
\end{aligned}$$ □

We now present the obvious partition regularity corollary. But note that with a little effort a stronger result (Theorem 15.10) follows.

Corollary 15.6. *Let $(S, +)$ be an infinite commutative semigroup and let A be a finite matrix with entries from ω which satisfies the first entries condition. If for each first entry c of A, cS is a central* set, then A is image partition regular over S.*

Proof. Let $S = \bigcup_{i=1}^{r} E_i$. Then some E_i is central in S. If S has a two sided identity, then Theorem 15.5 applies directly. Otherwise, adjoin an identity 0. Then, by Exercise 15.1.1, E_i is central in $S \cup \{0\}$. Further, given any first entry c of A, $c(S \cup \{0\})$ is central* in $S \cup \{0\}$ so again Theorem 15.5 applies. □

Some common, well behaved, semigroups fail to satisfy the hypothesis that cS is a central* set for each $c \in \mathbb{N}$. For example, in the semigroup $(\mathbb{N}, \cdot)$, $\{x^2 : x \in \mathbb{N}\}$ (the multiplicative analogue of $2S$) is not even a central set. (See Exercise 15.1.2.) Consequently $B = \mathbb{N} \backslash \{x^2 : x \in \mathbb{N}\}$ is a central* set and $A = (2)$ is a first entries matrix. But there does not exist $a \in \mathbb{N}$ with $a^2 \in B$.

To derive the stronger partition result, we need the following immediate corollary. For it, one needs only to recall that for any $n \in \mathbb{N}$, $n\mathbb{N}$ is central*. (In fact it is an IP* set; that is $n\mathbb{N}$ is a member of every idempotent in $(\beta\mathbb{N}, +)$ by Lemma 6.6.)

Corollary 15.7. *Any finite matrix with entries from ω which satisfies the first entries condition is image partition regular over $\mathbb{N}$.* □

Corollary 15.8. *Let A be a finite matrix with entries from ω which satisfies the first entries condition and let $r \in \mathbb{N}$. There exists $k \in \mathbb{N}$ such that whenever $\{1, 2, \ldots, k\}$ is r-colored, there exists $\vec{x} \in \{1, 2, \ldots, k\}^v$ such that the entries of $A\vec{x}$ are monochrome.*

Proof. This is a standard compactness argument using Corollary 15.7. (See Section 5.5 or the proof that (b) implies (a) in Theorem 15.30.) See also Exercise 15.1.3.   □

As we remarked after Definition 15.1, in the definition of image partition regularity we demand that the entries of $\vec{x}$ are all non-zero because that is the natural version for the classical applications. However, one may reasonably ask what happens if one weakens the conclusion. (As we shall see, the answer is "nothing".) Likewise, one may strengthen the definition by requiring that all entries of $A\vec{x}$ be non-zero. Again, we shall see that we get the same answer.

Definition 15.9. Let $(S, +)$ be a semigroup with identity 0, let $u, v \in \mathbb{N}$, and let A be a $u \times v$ matrix with entries from ω.

(a) The matrix A is *weakly image partition regular over S* if and only if whenever $r \in \mathbb{N}$ and $S = \bigcup_{i=1}^{r} E_i$, there exist $i \in \{1, 2, \ldots, r\}$ and $\vec{x} \in (S^v \backslash \{\vec{0}\})$ such that $A\vec{x} \in E_i{}^u$.

(b) The matrix A is *strongly image partition regular over S* if and only if whenever $r \in \mathbb{N}$ and $S \backslash \{0\} = \bigcup_{i=1}^{r} E_i$, there exist $i \in \{1, 2, \ldots, r\}$ and $\vec{x} \in (S \backslash \{0\})^v$ such that $A\vec{x} \in E_i{}^u$.

Theorem 15.10. *Let $(S, +)$ be a commutative semigroup with identity 0. The following statements are equivalent.*

(a) *Whenever A is a finite matrix with entries from ω which*

   satisfies the first entries condition, A is strongly image partition regular over S.

(b) *Whenever A is a finite matrix with entries from ω which*

   satisfies the first entries condition, A is image partition regular over S.

(c) *Whenever A is a finite matrix with entries from ω which*

   satisfies the first entries condition, A is weakly image partition regular over S.

(d) *For each $n \in \mathbb{N}$, $nS \neq \{0\}$.*

Proof. That (a) implies (b) and (b) implies (c) is trivial.

(c) implies (d). Let $n \in \mathbb{N}$ and assume that $nS = \{0\}$. Let

$$A = \begin{pmatrix} 1 & 1 \\ 1 & n \\ 0 & n \end{pmatrix}.$$

Then A satisfies the first entries condition. To see that A is not weakly image partition regular over S, let $E_1 = \{0\}$ and let $E_2 = S \backslash \{0\}$. Suppose we have some $\vec{x} = \begin{pmatrix} x_1 \\ x_2 \end{pmatrix} \in$

$S^2\backslash\{\vec{0}\}$ and some $i \in \{1, 2\}$ such that $A\vec{x} \in (E_i)^3$. Now $0 = nx_2$ so $i = 1$. Thus $x_1 = x_1 + nx_2 = 0$ so that $x_2 = x_1 + x_2 = 0$ and hence $\vec{x} = \vec{0}$, a contradiction.

(d) implies (a). Let A be a finite matrix with entries from ω which satisfies the first entries condition. To see that A is strongly image partition regular over S, let $r \in \mathbb{N}$ and let $S\backslash\{0\} = \bigcup_{i=1}^{r} E_i$. Pick by Corollary 15.8 some $k \in \mathbb{N}$ such that whenever $\{1, 2, \ldots, k\}$ is r-colored, there exists $\vec{x} \in (\{1, 2, \ldots, k\})^v$ such that the entries of $A\vec{x}$ are monochrome.

There exists $z \in S$ such that $\{z, 2z, 3z, \ldots, kz\} \cap \{0\} = \emptyset$. Indeed, otherwise one would have $k!S = \{0\}$. So pick such z and for $i \in \{1, 2, \ldots, r\}$, let $B_i = \{t \in \{1, 2, \ldots, k\} : tz \in E_i\}$. Pick $i \in \{1, 2, \ldots, r\}$ and $\vec{y} \in (\{1, 2, \ldots, k\})^v$ such that $A\vec{y} \in (B_i)^u$. Let $\vec{x} = \vec{y}z$. Then $A\vec{x} \in (E_i)^u$. $\square$

Exercise 15.1.1. Let $(S, +)$ be a semigroup without a two sided identity, and let C be central in S. Adjoin an identity 0 to S and prove that C is central in $S \cup \{0\}$.

Exercise 15.1.2. Prove that $\{x^2 : x \in \mathbb{N}\}$ is not piecewise syndetic (and hence not central) in $(\mathbb{N}, \cdot)$.

Exercise 15.1.3. Prove Corollary 15.8.

Exercise 15.1.4. Let $u, v \in \mathbb{N}$ and let A be a $u \times v$ matrix with entries from ω which is image partition regular over $\mathbb{N}$. Let T denote the set of ultrafilters $p \in \beta\mathbb{N}$ with the property that, for every $E \in p$, there exists $\vec{x} \in \mathbb{N}^v$ for which all the entries of $A\vec{x}$ are in E. Prove that T is a closed subsemigroup of $\beta\mathbb{N}$.

15.2 Kernel Partition Regular Matrices

If one has a group $(G, +)$ and a matrix C with integer entries, one can define C to be *kernel partition regular* over G if and only if whenever $G\backslash\{0\}$ is finitely colored there will exist $\vec{x}$ with monochrome entries such that $C\vec{x} = \vec{0}$. Thus in such a situation, one is saying that monochrome solutions to a given system of homogeneous linear equations can always be found. On the other hand, in an arbitrary semigroup we know that $-x$ may not mean anything, and so we generalize the definition.

Definition 15.11. Let $u, v \in \mathbb{N}$ and let C be a $u \times v$ matrix with entries from $\mathbb{Z}$. Then C^+ and C^- are the $u \times v$ matrices with entries from ω defined by $c_{i,j}^+ = (|c_{i,j}| + c_{i,j})/2$ and $c_{i,j}^- = (|c_{i,j}| - c_{i,j})/2$.

Thus, for example, if

$$C = \begin{pmatrix} 1 & -3 & 0 \\ 0 & -2 & 2 \end{pmatrix}, \quad \text{then} \quad C^+ = \begin{pmatrix} 1 & 0 & 0 \\ 0 & 0 & 2 \end{pmatrix} \quad \text{and} \quad C^- = \begin{pmatrix} 0 & 3 & 0 \\ 0 & 2 & 0 \end{pmatrix}.$$

Note that $C = C^+ - C^-$.

Definition 15.12. Let $(S, +)$ be a semigroup with identity 0, let $u, v \in \mathbb{N}$ and let C be a $u \times v$ matrix with entries from $\mathbb{Z}$. Then C is *kernel partition regular* over S if and only if whenever $r \in \mathbb{N}$ and $S \setminus \{0\} = \bigcup_{i=1}^{r} D_i$, there exist $i \in \{1, 2, \ldots, r\}$ and $\vec{x} \in (D_i)^v$ such that $C^+ \vec{x} = C^- \vec{x}$.

The condition which guarantees kernel partition regularity over most semigroups is known as the *columns condition*. It says that the columns of the matrix can be gathered into groups so that the sum of each group is a linear combination of columns from preceding groups (and in particular the sum of the first group is $\vec{0}$).

Definition 15.13. Let $u, v \in \mathbb{N}$, let C be a $u \times v$ matrix with entries from $\mathbb{Q}$, and let $\vec{c}_1, \vec{c}_2, \ldots, \vec{c}_v$ be the columns of C. Let $R = \mathbb{Z}$ or $R = \mathbb{Q}$. The matrix C satisfies the *columns condition* over R if and only if there exist $m \in \mathbb{N}$ and $I_1, I_2, \ldots, I_m$ such that

(1) $\{I_1, I_2, \ldots, I_m\}$ is a partition of $\{1, 2, \ldots, v\}$.

(2) $\sum_{i \in I_1} \vec{c}_i = \vec{0}$.

(3) If $m > 1$ and $t \in \{2, 3, \ldots, m\}$, let $J_t = \bigcup_{j=1}^{t-1} I_j$. Then there exist $\langle \delta_{t,i} \rangle_{i \in J_t}$ in R such that $\sum_{i \in I_t} \vec{c}_i = \sum_{i \in J_t} \delta_{t,i} \cdot \vec{c}_i$.

Observe that one can effectively check whether a given matrix satisfies the columns condition. (The problem is, however, NP complete because it implies the ability to determine whether a subset of a given set sums to 0.)

Lemma 15.14. *Suppose that C is a $u \times m$ matrix with entries from $\mathbb{Z}$ which satisfies the first entries condition. Let $k = \max\{|c_{i,j}| : (i, j) \in \{1, 2, \ldots, u\} \times \{1, 2, \ldots, m\}\} + 1$, and let E be the $m \times m$ matrix whose entry in row t and column j is*

$$e_{t,j} = \begin{cases} k^{j-t} & \text{if } t \leq j \\ 0 & \text{if } t > j. \end{cases}$$

Then CE is a matrix with entries from ω which also satisfies the first entries condition.

Proof. Let $D = CE$. To see that D satisfies the first entries condition and has entries from ω, let $i \in \{1, 2, \ldots, u\}$ and let $s = \min\{t \in \{1, 2, \ldots, m\} : c_{i,t} \neq 0\}$. Then for $j < s$, $d_{i,j} = 0$ and $d_{i,s} = c_{i,s}$. If $j > s$, then

$$\begin{aligned} d_{i,j} &= \sum_{t=1}^{j} c_{i,t} k^{j-t} \\ &= \sum_{t=s}^{j} c_{i,t} k^{j-t} \\ &\geq k^{j-s} - \sum_{t=s+1}^{j} |c_{i,t}| k^{j-t} \\ &\geq k^{j-s} - \sum_{t=s+1}^{j} (k-1) k^{j-t} \\ &= 1. \end{aligned}$$
$\square$

A connection between matrices satisfying the columns condition and those satisfying the first entries condition is provided by the following lemma.

Lemma 15.15. *Let $u, v \in \mathbb{N}$ and let C be a $u \times v$ matrix with entries from $\mathbb{Q}$ which satisfies the columns condition over $\mathbb{Q}$. There exist $m \in \{1, 2, \ldots, v\}$ and a $v \times m$ matrix B with entries from ω that satisfies the first entries conditions such that $CB = O$, where O is the $u \times m$ matrix with all zero entries. If C satisfies the columns condition over $\mathbb{Z}$, then the matrix B can be chosen so that its only first entry is 1.*

Proof. Pick $m \in \mathbb{N}$, $\langle I_t \rangle_{t=1}^{m}$, $\langle J_t \rangle_{t=2}^{m}$, and $\langle \langle \delta_{t,i} \rangle_{i \in J_t} \rangle_{t=2}^{m}$ as guaranteed by the columns condition for C. Let B' be the $v \times m$ matrix whose entry in row i and column t is given by

$$b'_{i,t} = \begin{cases} -\delta_{t,i} & \text{if } i \in J_t \\ 1 & \text{if } i \in I_t \\ 0 & \text{if } i \notin \bigcup_{j=1}^{t} I_j. \end{cases}$$

We observe that B' satisfies the first entries condition, with the first non-zero entry in each row being 1. We also observe that $CB' = O$. (Indeed, let $j \in \{1, 2, \ldots, u\}$ and $t \in \{1, 2, \ldots, m\}$. If $t = 1$, then $\Sigma_{i=1}^{v} c_{j,i} \cdot b'_{i,t} = \Sigma_{i \in I_1} c_{j,i} = 0$ and if $t > 1$, then $\Sigma_{i=1}^{v} c_{j,i} \cdot b'_{i,t} = \Sigma_{i \in J_t} -\delta_{t,i} \cdot c_{j,i} + \Sigma_{i \in I_t} c_{j,i} = 0$.)

We can choose a positive integer d for which dB' has entries in $\mathbb{Z}$. Then dB' also satisfies the first entries condition and the equation $C(dB') = O$. If all of the numbers $\delta_{i,j}$ are in $\mathbb{Z}$, let $d = 1$.

Let $k = \max\{|db'_{i,j}| : (i, j) \in \{1, 2, \ldots, v\} \times \{1, 2, \ldots, m\}\} + 1$, and let E be the $m \times m$ matrix whose entry in row i and column j is

$$e_{i,j} = \begin{cases} k^{j-i} & \text{if } i \leq j \\ 0 & \text{if } i > j. \end{cases}$$

By Lemma 15.14, $B = dB'E$ has entries in ω and satisfies the first entries condition. Clearly, $CB = O$.

Now, given $i \in \{1, 2, \ldots, v\}$, choose $s \in \{1, 2, \ldots, m\}$ such that $i \in I_s$. If $t < s$, then $b'_{i,t} = 0$ while if $t > j$, then $e_{t,j} = 0$. Thus if $j < s$, then $b_{i,j} = 0$ while if $j \geq s$, then $b_{i,j} = \Sigma_{t=s}^{j} db'_{i,t} k^{j-t}$. In particular the first nonzero entry of row i is d. $\square$

We see that, not only is the columns condition over $\mathbb{Q}$ sufficient for kernel partition regularity over most semigroups, but also that in many cases we can guarantee that solutions to the equations $C^+ \vec{x} = C^- \vec{x}$ can be found in any central set.

Theorem 15.16. *Let $(S, +)$ be an infinite commutative semigroup with identity 0, let $u, v \in \mathbb{N}$, and let C be a $u \times v$ matrix with entries from $\mathbb{Z}$.*

(a) If C satisfies the columns condition over $\mathbb{Z}$, then for any central subset D of S, there exists $\vec{x} \in D^v$ such that $C^+ \vec{x} = C^- \vec{x}$.

(b) If C satisfies the columns condition over $\mathbb{Q}$ and for each $d \in \mathbb{N}$, dS is a central*-set in S, then for any central subset D of S, there exists $\vec{x} \in D^v$ such that $C^+ \vec{x} = C^- \vec{x}$.

(c) If C satisfies the columns condition over $\mathbb{Q}$ and for each $d \in \mathbb{N}$, $dS \neq \{0\}$, then C is kernel partition regular over S.

Proof. (a) Pick a $v \times m$ matrix B as guaranteed for C by Lemma 15.15. Then the only first entry of B is 1 (and $1S = S$ is a central* set) so by Theorem 15.5 we may pick some $\vec{z} \in S^m$ such that $B\vec{z} \in D^v$. Let $\vec{x} = B\vec{z}$. Now $CB = O$ so $C^+B = C^-B$ (and all entries of C^+B and of C^-B are non-negative) so $C^+\vec{x} = C^+B\vec{z} = C^-B\vec{z} = C^-\vec{x}$.

(b) This proof is nearly identical to the proof of (a).

(c) Pick a matrix B as guaranteed for C by Lemma 15.15. Let $r \in \mathbb{N}$ and let $S\backslash\{0\} = \bigcup_{i=1}^{r} D_i$. By Theorem 15.10 choose a vector $\vec{z} \in (S\backslash\{0\})^m$ and $i \in \{1, 2, \ldots, r\}$ such that $B\vec{z} \in (D_i)^v$, and let $\vec{x} = B\vec{z}$. As above, we conclude that $C^+\vec{x} = C^+B\vec{z} = C^-B\vec{z} = C^-\vec{x}$. $\square$

Exercise 15.2.1. Note that the matrix $(2 \ -2 \ 1)$ satisfies the columns condition. Show that there is a set which is central in $(\mathbb{N}, \cdot)$ but contains no solution to the equation $x_1^2 \cdot x_3 = x_2^2$. (Hint: Consider Exercise 15.1.2.)

15.3 Kernel Partition Regularity Over $\mathbb{N}$ — Rado's Theorem

In this section we show that matrices satisfying the columns condition over $\mathbb{Q}$ are precisely those which are kernel partition regular over $(\mathbb{N}, +)$ (which is Rado's Theorem) and are also precisely those which are kernel partition regular over $(\mathbb{N}, \cdot)$.

Given a $u \times v$ matrix C with integer entries and given $\vec{x} \in \mathbb{N}^v$ and $\vec{y} \in \mathbb{N}^u$ we write $\vec{x}^C = \vec{y}$ to mean that for $i \in \{1, 2, \ldots, u\}$, $\prod_{j=1}^{v} x_j^{c_{i,j}} = y_i$.

Theorem 15.17. *Let $u, v \in \mathbb{N}$ and let C be a $u \times v$ matrix with entries from $\mathbb{Z}$. The following statements are equivalent.*

(a) *The matrix C is kernel partition regular over $(\mathbb{N}, +)$.*

(b) *The matrix C is kernel partition regular over $(\mathbb{N}, \cdot)$. That is, whenever $r \in \mathbb{N}$ and $\mathbb{N}\backslash\{1\} = \bigcup_{i=1}^{r} D_i$, there exist $i \in \{1, 2, \ldots, r\}$ and $\vec{x} \in (D_i)^v$ such that $\vec{x}^C = \vec{1}$.*

Proof. (a) implies (b). Let $r \in \mathbb{N}$ and let $\mathbb{N}\backslash\{1\} = \bigcup_{i=1}^{r} D_i$. For each $i \in \{1, 2, \ldots, r\}$, let $A_i = \{n \in \mathbb{N} : 2^n \in D_i\}$. Pick $i \in \{1, 2, \ldots, r\}$ and $\vec{y} \in (A_i)^v$ such that $C\vec{y} = \vec{0}$. For each $j \in \{1, 2, \ldots, u\}$, let $x_j = 2^{y_j}$. Then $\vec{x}^C = \vec{1}$.

(b) implies (a). Let $\langle p_i \rangle_{i=1}^{\infty}$ denote the sequence of prime numbers. Each $q \in \mathbb{Q}\backslash\{0\}$ can be expressed uniquely as $q = \prod_{i=1}^{\infty} p_i^{e_i}$, where $e_i \in \mathbb{Z}$ for every i. We define $\phi : \mathbb{Q}\backslash\{0\} \to \mathbb{Z}$ by putting $\phi(q) = \sum_{i=1}^{\infty} e_i$. (Thus if $q \in \mathbb{N}$, $\phi(q)$ is the length of the prime factorization of q.) We extend ϕ to a function $\psi : (\mathbb{Q}\backslash\{0\})^v \to \mathbb{Z}^v$ by putting $\psi(\vec{x})_i = \phi(x_i)$ for each $\vec{x} \in (\mathbb{Q}\backslash\{0\})^v$ and each $i \in \{1, 2, \ldots, v\}$. Since ϕ is a homomorphism from $(\mathbb{Q}\backslash\{0\}, \cdot)$ to $(\mathbb{Z}, +)$, it follows easily that $\psi(\vec{x}^C) = C\psi(\vec{x})$ for every $\vec{x} \in (\mathbb{Q}\backslash\{0\})^v$. Let $\{E_i\}_{i=1}^{k}$ be a finite partition of $\mathbb{N}$. Then $\{\phi^{-1}[E_i] \cap \mathbb{N}\}_{i=1}^{k}$ is a finite partition of $\mathbb{N}\backslash\{1\}$, because $\phi[\mathbb{N}\backslash\{1\}] \subseteq \mathbb{N}$. So pick $i \in \{1, 2, \ldots, k\}$ and

$\vec{x} \in (\phi^{-1}[E_i] \cap \mathbb{N})^v$ such that $\vec{x}^C = \vec{1}$. Then $\psi(\vec{x}) \in (E_i)^v$ and $C\psi(\vec{x}) = \psi(\vec{x}^C) = \psi(\vec{1}) = \vec{0}$. □

Definition 15.18. Let $u, v \in \mathbb{N}$, let C be a $u \times v$ matrix with entries from $\mathbb{Z}$, and let J and I be disjoint nonempty subsets of $\{1, 2, \ldots, v\}$. Denote the columns of C as $\vec{c}_1, \vec{c}_2, \ldots, \vec{c}_v$.

   (a) If there exist $\langle z_j \rangle_{j \in J}$ in $\mathbb{Q}$ such that $\sum_{j \in I} \vec{c}_j = \sum_{j \in J} z_j \vec{c}_j$, then $E(C, J, I) = \emptyset$.

   (b) If there do not exist $\langle z_j \rangle_{j \in J}$ in $\mathbb{Q}$ such that $\sum_{j \in I} \vec{c}_j = \sum_{j \in J} z_j \vec{c}_j$, then

$$E(C, J, I) = \{q : q \text{ is a prime and there exist } \langle z_j \rangle_{j \in J} \text{ in } \omega,$$
$$a \in \{1, 2, \ldots, q - 1\}, \ d \in \omega, \text{ and } \vec{y} \in \mathbb{Z}^u$$
$$\text{such that } \sum_{j \in J} z_j \vec{c}_j + aq^d \sum_{j \in I} \vec{c}_j = q^{d+1} \vec{y} \}.$$

Lemma 15.19. *Let $u, v \in \mathbb{N}$, let C be a $u \times v$ matrix with entries from $\mathbb{Z}$, and let J and I be disjoint nonempty subsets of $\{1, 2, \ldots, v\}$. Then $E(C, J, I)$ is finite.*

Proof. If (a) of Definition 15.18 applies, the result is trivial so we assume that (a) does not apply. Let $\vec{b} = \sum_{j \in I} \vec{c}_j$. Then $\vec{b}$ is not in the vector subspace of $\mathbb{Q}^u$ spanned by $\langle \vec{c}_j \rangle_{j \in J}$ so pick some $\vec{w} \in \mathbb{Q}^u$ such that $\vec{w} \cdot \vec{c}_j = 0$ for each $j \in J$ and $\vec{w} \cdot \vec{b} \neq 0$. By multiplying by a suitable member of $\mathbb{Z}$ we may assume that all entries of $\vec{w}$ are integers and that $\vec{w} \cdot \vec{b} > 0$. Let $s = \vec{w} \cdot \vec{b}$. We show now that if $q \in E(C, J, I)$, then q divides s.

Let $q \in E(C, J, I)$ and pick $\langle z_j \rangle_{j \in J}$ in ω, $a \in \{1, 2, \ldots, q - 1\}$, $d \in \omega$, and $\vec{y} \in \mathbb{Z}^u$ such that $\sum_{j \in J} z_j \vec{c}_j + aq^d \sum_{j \in I} \vec{c}_j = q^{d+1} \vec{y}$. Then

$$\sum_{j \in J} z_j \vec{c}_j + aq^d \vec{b} = q^{d+1} \vec{y}$$

so

$$\sum_{j \in J} z_j (\vec{w} \cdot \vec{c}_j) + aq^d (\vec{w} \cdot \vec{b}) = q^{d+1}(\vec{w} \cdot \vec{y}).$$

Since $\vec{w} \cdot \vec{c}_j = 0$ for each $j \in J$ we then have that $aq^d s = q^{d+1}(\vec{w} \cdot \vec{y})$ and hence $as = q(\vec{w} \cdot \vec{y})$. Since $a \in \{1, 2, \ldots, q - 1\}$, it follows that q divides s as claimed. □

The equivalence of (a) and (c) in the following theorem is Rado's Theorem. Note that Rado's Theorem provides an effective computation to determine whether a given system of linear homogeneous equations is kernel partition regular.

Theorem 15.20. *Let $u, v \in \mathbb{N}$ and let C be a $u \times v$ matrix with entries from $\mathbb{Z}$. The following statements are equivalent.*

   (a) *The matrix C is kernel partition regular over $(\mathbb{N}, +)$.*

   (b) *The matrix C is kernel partition regular over $(\mathbb{N}, \cdot)$. That is, whenever $r \in \mathbb{N}$ and $\mathbb{N} \setminus \{1\} = \bigcup_{i=1}^{r} D_i$, there exist $i \in \{1, 2, \ldots, r\}$ and $\vec{x} \in (D_i)^v$ such that $\vec{x}^C = \vec{1}$.*

   (c) *The matrix C satisfies the columns condition over $\mathbb{Q}$.*

Proof. The equivalence of (a) and (b) is Theorem 15.17.

(a) implies (c). By Lemma 15.19, whenever J and I are disjoint nonempty subsets of $\{1, 2, \ldots, v\}$, $E(C, J, I)$ is finite. Consequently, we may pick a prime q such that $q > \max\{|\sum_{j \in J} c_{i,j}| : i \in \{1, 2, \ldots, u\}$ and $\emptyset \neq J \subseteq \{1, 2, \ldots, v\}\}$ and whenever J and I are disjoint nonempty subsets of $\{1, 2, \ldots, v\}$, $q \notin E(C, J, I)$.

Given $x \in \mathbb{N}$, pick $a(x) \in \{1, 2, \ldots, q-1\}$ and $\ell(x)$ and $b(x)$ in ω such that $x = a(x) \cdot q^{\ell(x)} + b(x) \cdot q^{\ell(x)+1}$. That is, in the base q expansion of x, $\ell(x)$ is the number of rightmost zeros and $a(x)$ is the rightmost nonzero digit.

For each $a \in \{1, 2, \ldots, q-1\}$ let $A_a = \{x \in \mathbb{N}\setminus\{1\} : a(x) = a\}$. Pick $a \in \{1, 2, \ldots, q-1\}$ and $x_1, x_2, \ldots, x_v \in A_a$ such that $C\vec{x} = \vec{0}$.

Partition $\{1, 2, \ldots, v\}$ according to $\ell(x_j)$. That is, pick $m \in \mathbb{N}$, sets $I_1, I_2, \ldots, I_m$ and numbers $\ell_1 < \ell_2 < \cdots < \ell_m$ such that $\{I_1, I_2, \ldots, I_m\}$ is a partition of $\{1, 2, \ldots, v\}$ and for each $t \in \{1, 2, \ldots, m\}$ and each $j \in I_t$, $\ell(x_j) = \ell_t$. For $n \in \{2, 3, \ldots, m\}$, if any, let $J_n = \bigcup_{t=1}^{n-1} I_t$.

We now show that

(1) $\sum_{j \in I_1} \vec{c}_j = \vec{0}$ and

(2) given $n \in \{2, 3, \ldots, m\}$, $\sum_{j \in I_n} \vec{c}_j$ is a linear combination over $\mathbb{Q}$ of $\langle \vec{c}_j \rangle_{j \in J_n}$,

so that C satisfies the columns condition over $\mathbb{Q}$.

To establish (1), let $d = \ell_1$. Then for any $j \in I_1$ let $e_j = b(x_j)$ and note that $x_j = a \cdot q^d + e_j \cdot q^{d+1}$. For $j \in \{1, 2, \ldots, v\}\setminus I_1$ we have $\ell(x_j) > d$ so $x_j = e_j \cdot q^{d+1}$ for some $e_j \in \mathbb{N}$. Then

$$\vec{0} = \sum_{j=1}^{v} x_j \cdot \vec{c}_j$$
$$= \sum_{j \in I_1} a \cdot q^d \cdot \vec{c}_j + \sum_{j=1}^{v} e_j \cdot q^{d+1} \cdot \vec{c}_j.$$

Suppose that $\sum_{j \in I_1} \vec{c}_j \neq \vec{0}$ and pick some $i \in \{1, 2, \ldots, u\}$ such that $\sum_{j \in I_1} c_{i,j} \neq 0$. Then

$$0 = \sum_{j \in I_1} a \cdot q^d \cdot c_{i,j} + \sum_{j=1}^{v} e_j \cdot q^{d+1} \cdot c_{i,j}$$

so, since $q \nmid a$ we have $q \mid \sum_{j \in I_1} c_{i,j}$, contradicting our choice of q.

Now consider (2). Let $n \in \{2, 3, \ldots, m\}$ be given and let $d = \ell_n$. For $j \in I_n$, let $e_j = b(x_j)$ and note that $x_j = a \cdot q^d + e_j \cdot q^{d+1}$. For $j \in \bigcup_{i=n+1}^{m} I_i$, if any, pick $e_j \in \mathbb{N}$ such that $x_j = e_j \cdot q^{d+1}$. Thus

$$\vec{0} = \sum_{j=1}^{v} x_j \cdot \vec{c}_j = \sum_{j \in J_n} x_j \cdot \vec{c}_j + \sum_{j \in I_n} a \cdot q^d \cdot \vec{c}_j + \sum_{j \in \bigcup_{i=n}^{m} I_i} e_j \cdot q^{d+1} \cdot \vec{c}_j.$$

Let $\vec{y} = \sum_{j \in \bigcup_{i=n}^{m} I_i} (-e_j) \cdot \vec{c}_j$. Then

$$q^{d+1} \vec{y} = \sum_{j \in J_n} x_j \cdot \vec{c}_j + \sum_{j \in I_n} a \cdot q^d \cdot \vec{c}_j.$$

Since $q \notin E(C, J_n, I_n)$ we have that $\sum_{j \in I_n} \vec{c}_j$ is a linear combination of $\langle \vec{c}_j \rangle_{j \in J_n}$ as required.

(c) implies (a). This follows from Theorem 15.16(c). $\square$

To illustrate the use of Theorem 15.20 we establish the case $\ell = 4$ of van der Waerden's Theorem. That is, we show that whenever $\mathbb{N}$ is finitely colored we can get some $a, d \in \mathbb{N}$ with $a, a + d, a + 2d, a + 3d$ monochrome. So one lets $x_1 = a$, $x_2 = a + d$, $x_3 = a + 2d$, and $x_4 = a + 3d$. Then $x_2 - x_1 = d$ so $x_3 = x_1 + 2(x_2 - x_1)$ and $x_4 = x_1 + 3(x_2 - x_1)$. That is we have the equations

$$x_1 - 2x_2 + x_3 = 0$$
$$2x_1 - 3x_2 + x_4 = 0$$

so we are asking to show that the matrix

$$\begin{pmatrix} 1 & -2 & 1 & 0 \\ 2 & -3 & 0 & 1 \end{pmatrix}$$

is kernel partition regular. Indeed it satisfies the columns condition because the sum of its columns is $\vec{0}$.

But there is a problem! The assignment $x_1 = x_2 = x_3 = x_4 = 1$ (or any other number) solves the equations. That is, these equations allow $d = 0$. So we strengthen the problem by asking that d also be the same color as the terms of the progression. Let $x_1 = a, x_2 = d, x_3 = a + d, x_4 = a + 2d$, and $x_5 = a + 3d$. Then we get the equations

$$x_3 = x_1 + x_2$$
$$x_4 = x_1 + 2x_2$$
$$x_5 = x_1 + 3x_2$$

so we need to show that the matrix

$$\begin{pmatrix} 1 & 1 & -1 & 0 & 0 \\ 1 & 2 & 0 & -1 & 0 \\ 1 & 3 & 0 & 0 & -1 \end{pmatrix}$$

is kernel partition regular. For this, let $I_1 = \{1, 3, 4, 5\}$, $I_2 = \{2\}$, $\delta_{2,1} = 0$, $\delta_{2,3} = -1$, $\delta_{2,4} = -2$, and $\delta_{2,5} = -3$.

Exercise 15.3.1. Derive the finite version of the Finite Sums Theorem as a consequence of Rado's Theorem. That is, show that for each $n, r \in \mathbb{N}$, whenever $\mathbb{N} = \bigcup_{i=1}^{r} A_i$, there exist $i \in \{1, 2, \ldots, r\}$ and $\langle x_t \rangle_{t=1}^{n}$ such that $FS(\langle x_t \rangle_{t=1}^{n}) \subseteq A_i$. For example, the case $n = 3$ requires that one show that the matrix

$$\begin{pmatrix} 1 & 1 & 0 & -1 & 0 & 0 & 0 \\ 1 & 0 & 1 & 0 & -1 & 0 & 0 \\ 0 & 1 & 1 & 0 & 0 & -1 & 0 \\ 1 & 1 & 1 & 0 & 0 & 0 & -1 \end{pmatrix}$$

is kernel partition regular.

15.4 Image Partition Regularity Over $\mathbb{N}$

We concluded the last section with a verification of the length 4 version of van der Waerden's Theorem via Rado's Theorem. To do this we had to figure out a set of equations to solve and had to strengthen the problem. On the other hand, we have already seen that the original problem is naturally seen as one of determining that the matrix

$$\begin{pmatrix} 1 & 0 \\ 1 & 1 \\ 1 & 2 \\ 1 & 3 \end{pmatrix}$$

is image partition regular. (And the strengthened problem clearly asks that the matrix

$$\begin{pmatrix} 0 & 1 \\ 1 & 0 \\ 1 & 1 \\ 1 & 2 \\ 1 & 3 \end{pmatrix}$$

be image partition regular.)

Since many problems in Ramsey Theory are naturally stated as deciding whether certain matrices are image partition regular, it is natural to ask for a determination of precisely which matrices are image partition regular. We do this in this section.

Since we are now dealing with the semigroup $(\mathbb{N}, +)$, we extend Definition 15.1 to allow entries from $\mathbb{Q}$.

Definition 15.21. Let $u, v \in \mathbb{N}$, and let A be a $u \times v$ matrix with entries from $\mathbb{Q}$. Then A is *image partition regular over* $\mathbb{N}$ if and only if whenever $r \in \mathbb{N}$ and $\mathbb{N} = \bigcup_{i=1}^{r} E_i$, there exist $i \in \{1, 2, \ldots, r\}$ and $\vec{x} \in \mathbb{N}^v$ such that $A\vec{x} \in E_i{}^u$.

We need a preliminary lemma which deals with a specialized notion from linear algebra.

Definition 15.22. Let $u, v \in \mathbb{N}$, let $\vec{c}_1, \vec{c}_2, \ldots, \vec{c}_v$ be in $\mathbb{Q}^u$, and let $I \subseteq \{1, 2, \ldots, v\}$. The *I-restricted span* of $(\vec{c}_1, \vec{c}_2, \ldots, \vec{c}_v)$ is

$$\{\Sigma_{i=1}^{v} \alpha_i \cdot \vec{c}_i : \text{ each } \alpha_i \in \mathbb{R} \text{ and if } i \in I, \text{ then } \alpha_i \geq 0\}.$$

Lemma 15.23. *Let $u, v \in \mathbb{N}$, let $\vec{c}_1, \vec{c}_2, \ldots, \vec{c}_v$ be in $\mathbb{Q}^u$, and let $I \subseteq \{1, 2, \ldots, v\}$. Let S be the I-restricted span of $(\vec{c}_1, \vec{c}_2, \ldots, \vec{c}_v)$.*
   (a) S is closed in $\mathbb{R}^u$.
   (b) If $\vec{y} \in S \cap \mathbb{Q}^u$, then there exist $\delta_1, \delta_2, \ldots, \delta_v$ in $\mathbb{Q}$ such that $\vec{y} = \Sigma_{i=1}^{v} \delta_i \cdot \vec{c}_i$ and $\delta_i \geq 0$ for each $i \in I$.

Proof. (a) We proceed by induction on $|I|$ (for all v). If $I = \emptyset$, this is simply the assertion that any vector subspace of $\mathbb{R}^u$ is closed. So we assume $I \neq \emptyset$ and assume without loss of generality that $1 \in I$. Let T be the $(I \backslash \{1\})$-restricted span of $(\vec{c}_2, \vec{c}_3, \ldots, \vec{c}_v)$. By the induction hypothesis, T is closed.

To see that S is closed, let $\vec{b} \in c\ell\, S$. We show $\vec{b} \in S$. If $\vec{c}_1 = \vec{0}$, then $\vec{b} \in c\ell\, S = c\ell\, T = T = S$, so assume that $\vec{c}_1 \neq \vec{0}$. For each $n \in \mathbb{N}$, pick $\langle \alpha_i(n) \rangle_{i=1}^{v}$ such that $\alpha_i(n) \geq 0$ for each $i \in I$ and $\| \vec{b} - \sum_{i=1}^{v} \alpha_i(n) \cdot \vec{c}_i \| < 1/n$.

Assume first that $\{\alpha_1(n) : n \in \mathbb{N}\}$ is bounded. Pick a limit point δ of the sequence $\langle \alpha_1(n) \rangle_{n=1}^{\infty}$ and note that $\delta \geq 0$. We claim that $\vec{b} - \delta \cdot \vec{c}_1 \in T$. To see this we show that $\vec{b} - \delta \cdot \vec{c}_1 \in c\ell\, T$. So let $\epsilon > 0$ be given and pick $n > 2/\epsilon$ such that $|\alpha_1(n) - \delta| < \epsilon/(2\|\vec{c}_1\|)$. Then

$$\| \vec{b} - \delta \cdot \vec{c}_1 - \sum_{i=2}^{v} \alpha_i(n) \cdot \vec{c}_i \| \leq \| \vec{b} - \sum_{i=1}^{v} \alpha_i(n) \cdot \vec{c}_i \| + \| \alpha_1(n) \vec{c}_1 - \delta \vec{c}_1 \| < \epsilon.$$

Since $\vec{b} - \delta \cdot \vec{c}_1 \in T$, we have $\vec{b} \in \delta \cdot \vec{c}_1 + T \subseteq S$.

Now assume that $\{\alpha_1(n) : n \in \mathbb{N}\}$ is unbounded. To see that S is closed, it suffices to show that $-\vec{c}_1 \in T$. For then S is in fact the $(I \backslash \{1\})$-restricted span of $(\vec{c}_1, \vec{c}_2, \ldots, \vec{c}_v)$ which is closed by the induction hypothesis.

To see that $-\vec{c}_1 \in T$, we show that $-\vec{c}_1 \in c\ell\, T$. Let $\epsilon > 0$ be given and pick $n \in \mathbb{N}$ such that $\alpha_1(n) > (1 + \|\vec{b}\|)/\epsilon$. For $i \in \{2, 3, \ldots, v\}$ let $\delta_i = \alpha_i(n)/\alpha_1(n)$ and note that for $i \in I \backslash \{1\}$, $\delta_i \geq 0$. Then

$$\begin{aligned}
\| -\vec{c}_1 - \sum_{i=2}^{v} \delta_i \cdot \vec{c}_i \| &\leq \| \vec{b}/\alpha_1(n) - \vec{c}_1 - \sum_{i=2}^{v} (\alpha_i(n)/\alpha_1(n)) \cdot \vec{c}_i \| + \| \vec{b}/\alpha_1(n) \| \\
&= (1/\alpha_1(n)) \cdot \| \vec{b} - \sum_{i=1}^{v} \alpha_i(n) \cdot \vec{c}_i \| + \| \vec{b} \|/\alpha_1(n) \\
&< 1/(n \cdot \alpha_1(n)) + \| \vec{b} \|/\alpha_1(n) \\
&< (1 + \|\vec{b}\|)/\alpha_1(n) \\
&< \epsilon.
\end{aligned}$$

(b) Again we proceed by induction on $|I|$. The case $I = \emptyset$ is immediate, being merely the assertion that a rational vector in the linear span of some other rational vectors is actually in their linear rational span (which is true because one is solving linear equations with rational coefficients).

So assume $I \neq \emptyset$. Let $X = \{\vec{x} \in \mathbb{R}^v : \sum_{i=1}^{v} x_i \cdot \vec{c}_i = \vec{y}\}$. Thus X is an affine subspace of $\mathbb{R}^v$, and we are given (by the fact that $\vec{y} \in S$) that there is some $\vec{x} \in X$ such that $x_i \geq 0$ for all $i \in I$. Also (by the case $I = \emptyset$) there is some $\vec{z} \in X$ such that $z_i \in \mathbb{Q}$ for all $i \in \{1, 2, \ldots, v\}$. If $z_i \geq 0$ for all $i \in I$, then we are done, so assume there is some $i \in I$ such that $z_i < 0$. Given $i \in I$, $\{t \in [0, 1] : (1 - t)x_i + tz_i \geq 0\}$ is $[0, 1]$ if $z_i \geq 0$ and is $[0, x_i/(x_i - z_i)]$ if $z_i < 0$. Let t be the largest member of $[0, 1]$ such that the vector $\vec{w} = (1 - t) \cdot \vec{x} + t \cdot \vec{z}$ satisfies $w_i \geq 0$ for all $i \in I$. Then for some $i \in I$ we have that $w_i = 0$. We assume without loss of generality that $1 \in I$ and $w_1 = 0$. Then $\vec{y} = \sum_{i=2}^{v} w_i \cdot \vec{c}_i$ so $\vec{y}$ is in the $(I \backslash \{1\})$-restricted span of $(\vec{c}_2, \vec{c}_3, \ldots, \vec{c}_v)$ so by the induction hypothesis we may pick $\delta_2, \delta_3, \ldots, \delta_v$ in $\mathbb{Q}$ such that $\vec{y} = \sum_{i=2}^{v} \delta_i \cdot \vec{c}_i$ and $\delta_i \geq 0$ for each $i \in I \backslash \{1\}$. Letting $\delta_1 = 0$, we are done. $\square$

Theorem 15.24. *Let $u, v \in \mathbb{N}$ and let A be a $u \times v$ matrix with entries from $\mathbb{Q}$. The following statements are equivalent.*

(a) *A is image partition regular over $\mathbb{N}$.*

(b) *Let $\vec{c}_1, \vec{c}_2, \ldots, \vec{c}_v$ be the columns of A. There exist $s_1, s_2, \ldots, s_v \in \{x \in \mathbb{Q} : x > 0\}$ such that the matrix*

$$M = \begin{pmatrix} & & & & -1 & 0 & \cdots & 0 \\ & & & & 0 & -1 & \cdots & 0 \\ s_1 \cdot \vec{c}_1 & s_2 \cdot \vec{c}_2 & \cdots & s_v \cdot \vec{c}_v & \vdots & \vdots & \ddots & \vdots \\ & & & & 0 & 0 & \cdots & -1 \end{pmatrix}$$

is kernel partition regular over $\mathbb{N}$.

(c) *There exist $m \in \mathbb{N}$ and a $u \times m$ matrix B with entries from $\mathbb{Q}$ which satisfies the first entries condition such that given any $\vec{y} \in \mathbb{N}^m$ there is some $\vec{x} \in \mathbb{N}^v$ with $A\vec{x} = B\vec{y}$.*

(d) *There exist $m \in \mathbb{N}$ and a $u \times m$ matrix C with entries from $\mathbb{Z}$ which satisfies the first entries condition such that given any $\vec{y} \in \mathbb{N}^m$ there is some $\vec{x} \in \mathbb{N}^v$ with $A\vec{x} = C\vec{y}$.*

(e) *There exist $m \in \mathbb{N}$ and a $u \times m$ matrix D with entries from ω which satisfies the first entries condition such that given any $\vec{y} \in \mathbb{N}^m$ there is some $\vec{x} \in \mathbb{N}^v$ with $A\vec{x} = D\vec{y}$.*

(f) *There exist $m \in \mathbb{N}$, a $u \times m$ matrix E with entries from ω, and $c \in \mathbb{N}$ such that E satisfies the first entries condition, c is the only first entry of E, and given any $\vec{y} \in \mathbb{N}^m$ there is some $\vec{x} \in \mathbb{N}^v$ with $A\vec{x} = E\vec{y}$.*

Proof. (a) implies (b). Given any $p \in \mathbb{N}\backslash\{1\}$, we let the *start base p* coloring of $\mathbb{N}$ be the function

$$\sigma_p : \mathbb{N} \to \{1, 2, \ldots, p-1\} \times \{0, 1, \ldots, p-1\} \times \{0, 1\}$$

defined as follows: given $y \in \mathbb{N}$, write $y = \Sigma_{t=0}^n a_t p^t$, where each $a_t \in \{0, 1, \ldots, p-1\}$ and $a_n \neq 0$; if $n > 0$, $\sigma_p(y) = (a_n, a_{n-1}, i)$ where $i \equiv n \pmod 2$; if $n = 0$, $\sigma_p(y) = (a_0, 0, 0)$. (For example, given $p > 8$, if $x = 8320100$, $y = 503011$, $z = 834$, and $w = 834012$, all written in base p, then $\sigma_p(x) = \sigma_p(z) = (8, 3, 0)$, $\sigma_p(w) = (8, 3, 1)$, and $\sigma_p(y) = (5, 0, 1)$.)

Let $\vec{c}_1, \vec{c}_2, \ldots, \vec{c}_v$ be the columns of A and let $\vec{d}_1, \vec{d}_2, \ldots, \vec{d}_u$ denote the columns of the $u \times u$ identity matrix. Let B be the matrix

$$\begin{pmatrix} s_1 \cdot \vec{c}_1 & s_2 \cdot \vec{c}_2 & \cdots & s_v \cdot \vec{c}_v & -\vec{d}_1 & -\vec{d}_2 & \cdots & -\vec{d}_u \end{pmatrix}$$

where $s_1, s_2, \ldots, s_v$ are as yet unspecified positive rationals. Denote the columns of B by $\vec{b}_1, \vec{b}_2, \ldots, \vec{b}_{u+v}$. Then

$$\vec{b}_t = \begin{cases} s_t \cdot \vec{c}_t & \text{if } t \leq v \\ -\vec{d}_{t-v} & \text{if } t > v. \end{cases}$$

Given any $p \in \mathbb{N}\backslash\{1\}$ and any $x \in \mathbb{N}$, let $\gamma(p, x) = \max\{n \in \omega : p^n \leq x\}$. Now temporarily fix some $p \in \mathbb{N}\backslash\{1\}$. We obtain $m = m(p)$ and an ordered partition

$\big(D_1(p), D_2(p), \ldots, D_m(p)\big)$ of $\{1, 2, \ldots, u\}$ as follows. Pick $\vec{x} \in \mathbb{N}^v$ such that $A\vec{x}$ is monochrome with respect to the start base p coloring and let $\vec{y} = A\vec{x}$. Now divide up $\{1, 2, \ldots, u\}$ according to which of the y_i's start furthest to the left in their base p representation. That is, we get $D_1(p), D_2(p), \ldots, D_m(p)$ so that

(1) if $k \in \{1, 2, \ldots, m\}$ and $i, j \in D_k(p)$, then $\gamma(p, y_i) = \gamma(p, y_j)$, and

(2) if $k \in \{2, 3, \ldots, m\}$, $i \in D_k(p)$, and $j \in D_{k-1}(p)$, then $\gamma(p, y_j) > \gamma(p, y_i)$. We also observe that since $\sigma_p(y_i) = \sigma_p(y_j)$, we have that $\gamma(p, y_j) \equiv \gamma(p, y_i)$ (mod 2) and hence $\gamma(p, y_j) \geq \gamma(p, y_i) + 2$.

There are only finitely many ordered partitions of $\{1, 2, \ldots, u\}$. Therefore we may pick an infinite subset P of $\mathbb{N}\backslash\{1\}$, $m \in \mathbb{N}$, and an ordered partition $(D_1, D_2, \ldots, D_m)$ of $\{1, 2, \ldots, u\}$ so that for all $p \in P$, $m(p) = m$ and

$$\big(D_1(p), D_2(p), \ldots, D_m(p)\big) = (D_1, D_2, \ldots, D_m).$$

We shall utilize $(D_1, D_2, \ldots, D_m)$ to find $s_1, s_2, \ldots, s_v$ and a partition of $\{1, 2, \ldots, u + v\}$ as required for the columns condition.

We proceed by induction. First we shall find $E_1 \subseteq \{1, 2, \ldots, v\}$, specify $s_i \in \mathbb{Q}^+ = \{s \in \mathbb{Q} : s > 0\}$ for each $i \in E_1$, let $I_1 = E_1 \cup (v + D_1)$, and show that $\sum_{i \in I_1} \vec{b}_i = \vec{0}$. That is, we shall show that $\sum_{i \in E_1} s_i \cdot \vec{c}_i = \sum_{i \in D_1} \vec{d}_i$. In order to do this, we show that $\sum_{i \in D_1} \vec{d}_i$ is in the $\{1, 2, \ldots, v\}$-restricted span of $(\vec{c}_1, \vec{c}_2, \ldots, \vec{c}_v)$. For then by Lemma 15.23(b) one has $\sum_{i \in D_1} \vec{d}_i = \sum_{i=1}^{v} \alpha_i \cdot \vec{c}_i$, where each $\alpha_i \in \mathbb{Q}$ and each $\alpha_i \geq 0$. Let $E_1 = \{i \in \{1, 2, \ldots, v\} : \alpha_i > 0\}$ and for $i \in E_i$, let $s_i = \alpha_i$.

Let S be the $\{1, 2, \ldots, v\}$-restricted span of $(\vec{c}_1, \vec{c}_2, \ldots, \vec{c}_v)$. In order to show that $\sum_{i \in D_1} \vec{d}_i$ is in S it suffices, by Lemma 15.23(a), to show that $\sum_{i \in D_1} \vec{d}_i$ is in $c\ell S$. To this end, let $\epsilon > 0$ be given and pick $p \in P$ with $p > u/\epsilon$. Pick the $\vec{x} \in \mathbb{N}^v$ and $\vec{y} \in \mathbb{N}^u$ that we used to get $\big(D_1(p), D_2(p), \ldots, D_m(p)\big)$. That is, $A\vec{x} = \vec{y}$, $\vec{y}$ is monochrome with respect to the start base p coloring, and $(D_1, D_2, \ldots, D_m) = \big(D_1(p), D_2(p), \ldots, D_m(p)\big)$ is the ordered partition of $\{1, 2, \ldots, u\}$ induced by the starting positions of the y_i's. Pick γ so that for all $i \in D_1$, $\gamma(p, y_i) = \gamma$. Pick $(a, b, c) \in \{1, 2, \ldots, p-1\} \times \{0, 1, \ldots, p-1\} \times \{0, 1\}$ such that $\sigma_p(y_i) = (a, b, c)$ for all $i \in \{1, 2, \ldots, u\}$. Let $\ell = a + b/p$ and observe that $1 \leq \ell < p$. For $i \in D_1$, $y_i = a \cdot p^\gamma + b \cdot p^{\gamma-1} + z_i \cdot p^{\gamma-2}$ where $0 \leq z_i < p$, and hence $y_i/p^\gamma = \ell + z_i/p^2$; let $\lambda_i = z_i/p^2$ and note that $0 \leq \lambda_i < 1/p$. For $i \in \bigcup_{j=2}^{m} D_j$, we have $\gamma(p, y_i) \leq \gamma - 2$; let $\lambda_i = y_i/p^\gamma$ and note that $0 < \lambda_i < 1/p$.

Now $A\vec{x} = \vec{y}$ so

$$\textstyle\sum_{i=1}^{v} x_i \cdot \vec{c}_i = \vec{y} = \sum_{i=1}^{v} y_i \cdot \vec{d}_i = \sum_{i \in D_1} y_i \cdot \vec{d}_i + \sum_{j=2}^{m} \sum_{i \in D_j} y_i \cdot \vec{d}_i.$$

Thus $\sum_{i=1}^{v} (x_i/p^\gamma) \cdot \vec{c}_i = \sum_{i \in D_1} \ell \cdot \vec{d}_i + \sum_{i=1}^{u} \lambda_i \cdot \vec{d}_i$ and consequently

$$\| \textstyle\sum_{i \in D_1} \vec{d}_i - \sum_{i=1}^{v} \big(x_i/(\ell p^\gamma)\big) \cdot \vec{c}_i \| = \| \sum_{i=1}^{u} (\lambda_i/\ell) \cdot \vec{d}_i \| \leq \sum_{i=1}^{u} |\lambda_i/\ell| < u/p < \epsilon.$$

Since $\sum_{i=1}^{v} \big(x_i/(\ell p^\gamma)\big) \cdot \vec{c}_i$ is in S, we have that $\sum_{i \in D_i} \vec{d}_i \in c\ell S$ as required.

Now let $k \in \{2, 3, \ldots, m\}$ and assume we have chosen $E_1, E_2, \ldots, E_{k-1} \subseteq \{1, 2, \ldots, v\}$, $s_i \in \mathbb{Q}^+$ for $i \in \bigcup_{j=1}^{k-1} E_j$, and $I_j = E_j \cup (v + D_j)$ as required for the columns condition. Let $L_k = \bigcup_{j=1}^{k-1} E_j$ and let $M_k = \bigcup_{j=1}^{k-1} D_j$ and enumerate M_k in order as $q(1), q(2), \ldots, q(r)$. We claim that it suffices to show that $\Sigma_{i \in D_k} \vec{d}_i$ is in the $\{1, 2, \ldots, v\}$-restricted span of $(\vec{c}_1, \vec{c}_2, \ldots, \vec{c}_v, \vec{d}_{q(1)}, \vec{d}_{q(2)}, \ldots, \vec{d}_{q(r)})$, which we will again denote by S. Indeed, assume we have done this and pick by Lemma 15.23(b) $\alpha_1, \alpha_2, \ldots, \alpha_v$ in $\{x \in \mathbb{Q} : x \geq 0\}$ and $\delta_{q(1)}, \delta_{q(2)}, \ldots, \delta_{q(r)}$ in $\mathbb{Q}$ such that $\Sigma_{i \in D_k} \vec{d}_i = \Sigma_{i=1}^v \alpha_i \cdot \vec{c}_i + \Sigma_{i=1}^r \delta_{q(i)} \cdot \vec{d}_{q(i)}$. Let $E_k = \{i \in \{1, 2, \ldots, v\} \backslash L_k : \alpha_i > 0\}$ and for $i \in E_k$, let $s_i = \alpha_i$. Let $I_k = E_k \cup (v + D_k)$. Then

$$
\begin{aligned}
\Sigma_{i \in I_k} \vec{b}_i &= \Sigma_{i \in E_k} \alpha_i \cdot \vec{c}_i + \Sigma_{i \in D_k} -\vec{d}_i \\
&= \Sigma_{i \in E_k} \alpha_i \cdot \vec{c}_i + \Sigma_{i=1}^v -\alpha_i \cdot \vec{c}_i + \Sigma_{i=1}^r -\delta_{q(i)} \cdot \vec{d}_{q(i)}.
\end{aligned}
$$

Now, if $i \in \{1, 2, \ldots, v\}$ and $\alpha_i \neq 0$, then $i \in L_k \cup E_k$ so

$$
\Sigma_{i \in E_k} \alpha_i \cdot \vec{c}_i + \Sigma_{i=1}^v -\alpha_i \cdot \vec{c}_i + \Sigma_{i=1}^r -\delta_{q(i)} \cdot \vec{d}_{q(i)} = \Sigma_{i \in L_k} -\alpha_i \cdot \vec{c}_i + \Sigma_{i=1}^r -\delta_{q(i)} \cdot \vec{d}_{q(i)}
$$

and

$$
\begin{aligned}
\Sigma_{i \in L_k} -\alpha_i \cdot \vec{c}_i + \Sigma_{i=1}^r -\delta_{q(i)} \cdot \vec{d}_{q(i)} &= \Sigma_{i \in L_k} -\alpha_i \cdot \vec{c}_i + \Sigma_{i \in M_k} -\delta_i \cdot \vec{d}_i \\
&= \Sigma_{i \in L_k} (-\alpha_i/s_i) \cdot \vec{b}_i + \Sigma_{i \in M_k} \delta_i \cdot \vec{b}_{v+i}.
\end{aligned}
$$

Let $\beta_i = -\alpha_i/s_i$ if $i \in L_k$ and let $\beta_{v+i} = \delta_i$ if $i \in M_k$. Then we have

$$
\Sigma_{i \in \bigcup_{j=1}^{k-1} I_j} \beta_i \cdot \vec{b}_i = \Sigma_{i \in L_k} (-\alpha_i/s_i) \cdot \vec{b}_i + \Sigma_{i \in M_k} \delta_i \cdot \vec{b}_{v+i} = \Sigma_{i \in I_k} \vec{b}_i
$$

as required for the columns condition.

In order to show that $\Sigma_{i \in D_k} \vec{d}_i$ is in S, it suffices, by Lemma 15.23(a) to show that $\Sigma_{i \in D_k} \vec{d}_i$ is in $cl\, S$. To this end, let $\epsilon > 0$ be given and pick $p \in P$ with $p > u/\epsilon$. Pick the $\vec{x} \in \mathbb{N}^v$ and $\vec{y} \in \mathbb{N}^u$ that we used to get $(D_1(p), D_2(p), \ldots, D_m(p))$. Pick γ so that for all $i \in D_k$, $\gamma(p, y_i) = \gamma$. Pick $(a, b, c) \in \{1, 2, \ldots, p-1\} \times \{0, 1, \ldots, p-1\} \times \{0, 1\}$ such that $\sigma_p(y_i) = (a, b, c)$ for all $i \in \{1, 2, \ldots, u\}$. Let $\ell = a + b/p$. For $i \in D_k$, $y_i = a \cdot p^\gamma + b \cdot p^{\gamma-1} + z_i \cdot p^{\gamma-2}$ where $0 \leq z_i < p$, and hence $y_i/p^\gamma = \ell + z_i/p^2$; let $\lambda_i = z_i/p^2$ and note that $0 \leq \lambda_i < 1/p$. For $i \in \bigcup_{j=k+1}^m D_j$, we have $\gamma(p, y_i) \leq \gamma - 2$; let $\lambda_i = y_i/p^\gamma$ and note that $0 < \lambda_i < 1/p$. (Of course we have no control on the size of y_i/p^γ for $i \in M_k$.)

Now $A\vec{x} = \vec{y}$ so

$$
\begin{aligned}
\Sigma_{i=1}^v x_i \cdot \vec{c}_i &= \Sigma_{i=1}^u y_i \cdot \vec{d}_i \\
&= \Sigma_{i \in M_k} y_i \cdot \vec{d}_i + \Sigma_{i \in D_k} y_i \cdot \vec{d}_i + \Sigma_{j=k+1}^m \Sigma_{i \in D_j} y_i \cdot \vec{d}_i,
\end{aligned}
$$

where $\Sigma_{j=k+1}^m \Sigma_{i \in D_j} y_i \cdot \vec{d}_i = 0$ if $k = m$.

Thus $\Sigma_{i=1}^{v}(x_i/p^\gamma) \cdot \vec{c}_i = \Sigma_{i \in M_k}(y_i/p^\gamma) \cdot \vec{d}_i + \Sigma_{i \in D_k} \ell \cdot \vec{d}_i + \Sigma_{j=k}^{m} \Sigma_{i \in D_j} \lambda_i \cdot \vec{d}_i$. Consequently,

$$\| \Sigma_{i \in D_k} \vec{d}_i - \left(\Sigma_{i=1}^{v}\left(x_i/(\ell p^\gamma)\right) \cdot \vec{c}_i + \Sigma_{i \in M_k}\left(-y_i/(\ell p^\gamma)\right) \cdot \vec{d}_i\right)\|$$
$$= \| \Sigma_{j=k}^{m} \Sigma_{i \in D_j} (\lambda_i/\ell) \cdot \vec{d}_i \|$$
$$\leq \Sigma_{j=k}^{m} \Sigma_{i \in D_j} |\lambda_i/\ell|$$
$$< u/p$$
$$< \epsilon,$$

which suffices, since $x_i/(\ell p^\gamma) > 0$ for $i \in \{1, 2, \ldots, v\}$.

Having chosen $I_1, I_2, \ldots, I_m$, if $\{1, 2, \ldots, u + v\} = \bigcup_{j=1}^{m} I_j$, we are done. So assume $\{1, 2, \ldots, u + v\} \neq \bigcup_{j=1}^{m} I_j$. Let $I_{m+1} = \{1, 2, \ldots, u + v\} \backslash \bigcup_{j=1}^{m} I_j$. Now $\{1, 2, \ldots, u\} = \bigcup_{j=1}^{m} D_j$, so $\{-\vec{d}_1, -\vec{d}_2, \ldots, -\vec{d}_u\} \subseteq \{\vec{b}_i : i \in \bigcup_{j=1}^{m} I_j\}$ and hence we can write $\Sigma_{i \in I_{m+1}} \vec{b}_i$ as a linear combination of $\{\vec{b}_i : i \in \bigcup_{j=1}^{m} I_j\}$.

(b) implies (c). By Theorem 15.20 M satisfies the columns condition over $\mathbb{Q}$. Thus by Lemma 15.15, there exist some $m \in \{1, 2, \ldots, u + v\}$ and a $(u + v) \times m$ matrix F with entries from ω which satisfies the first entries condition such that $MF = O$, the $u \times m$ matrix whose entries are all zero. Let S denote the diagonal $v \times v$ matrix whose diagonal entries are $s_1, s_2, \ldots, s_v$. Then M can be written in block form as $\begin{pmatrix} AS & -I \end{pmatrix}$ and F can be written in block form as $\begin{pmatrix} G \\ H \end{pmatrix}$, where I denotes the $u \times u$ identity matrix and G and H denote $v \times m$ and $u \times m$ matrices respectively. We observe that G and H are first entries matrices and that $ASG = H$, because $MF = O$. We can choose $d \in \mathbb{N}$ such that all the entries of dSG are in ω. Let $B = dH$. Then B is a first entries matrix. Let $\vec{y} \in \mathbb{N}^m$ be given and let $\vec{x} = dSG\vec{y}$. Then $A\vec{x} = B\vec{y}$ as required.

(c) implies (d). Given B, let $d \in \mathbb{N}$ be a common multiple of the denominators in entries of B and let $C = Bd$. Given $\vec{y}$, let $\vec{z} = d\vec{y}$ and pick $\vec{x}$ such that $A\vec{x} = B\vec{z} = C\vec{y}$.

(d) implies (e). Let C be given as guaranteed by (d). By Lemma 15.14, there is an $m \times m$ matrix E for which $D = CE$ has entries from ω and also satisfies the first entries condition. Given $\vec{y} \in \mathbb{N}^m$, we define $\vec{z} \in \mathbb{N}^m$ by $\vec{z} = E\vec{y}$. Pick $\vec{x} \in \mathbb{N}^v$ such that $A\vec{x} = C\vec{z}$. Then $C\vec{z} = CE\vec{y} = D\vec{y}$.

(e) implies (f). Let D be given as guaranteed by (e), and for each $j \in \{1, 2, \ldots, m\}$, pick $w_j \in \mathbb{N}$ such that for any $i \in \{1, 2, \ldots, v\}$ if $j = \min\{t \in \{1, 2, \ldots, m\} : d_{i,t} \neq 0\}$, then $d_{i,j} = w_j$. (That is, w_j is the first entry associated with column j, if there are any first entries in column j.) Let c be a common multiple of $\{w_1, w_2, \ldots, w_m\}$. Define the $u \times m$ matrix E by, for $(i, j) \in \{1, 2, \ldots, u\} \times \{1, 2, \ldots, m\}$, $e_{i,j} = (c/w_j)d_{i,j}$. Now, given $\vec{y} \in \mathbb{N}^m$, we define $\vec{z} \in \mathbb{N}^m$ by, for $j \in \{1, 2, \ldots, m\}$, $z_j = (c/w_j)y_j$. Then $D\vec{z} = E\vec{y}$.

(f) implies (a). For each $c \in \mathbb{N}$, $c\mathbb{N}$ is a member of every idempotent in $\beta\mathbb{N}$ by Lemma 6.6, and is in particular a central* set. Thus by Corollary 15.6 E is image partition regular over $\mathbb{N}$. To see that A is image partition regular over $\mathbb{N}$, let $r \in \mathbb{N}$ and let $\mathbb{N} = \bigcup_{i=1}^{r} F_i$. Pick $i \in \{1, 2, \ldots, r\}$ and $\vec{y} \in \mathbb{N}^m$ such that $E\vec{y} \in F_i^u$. Then pick $\vec{x} \in \mathbb{N}^v$ such that $A\vec{x} = E\vec{y}$. $\square$

Statement (b) of Theorem 15.24 is a computable condition. We illustrate its use by determining whether the matrices

$$\begin{pmatrix} 3 & 1 \\ 2 & 3 \end{pmatrix} \quad \text{and} \quad \begin{pmatrix} 1 & -1 \\ 3 & 2 \\ 4 & 6 \end{pmatrix}$$

are image partition regular over $\mathbb{N}$.

Consider the matrix

$$\begin{pmatrix} 3s_1 & 1s_2 & -1 & 0 \\ 2s_1 & 3s\ 2 & 0 & -1 \end{pmatrix}$$

where s_1 and s_2 are positive rationals. One quickly sees that the only possible choice for a set I_1 of columns summing to $\vec{0}$ is $I_1 = \{1, 2, 3, 4\}$ and then, solving the equations

$$3s_1 + s_2 = 1$$
$$2s_1 + 3s_2 = 1$$

one gets $s_1 = 2/7$ and $s_2 = 1/7$ and one has established that

$$\begin{pmatrix} 3 & 1 \\ 2 & 3 \end{pmatrix}$$

is image partition regular.

Now consider the matrix

$$\begin{pmatrix} 1s_1 & -1s_2 & -1 & 0 & 0 \\ 3s_1 & 2s_2 & 0 & -1 & 0 \\ 4s_1 & 6s_2 & 0 & 0 & -1 \end{pmatrix}$$

where again s_1 and s_2 are positive rationals. By a laborious consideration of cases one sees that the only non zero choices for s_1 and s_2 which make this matrix satisfy the columns condition are $s_1 = 3/5$ and $s_2 = -2/5$. Consequently,

$$\begin{pmatrix} 1 & -1 \\ 3 & 2 \\ 4 & 6 \end{pmatrix}$$

is *not* image partition regular.

Exercise 15.4.1. Prove that the matrix

$$\begin{pmatrix} 2 & 0 & 0 \\ 4 & 1 & -9 \\ 2 & -2 & 3 \end{pmatrix}$$

is image partition regular.

Exercise 15.4.2. Let A be a $u \times v$ matrix with entries from ω. Prove that A is image partition regular over $(\mathbb{N}, +)$ if and only if A is image partition regular over $(\mathbb{N}, \cdot)$. (Hint: Consider the proof of Theorem 15.17.)

15.5 Matrices with Entries from Fields

In the general situation where one is dealing with arbitrary commutative semigroups, we restricted our coefficients to have entries from ω. In the case of $(\mathbb{N}, +)$, we allowed the entries of matrices to come from $\mathbb{Q}$. There is another natural setting in which the entries of a coefficient matrix can be allowed to come from other sets. This is the case in which the semigroup is a vector space over a field. We show here that the appropriate analogue of the first entries condition is sufficient for image partition regularity in this case, and that the appropriate analogue of the columns condition is sufficient for kernel partition regularity. We also show that in the case of a vector space over a *finite* field, the columns condition is also necessary for kernel partition regularity.

We begin by generalizing the notion of the first entries condition from Definition 15.2.

Definition 15.25. Let F be a field, let $u, v \in \mathbb{N}$, and let A be a $u \times v$ matrix with entries from F. Then A satisfies the *first entries condition over F* if and only if no row of A is $\vec{0}$ and whenever $i, j \in \{1, 2, \ldots, u\}$ and

$$k = \min\{t \in \{1, 2, \ldots, v\} : a_{i,t} \neq 0\} = \min\{t \in \{1, 2, \ldots, v\} : a_{j,t} \neq 0\},$$

then $a_{i,k} = a_{j,k}$. An element b of F is a *first entry* of A if and only if there is some row i of A such that $b = a_{i,k}$ where $k = \min\{t \in \{1, 2, \ldots, v\} : a_{i,t} \neq 0\}$, in which case b is the *first entry of A from column k*.

Notice that the notion of "first entries condition" from Definition 15.2 and the special case of Definition 15.25 in which $F = \mathbb{Q}$ are not exactly the same since there is no requirement in Definition 15.25 that the first entries be positive (as this has no meaning in many fields).

We now consider vector spaces V over arbitrary fields. We shall be dealing with vectors (meaning ordered v-tuples) each of whose entries is a vector (meaning a member of V). We shall use bold face for the members of V and continue to represent v-tuples by an arrow above the letter.

The following theorem is very similar to Theorem 15.5 and so is its proof. (The main difference is that given a vector space V over a field F and $d \in F \setminus \{0\}$, one has $dV = V$ so that dV is automatically central* in $(V, +)$.) Accordingly, we leave the proof as an exercise.

Theorem 15.26. *Let F be a field and let V be an infinite vector space over F. Let $u, v \in \mathbb{N}$ and let A be a $u \times v$ matrix with entries from F which satisfies the first entries condition over F. Let C be central in $(V, +)$. Then there exist sequences $\langle \mathbf{x}_{1,n} \rangle_{n=1}^{\infty}$, $\langle \mathbf{x}_{2,n} \rangle_{n=1}^{\infty}, \ldots, \langle \mathbf{x}_{v,n} \rangle_{n=1}^{\infty}$ in V such that for every $G \in \mathcal{P}_f(\mathbb{N})$, $\vec{\mathbf{x}}_G \in (V \setminus \{\mathbf{0}\})^v$ and $A\vec{\mathbf{x}}_G \in C^u$, where*

$$\vec{\mathbf{x}}_G = \begin{pmatrix} \sum_{n \in G} \mathbf{x}_{1,n} \\ \sum_{n \in G} \mathbf{x}_{2,n} \\ \vdots \\ \sum_{n \in G} \mathbf{x}_{v,n} \end{pmatrix}.$$

Proof. This is Exercise 15.5.1.   □

The assertion that V is an infinite vector space over F reduces, of course, to the assertion that either F is infinite and V is nontrivial or V is infinite dimensional over F.

Corollary 15.27. *Let F be a field and let V be an infinite vector space over F. Let $u, v \in \mathbb{N}$ and let A be a $u \times v$ matrix with entries from F which satisfies the first entries condition over F. Then A is strongly image partition regular over V. That is, whenever $r \in \mathbb{N}$ and $V \backslash \{\mathbf{0}\} = \bigcup_{i=1}^{r} E_i$, there exist $i \in \{1, 2, \ldots, r\}$ and $\vec{x} \in (V \backslash \{\mathbf{0}\})^v$ such that $A\vec{x} \in E_i{}^u$.*

Proof. We first observe that $\{\mathbf{0}\}$ is not central in $(V, +)$. To see this, note that by Corollary 4.33, V^* is an ideal of $(\beta V, +)$ so that all minimal idempotents are in V^*. Consequently one may choose $i \in \{1, 2, \ldots, r\}$ such that E_i is central in $(V, +)$. Pick, by Theorem 15.26 some $\vec{x} \in (V \backslash \{\mathbf{0}\})^v$ with $A\vec{x} \in E_i{}^u$.   □

Now we turn our attention to kernel partition regularity. We extend the definition of the columns condition to apply to matrices with entries from an arbitrary field.

Definition 15.28. Let F be a field, let $u, v \in \mathbb{N}$, let C be a $u \times v$ matrix with entries from F, and let $\vec{c}_1, \vec{c}_2, \ldots, \vec{c}_v$ be the columns of C. The matrix C satisfies the columns condition over F if and only if there exist $m \in \mathbb{N}$ and $I_1, I_2, \ldots, I_m$ such that

(1) $\{I_1, I_2, \ldots, I_m\}$ is a partition of $\{1, 2, \ldots, v\}$.
(2) $\sum_{i \in I_1} \vec{c}_i = \vec{0}$.
(3) If $m > 1$ and $t \in \{2, 3, \ldots, m\}$, let $J_t = \bigcup_{j=1}^{t-1} I_j$. Then there exist $\langle \delta_{t,i} \rangle_{i \in J_t}$ in F such that $\sum_{i \in I_t} \vec{c}_i = \sum_{i \in J_t} \delta_{t,i} \cdot \vec{c}_i$.

Note that Definitions 15.13 and 15.28 agree in the case that $F = \mathbb{Q}$.

Theorem 15.29. *Let F be a field, let V be an infinite vector space over F, let $u, v \in \mathbb{N}$, and let C be a $u \times v$ matrix with entries from F that satisfies the columns condition over F. Then C is kernel partition regular over V. That is, whenever $r \in \mathbb{N}$ and $V \backslash \{\mathbf{0}\} = \bigcup_{i=1}^{r} E_i$, there exist $i \in \{1, 2, \ldots, r\}$ and $\vec{x} \in E_i^v$ such that $C\vec{x} = \vec{0}$.*

Proof. Pick $m \in \mathbb{N}$, $I_1, I_2, \ldots, I_m$, and for $t \in \{2, 3, \ldots, m\}$, J_t and $\langle \delta_{t,i} \rangle_{i \in J_t}$ as guaranteed by the definition of the columns condition.

Define the $v \times m$ matrix B by, for $(i, t) \in \{1, 2, \ldots, v\} \times \{1, 2, \ldots, m\}$,

$$b_{i,t} = \begin{cases} -\delta_{t,i} & \text{if } i \in J_t \\ 1 & \text{if } i \in I_t \\ 0 & \text{if } i \notin \bigcup_{j=1}^{t} I_j. \end{cases}$$

We observe that B satisfies the first entries condition. We also observe that $CB = O$. (Indeed, let $j \in \{1, 2, \ldots, u\}$ and $t \in \{1, 2, \ldots, m\}$. If $t = 1$, then $\sum_{i=1}^{v} c_{j,i} \cdot b_{i,t} = \sum_{i \in I_1} c_{j,i} = 0$ and if $t > 1$ then $\sum_{i=1}^{v} c_{j,i} \cdot b_{i,t} = \sum_{i \in J_t} -\delta_{t,i} \cdot c_{j,i} + \sum_{i \in I_t} c_{j,i} = 0$.)

Now let $r \in \mathbb{N}$ and let $V\backslash\{0\} = \bigcup_{i=1}^{r} E_i$. Pick by Corollary 15.27 some $i \in \{1, 2, \ldots, r\}$ and $\vec{y} \in (V\backslash\{0\})^m$ such that $B\vec{y} \in E_i{}^v$. Let $\vec{x} = B\vec{y}$. Then $C\vec{x} = CB\vec{y} = O\vec{y} = \vec{0}$. □

We see that in the case that F is a finite field, we in fact have a characterization of kernel partition regularity.

Theorem 15.30. *Let F be a finite field, let $u, v \in \mathbb{N}$, and let C be a $u \times v$ matrix with entries from F. The following statements are equivalent.*

(a) *For each $r \in \mathbb{N}$ there is some $n \in \mathbb{N}$ such that whenever V is a vector space of dimension at least n over F and $V\backslash\{0\} = \bigcup_{i=1}^{r} E_i$ there exist some $i \in \{1, 2, \ldots, r\}$ and some $\vec{x} \in E_i{}^v$ such that $C\vec{x} = \vec{0}$.*

(b) *The matrix C satisfies the columns condition over F.*

Proof. (a) implies (b). Let $r = |F| - 1$ and pick $n \in \mathbb{N}$ such that whenever V is a vector space of dimension at least n over F and $V\backslash\{0\} = \bigcup_{i=1}^{r} E_i$ there exist some $i \in \{1, 2, \ldots, r\}$ and some $\vec{x} \in E_i{}^v$ such that $C\vec{x} = \vec{0}$.

Let $V = \bigtimes_{i=1}^{n} F$. Since in this case we will be working with v-tuples of n-tuples, let us establish our notation. Given $\vec{x} \in V^v$, we write

$$\vec{x} = \begin{pmatrix} x_1 \\ x_2 \\ \vdots \\ x_v \end{pmatrix}$$

and given $i \in \{1, 2, \ldots, v\}$, $x_i = \big(x_i(1), x_i(2), \ldots, x_i(n)\big)$.

We color V according to the value of its first nonzero coordinate. For each $x \in V\backslash\{0\}$, let $\gamma(x) = \min\{i \in \{1, 2, \ldots, n\} : x(i) \neq 0\}$. For each $\alpha \in F\backslash\{0\}$, let $E_\alpha = \{x \in V\backslash\{0\} : x(\gamma(x)) = \alpha\}$. Pick some $\alpha \in F\backslash\{0\}$ and some $\vec{x} \in E_\alpha{}^v$ such that $C\vec{x} = \vec{0}$.

Let $D = \big\{\gamma(x_i) : i \in \{1, 2, \ldots, v\}\big\}$ and let $m = |D|$. Enumerate D in increasing order as $\{d_1, d_2, \ldots, d_m\}$. For each $t \in \{1, 2, \ldots, m\}$, let $I_t = \{i \in \{1, 2, \ldots, v\} : \gamma(x_i) = d_t\}$. For $t \in \{2, 3, \ldots, m\}$, let $J_t = \bigcup_{j=1}^{t-1} I_j$ and for $i \in J_t$, let $\delta_{t,i} = -x_i(d_t) \cdot \alpha^{-1}$.

To see that these are as required for the columns condition, we first show that $\sum_{i \in I_1} \vec{c}_i = \vec{0}$. To this end, let $j \in \{1, 2, \ldots, u\}$ be given. We show that $\sum_{i \in I_1} c_{j,i} = 0$. Now $C\vec{x} = \vec{0}$ so $\sum_{i=1}^{v} c_{j,i} \cdot x_i = 0$ so in particular, $\sum_{i=1}^{v} c_{j,i} \cdot x_i(d_1) = 0$. Now if $i \in \{1, 2, \ldots, v\}\backslash I_1$, then $d_1 < \gamma(x_i)$ so $x_i(d_1) = 0$. Thus $0 = \sum_{i \in I_1} c_{j,i} \cdot x_i(d_1) = \alpha \cdot \sum_{i \in I_1} c_{j,i}$ so $\sum_{i \in I_1} c_{j,i} = 0$.

Now assume $m > 1$ and $t \in \{2, 3, \ldots, m\}$. To see that $\sum_{i \in I_t} \vec{c}_i = \sum_{i \in J_t} \delta_{t,i} \cdot \vec{c}_i$, we again let $j \in \{1, 2, \ldots, u\}$ be given and show that $\sum_{i \in I_t} c_{j,i} = \sum_{i \in J_t} \delta_{t,i} \cdot c_{j,i}$. Then $\sum_{i=1}^{v} c_{j,i} \cdot x_i = 0$ so in particular, $\sum_{i=1}^{v} c_{j,i} \cdot x_i(d_t) = 0$. If $i \in \{1, 2, \ldots, v\}\backslash(I_t \cup J_t)$, then $x_i(d_t) = 0$ so $0 = \sum_{i \in I_t} c_{j,i} \cdot x_i(d_t) + \sum_{i \in J_t} c_{j,i} \cdot x_i(d_t) = \sum_{i \in I_t} c_{j,i} \cdot \alpha + \sum_{i \in J_t} c_{j,i} \cdot$

$(-\delta_{t,i} \cdot \alpha)$. Thus $\alpha \cdot \Sigma_{i \in I_t} c_{j,i} = \alpha \cdot \Sigma_{i \in J_t} c_{j,i} \cdot \delta_{t,i}$ and so $\Sigma_{i \in I_t} c_{j,i} = \Sigma_{i \in J_t} c_{j,i} \cdot \delta_{t,i}$ as required.

(b) implies (a). We use a standard compactness argument. Let $r \in \mathbb{N}$ and suppose the conclusion fails. For each $n \in \mathbb{N}$ let $V_n = \{x \in \times_{i=1}^{\infty} F : \text{for each } i > n, x(i) = 0\}$. Then for each n, V_n is an n-dimensional vector space over F and $V_n \subseteq V_{n+1}$.

Now given any $n \in \mathbb{N}$, there is a vector space V of dimension at least n over F for which there exist $E_1, E_2, \ldots, E_r$ with $V \setminus \{\mathbf{0}\} = \bigcup_{i=1}^{r} E_i$ such that for each $i \in \{1, 2, \ldots, r\}$ and each $\vec{x} \in E_i{}^v$, $C\vec{x} \neq \vec{\mathbf{0}}$. The same assertion holds for any n-dimensional subspace of V as well. Thus we can assume that we have some $\varphi_n : V_n \setminus \{\mathbf{0}\} \to \{1, 2, \ldots, r\}$ such that for any $i \in \{1, 2, \ldots, r\}$ and any $\vec{x} \in (\varphi_n{}^{-1}[\{i\}])^v$, $C\vec{x} \neq \vec{\mathbf{0}}$.

Choose an infinite subset A_1 of $\mathbb{N}$ so that for $n, m \in A_1$ one has $\varphi_{n|V_1 \setminus \{\mathbf{0}\}} = \varphi_{m|V_1 \setminus \{\mathbf{0}\}}$. Inductively, given A_{t-1}, choose an infinite subset A_t of A_{t-1} with $\min A_t \geq t$ such that for $n, m \in A_t$ one has $\varphi_{n|V_t \setminus \{\mathbf{0}\}} = \varphi_{m|V_t \setminus \{\mathbf{0}\}}$. For each t pick $n(t) \in A_t$.

Let $V = \bigcup_{n=1}^{\infty} V_n$. Then V is an infinite vector space over F. For $x \in V \setminus \{\mathbf{0}\}$ pick the first t such that $x \in V_t$. Then $x \in V_t \subseteq V_{n(t)}$ because $n(t) \in A_t$ and $\min A_t \geq t$. Define $\varphi(x) = \varphi_{n(t)}(x)$. By Theorem 15.29 pick $i \in \{1, 2, \ldots, r\}$ and $\vec{x} \in (\varphi^{-1}[\{i\}])^v$ such that $C\vec{x} = \vec{\mathbf{0}}$. Pick $t \in \mathbb{N}$ such that $\{x_1, x_2, \ldots, x_v\} \subseteq V_t$.

We claim that for each $j \in \{1, 2, \ldots, v\}$ one has $\varphi_{n(t)}(x_j) = i$. To see this, let $j \in \{1, 2, \ldots, v\}$ be given and pick the least s such that $x_j \in V_s$. Then $n(s), n(t) \in A_s$ so $\varphi_{n(t)}(x_j) = \varphi_{n(s)}(x_j) = \varphi(x_j) = i$. Thus $\vec{x} \in (\varphi_{n(t)}{}^{-1}[\{i\}])^v$ and $C\vec{x} = \vec{\mathbf{0}}$, a contradiction. $\square$

Of course any field is a vector space over itself. Thus Corollary 15.27 implies that, if F is an infinite field, any matrix with entries from F which satisfies the first entries condition over F is strongly image partition regular over F.

Exercise 15.5.1. Prove Theorem 15.26 by suitably modifying the proof of Theorem 15.5.

Notes

The terminology "image partition regular" and "kernel partition regular" was suggested by W. Deuber. Matrices satisfying the first entries condition are based on Deuber's (m, p, c) sets [76].

The columns condition was introduced by Rado [206] and he showed there that a matrix is kernel partition regular over $(\mathbb{N}, +)$ if and only if it satisfies the columns condition over $\mathbb{Q}$. Other generalizations of this result were obtained in [206] and [207].

The proof that (a) and (c) are equivalent in Theorem 15.20 is based on Rado's original arguments [206]. It is shown in [162], a result of collaboration with W. Woan, that there are solutions to the system of equations $\vec{x}^C = \vec{1}$ in any central set in $(\mathbb{N}, \cdot)$ if and only if C satisfies the columns condition over $\mathbb{Z}$.

The proof given here of the sufficiency of the columns condition using the image partition regularity of matrices satisfying the first entries condition is based on Deuber's proof that the set of subsets of $\mathbb{N}$ containing solutions to all kernel partition regular matrices is partition regular [76].

The material from Section 15.4 is taken from [142], a result of collaboration with I. Leader, where several other characterizations of image partition regular matrices are given, including one which seems to be easier to verify. This characterization of the image partition regular matrices is quite recent. Whereas Rado's Theorem was proved in 1933 [206] and (m, p, c) sets (on which the first entries condition was based) were introduced in 1973 [76], the characterization of image partition regular matrices was not obtained until 1993 [142].

Theorems 15.26, 15.29, and 15.30 are from [21], written in collaboration with V. Bergelson and W. Deuber, except that there the field F was required to be countable and the vector space V was required to be of countable dimension. (At the time [21] was written, the Central Sets Theorem was only known to hold for countable commutative semigroups.)

Chapter 16

IP, IP*, Central, and Central* Sets

We saw in Chapter 14 that in any semigroup S, central sets have rich combinatorial content. And our introduction to the combinatorial applications of the algebraic structure of βS came through the Finite Products Theorem (Corollary 5.9). We shall see in this chapter that sets which intersect $FP(\langle x_n \rangle_{n=1}^{\infty})$ for every sequence $\langle x_n \rangle_{n=1}^{\infty}$ (that is, the *IP* sets*) have very rich combinatorial structure, especially in the semigroups $(\mathbb{N}, +)$ and $(\mathbb{N}, \cdot)$. Further, by means of the old and often studied combinatorial notion of *spectra* of numbers, we exhibit a large class of examples of these special sets in $(\mathbb{N}, +)$.

16.1 IP, IP*, Central, and Central* Sets
in Arbitrary Semigroups

Recall that we have defined a *central* set in a semigroup S as one which is a member of a minimal idempotent in βS (Definition 4.42). As one of our main concerns in this chapter is the presentation of examples, we begin by describing a class of sets that are central in an arbitrary semigroup.

Theorem 16.1. *Let S be a semigroup and for each $F \in \mathcal{P}_f(S)$, let $x_F \in S$. Then $\bigcup_{F \in \mathcal{P}_f(S)} (F \cdot x_F)$ is central in S.*

Proof. Let $A = \bigcup_{F \in \mathcal{P}_f(S)} (F \cdot x_F)$. For each $F \in \mathcal{P}_f(S)$, let

$$B_F = \{x_H : H \in \mathcal{P}_f(S) \text{ and } F \subseteq H\}.$$

Then $\{B_F : F \in \mathcal{P}_f(S)\}$ is a set of subsets of S with the finite intersection property so choose (by Theorem 3.8) some $p \in \beta S$ such that $\{B_F : F \in \mathcal{P}_f(S)\} \subseteq p$.

We claim that $\beta S \cdot p \subseteq \overline{A}$ for which it suffices, since ρ_p is continuous, to show that $S \cdot p \subseteq \overline{A}$. To this end, let $s \in S$. Then $B_{\{s\}} \subseteq s^{-1}A$ so $s^{-1}A \in p$ so by Theorem 4.12 $A \in sp$. By Corollary 2.6, there is a minimal idempotent $q \in \beta S \cdot p$ and so $A \in q$. $\square$

In the event that S is countable, the description given by Theorem 16.1 can be simplified.

Corollary 16.2. *Let $S = \{a_n : n \in \mathbb{N}\}$ be a countable semigroup and let $\langle x_n \rangle_{n=1}^{\infty}$ be a sequence in S. Then $\{a_n \cdot x_m : n, m \in \mathbb{N}$ and $n \leq m\}$ is central in S.*

Proof. This is Exercise 16.1.1.   □

We now introduce some terminology which is due to Furstenberg [98] and is commonly used in Topological Dynamics circles.

Definition 16.3. Let S be a semigroup. A subset A of S is an *IP set* if and only if there is a sequence $\langle x_n \rangle_{n=1}^{\infty}$ in S such that $FP(\langle x_n \rangle_{n=1}^{\infty}) \subseteq A$.

Actually, as defined by Furstenberg, an IP set is a set which can be written as $FP(\langle x_n \rangle_{n=1}^{\infty})$ for some sequence $\langle x_n \rangle_{n=1}^{\infty}$. We have modified the definition because we already have a notation for $FP(\langle x_n \rangle_{n=1}^{\infty})$ and because of the nice characterization of IP set obtained in Theorem 16.4 below.

The terminology may be remembered because of the intimate relationship between IP sets and idempotents. However, the origin as described in [98] is as an "infinite dimensional parallelepiped". To see the idea behind that term, consider the elements of $FP(\langle x_i \rangle_{i=1}^{3})$ which we have placed at seven of the vertices of a cube (adding an identity e at the origin).

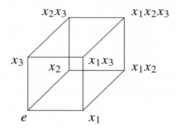

Theorem 16.4. *Let S be a semigroup and let A be a subset of S. Then A is an IP set if and only if there is some idempotent $p \in \beta S$ such that $A \in p$.*

Proof. This is a reformulation of Theorem 5.12.   □

In general, given a class $\mathcal{R}$ of subsets of a set S, one may define the class $\mathcal{R}*$ of sets that meet ever member of $\mathcal{R}$. We have already done so (in Definition 15.3) for central sets.

Definition 16.5. Let S be a semigroup and let $A \subseteq S$. Then A is an *IP* set* if and only if for every IP set $B \subseteq S$, $A \cap B \neq \emptyset$.

Recall from Lemma 15.4 that a set is a central* set if and only if it is a member of every minimal idempotent. A similar characterization is valid for IP* sets.

Theorem 16.6. *Let S be a semigroup and let $A \subseteq S$. The following statements are equivalent.*

(a) *A is an IP* set.*

(b) *A is a member of every idempotent of βS.*

(c) *$A \cap B$ is an IP set for every IP set B of S.*

Proof. (a) implies (b). Let p be an idempotent of βS and suppose that $A \notin p$. Then $S \backslash A \in p$ so by Theorem 5.8 there is a sequence $\langle x_n \rangle_{n=1}^{\infty}$ in S with $\mathrm{FP}(\langle x_n \rangle_{n=1}^{\infty}) \subseteq S \backslash A$. That is, $S \backslash A$ is an IP set which misses A, a contradiction.

(b) implies (c). Let B be an IP set and pick by Lemma 5.11 an idempotent p of βS such that $B \in p$. Then $A \cap B \in p$ so $A \cap B$ is an IP set by Theorem 5.8.

That (c) implies (a) is trivial. $\square$

As a trivial consequence of Lemma 15.4, Theorems 16.4 and 16.6, and the definition of central, one has that

$$\mathrm{IP^*} \Rightarrow \mathrm{central^*} \Rightarrow \mathrm{central} \Rightarrow \mathrm{IP}.$$

One also sees immediately the following:

Remark 16.7. *Let S be a semigroup and let A and B be subsets of S.*

(a) *If A and B are IP* sets, then $A \cap B$ is an IP* set.*

(b) *If A and B are central* sets, then $A \cap B$ is a central* set.*

We see now that in most reasonable semigroups, it is easy to produce a specific central* set which is not an IP* set.

Theorem 16.8. *Let S be a semigroup and assume that $\langle x_n \rangle_{n=1}^{\infty}$ is a sequence in S such that $\mathrm{FP}(\langle x_n \rangle_{n=1}^{\infty})$ is not piecewise syndetic. Then $S \backslash \mathrm{FP}(\langle x_n \rangle_{n=1}^{\infty})$ is a central* set which is not an IP* set.*

Proof. Trivially $S \backslash \mathrm{FP}(\langle x_n \rangle_{n=1}^{\infty})$ is not an IP* set. Since $\mathrm{FP}(\langle x_n \rangle_{n=1}^{\infty})$ is not piecewise syndetic, we have by Theorem 4.40 $\overline{\mathrm{FP}(\langle x_n \rangle_{n=1}^{\infty})} \cap K(\beta S) = \emptyset$ so for every minimal idempotent p, $S \backslash \mathrm{FP}(\langle x_n \rangle_{n=1}^{\infty}) \in p$. $\square$

Of course, since the notions of central and IP are characterized by membership in an idempotent, they are partition regular notions. In trivial situations (see Exercises 16.1.2 and 16.1.3) the notions of central* and IP* may also be partition regular.

Lemma 16.9. *Let S be a semigroup.*

(a) *The notion of IP* is partition regular in S if and only if βS has a unique idempotent.*

(b) *The notion of central* is partition regular in S if and only if $K(\beta S)$ has a unique idempotent.*

Proof. We establish (a) only, the other proof being very similar.

Necessity. Suppose that βS has two idempotents p and q and pick $A \in p \backslash q$. Then $A \cup (S \backslash A)$ is an IP* set while neither A nor $S \backslash A$ is an IP* set.

Sufficiency. Let p be the unique idempotent of βS. Then a subset A of S is an IP* set if and only if $A \in p$.   □

As a consequence of Lemma 16.9 we have the following.

Remark 16.10. *Let S be a semigroup. If the notion of IP* is partition regular in S, then so is the notion of central*.*

Exercise 16.1.3 shows that one may have the notion of central* partition regular when the notion of IP* is not.

We see now that in more civilized semigroups, the notions of IP* and central* are not partition regular. (Theorem 16.11 is in fact a corollary to Corollary 6.43 and Lemma 16.9, but it has a simple self contained proof, so we present it.)

Theorem 16.11. *Let S be an infinite weakly left cancellative semigroup. There exist disjoint central subsets of S. Consequently, neither the notions of central* nor IP* are partition regular in S.*

Proof. Let $\kappa = |S|$ and enumerate $\mathcal{P}_f(S)$ as $\langle F_\alpha \rangle_{\alpha < \kappa}$. Observe that whenever $B \subseteq S$ and $|B| < \kappa$ and $F \in \mathcal{P}_f(S)$ there is some $x \in S$ such that $(F \cdot x) \cap B = \emptyset$. (For each $u \in F$ and $v \in B$, $\{x \in S : ux = v\}$ is finite by the definition of weakly left cancellative so $|\bigcup_{u \in F} \bigcup_{v \in B} \{x \in S : ux = v\}| < \kappa$.)

Choose $x_0 \in S$ and choose $y_0 \in S$ such that $(F_0 \cdot y_0) \cap (F_0 \cdot x_0) = \emptyset$. Inductively, let $\alpha < \kappa$ and assume we have chosen $\langle x_\sigma \rangle_{\sigma < \alpha}$ and $\langle y_\sigma \rangle_{\sigma < \alpha}$ in S so that

$$\left(\bigcup_{\sigma < \alpha} F_\sigma \cdot x_\sigma \right) \cap \left(\bigcup_{\sigma < \alpha} F_\sigma \cdot y_\sigma \right) = \emptyset.$$

Let $B = \left(\bigcup_{\sigma < \alpha} F_\sigma \cdot x_\sigma \right) \cup \left(\bigcup_{\sigma < \alpha} F_\sigma \cdot y_\sigma \right)$ and choose $x_\alpha \in S$ such that $(F_\alpha \cdot x_\alpha) \cap B = \emptyset$ and choose $y_\alpha \in S$ such that $(F_\alpha \cdot y_\alpha) \cap (B \cup F_\alpha \cdot x_\alpha) = \emptyset$.

By Theorem 16.1 we have that $\left(\bigcup_{\sigma < \kappa} F_\sigma \cdot x_\sigma \right)$ and $\left(\bigcup_{\sigma < \kappa} F_\sigma \cdot y_\sigma \right)$ are disjoint central subsets of S.   □

In any semigroup, we see that IP* sets satisfy a significantly stronger combinatorial conclusion than that given by their definition.

Theorem 16.12. *Let S be a semigroup, let A be an IP* set in S, and let $\langle x_n \rangle_{n=1}^\infty$ be a sequence in S. There is a product subsystem $\langle y_n \rangle_{n=1}^\infty$ of $\langle x_n \rangle_{n=1}^\infty$ such that $FP(\langle y_n \rangle_{n=1}^\infty) \subseteq A$.*

Proof. Pick by Lemma 5.11 an idempotent $p \in \beta S$ such that $FP(\langle x_n \rangle_{n=m}^\infty) \in p$ for every $m \in \mathbb{N}$. Then $A \in p$ by Theorem 16.6, so Theorem 5.14 applies.   □

Exercise 16.1.1. Prove Corollary 16.2.

Exercise 16.1.2. Let S be a set and let $a \in S$. Define $xy = a$ for all x and y in S. Show that a subset A of S is an IP set if and only if $a \in A$ and consequently that the notions of IP* and central* are partition regular for this semigroup.

Exercise 16.1.3. Recall the semigroup $(\mathbb{N}, \wedge)$ where $n \wedge m = \min\{n, m\}$. (See Exercise 4.1.11.) Verify each of the following assertions.

(a) In $(\beta\mathbb{N}, \wedge)$ every element is idempotent and $K(\beta\mathbb{N}, \wedge) = \{1\}$.

(b) The IP sets in $(\mathbb{N}, \wedge)$ are the nonempty subsets of $\mathbb{N}$ and the central sets in $(\mathbb{N}, \wedge)$ are the subsets A of $\mathbb{N}$ with $1 \in A$.

(c) The only IP* set in $(\mathbb{N}, \wedge)$ is $\mathbb{N}$, while the central* sets in $(\mathbb{N}, \wedge)$ are the same as the central sets. Consequently, the notion of IP* is not partition regular in $(\mathbb{N}, \wedge)$ while the notion of central* is partition regular in $(\mathbb{N}, \wedge)$.

Exercise 16.1.4. Recall the semigroup $(\mathbb{N}, \vee)$ where $n \vee m = \max\{n, m\}$. (See Exercise 4.1.11.) Verify each of the following assertions.

(a) In $(\beta\mathbb{N}, \vee)$ every element is idempotent and $K(\beta\mathbb{N}, \vee) = \mathbb{N}^*$.

(b) The IP sets in $(\mathbb{N}, \vee)$ are the nonempty subsets of $\mathbb{N}$ and the central sets in $(\mathbb{N}, \vee)$ are the infinite subsets of $\mathbb{N}$.

(c) The only IP* set in $(\mathbb{N}, \vee)$ is $\mathbb{N}$, while the central* sets in $(\mathbb{N}, \vee)$ are the cofinite subsets of $\mathbb{N}$. Consequently, neither the notions of IP* nor the notions of central* are partition regular in $(\mathbb{N}, \vee)$.

16.2 IP* and Central Sets in $\mathbb{N}$

In this section we compare the additive and multiplicative structures of $\mathbb{N}$. We begin with a trivial observation.

Lemma 16.13. *Let* $n \in \mathbb{N}$. *Then* $\mathbb{N}n$ *is an IP* set in* $(\mathbb{N}, +)$.

Proof. This is an immediate consequence of Theorem 16.6 and Lemma 6.6. □

Now we establish that central sets in $(\mathbb{N}, +)$ have a richer structure than that guaranteed to an arbitrary commutative semigroup (and in fact richer than that possessed by central sets in $(\mathbb{N}, \cdot)$).

Theorem 16.14. *Let* B *be central in* $(\mathbb{N}, +)$, *let* A *be a* $u \times v$ *matrix with entries from* $\mathbb{Q}$.

(a) *If* A *is image partition regular, then there exists* $\vec{y} \in \mathbb{N}^v$ *such that all entries of* $A\vec{y}$ *are in* B.

(b) *If* A *is kernel partition regular, then there exists* $\vec{y} \in B^v$ *such that* $A\vec{y} = \vec{0}$.

Proof. (a) Pick by Theorem 15.24 some $m \in \mathbb{N}$ and a $u \times m$ matrix D satisfying the first entries condition with entries from ω such that for any $\vec{z} \in \mathbb{N}^m$ there is some $\vec{y} \in \mathbb{N}^v$ with $A\vec{y} = D\vec{z}$. By Lemma 16.13, for each $n \in \mathbb{N}$, $\mathbb{N}n$ is an IP* set and hence a central* set. Consequently, by Theorem 15.5, pick $\vec{z} \in \mathbb{N}^m$ such that all entries of $D\vec{z}$ are in B and pick $\vec{y} \in \mathbb{N}^v$ such that $A\vec{y} = D\vec{z}$.

(b) Let $d \in \mathbb{N}$ be a common multiple of all of the denominators of entries of A. Then dA is a matrix with entries from $\mathbb{Z}$ which is kernel partition regular. (If $\mathbb{N} = \bigcup_{i=1}^{r} C_i$, pick $i \in \{1, 2, \ldots, r\}$ and $\vec{x} \in C_i{}^v$ such that $A\vec{x} = \vec{0}$. Then $(dA)\vec{x} = \vec{0}$.) Thus by Rado's Theorem (Theorem 15.20) dA satisfies the columns condition over $\mathbb{Q}$ so, by Lemma 15.15, pick $m \in \mathbb{N}$ and a $v \times m$ matrix C with entries from ω which satisfies the first entries condition such that $dAC = O$. As in part (a), pick $\vec{x} \in \mathbb{N}^m$ such that all entries of $C\vec{x}$ are in B and let $\vec{y} = C\vec{x}$. Then $dA\vec{y} = dAC\vec{x} = O\vec{x} = \vec{0}$ so $A\vec{y} = \vec{0}$. $\square$

In fact, central sets in $(\mathbb{N}, +)$ not only contain images of all image partition regular matrices, but all finite sums choosing at most one term from each such image as well.

Definition 16.15. Let $(S, +)$ be a semigroup and let $\langle Y_n \rangle_{n=1}^{\infty}$ be a sequence of subsets of S. Then $FS(\langle Y_n \rangle_{n=1}^{\infty}) = \{\sum_{n \in F} a_n : F \in \mathcal{P}_f(\mathbb{N})$ and for all $n \in F$, $a_n \in Y_n\}$.

We define $FS(\langle Y_n \rangle_{n=1}^{m})$ and $FS(\langle Y_n \rangle_{n=m}^{\infty})$ analogously.

Theorem 16.16. *Let B be central in $(\mathbb{N}, +)$. Let $\langle A(n) \rangle_{n=1}^{\infty}$ enumerate the image partition regular matrices with entries from $\mathbb{Q}$ and for each n, let $m(n)$ be the number of columns of $A(n)$. There exists for each $n \in \mathbb{N}$ a choice of $\vec{x}(n) \in \mathbb{N}^{m(n)}$ such that, if Y_n is the set of entries of $A(n)\vec{x}(n)$, then $FS(\langle Y_n \rangle_{n=1}^{\infty}) \subseteq B$.*

Proof. Pick a minimal idempotent in $(\beta\mathbb{N}, +)$ such that $B \in p$. Pick by Theorem 16.14 some $\vec{x}(1) \in \mathbb{N}^{m(1)}$ such that all entries of $A(1)\vec{x}(1)$ are in $B^{\star} = \{a \in B : -a + B \in p\}$. Let Y_1 be the set of entries of $A(1)\vec{x}(1)$.

Inductively, let $n \in \mathbb{N}$ and assume that we have chosen $\vec{x}(k) \in \mathbb{N}^{m(k)}$ for each $k \in \{1, 2, \ldots, n\}$ so that, with Y_k as the set of entries of $A(k)\vec{x}(k)$, one has $FS(\langle Y_k \rangle_{k=1}^{n}) \subseteq B^{\star}$. By Lemma 4.14, for each $a \in B^{\star}$, $-a + B^{\star} \in p$ so

$$B^{\star} \cap \bigcap \{-a + B^* : a \in FS(\langle Y_k \rangle_{k=1}^{n})\} \in p$$

so pick $\vec{x}(n+1) \in \mathbb{N}^{m(n+1)}$ such that all entries of $A(n+1)\vec{x}(n+1)$ are in

$$B^{\star} \cap \bigcap \{-a + B^* : a \in FS(\langle Y_k \rangle_{k=1}^{n})\}.$$

Letting Y_{n+1} be the set of entries of $A(n+1)\vec{x}(n+1)$ one has $FS(\langle Y_k \rangle_{k=1}^{n+1}) \subseteq B^{\star}$. $\square$

A similar result applies to kernel partition regular matrices.

Theorem 16.17. *Let B be central in $(\mathbb{N}, +)$. Let $\langle A(n) \rangle_{n=1}^{\infty}$ enumerate the kernel partition regular matrices with entries from $\mathbb{Q}$ and for each n, let $m(n)$ be the number of columns of $A(n)$. There exists for each $n \in \mathbb{N}$ a choice of $\vec{x}(n) \in \mathbb{N}^{m(n)}$ such that $A(n)\vec{x}(n) = \vec{0}$ and, if Y_n is the set of entries of $\vec{x}(n)$, then $FS(\langle Y_n \rangle_{n=1}^{\infty}) \subseteq B$.*

Proof. This can be copied nearly verbatim from the proof of Theorem 16.16. $\square$

The contrast with Theorems 16.14, 16.16, and 16.17 in the case of sets central, in fact central* in $(\mathbb{N}, \cdot)$ is striking. Recall from Section 15.3 the notation $\vec{x}^A = \vec{1}$ which represents the multiplicative analogue of the equation $A\vec{x} = \vec{0}$.

Theorem 16.18. *Let $C = \mathbb{N} \backslash \{x^2 : x \in \mathbb{N}\}$, let $A = (2)$ and let $B = (2 \ \ -2 \ \ 1)$. Then C is central* in $(\mathbb{N}, \cdot)$, A is image partition regular in $(\mathbb{N}, \cdot)$, and B is kernel partition regular in $(\mathbb{N}, \cdot)$. However, for no $x \in \mathbb{N}$ is $x^2 \in C$ and for no $\vec{x} \in C^3$ is $\vec{x}^B = 1$.*

Proof. The reader was asked to show in Exercise 15.1.2 that $\{x^2 : x \in \mathbb{N}\}$ is not central in $(\mathbb{N}, \cdot)$. Consequently, its complement must be central*. Trivially A is image partition regular in $(\mathbb{N}, \cdot)$ and B satisfies the columns condition over $\mathbb{Q}$, so is kernel partition regular in $(\mathbb{N}, \cdot)$ by Theorem 15.20. $\qquad \square$

The following lemma establishes that multiplication preserves IP* sets in $(\mathbb{N}, +)$.

Lemma 16.19. *Let $A \subseteq \mathbb{N}$ and let $n \in \mathbb{N}$. The following statements are equivalent.*

(a) *A is an IP* set in $(\mathbb{N}, +)$.*

(b) *$n^{-1}A$ is an IP* set in $(\mathbb{N}, +)$.*

(c) *nA is an IP* set in $(\mathbb{N}, +)$.*

Proof. (a) implies (b). Assume that A is an IP* set, and let B be an IP set. Pick a sequence $\langle x_t \rangle_{t=1}^{\infty}$ such that $FS(\langle x_t \rangle_{t=1}^{\infty}) \subseteq B$. Then $FS(\langle nx_t \rangle_{t=1}^{\infty}) \cap A \neq \emptyset$ so $n^{-1}A \cap B \neq \emptyset$.

(b) implies (a). Assume that $n^{-1}A$ is an IP* set, and let B be an IP set. Pick a sequence $\langle x_t \rangle_{t=1}^{\infty}$ such that $FS(\langle x_t \rangle_{t=1}^{\infty}) \subseteq B$. By Lemma 16.13, $\mathbb{N}n$ is an IP* set in $(\mathbb{N}, +)$ so pick by Theorem 16.12 a sum subsystem $\langle y_t \rangle_{t=1}^{\infty}$ of $\langle x_t \rangle_{t=1}^{\infty}$ such that $FS(\langle y_t \rangle_{t=1}^{\infty}) \subseteq \mathbb{N}n$. In particular n divides y_t for each t. Consequently

$$FS(\langle \tfrac{y_t}{n} \rangle_{t=1}^{\infty}) \cap n^{-1}A \neq \emptyset$$

so that $FS(\langle y_t \rangle_{t=1}^{\infty}) \cap A \neq \emptyset$. Since $FS(\langle y_t \rangle_{t=1}^{\infty}) \subseteq FS(\langle x_t \rangle_{t=1}^{\infty}) \subseteq B$, we are done.

Since $n^{-1}(nA) = A$, the equivalence of (a) with (c) follows from the equivalence of (a) with (b). $\qquad \square$

We see now that IP* sets in $(\mathbb{N}, +)$ are guaranteed to have substantial multiplicative structure.

Theorem 16.20. *Let $\mathcal{S}$ be the set of finite sequences in $\mathbb{N}$ (including the empty sequence) and let $f : \mathcal{S} \to \mathbb{N}$. Let $\langle y_n \rangle_{n=1}^{\infty}$ be a sequence in $\mathbb{N}$ and let A be an IP* set in $(\mathbb{N}, +)$. Then there is a sum subsystem $\langle x_n \rangle_{n=1}^{\infty}$ of $\langle y_n \rangle_{n=1}^{\infty}$ such that whenever $F \in \mathcal{P}_f(\mathbb{N})$, $\ell = f(\langle x_k \rangle_{k=1}^{\min F - 1})$, and $t \in \{1, 2, \dots, \ell\}$, one has $t \cdot \Sigma_{n \in F} x_n \in A$.*

Proof. Pick by Lemma 5.11 some $p = p + p$ in $\beta\mathbb{N}$ such that $FS(\langle y_n \rangle_{n=m}^{\infty}) \in p$ for each $m \in \mathbb{N}$. Then by Lemma 16.19, we have for each $t \in \mathbb{N}$ that $t^{-1}A$ is an IP* set in $(\mathbb{N}, +)$ and hence is in p. Let $B_1 = FS(\langle y_n \rangle_{n=1}^{\infty}) \cap \bigcap_{t=1}^{f(\emptyset)} t^{-1}A$ and note that $B_1 \in p$. Pick $x_1 \in B_1^{\star}$ and pick $H_1 \in \mathcal{P}_f(\mathbb{N})$ such that $x_1 = \Sigma_{t \in H_1} y_t$.

Inductively, let $n \in \mathbb{N}$ and assume that we have chosen $\langle x_i \rangle_{i=1}^{n}$, $\langle H_i \rangle_{i=1}^{n}$, and $\langle B_i \rangle_{i=1}^{n}$ such that for each $i \in \{1, 2, \dots, n\}$:

(1) $x_i = \Sigma_{t \in H_i}\, y_t$,

(2) if $i > 1$, then $\min H_i > \max H_{i-1}$,

(3) $B_i \in p$,

(4) if $\emptyset \neq F \subseteq \{1, 2, \ldots, i\}$ and $m = \min F$, then $\Sigma_{j \in F}\, x_j \in B_m{}^\star$, and

(5) if $i > 1$, then $B_i \subseteq \bigcap_{t=1}^{f(\langle x_j \rangle_{j=1}^{i-1})} t^{-1}A$.

Only hypotheses (1), (3), and (4) apply at $n = 1$ and they hold trivially. Let $k = \max H_n + 1$. By assumption $FS(\langle y_t \rangle_{t=k}^{\infty}) \in p$. Again we have by Lemma 16.19 that for each $t \in \mathbb{N}$, $t^{-1}A$ is an IP* set in $(\mathbb{N}, +)$ and is thus in p.

For each $m \in \{1, 2, \ldots, n\}$ let

$$E_m = \{\Sigma_{j \in F}\, x_j : \emptyset \neq F \subseteq \{1, 2, \ldots, n\} \text{ and } m = \min F\}.$$

By hypothesis (4) we have for each $m \in \{1, 2, \ldots, n\}$ and each $a \in E_m$ that $a \in B_m{}^\star$ and hence, by Lemma 4.14, $-a + B_m{}^\star \in p$. Let

$$B_{n+1} = FS(\langle y_t \rangle_{t=k}^{\infty}) \cap \bigcap_{t=1}^{f(\langle x_j \rangle_{j=1}^{n})} t^{-1}A \cap \bigcap_{m=1}^{n} \bigcap_{a \in E_m} (-a + B_m{}^\star).$$

Then $B_{n+1} \in p$. Choose $x_{n+1} \in B_{n+1}{}^\star$. Since $x_{n+1} \in FS(\langle y_t \rangle_{t=k}^{\infty})$, choose $H_{n+1} \in \mathcal{P}_f(\mathbb{N})$ such that $\min H_{n+1} \geq k$ and $x_{n+1} = \Sigma_{t \in H_{n+1}}\, y_t$. Then hypotheses (1), (2), (3), and (5) are satisfied directly.

To verify hypothesis (4), let $\emptyset \neq F \subseteq \{1, 2, \ldots, n + 1\}$ and let $m = \min F$. If $n+1 \notin F$, the conclusion holds by hypothesis, so assume that $n+1 \in F$. If $F = \{n+1\}$, then $\Sigma_{j \in F}\, x_j = x_{n+1} \in B_{n+1}{}^\star$, so assume $F \neq \{n+1\}$ and let $G = F\backslash\{n+1\}$. Let $a = \Sigma_{j \in G}\, x_j$. Then $a \in E_m$ so $x_{n+1} \in -a + B_m{}^\star$ so $\Sigma_{j \in F}\, x_j = a + x_{n+1} \in B_m{}^\star$ as required.

We thus have that $\langle x_n \rangle_{n=1}^{\infty}$ is a sum subsystem of $\langle y_n \rangle_{n=1}^{\infty}$. To complete the proof, let $F \in \mathcal{P}_f(\mathbb{N})$, let $\ell = f(\langle x_k \rangle_{k=1}^{\min F - 1})$, and let $t \in \{1, 2, \ldots, \ell\}$. Let $m = \min F$. Then by hypotheses (4) and (5), $\Sigma_{j \in F}\, x_j \in B_m \subseteq t^{-1}A$ so $t \cdot \Sigma_{n \in F}\, x_n \in A$. $\qquad\square$

Corollary 16.21. *Let A be an IP* set in $(\mathbb{N}, +)$ and let $\langle y_n \rangle_{n=1}^{\infty}$ be a sequence in $\mathbb{N}$. There is a sum subsystem $\langle x_n \rangle_{n=1}^{\infty}$ of $\langle y_n \rangle_{n=1}^{\infty}$ such that*

$$FS(\langle x_n \rangle_{n=1}^{\infty}) \cup FP(\langle x_n \rangle_{n=1}^{\infty}) \subseteq A.$$

Proof. Let $\mathcal{S}$ be the set of finite sequences in $\mathbb{N}$. Define $f(\emptyset) = 1$ and given $\langle x_j \rangle_{j=1}^{n}$ define $f(\langle x_j \rangle_{j=1}^{n}) = \prod_{j=1}^{n} x_j$. Choose $\langle x_n \rangle_{n=1}^{\infty}$ as guaranteed by Theorem 16.20. Letting $t = 1$, one sees that $FS(\langle x_n \rangle_{n=1}^{\infty}) \subseteq A$. To see that $FP(\langle x_n \rangle_{n=1}^{\infty}) \subseteq A$, let $F \in \mathcal{P}_f(\mathbb{N})$ and let $n = \max F$. If $|F| = 1$, then $\prod_{j \in F} x_j = x_n \in A$, so assume $|F| > 1$ and let $G = F\backslash\{n\}$. Let $t = \prod_{j \in G} x_j$. Then $t \leq f(\langle x_j \rangle_{j=1}^{n-1})$ so $\prod_{j \in F} x_j = t \cdot x_n \in A$. $\square$

Theorem 16.20 and Corollary 16.21 establish that an IP* set in $(\mathbb{N}, +)$ must have substantial multiplicative structure. One may naturally ask whether similar results apply to central* or central sets. On the one hand, we shall see in Corollary 16.26 that sets which are central* in $(\mathbb{N}, +)$ have significant multiplicative structure, in fact are central in $(\mathbb{N}, \cdot)$. On the other hand, central sets in $(\mathbb{N}, +)$ need have no multiplicative structure at all (Theorem 16.27) while central sets in $(\mathbb{N}, \cdot)$ must have a significant amount of additive structure (Theorem 16.28).

Definition 16.22. $\mathbb{M} = \{p \in \beta\mathbb{N} : \text{for each } A \in p, \ A \text{ is central in } (\mathbb{N}, +)\}$.

As an immediate consequence of the definition of central we have the following.

Remark 16.23. $\mathbb{M} = c\ell \, E\big(K(\beta\mathbb{N}, +)\big)$.

Theorem 16.24. $\mathbb{M}$ *is a left ideal of* $(\beta\mathbb{N}, \cdot)$.

Proof. Let $p \in \mathbb{M}$ and let $q \in \beta\mathbb{N}$. To see that $qp \in \mathbb{M}$, let $B \in qp$ and pick $a \in \mathbb{N}$ such that $a^{-1}B \in p$. Then $a^{-1}B$ is central in $(\mathbb{N}, +)$ so pick, by Theorem 14.25 a decreasing sequence $\langle C_n \rangle_{n=1}^{\infty}$ such that

(i) for each $n \in \mathbb{N}$ and each $x \in C_n$, there is some $m \in \mathbb{N}$ such that $C_m \subseteq -x + C_n$ and

(ii) $\{C_n : n \in \mathbb{N}\}$ is collectionwise piecewise syndetic.

Then $\langle aC_n \rangle_{n=1}^{\infty}$ is a decreasing sequence of subsets of B and immediately one has that for each $n \in \mathbb{N}$ and each $x \in aC_n$, there is some $m \in \mathbb{N}$ such that $aC_m \subseteq -x + aC_n$.

It thus suffices by Theorem 14.25 to show that $\{aC_n : n \in \mathbb{N}\}$ is collectionwise piecewise syndetic. That is, using Lemma 14.20, that there exist sequences $\langle g(n) \rangle_{n=1}^{\infty}$ and $\langle x_n \rangle_{n=1}^{\infty}$ in $\mathbb{N}$ such that for all $n, m \in \mathbb{N}$ with $m \geq n$,

$$\{x_m + 1, x_m + 2, \ldots, x_m + m\} \subseteq \bigcup_{j=1}^{g(n)} (-j + \bigcap_{i=1}^{n} aC_i).$$

If $a = 1$, the conclusion is immediate from the fact that $\{C_n : n \in \mathbb{N}\}$ is collectionwise piecewise syndetic so assume that $a > 1$.

Since $\{C_n : n \in \mathbb{N}\}$ is collectionwise piecewise syndetic, choose sequences $\langle h(n) \rangle_{n=1}^{\infty}$ and $\langle y_n \rangle_{n=1}^{\infty}$ in $\mathbb{N}$ such that for all $n, m \in \mathbb{N}$ with $m \geq n$,

$$\{y_m + 1, y_m + 2, \ldots, y_m + m\} \subseteq \bigcup_{j=1}^{h(n)} (-j + \bigcap_{i=1}^{n} C_i).$$

For each $n \in \mathbb{N}$, let $x_n = ay_n$ and let $g(n) = ah(n) + a$. Given $n, m \in \mathbb{N}$ with $m \geq n$ and given $k \in \{1, 2, \ldots, m\}$, pick $r \in \{0, 1, \ldots, a - 1\}$ such that $k + r = al$ for some $l \in \{1, 2, \ldots, m\}$. Then for some $j \in \{1, 2, \ldots, h(n)\}$ we have $y_m + l \in -j + \bigcap_{i=1}^{n} C_i$ so that $ay_m + al \in -aj + \bigcap_{i=1}^{n} aC_i$ so $x_m + k \in -(aj + r) + \bigcap_{i=1}^{n} aC_i$. Since $aj + r < g(n)$ we are done. $\square$

We saw in Corollary 13.15 that $K(\beta\mathbb{N}, \cdot) \cap K(\beta\mathbb{N}, +) = \emptyset$. We pause to observe now that these objects are nonetheless close.

Corollary 16.25. $K(\beta\mathbb{N}, \cdot) \cap c\ell \, K(\beta\mathbb{N}, +) \neq \emptyset$.

Proof. By Theorem 16.24, $\mathbb{M}$ is a left ideal of $(\beta\mathbb{N}, \cdot)$ and consequently has nonempty intersection with $K(\beta\mathbb{N}, \cdot)$ while by Remark 16.23, $\mathbb{M} \subseteq c\ell\, K(\beta\mathbb{N}, +)$. $\square$

Corollary 16.26. *Let A be central* in $(\mathbb{N}, +)$. Then A is central in $(\mathbb{N}, \cdot)$.*

Proof. By Lemma 15.4, $E\big(K(\beta\mathbb{N}, +)\big) \subseteq \overline{A}$ so $\mathbb{M} \subseteq \overline{A}$. Since $\mathbb{M}$ is a left ideal of $(\beta\mathbb{N}, \cdot)$, $\mathbb{M} \cap E\big(K(\beta\mathbb{N}, \cdot)\big) \neq \emptyset$ by Corollary 2.6 so A is central in $(\mathbb{N}, \cdot)$. $\square$

Theorem 16.27. *There is a set $A \subseteq \mathbb{N}$ which is central in $(\mathbb{N}, +)$ such that for no y and z in $\mathbb{N}$ is $\{y, z, yz\} \subseteq A$. In particular for no $y \in \mathbb{N}$ is $\{y, y^2\} \subseteq A$.*

Proof. Let $x_1 = 2$ and inductively for $n \in \mathbb{N}$ choose $x_{n+1} \geq (x_n + n)^2$. Notice in particular that for each n, $x_n > n$. Let $A = \{x_n + k : n, k \in \mathbb{N} \text{ and } k \leq n\}$. By Corollary 16.2 A is central in $(\mathbb{N}, +)$. Suppose now we have $y \leq z$ in $\mathbb{N}$ with $\{y, z, yz\} \subseteq A$ and pick $n \in \mathbb{N}$ such that $z \in \{x_n + 1, x_n + 2, \ldots, x_n + n\}$. Then $y > 2$ so $yz > 2z > 2x_n > x_n + n$ so $yz \geq x_{n+1} + 1 > (x_n + n)^2 \geq z^2 \geq yz$, a contradiction. $\square$

We see now that sets central in $(\mathbb{N}, \cdot)$ must contain large finite additive structure.

Theorem 16.28. *Let $A \subseteq \mathbb{N}$ be central in $(\mathbb{N}, \cdot)$. For each $m \in \mathbb{N}$ there exists a finite sequence $\langle x_t \rangle_{t=1}^{m}$ such that $\mathrm{FS}(\langle x_t \rangle_{t=1}^{m}) \subseteq A$.*

Proof. Let

$$T = \{p \in \beta\mathbb{N} : \text{ for all } B \in p \text{ and all } m \in \mathbb{N} \text{ there}$$
$$\text{exists } \langle x_t \rangle_{t=1}^{m} \text{ with } \mathrm{FS}(\langle x_t \rangle_{t=1}^{m}) \subseteq B\}.$$

By Theorem 5.8, all idempotents in $(\beta\mathbb{N}, +)$ are in T so $T \neq \emptyset$. We claim that T is a two sided ideal of $(\beta\mathbb{N}, \cdot)$ so that $K(\beta\mathbb{N}, \cdot) \subseteq T$.

Let $p \in T$ and let $q \in \beta\mathbb{N}$. To see that $q \cdot p \in T$, let $B \in q \cdot p$ and let $m \in \mathbb{N}$. Pick $a \in \mathbb{N}$ such that $a^{-1}B \in p$ and pick $\langle x_t \rangle_{t=1}^{m}$ such that $\mathrm{FS}(\langle x_t \rangle_{t=1}^{m}) \subseteq a^{-1}B$. Then $\mathrm{FS}(\langle ax_t \rangle_{t=1}^{m}) \subseteq B$.

To see that $p \cdot q \in T$, let $B \in p \cdot q$ and let $m \in \mathbb{N}$. Let $C = \{a \in \mathbb{N} : a^{-1}B \in q\}$. Then $C \in p$ so pick $\langle x_t \rangle_{t=1}^{m}$ such that $\mathrm{FS}(\langle x_t \rangle_{t=1}^{m}) \subseteq C$. Let $E = \mathrm{FS}(\langle x_t \rangle_{t=1}^{m})$. Then E is finite, so $\bigcap_{a \in E} a^{-1}B \in q$ so pick $b \in \bigcap_{a \in E} a^{-1}B$. Then $\mathrm{FS}(\langle bx_t \rangle_{t=1}^{m}) \subseteq B$.

Since A is central in $(\mathbb{N}, \cdot)$, pick a minimal idempotent p in $(\beta\mathbb{N}, \cdot)$ such that $A \in p$. Then $p \in K(\beta\mathbb{N}, \cdot) \subseteq T$ so for each $m \in \mathbb{N}$ there exists a finite sequence $\langle x_t \rangle_{t=1}^{m}$ such that $\mathrm{FS}(\langle x_t \rangle_{t=1}^{m}) \subseteq A$. $\square$

It is natural, in view of the above theorem, to ask whether one can extend the conclusion to infinite sequences. We see now that one cannot.

Theorem 16.29. *There is a set $A \subseteq \mathbb{N}$ which is central in $(\mathbb{N}, \cdot)$ such that for no sequence $\langle y_n \rangle_{n=1}^{\infty}$ does one have $\mathrm{FS}(\langle y_n \rangle_{n=1}^{\infty}) \subseteq A$.*

Proof. Let $x_1 = 1$ and for $n \in \mathbb{N}$ pick some $x_{n+1} > nx_n$. Let

$$A = \{kx_n : n, k \in \mathbb{N} \text{ and } k \leq n\}.$$

By Corollary 16.2 A is central in $(\mathbb{N}, \cdot)$.

Suppose we have a sequence $\langle y_n \rangle_{n=1}^{\infty}$ with $\mathrm{FS}(\langle y_n \rangle_{n=1}^{\infty}) \subseteq A$. We may presume that the sequence $\langle y_n \rangle_{n=1}^{\infty}$ is increasing since in any event an increasing sum subsystem of $\langle y_n \rangle_{n=1}^{\infty}$ can be chosen. Pick m, n, and r in $\mathbb{N}$ such that $n < r$, $y_m \in \{x_n, 2x_n, 3x_n, \ldots, nx_n\}$, and $y_{m+1} \in \{x_r, 2x_r, 3x_r, \ldots, rx_r\}$.

Pick $k \in \{1, 2, \ldots, r\}$ such that $y_{m+1} = kx_r$. Then

$$kx_r < y_m + y_{m+1} \leq kx_r + nx_n < kx_r + x_r = (k+1)x_r$$

so $y_m + y_{m+1} \notin A$. $\square$

Given any central set A in $(\mathbb{N}, +)$, one has by definition that there is a minimal idempotent p in $(\beta\mathbb{N}, +)$ such that $A \in p$. Since $p = p + p$, one has that

$$\{x \in \mathbb{N} : -x + A \in p\} \in p$$

and so, in particular, $\{x \in \mathbb{N} : -x + A \text{ is central}\}$ is central. That is, every central set often translates down to a central set.

Recall that for any $p \in \beta\mathbb{N}$, $-p = (-1) \cdot p \in \beta\mathbb{Z}$. Since by Exercise 4.3.5, $-\mathbb{N}^*$ is a left ideal of $(\beta\mathbb{Z}, +)$, one cannot have $p = p + (-p)$ for any $p \in \mathbb{N}^*$. If one had $p = (-p) + p$ for some minimal idempotent p in $(\beta\mathbb{N}, +)$ one would have as above that for any $A \in p$, $\{x \in \mathbb{N} : x + A \in p\} \in p$ and hence $\{x \in \mathbb{N} : x + A \text{ is central}\}$ is central. However, according to Corollary 13.19 such an equation cannot hold. Nonetheless, we are able to establish that for any central set A in $(\mathbb{N}, +)$, $\{x \in \mathbb{N} : x + A \text{ is central}\}$ is central.

Theorem 16.30. *Let $p \in K(\beta\mathbb{N}, +)$ and let L be a minimal left ideal of $(\beta\mathbb{N}, +)$. There is an idempotent $q \in L$ such that $p = (-q) + p$.*

Proof. Let $T = \{q \in L : (-q) + p = p\}$. Given $q, r \in \beta\mathbb{N}$, $-(q+r) = (-q) + (-r)$ by Lemma 13.1, so if $q, r \in T$, then $q + r \in T$. Define $\ell : \beta\mathbb{Z} \to \beta\mathbb{Z}$ by $\ell(p) = -p$. (That is, in the semigroup $(\beta\mathbb{Z}, \cdot)$, $\ell = \lambda_{-1}$.) Then ℓ is continuous and $T = L \cap (\rho_p \circ \ell)^{-1}[\{p\}]$ so T is compact. Therefore, it suffices to show that $T \neq \emptyset$. For then, T is a compact subsemigroup of $\beta\mathbb{N}$, hence contains an idempotent by Theorem 2.5.

Since $p \in K(\beta\mathbb{N}, +)$, pick a minimal left ideal L' of $\beta\mathbb{N}$ such that $p \in L'$. Now $\mathbb{N}^*$ is a left ideal of $\beta\mathbb{Z}$ by Exercise 4.3.5 so by Lemma 1.43(c), L and L' are left ideals of $\beta\mathbb{Z}$. In particular $-L + p \subseteq L'$. We claim that $-L + p$ is a left ideal of $\beta\mathbb{N}$ so that $-L + p = L'$. To see this, let $q \in -L$ and let $r \in \beta\mathbb{N}$. We show that $r + q + p \in -L + p$. Now $-q \in L$ so $-r + -q \in L$. Since $-r + -q = -(r + q)$ by Lemma 13.1 we have $r + q \in -L$ so $r + q + p \in -L + p$. Since $-L + p = L'$ and $p \in L'$ we have $T \neq \emptyset$ as required. $\square$

Corollary 16.31. *Let $A \subseteq \mathbb{N}$ be central in $(\mathbb{N}, +)$. Then $\{x \in \mathbb{N} : x + A \text{ is central}\}$ is central.*

Proof. Pick a minimal idempotent p in $(\beta\mathbb{N}, +)$ such that $A \in p$ and pick, by Theorem 16.30, a minimal idempotent q in $(\beta\mathbb{N}, +)$ such that $p = (-q) + p$. Then $\{x \in \mathbb{N} : x + A \in p\} \in q$ and $\{x \in \mathbb{N} : x + A \in p\} \subseteq \{x \in \mathbb{N} : x + A \text{ is central}\}$. $\qquad\square$

We do not know whether every central* set often translates either up or down to another central* set. However we do have the following strong contrast with Corollary 16.31 for IP* sets.

Theorem 16.32. *There is an IP* set A of $(\mathbb{N}, +)$ such that for all $n \in \mathbb{N}$, neither $n + A$ nor $-n + A$ is an IP* set.*

Proof. Let $\langle D_n \rangle_{n \in \mathbb{Z}}$ be a sequence of pairwise disjoint infinite sets of positive even integers such that for each n, $\min D_n > |2n|$. For each $n \in \mathbb{Z}$ enumerate D_n in increasing order as $\langle a(n, k) \rangle_{k=1}^{\infty}$ and for each $k \in \mathbb{N}$, let $y_{n,k} = 2^{a(n,2k)} + 2^{a(n,2k-1)}$. For each $n \in \mathbb{Z}\setminus\{0\}$, let $B_n = \text{FS}(\langle y_{n,k} \rangle_{k=1}^{\infty})$. Let $C = \{n + z : n \in \mathbb{Z}\setminus\{0\}$ and $z \in B_n\}$. Notice that if $n \in \mathbb{Z}\setminus\{0\}$ and $z \in B_n$, then $|n| < z$ so $C \subseteq \mathbb{N}$. Let $A = \mathbb{N}\setminus C$. Then given any $n \in \mathbb{Z}\setminus\{0\}$, $(A + n) \cap B_{-n} = \emptyset$, so $A + n$ is not an IP* set.

We now claim that A is an IP* set, so suppose instead we have a sequence $\langle x_n \rangle_{n=1}^{\infty}$ with $\text{FS}(\langle x_n \rangle_{n=1}^{\infty}) \cap A = \emptyset$. That is, $\text{FS}(\langle x_n \rangle_{n=1}^{\infty}) \subseteq C$. By passing to a suitable sum subsystem we may presume that the sequence $\langle x_n \rangle_{n=1}^{\infty}$ is increasing.

We first observe that if $n \in \mathbb{Z}\setminus\{0\}$, $u \in n + B_n$, and $u = \Sigma_{t \in F} 2^t$, then $\max F \in D_n$. Indeed $u = n + z$ where $z \in B_n$ and $z = \Sigma_{t \in H} 2^t$ where $H \subseteq D_n$ and H has at least two members (because each $y_{n,k}$ has two binary digits). Thus, if $n < 0$, borrowing will not reach $\max H$. Since then $\min H > |2n|$ one has $\max F = \max H$.

We now show that one cannot have $n \in \mathbb{Z}\setminus\{0\}$ and $i < j$ with $\{x_i, x_j\} \subseteq n + B_n$, so suppose instead that we do. Now $x_j = \Sigma_{t \in F} 2^t$ where $\max F \in D_n$. Let $k = \max F$. Since $x_i < x_j$ we have that $x_j + x_i = \Sigma_{t \in H} 2^t$ where either $\max H = k$ or $\max H = k + 1$. But as we have just seen, given any member of C, the largest element of its binary support is even, so the latter case is impossible. The former case tells us that $x_j + x_i \in n + B_n$. But now $\{x_i - n, x_j - n, x_i + x_j - n\} \subseteq B_n$, so if $\ell = \min D_n$ we have that 2^ℓ divides each of $x_i - n$, $x_j - n$, and $x_i + x_j - n$, and hence 2^ℓ divides n, a contradiction.

Consequently, we may choose for each $i \in \mathbb{N}$ some $n(i) \in \mathbb{Z}\setminus\{0\}$ such that $x_i \in n(i) + B_{n(i)}$ and $n(i) \neq n(j)$ for $i \neq j$. Choose i such that $|n(i)| > x_1$. Then $x_i = \Sigma_{t \in F} 2^t$ and $x_i + x_1 = \Sigma_{t \in G} 2^t$ where $\max F = \max G$ and consequently, $x_i + x_1 \in n(i) + B_{n(i)}$. But now $x_1 = (x_i + x_1) - x_i$ is a difference of two members of $B_{n(i)}$ and is hence divisible by 2^ℓ, where $\ell = \min D_{n(i)}$. This is a contradiction because $\ell > |2n(i)| > x_1$. $\qquad\square$

Exercise 16.2.1. In the proof of Lemma 16.19 we used the fact that in the semigroup $(\mathbb{N}, +)$, one must have $n^{-1}(nA) = A$. Show that in $(\mathbb{N}, +)$ one need not have $n(n^{-1}A) = A$. Show in fact that $n(n^{-1}A) = A$ if and only if $A \subseteq \mathbb{N}n$.

Exercise 16.2.2. Prove that if $r \in \mathbb{N}$ and $\mathbb{N} = \bigcup_{i=1}^{r} C_i$, and for only one $i \in \{1, 2, \ldots, r\}$ is there a sequence $\langle y_n \rangle_{n=1}^{\infty}$ in $\mathbb{N}$ with $\mathrm{FS}(\langle y_n \rangle_{n=1}^{\infty}) \subseteq C_i$, then in fact for this i, C_i is an IP* set (so that there is a sequence $\langle x_n \rangle_{n=1}^{\infty}$ in $\mathbb{N}$ such that $\mathrm{FS}(\langle x_n \rangle_{n=1}^{\infty}) \cup \mathrm{FP}(\langle x_n \rangle_{n=1}^{\infty}) \subseteq C_i$).

16.3 IP* Sets in Weak Rings

In this section we extend Corollary 16.21 to apply to a much wider class, called "weak rings", obtaining a sequence and its finite sums and, depending on the precise hypotheses, either all products or almost all products in a given IP* set.

Definition 16.33. (a) A *left weak ring* is a triple $(S, +, \cdot)$ such that $(S, +)$ and $(S, \cdot)$ are semigroups and the left distributive law holds. That is, for all $x, y, z \in S$ one has $x \cdot (y + z) = x \cdot y + x \cdot z$.

(b) A *right weak ring* is a triple $(S, +, \cdot)$ such that $(S, +)$ and $(S, \cdot)$ are semigroups and the right distributive law holds. That is, for all $x, y, z \in S$ one has $(x + y) \cdot z = x \cdot z + y \cdot z$.

(c) A *weak ring* is a triple $(S, +, \cdot)$ which is both a left weak ring and a right weak ring.

In the above definition, we have followed the usual custom regarding order of operations. That is $x \cdot y + x \cdot z = (x \cdot y) + (x \cdot z)$.

Notice that neither of the semigroups $(S, +)$ nor $(S, \cdot)$ is assumed to be commutative. Of course all rings are weak rings. Other examples of weak rings include all subsets of $\mathbb{C}$ that are closed under both addition and multiplication.

Lemma 16.34. *Let $(S, +)$ be any semigroup and let $\cdot$ be the operation making $(S, \cdot)$ a right zero semigroup. Then $(S, +, \cdot)$ is left weak ring. If $(S, +)$ has at least one element which is not idempotent, then $(S, +, \cdot)$ is not a right weak ring.*

Proof. This is Exercise 16.3.2.   □

Analogously to Lemma 16.19, we have the following. Notice that it does not matter whether we define $a^{-1} A b^{-1}$ to be $a^{-1}(A b^{-1})$ or $(a^{-1} A) b^{-1}$. In either case, $y \in a^{-1} A b^{-1}$ if and only if $a y b \in A$.

Lemma 16.35. *Let S be a set, let $A \subseteq S$, and let $a, b \in S$.*

(a) *If $(S, +, \cdot)$ is a left weak ring and A is an IP* set in $(S, +)$, then $a^{-1} A$ is an IP* set in $(S, +)$.*

(b) *If $(S, +, \cdot)$ is a right weak ring and A is an IP* set in $(S, +)$, then $A b^{-1}$ is an IP* set in $(S, +)$.*

(c) *If $(S, +, \cdot)$ is a weak ring and A is an IP* set in $(S, +)$, then $a^{-1} A b^{-1}$ is an IP* set in $(S, +)$.*

Proof. It suffices to establish (a) since then (b) follows from a left-right switch and (c) follows from (a) and (b). So, let $\langle x_n \rangle_{n=1}^{\infty}$ be a sequence in S. Then

$$\mathrm{FS}(\langle a \cdot x_n \rangle_{n=1}^{\infty}) \cap A \neq \emptyset$$

so pick $F \in \mathcal{P}_f(\mathbb{N})$ such that $\sum_{n \in F} a \cdot x_n \in A$. Then, using the left distributive law we have that $\sum_{n \in F} x_n \in a^{-1}A$. $\square$

Recall that in $\mathrm{FP}(\langle x_n \rangle_{n=1}^{\infty})$, the products are taken in increasing order of indices.

Definition 16.36. Let $(S, \cdot)$ be a semigroup, let $\langle x_n \rangle_{n=1}^{\infty}$ be a sequence in S, and let $k \in \mathbb{N}$. Then $\mathrm{AP}(\langle x_n \rangle_{n=1}^{k})$ is the set of all products of terms of $\langle x_n \rangle_{n=1}^{k}$ in any order with no repetitions. Similarly $\mathrm{AP}(\langle x_n \rangle_{n=1}^{\infty})$ is the set of all products of terms of $\langle x_n \rangle_{n=1}^{\infty}$ in any order with no repetitions.

For example, with $k = 3$, we obtain the following:

$$\mathrm{AP}(\langle x_n \rangle_{n=1}^{3}) = \{x_1, x_2, x_3, x_1 x_2, x_1 x_3, x_2 x_3, x_2 x_1, x_3 x_2, x_3 x_1, x_1 x_2 x_3,$$
$$x_1 x_3 x_2, x_2 x_1 x_3, x_2 x_3 x_1, x_3 x_1 x_2, x_3 x_2 x_1\}.$$

Theorem 16.37. *Let $(S, +, \cdot)$ be a left weak ring, let A be an IP* set in $(S, +)$, and let $\langle y_n \rangle_{n=1}^{\infty}$ be any sequence in S. Then there exists a sum subsystem $\langle x_n \rangle_{n=1}^{\infty}$ of $\langle y_n \rangle_{n=1}^{\infty}$ such that if $m \geq 2$, $F \in \mathcal{P}_f(\mathbb{N})$ with $\min F \geq m$, and $b \in \mathrm{AP}(\langle x_n \rangle_{n=1}^{m-1})$, then $b \cdot \sum_{t \in F} x_t \in A$. In particular,*

$$\mathrm{FS}(\langle x_n \rangle_{n=1}^{\infty}) \cup \{b \cdot x_m : m \geq 2 \text{ and } b \in \mathrm{AP}(\langle x_n \rangle_{n=1}^{m-1})\} \subseteq A.$$

Proof. Pick by Lemma 5.11 some idempotent p of $(\beta S, +)$ with

$$p \in \bigcap_{m=1}^{\infty} c\ell \, \mathrm{FS}(\langle x_n \rangle_{n=m}^{\infty}).$$

Then by Lemma 16.35, we have for each $a \in S$ that $a^{-1}A$ is an IP* set in $(S, +)$ and hence is in p.

Let $B_1 = \mathrm{FS}(\langle y_n \rangle_{n=1}^{\infty})$ and note that $B_1 \in p$. Pick $x_1 \in B_1^{\star}$ and pick $H_1 \in \mathcal{P}_f(\mathbb{N})$ such that $x_1 = \sum_{t \in H_1} y_t$.

Inductively, let $n \in \mathbb{N}$ and assume that we have chosen $\langle x_i \rangle_{i=1}^{n}$, $\langle H_i \rangle_{i=1}^{n}$, and $\langle B_i \rangle_{i=1}^{n}$ such that for each $i \in \{1, 2, \ldots, n\}$:

(1) $x_i = \sum_{t \in H_i} y_t$,

(2) if $i > 1$, then $\min H_i > \max H_{i-1}$,

(3) $B_i \in p$,

(4) if $\emptyset \neq F \subseteq \{1, 2, \ldots, i\}$ and $m = \min F$, then $\sum_{j \in F} x_j \in B_m^{\star}$, and

(5) if $i > 1$, then $B_i \subseteq \bigcap \{a^{-1}A : a \in \mathrm{AP}(\langle x_t \rangle_{t=1}^{i-1})\}$.

Only hypotheses (1), (3), and (4) apply at $n = 1$ and they hold trivially. Let $k = \max H_n + 1$. By assumption $\mathrm{FS}(\langle y_t\rangle_{t=k}^\infty) \in p$. Again we have by Lemma 16.35 that for each $a \in S$, $a^{-1}A$ is an IP* set in $(S, +)$ and is thus in p.

For each $m \in \{1, 2, \ldots, n\}$ let

$$E_m = \{\textstyle\sum_{j\in F} x_j : \emptyset \neq F \subseteq \{1, 2, \ldots, n\} \text{ and } m = \min F\}.$$

By hypothesis (4) we have for each $m \in \{1, 2, \ldots, m\}$ and each $a \in E_m$ that $a \in B_m{}^\star$ and hence, by Lemma 4.14, $-a + B_m{}^\star \in p$. Let

$$B_{n+1} = \mathrm{FS}(\langle y_t\rangle_{t=k}^\infty) \cap \bigcap \{a^{-1}A : a \in \mathrm{AP}(\langle x_t\rangle_{t=1}^n)\} \cap \bigcap_{m=1}^n \bigcap_{a\in E_m} (-a + B_m{}^\star).$$

Then $B_{n+1} \in p$. Choose $x_{n+1} \in B_{n+1}{}^\star$. Since $x_{n+1} \in \mathrm{FS}(\langle y_t\rangle_{t=k}^\infty)$, choose $H_{n+1} \in \mathcal{P}_f(\mathbb{N})$ such that $\min H_{n+1} \geq k$ and $x_{n+1} = \sum_{t\in H_{n+1}} y_t$. Then hypotheses (1), (2), (3), and (5) are satisfied directly.

To verify hypothesis (4), let $\emptyset \neq F \subseteq \{1, 2, \ldots, n + 1\}$ and let $m = \min F$. If $n+1 \notin F$, the conclusion holds by hypothesis, so assume that $n+1 \in F$. If $F = \{n+1\}$, then $\sum_{j\in F} x_j = x_{n+1} \in B_{n+1}{}^\star$, so assume $F \neq \{n + 1\}$ and let $G = F\backslash\{n + 1\}$. Let $a = \sum_{j\in F} x_j$. Then $a \in E_m$ so $x_{n+1} \in -a + B_m{}^\star$ so $\sum_{j\in F} x_j = a + x_{n+1} \in B_m{}^\star$ as required.

We thus have that $\langle x_n\rangle_{n=1}^\infty$ is a sum subsystem of $\langle y_n\rangle_{n=1}^\infty$. To complete the proof, let $m \geq 2$, let $F \in \mathcal{P}_f(\mathbb{N})$ with $\min F \geq m$ and let $a \in \mathrm{AP}(\langle x_n\rangle_{n=1}^{m-1})$. Then by hypotheses (4) and (5), $\sum_{j\in F} x_j \in B_m \subseteq a^{-1}A$ so that $a \cdot \sum_{t\in F} x_t \in A$. $\qquad\square$

Observe that $\{b \cdot x_m : m \geq 2 \text{ and } b \in \mathrm{AP}(\langle x_n\rangle_{n=1}^{m-1})\}$ is the set of all products (without repetition) from $\langle x_n\rangle_{n=1}^\infty$ that have their largest index occurring on the right.

Notice that the proof of Theorem 16.37 is nearly identical to that of Theorem 16.20. Essentially the same proof establishes a much stronger conclusion in the event that one has a weak ring rather than just a left weak ring.

Theorem 16.38. *Let $(S, +, \cdot)$ be a weak ring, let A be an IP* set in $(S, +)$, and let $\langle y_n\rangle_{n=1}^\infty$ be any sequence in X. Then there exists a sum subsystem $\langle x_n\rangle_{n=1}^\infty$ of $\langle y_n\rangle_{n=1}^\infty$ in S such that $\mathrm{FS}(\langle x_n\rangle_{n=1}^\infty) \cup \mathrm{AP}(\langle x_n\rangle_{n=1}^\infty) \subseteq A$.*

Proof. Modify the proof of Theorem 16.37 by replacing induction hypothesis (5) with:
(5) if $i > 1$, then

$$B_i \subseteq \bigcap \{a^{-1}A : a \in \mathrm{AP}(\langle x_t\rangle_{t=1}^{i-1})\} \cap \bigcap \{Ab^{-1} : b \in \mathrm{AP}(\langle x_t\rangle_{t=1}^{i-1})\}$$
$$\cap \bigcap \{a^{-1}Ab^{-1} : a, b \in \mathrm{AP}(\langle x_t\rangle_{t=1}^{i-1})\}.$$

Then replace the definition of B_{n+1} with

$$B_{n+1} = \mathrm{FS}(\langle y_t\rangle_{t=k}^\infty) \cap \bigcap \{a^{-1}A : a \in \mathrm{AP}(\langle x_t\rangle_{t=1}^{i-1})\} \cap \bigcap \{Ab^{-1} : b \in \mathrm{AP}(\langle x_t\rangle_{t=1}^{i-1})\}$$
$$\cap \bigcap \{a^{-1}Ab^{-1} : a, b \in \mathrm{AP}(\langle x_t\rangle_{t=1}^{i-1})\} \cap \bigcap_{m=1}^n \bigcap_{a\in E_m} (-a + B_m{}^\star). \qquad\square$$

Theorems 16.37 and 16.38 raise the natural question of whether the stronger conclusion in fact holds in any left weak ring. The example of Lemma 16.34 is not a counterexample to this question because, in this left weak ring, given any sequence $\langle x_n \rangle_{n=1}^{\infty}$, one has $\text{AP}(\langle x_n \rangle_{n=1}^{\infty}) = \{x_n : n \in \mathbb{N}\}$.

Theorem 16.39. *Let S be the free semigroup on the two distinct letters a and b and let* $\hom(S, S)$ *be the set of homomorphisms from S to S. Let $\circ$ be the usual composition of functions and define an operation $\oplus$ on* $\hom(S, S)$ *as follows. Given $f, g \in \hom(S, S)$ and $u_1, u_2, \ldots, u_t \in \{a, b\}$,*

$$(f \oplus g)(u_1 u_2 \ldots u_t) = f(u_1){}^\frown g(u_1){}^\frown f(u_2){}^\frown g(u_2){}^\frown \ldots {}^\frown f(u_t){}^\frown g(u_t).$$

Then $(\hom(S, S), \oplus, \circ)$ *is a left weak ring and there exist an IP* set A in* $(\hom(S, S), \oplus)$ *and a sequence $\langle f_n \rangle_{n=1}^{\infty}$ in* $\hom(S, S)$ *such that no sum subsystem $\langle g_n \rangle_{n=1}^{\infty}$ of $\langle f_n \rangle_{n=1}^{\infty}$ has* $\text{AP}(\langle g_n \rangle_{n=1}^{\infty}) \subseteq A$.

Proof. The verification that $(\hom(S, S), \oplus, \circ)$ is a left weak ring is Exercise 16.3.3. Notice that in order to define a member of $\hom(S, S)$ it is enough to define its values at a and b. Define $f_1 \in \hom(S, S)$ by $f_1(a) = ab$ and $f_1(b) = b$ and define inductively for $n \in \mathbb{N}$, $f_{n+1} = f_n \oplus f_1$. Notice that for each $n \in \mathbb{N}$, $f_n(a) = (ab)^n$ and $f_n(b) = b^n$.

Let $A = \hom(S, S) \setminus \{f_r \circ f_s : r, s \in \mathbb{N} \text{ and } r > s\}$. Notice that, given $r, s \in \mathbb{N}$,

$$f_r(f_s(a)) = f_r((ab)^s) = (f_r(a) f_r(b))^s = ((ab)^r b^r)^s.$$

(We have used the fact that f_r is a homomorphism.) In particular, notice that if $f_r \circ f_s = f_m \circ f_n$, then $(r, s) = (m, n)$.

We claim that if $m, n, r, s, \ell, t \in \mathbb{N}$, $h = f_m \circ f_n$, $k = f_r \circ f_s$, and $h \oplus k = f_\ell \circ f_t$, then $\ell = m = r$ and $t = n + s$. Indeed,

$$((ab)^\ell b^\ell)^t = (h \oplus k)(a) = h(a){}^\frown k(a) = ((ab)^m b^m)^n ((ab)^r b^r)^s.$$

Suppose now that $\langle h_n \rangle_{n=1}^{\infty}$ is a sequence in $\hom(S, S)$ with

$$\text{FS}(\langle h_n \rangle_{n=1}^{\infty}) \subseteq \{f_r \circ f_s : r, s \in \mathbb{N} \text{ and } r > s\}.$$

Then, using the fact just established, there is some r and for each n some $s(n)$ such that $h_n = f_r \circ f_{s(n)}$. Then

$$h_1 \oplus h_2 \oplus \cdots \oplus h_r = f_r \circ f_t$$

where $t = \Sigma_{n=1}^{r} s(n) \geq r$, a contradiction. Thus A is an IP* set in $(\hom(S, S), +)$.

Finally, suppose that $\langle g_n \rangle_{n=1}^{\infty}$ is a sum subsystem of $\langle f_n \rangle_{n=1}^{\infty}$ with $\text{AP}(\langle g_n \rangle_{n=1}^{\infty}) \subseteq A$. Then $g_1 = \Sigma_{n \in H} f_n$ for some $H \in \mathcal{P}_f(\mathbb{N})$ so $g_1 = f_k$ where $k = \Sigma H$. Then $g_{k+1} = f_\ell$ for some $\ell > k$ and thus $g_{k+1} \circ g_1 = f_\ell \circ f_k \notin A$, a contradiction. $\square$

Exercise 16.3.1. Show that if $(S, +, \cdot)$ is a weak ring in which $(S, +)$ is commutative, $n \in \mathbb{N}$ and the operations on the $n \times n$ matrices with entries from S are defined as usual (so, for example, the entry in row i and column j of $A \cdot B$ is $\Sigma_{k=1}^{n} a_{i,k} \cdot b_{k,j}$), then these matrices form a weak ring. Give an example of a left weak ring $(S, +, \cdot)$ for which $(S, +)$ is commutative, but the 2×2 matrices over S do not form a left weak ring.

Exercise 16.3.2. Prove Lemma 16.34.

Exercise 16.3.3. Prove that $(\hom(S, S), \oplus, \circ)$ as described in Theorem 16.39 is a left weak ring.

Exercise 16.3.4. We know that the left weak ring $(\hom(S, S), \oplus, \circ)$ of Theorem 16.39 is not a right weak ring because it does not satisfy the conclusion of Theorem 16.38. Establish this fact directly by producing $f, g, h \in \hom(S, S)$ such that $(f \oplus g) \circ h \neq f \circ h \oplus g \circ h$.

16.4 Spectra and Iterated Spectra

Spectra of numbers are sets of the form $\{\lfloor n\alpha \rfloor : n \in \mathbb{N}\}$ or $\{\lfloor n\alpha + \gamma \rfloor : n \in \mathbb{N}\}$ where α and γ are positive reals. Sets of this form or of the form $\{\lfloor n\alpha \rfloor : n \in A\}$ or $\{\lfloor n\alpha + \gamma \rfloor : n \in A\}$ for some specified sets A have been extensively studied in number theory. (See the notes to this chapter for some references.) We are interested in these sets because they provide us with a valuable collection of rather explicit examples of IP* sets and central* sets in $(\mathbb{N}, +)$.

By the very nature of their definition, it is easy to give examples of IP sets. And anytime one explicitly describes a finite partition of $\mathbb{N}$ at least one cell must be a central set and it is often easy to identify which cells are central. The situation with respect to IP* sets and central* sets is considerably different however. We know from Theorem 16.8 that whenever $\langle x_n \rangle_{n=1}^\infty$ is a sequence in $\mathbb{N}$ such that $\mathrm{FS}(\langle x_n \rangle_{n=1}^\infty)$ is not piecewise syndetic, one has $\mathbb{N} \setminus \mathrm{FS}(\langle x_n \rangle_{n=1}^\infty)$ is a central* set which is not IP*. So, for example, $\{\sum_{t \in F} 2^t : F \in \mathcal{P}_f(\mathbb{N}) \text{ and some } t \in F \text{ is even}\}$ is central* but not IP* because it is $\mathbb{N} \setminus \mathrm{FS}(\langle 2^{2t-1} \rangle_{t=1}^\infty)$.

We also know from Lemma 16.13 that for each $n \in \mathbb{N}$, $\mathbb{N}n$ is an IP* set. But at this point, we would be nearly at a loss to come up with an IP* set which doesn't almost contain $\mathbb{N}n$ for some n. (The set of Theorem 16.32 is one such example.)

In this section we shall be utilizing some information obtained in Section 10.1. Recall that we defined there for any subsemigroup S of $(\mathbb{R}, +)$ and any positive real number α, functions $g_\alpha : S \to \mathbb{Z}$, $f_\alpha : S \to [-\frac{1}{2}, \frac{1}{2})$, and $h_\alpha : S \to \mathbb{T}$ by $g_\alpha(x) = \lfloor x\alpha + \frac{1}{2} \rfloor$, $f_\alpha(x) = x\alpha - g_\alpha(x)$, and $h_\alpha(x) = \pi(f_\alpha(x))$ where the circle group $\mathbb{T} = \mathbb{R}/\mathbb{Z}$ and π is the projection of $\mathbb{R}$ onto $\mathbb{T}$.

Definition 16.40. Let $\alpha > 0$ and let $0 < \gamma < 1$. The function $g_{\alpha,\gamma} : \mathbb{N} \to \omega$ is defined by $g_{\alpha,\gamma}(n) = \lfloor n\alpha + \gamma \rfloor$.

We denote by $\widetilde{g_{\alpha,\gamma}}$, the continuous extension of $g_{\alpha,\gamma}$ from $\beta\mathbb{N}$ to $\beta\omega$. Notice that $g_\alpha = g_{\alpha,\frac{1}{2}}$.

Recall that for $\alpha > 0$ we have defined $Z_\alpha = \{p \in \beta S : \widetilde{f_\alpha}(p) = 0\}$.

Lemma 16.41. *Let $\alpha > 0$, let $0 < \gamma < 1$, and let $p \in Z_\alpha$. Then $\widetilde{g_{\alpha,\gamma}}(p) = \widetilde{g_\alpha}(p)$.*

Proof. Let $\epsilon = \min\{\gamma, 1 - \gamma\}$ and let $A = \{n \in \mathbb{N} : -\epsilon < f_\alpha(n) < \epsilon\}$. Then $A \in p$ so it suffices to show that $g_{\alpha,\gamma}$ and g_α agree on A. So let $n \in A$ and let $m = g_\alpha(n)$. Then $m - \epsilon < n\alpha < m + \epsilon$ so

$$m \leq m - \epsilon + \gamma < n\alpha + \gamma < m + \epsilon + \gamma \leq m + 1$$

so $m = g_{\alpha,\gamma}(n)$. $\qquad\qquad\qquad\qquad\qquad\qquad\qquad\qquad\qquad\qquad\qquad\quad$ $\square$

As regards spectra, we see that "as you sew, so shall you reap".

Theorem 16.42. *Let $\alpha > 0$, let $0 < \gamma < 1$, and let $A \subseteq \mathbb{N}$.*
  (a) *If A is an IP* set, then $g_{\alpha,\gamma}[A]$ is an IP* set.*
  (b) *If A is a central* set, then $g_{\alpha,\gamma}[A]$ is a central* set.*
  (c) *If A is a central set, then $g_{\alpha,\gamma}[A]$ is a central set.*
  (d) *If A is an IP set, then $g_{\alpha,\gamma}[A]$ is an IP set.*

Proof. (a) By Theorem 16.6 we need to show that $g_{\alpha,\gamma}[A]$ is a member of every idempotent in $(\beta\mathbb{N}, +)$. So let p be an idempotent in $(\beta\mathbb{N}, +)$. Since $Z_{1/\alpha}$ is the kernel of a homomorphism by Lemma 10.3 we have that $p \in Z_{1/\alpha}$. Since, by Theorem 10.12, $\widetilde{g_{1/\alpha}}$ is an isomorphism from $Z_{1/\alpha}$ to Z_α, $\widetilde{g_{1/\alpha}}(p)$ is an idempotent and so $A \in \widetilde{g_{1/\alpha}}(p)$. By Lemma 16.41 $\widetilde{g_{\alpha,\gamma}}\big(\widetilde{g_{1/\alpha}}(p)\big) = \widetilde{g_\alpha}\big(\widetilde{g_{1/\alpha}}(p)\big)$ which is p by Theorem 10.12. Since $A \in \widetilde{g_{1/\alpha}}(p)$ and $\widetilde{g_{\alpha,\gamma}}\big(\widetilde{g_{1/\alpha}}(p)\big) = p$ we have that $g_{\alpha,\gamma}[A] \in p$ as required.

(b) Let p be a minimal idempotent of $(\beta\mathbb{N}, +)$. We need to show that $g_{\alpha,\gamma}[A] \in p$. Now $p \in K(\beta\mathbb{N}) \cap Z_{1/\alpha}$ and $K(\beta\mathbb{N}) \cap Z_{1/\alpha} = K(Z_{1/\alpha})$ by Theorem 1.65. Thus p is a minimal idempotent in $Z_{1/\alpha}$ so by Theorem 10.12 $\widetilde{g_{1/\alpha}}(p)$ is a minimal idempotent in Z_α. That is, $\widetilde{g_{1/\alpha}}(p) \in K(Z_\alpha) = K(\beta\mathbb{N}) \cap Z_\alpha$ so $\widetilde{g_{1/\alpha}}(p)$ is a minimal idempotent in $(\beta\mathbb{N}, +)$. Consequently, $A \in \widetilde{g_{1/\alpha}}(p)$ so as in part (a), $g_{\alpha,\gamma}[A] \in p$.

(c) Since A is central, pick a minimal idempotent p with $A \in p$. Then $p \in K(\beta\mathbb{N}) \cap Z_\alpha = K(Z_\alpha)$ so by Theorem 10.12, $\widetilde{g_\alpha}(p) \in K(Z_{1/\alpha}) = K(\beta\mathbb{N}) \cap Z_{1/\alpha}$ so $\widetilde{g_\alpha}(p)$ is a minimal idempotent and $g_{\alpha,\gamma}[A] \in \widetilde{g_{\alpha,\gamma}}(p) = \widetilde{g_\alpha}(p)$.

(d) Since A is an IP set, pick an idempotent p with $A \in p$. Then $p \in Z_\alpha$ so by Theorem 10.12, $\widetilde{g_\alpha}(p)$ is an idempotent and $g_{\alpha,\gamma}[A] \in \widetilde{g_{\alpha,\gamma}}(p) = \widetilde{g_\alpha}(p)$. $\qquad$ $\square$

In the event that $\alpha \geq 1$ we get an even stronger correspondence, because then $g_{\alpha,\gamma}$ is one to one.

Corollary 16.43. *Let $\alpha \geq 1$, let $0 < \gamma < 1$, and let $A \subseteq \mathbb{N}$.*
  (a) *A is an IP* set if and only if $g_{\alpha,\gamma}[A]$ is an IP* set.*
  (b) *A is a central* set if and only if $g_{\alpha,\gamma}[A]$ is a central* set.*
  (c) *A is a central set if and only if $g_{\alpha,\gamma}[A]$ is a central set.*
  (d) *A is an IP set if and only if $g_{\alpha,\gamma}[A]$ is an IP set.*

Proof. We establish (a) and (c). The proof of (b) is similar to that of (a) and the proof of (d) is similar to that of (c).

(a) The necessity is Theorem 16.42(a). Assume that $g_{\alpha,\gamma}[A]$ is an IP* set and suppose that A is not an IP* set. Pick an IP set B such that $A \cap B = \emptyset$. Since $g_{\alpha,\gamma}$ is one to one, $g_{\alpha,\gamma}[A] \cap g_{\alpha,\gamma}[B] = \emptyset$ while by Theorem 16.42(d), $g_{\alpha,\gamma}[B]$ is an IP set, a contradiction.

(c) The necessity is Theorem 16.42(c). Assume that $g_{\alpha,\gamma}[A]$ is a central set and suppose that A is not a central set. Then $\mathbb{N}\backslash A$ is a central* set so by Theorem 16.42(b), $g_{\alpha,\gamma}[\mathbb{N}\backslash A]$ is a central* set. But this is a contradiction because $g_{\alpha,\gamma}[A] \cap g_{\alpha,\gamma}[\mathbb{N}\backslash A] = \emptyset$. $\square$

Because for $\alpha > 1$ the output of $g_{\alpha,\gamma}$ is the same kind of set as the input, one may iterate the functions at will. Thus, for example, if A is a set which is central* but not IP* (such as one given by Theorem 16.8), then

$$\left\{ \left\lfloor \lfloor n\sqrt{10} + .2 \rfloor \pi + \frac{1}{\pi} \right\rfloor : n \in A \right\}$$

is a central* set which is not IP*.

Notice also that the description of a set $g_{\alpha,\gamma}[\mathbb{N}]$ is effective. That is, one only needs a sufficiently precise decimal approximation to α and γ in order to determine whether a specified number is a member of $g_{\alpha,\gamma}[\mathbb{N}]$.

Notes

The notions of IP, IP*, central, and central* are due to H. Furstenberg in [98] where central sets were defined in terms of notions from topological dynamics. See Chapter 19 for a proof of the equivalence of the notions of central.

Theorem 16.20 is from [120] where it had a purely combinatorial proof. Most of the remaining results of Section 16.2 are from [31], obtained in collaboration with V. Bergelson, except for Theorem 16.16 (which is from [78] and was obtained in collaboration with W. Deuber) and Theorem 16.30 which, while previously unpublished, is from an early draft of [33], obtained in collaboration with V. Bergelson and B. Kra.

The results of Section 16.3 are due to E. Terry in [234]. The left weak ring $(\hom(S, S), \oplus, \circ)$ is due to J. Clay in [65] where it is given as an example of a left nearring which is not a right nearring. (The notion of *nearring* is a stronger notion than that of a weak ring.)

The results of Section 16.4 are from [33], results of collaboration with V. Bergelson and B. Kra.

H. Furstenberg [98, Proposition 9.4] gives an explicit description of certain IP* sets in terms of topological dynamics, and other examples can be deduced from his paper [100] with B. Weiss.

Spectra of the form $\{\lfloor n\alpha + \gamma \rfloor : n \in \mathbb{N}\}$ were introduced by T. Skolem [224] and given the name *the γ-nonhomogenous spectrum of α* by R. Graham, S. Lin, and C. Lin in [109]. See the introduction to [33] for a brief history of the spectra $\{\lfloor n\alpha \rfloor : n \in \mathbb{N}\}$ and further references.

Chapter 17

Sums and Products

We saw in Chapter 16 that any IP* set in $(\mathbb{N}, +)$ has extensive multiplicative structure in addition to the additive structure that one would expect. In particular, by Corollary 16.21, if A is an IP* set in $(\mathbb{N}, +)$, then there is some sequence $\langle x_n \rangle_{n=1}^{\infty}$ in $\mathbb{N}$ such that $\mathrm{FS}(\langle x_n \rangle_{n=1}^{\infty}) \cup \mathrm{FP}(\langle x_n \rangle_{n=1}^{\infty}) \subseteq A$. On the other hand, IP* sets in $(\mathbb{N}, +)$ are not partition regular by Theorem 16.11 so this fact does not yield any results about finite partitions of $\mathbb{N}$. In fact, we shall show in Theorem 17.16 that there is a finite partition of $\mathbb{N}$ such that no cell contains all pairwise sums and products from the same sequence.

We saw in Corollary 5.22 that given any finite partition of $\mathbb{N}$, there must be one cell A and sequences $\langle x_n \rangle_{n=1}^{\infty}$ and $\langle y_n \rangle_{n=1}^{\infty}$ with $\mathrm{FS}(\langle x_n \rangle_{n=1}^{\infty}) \cup \mathrm{FP}(\langle y_n \rangle_{n=1}^{\infty}) \subseteq A$. We are concerned in this chapter with extensions of this result in various different directions.

17.1 Ultrafilters with Rich Additive and Multiplicative Structure

Recall that we have defined $\mathbb{M} = \{p \in \beta\mathbb{N} : \text{for all } A \in p, A \text{ is central in } (\mathbb{N}, +)\}$ and $\Delta = \{p \in \beta\mathbb{N} : \text{for all } A \in p, \overline{d}(A) > 0\}$.

Definition 17.1. A *combinatorially rich ultrafilter* is any $p \in \mathbb{M} \cap \Delta \cap K(\beta\mathbb{N}, \cdot)$ such that $p \cdot p = p$.

We shall see the reason for the name "combinatorially rich" in Theorem 17.3. First we observe that they exist.

Lemma 17.2. *There exists a combinatorially rich ultrafilter.*

Proof. By Theorem 6.79 we have that Δ is a left ideal of $(\beta\mathbb{N}, +)$ and so, by Corollary 2.6 contains an additive idempotent r which is minimal in $(\beta\mathbb{N}, +)$. By Remark 16.23 we have $r \in \mathbb{M}$ and consequently $\mathbb{M} \cap \Delta \neq \emptyset$.

Thus by Theorems 6.79 and 16.24, $\mathbb{M} \cap \Delta$ is a left ideal of $(\beta\mathbb{N}, \cdot)$ and hence contains a multiplicative idempotent which is minimal in $(\beta\mathbb{N}, \cdot)$. $\qquad \square$

Recall the notion of FP-tree introduced in Definition 14.23. We call the corresponding additive notion an *FS-tree*. Since any member of a combinatorially rich ultrafilter is
additively central, it must contain by Theorem 14.25 an FS-tree T such that $\{B_f : f \in T\}$
is collectionwise piecewise syndetic, and in particular each B_f is piecewise syndetic.
(Recall that B_f is the set of successors to the node f of T.)

Notice, however, that in an arbitrary central set none of the B_f's need have positive
upper density. (In fact, recall from Theorem 6.80 that $\mathbb{N}^* \backslash \Delta$ is a left ideal of $(\beta\mathbb{N}, +)$
and hence there are central sets in $(\mathbb{N}, +)$ with zero density.)

Theorem 17.3. *Let p be a combinatorially rich ultrafilter and let $C \in p$.*
   (a) *C is central in $(\mathbb{N}, +)$.*
   (b) *C is central in $(\mathbb{N}, \cdot)$.*
   (c) *Let $\langle A(n)\rangle_{n=1}^{\infty}$ enumerate the image partition regular matrices with entries from
$\mathbb{Q}$ and for each n, let $m(n)$ be the number of columns of $A(n)$. There exists for each
$n \in \mathbb{N}$ a choice of $\vec{x}(n) \in \mathbb{N}^{m(n)}$ such that, if Y_n is the set of entries of $A(n)\vec{x}(n)$, then
$\mathrm{FS}(\langle Y_n \rangle_{n=1}^{\infty}) \subseteq C$.*
   (d) *Let $\langle A(n)\rangle_{n=1}^{\infty}$ enumerate the kernel partition regular matrices with entries from
$\mathbb{Q}$ and for each n, let $m(n)$ be the number of columns of $A(n)$. There exists for each
$n \in \mathbb{N}$ a choice of $\vec{x}(n) \in \mathbb{N}^{m(n)}$ such that $A(n)\vec{x}(n) = \vec{0}$ and, if Y_n is the set of entries
of $\vec{x}(n)$, then $\mathrm{FS}(\langle Y_n \rangle_{n=1}^{\infty}) \subseteq C$.*
   (e) *There is an FS-tree T in C such that for each $f \in T$, $\overline{d}(B_f) > 0$.*
   (f) *There is an FP-tree T in C such that for each $f \in T$, $\overline{d}(B_f) > 0$.*

Proof. Conclusion (a) holds because $p \in \mathbb{M}$ while conclusion (b) holds because $p \cdot p = p \in K(\beta\mathbb{N}, \cdot)$.

Conclusions (c) and (d) follow from Theorems 16.16 and 16.17 respectively.

To verify conclusion (e), notice that by Lemma 14.24 there is an FS-tree T in A such
that for each $f \in T$, $B_f \in p$. Since each member of p has positive upper density the
conclusion follows.

Conclusion (f) follows in the same way. $\square$

As a consequence of conclusions (a) and (b) of Theorem 17.3 we have in particular
that any member of a combinatorially rich ultrafilter satisfies the conclusions of the
Central Sets Theorem (Theorem 14.11), phrased both additively and multiplicatively.

There is naturally a corresponding partition result. Notice that Corollary 17.4 applies
in particular when $C = \mathbb{N}$.

Corollary 17.4. *Let C be a central* set in $(\mathbb{N}, +)$, let $r \in \mathbb{N}$, and let $C = \bigcup_{i=1}^{r} C_i$.
There is some $i \in \{1, 2, \ldots, r\}$ such that each of the conclusions of Theorem 17.3 hold
with C_i replacing C.*

Proof. By Lemma 15.4 we have $\{p \in K(\beta\mathbb{N}, +) : p = p + p\} \subseteq \overline{C}$ so $\mathbb{M} \subseteq \overline{C}$. Pick
a combinatorially rich ultrafilter p. Then $C \in p$ so some $C_i \in p$. $\square$

17.2 Pairwise Sums and Products

We know from Corollary 16.21 that any IP* set in $(\mathbb{N}, +)$ contains $\text{FS}(\langle x_n \rangle_{n=1}^{\infty}) \cup \text{FP}(\langle x_n \rangle_{n=1}^{\infty})$ for some sequence in $\mathbb{N}$. It is natural to ask whether there is some partition analogue of this result. We give a strong negative answer to this question in this section. That is we produce a finite partition of $\mathbb{N}$ such that no cell contains the pairwise sums and products of any injective sequence. As a consequence, we see that the equation $p + p = p \cdot p$ has no solutions in $\mathbb{N}^*$.

Definition 17.5. Let $\langle x_n \rangle_{n=1}^{\infty}$ be a sequence in $\mathbb{N}$.
  (a) $\text{PS}(\langle x_n \rangle_{n=1}^{\infty}) = \{x_n + x_m : n, m \in \mathbb{N} \text{ and } n \neq m\}$.
  (b) $\text{PP}(\langle x_n \rangle_{n=1}^{\infty}) = \{x_n \cdot x_m : n, m \in \mathbb{N} \text{ and } n \neq m\}$.

The partition we use is based on the binary representation of an integer. Recall that for any $x \in \mathbb{N}$, $x = \Sigma_{t \in \text{supp}(x)} 2^t$.

Definition 17.6. Let $x \in \mathbb{N}$. Then

  (a) $a(x) = \max \text{supp}(x)$.
  (b) If $x \notin \{2^t : t \in \omega\}$, then $b(x) = \max(\text{supp}(x) \backslash \{a(x)\})$.
  (c) $c(x) = \max\big(\{-1, 0, 1, \ldots, a(x)\} \backslash \text{supp}(x)\big)$.
  (d) $d(x) = \min \text{supp}(x)$.
  (e) If $x \notin \{2^t : t \in \omega\}$, then $e(x) = \min(\text{supp}(x) \backslash \{d(x)\})$.

When x is written (without leading 0's) in binary, $a(x)$, $b(x)$, $c(x)$, $d(x)$, and $e(x)$ are respectively the positions of the leftmost 1, the next to leftmost 1, the leftmost 0, the rightmost 1, and the next to rightmost 1.

Remark 17.7. *Let $x \in \mathbb{N} \backslash \{2^t : t \in \mathbb{N}\}$ and let $k \in \omega$.*
  (a) $b(x) \geq k$ *if and only if* $x \geq 2^{a(x)} + 2^k$.
  (b) $b(x) \leq k$ *if and only if* $x < 2^{a(x)} + 2^{k+1}$.
  (c) $c(x) \geq k$ *if and only if* $x < 2^{a(x)+1} - 2^k$.
  (d) $c(x) \leq k$ *if and only if* $x \geq 2^{a(x)+1} - 2^{k+1}$.
  (e) $e(x) = k$ *if and only if* $k > d(x)$ *and there is some* $m \in \omega$ *such that* $x = 2^{k+1}m + 2^k + 2^{d(x)}$.

We now introduce some sets that will be used to define the partition that we are seeking.

Definition 17.8. (a)

$$A_0 = \{x \in \mathbb{N} : a(x) \text{ is even and } 2^{a(x)} < x < 2^{a(x)+\frac{1}{2}}\}.$$
$$A_1 = \{x \in \mathbb{N} : a(x) \text{ is even and } 2^{a(x)+\frac{1}{2}} < x < 2^{a(x)+1}\}.$$
$$A_2 = \{x \in \mathbb{N} : a(x) \text{ is odd and } 2^{a(x)} < x < 2^{a(x)+\frac{1}{2}}\}.$$
$$A_3 = \{x \in \mathbb{N} : a(x) \text{ is odd and } 2^{a(x)+\frac{1}{2}} < x < 2^{a(x)+1}\}.$$
$$A_4 = \{2^t : t \in \omega\}.$$

(b) For $i \in \{0, 1, 2\}$

$$B_i = \{x \in \mathbb{N}\backslash A_4 : x < 2^{a(x)+1}(1 - 2^{c(x)-a(x)})^{\frac{1}{2}} \text{ and } a(x) - c(x) \equiv i \pmod{3}\}$$
$$\cup \{x \in \mathbb{N}\backslash A_4 : x \geq 2^{a(x)+1}(1 - 2^{c(x)-a(x)})^{\frac{1}{2}} \text{ and } a(x) - c(x) \equiv i + 1 \pmod{3}\}.$$

(c) $\{C_0, C_1\}$ is any partition of $\mathbb{N}$ such that for all $k \in \mathbb{N}\backslash\{1\}$, $k + 1 \in C_0$ if and only if $2k \in C_1$.

Notice that a partition as specified in Definition 17.8(c) is easy to come by. Odd numbers may be assigned at will, and if the numbers less than $2k$ have been assigned, assign $2k$ to the cell which does not contain $k + 1$.

Remark 17.9. (a) *If $x, y \in A_0 \cup A_2$, then $a(xy) = a(x) + a(y)$.*
(b) *If $x, y \in A_1 \cup A_3$, then $a(xy) = a(x) + a(y) + 1$.*

Lemma 17.10. (a) *If $x, y \in A_0 \cup A_2$, then $a(xy) - b(xy) \leq a(x) - b(x)$.*
(b) *If $x, y \in A_1 \cup A_3$, then $a(xy) - c(xy) \leq a(x) - c(x)$.*

Proof. We use Remarks 17.7 and 17.9.
(a) We have $x \geq 2^{a(x)} + 2^{b(x)}$ and $y > 2^{a(y)}$ so that

$$xy > 2^{a(x)+a(y)} + 2^{b(x)+a(y)} = 2^{a(xy)} + 2^{a(xy)-a(x)+b(x)}$$

so $b(xy) \geq a(xy) - a(x) + b(x)$.
(b) We have $x < 2^{a(x)+1} - 2^{c(x)}$ and $y < 2^{a(y)+1}$ so that

$$xy < 2^{a(x)+a(y)+2} - 2^{a(y)+c(x)+1} = 2^{a(xy)+1} - 2^{a(xy)-a(x)+c(x)}$$

so $c(xy) \geq a(xy) - a(x) + c(x)$. $\square$

We are now ready to define a partition of $\mathbb{N}$. When we write $x \approx y \pmod{\mathcal{R}}$ we mean, of course, that x and y are elements of the same member of $\mathcal{R}$.

Definition 17.11. Define a partition $\mathcal{R}$ of $\mathbb{N}$ by specifying that A_4 and $2\mathbb{N} + 1$ are cells of $\mathcal{R}$ and for any $x, y \in \mathbb{N}\backslash((2\mathbb{N} + 1) \cup A_4)$, $x \approx y \pmod{\mathcal{R}}$ if and only if each of the following statements holds.

(1) For $i \in \{0, 1, 2\}$, $x \in B_i$ if and only if $y \in B_i$.
(2) For $i \in \{0, 1\}$, $d(x) \in C_i$ if and only if $d(y) \in C_i$.
(3) $a(x) - b(x) \leq d(x)$ if and only if $a(y) - b(y) \leq d(y)$.
(4) $a(x) - c(x) \leq d(x)$ if and only if $a(y) - c(y) \leq d(y)$.
(5) $a(x) - b(x) \equiv a(y) - b(y) \pmod{3}$.
(6) $a(x) \equiv a(y) \pmod{2}$.
(7) $e(x) \equiv e(y) \pmod{2}$.
(8) $x \equiv y \pmod{1}6$.

Notice that there is no sequence $\langle x_n \rangle_{n=1}^{\infty}$ with $\mathrm{PS}(\langle x_n \rangle_{n=1}^{\infty}) \subseteq A_4$ or with $\mathrm{PS}(\langle x_n \rangle_{n=1}^{\infty}) \subseteq 2\mathbb{N} + 1$.

Lemma 17.12. *Let $\langle x_n \rangle_{n=1}^{\infty}$ be a one-to-one sequence in $\mathbb{N}$. If $\mathrm{PS}(\langle x_n \rangle_{n=1}^{\infty}) \cup \mathrm{PP}(\langle x_n \rangle_{n=1}^{\infty})$ is contained in one cell of the partition $\mathcal{R}$, then $\{d(x_n) : n \in \mathbb{N}\}$ is unbounded.*

Proof. Suppose instead that $\{d(x_n) : n \in \mathbb{N}\}$ is bounded and pick $k \in \mathbb{N}$ such that for infinitely many n, $d(x_n) = k$. If $k = 0$, then we would have $\mathrm{PP}(\langle x_n \rangle_{n=1}^{\infty}) \subseteq 2\mathbb{N} + 1$ so $\mathrm{PS}(\langle x_n \rangle_{n=1}^{\infty}) \subseteq 2\mathbb{N} + 1$, which is impossible. Thus we may assume that $k \geq 1$.

If $k > 1$, then pick $n < r$ such that $d(x_n) = d(x_r) = k$ and either

$$k + 1 \in \mathrm{supp}(x_n) \cap \mathrm{supp}(x_r) \quad \text{or} \quad k + 1 \notin \mathrm{supp}(x_n) \cup \mathrm{supp}(x_r).$$

Then $d(x_n + x_r) = k + 1$ and $d(x_n x_r) = 2k$ so one can't have $i \in \{0, 1\}$ with $d(x_n + x_r), d(x_n x_r) \in C_i$.

Thus we must have that $k = 1$. Suppose first that for infinitely many n one has $d(x_n) = 1$ and $e(x_n) = 2$. Pick $n < r$ and $u, v \in \omega$ such that $u \equiv v \pmod 2$ and $x_n = 2 + 4 + 8u$ and $x_r = 2 + 4 + 8v$. Then $x_n + x_r = 12 + 16 \cdot (\frac{u+v}{2}) \equiv 12 \pmod{16}$ while $x_n x_r = 36 + 48(u + v) + 64uv \equiv 4 \pmod{16}$.

Consequently one has infinitely many n with $d(x_n) = 1$ and $e(x_n) > 2$. Pick $n < r$ and $u, v \in \omega$ such that $d(x_n) = d(x_r) = 1$, $3 \leq e(x_n) \leq e(x_r)$, $u \equiv v \pmod 2$, and

$$x_n = 2 + 2^{e(x_n)} + u \cdot 2^{e(x_n)+1} \quad \text{and} \quad x_r = 2 + 2^{e(x_r)} + v \cdot 2^{e(x_r)+1}.$$

Suppose first that $e(x_n) < e(x_r)$. Then

$$x_n + x_r = 4 + 2^{e(x_n)} + 2^{e(x_n)+1}\big(2^{e(x_r)-e(x_n)-1} + u + v \cdot 2^{e(x_r)-e(x_n)}\big)$$

and

$$\begin{aligned}
x_n x_r = 4 + 2^{e(x_n)+1} + 2^{e(x_n)+2}\big(u + 2^{e(x_r)-e(x_n)-1} + 2^{e(x_r)-2} \\
+ u \cdot 2^{e(x_r)-1} + v \cdot 2^{e(x_r)-e(x_n)} + v \cdot 2^{e(x_r)-1} + uv \cdot 2^{e(x_r)}\big)
\end{aligned}$$

so $e(x_n + x_r) = e(x_n)$ while $e(x_n x_r) = e(x_n) + 1$, a contradiction.

Consequently we have some $\ell > 2$ such that $e(x_n) = e(x_r) = \ell$. Then

$$x_n + x_r = 4 + 2^{\ell+1} + 2^{\ell+2}\Big(\frac{u+v}{2}\Big)$$

and

$$x_n x_r = 4 + 2^{\ell+2} + 2^{\ell+3}\Big(2^{\ell-3} + \frac{u+v}{2} + (u+v)2^{\ell-2} + uv2^{\ell-1}\Big)$$

so that $e(x_n + x_r) = \ell + 1$ while $e(x_n x_r) = \ell + 2$, again a contradiction. $\square$

As a consequence of Lemma 17.12, we know that if $\langle x_n \rangle_{n=1}^{\infty}$ is a one-to-one sequence with $\mathrm{PS}(\langle x_n \rangle_{n=1}^{\infty}) \cup \mathrm{PP}(\langle x_n \rangle_{n=1}^{\infty})$ contained in one cell of the partition $\mathcal{R}$, then we can assume that for each n, $a(x_n) < d(x_{n+1})$ and consequently, there is no mixing of the bits of x_n and x_m when they are added in binary.

Lemma 17.13. *Let $\langle x_n \rangle_{n=1}^{\infty}$ be a one-to-one sequence in $\mathbb{N}$. If $\mathrm{PS}(\langle x_n \rangle_{n=1}^{\infty}) \cup$ $\mathrm{PP}(\langle x_n \rangle_{n=1}^{\infty})$ is contained in one cell of the partition $\mathcal{R}$, then $\{n \in \mathbb{N} : x_n \in A_0\}$ is infinite or $\{n \in \mathbb{N} : x_n \in A_3\}$ is infinite*

Proof. One cannot have $\mathrm{PS}(\langle x_n \rangle_{n=1}^{\infty}) \subseteq A_4$ and one cannot have $\mathrm{PS}(\langle x_n \rangle_{n=1}^{\infty}) \subseteq 2\mathbb{N}+1$ so one has $a(x_n + x_r) \equiv a(x_n x_r) \pmod{2}$ whenever n and r are distinct members of $\mathbb{N}$. By the pigeon hole principle we may presume that we have some $i \in \{0, 1, 2, 3, 4\}$ such that $\{x_n : n \in \mathbb{N}\} \subseteq A_i$. If one had $i = 4$, then one would have $\mathrm{PP}(\langle x_n \rangle_{n=1}^{\infty}) \subseteq A_4$ and hence that $\mathrm{PS}(\langle x_n \rangle_{n=1}^{\infty}) \subseteq A_4$, which we have already noted is impossible.

By Lemma 17.12 we have that $\{d(x_n) : n \in \mathbb{N}\}$ is unbounded so pick $n \in \mathbb{N}$ such that $d(x_n) > a(x_1)$. Then $a(x_n + x_1) = a(x_n)$.

Suppose $i = 1$, that is $\{x_n : n \in \mathbb{N}\} \subseteq A_1$. Then by Remark 17.9, $a(x_n x_1) = a(x_n) + a(x_1) + 1$ so $a(x_n x_1)$ is odd while $a(x_n + x_1)$ is even.

Similarly, if $i = 2$, then $a(x_n x_1)$ is even while $a(x_n + x_1)$ is odd. □

Lemma 17.14. *Let $\langle x_n \rangle_{n=1}^{\infty}$ be a one-to-one sequence in $\mathbb{N}$. If $d(x_{n+1}) > a(x_n)$ for each n, $\mathrm{PS}(\langle x_n \rangle_{n=1}^{\infty}) \cup \mathrm{PP}(\langle x_n \rangle_{n=1}^{\infty})$ is contained in one cell of the partition $\mathcal{R}$, and $\{x_n : n \in \mathbb{N}\} \subseteq A_0$, then $\{a(x_n) - b(x_n) : n \in \mathbb{N}\}$ is bounded.*

Proof. Suppose instead that $\{a(x_n) - b(x_n) : n \in \mathbb{N}\}$ is unbounded. Pick $n \in \mathbb{N}$ such that $a(x_n) - b(x_n) > d(x_1)$. Then, using the fact that $d(x_n) > a(x_1)$ we have that

$$a(x_n + x_1) - b(x_n + x_1) = a(x_n) - b(x_n) > d(x_1) = d(x_n + x_1)$$

while by Lemma 17.10

$$a(x_n x_1) - b(x_n x_1) \leq a(x_1) - b(x_1) < d(x_n) \leq d(x_n x_1).$$ □

Lemma 17.15. *Let $\langle x_n \rangle_{n=1}^{\infty}$ be a one-to-one sequence in $\mathbb{N}$. If $d(x_{n+1}) > a(x_n)$ for each n, $\mathrm{PS}(\langle x_n \rangle_{n=1}^{\infty}) \cup \mathrm{PP}(\langle x_n \rangle_{n=1}^{\infty})$ is contained in one cell of the partition $\mathcal{R}$, and $\{x_n : n \in \mathbb{N}\} \subseteq A_3$, then $\{a(x_n) - c(x_n) : n \in \mathbb{N}\}$ is bounded.*

Proof. This is nearly identical to the proof of Lemma 17.14. □

Theorem 17.16. *There is no one-to-one sequence $\langle x_n \rangle_{n=1}^{\infty}$ in $\mathbb{N}$ such that $\mathrm{PS}(\langle x_n \rangle_{n=1}^{\infty}) \cup$ $\mathrm{PP}(\langle x_n \rangle_{n=1}^{\infty})$ is contained in one cell of the partition $\mathcal{R}$.*

Proof. Suppose instead we have such a sequence. By Lemmas 17.12, 17.13, 17.14, and 17.15 we may presume that for each $n \in \mathbb{N}$, $d(x_{n+1}) > a(x_n) + 1$ and either

(i) $\{x_n : n \in \mathbb{N}\} \subseteq A_0$ and $\{a(x_n) - b(x_n) : n \in \mathbb{N}\}$ is bounded, or

(ii) $\{x_n : n \in \mathbb{N}\} \subseteq A_3$ and $\{a(x_n) - c(x_n) : n \in \mathbb{N}\}$ is bounded.

Assume first that $\{x_n : n \in \mathbb{N}\} \subseteq A_0$ and $\{a(x_n) - b(x_n) : n \in \mathbb{N}\}$ is bounded. Pick some $k \in \mathbb{N}$ and $n < r$ in $\mathbb{N}$ such that $a(x_n) - b(x_n) = a(x_r) - b(x_r) = k$. Then $a(x_r + x_n) - b(x_r + x_n) = a(x_r) - b(x_r) = k$.

Also $2^{a(x_n)} + 2^{a(x_n)-k} \leq x_n$ and $2^{a(x_r)} + 2^{a(x_r)-k} \leq x_r$, so

$$2^{a(x_r x_n)} + 2^{a(x_r x_n)-k+1} = 2^{a(x_r)+a(x_n)} + 2^{a(x_r)+a(x_n)-k+1} < x_r x_n$$

so $b(x_r x_n) \geq a(x_r x_n) - k + 1$. And $x_n < 2^{a(x_n)} + 2^{a(x_n)-k+1}$ and $x_r < 2^{a(x_r)} + 2^{a(x_r)-k+1}$
so

$$\begin{aligned}
x_r x_n &< 2^{a(x_r)+a(x_n)} + 2^{a(x_r)+a(x_n)-k+2} + 2^{a(x_r)+a(x_n)-2k+2} \\
&= 2^{a(x_r x_n)} + 2^{a(x_r x_n)-k+2} + 2^{a(x_r x_n)-2k+2} \\
&< 2^{a(x_r x_n)} + 2^{a(x_r x_n)-k+3}
\end{aligned}$$

so $b(x_r x_n) \leq a(x_r x_n) - k + 2$. Thus $a(x_r x_n) - b(x_r x_n) \in \{k-1, k-2\}$ so $a(x_r x_n) - b(x_r x_n) \not\equiv a(x_r + x_n) - b(x_r + x_n) \pmod 3$, a contradiction.

Finally assume that $\{x_n : n \in \mathbb{N}\} \subseteq A_3$ and $\{a(x_n) - c(x_n) : n \in \mathbb{N}\}$ is bounded. We may presume that we have some $k \in \mathbb{N}$ such that $a(x_n) - c(x_n) = k$ for all $n \in \mathbb{N}$. By the pigeon hole principle, pick $n < r$ in $\mathbb{N}$ such that either

(a) $x_n < 2^{a(x_n)+1}(1 - 2^{-k})^{\frac{1}{2}}$ and $x_r < 2^{a(x_r)+1}(1 - 2^{-k})^{\frac{1}{2}}$ or

(b) $x_n \geq 2^{a(x_n)+1}(1 - 2^{-k})^{\frac{1}{2}}$ and $x_r \geq 2^{a(x_r)+1}(1 - 2^{-k})^{\frac{1}{2}}$.

In either case we have $a(x_r x_n) = a(x_r) + a(x_n) + 1$. Also in either case we have $x_n \geq 2^{a(x_n)+1} - 2^{a(x_n)-k+1}$ and $x_r \geq 2^{a(x_r)+1} - 2^{a(x_r)-k+1}$ so that

$$x_r x_n \geq 2^{a(x_r x_n)+1} - 2^{a(x_r x_n)-k+2} + 2^{a(x_r x_n)-2k+1} > 2^{a(x_r x_n)+1} - 2^{a(x_r x_n)-k+2}$$

and consequently $c(x_r x_n) \leq a(x_r x_n) - k + 1$.

Assume that $x_n < 2^{a(x_n)+1}(1 - 2^{-k})^{\frac{1}{2}}$ and $x_r < 2^{a(x_r)+1}(1 - 2^{-k})^{\frac{1}{2}}$. We have $a(x_r + x_n) - c(x_r + x_n) = a(x_r) - c(x_r) = k$. Since $x_r x_n < 2^{a(x_r x_n)+1}(1 - 2^{-k}) = 2^{a(x_r x_n)+1} - 2^{a(x_r x_n)+1-k}$ and hence $c(x_r x_n) \geq a(x_r x_n) - k + 1$, one has $c(x_r x_n) = a(x_r x_n) - k + 1$. But then $a(x_r x_n) - c(x_r x_n) \not\equiv a(x_r + x_n) - c(x_r + x_n) \pmod 2$, a contradiction.

Thus we must have that $x_n \geq 2^{a(x_n)+1}(1 - 2^{-k})^{\frac{1}{2}}$ and $x_r \geq 2^{a(x_r)+1}(1 - 2^{-k})^{\frac{1}{2}}$. Now $a(x_r + x_n) = a(x_r)$ so $x_r + x_n > 2^{a(x_r+x_n)+1}(1 - 2^{-k})^{\frac{1}{2}}$ and $a(x_r + x_n) - c(x_r + x_n) = a(x_r) - c(x_r) = k$, so picking $i \in \{0, 1, 2\}$ such that $i + 1 \equiv k \pmod 3$ we have that $x_r + x_n \in B_i$ and consequently $x_r x_n \in B_i$.

Now $x_r x_n \geq 2^{a(x_r x_n)+1}(1 - 2^{-k}) = 2^{a(x_r x_n)+1} - 2^{a(x_r x_n)+1-k}$ so $c(x_r x_n) \leq a(x_r x_n) - k$. By Lemma 17.10 $a(x_r x_n) - c(x_r x_n) \leq k$ so $a(x_r x_n) - c(x_r x_n) = k$.

Notice that $(1 - 2^{-k-1} - 2^{-2k-2}) < (1 - 2^{-k})^{\frac{1}{2}}$, a fact that may be verified by squaring both sides. Since $x_n < 2^{a(x_n)+1} - 2^{a(x_n)-k}$ and $x_r < 2^{a(x_r)+1} - 2^{a(x_r)-k}$ we have

$$\begin{aligned}
x_r x_n &< 2^{a(x_r x_n)+1} - 2^{a(x_r x_n)-k+1} + 2^{a(x_r x_n)-2k-1} \\
&< 2^{a(x_r x_n)+1} - 2^{a(x_r x_n)-k} - 2^{a(x_r x_n)-2k-1} \\
&= 2^{a(x_r x_n)+1}(1 - 2^{-k-1} - 2^{-2k-2}) \\
&< 2^{a(x_r x_n)+1}(1 - 2^{-k})^{\frac{1}{2}} \\
&= 2^{a(x_r x_n)+1}(1 - 2^{c(x_r x_n)-a(x_r x_n)})^{\frac{1}{2}}.
\end{aligned}$$

Thus since $x_r x_n \in B_i$ we have that $k = a(x_r x_n) - c(x_r x_n) \equiv i \equiv k - 1 \pmod 3$, a contradiction. □

Recall from Theorem 13.14 that if $p \in \beta\mathbb{N}$ and $\mathbb{N}n \in p$ for infinitely many n, then there do not exist q, r, and s in $\mathbb{N}^*$ such that $q \cdot p = r + s$. The following corollary has a much weaker conclusion, but applies to any $p \in \mathbb{N}^*$.

Corollary 17.17. *Let $p \in \mathbb{N}^*$. Then $p + p \neq p \cdot p$.*

Proof. Suppose instead that $p + p = p \cdot p$ and pick some $A \in \mathcal{R}$ such that $A \in p \cdot p$. Let $B = \{x \in \mathbb{N} : -x + A \in p\} \cap \{x \in \mathbb{N} : x^{-1}A \in p\}$ and pick $x_1 \in B$. Inductively, let $n \in \mathbb{N}$ and assume we have chosen $\langle x_t \rangle_{t=1}^n$. Pick

$$x_{n+1} \in \left(B \cap \bigcap_{t=1}^n (-x_t + A) \cap (x_t{}^{-1}A) \right) \backslash \{x_1, x_2, \ldots, x_n\}.$$

Then $\mathrm{PS}(\langle x_n \rangle_{n=1}^\infty) \cup \mathrm{PP}(\langle x_n \rangle_{n=1}^\infty) \subseteq A$, contradicting Theorem 17.16. □

Theorem 17.16 establishes that one cannot expect any sort of combined additive and multiplicative results from an infinite sequence in an arbitrary finite partition of $\mathbb{N}$. On the other hand, the following question remains wide open.

Question 17.18. *Let $r, n \in \mathbb{N}$. If $\mathbb{N} = \bigcup_{i=1}^r A_i$, must there exist $i \in \{1, 2, \ldots, r\}$ and a one-to-one sequence $\langle x_t \rangle_{t=1}^n$ such that $\mathrm{FS}(\langle x_t \rangle_{t=1}^n) \cup \mathrm{FP}(\langle x_t \rangle_{t=1}^n) \subseteq A_i$?*

We would conjecture strongly that the answer to Question 17.18 is "yes". However, the only nontrivial case for which it is known to be true is $n = r = 2$. (See the notes to this chapter).

17.3 Sums of Products

Shortly after the original (combinatorial) proof of the Finite Sums Theorem and its corollary, the Finite Products Theorem, Erdős asked [89] whether, given any finite partition of $\mathbb{N}$, there must exist one cell containing a sequence and all of its "multilinear combinations". We know, of course, that whatever the precise meaning of "multilinear combinations", the answer is "no". (Theorem 17.16.) We see in this section, however, that a certain regularity can be imposed on sums of products of a sequence.

Recall that, if $n \in \mathbb{N}$ and $p \in \beta\mathbb{N}$, then $n \cdot p$ is the product of n and p in the semigroup $(\beta\mathbb{N}, \cdot)$ which need not be the same as the sum of p with itself n times. (We already know from Theorem 13.18 that if $p \in \mathbb{N}^*$, then $p + p \neq 2 \cdot p$.) Consequently we introduce some notation for the sum of p with itself n times.

Definition 17.19. Let $p \in \beta\mathbb{N}$. Then $\sigma_1(p) = p$ and, given $n \in \mathbb{N}$, $\sigma_{n+1}(p) = \sigma_n(p) + p$.

Of course, if $p \in \mathbb{N}$, then for all n, $n \cdot p = \sigma_n(p)$. The question naturally arises as to whether it is possible to have $n \cdot p = \sigma_n(p)$ for some $n \in \mathbb{N}\setminus\{1\}$ and $p \in \mathbb{N}^*$. We shall see in Corollary 17.22 that it is not possible.

Lemma 17.20. *Let $n \in \mathbb{N}$, let $p \in \mathbb{N}^*$, let $A \in p$ and let $B \in \sigma_n(p)$. There is a one-to-one sequence $\langle x_t \rangle_{t=1}^{\infty}$ in A such that, for each $F \in [\mathbb{N}]^n$, $\Sigma_{t \in F}\, x_t \in B$.*

Proof. This is Exercise 17.3.1.   □

Theorem 17.21. *Let $n \in \mathbb{N}\setminus\{1\}$. There is a finite partition $\mathcal{R}$ of $\mathbb{N}$ such that there do not exist $A \in \mathcal{R}$ and a one-to-one sequence $\langle x_t \rangle_{t=1}^{\infty}$ such that*

(1) *for each $t \in \mathbb{N}$, $n \cdot x_t \in A$, and*
(2) *whenever $F \in [\mathbb{N}]^n$, $\Sigma_{t \in F}\, x_t \in A$.*

Proof. For $i \in \{0, 1, 2, 3\}$, let

$$A_i = \bigcup_{k=0}^{\infty} \left\{ x \in \mathbb{N} : n^{2k+i/2} \leq x < n^{2k+(i+1)/2} \right\}$$

and let $\mathcal{R} = \{A_0, A_1, A_2, A_3\}$. Suppose that one has a one-to-one sequence $\langle x_t \rangle_{t=1}^{\infty}$ and $i \in \{0, 1, 2, 3\}$ such that

(1) for each $t \in \mathbb{N}$, $n \cdot x_t \in A_i$, and
(2) whenever $F \in [\mathbb{N}]^n$, $\Sigma_{t \in F}\, x_t \in A_i$.

By the pigeon hole principle, we may presume that we have some $j \in \{0, 1, 2, 3\}$ such that $\{x_t : t \in \mathbb{N}\} \subseteq A_j$.

Now if $n^{2k+j/2} \leq x_t < n^{2k+(j+1)/2}$, then $n^{2k+(j+2)/2} \leq n \cdot x_t < n^{2k+(j+3)/2}$ so that $i \equiv j+2 \pmod 4$. On the other hand, for sufficiently large t, if $n^{2k+j/2} \leq x_t < n^{2k+(j+1)/2}$, then

$$n^{2k+j/2} \leq x_1 + x_2 + \cdots + x_{n-1} + x_t < n^{2k+(j+2)/2}$$

so that $i \not\equiv j+2 \pmod 4$, a contradiction.   □

Corollary 17.22. *Let $p \in \mathbb{N}^*$ and let $n \in \mathbb{N}\setminus\{1\}$. Then $n \cdot p \neq \sigma_n(p)$.*

Proof. Suppose that $n \cdot p = \sigma_n(p)$ and pick $A \in \mathcal{R}$ such that $A \in n \cdot p$. Then $n^{-1}A \in p$ so, by Lemma 17.20, choose a sequence $\langle x_t \rangle_{t=1}^{\infty}$ in $n^{-1}A$ such that for each $F \in [\mathbb{N}]^n$, $\Sigma_{t \in F}\, x_t \in A$. This contradicts Theorem 17.21.   □

Recall that, given $F, G \in \mathcal{P}_f(\mathbb{N})$ we write $F < G$ if and only if $\max F < \min G$.

Definition 17.23. Let $\langle x_t \rangle_{t=1}^{\infty}$ be a sequence in $\mathbb{N}$ and let $m \in \mathbb{N}$. Then

$$\mathrm{SP}_m(\langle x_t \rangle_{t=1}^{\infty}) = \left\{ \Sigma_{i=1}^{m} \Pi_{t \in F_i}\, x_t : F_1, F_2, \ldots, F_m \in \mathcal{P}_f(\mathbb{N}) \text{ and } F_1 < F_2 < \cdots < F_m \right\}.$$

Theorem 17.24. *Let* $p \cdot p = p \in \beta\mathbb{N}$, *let* $m \in \mathbb{N}$, *and let* $A \in \sigma_m(p)$. *Then there is a sequence* $\langle x_t \rangle_{t=1}^{\infty}$ *such that* $\mathrm{SP}_m(\langle x_t \rangle_{t=1}^{\infty}) \subseteq A$.

Proof. If $m = 1$, this is just the finite products theorem (Theorem 5.8). In the proof, we are dealing with two semigroups, so the notation $B^{\star}$ is ambiguous. We shall use it to refer to the semigroup $(\beta\mathbb{N}, \cdot)$, so that $B^{\star} = \{x \in B : x^{-1}B \in p\}$.

We do the $m = 2$ case separately because it lacks some of the complexity of the general theorem, so assume $A \in p + p$. Let $B_1 = \{x \in \mathbb{N} : -x + A \in p\}$. Then $B_1 \in p$ so $B_1^{\star} \in p$. Pick $x_1 \in B_1^{\star}$. Inductively, let $n \in \mathbb{N}$ and assume that we have chosen $\langle x_t \rangle_{t=1}^{n}$ in $\mathbb{N}$ and $B_1 \supseteq B_2 \supseteq \cdots \supseteq B_n$ in p such that, if $\emptyset \neq F \subseteq \{1, 2, \ldots, n\}$, $k = \min F$, and $\ell = \max F$, then

(1) $\prod_{t \in F} x_t \in B_k^{\star}$ and

(2) if $\ell < n$, then $B_{\ell+1} \subseteq (-\prod_{t \in F} x_t + A)$.

For $k \in \{1, 2, \ldots, n\}$, let $E_k = \{\prod_{t \in F} x_t : F \subseteq \{1, 2, \ldots, n\}$ and $k = \min F\}$. Let

$$B_{n+1} = B_n \cap \bigcap_{k=1}^{n} \bigcap_{a \in E_k} \left((-a + A) \cap a^{-1}B_k^{\star}\right).$$

Now, given $k \in \{1, 2, \ldots, n\}$ and $a \in E_k$, $a \in B_k^{\star} \subseteq B_1$ so $-a + A \in p$ and $a^{-1}B_k^{\star} \in p$ so $B_{n+1} \in p$. Choose $x_{n+1} \in B_{n+1}^{\star}$.

To verify the induction hypotheses, let $\emptyset \neq F \subseteq \{1, 2, \ldots, n+1\}$, $k = \min F$, and $\ell = \max F$. If $\ell < n$, then both hypotheses hold by assumption. Assume that $\ell = n$. Then (1) holds by assumption and $\prod_{t \in F} x_t \in E_k$ so (2) holds. Finally assume that $\ell = n + 1$. Then (2) is vacuous. If $k = n + 1$, then $\prod_{t \in F} x_t = x_{n+1} \in B_{n+1}^{\star}$. If $k \leq n$, then let $G = F \setminus \{k\}$. Then $\prod_{t \in G} x_t \in E_k$ so $x_{n+1} \in B_{n+1} \subseteq (\prod_{t \in G} x_t)^{-1}B_k^{\star}$ and thus (1) holds.

The construction being complete, let $F_1, F_2 \in \mathcal{P}_f(\mathbb{N})$ with $F_1 < F_2$. Let $k = \min F_2$ and let $\ell = \max F_1$. Then

$$\prod_{t \in F_2} x_t \in B_k^{\star} \subseteq B_{\ell+1} \subseteq (-\prod_{t \in F_1} x_t + A).$$

This completes the proof in the case $m = 2$.

Now assume that $m \geq 3$. Let $B_1 = \{x \in \mathbb{N} : -x + A \in \sigma_{m-1}(p)\}$ and choose $x_1 \in B_1^{\star}$.

Inductively, let $n \in \mathbb{N}$ and assume that we have chosen $\langle x_k \rangle_{k=1}^{n}$ in $\mathbb{N}$ and $\langle B_k \rangle_{k=1}^{n}$ in p so that for each $r \in \{1, 2, \ldots, n\}$ each of the following statements holds.

(I) If $\emptyset \neq F \subseteq \{1, 2, \ldots, r\}$ and $k = \min F$, then $\prod_{t \in F} x_t \in B_k^{\star}$.

(II) If $r < n$, then $B_{r+1} \subseteq B_r$.

(III) If $\ell \in \{1, 2, \ldots, m - 1\}$, $F_1, F_2, \ldots, F_\ell \in \mathcal{P}_f(\{1, 2, \ldots, r\})$, and $F_1 < F_2 < \cdots < F_\ell$, then $-\sum_{i=1}^{\ell} \prod_{t \in F_i} x_t + A \in \sigma_{m-\ell}(p)$.

(IV) If $F_1, F_2, \ldots, F_{m-1} \in \mathcal{P}_f(\{1, 2, \ldots, r\})$, $F_1 < F_2 < \cdots < F_{m-1}$, and $r < n$, then $B_{r+1} \subseteq -\sum_{i=1}^{m-1} \prod_{t \in F_i} x_t + A$.

(V) If $\ell \in \{1, 2, \ldots, m - 2\}$, $F_1, F_2, \ldots, F_\ell \in \mathscr{P}_f(\{1, 2, \ldots, r\})$ with $F_1 < F_2 < \cdots < F_\ell$, and $r < n$, then

$$B_{r+1} \subseteq \{x \in \mathbb{N} : -x + (- \Sigma_{i=1}^{\ell} \Pi_{t \in F_i} x_t + A) \in \sigma_{m-\ell-1}(p)\}.$$

At $n = 1$, hypothesis (I) says that $x_1 \in B_1^{\star}$. Hypotheses (II), (IV), and (V) are vacuous, and hypothesis (III) says that $-x_1 + A \in \sigma_{m-1}(p)$.

For $\ell \in \{1, 2, \ldots, m - 1\}$, let

$$\mathscr{F}_\ell = \{(F_1, F_2, \ldots, F_\ell) : F_1, F_2, \ldots, F_\ell \in \mathscr{P}_f(\{1, 2, \ldots, n\})$$
$$\text{and } F_1 < F_2 < \cdots < F_\ell\}$$

and for $k \in \{1, 2, \ldots, n\}$, let

$$E_k = \{\Pi_{t \in F} x_t : \emptyset \neq F \subseteq \{1, 2, \ldots, n\} \text{ and } \min F = k\}.$$

Given $a \in E_k$, we have that $a \in B_k^{\star}$ by hypothesis (I) and so $a^{-1} B_k^{\star} \in p$ by Lemma 4.14. If $(F_1, F_2, \ldots, F_{m-1}) \in \mathscr{F}_{m-1}$, then by (III) we have

$$- \Sigma_{i=1}^{m-1} \Pi_{t \in F_i} x_t + A \in p.$$

If $\ell \in \{1, 2, \ldots, m - 2\}$ and $(F_1, F_2, \ldots, F_\ell) \in \mathscr{F}_\ell$ we have by hypothesis (III) that

$$\{x \in \mathbb{N} : -x + (- \Sigma_{i=1}^{\ell} \Pi_{t \in F_i} x_t + A) \in \sigma_{m-\ell-1}(p)\} \in p.$$

So we let

$$B_{n+1} = B_n \cap \bigcap_{k=1}^{n} \bigcap_{a \in E_k} a^{-1} B_k^{\star}$$

$$\cap \bigcap_{(F_1, F_2, \ldots, F_{m-1}) \in \mathscr{F}_{m-1}} (- \Sigma_{i=1}^{m-1} \Pi_{t \in F_i} x_t + A)$$

$$\cap \bigcap_{\ell=1}^{m-2} \bigcap_{(F_1, F_2, \ldots, F_\ell) \in \mathscr{F}_\ell} \{x \in \mathbb{N} : -x + (- \Sigma_{i=1}^{\ell} \Pi_{t \in F_i} x_t + A) \in \sigma_{m-\ell-1}(p)\}$$

and notice that $B_{n+1} \in p$.

For simplicity of notation we are taking $\bigcap \emptyset = \mathbb{N}$ in the above. Thus, for example, if $n = m - 3$, then $\mathscr{F}_{m-2} = \mathscr{F}_{m-1} = \emptyset$ and so

$$B_{n+1} = B_n \cap \bigcap_{k=1}^{n} \bigcap_{a \in E_k} a^{-1} B_k^{\star}$$

$$\cap \bigcap_{\ell=1}^{m-3} \bigcap_{(F_1, F_2, \ldots, F_\ell) \in \mathscr{F}_\ell} \{x \in \mathbb{N} : -x + (- \Sigma_{i=1}^{\ell} \Pi_{t \in F_i} x_t + A) \in \sigma_{m-\ell-1}(p)\}.$$

Choose $x_{n+1} \in B_{n+1}^{\star}$. To verify hypothesis (I), let $\emptyset \neq F \subseteq \{1, 2, \ldots, n + 1\}$ and let $k = \min F$. If $n + 1 \notin F$, then (I) holds by assumption, so assume that $n + 1 \in F$. If $k = n + 1$, we have $x_{n+1} \in B_{n+1}^{\star}$ directly, so assume that $k < n + 1$ and let $G = F \backslash \{n + 1\}$. Then $\Pi_{t \in G} x_t \in E_k$ so $x_{n+1} \in (\Pi_{t \in G} x_t)^{-1} B_k^{\star}$.

Hypothesis (II) holds trivially and hypotheses (IV) and (V) hold directly.

To verify hypothesis (III), let $\ell \in \{1, 2, \ldots, m - 1\}$ and let $F_1, F_2, \ldots, F_\ell \in \mathscr{P}_f(\{1, 2, \ldots, n + 1\})$ with $F_1 < F_2 < \cdots < F_\ell$. If $\ell = 1$, then by hypothesis

(I) and (II), $\prod_{t \in F_1} x_t \in B_1^\star \subseteq B_1$ so $-\prod_{t \in F_1} x_t + A \in \sigma_{m-1}(p)$. So assume that $\ell > 1$. Let $k = \min F_\ell$ and let $j = \max F_{\ell-1}$. Then by hypotheses (I) and (II), $\prod_{t \in F_\ell} x_t \in B_k^\star \subseteq B_{j+1}$ and by hypothesis (V) at $r = j$,

$$B_{j+1} \subseteq \{x \in \mathbb{N} : -x + (-\textstyle\sum_{i=1}^{\ell-1} \prod_{t \in F_i} x_t + A) \in \sigma_{m-\ell}(p)\}.$$

The induction being complete, we have that whenever $F_1, F_2, \ldots, F_m \in \mathcal{P}_f(\mathbb{N})$ with $F_1 < F_2 < \cdots < F_m$, if $k = \min F_m$ and $r = \max F_{m-1}$, then by (I), (II), and (IV),

$$\prod_{t \in F_m} x_t \in B_k^\star \subseteq B_{r+1} \subseteq -\textstyle\sum_{i=1}^{m-1} \prod_{t \in F_i} x_t + A$$

and thus $\sum_{i=1}^{m} \prod_{t \in F_i} x_t \in A$. $\qquad\square$

Corollary 17.25. *Let $r, m \in \mathbb{N}$ and let $\mathbb{N} = \bigcup_{j=1}^{r} A_j$. Then there exist $j \in \{1, 2, \ldots, r\}$ and a sequence $\langle x_t \rangle_{t=1}^{\infty}$ such that $\mathrm{SP}_m(\langle x_t \rangle_{t=1}^{\infty}) \subseteq A_j$.*

Proof. Pick any idempotent p in $(\beta\mathbb{N}, \cdot)$ and pick $j \in \{1, 2, \ldots, r\}$ such that $A_j \in \sigma_m(p)$. $\qquad\square$

There is a partial converse to Theorem 17.24. In this converse, the meaning of $\mathrm{SP}_n(\langle x_t \rangle_{t=k}^{\infty})$ should be obvious. Notice that one does not require that $p \cdot p = p$.

Theorem 17.26. *Let $\langle x_t \rangle_{t=1}^{\infty}$ be a sequence in $\mathbb{N}$. If $p \in \bigcap_{k=1}^{\infty} \overline{\mathrm{FP}(\langle x_t \rangle_{t=k}^{\infty})}$, then for all n and k in $\mathbb{N}$, $\mathrm{SP}_n(\langle x_t \rangle_{t=k}^{\infty}) \in \sigma_n(p)$.*

Proof. We proceed by induction on n, the case $n = 1$ holding by assumption. So let $n \in \mathbb{N}$ and assume that for each $k \in \mathbb{N}$, $\mathrm{SP}_n(\langle x_t \rangle_{t=k}^{\infty}) \in \sigma_n(p)$.

Let $k \in \mathbb{N}$. We claim that

$$\mathrm{SP}_n(\langle x_t \rangle_{t=k}^{\infty}) \subseteq \{a \in \mathbb{N} : -a + \mathrm{SP}_{n+1}(\langle x_t \rangle_{t=k}^{\infty}) \in p\}$$

so that $\mathrm{SP}_{n+1}(\langle x_t \rangle_{t=k}^{\infty}) \in \sigma_n(p) + p = \sigma_{n+1}(p)$. So let $a \in \mathrm{SP}_n(\langle x_t \rangle_{t=k}^{\infty})$ and pick $F_1 < F_2 < \cdots < F_n$ in $\mathcal{P}_f(\mathbb{N})$ such that $\min F_1 \geq k$ and $a = \sum_{i=1}^{n} \prod_{t \in F_i} x_t$ and let $\ell = \max F_n + 1$. Then

$$\mathrm{FP}(\langle x_t \rangle_{t=\ell}^{\infty}) \subseteq -a + \mathrm{SP}_{n+1}(\langle x_t \rangle_{t=k}^{\infty})$$

so that $-a + \mathrm{SP}_{n+1}(\langle x_t \rangle_{t=k}^{\infty}) \in p$. $\qquad\square$

We see now that one cannot necessarily expect to find $\mathrm{SP}_m(\langle x_t \rangle_{t=1}^{\infty})$ and $\mathrm{SP}_1(\langle x_t \rangle_{t=1}^{\infty})$ $(= \mathrm{FP}(\langle x_t \rangle_{t=1}^{\infty}))$ in one cell of a partition, indeed not even $\mathrm{SP}_m(\langle x_t \rangle_{t=1}^{\infty}) \cup \mathrm{SP}_1(\langle y_t \rangle_{t=1}^{\infty})$ with possibly different sequences $\langle x_t \rangle_{t=1}^{\infty}$ and $\langle y_t \rangle_{t=1}^{\infty}$. In fact much stronger conclusions are known — see the notes to this chapter.

Theorem 17.27. *Let $m \in \mathbb{N} \setminus \{1\}$. There is a finite partition $\mathcal{R}$ of $\mathbb{N}$ such that, given any $A \in \mathcal{R}$, there do not exist one-to-one sequences $\langle x_t \rangle_{t=1}^{\infty}$ and $\langle y_t \rangle_{t=1}^{\infty}$ with $\mathrm{SP}_m(\langle x_t \rangle_{t=1}^{\infty}) \subseteq A$ and $\mathrm{SP}_1(\langle y_t \rangle_{t=1}^{\infty}) \subseteq A$.*

Proof. For $x \in \mathbb{N}$, define $a(x)$ and $d(x)$ in ω by $m^{a(x)} \leq x < m^{a(x)+1}$ and $d(x) = \max\{t \in \omega : m^t | x\}$. Thus when x is written in base m, $a(x)$ is the position of the leftmost nonzero digit and $d(x)$ is the position of the rightmost nonzero digit. Let

$A_0 = \mathbb{N}\backslash\mathbb{N}m = \{x \in \mathbb{N} : d(x) = 0\}$
$A_1 = \{x \in \mathbb{N} : d(x) = 1\}$
$A_2 = \{m^t : t \in \mathbb{N} \text{ and } t > 1\}$
$A_3 = \{x \in \mathbb{N} : a(x) \text{ is even, } d(x) > 1 \text{ and } m^{a(x)} < x < m^{a(x)} + m^{a(x)-d(x)}\}$
$A_4 = \{x \in \mathbb{N} : a(x) \text{ is even, } d(x) > 1 \text{ and } m^{a(x)} + m^{a(x)-d(x)} \leq x\}$
$A_5 = \{x \in \mathbb{N} : a(x) \text{ is odd, } d(x) > 1 \text{ and } m^{a(x)+1} - m^{a(x)-d(x)} < x\}$
$A_6 = \{x \in \mathbb{N} : a(x) \text{ is odd, } d(x) > 1 \text{ and } m^{a(x)} < x \leq m^{a(x)+1} - m^{a(x)-d(x)}\}$.

Then $\{A_0, A_1, A_2, A_3, A_4, A_5, A_6\}$ is a partition of $\mathbb{N}$. Trivially, neither A_0 nor A_2 contains $\text{SP}_m(\langle x_t \rangle_{t=1}^\infty)$ and A_1 does not contain $\text{SP}_1(\langle x_t \rangle_{t=1}^\infty) = \text{FP}(\langle x_t \rangle_{t=1}^\infty)$ for any sequence $\langle x_t \rangle_{t=1}^\infty$ in $\mathbb{N}$.

We now claim that A_3 does not contain any $\text{FP}(\langle x_t \rangle_{t=1}^\infty)$, so suppose instead we have some one-to-one sequence $\langle x_t \rangle_{t=1}^\infty$ with $\text{FP}(\langle x_t \rangle_{t=1}^\infty) \subseteq A_3$. Then for each $t \in \mathbb{N}$, $d(x_t) > 1$ so pick a product subsystem $\langle y_t \rangle_{t=1}^\infty$ of $\langle x_t \rangle_{t=1}^\infty$ such that for each t, $a(y_t) < d(x_{t+1})$. Then $m^{a(y_1)} < y_1 < m^{a(y_1)+1}$ and $m^{a(y_2)} < y_2 < m^{a(y_2)+1}$ so $m^{a(y_1)+a(y_2)} < y_1 y_2 < m^{a(y_1)+a(y_2)+2}$ and consequently, $a(y_1 y_2) \in \{a(y_1)+a(y_2), a(y_1)+a(y_2)+1\}$. Since $a(y_1)$, $a(y_2)$, and $a(y_1 y_2)$ are even, $a(y_1 y_2) = a(y_1) + a(y_2)$.

Now $d(y_1 y_2) \geq d(y_1) + d(y_2)$ so

$$a(y_1 y_2) - a(y_1) + d(y_1) > a(y_1 y_2) - a(y_1) > a(y_1 y_2) - d(y_2) > a(y_1 y_2) - d(y_1 y_2).$$

Also, $y_1 \geq m^{a(y_1)} + m^{d(y_1)}$ and $y_2 > m^{a(y_2)}$ so

$$\begin{aligned} y_1 y_2 &> m^{a(y_1)+a(y_2)} + m^{a(y_2)+d(y_1)} \\ &= m^{a(y_1 y_2)} + m^{a(y_1 y_2)-a(y_1)+d(y_1)} \\ &> m^{a(y_1 y_2)} + m^{a(y_1 y_2)-d(y_1 y_2)} \end{aligned}$$

a contradiction.

Next we claim that A_4 does not contain any $\text{SP}_m(\langle x_t \rangle_{t=1}^\infty)$, so suppose that we have a one-to-one sequence $\langle x_t \rangle_{t=1}^\infty$ with $\text{SP}_m(\langle x_t \rangle_{t=1}^\infty) \subseteq A_4$. If for infinitely many t's we have $d(x_t) = 0$, we may choose $F \in [\mathbb{N}]^m$ such that for any $t, s \in F$ one has

$$x_t \equiv x_s \not\equiv m \pmod{m^2}.$$

Then $d(\sum_{t \in F} x_t) = 1$ so $\sum_{t \in F} x_t \in \text{SP}_m(\langle x_t \rangle_{t=1}^\infty)\backslash A_4$.

Thus we may assume that each $d(x_t) > 0$ and hence, by passing to a product subsystem as we did when discussing A_3, we may presume that for each t, $a(x_t) < d(x_{t+1})$, so that there is no carrying when x_t and x_{t+1} are added in base m arithmetic. (Notice that by Exercise 17.3.2, if $\langle y_t \rangle_{t=1}^\infty$ is a product subsystem of $\langle x_t \rangle_{t=1}^\infty$, then $\text{SP}_m(\langle y_t \rangle_{t=1}^\infty) \subseteq \text{SP}_m(\langle x_t \rangle_{t=1}^\infty)$.)

We may further assume that for each $t \geq m$, $a(x_t) \geq a(x_{m-1}) + d(x_1) + 2$. We claim that there is some $t \geq m$ such that $x_t < m^{a(x_t)} + m^{a(x_t)-d(x_1)-1}$. Suppose instead that for each $t \geq m$, $x_t \geq m^{a(x_t)}(1 + m^{-d(x_1)-1})$. Now $1 + m^{-d(x_1)-1} > 1$ so for some ℓ, $(1 + m^{-d(x_1)-1})^{\ell+1} > m$ and hence

$$\Pi_{t=m}^{m+\ell} x_t > m^{\sum_{t=m}^{m+\ell} a(x_t)+1}.$$

Pick the first k such that

$$\Pi_{t=m}^{k} x_t \geq m^{\sum_{t=m}^{k} a(x_t)+1}.$$

Since

$$\Pi_{t=m}^{k-1} x_t < m^{\sum_{t=m}^{k-1} a(x_t)+1} \text{ and } x_k < m^{a(x_k)+1}$$

we have that

$$\Pi_{t=m}^{k} x_t < m^{\sum_{t=m}^{k} a(x_t)+2}$$

and consequently $a(\Pi_{t=m}^{k} x_t) = \Sigma_{t=m}^{k} a(x_t) + 1$.

Since, for each t, $d(x_{t+1}) > a(x_t)$, we have that for each $s \in \{m, m+1, \ldots, k\}$, $a(x_1 + x_2 + \cdots + x_{m-1} + x_s) = a(x_s)$ and, since

$$x_1 + x_2 + \cdots + x_{m-1} + x_s \in SP_m(\langle x_t \rangle_{t=1}^{\infty}) \subseteq A_4,$$

$a(x_s)$ is even. Also $a(x_1 + x_2 + \cdots + x_{m-1} + \Pi_{t=m}^{k} x_t) = a(\Pi_{t=m}^{k} x_t)$, so $a(\Pi_{t=m}^{k} x_t)$ is even, a contradiction.

Thus we have some $t \geq m$ such that $x_t < m^{a(x_t)} + m^{a(x_t)-d(x_1)-1}$. Let $y = x_1 + x_2 + \cdots + x_{m-1} + x_t$. Then, $a(y) = a(x_t)$ and $d(y) = d(x_1)$. And because there is no carrying when these numbers are added in base m, we have

$$\begin{aligned} x_1 + x_2 + \cdots + x_{m-1} &< m^{a(x_{m-1})+1} \\ &\leq m^{a(x_t)-d(x_1)-1} \end{aligned}$$

so

$$\begin{aligned} y &< m^{a(x_t)} + m^{a(x_t)-d(x_1)-1} + m^{a(x_t)-d(x_1)-1} \\ &\leq m^{a(x_t)} + m^{a(x_t)-d(x_1)} \\ &= m^{a(y)} + m^{a(y)-d(y)} \end{aligned}$$

contradicting the fact that $y \in A_4$.

Now we claim that A_5 does not contain any $FP(\langle x_t \rangle_{t=1}^{\infty})$. So suppose instead that we have a one-to-one sequence $\langle x_t \rangle_{t=1}^{\infty}$ such that $FP(\langle x_t \rangle_{t=1}^{\infty}) \subseteq A_5$. As before, we may presume that for each t, $d(x_{t+1}) > a(x_t)$. Now $x_1 \leq m^{a(x_1)+1} - m^{d(x_1)}$ as can be seen by considering the base m expansion of x_1. Also, $x_2 < m^{a(x_2)+1}$ so that $x_1 x_2 < m^{a(x_1)+a(x_2)+2} - m^{a(x_2)+d(x_1)+1}$. Since also $x_1 x_2 > m^{a(x_1)+a(x_2)}$ and $a(x_1)$, $a(x_2)$, and $a(x_1 x_2)$ are all odd, we have $a(x_1 x_2) = a(x_1) + a(x_2) + 1$. Also

$$a(x_2) + d(x_1) + 1 = a(x_1 x_2) - a(x_1) + d(x_1) > a(x_1 x_2) - d(x_1 x_2)$$

because $d(x_1 x_2) \geq d(x_1) + d(x_2) > a(x_1) - d(x_1)$. Thus

$$x_1 x_2 < m^{a(x_1 x_2)+1} - m^{a(x_1 x_2)-d(x_1 x_2)},$$

contradicting the fact that $x_1 x_2 \in A_5$.

Finally we show that A_6 does not contain any $\mathrm{SP}_m(\langle x_t \rangle_{t=1}^\infty)$, so suppose instead that we have a one-to-one sequence $\langle x_t \rangle_{t=1}^\infty$ such that $\mathrm{SP}_m(\langle x_t \rangle_{t=1}^\infty) \subseteq A_6$. As in the consideration of A_4, we see that we cannot have $d(x_t) = 0$ for infinitely many t's and hence we can assume that for all t, $d(x_{t+1}) > a(x_t)$.

We claim that there is some $t \geq m$ such that $x_t > m^{a(x_t)+1} - m^{a(x_t)-d(x_1)}$. For suppose instead that for each $t \geq m$, $x_t \leq m^{a(x_t)+1}(1 - m^{-d(x_1)-1})$. Then for some ℓ, $(1 - m^{-d(x_1)-1})^{\ell+1} < 1/m$ so

$$\prod_{t=m}^{m+\ell} x_t < m^{\sum_{t=m}^{m+\ell} a(x_t)+\ell}.$$

Pick the first k such that

$$\prod_{t=m}^{k} x_t \leq m^{\sum_{t=m}^{k} a(x_t)+k-m}.$$

Then

$$\prod_{t=m}^{k-1} x_t > m^{\sum_{t=m}^{k-1} a(x_t)+k-1-m} \text{ and } x_k > m^{a(x_k)},$$

so

$$\prod_{t=m}^{k} x_t > m^{\sum_{t=m}^{k} a(x_t)+k-1-m}$$

and hence

$$a(x_1 + x_2 + \cdots + x_{m-1} + \prod_{t=m}^{k} x_t) = a(\prod_{t=m}^{k} x_t) = \sum_{t=m}^{k} a(x_t) + k - 1 - m.$$

Also for each $t \in \{m, m+1, \ldots, k\}$, $a(x_1 + x_2 + \cdots + x_{m-1} + x_t) = a(x_t)$ so each $a(x_t)$ as well as $\sum_{t=m}^{k} a(x_t) + k - 1 - m$ are odd, which is impossible.

Thus we have some $t \geq m$ such that $x_t > m^{a(x_t)+1} - m^{a(x_t)-d(x_1)}$. Let $y = x_1 + x_2 + \cdots + x_{m-1} + x_t$. Then $a(y) = a(x_t)$ and $d(y) = d(x_1)$ and $y > m^{a(x_t)+1} - m^{a(x_t)-d(x_1)} = m^{a(y)+1} - m^{a(y)-d(y)}$, a contradiction. $\square$

We see now that we can in fact assume that the partition of Theorem 17.27 has only two cells.

Corollary 17.28. *Let $m \in \mathbb{N}\backslash\{1\}$. There is a set $B \subseteq \mathbb{N}$ such that*

(1) *whenever $p \cdot p = p \in \beta\mathbb{N}$, $B \in p$,*
(2) *whenever $p \cdot p = p \in \beta\mathbb{N}$, $\mathbb{N}\backslash B \in \sigma_m(p)$,*
(3) *there is no one-to-one sequence $\langle x_t \rangle_{t=1}^\infty$ in $\mathbb{N}$ with $\mathrm{SP}_m(\langle x_t \rangle_{t=1}^\infty) \subseteq B$, and*
(4) *there is no one-to-one sequence $\langle x_t \rangle_{t=1}^\infty$ in $\mathbb{N}$ with $\mathrm{SP}_1(\langle x_t \rangle_{t=1}^\infty) \subseteq \mathbb{N}\backslash B$.*

Proof. Let $\mathcal{R}$ be the partition guaranteed by Theorem 17.27 and let

$$B = \bigcup\{A \in \mathcal{R} : \text{there exists } \langle x_t \rangle_{t=1}^\infty \text{ with } \mathrm{FP}(\langle x_t \rangle_{t=1}^\infty) \subseteq A\}.$$

To verify conclusion (1), let $p \cdot p = p \in \beta\mathbb{N}$ and pick $A \in \mathcal{R}$ such that $A \in p$. Then by Theorem 5.8 there is a sequence $\langle x_t \rangle_{t=1}^\infty$ such that $\mathrm{FP}(\langle x_t \rangle_{t=1}^\infty) \subseteq A$, so $A \subseteq B$ so $B \in p$.

To verify conclusion (2), let $p \cdot p = p \in \beta \mathbb{N}$ and pick $A \in \mathcal{R}$ such that $A \in \sigma_m(p)$. Then by Theorem 17.24 pick a sequence $\langle y_t \rangle_{t=1}^{\infty}$ with $\mathrm{SP}_m(\langle y_t \rangle_{t=1}^{\infty}) \subseteq A$. By Theorem 17.27 there does not exist a sequence $\langle x_t \rangle_{t=1}^{\infty}$ with $\mathrm{FP}(\langle x_t \rangle_{t=1}^{\infty}) \subseteq A$, so $A \cap B = \emptyset$.

To verify conclusion (3), suppose that one had a one-to-one sequence $\langle x_t \rangle_{t=1}^{\infty}$ with $\mathrm{SP}_m(\langle x_t \rangle_{t=1}^{\infty}) \subseteq B$. By Lemma 5.11, pick $p \cdot p = p \in \bigcap_{k=1}^{\infty} \overline{\mathrm{FP}(\langle x_t \rangle_{t=k}^{\infty})}$. Then by Theorem 17.26 $\mathrm{SP}_m(\langle x_t \rangle_{t=1}^{\infty}) \in \sigma_m(p)$ so that $B \in \sigma_m(p)$, contradicting conclusion (2).

To verify conclusion (4), suppose that one had a one-to-one sequence $\langle x_t \rangle_{t=1}^{\infty}$ with $\mathrm{FP}(\langle x_t \rangle_{t=1}^{\infty}) \subseteq \mathbb{N} \backslash B$. Again using Lemma 5.11, pick $p \cdot p = p \in \bigcap_{k=1}^{\infty} \overline{\mathrm{FP}(\langle x_t \rangle_{t=k}^{\infty})}$. Then $\mathrm{FP}(\langle x_t \rangle_{t=1}^{\infty}) \in p$ so $\mathbb{N} \backslash B \in p$, contradicting conclusion (1).   □

Exercise 17.3.1. Prove Lemma 17.20. (Hint: See the proofs of Corollary 17.17 and Theorem 17.24.)

Exercise 17.3.2. Prove that, if $m \in \mathbb{N}$ and $\langle y_t \rangle_{t=1}^{\infty}$ is a product subsystem of $\langle x_t \rangle_{t=1}^{\infty}$, then $\mathrm{SP}_m(\langle y_t \rangle_{t=1}^{\infty}) \subseteq \mathrm{SP}_m(\langle x_t \rangle_{t=1}^{\infty})$.

Exercise 17.3.3. Let $\langle y_t \rangle_{t=1}^{\infty}$. Modify the proof of Theorem 17.24 to show that if $p \cdot p = p \in \bigcap_{k=1}^{\infty} \overline{\mathrm{FP}(\langle y_t \rangle_{t=k}^{\infty})}$ and $A \in \sigma_m(p)$, then there is a product subsystem $\langle x_t \rangle_{t=1}^{\infty}$ of $\langle y_t \rangle_{t=k}^{\infty}$ such that $\mathrm{SP}_m(\langle x_t \rangle_{t=1}^{\infty}) \subseteq A$. (Hint: See the proof of Theorem 17.31)

Exercise 17.3.4. Let $n, r \in \mathbb{N}$ and let $\mathbb{N} = \bigcup_{i=1}^{r} A_i$. Prove that there is a function $f : \{1, 2, \ldots, n\} \rightarrow \{1, 2, \ldots, r\}$ and a sequence $\langle x_t \rangle_{t=1}^{\infty}$ such that for each $m \in \{1, 2, \ldots, n\}$, $\mathrm{SP}_m(\langle x_t \rangle_{t=1}^{\infty}) \subseteq A_{f(m)}$. (Hint: Use the results of Exercises 17.3.2 and 17.3.3.)

17.4 Linear Combinations of Sums —
Infinite Partition Regular Matrices

In this section we show that, given a finite sequence of coefficients, one can always find one cell of a partition containing the linear combinations of sums of a sequence which have the specified coefficients. We show further that the cell can depend on the choice of coefficients.

In many respects, the results of this section are similar to those of Section 17.3. However, the motivation is significantly different. In Section 15.4, we characterized the (finite) image partition regular matrices with entries from $\mathbb{Q}$ One may naturally extend the notion of image partition regularity to infinite dimensional matrices.

Definition 17.29. Let A be an $\omega \times \omega$ matrix with entries from $\mathbb{Q}$ such that each row of A has only finitely many nonzero entries. Then A is *image partition regular over* $\mathbb{N}$ if and only if whenever $r \in \mathbb{N}$ and $\mathbb{N} = \bigcup_{i=1}^{r} C_i$, there exist $i \in \{1, 2, \ldots, r\}$ and $\vec{x} \in \mathbb{N}^{\omega}$ such that all entries of $A\vec{x}$ are in C_i.

A simple example of an infinite partition regular matrix is

$$
A = \begin{pmatrix}
1 & 0 & 0 & 0 & \cdots \\
0 & 1 & 0 & 0 & \cdots \\
1 & 1 & 0 & 0 & \cdots \\
0 & 0 & 1 & 0 & \cdots \\
1 & 0 & 1 & 0 & \cdots \\
0 & 1 & 1 & 0 & \cdots \\
1 & 1 & 1 & 0 & \cdots \\
\vdots & \vdots & \vdots & \vdots & \ddots
\end{pmatrix}
$$

Here A is a finite sums matrix. That is, the assertion that all entries of $A\vec{x}$ are in C_i is the same as the assertion that $FS(\langle x_n \rangle_{n=1}^{\infty}) \subseteq C_i$. We establish here the image partition regularity of certain infinite matrices and provide a strong contrast to the situation with respect to finite image partition regular matrices.

For example, any central set C in $(\mathbb{N}, +)$ has the property that, given any finite image partition regular matrix A, there must exist $\vec{x}$ with all entries of $A\vec{x}$ in C. (Theorems 15.5, 15.24 and Lemma 16.13.) Consequently, given any finite partition of $\mathbb{N}$, some one cell must contain the entries of $A\vec{x}$ for every finite image partition regular matrix A.

By way of contrast, we shall establish here that a certain class of infinite matrices are image partition regular, and then show that there are two of these matrices, A and B, and a two cell partition of $\mathbb{N}$, neither cell of which contains all entries of $A\vec{x}$ and $B\vec{y}$ for any $\vec{x}$ and $\vec{y}$.

We call the systems we are studying "Milliken–Taylor" systems because of their relation to the Milliken–Taylor Theorem, which we shall prove in Chapter 18.

Definition 17.30. (a) $\mathbb{A} = \{\langle a_i \rangle_{i=1}^{m} \in \mathbb{N}^m : m \in \mathbb{N} \text{ and for all } i \in \{1, 2, \ldots, m-1\}, a_i \neq a_{i+1}\}$.

(b) Given $\vec{a} \in \mathbb{A}$ with length m and a sequence $\langle x_t \rangle_{t=1}^{\infty}$ in $\mathbb{N}$,

$$
[\text{MT}(\vec{a}, \langle x_t \rangle_{t=1}^{\infty}) = \{\textstyle\sum_{i=1}^{m} a_i \sum_{t \in F_i} x_t : F_1, F_2, \ldots, F_m \in \mathcal{P}_f(\mathbb{N})
$$
$$
\text{and } F_1 < F_2 < \cdots < F_m \}.
$$

The reason for requiring that $a_i \neq a_{i+1}$ in the definition of $\mathbb{A}$ is of course that $a \sum_{t \in F} x_t + a \sum_{t \in G} x_t = a \sum_{t \in F \cup G} x_t$ when $F < G$.

As a consequence of Corollary 17.33 below, whenever $\vec{a} \in \mathbb{A}$ and $\mathbb{N}$ is partitioned into finitely many cells, one cell must contain $\text{MT}(\vec{a}, \langle x_t \rangle_{t=1}^{\infty})$ for some sequence $\langle x_t \rangle_{t=1}^{\infty}$. This is the same as the assertion that a particular infinite matrix is image partition regular.

For example, if

$$A = \begin{pmatrix} 1 & 2 & 0 & 0 & \cdots \\ 1 & 0 & 2 & 0 & \cdots \\ 0 & 1 & 2 & 0 & \cdots \\ 1 & 1 & 2 & 0 & \cdots \\ 1 & 2 & 2 & 0 & \cdots \\ 1 & 0 & 0 & 2 & \cdots \\ 0 & 1 & 0 & 2 & \cdots \\ 1 & 1 & 0 & 2 & \cdots \\ \vdots & \vdots & \vdots & \vdots & \ddots \end{pmatrix}$$

and $\vec{x} = \langle x_t \rangle_{t=1}^{\infty}$, then $MT(\langle 1, 2 \rangle, \langle x_t \rangle_{t=1}^{\infty})$ is the set of entries of $A\vec{x}$. (The rows of A are all rows with entries from $\{0, 1, 2\}$ such that (1) only finitely many entries are non-zero, (2) at least one entry is 1, (3) at least one entry is 2, and (4) all occurrences of 1 come before any occurrences of 2.)

Theorem 17.31. *Let $\vec{a} = \langle a_1, a_2, \ldots, a_m \rangle \in \mathbb{A}$, let $\langle y_t \rangle_{t=1}^{\infty}$ be a sequence in $\mathbb{N}$, let $p + p = p \in \bigcap_{k=1}^{\infty} \overline{FS(\langle y_t \rangle_{t=k}^{\infty})}$, and let $A \in a_1 p + a_2 p + \cdots + a_m p$. There is a sum subsystem $\langle x_t \rangle_{t=1}^{\infty}$ of $\langle y_t \rangle_{t=1}^{\infty}$ such that $MT(\vec{a}, \langle x_t \rangle_{t=1}^{\infty}) \subseteq A$.*

Proof. Assume first that $m = 1$. Then $a_1^{-1} A \in p$ so by Theorem 5.14 there is a sum subsystem $\langle x_t \rangle_{t=1}^{\infty}$ of $\langle y_t \rangle_{t=1}^{\infty}$ such that $FS(\langle x_t \rangle_{t=1}^{\infty}) \subseteq a_1^{-1} A$. This says precisely that $MT(\vec{a}, \langle x_t \rangle_{t=1}^{\infty}) \subseteq A$.

Assume now that $m \geq 2$ and notice that

$$\{x \in \mathbb{N} : -x + A \in a_2 p + a_3 p + \cdots + a_m p\} \in a_1 p$$

so that

$$\{x \in \mathbb{N} : -a_1 x + A \in a_2 p + a_3 p + \cdots + a_m p\} \in p.$$

Let

$$B_1 = \{x \in \mathbb{N} : -a_1 x + A \in a_2 p + a_3 p + \cdots + a_m p\} \cap FS(\langle y_t \rangle_{t=1}^{\infty}).$$

Pick $x_1 \in B_1^{\star}$ and pick $H_1 \in \mathcal{P}_f(\mathbb{N})$ such that $x_1 = \Sigma_{t \in H_1} y_t$. (Here we are using the additive version of $B_1^{\star}$. That is, $B_1^{\star} = \{x \in B_1 : -x + B_1 \in p\}$.)

Inductively, let $n \in \mathbb{N}$ and assume that we have chosen $\langle x_k \rangle_{k=1}^{n}$ in $\mathbb{N}$, $\langle B_k \rangle_{k=1}^{n}$ in p, and $\langle H_k \rangle_{k=1}^{n}$ in $\mathcal{P}_f(\mathbb{N})$ so that for each $r \in \{1, 2, \ldots, n\}$:

(I) If $\emptyset \neq F \subseteq \{1, 2, \ldots, r\}$ and $k = \min F$, then $\Sigma_{t \in F} x_t \in B_k^{\star}$.

(II) If $r < n$, then $B_{r+1} \subseteq B_r$ and $H_r < H_{r+1}$.

(III) If $\ell \in \{1, 2, \ldots, m-1\}$, $F_1, F_2, \ldots, F_\ell \in \mathcal{P}_f(\{1, 2, \ldots, r\})$, and $F_1 < F_2 < \cdots < F_\ell$, then

$$- \Sigma_{i=1}^{\ell} a_i \, \Sigma_{t \in F_i} x_t + A \in a_{\ell+1} p + a_{\ell+2} p + \cdots + a_m p.$$

(IV) If $F_1, F_2, \ldots, F_{m-1} \in \mathcal{P}_f(\{1, 2, \ldots, r\})$, $F_1 < F_2 < \cdots < F_{m-1}$, and $r < n$, then

$$B_{r+1} \subseteq a_m{}^{-1}(-\textstyle\sum_{i=1}^{m-1} a_i \sum_{t \in F_i} x_t + A).$$

(V) If $\ell \in \{1, 2, \ldots, m-2\}$, $F_1, F_2, \ldots, F_\ell \in \mathcal{P}_f(\{1, 2, \ldots, r\})$, $F_1 < F_2 < \cdots < F_\ell$, and $r < n$, then

$$B_{r+1} \subseteq \{x \in \mathbb{N} : -a_{\ell+1}x + (-\textstyle\sum_{i=1}^{\ell} a_i \sum_{t \in F_i} x_t + A) \in a_{\ell+2}p + a_{\ell+3}p + \cdots + a_m p\}.$$

(VI) $x_r = \sum_{t \in H_r} y_t$.

At $n = 1$, hypotheses (I) and (VI) hold directly, hypotheses (II), (IV), and (V) are vacuous, and hypothesis (III) says that $-a_1 x_1 + A \in a_2 p + a_3 p + \cdots + a_m p$ which is true because $x_1 \in B_1$.

For $\ell \in \{1, 2, \ldots, m-1\}$, let

$$\mathcal{F}_\ell = \{(F_1, F_2, \ldots, F_\ell) : F_1, F_2, \ldots, F_\ell \in \mathcal{P}_f(\{1, 2, \ldots, n\})$$
$$\text{and } F_1 < F_2 < \cdots < F_\ell\}$$

and for $k \in \{1, 2, \ldots, n\}$, let

$$E_k = \{\textstyle\sum_{t \in F} x_t : \emptyset \neq F \subseteq \{1, 2, \ldots, n\} \text{ and } \min F = k\}.$$

Given $b \in E_k$, we have $b \in B_k{}^\star$ by hypothesis (I), and so $-b + B_k{}^\star \in p$. If $(F_1, F_2, \ldots, F_{m-1}) \in \mathcal{F}_{m-1}$, then by (III) we have $(-\sum_{i=1}^{m-1} a_i \sum_{t \in F_i} x_t + A) \in a_m p$ so that $a_m{}^{-1}(-\sum_{i=1}^{m-1} a_i \sum_{t \in F_i} x_t + A) \in p$. If $\ell \in \{1, 2, \ldots, m-2\}$ and $(F_1, F_2, \ldots, F_\ell) \in \mathcal{F}_\ell$ we have by (III) that

$$-\textstyle\sum_{i=1}^{\ell} a_i \sum_{t \in F_i} x_t + A \in a_{\ell+1}p + a_{\ell+2}p + \cdots + a_m p$$

so that

$$\{x \in \mathbb{N} : -a_{\ell+1}x + (-\textstyle\sum_{i=1}^{\ell} a_i \sum_{t \in F_i} x_t + A) \in a_{\ell+2}p + a_{\ell+3}p + \cdots + a_m p\} \in p.$$

Let $s = \max H_n + 1$. Then we have that $B_{n+1} \in p$, where

$$B_{n+1} = B_n \cap \mathrm{FS}(\langle y_t \rangle_{t=s}^\infty) \cap \bigcap_{k=1}^{n} \bigcap_{b \in E_k} (-b + B_k{}^\star) \cap$$
$$\bigcap_{(F_1, F_2, \ldots, F_{m-1}) \in \mathcal{F}_{m-1}} a_m{}^{-1}(-\textstyle\sum_{i=1}^{m-1} a_i \sum_{t \in F_i} x_t + A) \cap$$
$$\bigcap_{\ell=1}^{m-2} \bigcap_{(F_1, F_2, \ldots, F_\ell) \in \mathcal{F}_\ell} \{x \in \mathbb{N} : -a_{\ell+1}x + (-\textstyle\sum_{i=1}^{\ell} a_i \sum_{t \in F_i} x_t + A)$$
$$\in a_{\ell+2}p + a_{\ell+3}p + \cdots + a_m p\}.$$

Here again we use the convention that $\bigcap \emptyset = \mathbb{N}$ so, for example, if $m = 2$, then

$$B_{n+1} = B_n \cap \mathrm{FS}(\langle y_t \rangle_{t=s}^\infty) \cap \bigcap_{k=1}^{n} \bigcap_{b \in E_k} (-b + B_k{}^\star) \cap$$
$$\bigcap_{\emptyset \neq F \subseteq \{1, 2, \ldots, n\}} a_2{}^{-1}(-a_1 \textstyle\sum_{t \in F} x_t + A).$$

Choose $x_{n+1} \in B_{n+1}{}^\star$ and pick $H_{n+1} \subseteq \{s, s+1, s+2, \ldots\}$ such that $x_{n+1} = \Sigma_{t \in H_{n+1}} y_t$.

Hypothesis (I) can be verified as in the second proof of Theorem 5.8 and hypothesis (II) holds trivially.

Hypotheses (IV), (V) and (VI) hold directly.

To verify hypothesis (III), let $\ell \in \{1, 2, \ldots, m-1\}$ and let $F_1 < F_2 < \cdots < F_\ell$ in $\mathcal{P}_f(\{1, 2, \ldots, n+1\})$. If $\ell = 1$, we have by hypotheses (I) and (II) that $\Sigma_{t \in F_1} x_t \in B_1$ so that $-a_1 \Sigma_{t \in F_1} x_t + A \in a_2 p + a_3 p + \cdots a_m p$ as required. So assume that $\ell > 1$ and let $k = \min F_\ell$ and $j = \max F_{\ell-1}$. Then

$$\Sigma_{t \in F_\ell} x_t \in B_k{}^\star \subseteq B_{j+1} \subseteq \{x \in \mathbb{N} : -a_\ell x + (- \Sigma_{i=1}^{\ell-1} a_i \Sigma_{t \in F_i} x_t + A)$$
$$\in a_{\ell+1} p + a_{\ell+2} p + \cdots + a_m p\}$$

(by hypothesis (V) at $r = j$) so

$$-(\Sigma_{i=1}^{\ell} a_i \Sigma_{t \in F_i} x_t) + A \in a_{\ell+1} p + a_{\ell+2} p + \cdots + a_m p$$

as required.

The induction being complete, let $F_1 < F_2 < \cdots < F_m$ in $\mathcal{P}_f(\mathbb{N})$ and let $k = \min F_m$ and $j = \max F_{m-1}$. Then

$$\Sigma_{t \in F_m} x_t \in B_k{}^\star \subseteq B_{j+1} \subseteq a_m{}^{-1}(- \Sigma_{i=1}^{m-1} a_i \Sigma_{t \in F_i} x_t + A)$$

so $\Sigma_{i=1}^{m} a_i \Sigma_{t \in F_i} x_t \in A$. $\square$

Just as Theorem 17.26 was a partial converse to Theorem 17.24, we now obtain a partial converse to Theorem 17.31. Notice that p is not required to be an idempotent.

Theorem 17.32. *Let $\vec{a} = \langle a_1, a_2, \ldots, a_m \rangle \in \mathbb{A}$, let $\langle x_t \rangle_{t=1}^{\infty}$ be a sequence in $\mathbb{N}$, and let $p \in \bigcap_{k=1}^{\infty} \overline{\mathrm{FS}(\langle x_t \rangle_{t=k}^{\infty})}$. Then $\mathrm{MT}(\vec{a}, \langle x_t \rangle_{t=1}^{\infty}) \in a_1 p + a_2 p + \cdots + a_m p$.*

Proof. We show, by downward induction on $\ell \in \{1, 2, \ldots, m\}$ that for each $k \in \mathbb{N}$,

$$\{\Sigma_{i=\ell}^{m} a_i \Sigma_{t \in F_i} x_t : F_\ell, F_{\ell+1}, \ldots, F_m \in \mathcal{P}_f(\{k, k+1, \ldots\}) \text{ and } F_\ell < F_{\ell+1} < \cdots < F_m\}$$

$$\in a_\ell p + a_{\ell+1} p + \cdots + a_m p.$$

For $\ell = m$, we have $\mathrm{FS}(\langle x_t \rangle_{t=k}^{\infty}) \in p$ so that $\mathrm{FS}(\langle a_m x_t \rangle_{t=k}^{\infty}) \in a_m p$. So let $\ell \in \{1, 2, \ldots, m-1\}$ and assume that the assertion is true for $\ell + 1$. Let $k \in \mathbb{N}$ and let

$$A = \{\Sigma_{i=\ell}^{m} a_i \Sigma_{t \in F_i} x_t : F_\ell, F_{\ell+1}, \ldots, F_m \in \mathcal{P}_f(\{k, k+1, \ldots\})$$
$$\text{and } F_\ell < F_{\ell+1} < \cdots < F_m\}.$$

We show that

$$\mathrm{FS}(\langle a_\ell x_t \rangle_{t=k}^{\infty}) \subseteq \{x \in \mathbb{N} : -x + A \in a_{\ell+1} p + a_{\ell+2} p + \cdots + a_m p\}.$$

So let $b \in \mathrm{FS}(\langle a_\ell x_t \rangle_{t=k}^\infty)$ and pick $F_\ell \in \mathcal{P}_f(\{k, k+1, \ldots\})$ such that $b = a_\ell \sum_{t \in F_\ell} x_t$. Let $r = \max F_\ell + 1$. Then

$$\{\textstyle\sum_{i=\ell+1}^m a_i \sum_{t \in F_i} x_t : F_{\ell+1}, F_{\ell+2}, \ldots, F_m \in \mathcal{P}_f(\{r, r+1, \ldots\})$$

$$\text{and } F_{\ell+1} < F_{\ell+2} < \cdots < F_m\} \subseteq -b + A$$

so $-b + A \in a_{\ell+1}p + a_{\ell+2}p + \cdots + a_m p$ as required. $\qquad\square$

We see now that Milliken–Taylor systems are themselves partition regular.

Corollary 17.33. *Let $\vec{a} \in \mathbb{A}$, let $B \subseteq \mathbb{N}$, let $\langle x_t \rangle_{t=1}^\infty$ be a sequence in $\mathbb{N}$, and assume that $\mathrm{MT}(\vec{a}, \langle x_t \rangle_{t=1}^\infty) \subseteq B$. If $r \in \mathbb{N}$ and $B = \bigcup_{i=1}^r B_i$, then there exist $i \in \{1, 2, \ldots, r\}$ and a sum subsystem $\langle y_t \rangle_{t=1}^\infty$ of $\langle x_t \rangle_{t=1}^\infty$ such that $\mathrm{MT}(\vec{a}, \langle y_t \rangle_{t=1}^\infty) \subseteq B_i$.*

Proof. Pick by Lemma 5.11 some $p + p = p \in \bigcap_{k=1}^\infty \overline{\mathrm{FS}(\langle x_t \rangle_{t=k}^\infty)}$. By Theorem 17.32, $\mathrm{MT}(\vec{a}, \langle x_t \rangle_{t=1}^\infty) \in a_1 p + a_2 p + \cdots + a_m p$, where $\vec{a} = \langle a_1, a_2, \ldots, a_m \rangle$. Pick $i \in \{1, 2, \ldots, r\}$ such that $B_i \in a_1 p + a_2 p + \cdots + a_m p$, and apply Theorem 17.31. $\quad\square$

We observe that certain different members of $\mathbb{A}$ always have Milliken–Taylor systems in the same cell of a partition.

Theorem 17.34. *Let $\vec{a}, \vec{b} \in \mathbb{A}$, and assume that there is a positive rational s such that $\vec{a} = s\vec{b}$. Then whenever $r \in \mathbb{N}$ and $\mathbb{N} = \bigcup_{i=1}^r A_i$, there exist $i \in \{1, 2, \ldots, r\}$ and sequences $\langle x_t \rangle_{t=1}^\infty$ and $\langle y_t \rangle_{t=1}^\infty$ such that $\mathrm{MT}(\vec{a}, \langle x_t \rangle_{t=1}^\infty) \subseteq A_i$ and $\mathrm{MT}(\vec{b}, \langle y_t \rangle_{t=1}^\infty) \subseteq A_i$.*

Proof. Assume that $s = \frac{m}{n}$ where $m, n \in \mathbb{N}$ and let $\vec{c} = m\vec{b}$. Pick by Corollary 17.33 some $i \in \{1, 2, \ldots, r\}$ and a sequence $\langle z_t \rangle_{t=1}^\infty$ such that $\mathrm{MT}(\vec{c}, \langle z_t \rangle_{t=1}^\infty) \subseteq A_i$. For each $t \in \mathbb{N}$, let $x_t = nz_t$ and $y_t = mz_t$. Then $\mathrm{MT}(\vec{a}, \langle x_t \rangle_{t=1}^\infty) = \mathrm{MT}(\vec{b}, \langle y_t \rangle_{t=1}^\infty) = \mathrm{MT}(\vec{c}, \langle z_t \rangle_{t=1}^\infty)$. $\quad\square$

We show now, as we promised earlier, that we can separate certain Milliken–Taylor systems.

Theorem 17.35. *There is a set $A \subseteq \mathbb{N}$ such that for any sequence $\langle x_t \rangle_{t=1}^\infty$ in $\mathbb{N}$, $\mathrm{MT}(\langle 1, 2 \rangle, \langle x_t \rangle_{t=1}^\infty) \not\subseteq A$ and $\mathrm{MT}(\langle 1, 4 \rangle, \langle x_t \rangle_{t=1}^\infty) \not\subseteq \mathbb{N} \backslash A$.*

Proof. For $x \in \mathbb{N}$, let $\mathrm{supp}(x)$ be the binary support of x so that $x = \sum_{t \in \mathrm{supp}(x)} 2^t$. For each $x \in \mathbb{N}$, let $a(x) = \max \mathrm{supp}(x)$ and let $d(x) = \min \mathrm{supp}(x)$. We also define a function z so that $z(x)$ counts the number of blocks of zeros of odd length interior to the binary expansion of x, that is between $a(x)$ and $d(x)$. Thus, for example, if in binary, $x = 1011001010001100000$, then $z(x) = 3$. (The blocks of length 1 and 3 were counted, but the block of length 5 is not counted because it is not interior to the expansion of x.)

Let $A = \{x \in \mathbb{N} : z(x) \equiv 0 \pmod{4} \text{ or } z(x) \equiv 3 \pmod{4}\}$. Suppose that we have a sequence $\langle x_t \rangle_{t=1}^{\infty}$ such that either

$$\text{MT}(\langle 1, 2 \rangle, \langle x_t \rangle_{t=1}^{\infty}) \subseteq A \quad \text{or} \quad \text{MT}(\langle 1, 4 \rangle, \langle x_t \rangle_{t=1}^{\infty}) \subseteq \mathbb{N} \backslash A.$$

Pick a sum subsystem $\langle y_t \rangle_{t=1}^{\infty}$ of $\langle x_t \rangle_{t=1}^{\infty}$ such that for each t, $d(y_{t+1}) > a(y_t)$. (See Exercise 17.4.1.) For $i \in \{0, 1, 2, 3\}$ let

$$B_i = \{x \in \mathbb{N} : z(x) \equiv i \pmod{4}\}.$$

By Corollary 17.33 we may presume that if

$$\text{MT}(\langle 1, 2 \rangle, \langle x_t \rangle_{t=1}^{\infty}) \subseteq A,$$

then

$$\text{MT}(\langle 1, 2 \rangle, \langle y_t \rangle_{t=1}^{\infty}) \subseteq B_0 \quad \text{or} \quad \text{MT}(\langle 1, 2 \rangle, \langle y_t \rangle_{t=1}^{\infty}) \subseteq B_3$$

and, if

$$\text{MT}(\langle 1, 4 \rangle, \langle x_t \rangle_{t=1}^{\infty}) \subseteq \mathbb{N} \backslash A,$$

then

$$\text{MT}(\langle 1, 4 \rangle, \langle y_t \rangle_{t=1}^{\infty}) \subseteq B_1 \quad \text{or} \quad \text{MT}(\langle 1, 4 \rangle, \langle y_t \rangle_{t=1}^{\infty}) \subseteq B_2.$$

By the pigeon hole principle we may presume that for $t, k \in \mathbb{N}$,

$$a(y_t) \equiv a(y_k) \pmod{2}, \quad d(y_t) \equiv d(y_k) \pmod{2}, \quad \text{and} \quad z(y_t) \equiv z(y_k) \pmod{4}.$$

Assume first that for $t \in \mathbb{N}$, $a(y_t) \equiv d(y_t) \pmod{2}$. Now if $\text{MT}(\langle 1, 2 \rangle, \langle y_t \rangle_{t=1}^{\infty}) \subseteq A$, we have $y_1 + y_2 + 2y_3$ and $y_2 + 2y_3$ are in the same B_i so

$$z(y_1 + y_2 + 2y_3) \equiv z(y_2 + 2y_3) \pmod{4}.$$

Since $a(y_1) \equiv d(y_2) \pmod{2}$ there is an odd block of zeros between y_1 and y_2 in $y_1 + y_2 + 2y_3$, so $z(y_1 + y_2 + 2y_3) = z(y_1) + 1 + z(y_2 + 2y_3)$ and consequently $z(y_1) \equiv 3 \pmod{4}$. But $d(2y_3) = d(y_3) + 1 \not\equiv a(y_2) \pmod{2}$ so $z(y_2 + 2y_3) = z(y_2) + z(y_3) \equiv 3 + 3 \equiv 2 \pmod{4}$, a contradiction to the fact that $\text{MT}(\langle 1, 2 \rangle, \langle y_t \rangle_{t=1}^{\infty}) \subseteq A$.

If $\text{MT}(\langle 1, 4 \rangle, \langle y_t \rangle_{t=1}^{\infty}) \subseteq \mathbb{N} \backslash A$, we have as above that

$$z(y_1 + y_2 + 4y_3) \equiv z(y_2 + 4y_3) \pmod{4}$$

and $z(y_1 + y_2 + 4y_3) = z(y_1) + 1 + z(y_2 + 4y_3)$ so that $z(y_1) \equiv 3 \pmod{4}$. Since $d(4y_3) = d(y_3) + 2 \equiv a(y_2) \pmod{2}$ we have $z(y_2 + 4y_3) = z(y_2) + 1 + z(y_3) \equiv 3 \pmod{4}$, a contradiction to the fact that $\text{MT}(\langle 1, 4 \rangle, \langle y_t \rangle_{t=1}^{\infty}) \subseteq \mathbb{N} \backslash A$.

Now assume that for $t \in \mathbb{N}$, $a(y_t) \not\equiv d(y_t) \pmod{2}$. If $\text{MT}(\langle 1, 2 \rangle, \langle y_t \rangle_{t=1}^{\infty}) \subseteq A$ we have $z(y_1 + y_2 + 2y_3) \equiv z(y_2 + 2y_3) \pmod{4}$ and, since $d(y_2) \not\equiv a(y_1) \pmod{2}$, $z(y_1 + y_2 + 2y_3) = z(y_1) + z(y_2 + 2y_3)$ so that $z(y_1) \equiv 0 \pmod{4}$. Since

$$d(2y_3) = d(y_3) + 1 \equiv a(y_2) \pmod{2}$$

we have that $z(y_2 + 2y_3) = z(y_2) + 1 + z(y_3) \equiv 1 \pmod{4}$, a contradiction.

If $\mathrm{MT}(\langle 1, 4 \rangle, \langle y_t \rangle_{t=1}^{\infty}) \subseteq \mathbb{N} \backslash A$, we have as above that $z(y_1) \equiv 0 \pmod{4}$ and since $d(4y_3) = d(y_3) + 2 \not\equiv a(y_2) \pmod{2}$, $z(y_2 + 4y_3) = z(y_2) + z(y_3) \equiv 0 \pmod{4}$, a contradiction. $\qquad\square$

As a consequence of Theorem 17.35, we have the following contrast between infinite and finite image partition regular matrices.

Corollary 17.36. *There exist infinite image partition regular matrices B and C and a two cell partition of $\mathbb{N}$, neither cell of which contains all of the entries of $B\vec{x}$ and of $C\vec{y}$ for any $\vec{x}, \vec{y} \in \mathbb{N}^{\omega}$.*

Proof. The matrices for the Milliken–Taylor systems $\mathrm{MT}(\langle 1, 2 \rangle, \langle x_t \rangle_{t=1}^{\infty})$ and $\mathrm{MT}(\langle 1, 4 \rangle, \langle y_t \rangle_{t=1}^{\infty})$ are image partition regular. $\qquad\square$

In Section 15.4, we found several characterizations of finite image partition regular matrices.

Question 17.37. *Can one find a reasonable combinatorial characterization of infinite image partition regular matrices?*

Exercise 17.4.1. In Exercise 5.2.2, the reader was asked to show combinatorially that given any sequence $\langle x_t \rangle_{t=1}^{\infty}$ there is a sum subsystem $\langle y_t \rangle_{t=1}^{\infty}$ of $\langle x_t \rangle_{t=1}^{\infty}$ such that whenever $2^k < y_t$, one has $2^{k+1} | y_{t+1}$. Use the fact that, by Lemma 6.8, given any $p + p = p \in \bigcap_{k=1}^{\infty} \overline{\mathrm{FS}(\langle x_t \rangle_{t=k}^{\infty})}$, one has $p \in \mathbb{H}$, to show that there is a sum subsystem $\langle y_t \rangle_{t=1}^{\infty}$ of $\langle x_t \rangle_{t=1}^{\infty}$ such that for each $t \in \mathbb{N}$, $a(y_t) + 2 \leq d(y_{t+1})$.

Exercise 17.4.2. Prove that if H is a finite subset of $\mathbb{A}$, $r \in \mathbb{N}$, and $\mathbb{N} = \bigcup_{i=1}^{r} A_i$, then there exist a function $f : H \to \{1, 2, \ldots, r\}$ and a sequence $\langle x_t \rangle_{t=1}^{\infty}$ such that for each $\vec{a} \in H$, $\mathrm{MT}(\vec{a}, \langle x_t \rangle_{t=1}^{\infty}) \subseteq A_{f(\vec{a})}$. (Hint: Use Corollary 17.33.)

17.5 Sums and Products in $(0, 1)$ — Measurable and Baire Partitions

We show in this section that if one has a finite partition of the real interval $(0, 1)$ each member of which is Lebesgue measurable, or each member of which is a Baire set, then one cell will contain $\mathrm{FS}(\langle x_n \rangle_{n=1}^{\infty}) \cup \mathrm{FP}(\langle x_n \rangle_{n=1}^{\infty})$ for some sequence $\langle x_n \rangle_{n=1}^{\infty}$ in $(0, 1)$. (The terminology "Baire set" has different meanings in the literature. We take it to mean members of the smallest σ-algebra containing the open sets and the nowhere dense sets.) In fact, we show that certain sums of products and products of sums can also be found in the specified cell.

We use the algebraic structure of the subset 0^+ of $\beta(0, 1)_d$ which we discussed in Section 13.3. The results that we obtain are topological in nature, so it is natural to wonder how one can obtain them starting with the discrete topology on $(0, 1)$. The answer is that we use certain topologically characterized subsets of 0^+. Recall that a subset A of $\mathbb{R}$ is *meager* (or *first category*) if and only if A is the countable union of nowhere dense sets.

Definition 17.38. (a) A set $A \subseteq (0, 1)$ is *Baire large* if and only if for every $\epsilon > 0$, $A \cap (0, \epsilon)$ is not meager.

(b) $\mathcal{B} = \{p \in 0^+ : \text{for all } A \in p, A \text{ is Baire large}\}$.

Lemma 17.39. $\mathcal{B}$ *is a left ideal of* $(0^+, \cdot)$.

Proof. Since the union of finitely many meager sets is meager, one sees easily that whenever $(0, 1)$ is partitioned into finitely many sets, one of them is Baire large.

Consequently, by Theorem 5.7, it follows that $\{p \in \beta(0, 1)_d : \text{for all } A \in p, A \text{ is Baire large}\} \neq \emptyset$. On the other hand, if $p \in \beta(0, 1)_d \backslash 0^+$ one has some $\epsilon > 0$ such that $(\epsilon, 1) \in p$ and $(\epsilon, 1)$ is not Baire large. Thus $\mathcal{B} \neq \emptyset$. To see that $\mathcal{B}$ is a left ideal of 0^+, let $p \in \mathcal{B}$ and $q \in 0^+$ and let $A \in q \cdot p$. Pick $x \in (0, 1)$ such that $x^{-1}A \in p$. Given $\epsilon > 0$, $x^{-1}A \cap (0, \epsilon)$ is not meager so $x(x^{-1}A \cap (0, \epsilon))$ is not meager and $x(x^{-1}A \cap (0, \epsilon)) \subseteq A \cap (0, \epsilon)$. $\square$

The following result, and its measurable analogue Theorem 17.46, are the ones that allow us to obtain the combined additive and multiplicative results. Recall that a subset of $\mathbb{R}$ is a Baire set if and only if it can be written as the symmetric difference of an open set and a meager set. (Or do Exercise 17.5.1.)

Theorem 17.40. *Let p be a multiplicative idempotent in $\mathcal{B}$ and let A be a Baire set which is a member of p. Then $\{x \in A : x^{-1}A \in p \text{ and } -x + A \in p\} \in p$.*

Proof. Let $B = \{x \in A : x^{-1}A \in p\}$. Then since $p = p \cdot p$, $B \in p$. Also A is a Baire set so pick an open set U and a meager set M such that $A = U \triangle M$. Now $M \backslash U$ is meager so $M \backslash U \notin p$ so $U \backslash M \in p$. We claim that

$$(U \backslash M) \cap B \subseteq \{x \in A : x^{-1}A \in p \text{ and } -x + A \in p\}.$$

So let $x \in (U \backslash M) \cap B$ and pick $\epsilon > 0$ such that $(x, x + \epsilon) \subseteq U$. To see that $-x + A \in p$, we observe that $(0, 1) \backslash (-x + A)$ is not Baire large. Indeed one has $((0, 1) \backslash (-x + A)) \cap (0, \epsilon) \subseteq -x + M$, a meager set. (Given $y \in (0, \epsilon)$, $x + y \in U$ so, if $x + y \notin A$, then $x + y \in M$.) $\square$

Now we turn our attention to deriving the measurable analogue of Theorem 17.40. We denote by $\mu(A)$ the Lebesgue measure of the measurable set A, and by $\mu^*(B)$ the outer Lebesgue measure of an arbitrary set B. We assume familiarity with the basic facts of Lebesgue measure as encountered in any standard introductory analysis course. The notion corresponding to "Baire large" is that of having positive upper density near 0. We use the same notation for upper density near 0 as we used for upper density of sets of integers.

Definition 17.41. Let $A \subseteq (0, 1)$.

(a) The *upper density near* 0 *of* A, $\overline{d}(A)$ is defined by

$$\overline{d}(A) = \lim_{h \downarrow 0} \sup \frac{\mu^*(A \cap (0, h))}{h}.$$

(b) A point $x \in (0, 1)$ is a *density point* of A if and only if

$$\lim_{h \downarrow 0} \frac{\mu^*(A \cap (x - h, x + h))}{2h} = 1.$$

The following result, while quite well known, is not a part of standard introductory analysis courses so we include a proof.

Theorem 17.42 (Lebesgue Density Theorem). *Let* A *be a measurable subset of* $(0, 1)$. *Then* $\mu(\{x \in A : x \text{ is not a density point of } A\}) = 0$.

Proof. If x is not a density point of A, then there is some $\epsilon > 0$ such that

$$\liminf_{h \downarrow 0} \frac{\mu(A \cap (x - h, x + h))}{2h} < 1 - \epsilon.$$

Since the union of countably many sets of measure 0 is again of measure 0, it suffices to let $\epsilon > 0$ be given, let

$$B = \{x \in A : \liminf_{h \downarrow 0} \frac{\mu(A \cap (x - h, x + h))}{2h} < 1 - \epsilon\}$$

and show that $\mu(B) = 0$. Suppose instead that $\mu^*(B) > 0$. Pick an open set U such that $B \subseteq U$ and $\mu(U) < \frac{\mu^*(B)}{1 - \epsilon}$.

We first claim that

(†) if $\langle I_n \rangle_{n \in F}$ is a (finite or countably infinite) indexed family of pairwise disjoint intervals contained in U and for each $n \in F$, $\mu(A \cap I_n) < (1 - \epsilon)\mu(I_n)$, then $\mu^*(B \setminus \bigcup_{n \in F} I_n) > 0$.

To see this, since in general $\mu^*(C \cup D) \leq \mu^*(C) + \mu^*(D)$, it suffices to show that $\mu^*(B \cap \bigcup_{n \in F} I_n) < \mu^*(B)$. And indeed

$$\begin{aligned} \mu^*(B \cap \bigcup_{n \in F} I_n) &\leq \mu(A \cap \bigcup_{n \in F} I_n) \\ &= \Sigma_{n \in F} \, \mu(A \cap I_n) \\ &< \Sigma_{n \in F} (1 - \epsilon)\mu(I_n) \\ &\leq (1 - \epsilon)\mu(U) \\ &< \mu^*(B). \end{aligned}$$

Now choose $x_1 \in B$ and $h_1 > 0$ such that $[x_1 - h_1, x_1 + h_1] \subseteq U$ and

$$\mu(A \cap (x_1 - h_1, x_1 + h_1)) < (1 - \epsilon)2h_1.$$

Inductively, let $n \in \mathbb{N}$ and assume we have chosen $x_1, x_2, \ldots, x_n$ in B and positive $h_1, h_2, \ldots, h_n$ such that each $[x_i - h_i, x_i + h_i] \subseteq U$, each $\mu\big(A \cap (x_i - h_i, x_i + h_i)\big) < (1 - \epsilon)2h_i$, and $[x_i - h_i, x_i + h_i] \cap [x_j - h_j, x_j + h_j] = \emptyset$ for $i \neq j$.

Let

$$d_n = \sup\{h : \text{there exist } x \in B \text{ such that } [x - h, x + h] \subseteq U,$$

$[x - h, x + h] \cap \bigcup_{i=1}^{n}[x_i - h_i, x_i + h_i] = \emptyset$, and $\mu\big(A \cap (x - h, x + h)\big) < (1 - \epsilon)2h\}$.
Notice that $d_n > 0$. Indeed, by (†) one may pick $x \in B \backslash \bigcup_{i=1}^{n}[x_i - h_i, x_i + h_i]$ and then pick $\delta > 0$ such that $(x - \delta, x + \delta) \cap \bigcup_{i=1}^{n}[x_i - h_i, x_i + h_i] = \emptyset$ and $(x - \delta, x + \delta) \subseteq U$. Then, since $x \in B$, one may pick positive $h < \delta$ such that $\mu\big(A \cap (x - h, x + h)\big) < (1 - \epsilon)2h$.

Pick $x_{n+1} \in B$ and $h_{n+1} > \frac{d_n}{2}$ such that $[x_{n+1} - h_{n+1}, x_{n+1} + h_{n+1}] \subseteq U$,

$$[x_{n+1} - h_{n+1}, x_{n+1} + h_{n+1}] \cap \bigcup_{i=1}^{n}[x_i - h_i, x_i + h_i] = \emptyset,$$

and $\mu\big(A \cap (x_{n+1} - h_{n+1}, x_{n+1} + h_{n+1})\big) < (1 - \epsilon)2h_{n+1}$.

The inductive construction being complete, let $C = B \backslash \bigcup_{n=1}^{\infty}[x_n - h_n, x_n + h_n]$. Then by (†) $\mu^*(C) > 0$. Also, since $\langle [x_n - h_n, x_n + h_n] \rangle_{n=1}^{\infty}$ is a pairwise disjoint collection contained in $(0, 1)$, $\sum_{n=1}^{\infty} h_n$ converges, so pick $m \in \mathbb{N}$ such that

$$\sum_{n=m+1}^{\infty} h_n < \mu^*(C)/6.$$

Then $\mu(\bigcup_{n=m+1}^{\infty}[x_n - 3h_n, x_n + 3h_n]) < \mu^*(C)$ so pick $x \in C \backslash (\bigcup_{n=m+1}^{\infty}[x_n - 3h_n, x_n + 3h_n])$.

Then $x \in B \backslash \bigcup_{n=1}^{m}[x_n - h_n, x_n + h_n]$ so pick $h > 0$ such that $[x - h, x + h] \subseteq U$, $[x - h, x + h] \cap \bigcup_{n=1}^{m}[x_n - h_n, x_n + h_n] = \emptyset$, and $\mu\big(A \cap (x - h, x + h)\big) < (1 - \epsilon)2h$. Now there is some k such that $[x - h, x + h] \cap \bigcup_{n=1}^{k}[x_n - h_n, x_n + h_n] \neq \emptyset$ (since otherwise each $d_k \geq h$ and consequently $\sum_{n=1}^{\infty} h_n$ diverges) so pick the first k such that $[x - h, x + h] \cap [x_k - h_k, x_k + h_k] \neq \emptyset$ and notice that $k > m$. Then $h_k \geq \frac{d_{k-1}}{2} \geq \frac{h}{2}$ so that $|x - x_k| \leq h + h_k \leq 3h_k$ and hence $x \in [x_k - 3h_k, x_k + 3h_k]$, contradicting the fact that $x \notin \bigcup_{n=m+1}^{\infty}[x_n - 3h_n, x_n + 3h_n]$. $\qquad\square$

We also need another basic result about measurable sets.

Lemma 17.43. *Let A be a measurable subset of $(0, 1)$ such that $\overline{d}(A) > 0$. There exists $B \subseteq A$ such that $B \cup \{0\}$ is compact and $\overline{d}(A \backslash B) = 0$.*

Proof. For each $n \in \mathbb{N}$, let $A_n = A \cap (1/2^n, 1/2^{n-1})$ and let $T = \{n \in \mathbb{N} : \mu(A_n) > 0\}$. As is well known, given any bounded measurable set C and any $\epsilon > 0$ there is a compact subset D of C with $\mu(D) > \mu(C) - \epsilon$. Thus for each $n \in T$, pick compact $B_n \subseteq A_n$ with $\mu(B_n) > \mu(A_n) - \frac{1}{4^{n+1}}$. For $n \in \mathbb{N} \backslash T$, if any, let $B_n = \emptyset$. Let $B = \bigcup_{n=1}^{\infty} B_n$. Then $B \cup \{0\}$ is compact.

Suppose now that $\overline{d}(A \backslash B) = \alpha > 0$. Pick $m \in \mathbb{N}$ such that $\frac{1}{3 \cdot 2^m} < \alpha$. Pick $x < 1/2^m$ such that $\mu\big((A \backslash B) \cap (0, x)\big)/x > \frac{1}{3 \cdot 2^m}$. Pick $n \in \mathbb{N}$ with $1/2^n \leq x < 1/2^{n-1}$ and note that $n > m$. Then

$$\mu\big((A \backslash B) \cap (0, x)\big) \leq \sum_{k=n}^{\infty} \mu(A_n \backslash B_n) < \sum_{k=n}^{\infty} \frac{1}{4^{k+1}} = \frac{1}{3 \cdot 4^n}$$

and $x \geq \frac{1}{2^n}$ so

$$\mu\big((A \backslash B) \cap (0, x)\big)/x < \frac{1}{3 \cdot 2^n} < \frac{1}{3 \cdot 2^m},$$

a contradiction. $\qquad\square$

Definition 17.44. $\mathcal{L} = \{p \in 0^+ : \text{for all } A \in p, \overline{d}(A) > 0\}$.

Lemma 17.45. $\mathcal{L}$ *is a left ideal of* $(0^+, \cdot)$.

Proof. It is an easy exercise to show that if $\overline{d}(A \cup B) > 0$ then either $\overline{d}(A) > 0$ or $\overline{d}(B) > 0$. Consequently, by Theorem 3.11, it follows that $\{p \in \beta(0, 1)_d : \text{for all } A \in p, \overline{d}(A) > 0\} \neq \emptyset$. On the other hand, if $p \in \beta(0, 1)_d \backslash 0^+$ one has some $\epsilon > 0$ such that $(\epsilon, 1) \in p$ and $\overline{d}\big((\epsilon, 1)\big) = 0$. Thus $\mathcal{L} \neq \emptyset$. Let $p \in \mathcal{L}$ and let $q \in 0^+$. To see that $q \cdot p \in \mathcal{L}$, let $A \in q \cdot p$ and pick x such that $x^{-1}A \in p$. Another easy exercise establishes that $\overline{d}(A) = \overline{d}(x^{-1}A) > 0$. $\qquad\square$

Theorem 17.46. *Let p be a multiplicative idempotent in $\mathcal{L}$ and let A be a measurable member of p. Then $\{x \in A : x^{-1}A \in p \text{ and } -x + A \in p\} \in p$.*

Proof. Let $B = \{x \in A : x^{-1}A \in p\}$. Then $B \in p$ since $p = p \cdot p$. Let $C = \{y \in A : y$ is not a density point of $A\}$. By Theorem 17.42, $\mu(C) = 0$. Consequently since $p \in \mathcal{L}$, $C \notin p$ so $B \backslash C \in p$. We claim that $B \backslash C \subseteq \{x \in A : x^{-1}A \in p \text{ and } -x + A \in p\}$. Indeed, given $x \notin C$ one has 0 is a density point of $-x + A$ so by an easy computation, $\overline{d}\big((0, 1) \backslash (-x + A)\big) = 0$ so $(0, 1) \backslash (-x + A) \notin p$ so $-x + A \in p$. $\qquad\square$

Now let us define the kind of combined additive and multiplicative structures we obtain.

Definition 17.47. Let $\langle x_n \rangle_{n=1}^{\infty}$ be a sequence in $(0, 1)$. We define $\text{FSP}(\langle x_n \rangle_{n=1}^{\infty})$ and $\sigma : \text{FSP}(\langle x_n \rangle_{n=1}^{\infty}) \to \mathcal{P}\big(\mathcal{P}_f(\mathbb{N})\big)$ inductively to consist of only those objects obtainable by iteration of the following:

(1) If $m \in \mathbb{N}$, then $x_m \in \text{FSP}(\langle x_n \rangle_{n=1}^{\infty})$ and $\{m\} \in \sigma(x_m)$.
(2) If $x \in \text{FSP}(\langle x_n \rangle_{n=1}^{\infty})$, $m \in \mathbb{N}$, $F \in \sigma(x)$, and $\min F > m$, then $\{x_m \cdot x, x_m + x\} \subseteq \text{FSP}(\langle x_n \rangle_{n=1}^{\infty})$, $F \cup \{m\} \in \sigma(x_m \cdot x)$, and $F \cup \{m\} \in \sigma(x_m + x)$.

For example, if $z = x_3 + x_5 \cdot x_7 \cdot (x_8 + x_{10} \cdot x_{11})$, then $z \in \text{FSP}(\langle x_n \rangle_{n=1}^{\infty})$ and $\{3, 5, 7, 8, 10, 11\} \in \sigma(z)$. (Of course, it is also possible that $z = x_4 + x_{12} \cdot x_{13}$, in which case also $\{4, 12, 13\} \in \sigma(z)$.) Note also that $(x_3 + x_5) \cdot x_7$ is not, on its face, a member of $\text{FSP}(\langle x_n \rangle_{n=1}^{\infty})$. Notice that trivially $\text{FS}(\langle x_n \rangle_{n=1}^{\infty}) \cup \text{FP}(\langle x_n \rangle_{n=1}^{\infty}) \subseteq \text{FSP}(\langle x_n \rangle_{n=1}^{\infty})$.

The proof of the following theorem is reminiscent of the first proof of Theorem 5.8.

Theorem 17.48. *Let p be a multiplicative idempotent in $\mathcal{B}$ and let A be a Baire set which is a member of p. Then there is a sequence $\langle x_n \rangle_{n=1}^{\infty}$ in $(0, 1)$ such that $\text{FSP}(\langle x_n \rangle_{n=1}^{\infty}) \subseteq A$*

Proof. Let $A_1 = A$. By Theorem 17.40, $\{x \in A_1 : x^{-1}A_1 \in p$ and $-x + A_1 \in p\} \in p$ so pick $x_1 \in A_1$ such that $x_1^{-1}A_1 \in p$ and $-x_1 + A_1 \in p$. Let

$$A_2 = A_1 \cap x_1^{-1}A_1 \cap (-x_1 + A_1).$$

Then $A_2 \in p$ and, since multiplication by x_1^{-1} and addition of $-x_1$ are homeomorphisms, A_2 is a Baire set.

Inductively, let $n \in \mathbb{N} \setminus \{1\}$ and assume that we have chosen $\langle x_t \rangle_{t=1}^{n-1}$ and $\langle A_t \rangle_{t=1}^{n}$ so that A_n is a Baire set which is a member of p. Again invoking Theorem 17.40, one has that $\{x \in A_n : x^{-1}A_n \in p$ and $-x + A_n \in p\} \in p$ so pick $x_n \in A_n$ such that $x_n^{-1}A_n \in p$ and $-x_n + A_n \in p$. Let $A_{n+1} = A_n \cap x_n^{-1}A_n \cap (-x_n + A_n)$. Then A_{n+1} is a Baire set which is a member of p.

The induction being complete, we show that $\text{FSP}(\langle x_n \rangle_{n=1}^{\infty}) \subseteq A$ by establishing the following stronger assertion: If

$$z \in \text{FSP}(\langle x_n \rangle_{n=1}^{\infty}), \; F \in \sigma(z), \text{ and } m = \min F,$$

then $z \in A_m$. Suppose instead that this conclusion fails, and choose

$$z \in \text{FSP}(\langle x_n \rangle_{n=1}^{\infty}), \; F \in \sigma(z), \text{ and } m = \min F$$

such that $z \notin A_m$ and $|F|$ is as small as possible among all such counterexamples. Now if $F = \{m\}$, then $z = x_m \in A_m$, so we must have $|F| > 1$. Let $G = F \setminus \{m\}$ and pick $y \in \text{FSP}(\langle x_n \rangle_{n=1}^{\infty})$ such that $G \in \sigma(y)$ and either $z = x_m + y$ or $z = x_m \cdot y$. Let $r = \min G$. Since $|G| < |F|$, we have that

$$y \in A_r \subseteq A_{m+1} \subseteq x_m^{-1}A_m \cap (-x_m + A_m)$$

and hence $x_m + y \in A_m$ and $x_m \cdot y \in A_m$, contradicting the choice of z. □

Corollary 17.49. *Let $r \in \mathbb{N}$ and let $(0, 1) = \bigcup_{i=1}^{r} A_i$. If each A_i is a Baire set, then there exist $i \in \{1, 2, \ldots, r\}$ and a sequence $\langle x_n \rangle_{n=1}^{\infty}$ in $(0, 1)$ such that $\text{FSP}(\langle x_n \rangle_{n=1}^{\infty}) \subseteq A_i$.*

Proof. By Lemma 17.39, $\mathscr{B}$ is a left ideal of $(0^+, \cdot)$, which is a compact right topological semigroup by Lemma 13.29. Thus, by Corollary 2.6 we may pick an idempotent $p \in \mathscr{B}$. Pick $i \in \{1, 2, \ldots, r\}$ such that $A_i \in p$ and apply Theorem 17.48. □

We have similar, but stronger, results for measurable sets.

Theorem 17.50. *Let p be a multiplicative idempotent in $\mathscr{L}$ and let A be a measurable set which is a member of p. Then there is a sequence $\langle x_n \rangle_{n=1}^{\infty}$ in $(0, 1)$ such that $c\ell \, \text{FSP}(\langle x_n \rangle_{n=1}^{\infty}) \subseteq A \cup \{0\}$*

Proof. Since $A \in p$, $\overline{d}(A) > 0$. By Lemma 17.43, pick $B \subseteq A$ such that $B \cup \{0\}$ is compact and $\overline{d}(A \setminus B) = 0$. Then $A \setminus B \notin p$ so $B \in p$.

Now, proceeding exactly as in the proof of Theorem 17.48, invoking Theorem 17.46 instead of Theorem 17.40, one obtains a sequence $\langle x_n \rangle_{n=1}^{\infty}$ such that $\text{FSP}(\langle x_n \rangle_{n=1}^{\infty}) \subseteq B$ and consequently $c\ell\,\text{FSP}(\langle x_n \rangle_{n=1}^{\infty}) \subseteq B \cup \{0\} \subseteq A \cup \{0\}$.   □

Notice that if $c\ell\,\text{FSP}(\langle x_n \rangle_{n=1}^{\infty}) \subseteq A \cup \{0\}$, then in particular $c\ell\,\text{FS}(\langle x_n \rangle_{n=1}^{\infty}) \subseteq A \cup \{0\}$ and hence, if F is any subset of $\mathbb{N}$, finite or infinite, one has $\Sigma_{n \in F}\, x_n \in A$.

Corollary 17.51. *Let $r \in \mathbb{N}$ and let $(0, 1) = \bigcup_{i=1}^{r} A_i$. If each A_i is a measurable set, then there exist $i \in \{1, 2, \ldots, r\}$ and a sequence $\langle x_n \rangle_{n=1}^{\infty}$ in $(0, 1)$ such that $c\ell\,\text{FSP}(\langle x_n \rangle_{n=1}^{\infty}) \subseteq A_i \cup \{0\}$.*

Proof. By Lemma 17.45, $\mathcal{L}$ is a left ideal of $(0^+, \cdot)$, which is a compact right topological semigroup by Lemma 13.29. Thus, by Corollary 2.6 we may pick an idempotent $p \in \mathcal{L}$. Pick $i \in \{1, 2, \ldots, r\}$ such that $A_i \in p$ and apply Theorem 17.50.   □

Exercise 17.5.1. Let X be a topological space and let $\mathcal{B} = \{U \bigtriangleup M : U$ is open in X and M is meager in $X\}$. Show that $\mathcal{B}$ is the set of Baire sets in X.

Notes

The material in Section 17.1 is from [25], [27], and [31] (results obtained in collaboration with V. Bergelson).

The material in Section 17.2 is from [127]. (In [127], some pains were taken to reduce the number of cells of the partition so that, in lieu of the 4610 cells of the partition we used, only 7 were needed.)

It is a result of R. Graham (presented in [120, Theorem 4.3]) that whenever $\{1, 2, \ldots, 252\} = A_1 \cup A_2$, there must be some $x \neq y$ and some $i \in \{1, 2\}$ with $\{x, y, x + y, x \cdot y\} \subseteq A_i$, and consequently the answer to Question 17.18 is "yes" for $n = r = 2$.

Theorem 17.21 is from [119] in the case $n = 2$. It is shown in [119] that the partition $\mathcal{R}$ can in fact have 3 cells. It is a still unsolved problem of J. Owings [188] as to whether there is a two cell partition of $\mathbb{N}$ such that neither cell contains $\{x_n + x_m : n, m \in \mathbb{N}\}$ for any one-to-one sequence $\langle x_t \rangle_{t=1}^{\infty}$ (where, of course, one specifically does allow $n = m$).

Theorem 17.24 is due to G. Smith [226] and Theorem 17.27 is a special case of a much more general result from [226], namely that *given any $m \neq n$ in $\mathbb{N}$, there is a two cell partition of $\mathbb{N}$ neither cell of which contains both $\text{SP}_n(\langle x_t \rangle_{t=1}^{\infty})$ and $\text{SP}_m(\langle y_t \rangle_{t=1}^{\infty})$ for any sequences $\langle x_t \rangle_{t=1}^{\infty}$ and $\langle y_t \rangle_{t=1}^{\infty}$.*

The results of Section 17.4 are from [79], which is a result of collaboration with W. Deuber, H. Lefmann, and I. Leader. In fact a result much stronger than Theorem 17.35 is proved in [79]. That is, *if $\vec{a}$ and $\vec{b}$ are elements of $\mathbb{A}$, neither of which is a rational multiple of the other, then there is a subset A of $\mathbb{N}$ such that for no sequence $\langle x_t \rangle_{t=1}^{\infty}$ in $\mathbb{N}$, is either $\text{MT}(\vec{a}, \langle x_t \rangle_{t=1}^{\infty}) \subseteq A$ or $\text{MT}(\vec{b}, \langle x_t \rangle_{t=1}^{\infty}) \subseteq \mathbb{N} \backslash A$.*

Most of the material in Section 17.5 is from [35], results obtained in collaboration with V. Bergelson and I. Leader. The proof of the Lebesgue Density Theorem (Theorem

17.42) is from [189]. The idea of considering Baire or measurable partitions arises from research of Plewik, Prömel, and Voigt: Given a sequence $\langle t_n \rangle_{n=1}^{\infty}$ in $(0, 1]$, such that $\Sigma_{n=1}^{\infty} t_n$ converges, define the set of *all sums* of the sequence by $AS(\langle t_n \rangle_{n=1}^{\infty}) = \{\Sigma_{n \in F} t_n : \emptyset \neq F \subseteq \mathbb{N}\}$. In [197], Prömel and Voigt considered the question: If $(0, 1] = \bigcup_{i=1}^{r} A_i$, must there exist $i \in \{1, 2, \ldots, r\}$ and a sequence $\langle t_n \rangle_{n=1}^{\infty}$ in $(0, 1]$ with $AS(\langle t_n \rangle_{n=1}^{\infty}) \subseteq A_i$? As they pointed out, one easily sees (using the Axiom of Choice) that the answer is "no" by a standard diagonalization argument. They showed, however that if one adds the requirement that each A_i has the property of Baire, then the answer becomes "yes". In [195] Plewik and Voigt reached the same conclusion in the event that each A_i is assumed to be Lebesgue measurable. A unified and simplified proof of both results was presented in [36], a result of collaboration with V. Bergelson and B. Weiss.

Chapter 18

Multidimensional Ramsey Theory

Several results in Ramsey Theory, including Ramsey's Theorem itself, naturally apply to more than one "dimension", suitably interpreted. That is, while van der Waerden's Theorem and the Finite Sums Theorem, for example, deal with colorings of the elements of $\mathbb{N}$, Ramsey's Theorem deals with colorings of finite subsets of $\mathbb{N}$, which can be identified with points in Cartesian products.

18.1 Ramsey's Theorem and Generalizations

In this section we present a proof of Ramsey's Theorem which utilizes an arbitrary nonprincipal ultrafilter. Then we adapt that proof to obtain some generalizations, including the Milliken–Taylor Theorem. The main feature of the adaptation is that we utilize ultrafilters with special properties. (For example, to obtain the Milliken–Taylor Theorem, we use an idempotent in place of the arbitrary nonprincipal ultrafilter.)

While many of the applications are algebraic, the basic tools are purely set theoretic.

Lemma 18.1. *Let S be a set, let $p \in S^*$, let $k, r \in \mathbb{N}$, and let $[S]^k = \bigcup_{i=1}^r A_i$. For each $i \in \{1, 2, \ldots, r\}$, each $t \in \{1, 2, \ldots, k\}$, and each $E \in [S]^{t-1}$, define $B_t(E, i)$ by downward induction on t:*

(1) *For $E \in [S]^{k-1}$, $B_k(E, i) = \{y \in S \backslash E : E \cup \{y\} \in A_i\}$.*
(2) *For $t \in \{1, 2, \ldots, k - 1\}$ and $E \in [S]^{t-1}$,*

$$B_t(E, i) = \{y \in S \backslash E : B_{t+1}(E \cup \{y\}, i) \in p\}.$$

Then for each $t \in \{1, 2, \ldots, k\}$ and each $E \in [S]^{t-1}$, $S \backslash E = \bigcup_{i=1}^r B_t(E, i)$.

Proof. We proceed by downward induction on t. If $t = k$, then for each $y \in S \backslash E$, $E \cup \{y\} \in A_i$ for some i.

So let $t \in \{1, 2, \ldots, k - 1\}$ and let $E \in [S]^{t-1}$. Then given $y \in S \backslash E$, one has by the induction hypothesis that $S \backslash (E \cup \{y\}) = \bigcup_{i=1}^r B_{t+1}(E \cup \{y\}, i)$ so for some i, $B_{t+1}(E \cup \{y\}, i) \in p$. □

Theorem 18.2 (Ramsey's Theorem). *Let S be an infinite set and let $k, r \in \mathbb{N}$. If $[S]^k = \bigcup_{i=1}^r A_i$, then there exist $i \in \{1, 2, \ldots, r\}$ and an infinite subset C of S with $[C]^k \subseteq A_i$.*

Proof. If $k = 1$, this is just the pigeon hole principle, so assume that $k \geq 2$.

Let p be any nonprincipal ultrafilter on S. Define $B_t(E, i)$ as in the statement of Lemma 18.1. Then $S = \bigcup_{i=1}^r B_1(\emptyset, i)$ so pick $i \in \{1, 2, \ldots, r\}$ such that $B_1(\emptyset, i) \in p$. Pick $x_1 \in B_1(\emptyset, i)$ (so that $B_2(\{x_1\}, i) \in p$.

Inductively, let $n \in \mathbb{N}$ and assume that $\langle x_m \rangle_{m=1}^n$ has been chosen so that whenever $t \in \{1, 2, \ldots, k-1\}$ and $m_1 < m_2 < \cdots < m_t \leq n$, one has

$$B_{t+1}(\{x_{m_1}, x_{m_2}, \ldots, x_{m_t}\}, i) \in p.$$

Choose

$$x_{n+1} \in (B_1(\emptyset, i) \setminus \{x_1, x_2, \ldots, x_n\})$$
$$\cap \bigcap \{B_{t+1}(\{x_{m_1}, x_{m_2}, \ldots, x_{m_t}\}, i) : t \in \{1, 2, \ldots, k-1\}$$
$$\text{and } m_1 < m_2 < \cdots < m_t \leq n\}.$$

To see that $B_{t+1}(\{x_{m_1}, x_{m_2}, \ldots, x_{m_t}\}, i) \in p$ whenever $t \in \{1, 2, \ldots, k-1\}$ and $m_1 < m_2 < \cdots < m_t \leq n + 1$, let such t and $m_1, m_2, \ldots, m_t$ be given. If $m_t \leq n$, then the induction hypothesis applies, so assume that $m_t = n + 1$. If $t = 1$, the conclusion holds because $x_{n+1} \in B_1(\emptyset, i)$. If $t > 1$, the conclusion holds because $x_{n+1} \in B_t(\{x_{m_1}, x_{m_2}, \ldots, x_{m_{t-1}}\}, i)$.

The sequence $\langle x_n \rangle_{n=1}^\infty$ having been chosen, let $m_1 < m_2 < \cdots < m_k$. Then $x_{m_k} \in B_k(\{x_{m_1}, x_{m_2}, \ldots, x_{m_{k-1}}\}, i)$ so $\{x_{m_1}, x_{m_2}, \ldots, x_{m_k}\} \in A_i$. $\square$

In order to establish some generalizations of Ramsey's Theorem, we introduce some special notation.

Definition 18.3. Let S be a set, let $k \in \mathbb{N}$, and let $\varphi : S \to \mathbb{N}$.

(a) If $\langle D_n \rangle_{n=1}^\infty$ is a sequence of subsets of S, then

$$[\langle D_n \rangle_{n=1}^\infty, \varphi]_<^k = \big\{ \{x_1, x_2, \ldots, x_k\} : x_i \neq x_j \text{ for } i \neq j \text{ in } \{1, 2, \ldots, k\}$$
$$\text{and there exist } m_1 < m_2 < \cdots < m_k \text{ in } \mathbb{N} \text{ such that}$$
$$\text{for each } j \in \{1, 2, \ldots, k\}, x_j \in D_{m_j} \text{ and for each}$$
$$j \in \{2, 3, \ldots, k\}, m_j > \varphi(x_{j-1})\big\}.$$

(b) If $t \in \mathbb{N}$ and $\langle D_n \rangle_{n=1}^t$ is a sequence of subsets of S, then

$$[\langle D_n \rangle_{n=1}^t, \varphi]_<^k = \big\{ \{x_1, x_2, \ldots, x_k\} : x_i \neq x_j \text{ for } i \neq j \text{ in } \{1, 2, \ldots, k\}$$
$$\text{and there exist } m_1 < m_2 < \cdots < m_k \text{ in } \{1, 2, \ldots, t\}$$
$$\text{such that for each } j \in \{1, 2, \ldots, k\}, x_j \in D_{m_j} \text{ and}$$
$$\text{for each } j \in \{2, 3, \ldots, k\}, m_j > \varphi(x_{j-1})\big\}.$$

For example, if $S = \mathbb{N}$, $\varphi(n) = n$, and $D_n = \{m \in \mathbb{N} : m \geq n\}$ for each $n \in \mathbb{N}$, then

$$[\langle D_n \rangle_{n=1}^\infty, \varphi]_<^k = \big\{\{x_1, x_2, \ldots, x_k\} \subseteq \mathbb{N} : x_1 < x_2 < \cdots < x_k\big\} = [\mathbb{N}]^k.$$

Theorem 18.4. *Let S be an infinite set, let $p \in S^*$, let $k, r \in \mathbb{N}$, and assume that $[S]^k = \bigcup_{i=1}^{r} A_i$. Let $\varphi : S \to \mathbb{N}$ and assume that there is some $C \in p$ on which φ is finite-to-one. Then there exist $i \in \{1, 2, \ldots, r\}$ and a sequence $\langle D_n \rangle_{n=1}^{\infty}$ of members of p such that $D_{n+1} \subseteq D_n$ for each n and $[\langle D_n \rangle_{n=1}^{\infty}, \varphi]_{<}^{k} \subseteq A_i$.*

Proof. First assume that $k = 1$. Pick $i \in \{1, 2, \ldots, r\}$ such that

$$E = \{x \in S : \{x\} \in A_i\} \in p$$

and let $D_n = E$ for each n. Then $[\langle D_n \rangle_{n=1}^{\infty}, \varphi]_{<}^{1} = A_i$.

Now assume that $k > 1$ and let $B_t(E, i)$ be defined for each $i \in \{1, 2, \ldots, r\}$, each $t \in \{1, 2, \ldots, k\}$, and each $E \in [S]^{t-1}$ as in Lemma 18.1. Then by Lemma 18.1, $S = \bigcup_{i=1}^{r} B_1(\emptyset, i)$, so pick $i \in \{1, 2, \ldots, r\}$ such that $B_1(\emptyset, i) \in p$.

For each $\ell \in \mathbb{N}$, let $H_\ell = \{x \in S : \varphi(x) \leq \ell\}$. Then by assumption, for each ℓ, $H_\ell \cap C$ is finite. Let $D_1 = C \cap B_1(\emptyset, i)$. Inductively let $n \in \mathbb{N}$ and assume that we have chosen $\langle D_m \rangle_{m=1}^{n}$ in p such that for each $m \in \{1, 2, \ldots, n\}$:

(1) If $m < n$, then $D_{m+1} \subseteq D_m$.

(2) If $m > 1$, then for each $t \in \{1, 2, \ldots, k-1\}$ and each $E \in [\langle D_j \cap H_{m-1} \rangle_{j=1}^{m-1}, \varphi]_{<}^{t}$, $D_m \subseteq B_{t+1}(E, i)$.

Notice that both hypotheses are vacuous when $n = 1$.

We now claim that if $t \in \{1, 2, \ldots, k-1\}$ and $E \in [\langle D_j \cap H_n \rangle_{j=1}^{n}, \varphi]_{<}^{t}$, then $B_{t+1}(E, i) \in p$. So let such t and E be given. Then there exist $x_1, x_2, \ldots, x_t$ in S and $m_1 < m_2 < \cdots < m_t$ in $\{1, 2, \ldots, n\}$ such that $E = \{x_1, x_2, \ldots, x_t\}$, $x_j \in D_{m_j} \cap H_n$ for each $j \in \{1, 2, \ldots, t\}$, and for each $j \in \{2, 3, \ldots, t\}$, $m_j > \varphi(x_{j-1})$.

If $t = 1$, we have $x_1 \in D_{m_1} \subseteq D_1 \subseteq B_1(\emptyset, i)$, so $B_2(\{x_1\}, i) \in p$. So now assume that $t > 1$ and let $F = \{x_1, x_2, \ldots, x_{t-1}\}$. We claim that $F \in [\langle D_j \cap H_{m_t-1} \rangle_{j=1}^{m_t-1}, \varphi]_{<}^{t-1}$. Indeed, one only needs to verify that for each $j \in \{1, 2, \ldots, t-1\}$, $x_j \in H_{m_t-1}$. To see this, notice that $\varphi(x_j) < m_{j+1} \leq m_t$.

Since $F \in [\langle D_j \cap H_{m_t-1} \rangle_{j=1}^{m_t-1}, \varphi]_{<}^{t-1}$, we have by hypothesis (2) that

$$x_t \in D_{m_t} \subseteq B_t(F, i)$$

so that $B_{t+1}(E, i) \in p$ as required.

Now for each $t \in \{1, 2, \ldots, k-1\}$, $[\langle D_j \cap H_n \rangle_{j=1}^{n}, \varphi]_{<}^{t}$ is finite (because $C \cap H_n$ is finite), so we may choose $D_{n+1} \in p$ such that $D_{n+1} \subseteq D_n$ and for each $t \in \{1, 2, \ldots, k-1\}$ and for each $E \in [\langle D_j \cap H_n \rangle_{j=1}^{n}, \varphi]_{<}^{t}$, $D_{n+1} \subseteq B_{t+1}(E, i)$.

The construction being complete, we now only need to show that $[\langle D_n \rangle_{n=1}^{\infty}, \varphi]_{<}^{k} \subseteq A_i$. So let $E \in [\langle D_n \rangle_{n=1}^{\infty}, \varphi]_{<}^{k}$ and pick $x_1, x_2, \ldots, x_k$, and $m_1 < m_2 < \cdots < m_k$ such that for each $j \in \{1, 2, \ldots, k\}$, $x_j \in D_{m_j}$, and for each $j \in \{2, 3, \ldots, k\}$, $m_j > \varphi(x_{j-1})$, and $E = \{x_1, x_2, \ldots, x_k\}$.

Let $F = \{x_1, x_2, \ldots, x_{k-1}\}$. Then for each $j \in \{1, 2, \ldots, k-1\}$, $\varphi(x_j) < m_{j+1} \leq m_k$ so $F \in [\langle D_j \cap H_{m_k-1} \rangle_{j=1}^{m_k-1}, \varphi]_{<}^{k-1}$ so $x_k \in D_{m_k} \subseteq B_k(F, i)$, so $E \in A_i$ as required. $\square$

As the first application of Theorem 18.4, consider the following theorem which is a common generalization of Ramsey's Theorem and van der Waerden's Theorem (Corollary 14.2). To prove this generalization, we use an ultrafilter which has the property that every member contains an arbitrarily long arithmetic progression.

Theorem 18.5. *Let $k, r \in \mathbb{N}$ and let $[\mathbb{N}]^k = \bigcup_{i=1}^{r} A_i$. Then there exist sequences $\langle a_n \rangle_{n=1}^{\infty}$ and $\langle d_n \rangle_{n=1}^{\infty}$ in $\mathbb{N}$ such that whenever $n_1 < n_2 < \cdots < n_k$ and for each $j \in \{1, 2, \ldots, k\}$, $t_j \in \{0, 1, \ldots, n_j\}$, one has*

$$\{a_{n_1} + t_1 d_{n_1}, a_{n_2} + t_2 d_{n_2}, \ldots, a_{n_k} + t_k d_{n_k}\} \in A_i.$$

Proof. Let $\varphi : \mathbb{N} \to \mathbb{N}$ be the identity map and pick $p \in \mathcal{AP}$ (which is nonempty by Theorem 14.5). Pick $i \in \{1, 2, \ldots, r\}$ and a sequence $\langle D_n \rangle_{n=1}^{\infty}$ of members of p as guaranteed by Theorem 18.4.

Pick a_1 and d_1 such that $\{a_1, a_1 + d_1\} \subseteq D_1$ and let $\ell(1) = 1$. Inductively assume that $\langle a_m \rangle_{m=1}^{n}$, $\langle d_m \rangle_{m=1}^{n}$, and $\langle \ell(m) \rangle_{m=1}^{n}$ have been chosen so that for each $m \in \{1, 2, \ldots, n\}$, $\{a_m, a_m + d_m, a_m + 2d_m, \ldots, a_m + m d_m\} \subseteq D_{\ell(m)}$. Let $\ell(n+1) = \max\{\ell(n)+1, a_n + n d_n + 1\}$ and pick a_{n+1} and d_{n+1} such that

$$\{a_{n+1}, a_{n+1} + d_{n+1}, a_{n+1} + 2d_{n+1}, \ldots, a_{n+1} + (n+1)d_{n+1}\} \subseteq D_{\ell(n+1)}.$$

Assume now that $n_1 < n_2 < \cdots < n_k$ and for each $j \in \{1, 2, \ldots, k\}$ $t_j \in \{0, 1, \ldots, n_j\}$. Then

$$\{a_{n_1} + t_1 d_{n_1}, a_{n_2} + t_2 d_{n_2}, \ldots, a_{n_k} + t_k d_{n_k}\} \in [\langle D_n \rangle_{n=1}^{\infty}, \varphi]_<^k \subseteq A_i. \qquad \square$$

The case $k = 2$ of Theorem 18.5 is illustrated in Figure 18.1, where pairs of integers are identified with points below the diagonal. In this figure arithmetic progressions of length 2, 3, and 4 are indicated and the points which are guaranteed to be monochrome are circled. It is the content of Exercise 18.1.1 to show that Theorem 18.5 cannot be extended in the case $k = 2$ to require that pairs from the same arithmetic progression be included.

A significant generalization of Ramsey's Theorem, which has often been applied to obtain other results, is the Milliken–Taylor Theorem. To state it, we introduce some notation similar to that introduced in Definition 18.3. We state it in three forms, depending on whether the operation of the semigroup is written as "+", "$\cdot$", or "$\cup$".

Definition 18.6. (a) Let $(S, +)$ be a semigroup, let $\langle x_n \rangle_{n=1}^{\infty}$ be a sequence in S, and let $k \in \mathbb{N}$. Then

$$[FS(\langle x_n \rangle_{n=1}^{\infty})]_<^k = \big\{ \{\textstyle\sum_{t \in H_1} x_t, \sum_{t \in H_2} x_t, \ldots, \sum_{t \in H_k} x_t\} :$$
$$\text{for each } j \in \{1, 2, \ldots, k\},\ H_t \in \mathcal{P}_f(\mathbb{N})$$
$$\text{and if } j < k, \text{ then } \max H_j < \min H_{j+1}\big\}.$$

(b) Let $(S, \cdot)$ be a semigroup, let $\langle x_n \rangle_{n=1}^{\infty}$ be a sequence in S, and let $k \in \mathbb{N}$. Then

$$[FP(\langle x_n \rangle_{n=1}^{\infty})]_<^k = \big\{ \{\textstyle\prod_{t \in H_1} x_t, \prod_{t \in H_2} x_t, \ldots, \prod_{t \in H_k} x_t\} :$$
$$\text{for each } j \in \{1, 2, \ldots, k\},\ H_t \in \mathcal{P}_f(\mathbb{N})$$
$$\text{and if } j < k, \text{ then } \max H_j < \min H_{j+1}\big\}.$$

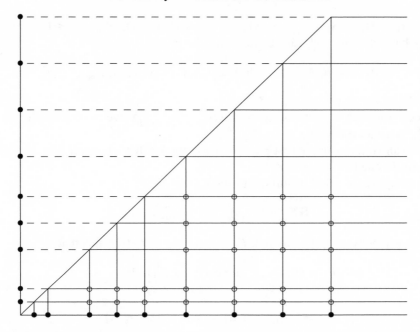

Figure 18.1: Monochrome Products of Arithmetic Progressions

(c) Let $\langle F_n \rangle_{n=1}^{\infty}$ be a sequence in the semigroup $(\mathcal{P}_f(\mathbb{N}), \cup)$ and let $k \in \mathbb{N}$. Then

$$[\text{FU}(\langle F_n \rangle_{n=1}^{\infty})]_<^k = \big\{ \{\textstyle\bigcup_{t \in H_1} F_t, \bigcup_{t \in H_2} F_t, \ldots, \bigcup_{t \in H_k} F_t\} :$$
$$\text{for each } j \in \{1, 2, \ldots, k\},\ H_t \in \mathcal{P}_f(\mathbb{N})$$
$$\text{and if } j < k, \text{ then } \max H_j < \min H_{j+1} \big\}.$$

The Milliken–Taylor Theorem is proved by replacing the arbitrary nonprincipal ultrafilter used in the proof of Theorem 18.2 by an idempotent.

Theorem 18.7 (Milliken–Taylor Theorem — Version 1). *Let $k, r \in \mathbb{N}$, let $\langle x_n \rangle_{n=1}^{\infty}$ be a sequence in $\mathbb{N}$, and assume that $[\mathbb{N}]^k = \bigcup_{i=1}^{r} A_i$. Then there exist $i \in \{1, 2, \ldots, r\}$ and a sum subsystem $\langle y_n \rangle_{n=1}^{\infty}$ of $\langle x_n \rangle_{n=1}^{\infty}$ such that*

$$[\text{FS}(\langle y_n \rangle_{n=1}^{\infty})]_<^k \subseteq A_i.$$

Proof. By passing to a suitable sum subsystem if necessary, we may assume that $\langle x_n \rangle_{n=1}^{\infty}$ satisfies uniqueness of finite sums. (That is, if $F, G \in \mathcal{P}_f(\mathbb{N})$ and $\Sigma_{n \in F} x_n = \Sigma_{n \in G} x_n$, then $F = G$.) Define $\varphi : \mathbb{N} \to \mathbb{N}$ so that for each $F \in \mathcal{P}_f(\mathbb{N})$, $\varphi(\Sigma_{n \in F} x_n) = \max F$, defining φ arbitrarily on $\mathbb{N} \setminus \text{FS}(\langle x_n \rangle_{n=1}^{\infty})$. Notice that φ is finite-to-one on $\text{FS}(\langle x_n \rangle_{n=1}^{\infty})$.

Pick by Lemma 5.11 an idempotent $p \in \beta\mathbb{N}$ such that $\text{FS}(\langle x_m \rangle_{m=n}^{\infty}) \in p$ for every $n \in \mathbb{N}$. Pick by Theorem 18.4 some $i \in \{1, 2, \ldots, r\}$ and a sequence $\langle D_n \rangle_{n=1}^{\infty}$ of members of p such that $D_{n+1} \subseteq D_n$ for each n and $[\langle D_n \rangle_{n=1}^{\infty}, \varphi]_<^k \subseteq A_i$.

Let $B_1 = D_1 \cap \mathrm{FS}(\langle x_m \rangle_{m=1}^\infty)$ and pick $y_1 \in B_1{}^\star$. Inductively, let $n \in \mathbb{N}$ and assume that we have chosen $\langle y_t \rangle_{t=1}^n$ in $\mathbb{N}$ and $\langle B_t \rangle_{t=1}^n$ in p. Let

$$B_{n+1} = B_n \cap (-y_n + B_n) \cap D_{\varphi(y_n)+1} \cap \mathrm{FS}(\langle x_m \rangle_{m=\varphi(y_n)+1}^\infty)$$

and pick $y_{n+1} \in B_{n+1}{}^\star$.

Since, for each $n \in \mathbb{N}$, $y_{n+1} \in \mathrm{FS}(\langle x_m \rangle_{m=\varphi(y_n)+1}^\infty)$, we have immediately that $\langle y_n \rangle_{n=1}^\infty$ is a sum subsystem of $\langle x_n \rangle_{n=1}^\infty$.

One can verify as in the first proof of Theorem 5.8 that whenever $F \in \mathcal{P}_f(\mathbb{N})$ and $m = \min F$, one has $\sum_{n \in F} y_n \in B_m$.

To complete the proof, it suffices to show that

$$[\mathrm{FS}(\langle y_n \rangle_{n=1}^\infty)]_<^k \subseteq [\langle D_n \rangle_{n=1}^\infty, \varphi]_<^k.$$

So, let $a \in [\mathrm{FS}(\langle y_n \rangle_{n=1}^\infty)]_<^k$ and pick $F_1, F_2, \ldots, F_k \in \mathcal{P}_f(\mathbb{N})$ such that

$$\max F_j < \min F_{j+1} \quad \text{for each} \quad j \in \{1, 2, \ldots, k-1\}$$

and

$$a = \{\textstyle\sum_{t \in F_1} y_t, \sum_{t \in F_2} y_t, \ldots, \sum_{t \in F_k} y_t\}.$$

For each $j \in \{1, 2, \ldots, k\}$, let $t_j = \min F_j$, let $\ell_j = \max F_j$, and let $m_j = \varphi(y_{t_j-1}) + 1$. Then $\sum_{t \in F_j} y_t \in B_{t_j} \subseteq D_{m_j}$. Further, if $j \in \{2, 3, \ldots, k\}$, then

$$\varphi(\textstyle\sum_{t \in F_{j-1}} y_t) = \varphi(y_{\ell_{j-1}}) \le \varphi(y_{t_j-1}) < m_j. \qquad \square$$

Corollary 18.8 (Milliken–Taylor Theorem — Version 2). *Let $k, r \in \mathbb{N}$ and let $[\mathcal{P}_f(\mathbb{N})]^k = \bigcup_{i=1}^r A_i$. Then there exist $i \in \{1, 2, \ldots, r\}$ and a sequence $\langle F_n \rangle_{n=1}^\infty$ in $\mathcal{P}_f(\mathbb{N})$ such that $\max F_n < \min F_{n+1}$ for all n and $[\mathrm{FU}(\langle F_n \rangle_{n=1}^\infty)]_<^k \subseteq A_i$.*

Proof. For each $i \in \{1, 2, \ldots, r\}$, let

$$B_i = \big\{\{\textstyle\sum_{t \in H_1} 2^{t-1}, \sum_{t \in H_2} 2^{t-1}, \ldots, \sum_{t \in H_k} 2^{t-1}\} : \{H_1, H_2, \ldots, H_k\} \in A_i\big\}.$$

Then $[\mathbb{N}]^k = \bigcup_{i=1}^r B_i$ so pick by Theorem 18.7, $i \in \{1, 2, \ldots, r\}$ and a sum subsystem $\langle y_n \rangle_{n=1}^\infty$ of $\langle 2^{n-1} \rangle_{n=1}^\infty$ such that $[\mathrm{FS}(\langle y_n \rangle_{n=1}^\infty)]_<^k \subseteq B_i$. Since $\langle y_n \rangle_{n=1}^\infty$ is a sum subsystem of $\langle 2^{n-1} \rangle_{n=1}^\infty$, pick for each n, some $F_n \in \mathcal{P}_f(\mathbb{N})$ such that $y_n = \sum_{t \in F_n} 2^{t-1}$ and $\max F_n < \min F_{n+1}$.

To see that $[\mathrm{FU}(\langle F_n \rangle_{n=1}^\infty)]_<^k \subseteq A_i$, let $H_1, H_2, \ldots, H_k \in \mathcal{P}_f(\mathbb{N})$ with $\max H_j < \min H_{j+1}$ for each $j \in \{1, 2, \ldots, k-1\}$. Then

$$\{\textstyle\sum_{t \in H_1} y_t, \sum_{t \in H_2} y_t, \ldots, \sum_{t \in H_k} y_t\} \in [\mathrm{FS}(\langle y_n \rangle_{n=1}^\infty)]_<^k \subseteq B_i$$

so there exist $\{K_1, K_2, \ldots, K_k\} \in A_i$ such that

$$\{\textstyle\sum_{t \in H_1} y_t, \sum_{t \in H_2} y_t, \ldots, \sum_{t \in H_k} y_t\} = \{\textstyle\sum_{t \in K_1} 2^{t-1}, \sum_{t \in K_2} 2^{t-1}, \ldots, \sum_{t \in K_k} 2^{t-1}\}.$$

Now, given $j \in \{1, 2, \ldots, k\}$, let $G_j = \bigcup_{n \in H_j} F_n$. Then $\Sigma_{t \in H_j} y_t = \Sigma_{t \in G_j} 2^{t-1}$ and if $j < k$, then $\max G_j < \min G_{j+1}$. Thus

$$\{\Sigma_{t \in K_1} 2^{t-1}, \Sigma_{t \in K_2} 2^{t-1}, \ldots, \Sigma_{t \in K_k} 2^{t-1}\}$$
$$= \{\Sigma_{t \in G_1} 2^{t-1}, \Sigma_{t \in G_2} 2^{t-1}, \ldots, \Sigma_{t \in G_k} 2^{t-1}\}$$

so $\{G_1, G_2, \ldots, G_k\} = \{K_1, K_2, \ldots, K_k\}$ (although, we don't know for example that $G_1 = K_1$) so that $\{G_1, G_2, \ldots, G_k\} \in A_i$ as required. $\square$

It is interesting to note that we obtain the following version of the Milliken–Taylor Theorem (which trivially implies the first two versions) as a corollary to Corollary 18.8, which was in turn a corollary to Theorem 18.7, but cannot prove it using the straight forward modification of the proof of Theorem 18.7 which simply replaces sums by products. The reason is that the function φ may not be definable if the sequence $\langle x_n \rangle_{n=1}^{\infty}$ in the semigroup $(S, \cdot)$ does not have a product subsystem which satisfies uniqueness of finite products.

Corollary 18.9 (Milliken–Taylor Theorem — Version 3). *Let $k, r \in \mathbb{N}$, let $(S, \cdot)$ be a semigroup, let $\langle x_n \rangle_{n=1}^{\infty}$ be a sequence in S, and assume that $[S]^k = \bigcup_{i=1}^r A_i$. Then there exist $i \in \{1, 2, \ldots, r\}$ and a product subsystem $\langle y_n \rangle_{n=1}^{\infty}$ of $\langle x_n \rangle_{n=1}^{\infty}$ such that $[FP(\langle y_n \rangle_{n=1}^{\infty})]_<^k \subseteq A_i$.*

Proof. For each $i \in \{1, 2, \ldots, r\}$, let

$$B_i = \left\{\{F_1, F_2, \ldots, F_k\} \in [\mathscr{P}_f(\mathbb{N})]^k : \{\Pi_{t \in F_1} x_t, \Pi_{t \in F_2} x_t, \ldots, \Pi_{t \in F_k} x_t\} \in A_i\right\}.$$

Then $[\mathscr{P}_f(\mathbb{N})]^k = \bigcup_{i=1}^r B_i$ so pick by Corollary 18.8 $i \in \{1, 2, \ldots, r\}$ and a sequence $\langle F_n \rangle_{n=1}^{\infty}$ in $\mathscr{P}_f(\mathbb{N})$ such that $\max F_n < \min F_{n+1}$ for all n and $[FU(\langle F_n \rangle_{n=1}^{\infty})]_<^k \subseteq B_i$.

For each $n \in \mathbb{N}$, let $y_n = \Pi_{t \in F_n} x_t$. Then $[FP(\langle y_n \rangle_{n=1}^{\infty})]_<^k \subseteq A_i$. (The verification of this assertion is Exercise 18.1.2.) $\square$

Exercise 18.1.1. Let

$$A_1 = \{\{x, y\} \in [\mathbb{N}]^2 : x < y < 2x\}$$

and let

$$A_2 = \{\{x, y\} \in [\mathbb{N}]^2 : y \geq 2x\}.$$

Then $[\mathbb{N}]^2 = A_1 \cup A_2$. Prove that there do not exist $i \in \{1, 2\}$ and sequences $\langle a_n \rangle_{n=1}^{\infty}$ and $\langle d_n \rangle_{n=1}^{\infty}$ in $\mathbb{N}$ such that

(1) whenever $n_1 < n_2$ and for each $j \in \{1, 2\}$ and $t_j \in \{1, 2, \ldots, n_j\}$ one has $\{a_{n_1} + t_1 d_{n_1}, a_{n_2} + t_2 d_{n_2}\} \in A_i$ and

(2) whenever $n \in \mathbb{N}$ and $1 \leq t < s \leq n$ one has $\{a_n + t d_n, a_n + s d_n\} \in A_i$.

Exercise 18.1.2. Complete the proof of Corollary 18.9 by verifying that

$$[FP(\langle y_n \rangle_{n=1}^{\infty})]_<^k \subseteq A_i.$$

Exercise 18.1.3. Apply Theorem 18.4 using a minimal idempotent in a countable semigroup S. See what kind of statements you can come up with. Do the same thing using a combinatorially rich ultrafilter in $\beta \mathbb{N}$. (See Section 17.1.) (This is an open ended exercise.)

18.2 IP* Sets in Product Spaces

We show in this section that IP* sets in any finite product of commutative semigroups contain products of finite subsystems of sequences in the coordinates. We first extend the notion of product subsystems to finite sequences.

Definition 18.10. Let $\langle y_n \rangle_{n=1}^{\infty}$ be a sequence in a semigroup S and let $k, m \in \mathbb{N}$. Then $\langle x_n \rangle_{n=1}^{m}$ is a *product subsystem* of $FP(\langle y_n \rangle_{n=k}^{\infty})$ if and only if there exists a sequence $\langle H_n \rangle_{n=1}^{m}$ in $\mathcal{P}_f(\mathbb{N})$ such that

(a) $\min H_1 \geq k$,

(b) $\max H_n < \min H_{n+1}$ for each $n \in \{1, 2, \ldots, m-1\}$, and

(c) $x_n = \prod_{t \in H_n} y_t$ for each $n \in \{1, 2, \ldots, m\}$.

The proof of the following theorem is analogous to the proof of Theorem 18.4.

Theorem 18.11. *Let $\ell \in \mathbb{N} \setminus \{1\}$ and for each $i \in \{1, 2, \ldots, \ell\}$ let S_i be a semigroup and let $\langle y_{i,n} \rangle_{n=1}^{\infty}$ be a sequence in S_i. Let $m, r \in \mathbb{N}$ and let $\times_{i=1}^{\ell} S_i = \bigcup_{j=1}^{r} C_j$. There exists $j \in \{1, 2, \ldots, r\}$ and for each $i \in \{1, 2, \ldots, \ell - 1\}$, there exists a product subsystem $\langle x_{i,n} \rangle_{n=1}^{m}$ of $FP(\langle y_{i,n} \rangle_{n=1}^{\infty})$ and there exists a product subsystem $\langle x_{\ell,n} \rangle_{n=1}^{\infty}$ of $FP(\langle y_{\ell,n} \rangle_{n=1}^{\infty})$ such that*

$$\left(\times_{i=1}^{\ell-1} FP(\langle x_{i,n} \rangle_{n=1}^{m}) \right) \times FP(\langle x_{\ell,n} \rangle_{n=1}^{\infty}) \subseteq C_j.$$

Proof. Given $i \in \{1, 2, \ldots, \ell\}$, pick by Lemma 5.11 an idempotent p_i with

$$p_i \in \bigcap_{k=1}^{\infty} c\ell_{\beta S_i} FP(\langle y_{i,n} \rangle_{n=k}^{\infty}).$$

For $(x_1, x_2, \ldots, x_{\ell-1}) \in \times_{i=1}^{\ell-1} S_i$ and $j \in \{1, 2, \ldots, r\}$, let

$$B_\ell(x_1, x_2, \ldots, x_{\ell-1}, j) = \{y \in S_\ell : (x_1, x_2, \ldots, x_{\ell-1}, y) \in C_j\}.$$

Now given $t \in \{2, 3, \ldots, \ell - 1\}$, assume that $B_{t+1}(x_1, x_2, \ldots, x_t, j)$ has been defined for each $(x_1, x_2, \ldots, x_t) \in \times_{i=1}^{t} S_i$ and each $j \in \{1, 2, \ldots, r\}$. Given

$$(x_1, x_2, \ldots, x_{t-1}) \in \times_{i=1}^{t-1} S_i \quad \text{and} \quad j \in \{1, 2, \ldots, r\},$$

let

$$B_t(x_1, x_2, \ldots, x_{t-1}, j) = \{y \in S_t : B_{t+1}(x_1, x_2, \ldots, x_{t-1}, y, j) \in p_{t+1}\}.$$

Finally, given that $B_2(x, j)$ has been defined for each $x \in S_1$ and each $j \in \{1, 2, \ldots, r\}$, let $B_1(j) = \{x \in S_1 : B_2(x, j) \in p_2\}$.

We show by downward induction on t that for each $t \in \{2, 3, \ldots, \ell\}$ and each $(x_1, x_2, \ldots, x_{t-1}) \in \times_{i=1}^{t-1} S_i$, $S_t = \bigcup_{j=1}^{r} B_t(x_1, x_2, \ldots, x_{t-1}, j)$. This is trivially true for $t = \ell$. Assume $t \in \{2, 3, \ldots, \ell - 1\}$ and the statement is true for $t + 1$. Let $(x_1, x_2, \ldots, x_{t-1}) \in \times_{i=1}^{t-1} S_i$. Given $y \in S_t$, one has that

$$S_{t+1} = \bigcup_{j=1}^{r} B_{t+1}(x_1, x_2, \ldots, x_{t-1}, y, j)$$

so one may pick $j \in \{1, 2, \ldots, r\}$ such that $B_{t+1}(x_1, x_2, \ldots, x_{t-1}, y, j) \in p_{t+1}$. Then $y \in B_t(x_1, x_2, \ldots, x_{t-1}, j)$.

Since for each $x \in S_1$, $S_2 = \bigcup_{j=1}^{r} B_2(x, j)$, one sees similarly that $S_1 = \bigcup_{j=1}^{r} B_1(j)$. Pick $j \in \{1, 2, \ldots, r\}$ such that $B_1(j) \in p_1$.

Pick by Theorem 5.14 a product subsystem $\langle x_{1,n} \rangle_{n=1}^{\infty}$ of $\mathrm{FP}(\langle y_{1,n} \rangle_{n=1}^{\infty})$ such that $\mathrm{FP}(\langle x_{1,n} \rangle_{n=1}^{\infty}) \subseteq B_1(j)$. Let $D_2 = \bigcap\{B_2(a, j) : a \in \mathrm{FP}(\langle x_{1,n} \rangle_{n=1}^{m})\}$. Since $\mathrm{FP}(\langle x_{1,n} \rangle_{n=1}^{m})$ is finite, we have $D_2 \in p_2$ so pick by Theorem 5.14 a product subsystem $\langle x_{2,n} \rangle_{n=1}^{\infty}$ of $\mathrm{FP}(\langle y_{2,n} \rangle_{n=1}^{\infty})$ such that $\mathrm{FP}(\langle x_{2,n} \rangle_{n=1}^{\infty}) \subseteq D_2$.

Let $t \in \{2, 3, \ldots, \ell - 1\}$ and assume $\langle x_{t,n} \rangle_{n=1}^{\infty}$ has been chosen. Let

$$D_{t+1} = \bigcap\{B_{t+1}(a_1, a_2, \ldots, a_t, j) : (a_1, a_2, \ldots, a_t) \in \times_{i=1}^{t} \mathrm{FP}(\langle x_{i,n} \rangle_{n=1}^{m})\}.$$

Then $D_{t+1} \in p_{t+1}$ so pick by Theorem 5.14 a product subsystem $\langle x_{t+1,n} \rangle_{n=1}^{\infty}$ of $\mathrm{FP}(\langle y_{t+1,n} \rangle_{n=1}^{\infty})$ such that $\mathrm{FP}(\langle x_{t+1,n} \rangle_{n=1}^{\infty}) \subseteq D_{t+1}$.

Then

$$\left(\times_{i=1}^{\ell-1} \mathrm{FP}(\langle x_{i,n} \rangle_{n=1}^{m}) \right) \times \mathrm{FP}(\langle x_{\ell,n} \rangle_{n=1}^{\infty}) \subseteq C_j$$

as required.   □

In contrast with Theorem 18.11, where the partition conclusion applied to products of arbitrary semigroups, we now restrict ourselves to products of commutative semigroups. We shall see in Theorem 18.16 that this restriction is necessary.

Definition 18.12. Let S be a semigroup, let $\langle y_n \rangle_{n=1}^{\infty}$ be a sequence in S, and let $m \in \mathbb{N}$. The sequence $\langle x_n \rangle_{n=1}^{m}$ is a *weak product subsystem* of $\langle y_n \rangle_{n=1}^{\infty}$ if and only if there exists a sequence $\langle H_n \rangle_{n=1}^{m}$ in $\mathcal{P}_f(\mathbb{N})$ such that $H_n \cap H_k = \emptyset$ when $n \neq k$ in $\{1, 2, \ldots, m\}$ and $x_n = \prod_{t \in H_n} y_t$ for each $n \in \{1, 2, \ldots, m\}$.

Recall by way of contrast, that in a *product subsystem* one requires that $\max H_n < \min H_{n+1}$.

Lemma 18.13. *Let $\ell \in \mathbb{N}$ and for each $i \in \{1, 2, \ldots, \ell\}$, let S_i be a commutative semigroup and let $\langle y_{i,n}\rangle_{n=1}^{\infty}$ be a sequence in S_i. Let*

$$\mathcal{L} = \{p \in \beta(\times_{i=1}^{\ell} S_i) : \text{for each } A \in p \text{ and each } m, k \in \mathbb{N} \text{ there exist for each}$$
$$i \in \{1, 2, \ldots, \ell\} \text{ a weak product subsystem } \langle x_{i,n}\rangle_{n=1}^{m}$$
$$\text{of } \mathrm{FS}(\langle y_{i,n}\rangle_{n=k}^{\infty}) \text{ such that } \times_{i=1}^{\ell} \mathrm{FP}(\langle x_{i,n}\rangle_{n=1}^{m}) \subseteq A\}.$$

Then $\mathcal{L}$ is a compact subsemigroup of $\beta(\times_{i=1}^{\ell} S_i)$.

Proof. Since any product subsystem is also a weak product subsystem, we have by Theorems 18.11 and 5.7 that $\mathcal{L} \neq \emptyset$. Since $\mathcal{L}$ is defined as the set of ultrafilters all of whose members satisfy a given property, $\mathcal{L}$ is closed, hence compact. To see that $\mathcal{L}$ is a semigroup, let $p, q \in \mathcal{L}$, let $A \in p \cdot q$ and let $m, k \in \mathbb{N}$. Then $\{\vec{a} \in \times_{i=1}^{\ell} S_i : \vec{a}^{-1}A \in q\} \in p$ so choose for each $i \in \{1, 2, \ldots, \ell\}$ a weak product subsystem $\langle x_{i,n}\rangle_{n=1}^{m}$ of $\mathrm{FP}(\langle y_{i,n}\rangle_{n=k}^{\infty})$ such that

$$\times_{i=1}^{\ell} \mathrm{FP}(\langle x_{i,n}\rangle_{n=1}^{m}) \subseteq \{\vec{a} \in \times_{i=1}^{\ell} S_i : \vec{a}^{-1}A \in q\}.$$

Given $i \in \{1, 2, \ldots, \ell\}$ and $n \in \{1, 2, \ldots, m\}$, pick $H_{i,n} \in \mathcal{P}_f(\mathbb{N})$ with $\min H_{i,n} \geq k$ such that $x_{i,n} = \prod_{t \in H_{i,n}} y_{i,t}$ and if $1 \leq n < s \leq m$, then $H_{i,n} \cap H_{i,s} = \emptyset$.

Let $r = \max(\bigcup_{i=1}^{\ell} \bigcup_{n=1}^{m} H_{i,n}) + 1$ and let

$$B = \bigcap\{\vec{a}^{-1}A : \vec{a} \in \times_{i=1}^{\ell} \mathrm{FP}(\langle x_{i,n}\rangle_{n=1}^{m})\}.$$

Then $B \in q$ so choose for each $i \in \{1, 2, \ldots, \ell\}$ a weak product subsystem $\langle z_{i,n}\rangle_{n=1}^{m}$ of $\mathrm{FP}(\langle y_{i,n}\rangle_{n=r}^{\infty})$ such that $\times_{i=1}^{\ell} \mathrm{FP}(\langle z_{i,n}\rangle_{n=1}^{m}) \subseteq B$. Given $i \in \{1, 2, \ldots, \ell\}$ and $n \in \{1, 2, \ldots, m\}$, pick $K_{i,n} \in \mathcal{P}_f(\mathbb{N})$ with $\min K_{i,n} \geq r$ such that $z_{i,n} = \prod_{t \in K_{i,n}} y_{i,t}$ and if $1 \leq n < s \leq m$, then $K_{i,n} \cap K_{i,s} = \emptyset$.

For $i \in \{1, 2, \ldots, \ell\}$ and $n \in \{1, 2, \ldots, m\}$, let $L_{i,n} = H_{i,n} \cup K_{i,n}$. Then

$$\prod_{t \in L_{i,n}} y_{i,t} = \prod_{t \in H_{i,n}} y_{i,t} \cdot \prod_{t \in K_{i,n}} y_{i,t} = x_{i,t} \cdot z_{i,t}$$

and if $1 \leq n < s \leq m$, then $L_{i,n} \cap L_{i,s} = \emptyset$. Thus for each $i \in \{1, 2, \ldots, \ell\}$, $\langle x_{i,t} \cdot z_{i,t}\rangle_{n=1}^{m}$ is a weak product subsystem of $\mathrm{FP}(\langle y_{i,n}\rangle_{n=k}^{\infty})$.

Finally we claim that $\times_{i=1}^{\ell} \mathrm{FP}(\langle x_{i,n} \cdot z_{i,n}\rangle_{n=1}^{m}) \subseteq A$. To this end let $\vec{c} \in \times_{i=1}^{\ell} \mathrm{FP}(\langle x_{i,n} \cdot z_{i,n}\rangle_{n=1}^{m})$ be given. For each $i \in \{1, 2, \ldots, \ell\}$, pick $F_i \subseteq \{1, 2, \ldots, m\}$ such that $c_i = \prod_{t \in F_i}(x_{i,t} \cdot z_{i,t})$ and let $a_i = \prod_{t \in F_i} x_{i,t}$ and $b_i = \prod_{t \in F_i} z_{i,t}$. Then $\vec{b} \in \times_{i=1}^{\ell} \mathrm{FP}(\langle z_{i,n}\rangle_{n=1}^{m})$ so $\vec{b} \in B$. Since $\vec{a} \in \times_{i=1}^{\ell} \mathrm{FP}(\langle x_{i,n}\rangle_{n=1}^{m})$ one has that $\vec{b} \in \vec{a}^{-1}A$ so that $\vec{a} \cdot \vec{b} \in A$. Since each S_i is commutative, we have for each $i \in \{1, 2, \ldots, \ell\}$ that $\prod_{t \in F_i}(x_{i,t} \cdot z_{i,t}) = (\prod_{t \in F_i} x_{i,t}) \cdot (\prod_{t \in F_i} z_{i,t})$ so that $\vec{c} = \vec{a} \cdot \vec{b}$ as required. $\square$

Theorem 18.14. *Let* $\ell \in \mathbb{N}$ *and for each* $i \in \{1, 2, \ldots, \ell\}$, *let* S_i *be a commutative semigroup and let* $\langle y_{i,n} \rangle_{n=1}^{\infty}$ *be a sequence in* S_i. *Let* C *be an IP* set in* $\bigtimes_{i=1}^{\ell} S_i$ *and let* $m \in \mathbb{N}$. *Then for each* $i \in \{1, 2, \ldots, \ell\}$ *there is a weak product subsystem* $\langle x_{i,n} \rangle_{n=1}^{m}$ *of* $\mathrm{FP}(\langle y_{i,n} \rangle_{n=1}^{\infty})$ *such that* $\bigtimes_{i=1}^{\ell} \mathrm{FP}(\langle x_{i,n} \rangle_{n=1}^{m}) \subseteq C$.

Proof. Let $\mathcal{L}$ be as in Lemma 18.13. Then $\mathcal{L}$ is a compact subsemigroup of $\bigtimes_{i=1}^{\ell} S_i$ so by Theorem 2.5 there is an idempotent $p \in \mathcal{L}$. By Theorem 16.6, $C \in p$. Thus by the definition of $\mathcal{L}$ for each $i \in \{1, 2, \ldots, \ell\}$ there is a weak product subsystem $\langle x_{i,n} \rangle_{n=1}^{m}$ of $\mathrm{FP}(\langle y_{i,n} \rangle_{n=1}^{\infty})$ such that $\bigtimes_{i=1}^{\ell} \mathrm{FP}(\langle x_{i,n} \rangle_{n=1}^{m}) \subseteq C$. $\square$

Three natural questions are raised by Theorem 18.14.

(1) Can one obtain infinite weak product subsystems (defined in the obvious fashion) such that $\bigtimes_{i=1}^{\ell} \mathrm{FP}(\langle x_{i,n} \rangle_{n=1}^{\infty}) \subseteq C$?

(2) Can one replace "weak product subsystems" with "product subsystems"?

(3) Can one omit the requirement that the semigroups S_i be commutative?

We answer all three of these questions in the negative.

The first two questions are answered in Theorem 18.15 using the semigroup $(\mathbb{N} \times \mathbb{N}, +)$. Since the operation is addition we refer to "sum subsystems" rather than "product subsystems".

Theorem 18.15. *There is an IP* set* C *in* $\mathbb{N} \times \mathbb{N}$ *such that:*

(a) *There do not exist* $z \in \mathbb{N}$ *and a sequence* $\langle x_n \rangle_{n=1}^{\infty}$ *in* $\mathbb{N}$ *such that either* $\{z\} \times \mathrm{FS}(\langle x_n \rangle_{n=1}^{\infty}) \subseteq C$ *or* $\mathrm{FS}(\langle x_n \rangle_{n=1}^{\infty}) \times \{z\} \subseteq C$. *(In particular there do not exist infinite weak sum subsystems* $\langle x_{1,n} \rangle_{n=1}^{\infty}$ *and* $\langle x_{2,n} \rangle_{n=1}^{\infty}$ *of* $\mathrm{FS}(\langle 2^n \rangle_{n=1}^{\infty})$ *with* $\mathrm{FS}(\langle x_{1,n} \rangle_{n=1}^{\infty}) \times \mathrm{FS}(\langle x_{2,n} \rangle_{n=1}^{\infty}) \subseteq C$.)

(b) *There do not exist sum subsystems* $\langle x_n \rangle_{n=1}^{2}$ *and* $\langle y_n \rangle_{n=1}^{2}$ *of* $\mathrm{FS}(\langle 2^n \rangle_{n=1}^{\infty})$ *with* $\{x_1, x_2\} \times \{y_1, y_2\} \subseteq C$.

Proof. Let

$$C = (\mathbb{N} \times \mathbb{N}) \setminus \{(\textstyle\sum_{n \in F} 2^n, \sum_{n \in G} 2^n) : F, G \in \mathcal{P}(\omega)$$
$$\text{and } \max F < \min G \text{ or } \max G < \min F\}.$$

To see that C is an IP* set in $\mathbb{N} \times \mathbb{N}$, suppose instead that we have a sequence $\langle (x_n, y_n) \rangle_{n=1}^{\infty}$ in $\mathbb{N} \times \mathbb{N}$ with

$$\mathrm{FS}(\langle (x_n, y_n) \rangle_{n=1}^{\infty}) \subseteq \{(\textstyle\sum_{n \in F} 2^n, \sum_{n \in G} 2^n) : F, G \in \mathcal{P}(\omega)$$
$$\text{and } \max F < \min G \text{ or } \max G < \min F\}.$$

Pick F_1 and G_1 in $\mathcal{P}(\omega)$ such that $x_1 = \sum_{t \in F_1} 2^t$ and $y_1 = \sum_{t \in G_1} 2^t$. Let $k = \max(F_1 \cup G_1)$. Choose, by Lemma 14.16, $H \in \mathcal{P}_f(\mathbb{N})$ such that $\min H > 1$, $2^{k+1} | \sum_{n \in H} x_n$, and $2^{k+1} | \sum_{n \in H} y_n$. Pick $F', G' \in \mathcal{P}(\omega)$ such that $\sum_{n \in H} x_n = \sum_{n \in F'} 2^n$ and $\sum_{n \in H} y_n =$

$\Sigma_{n \in G'} 2^n$. Then $k + 1 \leq \min(F' \cup G')$, so $x_1 + \Sigma_{n \in H} x_n = \Sigma_{n \in F_1 \cup F'} 2^n$ and $y_1 + \Sigma_{n \in H} y_n = \Sigma_{n \in G_1 \cup G'} 2^n$. Also

$$\Sigma_{n \in \{1\} \cup H}(x_n, y_n) = (x_1 + \Sigma_{n \in H} x_n, y_1 + \Sigma_{n \in H} y_n)$$
$$\in \{(\Sigma_{n \in F} 2^n, \Sigma_{n \in G} 2^n) : F, G \in \mathcal{P}(\omega)$$
$$\text{and } \max F < \min G \text{ or } \max G < \min F\}.$$

Thus $k + 1 \leq \max(F_1 \cup F') < \min(G_1 \cup G') \leq k$ or $k + 1 \leq \max(G_1 \cup G') < \min(F_1 \cup F') \leq k$, a contradiction.

To establish (a), suppose that one has $z \in \mathbb{N}$ and a sequence $\langle x_n \rangle_{n=1}^{\infty}$ in $\mathbb{N}$ such that either $\{z\} \times \text{FS}(\langle x_n \rangle_{n=1}^{\infty}) \subseteq C$ or $\text{FS}(\langle x_n \rangle_{n=1}^{\infty}) \times \{z\} \subseteq C$ and assume without loss of generality that $\{z\} \times \text{FS}(\langle x_n \rangle_{n=1}^{\infty}) \subseteq C$. Pick $F \in \mathcal{P}(\omega)$ such that $z = \Sigma_{t \in F} 2^t$ and let $k = \max F$. Pick $H \in \mathcal{P}_f(\mathbb{N})$ such that $2^{k+1} | \Sigma_{n \in H} x_n$. Pick $G \in \mathcal{P}(\omega)$ such that $\Sigma_{n \in H} x_n = \Sigma_{t \in G} 2^t$. Then $\max F < \min G$ so $(z, \Sigma_{n \in H} x_n) \notin C$.

To establish (b) suppose that one has sum subsystems $\langle x_n \rangle_{n=1}^{2}$ and $\langle y_n \rangle_{n=1}^{2}$ of $\text{FS}(\langle 2^n \rangle_{n=1}^{\infty})$ with $\{x_1, x_2\} \times \{y_1, y_2\} \subseteq C$. Pick $F_1, G_1, F_2, G_2 \in \mathcal{P}(\omega)$ such that $x_1 = \Sigma_{n \in F_1} 2^n$, $x_2 = \Sigma_{n \in F_2} 2^n$, $y_1 = \Sigma_{n \in G_1} 2^n$, $y_2 = \Sigma_{n \in G_2} 2^n$, $\max F_1 < \min F_2$, and $\max G_1 < \min G_2$. Without loss of generality, $\max F_1 \geq \max G_1$. But then we have $\max G_1 \leq \max F_1 < \min F_2$ so $(x_2, y_1) \notin C$, a contradiction. □

A striking contrast is provided by Theorem 18.15(b) and the case $\ell = 2$ of Theorem 18.11. That is IP* sets are not in general partition regular. It is easy to divide most semigroups into two classes, neither of which is an IP* set. Consequently it is not too surprising when one finds a property that must be satisfied by an IP* set in a semigroup S (such as containing a sequence with all of its sums and products when $S = \mathbb{N}$) which need not be satisfied by any cell of a partition of S. In this case we have a property, namely containing $\text{FS}(\langle x_{1,n} \rangle_{n=1}^{m}) \times \text{FS}(\langle x_{2,n} \rangle_{n=1}^{\infty})$ for some sum subsystems of any given sequences, which must be satisfied by some cell of a partition of $\mathbb{N} \times \mathbb{N}$, but need not be satisfied by IP* sets in $\mathbb{N} \times \mathbb{N}$.

The third question raised by Theorem 18.14 is answered with an example very similar to that used in the proof of Theorem 18.15.

Theorem 18.16. *Let S be the free semigroup on the alphabet $\{y_1, y_2, y_3, \ldots\}$. There is an IP* set C in $S \times S$ such that there do not exist weak product subsystems $\langle x_n \rangle_{n=1}^{2}$ and $\langle w_n \rangle_{n=1}^{2}$ of $\text{FP}(\langle y_n \rangle_{n=1}^{\infty})$ such that $\{x_1, x_2\} \times \{w_1, w_2\} \subseteq C$.*

Proof. Let

$$C = (S \times S) \setminus \{(\Pi_{n \in F} y_n, \Pi_{n \in G} y_n) : F, G \in \mathcal{P}_f(\mathbb{N})$$
$$\text{and } \max F < \min G \text{ or } \max G < \min F\}.$$

To see that C is an IP* set, suppose one has a sequence $\langle (x_n, w_n) \rangle_{n=1}^{\infty}$ with

$$\text{FP}(\langle (x_n, w_n) \rangle_{n=1}^{\infty}) \subseteq \{(\Pi_{n \in F} y_n, \Pi_{n \in G} y_n) : F, G \in \mathcal{P}_f(\mathbb{N})$$
$$\text{and } \max F < \min G \text{ or } \max G < \min F\}.$$

Given any $i < j$ in $\mathbb{N}$, pick F_i, F_j, G_i, G_j, $F_{i,j}$, $G_{i,j} \in \mathcal{P}_f(\mathbb{N})$ such that $x_i = \prod_{n \in F_i} y_n$, $x_j = \prod_{n \in F_j} y_n$, $w_i = \prod_{n \in G_i} y_n$, $w_j = \prod_{n \in G_j} y_n$, $x_i x_j = \prod_{n \in F_{i,j}} y_n$, and $w_i w_j = \prod_{n \in G_{i,j}} y_n$. Since $x_i x_j = \prod_{n \in F_{i,j}} y_n$, we have that $\max F_i < \min F_j$ and $F_{i,j} = F_i \cup F_j$ and similarly $\max G_i < \min G_j$ and $G_{i,j} = G_i \cup G_j$. Thus we may pick $j \in \mathbb{N}$ such that $\max(F_1 \cup G_1) < \min(F_j \cup G_j)$. Then

$$(x_1 x_j, w_1 w_j) \notin \{(\textstyle\prod_{n \in F} y_n, \prod_{n \in G} y_n) : F, G \in \mathcal{P}_f(\mathbb{N})$$
$$\text{and } \max F < \min G \text{ or } \max G < \min F\},$$

a contradiction.

Now suppose we have weak product subsystems $\langle x_n \rangle_{n=1}^2$ and $\langle w_n \rangle_{n=1}^2$ of $\mathrm{FP}(\langle y_n \rangle_{n=1}^\infty)$ such that $\{x_1, x_2\} \times \{w_1, w_2\} \subseteq C$. Pick F_1, F_2, G_1, $G_2 \in \mathcal{P}_f(\mathbb{N})$ such that $x_1 = \prod_{n \in F_1} y_n$, $x_2 = \prod_{n \in F_2} y_n$, $w_1 = \prod_{n \in G_1} y_n$, and $w_2 = \prod_{n \in G_2} y_n$. Since $x_1 x_2 \in \mathrm{FP}(\langle y_n \rangle_{n=1}^\infty)$, we must have $\max F_1 < \min F_2$ and similarly $\max G_1 < \min G_2$. Without loss of generality, $\max F_1 \geq \max G_1$. But then we have $\max G_1 \leq \max F_1 < \min F_2$ so $(x_2, w_1) \notin C$, a contradiction.   □

18.3 Spaces of Variable Words

In this section we obtain some infinite dimensional (or infinite parameter) extensions of the Hales–Jewett Theorem (Corollary 14.8) including the "main lemma" to the proof of Carlson's Theorem. (We shall prove Carlson's Theorem itself in Section 18.4.) These extensions involve sequences of variable words. Recall that we have defined a variable word over an alphabet Ψ to be a word over $\Psi \cup \{v\}$ in which v actually occurs (where v is a "variable" not in Ψ).

We introduce new notation (differing from that previously used) for variable words and nonvariable words, since we shall be dealing simultaneously with both.

Definition 18.17. Let Ψ be a (possibly empty) set (alphabet).
   (a) $W(\Psi; v)$ is the set of variable words over Ψ.
   (b) $W(\Psi)$ is the set of (nonvariable) words over Ψ.

Notice that $W(\Psi; v) \cup W(\Psi) = W(\Psi \cup \{v\})$.

Definition 18.18. Let Ψ be a set and let $\langle s_n(v) \rangle_{n=1}^\infty$ be a sequence in $W(\Psi; v)$.
   (a) The sequence $\langle t_n(v) \rangle_{n=1}^\infty$ is a *variable reduction* of $\langle s_n(v) \rangle_{n=1}^\infty$ if and only if there exist an increasing sequence of integers $\langle n_i \rangle_{i=1}^\infty$ with $n_1 = 1$ and a sequence $\langle a_i \rangle_{i=1}^\infty$ in $\Psi \cup \{v\}$ such that for each i,

$$v \in \{a_{n_i}, a_{n_i+1}, \dots, a_{n_{i+1}-1}\}$$

and

$$t_i(v) = s_{n_i}(a_{n_i}) {}^\frown s_{n_i+1}(a_{n_i+1}) {}^\frown \cdots {}^\frown s_{n_{i+1}-1}(a_{n_{i+1}-1}).$$

(b) The sequence $\langle t_n(v)\rangle_{n=1}^{\infty}$ is a *variable extraction* of $\langle s_n(v)\rangle_{n=1}^{\infty}$ if and only if $\langle t_n(v)\rangle_{n=1}^{\infty}$ is a variable reduction of some subsequence of $\langle s_n(v)\rangle_{n=1}^{\infty}$.

Thus, a variable reduction of $\langle s_n(v)\rangle_{n=1}^{\infty}$ is any infinite sequence of variable words obtained from $\langle s_n(v)\rangle_{n=1}^{\infty}$ by replacing each $s_i(v)$ by one of its instances, dividing the resulting sequence (of words over $\Psi \cup \{v\}$) into (infinitely many) finite blocks of consecutive words, and concatenating the members of each block, with the additional requirement that each of the resulting words has an occurrence of v. For example, if $\langle s_n(v)\rangle_{n=1}^{\infty}$ were the sequence $(avbv, v, bav, abvb, vavv, \ldots)$ then $\langle t_n(v)\rangle_{n=1}^{\infty}$ might be $(aabavbab, abvbbabb, \ldots)$.

Definition 18.19. Let Ψ be a set and let $\langle s_n(v)\rangle_{n=1}^{\infty}$ be a sequence in $W(\Psi; v)$.
   (a) The word w is a *reduced word* of $\langle s_n(v)\rangle_{n=1}^{\infty}$ if and only if there exist $k \in \mathbb{N}$ and $a_1, a_2, \ldots, a_k$ in $\Psi \cup \{v\}$ such that $w = s_1(a_1) {}^\frown s_2(a_2) {}^\frown \cdots {}^\frown s_k(a_k)$.
   (b) A *variable reduced word* of $\langle s_n(v)\rangle_{n=1}^{\infty}$ is a variable word which is a reduced word of $\langle s_n(v)\rangle_{n=1}^{\infty}$.
   (c) A *constant reduced word* of $\langle s_n(v)\rangle_{n=1}^{\infty}$ is a reduced word of $\langle s_n(v)\rangle_{n=1}^{\infty}$ which is not a variable reduced word.

Reduced words, variable reduced words and constant reduced words of a finite sequence $\langle s_n(v)\rangle_{n=1}^{k}$ are defined analogously.
   Notice that a word is a variable reduced word of $\langle s_n(v)\rangle_{n=1}^{\infty}$ if and only if it occurs as the first term in some variable reduction of $\langle s_n(v)\rangle_{n=1}^{\infty}$.

Definition 18.20. Let Ψ be a set and let $\langle s_n(v)\rangle_{n=1}^{\infty}$ be a sequence in $W(\Psi; v)$.
   (a) The word w is an *extracted word* of $\langle s_n(v)\rangle_{n=1}^{\infty}$ if and only if w is a reduced word of some subsequence of $\langle s_n(v)\rangle_{n=1}^{\infty}$.
   (b) A *variable extracted word* of $\langle s_n(v)\rangle_{n=1}^{\infty}$ is a variable word which is an extracted word of $\langle s_n(v)\rangle_{n=1}^{\infty}$.
   (c) A *constant extracted word* of $\langle s_n(v)\rangle_{n=1}^{\infty}$ is an extracted word of $\langle s_n(v)\rangle_{n=1}^{\infty}$ which is not a variable extracted word.

Notice that w is an extracted word of $\langle s_n(v)\rangle_{n=1}^{\infty}$ if and only if there exist $k \in \mathbb{N}$, $n_1 < n_2 < \cdots < n_k$ in $\mathbb{N}$, and $a_1, a_2, \ldots, a_k \in \Psi \cup \{v\}$ such that

$$w = s_{n_1}(a_1) {}^\frown s_{n_2}(a_2) {}^\frown \cdots {}^\frown s_{n_k}(a_k).$$

Extracted words are defined in the analogous way for a finite sequence $\langle s_n(v)\rangle_{n=1}^{k}$.
   We now introduce some special notation which will be used in the proof of the main theorem of this section. The notions defined depend on the choice of the alphabet Ψ, but the notation does not reflect this dependence.

Definition 18.21. Let Ψ be a set, let $k \in \mathbb{N} \cup \{\infty\}$ and let $\langle s_n(v)\rangle_{n=1}^{k}$ be a sequence in $W(\Psi; v)$. Let S be the semigroup $(W(\Psi \cup \{v\}), {}^\frown)$.
   (a) For each $m \in \mathbb{N}$, $A(\langle s_n(v)\rangle_{n=m}^{k})$ is the set of extracted words of $\langle s_n(v)\rangle_{n=m}^{k}$.
   (b) For each $m \in \mathbb{N}$, $V(\langle s_n(v)\rangle_{n=m}^{k})$ is the set of variable extracted words of $\langle s_n(v)\rangle_{n=m}^{k}$.

(c) For each $m \in \mathbb{N}$, $C(\langle s_n(v)\rangle_{n=m}^k)$ is the set of constant extracted words of $\langle s_n(v)\rangle_{n=m}^k$.

(d) $\mathcal{A}(\langle s_n(v)\rangle_{n=1}^\infty) = \bigcap_{m=1}^\infty c\ell_{\beta S}\ A(\langle s_n(v)\rangle_{n=m}^\infty)$.

(e) $\mathcal{V}(\langle s_n(v)\rangle_{n=1}^\infty) = \bigcap_{m=1}^\infty c\ell_{\beta S}\ V(\langle s_n(v)\rangle_{n=m}^\infty)$.

(f) $\mathcal{C}(\langle s_n(v)\rangle_{n=1}^\infty) = \bigcap_{m=1}^\infty c\ell_{\beta S}\ C(\langle s_n(v)\rangle_{n=m}^\infty)$.

Notice that $\mathcal{A}(\langle s_n(v)\rangle_{n=1}^\infty) = \mathcal{V}(\langle s_n(v)\rangle_{n=1}^\infty) \cup \mathcal{C}(\langle s_n(v)\rangle_{n=1}^\infty)$. Notice also that if $\Psi = \emptyset$, then each $C(\langle s_n(v)\rangle_{n=m}^k) = \emptyset$.

Lemma 18.22. *Let Ψ be a set, let $\langle s_n(v)\rangle_{n=1}^\infty$ be a sequence in $W(\Psi; v)$, and let S be the semigroup $(W(\Psi \cup \{v\}), \frown)$. Then $\mathcal{A}(\langle s_n(v)\rangle_{n=1}^\infty)$ is a compact subsemigroup of βS, and $\mathcal{V}(\langle s_n(v)\rangle_{n=1}^\infty)$ is an ideal of $\mathcal{A}(\langle s_n(v)\rangle_{n=1}^\infty)$. If $\Psi \neq \emptyset$, then $\mathcal{C}(\langle s_n(v)\rangle_{n=1}^\infty)$ is a compact subsemigroup of βS.*

Proof. We use Lemma 14.9 with $J = \{1\}$, $D = \mathbb{N}$, and $T_m = S$ for each $m \in \mathbb{N}$.

To see that $\mathcal{A}(\langle s_n(v)\rangle_{n=1}^\infty)$ is a subsemigroup of βS and $\mathcal{V}(\langle s_n(v)\rangle_{n=1}^\infty)$ is an ideal of $\mathcal{A}(\langle s_n(v)\rangle_{n=1}^\infty)$, for each $m \in \mathbb{N}$ let $E_m = A(\langle s_n(v)\rangle_{n=m}^\infty)$ and let $I_m = V(\langle s_n(v)\rangle_{n=m}^\infty)$. In order to invoke Lemma 14.9 we need to show that

(1) for each $m \in \mathbb{N}$, $I_m \subseteq E_m \subseteq T_m$,

(2) for each $m \in \mathbb{N}$ and each $w \in I_m$, there is some $k \in \mathbb{N}$ such that $w^\frown E_k \subseteq I_m$, and

(3) for each $m \in \mathbb{N}$ and each $w \in E_m \backslash I_m$, there is some $k \in \mathbb{N}$ such that $w^\frown E_k \subseteq E_m$ and $w^\frown I_k \subseteq I_m$.

Statement (1) is trivial. To verify statement (2), let $m \in \mathbb{N}$ and $w \in I_m$ be given. Pick $t \in \mathbb{N}$, $n_1 < n_2 < \cdots < n_t$ in $\mathbb{N}$ with $n_1 \geq m$, and $a_1, a_2, \ldots, a_t \in \Psi \cup \{v\}$ (with at least one $a_i = v$) such that

$$w = s_{n_1}(a_1)^\frown s_{n_2}(a_2)^\frown \cdots \frown s_{n_t}(a_t).$$

Let $k = n_t + 1$. Then $w^\frown E_k \subseteq I_m$ as required. The verification of statement (3) is identical except that $a_1, a_2, \ldots, a_t \in \Psi$.

Now assume that $\Psi \neq \emptyset$. To show that $\mathcal{C}(\langle s_n(v)\rangle_{n=1}^\infty)$ is a subsemigroup it suffices by Theorem 4.20 to show that for each $m \in \mathbb{N}$ and each $w \in C(\langle s_n(v)\rangle_{n=m}^\infty)$ there is some $k \in \mathbb{N}$ such that $w^\frown C(\langle s_n(v)\rangle_{n=k}^\infty) \subseteq C(\langle s_n(v)\rangle_{n=m}^\infty)$. The verification of this assertion is nearly identical to the verification of statement (2) above. $\square$

By now we are accustomed to using idempotents to obtain Ramsey-theoretic results. In the following theorem, we utilize two related idempotents in the case $\Psi \neq \emptyset$.

Theorem 18.23. *Let Ψ be a finite set and let $\langle s_n(v)\rangle_{n=1}^\infty$ be a sequence in $W(\Psi; v)$. Let $\mathcal{F}$ and $\mathcal{G}$ be finite partitions of $W(\Psi; v)$ and $W(\Psi)$ respectively. There exist $V \in \mathcal{F}$ and $C \in \mathcal{G}$ and a variable extraction $\langle w_n(v)\rangle_{n=1}^\infty$ of $\langle s_n(v)\rangle_{n=1}^\infty$ such that all variable extracted words of $\langle w_n(v)\rangle_{n=1}^\infty$ are in V and all constant extracted words of $\langle w_n(v)\rangle_{n=1}^\infty$ are in C.*

Proof. Assume first that $\Psi = \emptyset$. Then $W(\psi; v) = \{v^k : k \in \mathbb{N}\}$ and $W(\Psi) = \{\emptyset\}$. Given $n \in \mathbb{N}$, pick $k(n) \in \mathbb{N}$ such that $s_n(v) = v^{k(n)}$. For each $V \in \mathcal{F}$ let $B(V) = \{k \in \mathbb{N} : v^k \in V\}$. Pick by Corollary 5.15 $V \in \mathcal{F}$ and a sum subsystem $\langle \ell(n) \rangle_{n=1}^{\infty}$ of $\langle k(n) \rangle_{n=1}^{\infty}$ such that $FS(\langle \ell(n) \rangle_{n=1}^{\infty}) \subseteq B(V)$. For each $n \in \mathbb{N}$ let $w_n(v) = v^{\ell(n)}$. Then $\langle w_n(v) \rangle_{n=1}^{\infty}$ is a variable extraction of $\langle s_n(v) \rangle_{n=1}^{\infty}$ and the variable extracted words of $\langle w_n(v) \rangle_{n=1}^{\infty}$ correspond to the elements of $FS(\langle \ell(n) \rangle_{n=1}^{\infty})$ so that all variable extracted words of $\langle w_n(v) \rangle_{n=1}^{\infty}$ are in V. The conclusion about constant extracted words of $\langle w_n(v) \rangle_{n=1}^{\infty}$ holds vacuously.

Now assume that $\Psi \neq \emptyset$. By Lemma 18.22, $\mathcal{C}(\langle s_n(v) \rangle_{n=1}^{\infty})$ is a compact subsemigroup of βS, where S is the semigroup $(W(\Psi \cup \{v\}), \frown)$. Pick by Theorem 1.60 an idempotent p which is minimal in $\mathcal{C}(\langle s_n(v) \rangle_{n=1}^{\infty})$. Then $p \in \mathcal{A}(\langle s_n(v) \rangle_{n=1}^{\infty})$ so pick, again by Theorem 1.60, an idempotent q which is minimal in $\mathcal{A}(\langle s_n(v) \rangle_{n=1}^{\infty})$ such that $q \leq p$. Since, by Lemma 18.22, $\mathcal{V}(\langle s_n(v) \rangle_{n=1}^{\infty})$ is an ideal of $\mathcal{A}(\langle s_n(v) \rangle_{n=1}^{\infty})$, we have that $q \in \mathcal{V}(\langle s_n(v) \rangle_{n=1}^{\infty})$.

Now, for each $a \in \Psi$, define the evaluation map $\epsilon_a : W(\Psi \cup \{v\}) \to W(\Psi)$ as follows. Given $w \in W(\Psi \cup \{v\})$, $\epsilon_a(w)$ is w with all occurrences of v (if any) replaced by a. Thus, if $w = w(v) \in W(\Psi; v)$, then $\epsilon_a(w(v)) = w(a)$, while if $w \in W(\Psi)$, then $\epsilon_a(w) = w$. Notice that ϵ_a is a homomorphism from $W(\Psi \cup \{v\})$ to $W(\Psi)$ and hence, by Corollary 4.22, $\widetilde{\epsilon}_a$ is a homomorphism from $\beta(W(\Psi \cup \{v\}))$ to $\beta(W(\Psi))$.

We claim that for each $a \in \Psi$, $\widetilde{\epsilon}_a(q) = p$. Since $\widetilde{\epsilon}_a$ is a homomorphism, we have that $\widetilde{\epsilon}_a(q) \leq \widetilde{\epsilon}_a(p)$. Since ϵ_a is equal to the identity on $W(\psi)$, we have $\widetilde{\epsilon}_a(p) = p$ and consequently $\widetilde{\epsilon}_a(q) \leq p$. Since p is minimal in $\mathcal{C}(\langle s_n(v) \rangle_{n=1}^{\infty})$, to verify that $\widetilde{\epsilon}_a(q) = p$ we need only show that $\widetilde{\epsilon}_a(q) \in \mathcal{C}(\langle s_n(v) \rangle_{n=1}^{\infty})$.

To see that $\widetilde{\epsilon}_a(q) \in \mathcal{C}(\langle s_n(v) \rangle_{n=1}^{\infty})$, let $m \in \mathbb{N}$ be given. Now $V(\langle s_n(v) \rangle_{n=m}^{\infty}) \in q$ so $\epsilon_a[V(\langle s_n(v) \rangle_{n=m}^{\infty})] \in \widetilde{\epsilon}_a(q)$ by Lemma 3.30. It is routine to verify that

$$\epsilon_a[V(\langle s_n(v) \rangle_{n=m}^{\infty})] \subseteq C(\langle s_n(v) \rangle_{n=m}^{\infty})$$

so $C(\langle s_n(v) \rangle_{n=m}^{\infty}) \in \widetilde{\epsilon}_a(q)$ as required.

We shall, according to our usual custom, use $\frown$ for the binary operation on βS which extends the operation $\frown$ on S.

Let $V \in \mathcal{F} \cap q$ and let $C \in \mathcal{G} \cap p$. We put $V^{\sharp} = \{y \in V : y \frown p \in \overline{V} \text{ and } y \frown q \in \overline{V}\}$ and $C^{\sharp} = \{x \in C : x \frown p \in \overline{C} \text{ and } x \frown q \in \overline{V}\}$. Since $q \frown q = p \frown q = q \frown p = q$ and $p \frown p = p$, it follows that $V^{\sharp} \in q$ and $C^{\sharp} \in p$.

We claim that for each $y \in V^{\sharp}$ and each $x \in C^{\sharp}$, $y^{-1}(V^{\sharp}) \in q$, $y^{-1}(V^{\sharp}) \in p$, $x^{-1}(V^{\sharp}) \in q$, and $x^{-1}(C^{\sharp}) \in p$. To see this, let $y \in V^{\sharp}$ and $x \in C^{\sharp}$ be given. To see that $y^{-1}(V^{\sharp}) \in q$, we observe that $\lambda_y^{-1} \rho_p^{-1}[\overline{V}]$ is a neighborhood of q in βS, and so is $\lambda_y^{-1} \rho_q^{-1}[\overline{V}]$. Also $y^{-1}V \in q$. Since $y^{-1}(V^{\sharp}) = y^{-1}V \cap \lambda_y^{-1} \rho_p^{-1}[\overline{V}] \cap \lambda_y^{-1} \rho_q^{-1}[\overline{V}]$, it follows that $y^{-1}(V^{\sharp}) \in q$. Likewise, since $p \in \lambda_y^{-1} \rho_p^{-1}[\overline{V}] \cap \lambda_y^{-1} \rho_q^{-1}[\overline{V}]$, and $y^{-1}V \in p$, $y^{-1}(V^{\sharp}) \in p$. Similarly $x^{-1}(V^{\sharp}) = x^{-1}V \cap \lambda_x^{-1} \rho_p^{-1}[\overline{V}] \cap \lambda_x^{-1} \rho_q^{-1}[\overline{V}] \in q$ and $x^{-1}(C^{\sharp}) = x^{-1}C \cap \lambda_x^{-1} \rho_p^{-1}[\overline{C}] \cap \lambda_x^{-1} \rho_q^{-1}[\overline{V}] \in p$.

If $x, y \in A(\langle s_n(v) \rangle_{n=1}^{\infty})$, we shall write $x << y$ if there exists $m \in \mathbb{N}$ for which $x \in A(\langle s_n(v) \rangle_{n=1}^{m})$ and $y \in A(\langle s_n(v) \rangle_{n=m+1}^{\infty})$. We observe that, for any given $x \in A(\langle s_n(v) \rangle_{n=1}^{\infty})$, $\{y \in A(\langle s_n(v) \rangle_{n=1}^{\infty}) : x << y\} \in q$.

We now inductively construct a variable extraction $\langle w_n(v)\rangle_{n=1}^{\infty}$ of $\langle s_n(v)\rangle_{n=1}^{\infty}$ for which $V(\langle w_n(v)\rangle_{n=1}^{\infty}) \subseteq V^\sharp$ and $C(\langle w_n(v)\rangle_{n=1}^{\infty}) \subseteq C^\sharp$.

We first choose $w_1(v) \in V^\sharp \cap \bigcap_{a \in \Psi} \epsilon_a^{-1}[C^\sharp]$. We note that this is possible because $\epsilon_a^{-1}[C^\sharp] \in q$ for every $a \in \Psi$.

We then let $n \in \mathbb{N}$ and assume that we have chosen $w_i(v)$ for each $i \in \{1, 2, \ldots, n\}$ so that $w_i(v) << w_{i+1}(v)$ for every $i \in \{1, 2, \ldots, n-1\}$, $V(\langle w_i(v)\rangle_{i=1}^{n}) \subseteq V^\sharp$ and $C(\langle w_i(v)\rangle_{i=1}^{n}) \subseteq C^\sharp$. We can then choose $w_{n+1}(v)$ satisfying the following conditions.

(i) $w_{n+1}(v) \in V^\sharp \cap \bigcap_{a \in \Psi} \epsilon_a^{-1}[C^\sharp]$,

(ii) $w_n(v) << w_{n+1}(v)$,

(iii) $w_{n+1}(v) \in y^{-1}(V^\sharp)$ for every $y \in V(\langle w_i(v)\rangle_{i=1}^{n})$,

(iv) $w_{n+1}(v) \in \epsilon_a^{-1}[y^{-1}(V^\sharp)]$ for every $y \in V(\langle w_i(v)\rangle_{i=1}^{n})$ and every $a \in \Psi$,

(v) $w_{n+1}(v) \in x^{-1}(V^\sharp)$ for every $x \in C(\langle w_i(v)\rangle_{i=1}^{n})$, and

(vi) $w_{n+1}(v) \in \epsilon_a^{-1}[x^{-1}(C^\sharp)]$ for every $x \in C(\langle w_i(v)\rangle_{i=1}^{n})$ and every $a \in \Psi$.

This choice is possible, because each of these conditions for $w_{n+1}(v)$ is satisfied by all the elements of some member of q.

To see that $V(\langle w_i(v)\rangle_{i=1}^{n+1}) \subseteq V^\sharp$, let $u \in V(\langle w_i(v)\rangle_{i=1}^{n+1})$ and pick k and $m_1 < m_2 < \cdots < m_k$ in $\mathbb{N}$ and $a_1, a_2, \ldots, a_k \in \Psi \cup \{v\}$ such that some $a_i = v$ and

$$u = w_{m_1}(a_1)^\frown w_{m_2}(a_2)^\frown \cdots {}^\frown w_{m_k}(a_k).$$

If $m_k < n+1$, then $u \in V^\sharp$ by the induction hypothesis so assume that $m_k = n+1$. If $k = 1$, then $u \in V^\sharp$ by condition (i), so assume that $k > 1$. If $a_k \in \Psi$, then $u \in V^\sharp$ by condition (iv). If $a_k = v$ and $\{a_1, a_2, \ldots, a_{k-1}\} \subseteq \Psi$, then $u \in V^\sharp$ by condition (v). If $a_k = v$ and $v \in \{a_1, a_2, \ldots, a_{k-1}\}$, then $u \in V^\sharp$ by condition (iii).

That $C(\langle w_i(v)\rangle_{i=1}^{n+1}) \subseteq C^\sharp$ follows from the induction hypotheses and conditions (i) and (vi).

Finally, notice that $\langle w_n(v)\rangle_{n=1}^{\infty}$ is a variable extraction of $\langle s_n(v)\rangle_{n=1}^{\infty}$ by conditions (i) and (ii). $\qquad\square$

The following corollary includes the "main lemma" to the proof of Carlson's Theorem.

Corollary 18.24. *Let Ψ be a finite set and let $\langle s_n(v)\rangle_{n=1}^{\infty}$ be a sequence in $W(\Psi; v)$. Let $\mathcal{F}$ and $\mathcal{G}$ be finite partitions of $W(\Psi; v)$ and $W(\Psi)$ respectively. There exist $V \in \mathcal{F}$ and $C \in \mathcal{G}$ and a variable reduction $\langle t_n(v)\rangle_{n=1}^{\infty}$ of $\langle s_n(v)\rangle_{n=1}^{\infty}$ such that all variable reduced words of $\langle t_n(v)\rangle_{n=1}^{\infty}$ are in V and all constant reduced words of $\langle t_n(v)\rangle_{n=1}^{\infty}$ are in C.*

Proof. Define $\varphi : W(\Psi \cup \{v\}) \to W(\Psi \cup \{v\})$ by

$$\varphi(a_1 a_2 \ldots a_k) = s_1(a_1)^\frown s_2(a_2)^\frown \cdots {}^\frown s_k(a_k)$$

where $a_1, a_2, \ldots, a_k$ are in $\Psi \cup \{v\}$. Note that $\varphi[W(\Psi; v)] \subseteq W(\Psi; v)$ and $\varphi[W(\Psi)] \subseteq W(\Psi)$. Let $\mathcal{H} = \{\varphi^{-1}[B] : B \in \mathcal{F}\}$ and $\mathcal{K} = \{\varphi^{-1}[B] : B \in \mathcal{G}\}$. Then $\mathcal{H}$ and $\mathcal{K}$

are finite partitions of $W(\Psi; v)$ and $W(\Psi)$ respectively. Pick $V' \in \mathcal{H}$, $C' \in \mathcal{K}$, and a variable extraction $\langle w_n(v) \rangle_{n=1}^{\infty}$ of the sequence $\langle v, v, v, \ldots \rangle$ as guaranteed by Theorem 18.23. (That is, $\langle w_n(v) \rangle_{n=1}^{\infty}$ is simply a sequence of variable words, all of whose variable extracted words are in V' and all of whose constant extracted words are in C'.) Pick $V \in \mathcal{F}$ and $C \in \mathcal{G}$ such that $V' = \varphi^{-1}[V]$ and $C' = \varphi^{-1}[C]$.

For each $n \in \mathbb{N}$ pick $k_n \in \mathbb{N}$ and $a_{n,1}, a_{n,2}, \ldots, a_{n,k_n} \in \Psi \cup \{v\}$ such that

$$w_n = a_{n,1} a_{n,2} \ldots a_{n,k_n}.$$

Let $\ell_1 = 0$ and for each $n > 1$, let $\ell_n = \Sigma_{i=1}^{n-1} k_i$. For each $n \in \mathbb{N}$ let

$$t_n(v) = s_{\ell_n+1}(a_{n,1})^\frown s_{\ell_n+2}(a_{n,2})^\frown \cdots ^\frown s_{\ell_n+k_n}(a_{n,k_n})$$

and notice that $\langle t_n(v) \rangle_{n=1}^{\infty}$ is a variable reduction of $\langle s_n(v) \rangle_{n=1}^{\infty}$. Notice further that if $b_1, b_2, \ldots, b_n \in \Psi \cup \{v\}$, then

$$\varphi\big(w_1(b_1)^\frown w_2(b_2)^\frown \cdots ^\frown w_n(b_n)\big) = t_1(b_1)^\frown t_2(b_2)^\frown \cdots ^\frown t_n(b_n)$$

so if $b_1 b_2 \ldots b_n \in W(\Psi; v)$, then $t_1(b_1)^\frown t_2(b_2)^\frown \cdots ^\frown t_n(b_n) \in V$ and if $b_1 b_2 \ldots b_n \in W(\Psi)$, then $t_1(b_1)^\frown t_2(b_2)^\frown \cdots ^\frown t_n(b_n) \in C$. $\qquad\square$

Theorem 18.23 produces a sequence $\langle w_n(v) \rangle_{n=1}^{\infty}$ with a stronger homogeneity property than that in Corollary 18.24, but to obtain such a $\langle w_n(v) \rangle_{n=1}^{\infty}$ we must take an extraction of $\langle s_n(v) \rangle_{n=1}^{\infty}$; a reduction won't do. (To see why it won't do, consider $\langle s_n(v) \rangle_{n=1}^{\infty} = \langle av, bv, bv, bv, \ldots \rangle$ and partition the variable words according to whether the first letter is a or not.)

Exercise 18.3.1. Derive the Finite Sums Theorem (Corollary 5.10) as a consequence of Corollary 18.24. (Hint: Consider the function $f : W(\Psi; v) \to \mathbb{N}$ defined by the condition that $f(w)$ is the number of occurrences of v in w.)

18.4 Carlson's Theorem

In 1988 T. Carlson published [58] a theorem which has as corollaries many earlier Ramsey-theoretic results. We prove this theorem in this section.

Definition 18.25. Let Ψ be a (possibly empty) finite set and let $\mathcal{S}$ be the set of all infinite sequences in $W(\Psi; v)$. Denote the sequence $\langle s_n \rangle_{n=1}^{\infty}$ by $\vec{s}$. For $\vec{s} \in \mathcal{S}$, let

$$B_0(\vec{s}) = \{\vec{t} \in \mathcal{S} : \vec{t} \text{ is a variable reduction of } \vec{s}\},$$

and, for each $n \in \mathbb{N}$, let

$$B_n(\vec{s}) = \{\vec{t} \in B_0(\vec{s}) : \text{for each } i \in \{1, 2, \ldots, n\}, t_i = s_i\}.$$

Remark 18.26. *For every $\vec{s}, \vec{t} \in \mathcal{S}$ and $m, n \in \omega$, if $\vec{t} \in B_m(\vec{s})$ and if $m \leq n$, then $B_n(\vec{t}) \subseteq B_m(\vec{s})$.*

The following lemma is an easy consequence of Remark 18.26.

Lemma 18.27. *A topology can be defined on $\mathcal{S}$ by choosing $\{B_n(\vec{s}) : \vec{s} \in \mathcal{S}$ and $n \in \omega\}$ as a basis for the open sets.*

Proof. Let $\vec{s}, \vec{t} \in \mathcal{S}$, let $m, n \in \omega$, and let $\vec{u} \in B_m(\vec{s}) \cap B_n(\vec{t})$. Let $k = \max\{m, n\}$. Then by Remark 18.26 $B_k(\vec{u}) \subseteq B_m(\vec{s}) \cap B_n(\vec{t})$. $\qquad\qquad\qquad\qquad\qquad\qquad\qquad\square$

Definition 18.28. The *Ellentuck topology* on $\mathcal{S}$ is the topology with basis $\{B_n(\vec{s}) : \vec{s} \in \mathcal{S}$ and $n \in \omega\}$.

For the remainder of this section, we shall assume that $\mathcal{S}$ has the Ellentuck topology.

Definition 18.29. A set $X \subseteq \mathcal{S}$ is said to be *completely Ramsey* if and only if for every $n \in \omega$ and every $\vec{s} \in \mathcal{S}$, there is some $\vec{t} \in B_n(\vec{s})$ such that either $B_n(\vec{t}) \subseteq X$ or $B_n(\vec{t}) \cap X = \emptyset$.

Recall that we take the Baire sets in a topological space to be the σ-algebra generated by the open sets and the nowhere dense sets. (Other meanings for the term *Baire sets* also exist in the literature.) Recall also (see Exercise 17.5.1) that a set is a Baire set if and only if it can be written as the symmetric difference of an open set and a meager set, where the *meager* (or *first category*) sets are those sets that are the countable union of nowhere dense sets.

Carlson's Theorem (Theorem 18.44) is the assertion that a subset of $\mathcal{S}$ is completely Ramsey if and only if it is a Baire set (with respect to the Ellentuck topology). The proof in one direction is simple.

Theorem 18.30. *If X is a completely Ramsey subset of $\mathcal{S}$, then X is a Baire set. In fact X is the union of an open set and a nowhere dense set.*

Proof. Let X^o denote the interior of X. We shall show that X is a Baire set by showing that $X \setminus X^o$ is nowhere dense so that X is the union of an open set and a nowhere dense set.

If we assume the contrary, there exists $\vec{s} \in \mathcal{S}$ and $n \in \omega$ for which $B_n(\vec{s}) \subseteq c\ell_{\mathcal{S}}(X \setminus X^o)$. Since X is completely Ramsey, there exists $\vec{t} \in B_n(\vec{s})$ such that $B_n(\vec{t}) \subseteq X$ or $B_n(\vec{t}) \subseteq \mathcal{S} \setminus X$.

However, if $B_n(\vec{t}) \subseteq X$, then $B_n(\vec{t}) \subseteq X^o$ and hence $B_n(\vec{t}) \cap c\ell_{\mathcal{S}}(X \setminus X^o) = \emptyset$. This is a contradiction, because $B_n(\vec{t}) \subseteq B_n(\vec{s})$.

If $B_n(\vec{t}) \subseteq \mathcal{S} \setminus X$, then $B_n(\vec{t}) \cap c\ell_{\mathcal{S}}(X) = \emptyset$. This is again a contradiction, because $B_n(\vec{t}) \subseteq B_n(\vec{s}) \subseteq c\ell_{\mathcal{S}}(X)$. $\qquad\qquad\qquad\qquad\square$

Lemma 18.31. *The completely Ramsey subsets of $\mathcal{S}$ form a subalgebra of $\mathcal{P}(\mathcal{S})$.*

Proof. It is immediate from the definition of a completely Ramsey set that a subset of $\mathcal{S}$ is completely Ramsey if and only if its complement is. So it is sufficient to show that the completely Ramsey subsets of $\mathcal{S}$ are closed under finite unions.

To this end let X and Y be completely Ramsey subsets of $\mathcal{S}$. To see that $X \cup Y$ is completely Ramsey, let $\vec{s} \in \mathcal{S}$ and $n \in \omega$ be given. Pick $\vec{t} \in B_n(\vec{s})$ such that $B_n(\vec{t}) \subseteq X$ or $B_n(\vec{t}) \subseteq \mathcal{S}\backslash X$. If $B_n(\vec{t}) \subseteq X$, then $B_n(\vec{t}) \subseteq X \cup Y$ and we are done, so assume that $B_n(\vec{t}) \subseteq \mathcal{S}\backslash X$. Since Y is completely Ramsey, pick $\vec{u} \in B_n(\vec{t})$ such that $B_n(\vec{u}) \subseteq Y$ or $B_n(\vec{u}) \subseteq \mathcal{S}\backslash Y$. Now $B_n(\vec{u}) \subseteq B_n(\vec{t})$ and so if $B_n(\vec{u}) \subseteq Y$, then $B_n(\vec{u}) \subseteq X \cup Y$ and if $B_n(\vec{u}) \subseteq \mathcal{S}\backslash Y$, then $B_n(\vec{u}) \subseteq \mathcal{S}\backslash(X \cup Y)$.   □

Remark 18.32. *Let $\langle \vec{t}^m \rangle_{m=1}^\infty$ be a sequence in $\mathcal{S}$ with the property that $\vec{t}^{m+1} \in B_{m+1}(\vec{t}^m)$ for every $m \in \mathbb{N}$, and let $\vec{t} = \langle t^1{}_1, t^2{}_2, t^3{}_3, \ldots \rangle$. Then $\vec{t} \in B_m(\vec{t}^m)$ for every $m \in \omega$.*

We now need to introduce some more notation.

Definition 18.33. Let $n \in \omega$.

(a) If $\vec{s} \in \mathcal{S}$ and $n > 0$, then $\vec{s}_{|n} = \langle s_1, s_2, \ldots, s_n \rangle$ and $\vec{s}_{|0} = \emptyset$.

(b) If $\vec{s} \in \mathcal{S}$, then $\mathrm{tail}_n(\vec{s}) = \langle s_{n+1}, s_{n+2}, \ldots \rangle$.

(c) If $\vec{s} \in W(\Psi; v)^n$, then $|\vec{s}| = n$.

(d) If $\vec{s}$ is a finite sequence and $\vec{t}$ is a finite or infinite sequence, then $\langle \vec{s}, \vec{t} \rangle$ will denote the sequence in which $\vec{s}$ is followed by $\vec{t}$. (Thus, if $\vec{s} = \langle s_1, s_2, \ldots, s_m \rangle$ and $\vec{t} = \langle t_1, t_2, \ldots \rangle$, then $\langle \vec{s}, \vec{t} \rangle = \langle s_1, s_2, \ldots, s_m, t_1, t_2, \ldots \rangle$.)

(e) A subset X of $\mathcal{S}$ is said to be *almost dense* in $B_n(\vec{s})$ if $X \cap B_n(\vec{t}) \neq \emptyset$ whenever $\vec{t} \in B_n(\vec{s})$.

Notice that the notation $\vec{s}_{|n}$ differs slightly from our formal viewpoint since formally $n = \{0, 1, \ldots, n-1\}$ and $\{0, 1, \ldots, n-1\}$ is not a subset of the domain of the function $\vec{s}$.

Definition 18.34. We define a condition $(*)$ on a sequence $\langle T^n \rangle_{n\in\omega}$ of subsets of $\mathcal{S}$ as follows:

$(*)$ For every $n \in \omega$ and every $\vec{s} \in \mathcal{S}$

(a) $B_n(\vec{s}) \cap T^n \neq \emptyset$ and

(b) if $\vec{s} \in T^n$, then $B_n(\vec{s}) \subseteq T^n$.

We make some simple observations which will be useful in the proof of Carlson's Theorem.

Remark 18.35. *Let $\vec{s} \in \mathcal{S}$.*

(a) *If X is almost dense in $B_n(\vec{s})$, then X is almost dense in $B_n(\vec{t})$ for every $\vec{t} \in B_n(\vec{s})$.*

(b) *If $X \cap B_n(\vec{s}) = \emptyset$, then $X \cap B_n(\vec{t}) = \emptyset$ for every $\vec{t} \in B_n(\vec{s})$.*

(c) *For any $\vec{t} \in \mathcal{S}$, $\vec{t} \in B_n(\vec{s})$ if and only if $\vec{t}_{|n} = \vec{s}_{|n}$ and $\mathrm{tail}_n(\vec{t}) \in B_0\big(\mathrm{tail}_n(\vec{s})\big)$.*

Lemma 18.36. *Let* $\langle T^n \rangle_{n \in \omega}$ *be a sequence of subsets of* $\mathcal{S}$ *which satisfies* (∗) *and let* $\vec{s} \in \mathcal{S}$. *Then for each* $k \in \omega$, *there exists* $\vec{t} \in B_k(\vec{s})$ *such that, for every* $n > k$ *in* $\mathbb{N}$ *and every variable reduction* $\vec{r}$ *of* $\vec{t}_{|n}$, *we have* $\langle \vec{r}, \mathrm{tail}_n(\vec{t}) \rangle \in T^{|\vec{r}|}$.

Proof. We first show that

(†) for every $\vec{t} \in \mathcal{S}$ and every $n \in \mathbb{N}$, there exists $\vec{\sigma} \in B_n(\vec{t})$ such that $\langle \vec{r}, \mathrm{tail}_n(\vec{\sigma}) \rangle \in T^{|\vec{r}|}$ for every variable reduction $\vec{r}$ of $\vec{t}_{|n}$.

We enumerate the finite set of variable reductions of $\vec{t}_{|n}$ as $\{\vec{r}^{\,1}, \vec{r}^{\,2}, \ldots, \vec{r}^{\,m}\}$.

We put $\vec{\tau}^{\,0} = \mathrm{tail}_n(\vec{t})$ and then inductively choose $\vec{\tau}^{\,1}, \vec{\tau}^{\,2}, \ldots, \vec{\tau}^{\,m} \in \mathcal{S}$ so that the following properties hold for each $i \in \{1, 2, \ldots, m\}$:

(i) $\vec{\tau}^{\,i} \in B_0(\vec{\tau}^{\,i-1})$ and

(ii) $\langle \vec{r}^{\,i}, \vec{\tau}^{\,i} \rangle \in T^{|\vec{r}^{\,i}|}$.

Suppose that $i \in \{0, 1, 2, \ldots, m-1\}$ and that we have chosen $\vec{\tau}^{\,i}$. Let $l = |\vec{r}^{\,i+1}|$. By condition (a) of (∗), we can choose $\vec{u} \in B_l(\langle \vec{r}^{\,i+1}, \vec{\tau}^{\,i} \rangle) \cap T^l$. Let $\vec{\tau}^{\,i+1} = \mathrm{tail}_l(\vec{u})$. Then $\vec{\tau}^{\,i+1} \in B_0(\vec{\tau}^{\,i})$ and $\langle \vec{r}^{\,i+1}, \vec{\tau}^{\,i+1} \rangle = \vec{u} \in T^{|\vec{r}^{\,i+1}|}$.

Having chosen $\vec{\tau}^{\,m}$, we put $\vec{\sigma} = \langle \vec{t}_{|n}, \vec{\tau}^{\,m} \rangle$. Since $\vec{\tau}^{\,m} \in B_0(\mathrm{tail}_n(\vec{t}))$, we have $\vec{\sigma} \in B_n(\vec{t})$.

Now let $\vec{r}$ be a variable reduction of $\vec{t}_{|n}$. Then $\vec{r} = \vec{r}^{\,i}$ for some $i \in \{1, 2, \ldots, m\}$. We note that $\vec{\tau}^{\,m} \in B_0(\vec{\tau}^{\,i})$, and hence that $\langle \vec{r}^{\,i}, \vec{\tau}^{\,m} \rangle \in B_{|\vec{r}^{\,i}|}(\langle \vec{r}^{\,i}, \vec{\tau}^{\,i} \rangle)$. So $\langle \vec{r}^{\,i}, \vec{\tau}^{\,m} \rangle = \langle \vec{r}^{\,i}, \mathrm{tail}_n(\vec{\sigma}) \rangle \in T^{|\vec{r}^{\,i}|}$, by condition (b) of (∗). Thus (†) is established.

Now let $k \in \omega$ be given. By (†) we can inductively define elements $\vec{t}^{\,0}, \vec{t}^{\,1}, \vec{t}^{\,2}, \ldots$ of $\mathcal{S}$ with the following properties:

(i) $\vec{t}^{\,0} = \vec{t}^{\,1} = \cdots = \vec{t}^{\,k} = \vec{s}$,

(ii) $\vec{t}^{\,n} \in B_n(\vec{t}^{\,n-1})$ for every $n > k$, and

(iii) for every $n > k$ and every variable reduction $\vec{r}$ of $\vec{t}^{\,n-1}_{|n}$, $\langle \vec{r}, \mathrm{tail}_n(\vec{t}^{\,n}) \rangle \in T^{|\vec{r}|}$.

Let $\vec{t} = \langle t^1{}_1, t^2{}_2, t^3{}_3, \ldots \rangle$. Then $\vec{t} \in B_k(\vec{s})$. Let $n > k$ be given and let $\vec{r}$ be a variable reduction of $\vec{t}_{|n}$. Now $\vec{t}_{|n} = \vec{t}^{\,n-1}{}_{|n}$ so $\vec{r}$ is a variable reduction of $\vec{t}^{\,n-1}{}_{|n}$ and thus $\langle \vec{r}, \mathrm{tail}_n(\vec{t}^{\,n}) \rangle \in T^{|\vec{r}|}$. Also $\langle \vec{r}, \mathrm{tail}_n(\vec{t}) \rangle \in B_{|\vec{r}|}(\langle \vec{r}, \mathrm{tail}_n(\vec{t}^{\,n}) \rangle)$ and so by condition (b) of (∗), $\langle \vec{r}, \mathrm{tail}_n(\vec{t}) \rangle \in T^{|\vec{r}|}$. □

Corollary 18.37. *Let* $\vec{s} \in \mathcal{S}$, $k \in \omega$, *and* $X \subseteq \mathcal{S}$. *Then there exists* $\vec{t} \in B_k(\vec{s})$ *such that, for every* $n > k$ *and every variable reduction* $\vec{r}$ *of* $\vec{t}_{|n}$, *either* X *is almost dense in* $B_{|\vec{r}|}(\langle \vec{r}, \mathrm{tail}_n(\vec{t}) \rangle)$ *or else* $X \cap B_{|\vec{r}|}(\langle \vec{r}, \mathrm{tail}_n(\vec{t}) \rangle) = \emptyset$.

Proof. For each $n \in \omega$ we define $T^n \subseteq \mathcal{S}$ by stating that $\vec{\tau} \in T^n$ if and only if either X is almost dense in $B_n(\vec{\tau})$ or $X \cap B_n(\vec{\tau}) = \emptyset$. We shall show that $\langle T^n \rangle_{n \in \omega}$ satisfies (∗), and our claim will then follow from Lemma 18.36.

Let $n \in \omega$ and $\vec{\tau} \in \mathcal{S}$. If X is almost dense in $B_n(\vec{\tau})$, then $\vec{\tau} \in B_n(\vec{\tau}) \cap T^n$. Otherwise there exists $\vec{u} \in B_n(\vec{\tau})$ for which $B_n(\vec{u}) \cap X = \emptyset$ and so $\vec{u} \in B_n(\vec{\tau}) \cap T^n$. Thus $\langle T^n \rangle_{n \in \omega}$ satisfies condition (a) of (∗).

To see that $\langle T^n \rangle_{n \in \omega}$ satisfies condition (b) of $(*)$, let $n \in \omega$, let $\vec{\tau} \in T^n$ and let $\vec{u} \in B_n(\vec{\tau})$. Then $B_n(\vec{u}) \subseteq B_n(\vec{\tau})$. Consequently, if X is almost dense in $B_n(\vec{\tau})$, then X is almost dense in $B_n(\vec{u})$ and if $X \cap B_n(\vec{\tau}) = \emptyset$, then $X \cap B_n(\vec{u}) = \emptyset$. $\qquad\square$

Definition 18.38. (a) Let $\vec{t} \in \mathcal{S}$. then $R(\vec{t}) = \{w \in W(\Psi; v) : w$ is a variable reduced word of $\vec{t}\}$.

(b) Let $\vec{t} \in \mathcal{S}$ and let $w \in R(\vec{t})$. Then $N(w, \vec{t})$ is that $l \in \mathbb{N}$ for which there exist $a_1, a_2, \ldots, a_l \in \Psi \cup \{v\}$ such that $w = t_1(a_1)^\frown t_2(a_2)^\frown \cdots ^\frown t_l(a_l)$.

Remark 18.39. Let $\vec{s} \in \mathcal{S}$, $\vec{t} \in B_0(\vec{s})$ and $w \in R(\vec{t})$. Then $\langle w, \mathrm{tail}_{N(w,\vec{t})}(\vec{t}) \rangle \in B_0(\langle w, \mathrm{tail}_{N(w,\vec{s})}(\vec{s}) \rangle)$.

Lemma 18.40. Let $\vec{s} \in \mathcal{S}$, $k \in \omega$ and $X \subseteq \mathcal{S}$. Then there exists $\vec{t} \in B_k(\vec{s})$ such that, for every $n \in \mathbb{N}$ satisfying $n > k$ and every variable reduction $\vec{r}$ of $\vec{t}_{|n}$, either

(1) for every $w \in R(\mathrm{tail}_n(\vec{t}))$, X is almost dense in $B_{|\vec{r}|+1}(\langle \vec{r}, w, \mathrm{tail}_{N(w,\mathrm{tail}_n(\vec{t}))+n}(\vec{t}) \rangle$

or

(2) for every $w \in R(\mathrm{tail}_n(\vec{t}))$, $X \cap B_{|\vec{r}|+1}(\langle \vec{r}, w, \mathrm{tail}_{N(w,\mathrm{tail}_n(\vec{t}))+n}(\vec{t}) \rangle) = \emptyset$.

Proof. We define a sequence $\langle T^n \rangle_{n \in \omega}$ of subsets of $\mathcal{S}$ by stating that $\vec{\tau} \in T^n$ if and only if either $X \cap B_{n+1}(\langle \vec{\tau}_{|n}, w, \mathrm{tail}_{N(w,\mathrm{tail}_n(\vec{\tau}))+n}(\vec{\tau}) \rangle) = \emptyset$ for every $w \in R(\mathrm{tail}_n(\vec{\tau}))$, or else X is almost dense in $B_{n+1}(\langle \vec{\tau}_{|n}, w, \mathrm{tail}_{N(w,\mathrm{tail}_n(\vec{\tau}))+n}(\vec{\tau}) \rangle$ for every $w \in R(\mathrm{tail}_n(\vec{\tau}))$. We shall show that $\langle T^n \rangle_{n \in \omega}$ satisfies $(*)$. The claim then follows from Lemma 18.36.

Let $n \in \omega$ and $\vec{\tau} \in \mathcal{S}$. We shall show that T^n and $\vec{\tau}$ satisfy conditions (a) and (b) of $(*)$. By Corollary 18.37, there exists $\vec{u} \in B_n(\vec{\tau})$ such that, for every $m > n$ in $\mathbb{N}$ and every variable reduction $\vec{r}$ of $\vec{u}_{|m}$, $X \cap B_{|\vec{r}|}(\langle \vec{r}, \mathrm{tail}_m(\vec{u}) \rangle) = \emptyset$ or else X is almost dense in $B_{|\vec{r}|}(\langle \vec{r}, \mathrm{tail}_m(\vec{u}) \rangle)$. Let

$$V_0 = \{w \in R(\mathrm{tail}_n(\vec{u})) : X \cap B_{n+1}(\langle \vec{u}_{|n}, w, \mathrm{tail}_{N(w,\mathrm{tail}_n(\vec{u}))+n}(\vec{u}) \rangle) = \emptyset\},$$
$$V_1 = R(\mathrm{tail}_n(\vec{u})) \backslash V_0, \text{ and}$$
$$V_2 = W(\Psi; v) \backslash R(\mathrm{tail}_n(\vec{u})).$$

We note that, if $w \in R(\mathrm{tail}_n(\vec{u}))$, then $w \in V_1$ if and only if X is almost dense in $B_{n+1}(\langle \vec{u}_{|n}, w, \mathrm{tail}_{N(w,\mathrm{tail}_n(\vec{u}))+n}(\vec{u}) \rangle)$.

By Corollary 18.24, there exists $i \in \{0, 1, 2\}$ and $\vec{\sigma} \in B_0(\mathrm{tail}_n(\vec{u}))$ such that $R(\vec{\sigma}) \subseteq V_i$. Since $R(\vec{\sigma}) \subseteq R(\mathrm{tail}_n(\vec{u}))$, $i \neq 2$.

Let $\vec{\rho} = \langle \vec{u}_{|n}, \vec{\sigma} \rangle$. Then $\vec{\rho} \in B_n(\vec{\tau}) \cap T^n$. So T^n and $\vec{\tau}$ satisfy condition (a) of $(*)$.

To verify that T^n and $\vec{\tau}$ satisfy condition (b) of $(*)$, assume that $\vec{\tau} \in T^n$ and let $\vec{v} \in B_n(\vec{\tau})$. Then $\mathrm{tail}_n(\vec{v}) \in B_0(\mathrm{tail}_n(\vec{\tau}))$ and $R(\mathrm{tail}_n(\vec{v})) \subseteq R(\mathrm{tail}_n(\vec{\tau}))$. If $w \in R(\mathrm{tail}_n(\vec{v}))$, then

$$\langle \vec{v}_{|n}, w, \mathrm{tail}_{N(w,\mathrm{tail}_n(\vec{v}))+n}(\vec{v}) \rangle \in B_{n+1}(\langle \vec{\tau}_{|n}, w, \mathrm{tail}_{N(w,\mathrm{tail}_n(\vec{\tau}))+n}(\vec{\tau}) \rangle)$$

so that

$$B_{n+1}(\langle \vec{v}_{|n}, w, \mathrm{tail}_{N(w,\mathrm{tail}_n(\vec{v}))+n}(\vec{v}) \rangle) \subseteq B_{n+1}(\langle \vec{\tau}_{|n}, w, \mathrm{tail}_{N(w,\mathrm{tail}_n(\vec{\tau}))+n}(\vec{\tau}) \rangle).$$

It follows easily that $\vec{v} \in T^n$. $\qquad\square$

Lemma 18.41. *Every closed subset X of $\mathcal{S}$ is completely Ramsey.*

Proof. Let $\vec{s} \in \mathcal{S}$ and $k \in \omega$. We need to show that there exists $\vec{\sigma} \in B_k(\vec{s})$ for which $B_k(\vec{\sigma}) \subseteq X$ or $B_k(\vec{\sigma}) \cap X = \emptyset$. We suppose, on the contrary, that no such element $\vec{\sigma}$ exists. So X is almost dense in $B_k(\vec{s})$.

Let $\vec{t} \in B_k(\vec{s})$ be the element guaranteed by Lemma 18.40. We shall show that, for every $\vec{u} \in B_k(\vec{t})$ and every $n \geq k$ in ω, X is almost dense in $B_n(\vec{u})$. This is true if $n = k$. We shall assume that it is true for n and deduce that it is also true for $n + 1$.

Suppose then that $\vec{u} \in B_k(\vec{t})$ and let $m \in \mathbb{N}$ be the integer for which $\vec{u}_{|n}$ is a variable reduction of $\vec{t}_{|m}$. By our inductive assumption, there exists $\vec{x} \in X \cap B_n(\vec{u})$. Let $w = x_{n+1} \in R\big(\mathrm{tail}_m(\vec{t})\big)$. Since $\vec{x} \in B_{n+1}(\langle \vec{u}_{|n}, w, \mathrm{tail}_{N(w,\mathrm{tail}_m(\vec{t}))+m}(\vec{t}) \rangle) \cap X$, it follows from our choice of $\vec{t}$ that X is almost dense in $B_{n+1}(\langle \vec{u}_{|n}, w', \mathrm{tail}_{N(w',\mathrm{tail}_m(\vec{t}))+m}(\vec{t}) \rangle)$ for every $w' \in R(\mathrm{tail}_m(\vec{t}))$. In particular, this holds if we put $w' = u_{n+1}$. We then have $\vec{u} \in B_{n+1}(\langle \vec{u}_{|n}, w', \mathrm{tail}_{N(w',\mathrm{tail}_m(\vec{t}))+m}(\vec{t}) \rangle)$, and so X is almost dense in $B_{n+1}(\vec{u})$.

We have thus shown that, for every $\vec{u} \in B_k(\vec{t})$ and every $n > k$ in ω, $B_n(\vec{u}) \cap X \neq \emptyset$. Since X is closed, this implies that $\vec{u} \in X$. Thus $B_k(\vec{t}) \subseteq X$, a contradiction. $\square$

Lemma 18.42. *Let $\langle F_n \rangle_{n=0}^{\infty}$ be an increasing sequence of closed nowhere dense subsets of $\mathcal{S}$ and let $N = \bigcup_{n=0}^{\infty} F_n$. Then N is nowhere dense in $\mathcal{S}$.*

Proof. For each $n \in \omega$, let $T^n = \{\vec{\tau} \in \mathcal{S} : B_n(\vec{\tau}) \cap F_n = \emptyset\}$. Since each F_n is completely Ramsey (by Lemma 18.41) and nowhere dense, $\langle T^n \rangle_{n \in \omega}$ satisfies condition (a) of $(*)$. It clearly satisfies condition (b) of $(*)$.

To see that N is nowhere dense in $\mathcal{S}$, suppose instead that we have some $\vec{s}$ in the interior of $c\ell\, N$. Pick $k \in \omega$ such that $B_k(\vec{s}) \subseteq c\ell\, N$. By Lemma 18.36, there exists $\vec{t} \in B_k(\vec{s})$ such that, for every $m > k$ in ω and every variable reduction $\vec{r}$ of $\vec{t}_{|m}$, we have $\langle \vec{r}, \mathrm{tail}_m(\vec{t}) \rangle \in T^{|\vec{r}|}$.

Then $\vec{t} \in c\ell\, N$ so pick $\vec{u} \in B_k(\vec{t}) \cap N$ and pick $n > k$ such that $\vec{u} \in F_n$. Pick $m > k$ such that $\vec{u}_{|n}$ is a variable reduction of $\vec{t}_{|m}$. Then $\vec{u} \in B_n(\langle \vec{u}_{|n}, \mathrm{tail}_m(\vec{t}) \rangle)$ and $\langle \vec{u}_{|n}, \mathrm{tail}_m(\vec{t}) \rangle \in T^n$ so $\vec{u} \notin F_n$, a contradiction. $\square$

Corollary 18.43. *If N is a meager subset of $\mathcal{S}$, then N is completely Ramsey.*

Proof. We have that N is the union of an increasing sequence $\langle X_n \rangle_{n=0}^{\infty}$ of nowhere dense subsets of $\mathcal{S}$. Let $X = c\ell\left(\bigcup_{n=0}^{\infty} \overline{X_n}\right)$. Then X is nowhere dense, by Lemma 18.42, and X is completely Ramsey, by Lemma 18.41. It follows that, for each $\vec{s} \in \mathcal{S}$ and $n \in \omega$, there exists $\vec{t} \in B_n(\vec{s})$ for which $B_n(\vec{t}) \cap X = \emptyset$. Since $N \subseteq X$, this implies that $B_n(\vec{t}) \cap N = \emptyset$. $\square$

Theorem 18.44 (Carlson's Theorem). *A subset of $\mathcal{S}$ is completely Ramsey if and only if it is Baire.*

Proof. By Theorem 18.30, every completely Ramsey subset of $\mathcal{S}$ is Baire. Now assume that X is a Baire set and pick an open set U and a meager set M such that $X = U \triangle M$. By Lemmas 18.41 and 18.31, U is completely Ramsey and by Corollary 18.43, M

is completely Ramsey. Applying Lemma 18.31 again, one has that X is completely Ramsey. $\square$

As an amusing consequence of Theorems 18.30 and 18.44 one sees that in $\mathcal{S}$, every Baire set is the union of an open set and a nowhere dense set.

Notice that one has an immediate partition corollary to Carlson's Theorem.

Corollary 18.45. *Let $r \in \mathbb{N}$ and assume that $\mathcal{S} = \bigcup_{i=1}^{r} X_i$. If each X_i is a Baire set, then for every $n \in \omega$ and every $\vec{s} \in \mathcal{S}$, there exist some $i \in \{1, 2, \ldots, r\}$ and some $\vec{t} \in B_n(\vec{s})$ such that $B_n(\vec{t}) \subseteq X_i$.*

Proof. Let $n \in \omega$ and $\vec{s} \in \mathcal{S}$ and suppose that for each $i \in \{1, 2, \ldots, r\}$ and each $\vec{t} \in B_n(\vec{s})$ one does not have $B_n(\vec{t}) \subseteq X_i$. Pick $\vec{t}^{\,1} \in B_n(\vec{s})$ as guaranteed by the fact that X_1 is completely Ramsey. (So $B_n(\vec{t}^{\,1}) \cap X_1 = \emptyset$.) For $i \in \{1, 2, \ldots, r-1\}$, assume that $\vec{t}^{\,i}$ has been chosen and pick $\vec{t}^{\,i+1} \in B_n(\vec{t}^{\,i})$ as guaranteed by the fact that X_{i+1} is completely Ramsey. (So $B_n(\vec{t}^{\,i+1}) \cap X_{i+1} = \emptyset$.) Then $B_n(\vec{t}^{\,r}) \subseteq \emptyset$, a contradiction. $\square$

An extensive array of theorems in Ramsey Theory are a consequence of Carlson's Theorem. For example, Ellentuck's Theorem [82] is the special case of Carlson's Theorem which has the alphabet $\Psi = \emptyset$.

We have already seen in Exercise 18.3.1 that the Finite Sums Theorem is derivable from the "main lemma" to Carlson's Theorem. To illustrate a more typical application, we shall show how the Milliken–Taylor Theorem (Theorem 18.7) is derivable from Carlson's Theorem. As is common in such applications, one only uses the fact that *open* sets are completely Ramsey.

Corollary 18.46 (Milliken–Taylor Theorem). *Let $k, r \in \mathbb{N}$, let $\langle x_n \rangle_{n=1}^{\infty}$ be a sequence in $\mathbb{N}$, and assume that $[\mathbb{N}]^k = \bigcup_{i=1}^{r} A_i$. Then there exist $i \in \{1, 2, \ldots, r\}$ and a sum subsystem $\langle y_n \rangle_{n=1}^{\infty}$ of $\langle x_n \rangle_{n=1}^{\infty}$ such that*

$$[\mathrm{FS}(\langle y_n \rangle_{n=1}^{\infty})]_{<}^{k} \subseteq A_i.$$

Proof. Let $\Psi = \{a\}$ and define the function $f : W(\Psi; v) \to \mathbb{N}$ by letting $f(w)$ be the number of occurrences of v in w.

For each $i \in \{1, 2, \ldots, r\}$ let

$$X_i = \{\vec{s} \in \mathcal{S} : \{f(s_1), f(s_2), \ldots, f(s_k)\} \in A_i\}$$

and let

$$X_{r+1} = \{\vec{s} \in \mathcal{S} : |\{f(s_1), f(s_2), \ldots, f(s_k)\}| < k\}.$$

Then $\mathcal{S} = \bigcup_{i=1}^{r+1} X_i$. Now, given $i \in \{1, 2, \ldots, r+1\}$ and $\vec{s} \in X_i$, one has that $B_k(\vec{s}) \subseteq X_i$ so X_i is open.

For each $n \in \mathbb{N}$, let $s_n = v^{x_n}$, so that $f(s_n) = x_n$. Pick by Corollary 18.45 some $i \in \{1, 2, \ldots, r+1\}$ and some $\vec{t} \in B_0(\vec{s})$ such that $B_0(\vec{t}) \subseteq X_i$. Since $B_0(\vec{t}) \subseteq X_i$, $i \neq r+1$. For each $n \in \mathbb{N}$, let $y_n = f(t_n)$. Then $\langle y_n \rangle_{n=1}^{\infty}$ is a sum subsystem of $\langle x_n \rangle_{n=1}^{\infty}$.

To see that $[\mathrm{FS}(\langle y_n \rangle_{n=1}^{\infty})]_{<}^{k} \subseteq A_i$, let $H_1, H_2, \ldots, H_k \in \mathcal{P}_f(\mathbb{N})$ be given such that $\max H_j < \min H_{j+1}$ for each $j \in \{1, 2, \ldots, k-1\}$. For each $j \in \{1, 2, \ldots, k\}$, let $\ell_j = \max H_j$ and let $\ell_0 = 0$. For each $n \in \{1, 2, \ldots, \ell_k\}$, let

$$b_n = \begin{cases} a & \text{if } n \notin \bigcup_{i=1}^{k} H_i \\ v & \text{if } n \in \bigcup_{i=1}^{k} H_i \end{cases}$$

and for $j \in \{1, 2, \ldots, k\}$, let

$$w_j = t_{\ell_{j-1}+1}(b_{\ell_{j-1}+1}) ^\frown t_{\ell_{j-1}+2}(b_{\ell_{j-1}+2}) ^\frown \cdots ^\frown t_{\ell_j}(b_{\ell_j}).$$

For $j > k$, let $w_j = t_{l_k+j-k}$. Then $\vec{w} \in B_0(\vec{t})$ so $\{f(w_1), f(w_2), \ldots, f(w_k)\} \in A_i$. Since

$$\{f(w_1), f(w_2), \ldots, f(w_k)\} = \{\Sigma_{n \in H_1} y_n, \Sigma_{n \in H_2} y_n, \ldots, \Sigma_{n \in H_k} y_n\},$$

we are done.   □

Notes

The ultrafilter proof of Ramsey's Theorem (Theorem 18.2) is by now classical. See [67, p. 39] for a discussion of its origins. Version 1 of the Milliken–Taylor Theorem is essentially the version proved by K. Milliken [183] while Version 2 is that proved by A. Taylor [233]. The rest of the results of Section 18.1 are from [26] and were obtained in collaboration with V. Bergelson. The results of Section 18.2 are from [32], a result of collaboration with V. Bergelson.

Theorem 18.23 and Corollary 18.24 are from [20], a result of collaboration with V. Bergelson and A. Blass. The part of Corollary 18.24 that corresponds to $W(\Psi)$ is [28, Corollary 3.7], a result of collaboration with V. Bergelson. It is a modification of a result of T. Carlson and S. Simpson [59, Theorem 6.3]. The part of Corollary 18.24 that refers to $W(\Psi; v)$ is due to T. Carlson.

Carlson's Theorem is of course due to T. Carlson and is from [58]. The Ellentuck topology on the set of sequences of variable words was introduced by T. Carlson and S. Simpson in [59, Section 6]. It is analogous to a topology on the set of infinite subsets of ω which was introduced by E. Ellentuck in [82].

Given a finite alphabet Ψ, call L an *infinite dimensional subspace* of $W(\Psi)$ if and only if there is a sequence $\langle s_n(v) \rangle_{n=1}^{\infty}$ in $W(\Psi; v)$ such that L is the set of all constant reduced words of $\langle s_n(v) \rangle_{n=1}^{\infty}$. Similarly, given a finite sequence $\langle w_n(v) \rangle_{n=1}^{d}$ in $W(\Psi; v)$, call L a *d-dimensional subspace* of $W(\Psi)$ if and only if

$$L = \{w_1(a_1) ^\frown w_2(a_2) ^\frown \cdots ^\frown w_d(a_d) : a_1, a_2, \ldots, a_d \in \Psi\}.$$

It is a result of H. Furstenberg and Y. Katznelson [99, Theorem 3.1] that whenever the collection of all d-dimensional subspaces of $W(\Psi)$ are partitioned into finitely many pieces, there exists an infinite dimensional subspace of $W(\Psi)$ all of whose d-dimensional subspaces lie in the same cell of the partition. This result can be established by methods similar to those of Section 18.3. (See [20].)

Part IV

Connections With Other Structures

Chapter 19

Relations With Topological Dynamics

We have already seen that the notions of *syndetic* and *piecewise syndetic*, which have their origins in topological dynamics, are important in the theory of βS for a discrete semigroup S. We have also remarked that the notion of *central*, which is very important in the theory of βS and has a very simple algebraic definition, originated in topological dynamics. In this chapter we investigate additional relations between these theories. In particular, we establish the equivalence of the algebraic and dynamical definitions of "central".

19.1 Minimal Dynamical Systems

The most fundamental notion in the study of topological dynamics is that of *dynamical system*. We remind the reader that we take all hypothesized topological spaces to be Hausdorff.

Definition 19.1. A *dynamical system* is a pair $(X, \langle T_s \rangle_{s \in S})$ such that

(1) X is a compact topological space (called the *phase space* of the system);

(2) S is a semigroup;

(3) for each $s \in S$, T_s is a continuous function from X to X; and

(4) for all $s, t \in S$, $T_s \circ T_t = T_{st}$.

If $(X, \langle T_s \rangle_{s \in S})$ is a dynamical system, one says that the semigroup S *acts on* X via $\langle T_s \rangle_{s \in S}$.

We observe that we are indeed familiar with certain dynamical systems.

Remark 19.2. *Let S be a discrete semigroup. Then $(\beta S, \langle \lambda_s \rangle_{s \in S})$ is a dynamical system.*

Definition 19.3. Let $(X, \langle T_s \rangle_{s \in S})$ be a dynamical system. A subset Y of X is *invariant* if and only if for every $s \in S$, $T_s[Y] \subseteq Y$.

Lemma 19.4. *Let S be a semigroup. The closed invariant subsets of the dynamical system $(\beta S, \langle \lambda_s \rangle_{s \in S})$ are precisely the closed left ideals of βS.*

Proof. Given a left ideal L of βS, $p \in L$, and $s \in S$, one has $\lambda_s(p) = sp \in L$ so left ideals are invariant.

Given a closed invariant subset Y of βS and $p \in Y$ one has that $(\beta S)p = c\ell(Sp) \subseteq c\ell Y = Y$.   □

Definition 19.5. The dynamical system $(X, \langle T_s \rangle_{s \in S})$ is *minimal* if and only if there are no nonempty closed proper invariant subsets of X.

A simple application of Zorn's Lemma shows that, given any dynamical system $(X, \langle T_s \rangle_{s \in S})$, X contains a minimal nonempty closed invariant subset Y, and consequently $(Y, \langle T'_s \rangle_{s \in S})$ is a minimal dynamical system, where T'_s is the restriction of T_s to Y.

Lemma 19.6. *Let S be a semigroup. The minimal closed invariant subsets of the dynamical system $(\beta S, \langle \lambda_s \rangle_{s \in S})$ are precisely the minimal left ideals of βS.*

Proof. This is an immediate consequence of Lemma 19.4 and the fact that minimal left ideals of βS are closed (Corollary 2.6).   □

Of course, a given semigroup can act on many different topological spaces. For example, if X is any topological space, f is any continuous function from X to X, and for each $n \in \mathbb{N}$, $T_n = f^n$, then $(X, \langle T_n \rangle_{n \in \mathbb{N}})$ is a dynamical system. And, since any dynamical system contains a minimal dynamical system, a given semigroup may act minimally on many different spaces.

Definition 19.7. Let S be a semigroup. Then $(X, \langle T_s \rangle_{s \in S})$ is a *universal minimal dynamical system for S* if and only if $(X, \langle T_s \rangle_{s \in S})$ is a minimal dynamical system and, whenever $(Y, \langle R_s \rangle_{s \in S})$ is a minimal dynamical system, there is a continuous function φ from X onto Y such that for each $s \in S$ one has that $R_s \circ \varphi = \varphi \circ T_s$. (That is, given any $s \in S$, the diagram

$$
\begin{array}{ccc}
X & \xrightarrow{\;\;T_s\;\;} & X \\
\varphi \downarrow & & \downarrow \varphi \\
Y & \xrightarrow{\;\;R_s\;\;} & Y
\end{array}
$$

commutes.)

Theorem 19.8. *Let S be a semigroup, let L be a minimal left ideal of βS, and for each $s \in S$, let λ'_s be the restriction of λ_s to L. Then $(L, \langle \lambda'_s \rangle_{s \in S})$ is a universal minimal dynamical system for S.*

Proof. Let $(Y, \langle R_s \rangle_{s \in S})$ be a minimal dynamical system and fix $y \in Y$. Define $g : S \to Y$ by $g(s) = R_s(y)$ and let $\varphi = \widetilde{g}_{|L}$. Then immediately we have that φ is a continuous function from L to Y. To complete the proof it suffices to show that for all $s \in S$ and all $p \in L$, $\varphi(\lambda'_s(p)) = R_s(\varphi(p))$. (For then $R_s[\varphi[L]] \subseteq \varphi[L]$ so $\varphi[L]$ is invariant and thus by minimality $\varphi[L] = Y$.)

To see that $\varphi(\lambda'_s(p)) = R_s(\varphi(p))$ for all $s \in S$ and all $p \in L$, it suffices to show that for each $s \in S$, $\widetilde{g} \circ \lambda_s = R_s \circ \widetilde{g}$ on βS for which it in turn suffices to show that $\widetilde{g} \circ \lambda_s = R_s \circ \widetilde{g}$ on S. So let $t \in S$ be given. Then

$$g(\lambda_s(t)) = g(st) = R_{st}(y) = R_s(R_t(y)) = R_s(g(t)).$$

$\square$

We now wish to show that $(L, \langle \lambda'_s \rangle_{s \in S})$ is the *unique* universal minimal dynamical system for S. For this we need the following lemma which is interesting in its own right.

Lemma 19.9. *Let S be a semigroup and let L and L' be minimal left ideals of βS. If φ is a continuous function from L to L' such that $\lambda_s \circ \varphi = \varphi \circ \lambda'_s$ for all $s \in S$ (where λ'_s is the restriction of λ_s to L), then there is some $p \in L'$ such that φ is the restriction of ρ_p to L.*

Proof. Pick by Corollary 2.6 some idempotent $q \in L$ and let $p = \varphi(q)$. Then ρ_p and $\varphi \circ \rho_q$ are continuous functions from βS to L'. Further, given any $s \in S$,

$$\rho_p(s) = sp = s\varphi(q) = (\lambda_s \circ \varphi)(q) = (\varphi \circ \lambda'_s)(q) = \varphi(sq) = (\varphi \circ \rho_q)(s)$$

and thus ρ_p and $\varphi \circ \rho_q$ are continuous functions agreeing on S and are therefore equal. In particular, $\rho_{p|L} = (\varphi \circ \rho_q)_{|L}$. Now, by Lemma 1.30 and Theorem 1.59, q is a right identity for L, so $(\varphi \circ \rho_q)_{|L} = \varphi$ and thus $\varphi = \rho_{p|L}$ as required. $\square$

The following theorem is in many respects similar to Remark 3.26 which asserted the uniqueness of the Stone–Čech compactification. However, unlike that remark, the proof of Theorem 19.10 is not trivial. That is, given another universal minimal dynamical system $(X, \langle T_s \rangle_{s \in S})$ and given $s \in S$, one obtains the following diagram

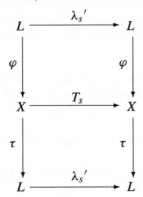

and would like to use it to show that φ is one to one. However, if for example S is a left zero semigroup, so that βS is also a left zero semigroup, then $\lambda_s(p) = \lambda_s(q)$ for all p and q in βS, and the above diagram is of no use in showing that φ is one to one.

Theorem 19.10. *Let S be a semigroup and let L be a minimal left ideal of βS. Then, up to a homeomorphism respecting the action of S, $(L, \langle \lambda'_s \rangle_{s \in S})$ is the unique universal minimal dynamical system for S. That is, given any universal minimal dynamical system $(X, \langle T_s \rangle_{s \in S})$ for S there is a homeomorphism $\varphi : L \to X$ such that $\varphi \circ \lambda'_s = T_s \circ \varphi$ for every $s \in S$.*

Proof. Pick $\varphi : L \to X$ as guaranteed by the fact that $(L, \langle \lambda'_s \rangle_{s \in S})$ is a universal minimal dynamical system for S and pick $\tau : X \to L$ as guaranteed by the fact that $(X, \langle T_s \rangle_{s \in S})$ is a universal minimal dynamical system for S.

Then one has immediately that $\varphi \circ \lambda'_s = T_s \circ \varphi$ for every $s \in S$ and that φ is a continuous function on a compact space to a Hausdorff space and is thus closed. Therefore, it suffices to show that φ is one to one. Now $\tau \circ \varphi : L \to L$ and for each $s \in S$,

$$\lambda_s \circ (\tau \circ \varphi) = \lambda'_s \circ (\tau \circ \varphi) = (\tau \circ \varphi) \circ \lambda'_s$$

so by Lemma 19.9 there is some $p \in L$ such that $\tau \circ \varphi$ is the restriction of ρ_p to L. Since, by Theorem 2.11(c), the restriction of ρ_p to L is one to one, we are done.   □

19.2 Enveloping Semigroups

We saw in Theorem 2.29 that if X is a topological space and $\mathcal{T}$ is a semigroup contained in $^X X$ (with the product topology) and each member of $\mathcal{T}$ is continuous, then the closure of $\mathcal{T}$ in $^X X$ is a semigroup under $\circ$, called the *enveloping semigroup* of $\mathcal{T}$. In particular (and this is the origin of the notion of enveloping semigroup), if $(X, \langle T_s \rangle_{s \in S})$ is a dynamical system, then the closure of $\{ T_s : s \in S \}$ in $^X X$ is a semigroup which is referred to as the enveloping semigroup of the dynamical system.

Theorem 19.11. *Let $(X, \langle T_s \rangle_{s \in S})$ be a dynamical system and define $\varphi : S \to {}^X X$ by $\varphi(s) = T_s$. Then $\widetilde{\varphi}$ is a continuous homomorphism from βS onto the enveloping semigroup of $(X, \langle T_s \rangle_{s \in S})$.*

Proof. Immediately one has that $\widetilde{\varphi}$ is continuous and that $\widetilde{\varphi}[\beta S] = c\ell\{T_s : s \in S\}$. Also by Theorem 2.2, each T_x is in the topological center of $^X X$ so by Corollary 4.22, φ is a homomorphism.   □

The following notation will be convenient in the next section.

Definition 19.12. Let $(X, \langle T_s \rangle_{s \in S})$ be a dynamical system and define $\varphi : S \to {}^X X$ by $\varphi(s) = T_s$. For each $p \in \beta S$, let $T_p = \widetilde{\varphi}(p)$.

As an immediate consequence of Theorem 19.11 we have the following. Be cautioned however that T_p is usually not continuous, so $(X, \langle T_p \rangle_{p \in \beta S})$ is usually not a dynamical system.

Remark 19.13. *Let $(X, \langle T_s \rangle_{s \in S})$ be a dynamical system and let $p, q \in \beta S$. Then $T_p \circ T_q = T_{pq}$ and for each $x \in X$, $T_p(x) = p\text{-}\lim_{s \in S} T_s(x)$.*

How close the enveloping semigroup comes to being a copy of βS can be viewed as a measure of the complexity of the action $\langle T_s \rangle_{s \in S}$ of S on X. With certain weak cancellation requirements on S, we see that there is one dynamical system for which βS is guaranteed to be the enveloping semigroup.

Lemma 19.14. *Let Q be a semigroup, let S be a subsemigroup of Q, and let $\Omega = {}^Q\{0, 1\}$ with the product topology. For each $s \in S$, define $T_s : \Omega \to \Omega$ by $T_s(f) = f \circ \rho_s$. Then $(\Omega, \langle T_s \rangle_{s \in S})$ is a dynamical system. Furthermore, for each $p \in \beta S$, $f \in \Omega$, and $s \in S$,*

$$\left(T_p(f) \right)(s) = 1 \Leftrightarrow \{ t \in S : f(t) = 1 \} \in sp.$$

Proof. Let $s \in S$. To see that T_s is continuous it is enough to show that $\pi_t \circ T_s$ is continuous for each $t \in Q$. So, let $t \in Q$ and let $f \in \Omega$. Then

$$(\pi_t \circ T_s)(f) = \pi_t(f \circ \rho_s) = (f \circ \rho_s)(t) = f(ts) = \pi_{ts}(f).$$

That is to say that $\pi_t \circ T_s = \pi_{ts}$ and is therefore continuous.

Now let $s, t \in S$. To see that $T_s \circ T_t = T_{st}$, let $f \in \Omega$. Then

$$(T_s \circ T_t)(f) = T_s\left(T_t(f) \right) = T_s(f \circ \rho_t) = f \circ \rho_t \circ \rho_s = f \circ \rho_{st} = T_{st}(f).$$

Finally, let $p \in \beta S$, $f \in \Omega$ and $s \in S$. Then

$$
\begin{aligned}
\left(T_p(f) \right)(s) = 1 &\Leftrightarrow \left(p\text{-}\lim_{u \in S} T_u(f) \right)(s) = 1 \\
&\Leftrightarrow p\text{-}\lim_{u \in S} \left(T_u(f) \right)(s) = 1 \\
&\Leftrightarrow p\text{-}\lim_{u \in S} f(su) = 1 \\
&\Leftrightarrow \widetilde{f}(sp) = 1 \\
&\Leftrightarrow \{ t \in S : f(t) = 1 \} \in sp. \qquad \square
\end{aligned}
$$

Notice that by Lemma 8.1, the hypotheses of the following theorem hold whenever S has any left cancelable element. In particular, they hold in the important cases in which S is $\mathbb{N}$ or $\mathbb{Z}$.

Theorem 19.15. *Let S be a semigroup, let $\Omega = {}^S\{0, 1\}$, and for each $s \in S$, define $T_s : \Omega \to \Omega$ by $T_s(f) = f \circ \rho_s$. If for every pair of distinct elements p and q of βS there is some $s \in S$ such that $sp \neq sq$, then βS is topologically and algebraically isomorphic to the enveloping semigroup of $(\Omega, \langle T_s \rangle_{s \in S})$.*

Proof. By Lemma 19.14, $(\Omega, \langle T_s \rangle_{s \in S})$ is a dynamical system. Define $\varphi : S \to {}^{\Omega}\Omega$ by $\varphi(s) = T_s$. Then by Theorem 19.11, $\widetilde{\varphi}$ is a continuous homomorphism from βS onto the enveloping semigroup of $(\Omega, \langle T_s \rangle_{s \in S})$. Thus we need only show that $\widetilde{\varphi}$ is one to one.

So let p and q be distinct members of βS and pick $s \in S$ such that $sp \neq sq$. Pick $A \in sp \backslash sq$ and let χ_A be the characteristic function of A. Then by Lemma 19.14, $\big(T_p(\chi_A)\big)(s) = 1$ and $\big(T_q(\chi_A)\big)(s) = 0$. Therefore $\widetilde{\varphi}(p) = T_p \neq T_q = \widetilde{\varphi}(q)$. $\square$

We have seen that if L is a closed left ideal of βS and λ'_s is the restriction of λ_s to L, then $(L, \langle \lambda'_s \rangle_{s \in S})$ is a dynamical system. We know further (by Theorems 19.8 and 19.10) that if L is a minimal left ideal of βS, then $(L, \langle \lambda'_s \rangle_{s \in S})$ is the universal minimal dynamical system for S. It is a natural question as to whether βS can be the enveloping semigroup of $(L, \langle \lambda'_s \rangle_{s \in S})$.

Lemma 19.16. *Let S be a semigroup and let L be a closed left ideal of βS and for each $s \in S$, let λ'_s be the restriction of λ_s to L. Define $\varphi : S \to {}^{L}L$ by $\varphi(s) = \lambda'_s$. Then for all $p \in \beta S$, $\widetilde{\varphi}(p) = \lambda_{p|L}$.*

Proof. Let $p \in \beta S$, let $q \in L$, and let $T_p - \widetilde{\varphi}(p)$. Then by Remark 19.13, $\widetilde{\varphi}(p)(q) = T_p(q) = p\text{-}\lim_{s \in S} sq = pq$. $\square$

Notice that if L contains any element which is right cancelable in βS then the hypotheses of the following theorem are satisfied. (See Chapter 8 for characterizations of right cancelability.) If S is a countable semigroup which can be embedded in a group and if $L \not\subseteq K(\beta S)$, then L does contain an element which is right cancelable in βS (by Theorem 6.56). So, in this case, the enveloping semigroup of $(L, \langle \lambda_s \rangle_{s \in S})$ is topologically and algebraically isomorphic to βS.

Theorem 19.17. *Let S be a semigroup, let L be a closed left ideal of βS, and for each $s \in S$, let λ'_s be the restriction of λ_s to L. Then βS is topologically and algebraically isomorphic to the enveloping semigroup of $(L, \langle \lambda_s \rangle_{s \in S})$ via a map which takes s to λ'_s if and only if whenever q and r are distinct elements of βS, there is some $p \in L$ such that $qp \neq rp$.*

Proof. Define $\varphi : S \to {}^{L}L$ by $\varphi(s) = \lambda'_s$. By Theorem 19.11 $\widetilde{\varphi}$ is a continuous homomorphism onto the enveloping semigroup of $(L, \langle \lambda_s \rangle_{s \in S})$ (and is the only continuous function extending φ). The conclusion now follows from Lemma 19.16. $\square$

If L is a minimal left ideal of βS, we have the following superficially weaker condition.

Theorem 19.18. *Let S be a semigroup, let L be a minimal left ideal of βS, and for each $s \in S$, let λ'_s be the restriction of λ_s to L. Then βS is topologically and algebraically isomorphic to the enveloping semigroup of $(L, \langle \lambda_s \rangle_{s \in S})$ via a map which takes s to λ'_s if and only if whenever q and r are distinct elements of βS, there is some $p \in K(\beta S)$ such that $qp \neq rp$.*

Proof. Since $L \subseteq K(\beta S)$, the necessity follows immediately from Theorem 19.17. For the sufficiency, let q and r be distinct members of βS and pick some $p \in K(\beta S)$ such that $qp \neq rp$. Pick by Theorem 2.8 some minimal left ideal L' of βS such that $p \in L'$ and pick some $z \in L$. By Theorem 2.11(c), the restriction of ρ_z to L' is one to one and qp and rp are in L' so $qpz \neq rpz$. Since $pz \in L$, Theorem 19.17 applies. □

We conclude this section by showing that $\beta \mathbb{N}$ is the enveloping semigroup of certain natural actions of $\mathbb{N}$ on the circle group $\mathbb{T}$, which we take here to be the quotient $\mathbb{R}/\mathbb{Z}$ under addition.

Theorem 19.19. *Let $a \in \{2, 3, 4, \ldots\}$ and for each $n \in \mathbb{N}$ define $T_n : \mathbb{T} \to \mathbb{T}$ by $T_n(\mathbb{Z} + x) = \mathbb{Z} + a^n x$. Then $(\mathbb{T}, \langle T_n \rangle_{n \in \mathbb{N}})$ is a dynamical system. If $\varphi : \mathbb{N} \to {}^{\mathbb{T}}\mathbb{T}$ is defined by $\varphi(n) = T_n$, then $\widetilde{\varphi}$ is a continuous isomorphism from $\beta \mathbb{N}$ onto the enveloping semigroup of $(\mathbb{T}, \langle T_n \rangle_{n \in \mathbb{N}})$.*

Proof. Since $a \in \mathbb{N}$, the functions T_n are well defined. Let $n \in \mathbb{N}$, let $\ell : \mathbb{R} \to \mathbb{R}$ be multiplication by a^n, and let $\pi : \mathbb{R} \to \mathbb{T}$ be the projection map. Then $T_n \circ \pi = \pi \circ \ell$ so T_n is continuous.

Trivially, if $n, m \in \mathbb{N}$, then $T_n \circ T_m = T_{n+m}$. Thus $(\mathbb{T}, \langle T_n \rangle_{n \in \mathbb{N}})$ is a dynamical system.

To complete the proof it suffices, by Theorem 19.11, to show that $\widetilde{\varphi}$ is one to one. To this end, let p and q be distinct members of $\beta \mathbb{N}$. Pick $B \in \{2\mathbb{N}, 2\mathbb{N} - 1\}$ such that $B \in p$ and pick $A \in p \backslash q$. Define $x \in (0, 1)$ by

$$x = \Sigma_{j=1}^{\infty} \alpha_j a^{-j} \quad \text{where} \quad \alpha_j = \begin{cases} 1 & \text{if } j - 1 \in A \cap B \\ 0 & \text{if } j - 1 \notin A \cap B. \end{cases}$$

Let

$$C = \{\mathbb{Z} + t : \frac{1}{a} \leq t \leq \frac{1}{a} + \frac{2}{a^3}\} \quad \text{and let} \quad D = \{\mathbb{Z} + t : 0 \leq t \leq \frac{1}{a^2} + \frac{2}{a^4}\}.$$

Notice that (since $a^2 + 2 < a^3$), $C \cap D = \emptyset$. We claim that $\widetilde{\varphi}(p)(x) \in C$ and $\widetilde{\varphi}(q)(x) \in D$. To see this it suffices to show, since C and D are closed, that for each $n \in \mathbb{N}$, if $n \in A \cap B$, then $T_n(x) \in C$, and if $n \notin A \cap B$, then $T_n(x) \in D$.

So let $n \in \mathbb{N}$ be given. Then

$$T_n(x) = \mathbb{Z} + \Sigma_{j=1}^{\infty} \alpha_j a^{n-j} = \mathbb{Z} + \Sigma_{j=n+1}^{\infty} \alpha_j a^{n-j}$$

since $\alpha_j a^{n-j} \in \mathbb{Z}$ if $j \leq n$. Now assume that $n \in A \cap B$. Then $\alpha_{n+1} = 1$ and (because $n + 1 \notin B$) $\alpha_{n+2} = 0$. Thus $\Sigma_{j=n+1}^{\infty} \alpha_j a^{n-j} = a^{-1} + \Sigma_{j=n+3}^{\infty} \alpha_j a^{n-j}$ so

$$\frac{1}{a} \leq \Sigma_{j=n+1}^{\infty} \alpha_j a^{n-j} \leq \frac{1}{a} + \frac{2}{a^3}$$

and hence $T_n(x) \in C$ as claimed.

Next assume that $n \notin A \cap B$. Then $\alpha_{n+1} = 0$ and either $\alpha_{n+2} = 0$ or $\alpha_{n+3} = 0$. Consequently

$$0 \leq \Sigma_{j=n+1}^{\infty} \alpha_j a^{n-j} \leq \frac{1}{a^2} + \frac{2}{a^4}$$

and hence $T_n(x) \in D$ as claimed.   □

If one views the circle group as $\{z \in \mathbb{C} : |z| = 1\}$, then Theorem 19.19 says that the enveloping semigroup of the system $(\mathbb{T}, \langle f^n \rangle_{n \in \mathbb{N}})$ is isomorphic to $\beta\mathbb{N}$, where $f(z) = z^a$.

19.3 Dynamically Central Sets

The term "central set" was originally defined by Furstenberg [98, Definition 8.3] for subsets of $\mathbb{N}$. Sets defined algebraically as elements of minimal idempotents were called "central" because many of the same theorems could be proved about them. The original definition (restricted to the semigroup $\mathbb{N}$ and applied only to metric dynamical systems) is as follows. We add the modifier "dynamically" because it is by no means obvious that the concepts are the same, even for subsets of $\mathbb{N}$.

Definition 19.20. Let S be a semigroup. A set $C \subseteq S$ is *dynamically central* if and only if there exist a dynamical system $(X, \langle T_s \rangle_{s \in S})$, points x and y in X, and a neighborhood U of y such that

(1) y is a uniformly recurrent point of X,

(2) x and y are proximal, and

(3) $C = \{s \in S : T_s(x) \in U\}$.

The terms "uniformly recurrent" and "proximal" are notions from topological dynamics, defined as follows.

Definition 19.21. Let $(X, \langle T_s \rangle_{s \in S})$ be a dynamical system.

(a) A point $y \in X$ is *uniformly recurrent* if and only if for every neighborhood U of y, $\{s \in S : T_s(y) \in U\}$ is syndetic.

(b) Points x and y of X are *proximal* if and only if there is a net $\langle s_\iota \rangle_{\iota \in I}$ in S such that the nets $\langle T_{s_\iota}(x) \rangle_{\iota \in I}$ and $\langle T_{s_\iota}(y) \rangle_{\iota \in I}$ converge to the same point of X.

The definition of "proximal" as standardly given for a dynamical system with a metric phase space (X, d) says that there is some sequence $\langle s_n \rangle_{n=1}^{\infty}$ such that $\lim_{n \to \infty} d\big(T_{s_n}(x), T_{s_n}(y)\big) = 0$. It is easy to see that this definition is equivalent to the one given above.

The characterization of proximality provided by the following lemma will be very convenient for us.

Lemma 19.22. *Let $(X, \langle T_s \rangle_{s \in S})$ be a dynamical system and let $x, y \in S$. Then x and y are proximal if and only if there exists a point $p \in \beta S$ such that $T_p(x) = T_p(y)$.*

Proof. Notice that if $p \in \beta S$, $\langle s_\iota \rangle_{\iota \in I}$ is a net in S which converges to p, and $z \in X$, then $T_p(z) = \lim_{\iota \in I} T_{s_\iota}(z)$. Thus if $p \in \beta S$ and $T_p(x) = T_p(y)$, then picking any net $\langle s_\iota \rangle_{\iota \in I}$ in S which converges to p, one has $\lim_{\iota \in I} T_{s_\iota}(x) = \lim_{\iota \in I} T_{s_\iota}(y)$.

Conversely, assume that one has a net $\langle s_\iota \rangle_{\iota \in I}$ in S such that $\lim_{\iota \in I} T_{s_\iota}(x) = \lim_{\iota \in I} T_{s_\iota}(y)$. Let p be any limit point of $\langle s_\iota \rangle_{\iota \in I}$ in βS. By passing to a subnet we may presume that $\langle s_\iota \rangle_{\iota \in I}$ converges to p. Then $T_p(x) = \lim_{\iota \in I} T_{s_\iota}(x) = \lim_{\iota \in I} T_{s_\iota}(y) = T_p(y)$. □

In order to establish the equivalence of the notions of "central" we investigate relationships between some algebraic and dynamical notions.

Theorem 19.23. *Let $(X, \langle T_s \rangle_{s \in S})$ be a dynamical system, let L be a minimal left ideal of βS, and let $x \in X$. The following statements are equivalent.*

(a) *The point x is a uniformly recurrent point of $(X, \langle T_s \rangle_{s \in S})$.*

(b) *There exists $u \in L$ such that $T_u(x) = x$.*

(c) *There exists an idempotent $e \in L$ such that $T_e(x) = x$.*

(d) *There exists $y \in X$ and an idempotent $e \in L$ such that $T_e(y) = x$.*

Proof. (a) implies (b). Choose any $v \in L$. Let $\mathcal{N}$ be the set of neighborhoods of x in X. For each $U \in \mathcal{N}$ let $B_U = \{s \in S : T_s(x) \in U\}$. Since x is uniformly recurrent, each B_U is syndetic and so there exists a finite set $F_U \subseteq S$ such that $S = \bigcup_{t \in F_U} (t^{-1} B_U)$. Pick $t_U \in F_U$ such that $t_U^{-1} B_U \in v$ so that $t_U v \in \overline{B_U}$. Let u be a limit point of the net $\langle t_U v \rangle_{U \in \mathcal{N}}$ in βS and notice that $u \in L$. We claim that $T_u(x) = x$. By Remark 19.13, it suffices to show that $B_V \in u$ for each $V \in \mathcal{N}$. So let $V \in \mathcal{N}$ and suppose that $B_V \notin u$. Pick $U \subseteq V$ such that $t_U v \in \overline{S \setminus B_V}$. Then $t_U v \in \overline{B_U} \subseteq \overline{B_V}$, a contradiction.

(b) implies (c). Since u belongs to a group contained in L (by Theorem 2.8), there is an idempotent $e \in L$ for which $eu = u$. So (using Remark 19.13) $T_e(x) = T_e\big(T_u(x)\big) = T_{eu}(x) = T_u(x) = x$.

That (c) implies (d) is trivial.

(d) implies (a). We note that $T_e(x) = T_e\big(T_e(y)\big) = T_{ee}(y) = T_e(y) = x$. Let U be a neighborhood of x. We need to show that $\{s \in S : T_s(x) \in U\}$ is syndetic. Pick a neighborhood V of x such that $c\ell\, V \subseteq U$ and let $A = \{s \in S : T_s(x) \in V\}$. Since $x = e\text{-}\lim_{s \in S} T_s(x)$ by Remark 19.13, $A \in e$. Let $B = \{s \in S : se \in \overline{A}\}$. Then by Theorem 4.39, B is syndetic. We claim that $B \subseteq \{s \in S : T_s(x) \in U\}$. Indeed, if $s \in B$, then $T_s(x) = T_s\big(T_e(x)\big) = T_{se}(x) \in \overline{\{T_t(x) : t \in A\}} \subseteq \overline{V} \subseteq U$. □

Theorem 19.24. *Let $(X, \langle T_s \rangle_{s \in S})$ be a dynamical system and let $x \in X$. Then there is a uniformly recurrent point $y \in c\ell\{T_s(x) : s \in S\}$ such that x and y are proximal.*

Proof. Let L be any minimal left ideal of βS and pick an idempotent $u \in L$. Let $y = T_u(x)$. Then trivially $y \in c\ell\{T_s(x) : s \in S\}$. By Theorem 19.23 y is a uniformly recurrent point of $(X, \langle T_s \rangle_{s \in S})$. By Remark 19.13 we have that $T_u(y) = T_u(T_u(x)) = T_{uu}(x) = T_u(x)$ so by Lemma 19.22 x and y are proximal.   □

Theorem 19.25. *Let $(X, \langle T_s \rangle_{s \in S})$ be a dynamical system and let $x, y \in X$. If x and y are proximal, then there is a minimal left ideal L of βS such that $T_u(x) = T_u(y)$ for all $u \in L$.*

Proof. By Lemma 19.22, $\{ p \in \beta S : T_p(x) = T_p(y) \} \neq \emptyset$. It is a left ideal in βS, because, for every $p, q \in \beta S$, $T_p(x) = T_p(y)$ implies that $T_{qp}(x) = T_q(T_p(x)) = T_q(T_p(y)) = T_{qp}(y)$.   □

Theorem 19.26. *Let $(X, \langle T_s \rangle_{s \in S})$ be a dynamical system and let $x, y \in X$. There is a minimal idempotent u in βS such that $T_u(x) = y$ if and only if x and y are proximal and y is uniformly recurrent.*

Proof. Necessity. Since u is minimal, there is a minimal left ideal L of βS such that $u \in L$. Thus by Theorem 19.23 y is uniformly recurrent. By Remark 19.13 $T_u(y) = T_u(T_u(x)) = T_{uu}(x) = T_u(x)$ so by Lemma 19.22 x and y are proximal.

Sufficiency. Pick by Theorem 19.25 a minimal left ideal L of βS such that $T_u(x) = T_u(y)$ for all $u \in L$. Pick by Theorem 19.23 an idempotent $u \in L$ such that $T_u(y) = y$.□

We are now prepared to establish the equivalence of the notions of central.

Theorem 19.27. *Let S be a semigroup and let $B \subseteq S$. Then B is central if and only if B is dynamically central.*

Proof. Necessity. Let $Q = S \cup \{e\}$ where e is a new identity adjoined to S (even if S already has an identity). Let $\Omega = {}^Q\{0, 1\}$ and for $s \in S$ define $T_s : \Omega \to \Omega$ by $T_s(f) = f \circ \rho_s$. Then by Lemma 19.14, $(\Omega, \langle T_s \rangle_{s \in S})$ is a dynamical system.

Let $x = \chi_B$, the characteristic function of $B \subseteq Q$. Pick a minimal idempotent u in βS such that $B \in u$ and let $y = T_u(x)$. Then by Theorem 19.26 y is uniformly recurrent and x and y are proximal.

Now let $U = \{z \in \Omega : z(e) = y(e)\}$. Then U is a neighborhood of y in Ω. We note that $y(e) = 1$. Indeed, $y = T_u(x)$ so $\{s \in S : T_s(x) \in U\} \in u$ so choose some $s \in B$ such that $T_s(x) \in U$. Then $y(e) = T_s(x)(e) = x(es) = 1$. Thus given any $s \in S$,

$$
\begin{aligned}
s \in B \quad &\Leftrightarrow \quad x(s) = 1 \\
&\Leftrightarrow \quad T_s(x)(e) = 1 \\
&\Leftrightarrow \quad T_s(x) \in U.
\end{aligned}
$$

Sufficiency. Choose a dynamical system $(X, \langle T_s \rangle_{s \in S})$, points $x, y \in X$, and a neighborhood U of y such that x and y are proximal, y is uniformly recurrent, and $B = \{s \in S : T_s(x) \in U\}$. Choose by Theorem 19.26 a minimal idempotent u in βS such that $T_u(x) = y$. Then $B \in u$.   □

It is not obvious from the definition that the notion of "dynamically central" is closed under passage to supersets. (If $B = \{s \in S : T_s(x) \in U\}$ and $B \subseteq C$, one can let $V = U \cup \{T_s(x) : s \in C\}$. But there is no reason to believe that $\{s \in S : T_s(x) \in V\} \subseteq C$.)

Corollary 19.28. *Let S be a semigroup and let $B \subseteq C \subseteq S$. If B is dynamically central, then C is dynamically central.*

Proof. This follows from Theorem 19.27 and the fact that supersets of central sets are central. □

Exercise 19.3.1. Let $(X, \langle T_s \rangle_{s \in S})$ be a dynamical system and let $x \in X$. The point $x \in X$ is *periodic* if and only if there is some $s \in S$ which is not an identity such that $T_s(x) = x$. The point $x \in X$ is *recurrent* if and only if every neighborhood of x contains a point of the form $T_s(x)$ for some $s \in S$ other than an identity.

Assume that x is not a periodic point of X and that S^* is a subsemigroup of βS. (See Theorem 4.28.) Prove that the following statements are equivalent:

(i) x is recurrent.

(ii) $T_p(x) = x$ for some $p \in S^*$.

(iii) $T_e(x) = x$ for some idempotent $e \in S^*$.

(iv) Given a neighborhood U of x, there exists a sequence $\langle s_n \rangle_{n=1}^{\infty}$ of distinct elements of S such that $T_s(x) \in U$ for every $s \in \mathrm{FP}(\langle s_n \rangle_{n=1}^{\infty})$.

19.4 Dynamically Generated IP* Sets

In this section we consider certain dynamically defined sets that are always IP* sets. In contrast with the results of Section 19.3, we see that the notion of a "dynamical IP* set" is not equivalent to that of "IP* set", but is rather significantly stronger.

The dynamical notion with which we shall be concerned in this section is that of a *measure preserving system*.

Definition 19.29. (a) A *measure space* is a triple $(X, \mathcal{B}, \mu)$, where X is a set, $\mathcal{B}$ is a σ-algebra of subsets of X, and μ is a countably additive measure on $\mathcal{B}$ with $\mu(X)$ finite.

(b) Given a measure space $(X, \mathcal{B}, \mu)$, a function $T : X \to X$ is a *measure preserving transformation* if and only if for all $B \in \mathcal{B}$, $T^{-1}[B] \in \mathcal{B}$ and $\mu(T^{-1}[B]) = \mu(B)$.

(c) Given a semigroup S and a measure space $(X, \mathcal{B}, \mu)$, a *measure preserving action* of S on X is an indexed family $\langle T_s \rangle_{s \in S}$ such that each T_s is a measure preserving transformation of X and $T_s \circ T_t = T_{st}$ for all $s, t \in S$. It is also required that if S has an identity e, then T_e is the identity function on X.

(d) A *measure preserving system* is a quadruple $(X, \mathcal{B}, \mu, \langle T_s \rangle_{s \in S})$ such that $(X, \mathcal{B}, \mu)$ is a measure space and $\langle T_s \rangle_{s \in S}$ is a measure preserving action of S on X.

We shall need a preliminary result about the semigroup $(\mathcal{P}_f(\mathbb{N}), \cup)$.

Definition 19.30. Let $(X, \mathcal{B}, \mu)$ be a measure space. A *monotone action* of $\mathcal{P}_f(\mathbb{N})$ on X is an indexed family $\langle T_F \rangle_{F \in \mathcal{P}_f(\mathbb{N})}$ such that

(a) each T_F is a measure preserving transformation of X and

(b) if $F, K \in \mathcal{P}_f(\mathbb{N})$ and $\max F < \min K$, then $T_F \circ T_K = T_{F \cup K}$.

Notice that, because of the "$\max F < \min K$" requirement, a monotone action of $\mathcal{P}_f(\mathbb{N})$ on X need not be a measure preserving action of $\mathcal{P}_f(\mathbb{N})$ on X.

Lemma 19.31. *Let* $\langle T_F \rangle_{F \in \mathcal{P}_f(\mathbb{N})}$ *be a monotone action of* $\mathcal{P}_f(\mathbb{N})$ *on the measure space* $(X, \mathcal{B}, \mu)$ *and let* $A \in \mathcal{B}$ *with* $\mu(A) > 0$. *Then for each* $k \in \mathbb{N}$ *there exists* $H \in \mathcal{P}_f(\mathbb{N})$ *with* $\min H > k$ *such that* $\mu(A \cap T_H{}^{-1}[A]) > 0$.

Proof. Pick $m \in \mathbb{N}$ such that $\mu(A) > \mu(X)/m$. For $n \in \{1, 2, \ldots, m\}$, let $F_n = \{k + n, k + n + 1, \ldots, k + m\}$. Then each $\mu(T_{F_n}{}^{-1}[A]) = \mu(A) > \mu(X)/m$ so pick $n, \ell \in \{1, 2, \ldots, m\}$ such that $n < \ell$ and $\mu(T_{F_n}{}^{-1}[A] \cap T_{F_\ell}{}^{-1}[A]) > 0$. Let $H = \{k+n, k+n+1, \ldots, k+\ell-1\}$. Then $T_H \circ T_{F_\ell} = T_{F_n}$ so $T_{F_n}{}^{-1}[A] = T_{F_\ell}{}^{-1}[T_H{}^{-1}[A]]$ so

$$
\begin{aligned}
\mu(T_H{}^{-1}[A] \cap A) &= \mu(T_{F_\ell}{}^{-1}[T_H{}^{-1}[A] \cap A]) \\
&= \mu(T_{F_\ell}{}^{-1}[T_H{}^{-1}[A]] \cap T_{F_\ell}{}^{-1}[A]) \\
&= \mu(T_{F_n}{}^{-1}[A] \cap T_{F_\ell}{}^{-1}[A]) > 0. \qquad \square
\end{aligned}
$$

Lemma 19.32. *Let* $(X, \mathcal{B}, \mu, \langle T_s \rangle_{s \in S})$ *be a measure preserving system and let* $A \in \mathcal{B}$ *satisfy* $\mu(A) > 0$. *For every sequence* $\langle s_n \rangle_{n=1}^{\infty}$ *and every* $k \in \mathbb{N}$, *there exists* $F \in \mathcal{P}_f(\mathbb{N})$ *such that* $k < \min F$ *and* $\mu(A \cap T_s^{-1}[A]) > 0$, *where* $s = \prod_{n \in F} s_n$.

Proof. For $F \in \mathcal{P}_f(\mathbb{N})$, let $R_F = T_{\prod_{n \in F} s_n}$. Then $\langle R_F \rangle_{F \in \mathcal{P}_f(\mathbb{N})}$ is a monotone action of $\mathcal{P}_f(\mathbb{N})$ on X. So we can choose a set F with the required properties by Lemma 19.31. $\qquad \square$

Theorem 19.33. *Let* $(X, \mathcal{B}, \mu, \langle T_s \rangle_{s \in S})$ *be a measure preserving system and let* $C \subseteq S$. *If there is some* $A \in \mathcal{B}$ *with* $\mu(A) > 0$ *such that* $\{s \in S : \mu(A \cap T_s^{-1}[A]) > 0\} \subseteq C$, *then* C *is an IP* set.*

Proof. This follows immediately from Lemma 19.32. $\qquad \square$

Definition 19.34. Let S be a semigroup. A subset C of S is a *dynamical IP* set* if and only if there exist a measure preserving system $(X, \mathcal{B}, \mu, \langle T_s \rangle_{s \in S})$ and an $A \in \mathcal{B}$ with $\mu(A) > 0$ such that $\{s \in S : \mu(A \cap T_s^{-1}[A]) > 0\} \subseteq C$.

Recall that by Theorem 16.32, there is an IP* set B in $(\mathbb{N}, +)$ such that for each $n \in \mathbb{N}$, neither $n + B$ nor $-n + B$ is an IP* set. Consequently, the following simple result shows that not every IP* set is a dynamical IP* set.

Theorem 19.35. *Let* B *be a dynamical IP* set in* $(\mathbb{N}, +)$. *There is a dynamical IP* set* $C \subseteq B$ *such that for each* $n \in C$, $-n + C$ *is a dynamical IP* set (and hence* $-n + B$ *is a dynamical IP* set).*

Proof. Pick a measure space $(X, \mathcal{B}, \mu)$, a measure preserving action $\langle T_n \rangle_{n \in \mathbb{N}}$ of $\mathbb{N}$ on X, and a set $A \in \mathcal{B}$ such that $\mu(A) > 0$ and $\{n \in \mathbb{N} : \mu(A \cap T_n^{-1}[A]) > 0\} \subseteq B$. Let $C = \{n \in \mathbb{N} : \mu(A \cap T_n^{-1}[A]) > 0\}$. To see that C is as required, let $n \in C$ and let $D = A \cap T_n^{-1}[A]$. We claim that $\{m \in \mathbb{N} : \mu(D \cap T_m^{-1}[D]) > 0\} \subseteq -n + C$. To this end, let $m \in \mathbb{N}$ such that $\mu(D \cap T_m^{-1}[D]) > 0$. Then

$$\begin{aligned}
D \cap T_m^{-1}[D] &= A \cap T_n^{-1}[A] \cap T_m^{-1}\big[A \cap T_n^{-1}[A]\big] \\
&\subseteq A \cap T_m^{-1}\big[T_n^{-1}[A]\big] \\
&= A \cap (T_n \circ T_m)^{-1}[A] \\
&= A \cap T_{n+m}^{-1}[A]
\end{aligned}$$

so $n + m \in C$. $\qquad\square$

Recall from Theorem 18.15 that an IP* set in $\mathbb{N} \times \mathbb{N}$ need not contain $\{w\} \times \text{FS}(\langle x_n \rangle_{n=1}^\infty)$ for any sequence $\langle x_n \rangle_{n=1}^\infty$. In the final result of this section we show that a dynamical IP* set in the product of two semigroups with identities must contain sets of the form $\text{FP}(\langle x_n \rangle_{n=1}^\infty) \times \text{FP}(\langle y_n \rangle_{n=1}^\infty)$. Indeed, it has a much stronger property: given any sequences $\langle w_n \rangle_{n=1}^\infty$ and $\langle z_n \rangle_{n=1}^\infty$ one can choose infinite product subsystems $\langle x_n \rangle_{n=1}^\infty$ of $\langle w_n \rangle_{n=1}^\infty$ and $\langle y_n \rangle_{n=1}^\infty$ of $\langle z_n \rangle_{n=1}^\infty$ with $\text{FP}(\langle x_n \rangle_{n=1}^\infty) \times \text{FP}(\langle y_n \rangle_{n=1}^\infty)$ contained in the given dynamical IP* set. More than this, they can be chosen in a parallel fashion. That is if $x_n = \prod_{t \in H_n} w_t$, then $y_n = \prod_{t \in H_n} z_t$.

We only discuss the product of two semigroups for simplicity and because the generalization to arbitrary finite products is straightforward.

In the proof of Theorem 19.36 we utilize product measure spaces. In doing so we shall use the customary notation. If $(X, \mathcal{B}, \mu)$ is a measure space and $m \in \mathbb{N}$, then $(X^m, \mathcal{B}^m, \mu^m)$ denotes the measure space defined as follows: X^m is the usual Cartesian product, $\mathcal{B}^m$ denotes the σ-algebra generated by the sets $\times_{i=1}^m A_i$ where $A_i \in \mathcal{B}$ for each $i \in \{1, 2, \ldots, m\}$, and μ^m denotes the countably additive measure on $\mathcal{B}^m$ such that $\mu^m(\times_{i=1}^m A_i) = \prod_{i=1}^m \mu(A_i)$ for every sequence $\langle A_i \rangle_{i=1}^m$ in $\mathcal{B}$.

Any transformation $T : X \to X$ defines a transformation $\Pi_m T : X^m \to X^m$ for which $\Pi_m T(x_1, x_2, \ldots, x_m) = \big(T(x_1), T(x_2), \ldots, T(x_m)\big)$. If T is measure preserving, so is $\Pi_m T$, because $(\Pi_m T)^{-1}[\times_{i=1}^m A_i] = \times_{i=1}^m (T^{-1}[A_i])$ for every $A_1, A_2, \ldots, A_m \subseteq X$.

Theorem 19.36. *Let S_1 and S_2 be semigroups with identities and let C be a dynamical IP* set in $S_1 \times S_2$. Let $\langle w_n \rangle_{n=1}^\infty$ be a sequence in S_1 and let $\langle z_n \rangle_{n=1}^\infty$ be a sequence in S_2. There exists a sequence $\langle H_n \rangle_{n=1}^\infty$ in $\mathcal{P}_f(\mathbb{N})$ such that*

(a) *for each n, $\max H_n < \min H_{n+1}$ and*

(b) *if for each n, $x_n = \prod_{t \in H_n} w_t$ and $y_n = \prod_{t \in H_n} z_t$, then*

$$\text{FP}(\langle x_n \rangle_{n=1}^\infty) \times \text{FP}(\langle y_n \rangle_{n=1}^\infty) \subseteq C.$$

Proof. Let $S = S_1 \times S_2$. Pick a measure space $(X, \mathcal{B}, \mu)$, a measure preserving action $\langle T_{(s,t)} \rangle_{(s,t) \in S_1 \times S_2}$ of S on X, and $B \in \mathcal{B}$ with $\mu(B) > 0$ such that,

$$D = \{(s, t) \in S : \mu(B \cap T_{(s,t)}^{-1}[B]) > 0\} \subseteq C.$$

Let e and f denote the identities of S_1 and S_2 respectively. For each $\sigma = (s, t) \in S$, we define $U_\sigma : X^3 \to X^3$ by putting $U_\sigma(x_1, x_2, x_3) = \left(T_{(s,f)}(x_1), T_{(e,t)}(x_2), T_{(s,t)}(x_3) \right)$. Since $U_\sigma^{-1}[A_1 \times A_2 \times A_3] = T_{(s,f)}^{-1}[A_1] \times T_{(e,t)}^{-1}[A_2] \times T_{(s,t)}^{-1}[A_3]$ for every $A_1, A_2, A_3 \in \mathcal{B}$, it follows that $(X^3, \mathcal{B}^3, \mu^3, \langle U_\sigma \rangle_{\sigma \in S})$ is a measure preserving system.

For each $n \in \mathbb{N}$, we define $\sigma_n \in S$ by $\sigma_n = (w_n, z_n)$.

We shall inductively choose a sequence $\langle H_n \rangle_{n=1}^\infty$ satisfying (a) and the following condition, which is clearly stronger than (b):

(c) if $x_i = \prod_{k \in H_i} w_k$ and $y_i = \prod_{k \in H_i} z_k$, then $\{(x, f), (e, y), (x, y)\} \subseteq D$ whenever $(x, y) \in \mathrm{FP}(\langle x_i \rangle_{i=1}^n) \times \mathrm{FP}(\langle y_i \rangle_{i=1}^n)$.

We first apply Lemma 19.32 to the measure preserving system $(X^3, \mathcal{B}^3, \mu^3, \langle U_\sigma \rangle_{\sigma \in S})$ and the set $A = B^3$. This allows us to choose $H_1 \in \mathcal{P}_f(\mathbb{N})$ such that $\mu^3(A \cap U_\sigma^{-1}[A]) > 0$, where $\sigma = \prod_{i \in H_1} \sigma_i = (x_1, y_1)$. This means that

$$\mu^3 \left(B^3 \cap (T_{(x_1, f)}^{-1}[B] \times T_{(e, y_1)}^{-1}[B] \times T_{(x_1, y_1)}^{-1}[B]) \right) > 0.$$

So H_1 satisfies (c).

We now suppose that $H_1, H_2, \ldots, H_n$ have been chosen.

Let $m = |\mathrm{FP}(\langle x_i \rangle_{i=1}^n) \times \mathrm{FP}(\langle y_i \rangle_{i=1}^n)|$. We apply Lemma 19.32 again, this time to the measure preserving system $\left((X^3)^{3m}, (\mathcal{B}^3)^{3m}, (\mu^3)^{3m}, \langle \prod_{3m} U_\sigma \rangle_{\sigma \in S} \right)$, and the set

$$A = \underset{}{\times} \left\{ (B \cap T_{(x,f)}^{-1}[B])^3 \times (B \cap T_{(e,y)}^{-1}[B])^3 \times (B \cap T_{(x,y)}^{-1}[B])^3) : \right.$$
$$\left. (x, y) \in \mathrm{FP}(\langle x_i \rangle_{i=1}^n) \times \mathrm{FP}(\langle y_i \rangle_{i=1}^n) \right\}.$$

This allows us to choose $H_{n+1} \in \mathcal{P}_f(\mathbb{N})$ such that $\max(H_n) < \min(H_{n+1})$ and $(\mu^3)^{3m}(A \cap (\prod_{3m} U_\sigma)^{-1}[A]) > 0$, where $\sigma = \prod_{i \in H_{n+1}} \sigma_i = (x_{n+1}, y_{n+1})$.

This implies that, for every $(x, y) \in \mathrm{FP}(\langle x_i \rangle_{i=1}^n) \times \mathrm{FP}(\langle y_i \rangle_{i=1}^n)$, we have

$$\mu(B \cap T_\alpha^{-1}[B] \cap T_\beta^{-1}[B \cap T_\alpha^{-1}[B]]) = \mu(B \cap T_\alpha^{-1}[B] \cap T_\beta^{-1}[B] \cap T_{\alpha\beta}^{-1}[B]) > 0$$

whenever $\alpha \in \{(x, f), (e, y), (x, y)\}$ and $\beta \in \{(x_{n+1}, f), (e, y_{n+1}), (x_{n+1}, y_{n+1})\}$.

We can therefore deduce each of the following statements:

(1) $(x_{n+1}, f) \in D$ and $(xx_{n+1}, f) \in D$ (putting $\alpha = (x, f)$ and $\beta = (x_{n+1}, f)$).

(1') $(e, y_{n+1}) \in D$ and $(e, yy_{n+1})) \in D$ (putting $\alpha = (e, y)$ and $\beta = (e, y_{n+1})$).

(2) $(x_{n+1}, y) \in D$ (putting $\alpha = (e, y)$ and $\beta = (x_{n+1}, f)$).

(2') $(x, y_{n+1}) \in D$ (putting $\alpha = (x, f)$ and $\beta = (e, y_{n+1})$).

(3) $(xx_{n+1}, y) \in D$ (putting $\alpha = (x, y)$ and $\beta = (x_{n+1}, f)$).

(3') $(x, yy_{n+1}) \in D$ (putting $\alpha = (x, y)$ and $\beta = (e, y_{n+1})$).

(4) $(x_{n+1}, yy_{n+1}) \in D$ (putting $\alpha = (e, y)$ and $\beta = (x_{n+1}, y_{n+1})$).

(4') $(xx_{n+1}, y) \in D$ (putting $\alpha = (x, f)$ and $\beta = (x_{n+1}, y_{n+1})$).

(5) $(x_{n+1}, y_{n+1}) \in D$ and $(xx_{n+1}, yy_{n+1}) \in D$ (putting $\alpha = (x, y)$ and $\beta = (x_{n+1}, y_{n+1})$).

It is now easy to check that our inductive assumption extends to $H_1, H_2, \ldots, H_{n+1}$. So the sequence $\langle H_n \rangle_{n=1}^\infty$ can be chosen inductively. $\qquad \square$

Notes

The notion of "dynamical system" is often defined only for compact metric spaces. The greater generality that we have chosen (which is essential if one is going to take βS as the phase space of a dynamical system) is also common in the literature of dynamical systems. General references for topological dynamics include the books by R. Ellis [86] and J. Auslander [5].

Theorem 19.8 and Lemma 19.9 are due to B. Balcar and F. Franek in [12] as is the proof that we give of Theorem 19.10. Theorem 19.10 is proved by R. Ellis [86] in the case that S is a group.

Theorem 19.15 is due to S. Glasner in [105], where it is stated in the case that S is a countable abelian group. Theorems 19.17 and 19.18 are from [141], a result of collaboration with J. Lawson and A. Lisan. Theorem 19.19 is due to W. Ruppert in [219]. S. Glasner has recently published [106] a proof that the enveloping semigroup of a minimal left ideal in $\beta\mathbb{Z}$ is not topologically and algebraically isomorphic to $\beta\mathbb{Z}$ via a map taking n to λ'_n. As a consequence, the condition of Theorem 19.18 does not hold in $\beta\mathbb{Z}$.

The equivalence of the notions of "central" and "dynamically central" was established in the case in which S is countable and the phase space X is metric in [27], a result of collaboration with V. Bergelson with the assistance of B. Weiss. The equivalence of these notions in the general case is a result of S. Hong-ting and Y. Hong-wei in [164].

Theorem 19.24 is due to J. Auslander [4] and R. Ellis [85].

Most of the results of Section 19.4 are from an early draft of [32], results obtained in collaboration with V. Bergelson. Lemma 19.31 is a modification of standard results about Poincaré recurrence. Theorem 19.33 for the case $S = \mathbb{Z}$ is due to H. Furstenberg [98].

Chapter 20

Density — Connections with Ergodic Theory

As we have seen, many results in Ramsey Theory assert that, given a finite partition of some set, one cell of the partition must contain a specified kind of structure. One may ask instead that such a structure be found in any "large" set.

For example, van der Waerden's Theorem (Corollary 14.3) says that whenever $\mathbb{N}$ is divided into finitely many classes, one of these contains arbitrarily long arithmetic progressions. Szemerédi's Theorem says that whenever A is a subset of $\mathbb{N}$ with positive upper density, A contains arbitrarily long progressions.

Szemerédi's original proof of this theorem [232] was elementary, but very long and complicated. Subsequently, using ergodic theory, H. Furstenberg [97] provided a shorter proof of this result. This proof used his "correspondence principle" which can be viewed as a device for translating some problems involving sets of positive density in $\mathbb{N}$ into problems involving measure preserving systems, the primary object of study in ergodic theory.

We present in Section 20.2 a proof of Furstenberg's Correspondence Principle using the notion of p-limit and in Section 20.3 a strong density version of the Finite Sums Theorem obtained using the algebraic structure of $(\beta\mathbb{N}, +)$.

20.1 Upper Density and Banach Density

We have already dealt with the notion of ordinary upper density of a subset A of $\mathbb{N}$, which was defined by

$$\overline{d}(A) = \lim_{n \to \infty} \sup \frac{|A \cap \{1, 2, \ldots, n\}|}{n}.$$

Another notion of density that is more useful in the context of ergodic theory is that of *Banach density* which we define here in a quite general context.

Definition 20.1. Let $(S, \cdot)$ be a countable semigroup which has been enumerated as $\langle s_n \rangle_{n=1}^{\infty}$ and let $A \subseteq S$. The *right Banach density* of A is

$$d_r^*(A) = \sup\{\alpha \in \mathbb{R} : \text{ for all } n \in \mathbb{N} \text{ there exist } m \geq n \text{ and } x \in S$$

$$\text{such that } \frac{|A \cap \{s_1, s_2, \ldots, s_m\} \cdot x|}{m} \geq \alpha\}$$

and the *left Banach density* of A is

$$d_l^*(A) = \sup\{\alpha \in \mathbb{R} : \text{ for all } n \in \mathbb{N} \text{ there exist } m \geq n \text{ and } x \in S$$

$$\text{such that } \frac{|A \cap x \cdot \{s_1, s_2, \ldots, s_m\}|}{m} \geq \alpha\}.$$

If S is commutative, one has $d_r^* = d_l^*$ and we simply write $d^*(A)$ for the Banach density of A.

Notice that the Banach density of a subset A of S depends on the fixed enumeration of S. For example, in the semigroup $(\mathbb{N}, +)$, if $\mathbb{N}$ is enumerated as

$$1, 2, 3, 4, 6, 5, 8, 10, 12, 14, 7, 16, 18, 20, 22, 24, 26, 28, 30, 9, 32, \ldots$$

then $d^*(2\mathbb{N}) = 1$, while with respect to the usual enumeration of $\mathbb{N}$, $d^*(2\mathbb{N}) = \frac{1}{2}$. When working with the semigroup $(\mathbb{N}, +)$ we shall always assume that it has its usual enumeration so that

$$d^*(A) = \sup\{\alpha \in \mathbb{R} : \text{ for all } n \in \mathbb{N} \text{ there exist } m \geq n \text{ and } x \in \mathbb{N}$$

$$\text{such that } \frac{|A \cap \{x+1, x+2, \ldots, x+m\}|}{m} \geq \alpha\}.$$

Lemma 20.2. *Let $(S, \cdot)$ be a countable semigroup which has been enumerated as $\langle s_n \rangle_{n=1}^{\infty}$ and let $A, B \subseteq S$. Then $d_r^*(A \cup B) \leq d_r^*(A) + d_r^*(B)$ and $d_l^*(A \cup B) \leq d_l^*(A) + d_l^*(B)$. If S is right cancellative, then $d_r^*(S) = 1$ and if S is left cancellative, then $d_l^*(S) = 1$*

Proof. This is Exercise 20.1.1. $\square$

Recall that we have defined $\Delta = \{p \in \beta\mathbb{N} : \text{ for all } A \in p, \ \overline{d}(A) > 0\}$. We introduce in a more general context the similar sets defined in terms of d_r^* and d_l^*.

Definition 20.3. Let $(S, \cdot)$ be a countable semigroup which has been enumerated as $\langle s_n \rangle_{n=1}^{\infty}$. Then $\Delta_r^*(S, \cdot) = \{p \in \beta S : \text{ for all } A \in p, \ d_r^*(A) > 0\}$ and $\Delta_l^*(S, \cdot) = \{p \in \beta S : \text{ for all } A \in p, \ d_l^*(A) > 0\}$.

Again, if S is commutative we write $\Delta^*(S, \cdot)$ for $\Delta_r^*(S, \cdot) = \Delta_l^*(S, \cdot)$.

Lemma 20.4. *Let $(S, \cdot)$ be a countable semigroup which has been enumerated as $\langle s_n \rangle_{n=1}^{\infty}$ and let $A \subseteq S$.*
(a) *If $d_r^*(A) > 0$, then $\overline{A} \cap \Delta_r^*(S, \cdot) \neq \emptyset$.*
(b) *If $d_l^*(A) > 0$, then $\overline{A} \cap \Delta_l^*(S, \cdot) \neq \emptyset$.*

Proof. We establish (a) only. By Theorem 3.11 it suffices to show that if $\mathcal{F}$ is a finite nonempty subset of $\mathcal{P}(S)$ and $d_r^*(\bigcup \mathcal{F}) > 0$, then for some $B \in \mathcal{F}$, $d_r^*(B) > 0$. This follows from Lemma 20.2. $\square$

Recall from Theorems 6.79 and 6.80 that both Δ and $\mathbb{N}^* \setminus \Delta$ are left ideals of $(\beta \mathbb{N}, +)$ and of $(\beta \mathbb{N}, \cdot)$. By way of contrast, we see in the following theorems that $S^* \setminus \Delta_r^*$ is far from being a left ideal of βS and $S^* \setminus \Delta_l^*$ is far from being a right ideal of βS.

Theorem 20.5. *Let $(S, \cdot)$ be a countable right cancellative semigroup which has been enumerated as $\langle s_n \rangle_{n=1}^{\infty}$. Then $\Delta_r^*(S, \cdot)$ is a right ideal of βS.*

Proof. One has that $\Delta_r^*(S, \cdot) \neq \emptyset$ by Lemma 20.4 and the fact from Lemma 20.2 that $d_r^*(S) > 0$. Let $p \in \Delta_r^*(S, \cdot)$, let $q \in \beta S$, and let $A \in p \cdot q$. We need to show that $d_r^*(A) > 0$. Let $B = \{y \in S : y^{-1} A \in q\}$. Then $B \in p$ so $d_r^*(B) > 0$. Pick $\alpha > 0$ such that for all $n \in \mathbb{N}$ there exist $m \geq n$ and $x \in S$ such that $\dfrac{|B \cap \{s_1, s_2, \ldots, s_m\} \cdot x|}{m} \geq \alpha$.

To see that $d_r^*(A) \geq \alpha$, let $n \in \mathbb{N}$ and pick $m \geq n$ and $x \in S$ such that $\dfrac{|B \cap \{s_1, s_2, \ldots, s_m\} \cdot x|}{m} \geq \alpha$. Let $C = B \cap \{s_1, s_2, \ldots, s_m\} \cdot x$ and pick $z \in \bigcap_{y \in C} y^{-1} A$. Then, since S is right cancellative, $\rho_{z|C}$ is a one-to-one function from C to $A \cap \{s_1, s_2, \ldots, s_m\} \cdot xz$ so $\frac{|A \cap \{s_1, s_2, \ldots, s_m\} \cdot xz|}{m} \geq \alpha$. $\qquad \square$

Theorem 20.6. *Let $(S, \cdot)$ be a countable left cancellative semigroup which has been enumerated as $\langle s_n \rangle_{n=1}^{\infty}$. Then $\Delta_l^*(S, \cdot)$ is a left ideal of βS.*

Proof. Let $p \in \Delta_l^*(S, \cdot)$, let $q \in \beta S$, and let $A \in q \cdot p$. We need to show that $d_l^*(A) > 0$. Now $\{y \in S : y^{-1} A \in p\} \in q$ so pick $y \in S$ such that $y^{-1} A \in p$. Then $d_l^*(y^{-1} A) > 0$ so pick $\alpha > 0$ such that for all $n \in \mathbb{N}$ there exist $m \geq n$ and $x \in S$ such that $\dfrac{|y^{-1} A \cap x \cdot \{s_1, s_2, \ldots, s_m\}|}{m} \geq \alpha$. Now given m and x, since S is left cancellative, λ_y is a one-to-one function from $y^{-1} A \cap x \cdot \{s_1, s_2, \ldots, s_m\}$ to $A \cap yx \cdot \{s_1, s_2, \ldots, s_m\}$, and consequently, $d_l^*(A) \geq \alpha$. $\qquad \square$

As a consequence of Theorems 20.5 and 20.6, if S is commutative and cancellative, then $\Delta^*(S)$ is an ideal of βS.

Theorem 20.7. *Let $(S, \cdot)$ be a countable right cancellative semigroup which has been enumerated as $\langle s_n \rangle_{n=1}^{\infty}$. For any $A \subseteq S$, if A is piecewise syndetic, then there is some $G \in \mathcal{P}_f(S)$ such that $d_r^*\big(\bigcup_{t \in G} t^{-1} A \big) = 1$. If $(S, \cdot) = (\mathbb{N}, +)$, then the converse holds. Consequently,*

$$c\ell \, K(\beta \mathbb{N}, +) = \{p \in \beta \mathbb{N} : \text{for all } A \in p \text{ there exists } G \in \mathcal{P}_f(\mathbb{N})$$

$$\text{such that } d^*\big(\bigcup_{t \in G} -t + A \big) = 1\}.$$

Proof. Let $A \subseteq S$, assume that A is piecewise syndetic, and pick $G \in \mathcal{P}_f(S)$ such that for each $F \in \mathcal{P}_f(S)$ there exists $x \in S$ with $F \cdot x \subseteq \bigcup_{t \in G} t^{-1} A$. To see that $d_r^*\big(\bigcup_{t \in G} t^{-1} A \big) = 1$, let $n \in \mathbb{N}$ be given and pick $x \in S$ such that

$$\{s_1, s_2, \ldots, s_n\} \cdot x \subseteq \bigcup_{t \in G} t^{-1} A.$$

Then

$$\left| \bigcup_{t \in G} t^{-1} A \cap \{s_1, s_2, \dots, s_n\} \cdot x \right| = n.$$

Now assume that $(S, \cdot) = (\mathbb{N}, +)$ and we have $G \in \mathcal{P}_f(\mathbb{N})$ such that $d^*\left(\bigcup_{t \in G} -t + A\right) = 1$. Let $F \in \mathcal{P}_f(\mathbb{N})$ be given and pick $k \in \mathbb{N}$ such that $F \subseteq \{1, 2, \dots, k\}$. We claim that there is some $x \in \mathbb{N}$ such that $\{x+1, x+2, \dots, x+k\} \subseteq \left(\bigcup_{t \in G} -t + A\right)$. To see this, pick some $m > 2k^2$ and some $y \in \mathbb{N}$ such that

$$\left| \{y+1, y+2, \dots, y+m\} \cap \left(\bigcup_{t \in G} -t + A \right) \right| > \left(1 - \tfrac{1}{2k}\right)m$$

and pick $v \in \mathbb{N}$ such that $vk < m \le (v+1)k$, noting that $v \ge 2k$. If for some $i \in \{0, 1, \dots, v-1\}$ one has $\{y+ik+1, y+ik+2, \dots, y+(i+1)k\} \subseteq \bigcup_{t \in G} -t+A$, then we are done, so suppose instead that for each $i \in \{0, 1, \dots, v-1\}$,

$$\left| \{y+ik+1, y+ik+2, \dots, y+(i+1)k\} \cap \left(\bigcup_{t \in G} -t + A \right) \right| \le k-1.$$

Then

$$
\begin{aligned}
\left(1 - \tfrac{1}{2k}\right)m &< \left| \{y+1, y+2, \dots, y+m\} \cap \left(\bigcup_{t \in G} -t + A \right) \right| \\
&= \Sigma_{i=0}^{v-1} \left| \{y+ik+1, y+ik+2, \dots, y+(i+1)k\} \cap \left(\bigcup_{t \in G} -t + A \right) \right| \\
&\quad + \left| \{y+vk+1, y+vk+2, \dots, y+m\} \cap \left(\bigcup_{t \in G} -t + A \right) \right| \\
&\le v(k-1) + k
\end{aligned}
$$

so, since $m > vk$, one has that $\left(1 - \tfrac{1}{2k}\right)vk < v(k-1)+k$ so that $2k > v$, a contradiction. The final conclusion now follows from Corollary 4.41.   □

The following result reflects the interaction of addition and multiplication in $\beta\mathbb{N}$ which was treated in Chapter 13.

Theorem 20.8. *The set $\Delta^*(\mathbb{N}, +)$ is a left ideal of $(\beta\mathbb{N}, \cdot)$.*

Proof. Let $p \in \Delta^*(\mathbb{N}, +)$ and let $q \in \beta\mathbb{N}$. To see that $q \cdot p \in \Delta^*(\mathbb{N}, +)$, let $A \in q \cdot p$. Then $\{x \in \mathbb{N} : x^{-1}A \in p\} \in q$ so pick x such that $x^{-1}A \in p$. Then $d^*(x^{-1}A) > 0$ (where here, of course, we are referring to the additive version of d^*) so pick $\alpha > 0$ such that for all $n \in \mathbb{N}$ there exist $m \ge n$ and $y \in \mathbb{N}$ such that $\dfrac{|x^{-1}A \cap \{y+1, y+2, \dots, y+m\}|}{m} \ge \alpha$. Then given such n, m, and y, one has that

$$\frac{|A \cap \{xy+1, xy+2, \dots, xy+xm\}|}{m} \ge \alpha$$

so that $d^*(A) \ge \dfrac{\alpha}{x}$.   □

Recall that, given a subset A of a semigroup $(S, \cdot)$ that we have defined a tree T in A whose nodes are functions. Recall further that given a node f of T, the set of successors to f is denoted by B_f and that T is an *FP-tree* provided that for each node

f of T, B_f consists of all finite products of entries on paths extending f which occur after f. (By a path in T we mean a function $g : \omega \to A$ such that for each $n \in \omega$, the restriction of g to $\{0, 1, \ldots, n-1\}$ is in T.)

Theorem 20.9. *Let $(S, \cdot)$ be a countable cancellative semigroup which has been enumerated as $\langle s_n \rangle_{n=1}^{\infty}$. Let $r \in \mathbb{N}$ and let $S = \bigcup_{i=1}^{r} A_i$. There exist $i \in \{1, 2, \ldots, r\}$ and a tree T in A_i such that for each path g in T, $\mathrm{FP}(\langle g(n) \rangle_{n=0}^{\infty}) \subseteq A_i$, and for each node f of T, $d_r^*(B_f) > 0$ and $d_l^*(B_f) > 0$.*

Proof. By Theorems 20.5 and 20.6, $\Delta_r^*(S, \cdot)$ is a right ideal of βS and $\Delta_l^*(S, \cdot)$ is a left ideal of βS. By Corollary 2.6 and Theorem 2.7 pick an idempotent $p \in \Delta_r^*(S, \cdot) \cap \Delta_l^*(S, \cdot)$ and pick $i \in \{1, 2, \ldots, r\}$ such that $A_i \in p$. Then by Lemma 14.24 there is an FP-tree T in A_i such that for each $f \in T$, $B_f \in p$. Thus, in particular, for each $f \in T$, $d_r^*(B_f) > 0$ and $d_l^*(B_f) > 0$. $\square$

The following theorem is not a corollary to Theorem 20.9 because $d^*(A) > 0$ does not imply that $\overline{d}(A) > 0$. The proof is essentially the same, however, so we leave that proof as an exercise.

Theorem 20.10 is an interesting example of the strength of combinatorial results obtainable using idempotents in $\beta\mathbb{N}$. Although many of the Ramsey-theoretic results which we have presented were first obtained by elementary methods or have subsequently been given elementary proofs, we doubt that an elementary proof of Theorem 20.10 will be found in the near future.

Theorem 20.10. *Let $r \in \mathbb{N}$ and let $\mathbb{N} = \bigcup_{i=1}^{r} A_i$. There exist $i \in \{1, 2, \ldots, r\}$ and a tree T in A_i such that for each path g in T, $\mathrm{FS}(\langle g(n) \rangle_{n=0}^{\infty}) \subseteq A_i$, and for each node f of T, $\overline{d}(B_f) > 0$.*

Proof. This is Exercise 20.1.2. $\square$

The following lemma will be needed in the next section.

Lemma 20.11. *Let $A \subseteq \mathbb{N}$ such that $d^*(A) = \alpha > 0$. Then there exists some sequence $\langle I_k \rangle_{k=1}^{\infty}$ of intervals in $\mathbb{N}$ such that $\lim_{k \to \infty} |I_k| = \infty$ and*

$$\lim_{k \to \infty} \frac{|A \cap I_k|}{|I_k|} = \alpha.$$

Proof. This is Exercise 20.1.3. $\square$

Exercise 20.1.1. Prove Lemma 20.2.

Exercise 20.1.2. Using the fact that Δ is a left ideal of $(\beta\mathbb{N}, +)$ (Theorem 6.79), prove Theorem 20.10.

Exercise 20.1.3. Prove Lemma 20.11.

20.2 The Correspondence Principle

Recall that a *measure preserving system* is a quadruple $(X, \mathcal{B}, \mu, \langle T_s \rangle_{s \in S})$ where $(X, \mathcal{B}, \mu)$ is a measure space and $\langle T_s \rangle_{s \in S}$ is a measure preserving action of a semigroup S on X.

If the semigroup is $(\mathbb{N}, +)$, the action $\langle T_n \rangle_{n \in \mathbb{N}}$ is generated by a single function T_1 and we put $T = T_1$ and refer to the *measure preserving system* $(X, \mathcal{B}, \mu, T)$.

Theorem 20.12 (Furstenberg's Correspondence Principle). *Let $A \subseteq \mathbb{N}$ with $d^*(A) > 0$. There exist a measure preserving system $(X, \mathcal{B}, \mu, T)$ (in which X is a compact metric space and T is a homeomorphism from X onto X) and a set $A' \in \mathcal{B}$ such that*

(1) $\mu(A') = d^*(A)$ *and*

(2) *for every $F \in \mathcal{P}_f(\mathbb{N})$, $d^*\big(A \cap \bigcap_{n \in F} (-n + A)\big) \geq \mu(A' \cap \bigcap_{n \in F} T^{-n}[A'])$.*

Proof. Let $\Omega = {}^{\mathbb{Z}}\{0, 1\}$ with the product topology and let $T : \Omega \to \Omega$ be the shift defined by $T(x)(n) = x(n + 1)$. (By Exercise 20.2.1 one has in fact that Ω is a metric space.) Let ξ be the characteristic function of A (viewed as a subset of $\mathbb{Z}$). Let $X = c\ell\{T^n(\xi) : n \in \mathbb{Z}\}$, the orbit closure of ξ. Then X is a compact metric space and (the restriction of) T is a homeomorphism from X onto X.

For each $n \in \mathbb{Z}$, let $D_n = X \cap \pi_n^{-1}[\{1\}] = \{\delta \in X : \delta(n) = 1\}$ and notice that D_n is clopen in X. Let $\mathcal{A}$ be the Boolean algebra of sets generated by $\{D_n : n \in \mathbb{Z}\}$ and let $\mathcal{B}$ be the σ-algebra generated by $\{D_n : n \in \mathbb{Z}\}$. Notice that for each $n \in \mathbb{Z}$, $T[D_n] = D_{n-1}$ and $T^{-1}[D_n] = D_{n+1}$. Consequently, if $B \in \mathcal{B}$, then $T^{-1}[B] \in \mathcal{B}$ (and $T[B] \in \mathcal{B}$).

Define $\varphi : \mathcal{P}(X) \to \mathcal{P}(\mathbb{N})$ by $\varphi(B) = \{n \in \mathbb{N} : T^n(\xi) \in B\}$. Let $\alpha = d^*(A)$ and pick by Lemma 20.11 a sequence $\langle I_k \rangle_{k=1}^{\infty}$ of intervals in $\mathbb{N}$ such that $\lim_{k \to \infty} |I_k| = \infty$ and

$$\lim_{k \to \infty} \frac{|A \cap I_k|}{|I_k|} = \alpha.$$

Pick any $p \in \mathbb{N}^*$ and define $\nu : \mathcal{P}(X) \to [0, 1]$ by

$$\nu(B) = p\text{-}\lim_{k \in \mathbb{N}} \frac{|\varphi(B) \cap I_k|}{|I_k|}.$$

Notice that $\nu(X) = 1$. Further, given any $B, C \subseteq X$, one has $\varphi(B \cap C) = \varphi(B) \cap \varphi(C)$ (so that, in particular, if $B \cap C = \emptyset$, then $\varphi(B) \cap \varphi(C) = \emptyset$) and $\varphi(B \cup C) = \varphi(B) \cup \varphi(C)$. Thus, if B and C are disjoint subsets of X, then

$$\begin{aligned}
\nu(B \cup C) &= p\text{-}\lim_{k \in \mathbb{N}} \frac{|\big(\varphi(B) \cup \varphi(C)\big) \cap I_k|}{|I_k|} \\
&= p\text{-}\lim_{k \in \mathbb{N}} \frac{|\varphi(B) \cap I_k| + |\varphi(C) \cap I_k|}{|I_k|} \\
&= \nu(B) + \nu(C).
\end{aligned}$$

Next we claim that for any $B \subseteq X$, $\nu(T^{-1}[B]) = \nu(B)$. Indeed, $\varphi(T^{-1}[B]) = \left(-1 + \varphi(B)\right) \cap \mathbb{N}$ so

$$\nu(T^{-1}[B]) = p\text{-}\lim_{k \in \mathbb{N}} \frac{\left|\left(-1 + \varphi(B)\right) \cap I_k\right|}{|I_k|} = p\text{-}\lim_{k \in \mathbb{N}} \frac{|\varphi(B) \cap I_k|}{|I_k|} = \nu(B)$$

because for any $k \in \mathbb{N}$, $\left|\left(-1 + \varphi(B)\right) \cap I_k\right|$ and $|\varphi(B) \cap I_k|$ differ by at most 1.

Thus ν is finitely additive and T-invariant on $\mathcal{P}(X)$ and hence in particular on $\mathcal{A}$. Further all members of $\mathcal{A}$ are clopen in X, so if one has a sequence $\langle B_n \rangle_{n=1}^{\infty}$ in $\mathcal{A}$ such that each $B_{n+1} \subseteq B_n$ and $\bigcap_{n=1}^{\infty} B_n = \emptyset$, then since X is compact there is some k such that for all $n \geq k$, $B_n = \emptyset$ and hence $\lim_{n \to \infty} \nu(B_n) = 0$.

Therefore, we are in a position to apply Hopf's Extension Theorem. (This is a standard result in first year analysis courses. See for example [115, Exercise 10.37] or [14, Satz 3.2].) The finitely additive measure ν on $\mathcal{A}$ can be extended to a countably additive measure μ on $\mathcal{B}$ where for each $B \in \mathcal{B}$,

$$\mu(B) = \inf\{\Sigma_{C \in \mathcal{G}} \, \nu(C) : \mathcal{G} \subseteq \mathcal{A}, \, |\mathcal{G}| \leq \omega, \text{ and } B \subseteq \bigcup \mathcal{G}\}.$$

Using this description of $\mu(B)$ and the fact that for $C \in \mathcal{A}$, $\nu(T^{-1}[C]) = \nu(C)$, one sees immediately that for all $B \in \mathcal{B}$, $\mu(T^{-1}[B]) = \mu(B)$. Thus we have that $(X, \mathcal{B}, \mu, T)$ is a measure preserving system.

Let $A' = D_0 = \{\delta \in X : \delta(0) = 1\}$ and observe that $\varphi(A') = A$ and for each $n \in \mathbb{N}$, $T^{-n}[A'] = D_n$ and $\varphi(D_n) = (-n + A) \cap \mathbb{N}$. Thus

$$\mu(A') = \nu(A') = p\text{-}\lim_{k \in \mathbb{N}} \frac{|\varphi(A') \cap I_k|}{|I_k|} = p\text{-}\lim_{k \in \mathbb{N}} \frac{|A \cap I_k|}{|I_k|} = \alpha$$

and given any $F \in \mathcal{P}_f(\mathbb{N})$,

$$\begin{aligned}
\mu(A' \cap \bigcap_{n \in F} T^{-n}[A']) &= \mu(D_0 \cap \bigcap_{n \in F} D_n) \\
&= \nu(D_0 \cap \bigcap_{n \in F} D_n) \\
&= p\text{-}\lim_{k \in \mathbb{N}} \frac{|\varphi(D_0) \cap \bigcap_{n \in F} \varphi(D_n) \cap I_k|}{|I_k|} \\
&= p\text{-}\lim_{k \in \mathbb{N}} \frac{|A \cap \bigcap_{n \in F} (-n + A) \cap I_k|}{|I_k|} \\
&\leq d^*(\bigcap_{n \in F} A \cap (-n + A)).
\end{aligned}$$

$\square$

When proving Szemerédi's Theorem, one wants to show that, given a set A with $d^*(A) > 0$ and $\ell \in \mathbb{N}$, there is some $d \in \mathbb{N}$ such that

$$\{a \in \mathbb{N} : \{a, a + d, a + 2d, \ldots, a + \ell d\} \subseteq A\} \neq \emptyset.$$

As is typical of ergodic theoretic proofs of combinatorial facts, one establishes this fact by showing that the set is in fact large. The proof of the following theorem requires extensive background development, so we do not present it here.

Theorem 20.13. *Let $(X, \mathcal{B}, \mu)$ be a measure space, let $T_1, T_2, \ldots, T_\ell$ be commuting measure preserving transformations of $(X, \mathcal{B}, \mu)$, and let $B \in \mathcal{B}$ with $\mu(B) > 0$. Then there exists $d \in \mathbb{N}$ such that $\mu(B \cap \bigcap_{i=1}^{\ell} T_i^{-d}[B]) > 0$.*

Proof. [98, Theorem 7.15]. □

Corollary 20.14 (Szemerédi's Theorem). *Let $A \subseteq \mathbb{N}$ with $d^*(A) > 0$. Then A contains arbitrarily long arithmetic progressions.*

Proof. Pick by Theorem 20.12 a measure preserving system $(X, \mathcal{B}, \mu, T)$ and a set $A' \in \mathcal{B}$ such that

(1) $\mu(A') = d^*(A)$ and

(2) for every $F \in \mathcal{P}_f(\mathbb{N})$,

$$d^*\left(A \cap \bigcap_{n \in F} (-n + A)\right) \geq \mu(A' \cap \bigcap_{n \in F} T^{-n}[A']).$$

Let $\ell \in \mathbb{N}$ be given and for each $i \in \{1, 2, \ldots, \ell\}$, let $T_i = T^i$. Notice that $T_1, T_2, \ldots, T_\ell$ are measure preserving transformations of $(X, \mathcal{B}, \mu)$ that commute with each other. Pick by Theorem 20.13 some $d \in \mathbb{N}$ such that $\mu(A' \cap \bigcap_{i=1}^{\ell} T_i^{-d}[A']) > 0$ and notice that for each i, $T_i^{-d} = T^{-id}$. Let $F = \{d, 2d, 3d, \ldots, \ell d\}$. Then

$$d^*\left(A \cap \bigcap_{n \in F} (-n + A)\right) \geq \mu(A' \cap \bigcap_{n \in F} T^{-n}[A']) > 0$$

so $A \cap \bigcap_{n \in F} (-n + A) \neq \emptyset$ so pick $a \in A \cap \bigcap_{n \in F} (-n + A)$. Then $\{a, a + d, a + 2d, \ldots, a + \ell d\} \subseteq A$. □

Exercise 20.2.1. Let $\Omega = {}^{\mathbb{Z}}\{0, 1\}$ and for $\xi \neq \eta$ in Ω, define $\rho(\xi, \eta) = \frac{1}{k+1}$ where $k = \min\{|t| : t \in \mathbb{Z} \text{ and } \xi(t) \neq \eta(t)\}$ (and of course $\rho(\xi, \xi) = 0$). Prove that ρ is a metric on Ω and the metric and product topologies on Ω agree.

20.3 A Density Version of the Finite Sums Theorem

The straightforward density version of the Finite Sums Theorem (which would assert that any set $A \subseteq \mathbb{N}$ such that $\overline{d}(A) > 0$ — or perhaps $d^*(A) > 0$ — would contain $FS(\langle x_n \rangle_{n=1}^{\infty})$ for some sequence $\langle x_n \rangle_{n=1}^{\infty}$) is obviously false. Consider $A = 2\mathbb{N} + 1$. However, a consideration of the proof of Szemerédi's Theorem via ergodic theory reminds us that the proof was obtained by showing that a set which was only required to be nonempty was in fact large.

Viewed in this way, the Finite Sums Theorem says that whenever $r \in \mathbb{N}$ and $\mathbb{N} = \bigcup_{i=1}^{r} A_i$, there exist $i \in \{1, 2, \ldots, r\}$ and a sequence $\langle x_n \rangle_{n=1}^{\infty}$ such that for each n, $x_{n+1} \in A_i \cap \bigcap\{-\sum_{t \in F} x_t + A_i : \emptyset \neq F \subseteq \{1, 2, \ldots, n\}\}$. In this section we show that not only this set, but indeed many of its subsets, can be made to have positive upper density and specify how big these sets can be made to be.

Lemma 20.15. *Let $A \subseteq \mathbb{N}$ such that $\overline{d}(A) > 0$ and let B be an infinite subset of $\mathbb{N}$. For every $\epsilon > 0$ there exist $x < y$ in B such that $\overline{d}\big(A \cap (-(y-x) + A)\big) \geq \overline{d}(A)^2 - \epsilon$.*

Proof. Let $a = \overline{d}(A)$ and pick a sequence $\langle x_n \rangle_{n=1}^{\infty}$ in $\mathbb{N}$ such that $\lim_{n \to \infty} x_n = \infty$ and

$$\lim_{n \to \infty} \frac{|A \cap \{1, 2, \ldots, x_n\}|}{x_n} = a.$$

Notice that for every $t \in \mathbb{N}$, $\displaystyle\lim_{n \to \infty} \frac{|(-t + A) \cap \{1, 2, \ldots, x_n\}|}{x_n} = a$.

Enumerate B in increasing order as $\langle y_i \rangle_{i=1}^{\infty}$. Let $\epsilon > 0$ be given. If $\epsilon \geq a^2$, the conclusion is trivial, so assume that $\epsilon < a^2$ and let $b = \sqrt{a^2 - \epsilon/2}$. Pick $k \in \mathbb{N}$ such that $k > \dfrac{4a}{\epsilon}$ and pick $\ell \in \mathbb{N}$ such that for all $i \in \{1, 2, \ldots, k\}$ and all $n \geq \ell$,

$$b < \frac{|(-y_i + A) \cap \{1, 2, \ldots, x_n\}|}{x_n} < 2a.$$

It suffices to show that for some pair (i, j) with $1 \leq i < j \leq k$, one has

$$\overline{d}\big((-y_i + A) \cap (-y_j + A)\big) \geq a^2 - \epsilon$$

since $\overline{d}\big((-y_i + A) \cap (-y_j + A)\big) = \overline{d}\big(A \cap (-(y_j - y_i) + A)\big)$. Suppose instead that this conclusion fails. Then in particular for each pair (i, j) with $1 \leq i < j \leq k$, there is some $v_{i,j} \in \mathbb{N}$ such that for all $n \geq v_{i,j}$

$$\frac{|(-y_i + A) \cap (-y_j + A) \cap \{1, 2, \ldots, x_n\}|}{x_n} < a^2 - \epsilon.$$

Let $n = \max(\{\ell\} \cup \{v_{i,j} : 1 \leq i < j \leq k\})$ and let $m = x_n$. For each $i \in \{1, 2, \ldots, k\}$, let $A_i = (-y_i + A) \cap \{1, 2, \ldots, m\}$ and for any $C \subseteq \{1, 2, \ldots, m\}$, let $\mu(C) = \dfrac{|C|}{m}$. Then for each $i \in \{1, 2, \ldots, k\}, b < \mu(A_i) < 2a$, and for $1 \leq i < j \leq k, \mu(A_i \cap A_j) < a^2 - \epsilon$. Notice also that for $C \subseteq \{1, 2, \ldots, k\}$,

$$\mu(C) = \frac{\sum_{t=1}^{m} \chi_C(t)}{m}.$$

We claim that it suffices to show that

$$k^2 b^2 < \sum_{i=1}^{k} \mu(A_i) + 2 \sum_{i=1}^{k-1} \sum_{j=i+1}^{k} \mu(A_i \cap A_j).$$

Indeed, once we have done so, we have $k^2 b^2 < 2ka + k(k-1)(a^2 - \epsilon)$ so that

$$2a > k\big(\epsilon - (a^2 - b^2)\big) + a^2 - \epsilon > k\big(\epsilon - (a^2 - b^2)\big) = \frac{k\epsilon}{2} > 2a$$

a contradiction.

To establish the desired inequality, observe first that

$$kb < \Sigma_{i=1}^{k} \mu(A_i) = \frac{\Sigma_{i=1}^{k} \Sigma_{t=1}^{m} \chi_{A_i}(t)}{m}$$

so that

(1)
$$k^2 b^2 < \frac{\left(\Sigma_{t=1}^{m} \Sigma_{i=1}^{k} \chi_{A_i}(t)\right)^2}{m^2}.$$

Next one establishes by a routine induction on m that for any function $f :$ $\{1, 2, \ldots, m\} \to \mathbb{R}$,

$$m \Sigma_{t=1}^{m} f^2(t) - \left(\Sigma_{t=1}^{m} f(t)\right)^2 = \Sigma_{t=2}^{m} \Sigma_{s=1}^{t-1} \left(f(t) - f(s)\right)^2$$

and thus $m \Sigma_{t=1}^{m} f^2(t) \geq \left(\Sigma_{t=1}^{m} f(t)\right)^2$. Then in particular

$$\left(\Sigma_{t=1}^{m} \Sigma_{i=1}^{k} \chi_{A_i}(t)\right)^2 \leq m \Sigma_{t=1}^{m} \left(\Sigma_{i=1}^{k} \chi_{A_i}(t)\right)^2$$

and thus by (1) we have

(2)
$$k^2 b^2 < \frac{\Sigma_{t=1}^{m} \left(\Sigma_{i=1}^{k} \chi_{A_i}(t)\right)^2}{m}.$$

Now, for any t,

$$\left(\Sigma_{i=1}^{k} \chi_{A_i}(t)\right)^2 = \Sigma_{i=1}^{k} \chi_{A_i}^2(t) + 2 \Sigma_{i=1}^{k-1} \Sigma_{j=i+1}^{k} \chi_{A_i}(t) \chi_{A_j}(t)$$
$$= \Sigma_{i=1}^{k} \chi_{A_i}(t) + 2 \Sigma_{i=1}^{k-1} \Sigma_{j=i+1}^{k} \chi_{A_i \cap A_j}(t)$$

so that

$$\frac{\Sigma_{t=1}^{m} \left(\Sigma_{i=1}^{k} \chi_{A_i}(t)\right)^2}{m} = \frac{\Sigma_{i=1}^{k} \Sigma_{t=1}^{m} \chi_{A_i}(t)}{m} + \frac{2 \Sigma_{i=1}^{k-1} \Sigma_{j=i+1}^{k} \Sigma_{t=1}^{m} \chi_{A_i \cap A_j}(t)}{m}$$
$$= \Sigma_{i=1}^{k} \mu(A_i) + 2 \Sigma_{i=1}^{k-1} \Sigma_{j=i+1}^{k} \mu(A_i \cap A_j).$$

Now, using (2) the required inequality is established.   □

Notice that in Lemma 20.15, one cannot do better than $\overline{d}(A)^2$. Indeed, choose a set $A \subseteq \mathbb{N}$ by randomly assigning (with probability $\frac{1}{2}$) each $n \in \mathbb{N}$ to A or its complement. Then for any $t \in \mathbb{N}$, $\overline{d}\left(A \cap (-t + A)\right) = \frac{1}{4} = \overline{d}(A)^2$.

Lemma 20.16. *Let $p \in \Delta$ such that $p + p = p$, let $A \in p$, and let $\epsilon > 0$. Then*

$$\{x \in A : (-x + A) \in p \text{ and } \overline{d}\left(A \cap (-x + A)\right) > \overline{d}(A)^2 - \epsilon\} \in p.$$

Proof. Let $B = \{x \in \mathbb{N} : \overline{d}(A \cap (-x + A)) \geq \overline{d}(A)^2 - \epsilon/2\}$. Since by Theorem 4.12, $A^* = \{x \in A : (-x + A) \in p\} \in p$, it suffices to show that $B \in p$. Suppose instead that $\mathbb{N} \backslash B \in p$ and pick by Theorem 5.8 a sequence $\langle x_n \rangle_{n=1}^{\infty}$ in $\mathbb{N}$ such that $\mathrm{FS}(\langle x_n \rangle_{n=1}^{\infty}) \subseteq \mathbb{N} \backslash B$. Let $C = \{\sum_{t=1}^{n} x_t : n \in \mathbb{N}\}$ and pick by Lemma 20.15 some $n < m$ such that $\sum_{t=1}^{m} x_t - \sum_{t=1}^{n} x_t \in B$. Since

$$\textstyle\sum_{t=1}^{m} x_t - \sum_{t=1}^{n} x_t = \sum_{t=n+1}^{m} x_t \in \mathrm{FS}(\langle x_t \rangle_{t=1}^{\infty}),$$

this is a contradiction. □

The proof of the following theorem is reminiscent of the proof of Lemma 14.24.

Theorem 20.17. *Let $p \in \Delta$, let $A \in p$ and let $0 < \delta < \overline{d}(A)$. There is a tree T in A such that*

(1) *for each path g of T*

    (a) $\mathrm{FS}(\langle g(t) \rangle_{t=1}^{\infty}) \subseteq A$ *and*

    (b) *for each $F \in \mathcal{P}_f(\omega)$, $\overline{d}(A \cap \bigcap\{-y + A : y \in \mathrm{FS}(\langle g(t) \rangle_{t \in F})\}) > \delta^{2^{|F|}}$*

and

(2) *for each $f \in T$, $\overline{d}(B_f) > 0$.*

Proof. Given $n \in \omega$, $g : \{0, 1, \ldots, n-1\} \to A$, and $F \subseteq \{0, 1, \ldots, n-1\}$, let

$$D(F, g) = A \cap \bigcap\{-y + A : y \in \mathrm{FS}(\langle g(t) \rangle_{t \in F})\},$$
$$C(F, g) = \{x \in D(F, g) : -x + D(F, g) \in p$$
$$\text{and } \overline{d}(D(F, g) \cap (-x + D(F, g))) > \delta^{2^{|F|+1}}\},$$

and

$$B(g) = \bigcap\{C(F, g) : F \subseteq \{0, 1, \ldots, n-1\}\}.$$

Notice that, following our usual convention that $\bigcap \emptyset$ is whatever semigroup we are concerned with at the time, one has $D(\emptyset, g) = A$.

Let $T_0 = \{\emptyset\}$ and inductively for $n \in \omega$, let

$$T_{n+1} = \{g \cup \{(n, x)\} : g \in T_n, \ x \in B(g), \text{ and if } n > 0, x > g(n-1)\}.$$

Let $T = \bigcup_{n=0}^{\infty} T_n$. Then trivially T is a tree in A. (Given $x \in B(g)$, $x \in C(\emptyset, g) \subseteq D(\emptyset, g) = A$.) Notice that for $n \in \mathbb{N}$ and $g \in T_n$, $B_g = \{x \in B(g) : x > g(n-1)\}$.

We show first by induction on n that for all $g \in T_n$ and all $F \subseteq \{0, 1, \ldots, n-1\}$, $C(F, g) \in p$ and $\overline{d}(D(F, g)) > \delta^{2^{|F|}}$. Observe that

(*) if $k = \max F$ and h is the restriction of g to $\{0, 1, \ldots, k\}$, then $D(F, g) = D(F, h)$ and $C(F, g) = C(F, h)$.

To ground the induction assume that $n = 0$. Then $g = F = \emptyset$ and $D(F, g) = A$ so that $\overline{d}(D(F, g)) > \delta = \delta^{2^0}$. Let $\epsilon = \overline{d}(A)^2 - \delta^2$. Then

$$C(F, g) = \{x \in A : -x + A \in p \text{ and } \overline{d}(A \cap (-x + A)) > \overline{d}(A)^2 - \epsilon\}$$

so that by Lemma 20.16, $C(F, g) \in p$.

Now let $n \in \omega$ and assume that for all $g \in T_n$ and all $F \subseteq \{0, 1, \ldots, n - 1\}$, $C(F, g) \in p$ and $\overline{d}(D(F, g)) > \delta^{2^{|F|}}$. Let $g \in T_{n+1}$ and $F \subseteq \{0, 1, \ldots, n\}$ be given. By the observation (*), we may assume that $n \in F$. Let h be the restriction of g to $\{0, 1, \ldots, n - 1\}$, let $H = F \setminus \{n\}$, and let $x = g(n)$. Then $h \in T_n$ and $x \in B(h)$.

We now claim that

(**) $D(F, g) = D(H, h) \cap (-x + D(H, h))$.

To see that $D(H, h) \cap (-x + D(H, h)) \subseteq D(F, g)$, let $z \in D(H, h) \cap (-x + D(H, h))$. Then $z \in D(H, h) \subseteq A$. Let $y \in \mathrm{FS}(\langle g(t) \rangle_{t \in F})$ and pick $G \subseteq F$ such that $y = \Sigma_{t \in G} g(t)$. We show that $z \in -y + A$. If $n \notin G$, then $y \in \mathrm{FS}(\langle h(t) \rangle_{t \in H})$ and hence, since $z \in D(H, h)$, $z \in -y + A$. We thus assume that $n \in G$. If $G = \{n\}$, then $z \in -x + D(H, h) \subseteq -x + A = -y + A$ as required. We therefore assume that $K = G \setminus \{n\} \neq \emptyset$. Then $K \subseteq H$. Let $v = \Sigma_{t \in H} g(t) = \Sigma_{t \in H} h(t)$. Then $v \in \mathrm{FS}(\langle h(t) \rangle_{t \in H})$ so $x + z \in D(H, h) \subseteq -v + A$ and thus $v + x + z = y + z \in A$ as required.

To see that $D(F, g) \subseteq D(H, h) \cap (-x + D(H, h))$, let $z \in D(F, g)$. Since $\mathrm{FS}(\langle h(t) \rangle_{t \in H}) \subseteq \mathrm{FS}(\langle g(t) \rangle_{t \in F})$, $z \in D(H, h)$. Since $x = g(n) \in \mathrm{FS}(\langle g(t) \rangle_{t \in F})$, $x + z \in A$. To see that $x + z \in D(H, h)$, let $y \in \mathrm{FS}(\langle h(t) \rangle_{t \in H})$. Then $y + x \in \mathrm{FS}(\langle g(t) \rangle_{t \in F})$ so $z \in -(y + x) + A$ so $x + z \in -y + A$. Thus (**) is established.

Since $x \in B(h) \subseteq C(H, h)$, we have $-x + D(H, h) \in p$. Thus

$$D(H, h) \cap (-x + D(H, h)) \in p.$$

That is $D(F, g) \in p$. Now since $x \in C(H, h)$ and $D(F, g) = D(H, h) \cap (-x + D(H, h))$, $\overline{d}(D(F, g)) > \delta^{2^{|H|+1}} = \delta^{2^{|F|}}$. Let $\gamma = \overline{d}(D(F, g)) - \delta^{2^{|F|}}$ and let $\epsilon = 2\gamma\delta^{2^{|F|}} + \gamma^2$. Let

$$E = \{z \in D(F, g) : -z + D(F, g) \in p \text{ and}$$
$$\overline{d}(D(F, g) \cap (-z + D(F, g))) > \overline{d}(D(F, g))^2 - \epsilon\}.$$

By Lemma 20.16, $E \in p$. Since $\overline{d}(D(F, g))^2 - \epsilon = (\delta^{2^{|F|}} + \gamma)^2 - \epsilon = \delta^{2^{|F|+1}}$, we have that $E = C(F, g)$ so that $C(F, g) \in p$. The induction is complete.

Now let f be any path of T. We show by induction on $|F|$, using essentially the first proof of Theorem 5.8, that if $F \in \mathcal{P}_f(\omega)$, $n = \min F$, and h is the restriction of f to $\{0, 1, \ldots, n - 1\}$, then $\Sigma_{t \in F} f(t) \in D(\{0, 1, \ldots, n - 1\}, h)$. First assume that $F = \{n\}$, let g be the restriction of f to $\{0, 1, \ldots, n\}$, and let h be the restriction of f to $\{0, 1, \ldots, n - 1\}$. Then $g = h \cup \{(n, f(n))\}$ with

$$f(n) \in B(h) \subseteq C(\{0, 1, \ldots, n - 1\}, h) \subseteq D(\{0, 1, \ldots, n - 1\}, h)$$

as required.

Now assume that $|F| > 1$, let $G = F \setminus \{n\}$, and let $m = \min G$. Let h, g, and k be the restrictions of f to $\{0, 1, \ldots, n - 1\}$, $\{0, 1, \ldots, n\}$, and $\{0, 1, \ldots, m - 1\}$ respectively. Then

$$\Sigma_{t \in G} f(t) \in D(\{0, 1, \ldots, m - 1\}, k) \subseteq D(\{0, 1, \ldots, n\}, g)$$

and $D(\{0, 1, \ldots, n\}, g) = D(\{0, 1, \ldots, n-1\}, h) \cap \left(-f(n) + D(\{0, 1, \ldots, n-1\}, h)\right)$ by (**). Thus $\Sigma_{t \in G} f(t) + f(n) \in D(\{0, 1, \ldots, n-1\}, h)$ as required. In particular, conclusion (1)(a) holds.

To establish conclusion (1)(b), let $F \in \mathcal{P}_f(\omega)$, pick $n > \max F$, and let h be the restriction of f to $\{0, 1, \ldots, n-1\}$. The conclusion of (1)(b) is precisely the statement, proved above, that $\overline{d}\big(D(F, h)\big) > \delta^{2^{|F|}}$.

As we observed above, for $n \in \mathbb{N}$ and $g \in T_n$, $B_g = \{x \in B(g) : x > g(n-1)\}$ and thus $B_g \in p$ so, since $p \in \Delta$, $\overline{d}(B_g) > 0$. $\qquad\square$

Notes

The notion which we have called "Banach density" should perhaps be called "Polya density" since it appears (with reference to $\mathbb{N}$) explicitly in [196]. To avoid proliferation of terminology, we have gone along with Furstenberg [98] who says the notion is of the kind appearing in early works of Banach. The definition of Banach density in the generality that we use in Section 20.1 is based on the definition of "maximal density" by H. Umoh in [237].

It is easy to construct subsets A of $\mathbb{N}$ such that for all $n \in \mathbb{N}$, $\overline{d}(\bigcup_{t=1}^{n} -t + A) = \overline{d}(A) > 0$. On the other hand, it was shown in [126] that if $d^*(A) > 0$, then for each $\epsilon > 0$ there is some $n \in \mathbb{N}$ such that $d^*(\bigcup_{t=1}^{n} -t + A) > 1 - \epsilon$. Consequently, in view of Theorem 20.7, it would seem that $\Delta^*(\mathbb{N}, +)$ is not much larger than $c\ell\, K(\beta\mathbb{N}, +)$. On the other hand it was shown in [73], a result of collaboration with D. Davenport, that $c\ell\, K(\beta\mathbb{N}, +)$ is the intersection of closed ideals of the form $c\ell(\mathbb{N}^* + p)$ lying strictly between $c\ell\, K(\beta\mathbb{N}, +)$ and $\Delta^*(\mathbb{N}, +)$.

Theorem 20.13, one of three nonelementary results results used in this book that we do not prove, is due to H. Furstenberg.

Lemma 20.15 is due to V. Bergelson in [18]. Theorem 20.17 is from [24], a result of collaboration with V. Bergelson.

Chapter 21

Other Semigroup Compactifications

Throughout this book we have investigated the structure of βS for a discrete semigroup S. According to Theorem 4.8, βS is a maximal right topological semigroup containing S within its topological center.

In this chapter we consider semigroup compactifications that are maximal right topological, semitopological, or topological semigroups, or topological groups, defined not only for discrete semigroups, but in fact for any semigroup which is also a topological space.

21.1 The $\mathcal{LMC}$, $\mathcal{WAP}$, $\mathcal{AP}$, and $\mathcal{SAP}$ Compactifications

Let S be a semigroup which is also a topological space. We shall describe a method of associating S with a compact right topological semigroup defined by a universal property. The names for the compactifications that we introduce in this section are taken from [39] and [40]. These names come from those of the complex valued functions on S that extend continuously to the specified compactification. The four classes of spaces in which we are interested are defined by the following statements.

Definition 21.1. (a) $\Psi_1(C)$ is the statement "C is a compact right topological semigroup".

(b) $\Psi_2(C)$ is the statement "C is a compact semitopological semigroup".

(c) $\Psi_3(C)$ is the statement "C is a compact topological semigroup".

(d) $\Psi_4(C)$ is the statement "C is a compact topological group".

Lemma 21.2. *Let S be a set with $|S| = \kappa \geq \omega$, let X be a compact space, and let $f : S \to X$. If $f[S]$ is dense in X, then $|X| \leq 2^{2^\kappa}$.*

Proof. Give S the discrete topology. Then the continuous extension $\tilde{f}$ of f to βS takes βS onto X so that $|X| \leq |\beta S|$. By Theorem 3.58, $|\beta S| = 2^{2^\kappa}$. $\square$

In the following lemma we need to be concerned with the technicalities of set theory more than is our custom. One would like to define $F_i = \{(f, C) : \Psi_i(C), f \text{ is a}$

continuous homomorphism from S to C, and $f[S] \subseteq \Lambda(C)$}. However, there is no such set. (Its existence would lead quickly to Russell's Paradox.)

Notice that the requirements in (1) and (2) concerning the topological center are redundant if $i \neq 1$.

Lemma 21.3. *Let S be an infinite semigroup which is also a topological space and let $i \in \{1, 2, 3, 4\}$. There is a set F_i of ordered pairs (f, C) such that*

(1) *if $(f, C) \in F_i$, then $\Psi_i(C)$, f is a continuous homomorphism from S to C, and $f[S] \subseteq \Lambda(C)$, and*

(2) *given any D such that $\Psi_i(D)$ and given any continuous homomorphism $g : S \to D$ with $g[S] \subseteq \Lambda(D)$, there exist $(f, C) \in F_i$ and a continuous one-to-one homomorphism $\varphi : C \to D$ such that $\varphi \circ f = g$.*

Proof. Let $\kappa = |S|$ and fix a set X with $|X| = 2^{2^\kappa}$. Let

$$\mathscr{G} = \{(f, (C, \mathcal{T}, \cdot)) : C \subseteq X, \mathcal{T} \text{ is a topology on } C, \cdot \text{ is an associative}$$
$$\text{binary operation on } C, \text{ and } f : S \to C\}.$$

Note that if $C \subseteq X$, $\mathcal{T}$ is a topology on C, $\cdot$ is a binary operation on C, and $f : S \to C$, then $f \subseteq S \times C \subseteq S \times X$, $C \subseteq X$, $\mathcal{T} \subseteq \mathcal{P}(C) \subseteq \mathcal{P}(X)$, and $\cdot : C \times C \to C$ so that $\cdot \subseteq (C \times C) \times C \subseteq (X \times X) \times X$. Thus

$$\mathscr{G} \subseteq \mathcal{P}(S \times X) \times \big(\mathcal{P}(X) \times \mathcal{P}\big(\mathcal{P}(X)\big) \times \mathcal{P}((X \times X) \times X)\big)$$

so $\mathscr{G}$ is a set. (More formally, the axiom schema of separation applied to the set

$$\mathcal{P}(S \times X) \times \big(\mathcal{P}(X) \times \mathcal{P}\big(\mathcal{P}(X)\big) \times \mathcal{P}((X \times X) \times X)\big)$$

and the statement "$C \subseteq X$, $\mathcal{T}$ is a topology on C, $\cdot$ is an associative binary operation on C, and $f : S \to C$" guarantees the existence of a set $\mathscr{G}$ as we have defined it.

In defining $\mathscr{G}$ we have, out of necessity, departed from the custom of not specifically mentioning the topology or the operation when talking about a semigroup with a topology. In the definition of the set F_i we return to that custom, writing C instead of $(C, \mathcal{T}, \cdot)$. Let

$$F_i = \{(f, C) \in \mathscr{G} : \Psi_i(C), \ f \text{ is a continuous homomorphism from}$$
$$S \text{ to } C, \text{ and } f[S] \subseteq \Lambda(C)\}.$$

(Notice again that the requirement that $f[S] \subseteq \Lambda(C)$ is redundant if $i \neq 1$.)

Trivially F_i satisfies conclusion (1). Now let g and D be given such that $\Psi_i(D)$ and g is a continuous homomorphism from S to D. Let $B = c\ell \, g[S]$. By Exercise 2.3.2, since $g[S] \subseteq \Lambda(D)$, B is a subsemigroup of D. Further, if D is a topological group (as it is if $i = 4$), then by Exercise 2.2.3 B is a topological group. Since the continuity requirements of Ψ_i are hereditary, we have $\Psi_i(B)$.

Now by Lemma 21.2, $|B| \leq 2^{2^\kappa} = |X|$ so pick a one-to-one function $\tau : B \to X$ and let $C = \tau[B]$. Give C the topology and operation making τ an isomorphism and a homeomorphism from B onto C. Let $f = \tau \circ g$ and let $\varphi = \tau^{-1}$. Then $(f, C) \in F_i$, φ is a continuous one-to-one homomorphism from C to D, and $\varphi \circ f = g$.   □

Notice that one may have f and distinct C_1 and C_2 with $(f, C_1) \in F_i$ and $(f, C_2) \in F_i$. We can now establish the existence of universal semigroup compactifications with respect to each of the statements Ψ_i.

Theorem 21.4. *Let S be an infinite semigroup which is also a topological space and let $i \in \{1, 2, 3, 4\}$. There exist a pair $(\eta_i, \gamma_i S)$ such that*

(1) $\Psi_i(\gamma_i S)$,

(2) η_i *is a continuous homomorphism from S to $\gamma_i S$,*

(3) $\eta_i[S]$ *is dense in $\gamma_i S$,*

(4) $\eta_i[S] \subseteq \Lambda(\gamma_i S)$, *and*

(5) *given any D such that $\Psi_i(D)$ and any continuous homomorphism $g : S \to D$ with $g[S] \subseteq \Lambda(D)$, there exists a continuous homomorphism $\mu : \gamma_i S \to D$ such that $\mu \circ \eta_i = g$.*

So the following diagram commutes:

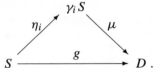

Proof. Pick a set F_i as guaranteed by Lemma 21.3 and let $T = \bigtimes_{(f,C) \in F_i} C$. Define $\eta_i : S \to T$ by $\eta_i(s)(f, C) = f(s)$. Let $\gamma_i S = c\ell_T \eta_i[S]$. By Theorem 2.22 T (with the product topology and coordinatewise operations) is a compact right topological semigroup and $\eta_i[S] \subseteq \Lambda(T)$. If $i \in \{2, 3, 4\}$ one easily verifies the remaining requirements needed to establish that $\Psi_i(T)$. By Exercise 2.3.2 $\gamma_i S$ is a subsemigroup of T and, if $i = 4$, by Exercise 2.2.3 $\gamma_i S$ is a topological group. Consequently $\Psi_i(\gamma_i S)$.

For each $(f, C) \in F_i$, $\pi_{(f,C)} \circ \eta_i = f$ so η_i is continuous. Given $s, t \in S$ and $(f, C) \in F_i$,

$$\big(\eta_i(s)\eta_i(t)\big)(f, C) = \big(\eta_i(s)(f, C)\big)\big(\eta_i(t)(f, C)\big) = f(s)f(t) = f(st) = \eta_i(st)(f, C)$$

so η_i is a homomorphism.

Trivially $\eta_i[S]$ is dense in $\gamma_i S$ and we have already seen that $\eta_i[S] \subseteq \Lambda(T)$ so that $\eta_i[S] \subseteq \Lambda(\gamma_i S)$.

Finally, let D be given such that $\Psi_i(D)$ and let $g : S \to D$ be a continuous homomorphism. Pick by Lemma 21.3 some $(f, C) \in F_i$ and a continuous one-to-one homomorphism $\varphi : C \to D$ such that $\varphi \circ f = g$. Let $\mu = \varphi \circ \pi_{(f,C)}$. Then, given $s \in S$,

$$(\mu \circ \eta_i)(s) = (\varphi \circ \pi_{(f,C)} \circ \eta_i)(s) = \varphi\big(\eta_i(s)(f, C)\big) = \varphi\big(f(s)\big) = g(s). \qquad □$$

We now observe that the compactifications whose existence is guaranteed by Theorem 21.4 are essentially unique.

Theorem 21.5. *Let S be an infinite semigroup which is also a topological space and let $i \in \{1, 2, 3, 4\}$. Let $(\eta_i, \gamma_i S)$ be as guaranteed by Theorem* 21.4. *Assume that also the pair (φ, T) satisfies*

(1) $\Psi_i(T)$,

(2) *φ is a continuous homomorphism from S to T,*

(3) *$\varphi[S]$ is dense in T,*

(4) *$\varphi[S] \subseteq \Lambda(T)$, and*

(5) *given any D such that $\Psi_i(D)$ and any continuous homomorphism $g : S \to D$ with $g[S] \subseteq \Lambda(D)$, there exists a continuous homomorphism $\mu : T \to D$ such that $\mu \circ \varphi = g$.*

Then there is a function $\delta : \gamma_i S \to Y$ which is both an isomorphism and a homeomorphism such that δ takes $\eta_i[S]$ onto $\varphi[S]$ and $\delta \circ \eta_i = \varphi$.

Proof. Let $\delta : \gamma_i S \to T$ be as guaranteed by conclusion (5) of Theorem 21.4 for T and φ and let $\mu : T \to \gamma_i S$ be as guaranteed by conclusion (5) above for $\gamma_i S$ and η_i. Then δ and μ are continuous homomorphisms, $\delta \circ \eta_i = \varphi$, and $\mu \circ \varphi = \eta_i$.

Now $\delta[\gamma_i S]$ is a compact set containing $\varphi[S]$ which is dense in T so $\delta[\gamma_i S] = T$. Also $\mu \circ \delta \circ \eta_i = \mu \circ \varphi = \eta_i$ so $\mu \circ \delta$ agrees with the identity on the dense set $\eta_i[S]$ and thus $\mu \circ \delta$ is the identity on $\gamma_i S$. Consequently $\mu = \delta^{-1}$ so δ is a homeomorphism and an isomorphism. $\square$

As a consequence of Theorem 21.5 it is reasonable to speak of "the" $\mathcal{LMC}$-compactification, and so forth.

Definition 21.6. Let S be an infinite semigroup which is also a topological space and for each $i \in \{1, 2, 3, 4\}$ let $(\eta_i, \gamma_i S)$ be as guaranteed by Theorem 21.4.

(a) $(\eta_1, \gamma_1 S)$ is the $\mathcal{LMC}$-compactification of S and $\mathcal{LMC}(S) = \gamma_1 S$.

(b) $(\eta_2, \gamma_2 S)$ is the $\mathcal{WAP}$-compactification of S and $\mathcal{WAP}(S) = \gamma_2 S$.

(c) $(\eta_3, \gamma_3 S)$ is the $\mathcal{AP}$-compactification of S and $\mathcal{AP}(S) = \gamma_3 S$.

(d) $(\eta_4, \gamma_4 S)$ is the $\mathcal{SAP}$-compactification of S and $\mathcal{SAP}(S) = \gamma_4 S$.

Notice that by Theorem 4.8, if S is a discrete semigroup, then $(\iota, \beta S)$ is another candidate to be called "the" $\mathcal{LMC}$-compactification of S.

We remind the reader that semigroup compactifications need not be topological compactifications because the functions (in this case η_i) are not required to be embeddings. In Exercise 21.1.1, the reader is asked to show that if it is possible for η_i to be an embedding, or just one-to-one, then it is.

Exercise 21.1.1. Let S be an infinite semigroup which is also a topological space and let $i \in \{1, 2, 3, 4\}$. Let Y be a semigroup with topology such that $\Psi_i(Y)$ Let $\tau : S \to Y$ be a continuous homomorphism with $\tau[S] \subseteq \Lambda(Y)$.

(a) Prove that if τ is one-to-one, then so is η_i.

(b) Prove that if τ is an embedding, then so is η_i.

Exercise 21.1.2. If S is a group (which is also a topological space), show that $\mathcal{AP}(S) \approx \mathcal{SAP}(S)$. (Hint: Prove that $\mathcal{AP}(S)$ is a group and apply Ellis' Theorem (Corollary 2.39).)

Exercise 21.1.3. Let C denote any of the semigroups $\mathcal{WAP}(\mathbb{N})$, $\mathcal{AP}(\mathbb{N})$, or $\mathcal{SAP}(\mathbb{N})$. Let $x \in C$ and let m and n be distinct positive integers. Show that $m + x \neq n + x$.

Exercise 21.1.4. Show that each of the compactifications $\mathcal{WAP}(\mathbb{N})$, $\mathcal{AP}(\mathbb{N})$ and $\mathcal{SAP}(\mathbb{N})$ has 2^c elements. (Hint: By Kronecker's Theorem, $\mathbb{N}$ can be densely and homomorphically embedded in a product of $\mathfrak{c}$ copies of the unit circle.)

21.2 Right Topological Compactifications

The requirements in conclusions (4) and (5) of Theorem 21.4 referring to the topological center seem awkward. They are redundant except when $i = 1$, when the property being considered is that of being a right topological semigroup. We show in this section that without the requirement that $g[S] \subseteq \Lambda(D)$, there is no maximal right topological compactification of any infinite discrete weakly right cancellative semigroup in the sense of Theorem 21.4.

Theorem 21.7. *Let S be an infinite discrete weakly right cancellative semigroup and let κ be an infinite cardinal. There exist a compact right topological semigroup T and an injective homomorphism $\tau : S \to T$ such that any closed subsemigroup of T containing $\tau[S]$ has cardinality at least κ.*

Proof. Let ∞ be a point not in $\kappa \times S$ and let $T = (\kappa \times S) \cup \{\infty\}$. (Recall that the cardinal κ is an ordinal, so that $\kappa = \{\alpha : \alpha \text{ is an ordinal and } \alpha < \kappa\}$.) Define an operation $\cdot$ on T as follows.

$$(0, s) \cdot (t', s') = (t', ss'),$$
$$(t, s) \cdot (t', s') = (t' + t, s) \text{ if } t \neq 0, \text{ and}$$
$$(t, s) \cdot \infty = \infty \cdot (t, s) = \infty \cdot \infty = \infty,$$

where $t' + t$ is ordinal addition. We leave it as an exercise (Exercise 21.2.1) to verify that the operation $\cdot$ is associative.

Now fix an element $\sigma \in S$ and define a topology on T as follows.

(1) Each point of $\big(\kappa \times (S \setminus \{\sigma\})\big) \cup \{(0, \sigma)\}$ is isolated.

(2) Basic open neighborhoods of ∞ are of the form

$$\{\infty\} \cup (\{t'' : t' < t'' < \kappa\} \times S) \cup \big(\{t'\} \times (S \setminus F)\big)$$

where $t' < \kappa$ and F is a finite subset of S with $\sigma \in F$.

(3) If $t = t' + 1$, then basic open neighborhoods of (t, σ) are of the form

$$\{(t, \sigma)\} \cup (\{t'\} \times (S\backslash F))$$

where F is a finite subset of S with $\sigma \in F$.

(4) If t is a nonzero limit ordinal, then basic open neighborhoods of (t, σ) are of the form

$$\{(t, \sigma)\} \cup (\{t'' : t' < t'' < t\} \times S) \cup (\{t'\} \times (S\backslash F))$$

where $t' < t$ and F is a finite subset of S with $\sigma \in F$.

That is to say, a basis for the topology on T is

$$\mathcal{B} = \big\{\{(t, s)\} : t < \kappa \text{ and } s \in S\backslash\{\sigma\}\big\} \cup \big\{\{(0, \sigma)\}\big\}$$

$$\cup \big\{\{\infty\} \cup (\{t'' : t' < t'' < \kappa\} \times S) \cup (\{t'\} \times (S\backslash F)) :$$
$$t' < \kappa, \ F \in \mathcal{P}_f(S), \text{ and } \sigma \in F\big\}$$

$$\cup \big\{\{(t'+1, \sigma)\} \cup (\{t'\} \times (S\backslash F)) : t' < \kappa, \ F \in \mathcal{P}_f(S), \text{ and } \sigma \in F\big\}$$

$$\cup \big\{\{(t, \sigma)\} \cup (\{t'' : t' < t'' < t\} \times S) \cup (\{t'\} \times (S\backslash F)) :$$
$$t \text{ is a limit}, \ t' < t < \kappa, \ F \in \mathcal{P}_f(S), \text{ and } \sigma \in F\big\}.$$

The verification that $\mathcal{B}$ is a basis for a Hausdorff topology on T is Exercise 21.2.2.

To see that, with this topology, T is compact, let $\mathcal{U}$ be an open cover of T. Pick $U \in \mathcal{U}$ such that $\infty \in U$ and pick $t_1 < \kappa$ such that $\{t'' : t_1 < t'' < \kappa\} \times S \subseteq U$. If a finite subfamily of $\mathcal{U}$ covers $\{t : t \leq t_1\} \times S$, then we are done. So assume that no finite subfamily of $\mathcal{U}$ covers $\{t : t \leq t_1\} \times S$ and pick the least t_0 such that no finite subfamily of $\mathcal{U}$ covers $\{t : t \leq t_0\} \times S$. Pick $V \in \mathcal{U}$ such that $(t_0 + 1, \sigma) \in V$ and pick $F \in \mathcal{P}_f(S)$ such that $\{t_0\} \times (S\backslash F) \subseteq V$. Pick finite $\mathcal{G} \subseteq \mathcal{U}$ such that $\{t_0\} \times F \subseteq \bigcup \mathcal{G}$.

Assume first that $t_0 = t' + 1$ for some t' and pick finite $\mathcal{F} \subseteq \mathcal{U}$ such that $\{t : t \leq t'\} \times S \subseteq \bigcup \mathcal{F}$. Then $\mathcal{F} \cup \mathcal{G} \cup \{V\}$ covers $\{t : t \leq t_0\} \times S$, a contradiction.

Thus t_0 is a limit ordinal. If $t_0 = 0$, then $\mathcal{G} \cup \{V\}$ covers $\{0\} \times S$. Thus $t_0 \neq 0$. Pick $W \in \mathcal{U}$ such that $(t_0, \sigma) \in W$ and pick $t' < t_0$ such that $\{t'' : t' < t'' < t_0\} \times S \subseteq W$. Pick finite $\mathcal{F} \subseteq \mathcal{U}$ such that $\{t : t \leq t'\} \times S \subseteq \bigcup \mathcal{F}$. Then $\mathcal{F} \cup \mathcal{G} \cup \{V, W\}$ covers $\{t : t \leq t_0\} \times S$, a contradiction.

Thus T is compact as claimed. We verify now that T is a right topological semigroup. Since ρ_∞ is constant it is continuous. Let $(t', s') \in \kappa \times S$. Trivially $\rho_{(t', s')}$ is continuous at each isolated point of T, so we only need to show that $\rho_{(t', s')}$ is continuous at ∞ and at each point of $(\kappa\backslash\{0\}) \times \{\sigma\}$.

To see that $\rho_{(t', s')}$ is continuous at ∞, let U be a neighborhood of ∞ and pick $u < \kappa$ such that $\{t'' : u < t'' < \kappa\} \times S \subseteq U$. Let

$$W = \{\infty\} \cup (\{t'' : u + 1 < t'' < \kappa\} \times S) \cup (\{u + 1\} \times (S\backslash\{\sigma\})).$$

Then $\rho_{(t', s')}[W] \subseteq U$ since $t'' \leq t' + t''$ for all $t'' < \kappa$.

Next let $0 < t < \kappa$ and let U be a neighborhood of $(t, \sigma) \cdot (t', s') = (t' + t, \sigma)$. Assume first that $t = t_0 + 1$ and pick $F \in \mathcal{P}_f(S)$ with $\sigma \in F$ such that $\{t' + t_0\} \times (S\backslash F) \subseteq$

U. Let $G = \{s \in S : ss' \in F\}$ and notice that, since S is weakly right cancellative, G is finite. Let
$$W = \{(t, \sigma)\} \cup \big(\{t_0\} \times (S \backslash (F \cup G))\big).$$

Then $\rho_{(t', s')}[W] \subseteq U$. (If $t_0 = 0$ and $s \in S \backslash G$, then $ss' \notin F$.)

Now assume that t is a limit ordinal and pick $t_0 < t$ and $F \in \mathcal{P}_f(S)$ with $\sigma \in F$ such that

$$\{(t' + t, \sigma)\} \cup (\{t'' : t' + t_0 < t'' < t' + t\} \times S) \cup \big(\{t' + t_0\} \times (S \backslash F)\big) \subseteq U$$

and let
$$W = \{(t, \sigma)\} \cup (\{t'' : t_0 < t'' < t\} \times S) \cup \big(\{t_0\} \times (S \backslash F)\big).$$

Then $\rho_{(t', s')}[W] \subseteq U$.

Thus T is a compact right topological semigroup. Define $\tau : S \to T$ by $\tau(s) = (0, s)$. Trivially τ is an injective homomorphism.

Now let H be a closed subsemigroup of T containing $\tau[S]$. We claim that $\kappa \times \{\sigma\} \subseteq H$. Suppose not, and pick the least $t < \kappa$ such that $(t, \sigma) \notin H$. Notice that $t > 0$. Pick a neighborhood U of (t, σ) such that $U \cap H = \emptyset$. Then there exist some $t' < t$ and some $F \in \mathcal{P}_f(S)$ such that $\{t'\} \times (S \backslash F) \subseteq U$. Then $(t', \sigma) \in H$ so

$$(\{0\} \times S) \cdot (t', \sigma) = \{t'\} \times S\sigma \subseteq H.$$

Since S is weakly right cancellative, $S\sigma$ is infinite, so $(\{t'\} \times S\sigma) \cap U \neq \emptyset$, a contradiction. $\qquad\square$

The following corollary says in the strongest terms that without the requirement that $\tau[S] \subseteq \Lambda(T)$ (or some other requirement), there could not be a maximal right topological compactification of S. Notice that the corollary does not even demand any relationship between the topology on X and the semigroup operation, nor are any requirements, either algebraic or topological, placed on the function φ. (Although, presumably, the kind of result one would be looking for would have X as a compact right topological semigroup and would have φ as a homomorphism with $\varphi[S]$ dense in X.)

Corollary 21.8. *Let S be an infinite discrete weakly right cancellative semigroup, let X be a semigroup with a compact topology, and let $\varphi : S \to X$. There exist a compact right topological semigroup T and an injective homomorphism $\tau : S \to T$ such that there is no continuous homomorphism $\eta : X \to T$ with $\eta \circ \varphi = \tau$.*

Proof. Pick a cardinal $\kappa > |X|$ and let T and τ be as guaranteed by Theorem 21.7 for S and κ. Suppose one has a continuous homomorphism $\eta : X \to T$ such that $\eta \circ \varphi = \tau$. Then $\eta[X]$ is a closed subsemigroup of T containing $\tau[S]$ and thus $\kappa > |X| \geq \eta[X] \geq \kappa$, a contradiction. $\qquad\square$

Exercise 21.2.1. Prove that the operation on T defined in the proof of Theorem 21.7 is associative.

Exercise 21.2.2. Let $\mathcal{B}$ and T be as in the proof of Theorem 21.7. Prove that $\mathcal{B}$ is a basis for a Hausdorff topology on T.

21.3 Periodic Compactifications as Quotients

Given a semigroup S with topology, the $\mathcal{WAP}$-, $\mathcal{AP}$-, and $\mathcal{SAP}$-compactifications are all compact right topological semigroups that are equal to their topological centers and thus, by Theorem 4.8 each is a quotient of βS_d, where S_d denotes S with the discrete topology.

In this section we shall identify equivalence relations on βS_d that yield the $\mathcal{WAP}$- and $\mathcal{AP}$-compactifications. We shall use the explicit descriptions of these compactifications as quotients to characterize the continuous functions from a semitopological semigroup S to a compact semitopological or topological semigroup that extend to the $\mathcal{WAP}$- or $\mathcal{AP}$-compactifications of S. We shall also show that, if S is commutative, $\mathcal{SAP}(S)$ is embedded in $\mathcal{WAP}(S)$ as the smallest ideal of $\mathcal{WAP}(S)$.

We first give some simple well-known properties of quotient spaces defined by families of functions.

Definition 21.9. Let X be a compact space and let $\phi(f, Y)$ be a statement which implies that f is a continuous function mapping X to a space Y. We define an equivalence relation on X by stating that $x \equiv y$ if and only if $f(x) = f(y)$ for every f for which $\phi(f, Y)$ holds for some Y. Let $X/\equiv$ denote the quotient of X defined by this relation, and let $\pi : X \to X/\equiv$ denote the canonical mapping. We give $X/\equiv$ the quotient topology, in which a set U is open in $X/\equiv$ if and only if $\pi^{-1}[U]$ is open in X.

Of course, the space $X/\equiv$ depends on the statement ϕ, although we have not indicated this in the notation.

Lemma 21.10. *Let X be a compact space and let* $\overset{.}{\phi}(f, Y)$ *be a statement which implies that f is a continuous function mapping X to a space Y. Then $X/\equiv$ has the following properties:*

(1) *If $g : X \to Z$ is a continuous function and if $\hat{g} : X/\equiv \to Z$ is a function for which $g = \hat{g} \circ \pi$, then $\hat{g}$ is continuous.*

(2) *$X/\equiv$ is a compact Hausdorff space.*

(3) *A net $\langle \pi(x_\iota) \rangle_{\iota \in I}$ converges to $\pi(x)$ in $X/\equiv$ if and only if $\langle f(x_\iota) \rangle_{\iota \in I}$ converges to $f(x)$ in Y for every (f, Y) for which $\phi(f, Y)$ holds.*

Proof. (1) If V is an open subset of Z, then $\pi^{-1}\big[\hat{g}^{-1}[V]\big] = g^{-1}[V]$, which is open in X and so $\hat{g}^{-1}[V]$ is open in $X/\equiv$.

(2) Since $X/\equiv$ is the continuous image of a compact space, it is compact. To see that it is Hausdorff, suppose that $x, y \in X$ and that $\pi(x) \neq \pi(y)$. Then there is a pair

(f, Y) such that $\phi(f, Y)$ holds and $f(x) \neq f(y)$. For every $u, v \in X$, $\pi(u) = \pi(v)$ implies that $f(u) = f(v)$. So there is a function $\hat{f} : X/\!\equiv \,\to Y$ for which $f = \hat{f} \circ \pi$. By (1), $\hat{f}$ is continuous. If G and H are disjoint subsets of Y which are neighborhoods of $f(x)$ and $f(y)$ respectively, then $\hat{f}^{-1}[G]$ and $\hat{f}^{-1}[H]$ are disjoint subsets of $X/\!\equiv$ which are neighborhoods of $\pi(x)$ and $\pi(y)$ respectively. So $X/\!\equiv$ is Hausdorff.

(3) As in (2), for each (f, Y) for which $\phi(f, Y)$ holds, let $\hat{f} : X/\!\equiv \,\to Y$ denote the continuous function for which $f = \hat{f} \circ \pi$. The set $\{\hat{f}^{-1}[U] : \phi(f, Y)$ and U is open in $Y\}$ is a subbase for a topology τ on $X/\!\equiv$ which is coarser than the quotient topology and is therefore compact. However, we saw in (2) that, for every $x, y \in X$, $\pi(x) \neq \pi(y)$ implied that $\pi(x)$ and $\pi(y)$ had disjoint τ-neighborhoods. So τ is Hausdorff and is therefore equal to the quotient topology on $X/\!\equiv$.

Let $\langle x_\iota \rangle_{\iota \in I}$ be a net in X and let $x \in X$. If $\langle f(x_\iota) \rangle_{\iota \in I}$ converges to $f(x)$ in Y whenever $\phi(f, Y)$ holds, then $\langle \pi(x_\iota) \rangle_{\iota \in I}$ converges to $\pi(x)$ in the topology τ. So $\langle \pi(x_\iota) \rangle_{\iota \in I}$ converges to $\pi(x)$ in the quotient topology.

Now assume that $\langle \pi(x_\iota) \rangle_{\iota \in I}$ converges to $\pi(x)$ and let (f, Y) be given such that $\phi(f, Y)$ holds. If U is a neighborhood of $f(x)$, then $\hat{f}^{-1}[U]$ is a neighborhood of $\pi(x)$. Thus $\langle \pi(x_\iota) \rangle_{\iota \in I}$ is eventually in $\hat{f}^{-1}[U]$ and so $\langle f(x_\iota) \rangle_{\iota \in I}$ is eventually in U. $\square$

We recall that we defined in Section 13.4 a binary operation $\diamond$ on βS_d by putting $x \diamond y = \lim_{t \to y} \lim_{s \to x} st$, where s and t denote elements of S. Note that, in defining $x \diamond y$, we have reversed the order of the limits used in our definition of the semigroup $(\beta S_d, \cdot)$. The semigroup $(\beta S_d, \diamond)$ is a left topological semigroup while $(\beta S_d, \cdot)$, as we defined it, is a right topological semigroup.

The following lemma summarizes the basic information that will be used in our construction of the $\mathcal{WAP}$- and $\mathcal{AP}$-compactifications of a semitopological semigroup.

Lemma 21.11. *Let S be a semitopological semigroup and let $\theta(f, C)$ be a statement which implies that f is a continuous function f mapping S to a compact semitopological semigroup C. Let $\phi(g, C)$ be the statement that there exists f for which $\theta(f, C)$ holds and $g = \widetilde{f} : \beta S_d \to C$. Suppose that, for every (f, C) for which $\theta(f, C)$ is true, we have:*

(1) *$\theta(f \circ \lambda_a, C)$ and $\theta(f \circ \rho_a, C)$ hold for every $a \in S$ (where $\lambda_a, \rho_a : S \to S$) and*
(2) *for all $x, y \in \beta S_d$, $\widetilde{f}(x \cdot y) = \widetilde{f}(x \diamond y)$.*

Let $\beta S_d/\!\equiv$ and π be defined by the statement ϕ as described in Definition 21.9. We define $\cdot$ on $\beta S_d/\!\equiv$ by putting $\pi(x) \cdot \pi(y) = \pi(x \cdot y)$.

(a) *The operation $\cdot$ on $\beta S_d/\!\equiv$ is well defined.*
(b) *With this operation, $\beta S_d/\!\equiv$ is a compact semitopological semigroup and π is a continuous homomorphism from $(\beta S_d, \cdot)$ and from $(\beta S_d, \diamond)$.*
(c) *The restriction $\pi_{|S}$ is continuous.*

Proof. Notice that for $x, y \in \beta S_d$, we have $x \equiv y$ if and only if for every pair (f, C) satisfying $\theta(f, C)$, one has $\widetilde{f}(x) = \widetilde{f}(y)$.

(a) Given $a \in S$ and $x, x' \in \beta S_d$ with $x \equiv x'$, we claim that $x \cdot a \equiv x' \cdot a$ and $a \cdot x \equiv a \cdot x'$. To see this, let (f, C) be given satisfying $\theta(f, C)$. Then $\theta(f \circ \rho_a, C)$ and $\theta(f \circ \lambda_a, C)$ hold. Since $\widetilde{f \circ \rho_a} = \widetilde{f} \circ \widetilde{\rho}_a$ and $\widetilde{f \circ \lambda_a} = \widetilde{f} \circ \widetilde{\lambda}_a$ we have that $\widetilde{f}(x \cdot a) = \widetilde{f}(x' \cdot a)$ and $\widetilde{f}(a \cdot x) = \widetilde{f}(a \cdot x')$ as required.

Now assume that $x \equiv x'$ and $y \in \beta S_d$. We claim that $x \cdot y \equiv x' \cdot y$. Recall that we denote by ℓ_x the continuous function from βS_d to βS_d defined by $\ell_x(z) = x \diamond z$. Now for each $a \in S$ and each (f, C) for which $\theta(f, C)$ holds,

$$\widetilde{f} \circ \ell_x(a) = \widetilde{f}(x \diamond a) = \widetilde{f}(x \cdot a) = \widetilde{f}(x' \cdot a) = \widetilde{f}(x' \diamond a) = \widetilde{f} \circ \ell_{x'}(a)$$

and so $\widetilde{f} \circ \ell_x$ and $\widetilde{f} \circ \ell_{x'}$ are continuous functions agreeing on S, hence on βS_d. Thus $\widetilde{f}(x \cdot y) = \widetilde{f}(x' \cdot y)$. Similarly (using $\rho_x(z) = z \cdot x$), $y \cdot x \equiv y \cdot x'$.

Now let $x \equiv x'$ and $y \equiv y'$. Then $x \cdot y \equiv x' \cdot y \equiv x' \cdot y'$.

(b) By Lemma 21.10, $\beta S_d/\equiv$ is compact and Hausdorff. By (a), π is a homomorphism from $(\beta S_d, \cdot)$, and by (a) and (2), π is a homomorphism from $(\beta S_d, \diamond)$. By Exercise 2.2.2, since $\beta S_d/\equiv$ is the continuous homomorphic image of $(\beta S_d, \cdot)$, it is a right topological semigroup, and since it is the continuous homomorphic image of $(\beta S_d, \diamond)$, it is a left topological semigroup.

(c) To see that $\pi_{|S} : S \to \beta S_d/\equiv$ is continuous, let $s \in S$ and let $\langle s_\iota \rangle_{\iota \in I}$ be a net in S converging to s in S. To see that $\langle \pi(s_\iota) \rangle_{\iota \in I}$ converges to $\pi(s)$ it suffices by Lemma 21.10 to let (g, C) be given such that $\phi(g, C)$ holds and show that $\langle g(s_\iota) \rangle_{\iota \in I}$ converges to $g(s)$. Pick f such that $\theta(f, C)$ holds and $g = \widetilde{f}$. Then $f : S \to C$ is continuous so $\langle f(s_\iota) \rangle_{\iota \in I}$ converges to $f(s)$. That is, $\langle g(s_\iota) \rangle_{\iota \in I}$ converges to $g(s)$ as required. □

The numbering θ_2 and ϕ_2 in the following definition is intended to correspond to the function $\eta_2 : S \to \mathcal{WAP}(S)$. The statements θ_2 and ϕ_2 depend on the semigroup S, but the notation does not reflect this dependence.

Definition 21.12. Let S be a semitopological semigroup. $\theta_2(f, C)$ is the statement "C is a compact semitopological semigroup, f is a continuous function from S to C, and $\widetilde{f}(x \diamond y) = \widetilde{f}(x \cdot y)$ for every $x, y \in \beta S_d$". $\phi_2(g, C)$ is the statement "there exists (f, C) for which $\theta_2(f, C)$ holds and $g = \widetilde{f} : \beta S_d \to C$".

Lemma 21.13. *Let S be a semitopological semigroup. If $\theta_2(f, C)$ holds and $a \in S$, then $\theta_2(f \circ \lambda_a, C)$ and $\theta_2(f \circ \rho_a, C)$ hold as well (where $\lambda_a, \rho_a : S \to S$).*

Proof. Since λ_a is continuous (because S is a semitopological semigroup), we have that $f \circ \lambda_a$ is continuous. Notice also that $\widetilde{f \circ \lambda_a} = \widetilde{f} \circ \widetilde{\lambda}_a$. Now, let $x, y \in \beta S_d$. Then, since $a \in S$, $a \diamond x = a \cdot x$ and so

$$\widetilde{f \circ \lambda_a}(x \diamond y) = \widetilde{f}\big(a \diamond (x \diamond y)\big) = \widetilde{f}\big((a \diamond x) \diamond y\big) = \widetilde{f}\big((a \cdot x) \diamond y\big)$$
$$= \widetilde{f}\big((a \cdot x) \cdot y\big) = \widetilde{f}\big(a \cdot (x \cdot y)\big) = \widetilde{f \circ \lambda_a}(x \cdot y).$$

Similarly $\theta_2(f \circ \rho_a, C)$ holds. □

The reader may wonder why we now require S to be a semitopological semigroup, after pointing out in Section 21.1 that we were not demanding this. The reason is that we need this fact for the validity of Lemma 21.13.

Theorem 21.14. *Let S be a semitopological semigroup, let $\phi = \phi_2$ and let $\beta S_d/\equiv$ and $\pi : \beta S_d \to \beta S_d/\equiv$ be defined by Definition 21.9. Then $(\pi_{|S}, \beta S_d/\equiv)$ is a $\mathcal{WAP}$-compactification of S. That is:*

(a) $\Psi_2(\beta S_d/\equiv)$,

(b) $\pi_{|S}$ *is a continuous homomorphism from S to $\beta S_d/\equiv$,*

(c) $\pi[S]$ *is dense in $\beta S_d/\equiv$, and*

(d) *given any D such that $\Psi_2(D)$ and any continuous homomorphism $g : S \to D$, there exists a continuous homomorphism $\mu : \beta S_d/\equiv \to D$ such that $\mu \circ \pi_{|S} = g$.*

Proof. Conclusions (a), (b), and (c) are an immediate consequence of Lemmas 21.11 and 21.13.

Let g and D be given such that $\Psi_2(D)$ and g is a continuous homomorphism from S to D. We claim that $\theta_2(g, D)$ holds. So let $x, y \in \beta S_d$. Notice that $\widetilde{g}$ is a homomorphism from $(\beta S_d, \cdot)$ to D by Corollary 4.22. Thus

$$\widetilde{g}(x \diamond y) = \widetilde{g}(y\text{-}\lim_{t \in S} x\text{-}\lim_{s \in S} s \cdot t) = y\text{-}\lim_{t \in S} x\text{-}\lim_{s \in S} g(s) \cdot g(t) = \widetilde{g}(x) \cdot \widetilde{g}(y) = \widetilde{g}(x \cdot y).$$

Let $\mu : \beta S_d/\equiv \to D$ denote the function for which $\mu \circ \pi = \widetilde{g}$. Then μ is continuous, by Lemma 21.10, and is easily seen to be a homomorphism. $\square$

We now turn our attention to characterizations of functions that extend to $\mathcal{WAP}(S)$.

Theorem 21.15. *Let S be a semitopological semigroup, let C be a compact semitopological semigroup, and let f be a continuous function from S to C. There is a continuous function $g : \mathcal{WAP}(S) \to C$ for which $f = g \circ \eta_2$ if and only if $\widetilde{f}(x \cdot y) = \widetilde{f}(x \diamond y)$ for every $x, y \in \beta S_d$.*

Proof. Necessity. Pick a continuous function $g : \mathcal{WAP}(S) \to C$ such that $f = g \circ \eta_2$. Then

$$\begin{aligned}
\widetilde{f}(x \cdot y) &= x\text{-}\lim_{s \in S} y\text{-}\lim_{t \in S} f(s \cdot t) = x\text{-}\lim_{s \in S} y\text{-}\lim_{t \in S} g\big(\eta_2(s \cdot t)\big) \\
&= g\big(x\text{-}\lim_{s \in S} y\text{-}\lim_{t \in S} \eta_2(s) \cdot \eta_2(t)\big) = g\big(\widetilde{\eta_2}(x) \cdot \widetilde{\eta_2}(y)\big) \\
&= g\big(y\text{-}\lim_{t \in S} x\text{-}\lim_{s \in S} \eta_2(s) \cdot \eta_2(t)\big) \\
&= y\text{-}\lim_{t \in S} x\text{-}\lim_{s \in S} g\big(\eta_2(s \cdot t)\big) = y\text{-}\lim_{t \in S} x\text{-}\lim_{s \in S} f(s \cdot t) = \widetilde{f}(x \diamond y).
\end{aligned}$$

Sufficiency. Define $\equiv$ as in Theorem 21.14. Then by Theorems 21.5 and 21.14, it suffices to show that there exists a continuous function $g : \beta S_d/\equiv \to C$ for which $f = g \circ \pi_{|S}$. Since $\phi_2(\widetilde{f}, C)$ holds, there is a function $g : \beta S_d/\equiv \to C$ for which $\widetilde{f} = g \circ \pi$. By Lemma 21.10, g is continuous. $\square$

If S is commutative, we obtain a characterization in terms of $(\beta S_d, \cdot)$ alone.

Corollary 21.16. *Let S be a commutative semitopological semigroup, let C be a compact semitopological semigroup, and let f be a continuous function from S to C. There is a continuous function g : $\mathcal{WAP}(S) \rightarrow C$ for which $f = g \circ \eta_2$ if and only if $\widetilde{f}(x \cdot y) = \widetilde{f}(y \cdot x)$ for every $x, y \in \beta S_d$.*

Proof. This is an immediate consequence of Theorems 13.37 and 21.15.   □

Recall that $C(S)$ denotes the algebra of bounded continuous complex-valued functions defined on S.

Definition 21.17. Let S be a semitopological semigroup and let $f \in C(S)$.

(a) The function f is *weakly almost periodic* if and only if there is a continuous $g : \mathcal{WAP}(S) \rightarrow \mathbb{C}$ such that $g \circ \eta_2 = f$.

(b) The function f is *almost periodic* if and only if there is a continuous $g : \mathcal{AP}(S) \rightarrow \mathbb{C}$ such that $g \circ \eta_3 = f$.

Theorem 21.18. *Let S be an infinite semitopological semigroup and let $f \in C(S)$. The following statements are equivalent:*

(1) *The function f is weakly almost periodic.*

(2) *Whenever $x, y \in \beta S_d$, $\widetilde{f}(x \cdot y) = \widetilde{f}(x \diamond y)$.*

(3) *Whenever $\langle a_n \rangle_{n=1}^{\infty}$ and $\langle b_n \rangle_{n=1}^{\infty}$ are sequences in S and all indicated limits exist,*
$$\lim_{m \to \infty} \lim_{n \to \infty} f(a_m \cdot b_n) = \lim_{n \to \infty} \lim_{m \to \infty} f(a_m \cdot b_n).$$

Proof. By Theorem 21.15, (1) and (2) are equivalent. We shall show that (2) and (3) are equivalent.

(2) implies (3). Let $\langle a_n \rangle_{n=1}^{\infty}$ and $\langle b_n \rangle_{n=1}^{\infty}$ be sequences in S for which all the limits indicated in the expressions $\lim_{m \to \infty} \lim_{n \to \infty} f(a_m \cdot b_n)$ and $\lim_{n \to \infty} \lim_{m \to \infty} f(a_m \cdot b_n)$ exist. Let x and y be limit points in βS_d of the sequences $\langle a_n \rangle_{n=1}^{\infty}$ and $\langle b_n \rangle_{n=1}^{\infty}$ respectively. Then $\lim_{n \to \infty} \lim_{m \to \infty} f(a_m \cdot b_n) = \widetilde{f}(x \diamond y)$ and $\lim_{m \to \infty} \lim_{n \to \infty} f(a_m \cdot b_n) = \widetilde{f}(x \cdot y)$. So (2) implies that these double limits are equal.

(3) implies (2). Let $x, y \in \beta S_d$, let $k = \widetilde{f}(x \cdot y)$, and let $\ell = \widetilde{f}(x \diamond y)$. For each $n \in \mathbb{N}$ let $A_n = \{a \in S : |\widetilde{f}(a \cdot y) - k| < \frac{1}{n}\}$ and let $B_n = \{b \in S : |\widetilde{f}(x \cdot b) - \ell| < \frac{1}{n}\}$. Since $(\beta S_d, \cdot)$ is a right topological semigroup, $(\beta S_d, \diamond)$ is a left topological semigroup, and $x \cdot b = x \diamond b$ for $b \in S$, $A_n \in x$ and $B_n \in y$.

Choose $a_1 \in A_1$ and $b_1 \in B_1$. Inductively, let $r \in \mathbb{N}$ and assume that we have chosen $\{a_1, a_2, \ldots, a_r\}$ and $\{b_1, b_2, \ldots, b_r\}$ so that for all $m, n \in \{1, 2, \ldots, r\}$

(i) $a_m \in A_m$ and $b_n \in B_n$,

(ii) if $m > n$, then $|f(a_m \cdot b_n) - \ell| < \frac{1}{n}$,

(iii) if $n > m$, then $|f(a_m \cdot b_n) - k| < \frac{1}{m}$,

(iv) if $m > n$, then $|f(a_m \cdot b_n) - \widetilde{f}(x \cdot b_n)| < \frac{1}{m}$, and

(v) if $n > m$, then $|f(a_m \cdot b_n) - \widetilde{f}(a_m \cdot y)| < \frac{1}{n}$.

Now for each $n \in \{1, 2, \ldots, r\}$, $b_n \in B_n$ so $\{a \in S : |f(a \cdot b_n) - \ell| < \frac{1}{n}\} \in x$. Also, for each $n \in \{1, 2, \ldots, r\}$, $\{a \in S : |f(a \cdot b_n) - \widetilde{f}(x \cdot b_n)| < \frac{1}{r+1}\} \in x$ so pick a_{r+1} such that for each $n \in \{1, 2, \ldots, r\}$, $|f(a_{r+1} \cdot b_n) - \ell| < \frac{1}{n}$ and $|f(a_{r+1} \cdot b_n) - \widetilde{f}(x \cdot b_n)| < \frac{1}{r+1}$. Similarly, choose

$$b_{r+1} \in \bigcap_{m=1}^{r} \big(\{b \in S : |f(a_m \cdot b) - k| < \tfrac{1}{m}\} \cap \{b \in S : |f(a_m \cdot b) - \widetilde{f}(a_m \cdot y)| < \tfrac{1}{r+1}\}\big).$$

Now, given $n \in \mathbb{N}$, one has $\lim_{m \to \infty} f(a_m \cdot b_n) = \widetilde{f}(x \cdot b_n)$ and given $m \in \mathbb{N}$ one has $\lim_{n \to \infty} f(a_m \cdot b_n) = \widetilde{f}(a_m \cdot y)$. Also $\lim_{m \to \infty} \lim_{n \to \infty} f(a_m \cdot b_n) = k$ and $\lim_{n \to \infty} \lim_{m \to \infty} f(a_m \cdot b_n) = \ell$. So $\ell = k$. $\square$

Theorem 21.19. *Let S be a commutative semitopological semigroup. Then $\mathcal{WAP}(S)$ is also commutative. Furthermore, $(\eta_2, \mathcal{WAP}(S))$ is the maximal commutative semigroup compactification of S in the following sense: If (ν, T) is any commutative semigroup compactification of S, then there is a continuous homomorphism $\mu : \mathcal{WAP}(S) \to T$ for which $\nu = \mu \circ \eta_2$.*

Proof. By Theorems 21.5 and 21.14, we may assume that $\mathcal{WAP}(S) = \beta S_d / \equiv$ and $\eta_2 = \pi_{|S}$. Now given (f, C) such that $\theta_2(f, C)$, one has by Theorem 13.37 that for all $x, y \in \beta S_d$, $\widetilde{f}(x \cdot y) = \widetilde{f}(y \cdot x)$ and thus $\pi(x) \cdot \pi(y) = \pi(y) \cdot \pi(x)$.

Now, if (ν, T) is a commutative semigroup compactification of S, then T is a semitopological semigroup so there exists a continuous homomorphism $\mu : \mathcal{WAP}(S) \to T$ for which $\nu = \mu \circ \eta_2$. $\square$

Theorem 21.20. *Let S be an infinite commutative semitopological semigroup. Then $K\big(\mathcal{WAP}(S)\big)$ can be identified with $\mathcal{SAP}(S)$.*

Proof. Let $K = K\big(\mathcal{WAP}(S)\big)$. By Theorem 21.19 and Corollary 2.40, K is a compact topological group. Let e denote the unique minimal idempotent of $\mathcal{WAP}(S)$. We define a continuous homomorphism $\tau : \mathcal{WAP}(S) \to K$ by $\tau(x) = xe$ and define a continuous homomorphism $\mu : S \to K$ by $\mu = \tau \circ \eta_2$.

Let T be a compact topological group and let $\nu : S \to T$ be a continuous homomorphism. By the defining property of $\mathcal{WAP}(S)$, there is a continuous homomorphism $\phi : \mathcal{WAP}(S) \to T$ for which $\phi \circ \eta_2 = \nu$. We note that $\phi(e)$ is the identity of T and so $\phi_{|K} \circ \mu = \nu$. This shows that we can identify $\mathcal{SAP}(S)$ with K (by Theorem 21.5). $\square$

We now discuss some simple properties of $\mathcal{WAP}(\mathbb{N})$. We use Theorems 21.5 and 21.14 to identify $\mathcal{WAP}(\mathbb{N})$ with $\beta \mathbb{N} / \equiv$ and η_2 with $\pi_{|\mathbb{N}}$.

If $S = \mathbb{N} \cup \{\infty\}$, the one point compactification of $\mathbb{N}$, and $\infty + x = x + \infty = \infty$ for all $x \in S$, then S is a semitopological semigroup. Thus, by Exercise 21.1.1, $\eta_2 : \mathbb{N} \to \mathcal{WAP}(\mathbb{N})$ is an embedding. Consequently, we shall regard $\mathbb{N}$ as being a subspace of $\mathcal{WAP}(\mathbb{N})$.

Lemma 21.21. *Let x be in the interior in $\beta \mathbb{N}$ of $\beta \mathbb{N} \setminus (\mathbb{N}^* + \mathbb{N}^*)$. Then $|\pi(x)| = 1$.*

Proof. Let $y \in \beta\mathbb{N}\setminus\{x\}$. We can choose a subset A of $\mathbb{N}$ for which $x \in \overline{A}$, $y \notin \overline{A}$, and $\overline{A} \cap (\mathbb{N}^* + \mathbb{N}^*) = \emptyset$.

Let $f = \chi_A : \mathbb{N} \to \{0, 1\}$ so that $\widetilde{f} = \chi_{\overline{A}}$. We claim that $\widetilde{f}(p+q) = \widetilde{f}(q+p)$ for all $p, q \in \beta\mathbb{N}$. Indeed, if $p \in \mathbb{N}$ or $q \in \mathbb{N}$, then $p+q = q+p$, while if $p, q \in \mathbb{N}^*$, then $\widetilde{f}(p + q) = 0 = \widetilde{f}(q + p)$. Thus by Corollary 21.16, there is a continuous function $g : \mathcal{WAP} \to \{0, 1\}$ such that $f = g \circ \pi_{|\mathbb{N}}$. Then $\widetilde{f}$ and $g \circ \pi$ are continuous functions from $\beta\mathbb{N}$ to $\{0, 1\}$ agreeing on $\mathbb{N}$ and so $\widetilde{f} = g \circ \pi$. Since $\widetilde{f}(x) = 1$ and $\widetilde{f}(y) = 0$, $\pi(x) \neq \pi(y)$. That is, $y \notin \pi(x)$. $\qquad\square$

Theorem 21.22. *There is a dense open subset U of $\mathbb{N}^*$ with the property that $|\pi(x)| = 1$ for every $x \in U$.*

Proof. By Exercise 4.1.7, if $\langle x_n \rangle_{n=1}^{\infty}$ is a sequence in $\mathbb{N}$ for which $x_{n+1} - x_n \to \infty$ and if $A = \{x_n : n \in \mathbb{N}\}$, then $(c\ell_{\beta\mathbb{N}} A) \cap (\mathbb{N}^* + \mathbb{N}^*) = \emptyset$. Now any infinite sequence in $\mathbb{N}$ contains a subsequence with this property, and so there is a dense open subset U of $\mathbb{N}^*$ for which $U \cap (\mathbb{N}^* + \mathbb{N}^*) = \emptyset$. $\qquad\square$

Theorem 21.23. *If x is a P-point in $\mathbb{N}^*$, then $\pi(x)$ is P-point in $\mathcal{WAP}(\mathbb{N})\setminus\mathbb{N}$.*

Proof. Let U denote the interior of $\beta\mathbb{N}\setminus(\mathbb{N}^* + \mathbb{N}^*)$. We shall show that $x \in U$. Our claim will then follow from Lemma 21.21 and the observation that $\pi[U]$ is open in $\mathcal{WAP}(\mathbb{N})$ because $\pi^{-1}\big[\pi[U]\big] = U$.

For each $k \in \mathbb{N}$, we can choose a set $A_k \in x$ with the property that $|a - a'| \geq k$ whenever a and a' are distinct integers in A_k. (This can be seen from the fact that there exists $i \in \{0, 1, 2, \ldots, k-1\}$ for which $\mathbb{N}k + i \in x$.) Since x is a P-point in $\mathbb{N}^*$, there is a set $A \in x$ for which $A\setminus A_k$ is finite for every k. It follows from Exercise 4.1.7 that $\overline{A} \cap (\mathbb{N}^* + \mathbb{N}^*) = \emptyset$. $\qquad\square$

We now turn our attention to representing $\mathcal{AP}(S)$ as a quotient of βS_d.

Definition 21.24. Let S be a semitopological semigroup. $\theta_3(f, C)$ is the statement "C is a compact semitopological semigroup, f is a continuous function from S to C, and there is a continuous function $f^* : \beta S_d \times \beta S_d \to C$ such that for all $s, t \in S$, $f(s \cdot t) = f^*(s, t)$". $\phi_3(g, C)$ is the statement "there exists (f, C) for which $\theta_3(f, C)$ holds and $g = \widetilde{f} : \beta S_d \to C$".

We show next that $\theta_3(f, C)$ implies $\theta_2(f, C)$.

Lemma 21.25. *Let S be a semitopological semigroup. If $\theta_3(f, C)$ holds and $a \in S$, then $\theta_3(f \circ \rho_a, C)$ and $\theta_3(f \circ \lambda_a, C)$ hold as well (where $\rho_a, \lambda_a : S \to S$). Furthermore, if f^* is as guaranteed by this statement, then for all $x, y \in \beta S_d$, $\widetilde{f}(x \cdot y) = f^*(x, y) = \widetilde{f}(x \diamond y)$.*

Proof. Suppose that $\theta_3(f, C)$ holds and that $a \in S$. Let $g : \beta S_d \times \beta S_d \to C$ be defined by $g(x, y) = f^*(ax, y)$. Then g is continuous and $g(s, t) = (f \circ \lambda_a)(s \cdot t)$ for every $s, t \in S$. So $\theta_3(f \circ \lambda_a, C)$ holds. Similarly, $\theta_3(f \circ \rho_a, C)$ holds.

Let $x, y \in \beta S_d$. Then

$$\widetilde{f}(x \cdot y) = x\text{-}\lim_{s \in S} y\text{-}\lim_{t \in S} f(s \cdot t) = x\text{-}\lim_{s \in S} y\text{-}\lim_{t \in S} f^*(s, t) = f^*(x, y)$$
$$= y\text{-}\lim_{t \in S} x\text{-}\lim_{s \in S} f^*(s, t) = y\text{-}\lim_{t \in S} x\text{-}\lim_{s \in S} f(s \cdot t) = \widetilde{f}(x \diamond y).$$

$\square$

For the remainder of this section, we reuse the notations $\equiv$ and π for a different quotient.

Theorem 21.26. *Let S be a semitopological semigroup. Let $\beta S_d/\equiv$ and π be defined by Definition 21.9 with $\phi = \phi_3$. For each $x, y \in \beta S_d$, let $\pi(x) \cdot \pi(y) = \pi(x \cdot y)$. Then $(\pi_{|S}, \beta S_d/\equiv)$ is an $\mathcal{AP}$-compactification of S. That is,*

(a) $\Psi_3(\beta S_d/\equiv)$,

(b) *$\pi_{|S}$ is a continuous homomorphism from S to $\beta S_d/\equiv$,*

(c) *$\pi[S]$ is dense in $\beta S_d/\equiv$, and*

(d) *given any D such that $\Psi_3(D)$ and any continuous homomorphism $g : S \to D$, there exists a continuous homomorphism $\mu : \beta S_d/\equiv \to D$ such that $\mu \circ \pi_{|S} = g$.*

Proof. It follows immediately from Lemmas 21.11 and 21.25 that π is well-defined. Conclusions (b) and (c) are also immediate from these lemmas, as is the fact that $(\beta S_d/\equiv, \cdot)$ is a compact semitopological semigroup. To complete the verification of (a), we need to show that multiplication in $\beta S_d/\equiv$ is jointly continuous.

Suppose that $x, y \in \beta S_d$ and that $\langle x_\iota \rangle_{\iota \in I}$ and $\langle y_\iota \rangle_{\iota \in I}$ are nets in βS_d for which $\langle \pi(x_\iota) \rangle_{\iota \in I}$ and $\langle \pi(y_\iota) \rangle_{\iota \in I}$ converge to $\pi(x)$ and $\pi(y)$ respectively in $\beta S_d/\equiv$. We claim that $\langle \pi(x_\iota) \cdot \pi(y_\iota) \rangle_{\iota \in I}$ converges to $\pi(x) \cdot \pi(y)$, that is that $\langle \pi(x_\iota \cdot y_\iota) \rangle_{\iota \in I}$ converges to $\pi(x \cdot y)$. By passing to a subnet, we may presume that $\langle x_\iota \rangle_{\iota \in I}$ and $\langle y_\iota \rangle_{\iota \in I}$ converge in βS_d to limits u and v respectively. Then $\pi(u) = \pi(x)$ and $\pi(v) = \pi(y)$.

To see that $\langle \pi(x_\iota \cdot y_\iota) \rangle_{\iota \in I}$ converges to $\pi(x \cdot y) = \pi(u \cdot v)$, it suffices by Lemma 21.10 to show that $\langle \widetilde{f}(x_\iota \cdot y_\iota) \rangle_{\iota \in I}$ converges to $\widetilde{f}(u \cdot v)$ for every (f, C) for which $\theta_3(f, C)$ holds. So assume that $\theta_3(f, C)$ holds and let f^* be the function guaranteed by this statement. Then $\langle f^*(x_\iota, y_\iota) \rangle_{\iota \in I}$ converges to $f^*(u, v)$ and so by Lemma 21.25, $\langle \widetilde{f}(x_\iota \cdot y_\iota) \rangle_{\iota \in I}$ converges to $\widetilde{f}(u \cdot v)$ as required.

To verify (d), let g and D be given such that $\Psi_3(D)$ and g is a continuous homomorphism from S to D. We claim that $\theta_3(g, D)$ holds. To see this, define $g^* : \beta S_d \times \beta S_d \to D$ by $g^*(x, y) = \widetilde{g}(x) \cdot \widetilde{g}(y)$. Then g^* is continuous and, given $s, t \in S$, one has $g^*(s, t) = g(s) \cdot g(t) = g(s \cdot t)$.

Let $\mu : \beta S_d/\equiv \to D$ be the function for which $\mu \circ \pi = \widetilde{g}$. Then μ is continuous (by Lemma 21.10) and is easily seen to be a homomorphism. $\square$

Theorem 21.27. *Let S be a semitopological semigroup, let C be a compact topological semigroup, and let f be a continuous function from S to C. There is a continuous function $g : \mathcal{AP}(S) \to C$ for which $f = g \circ \eta_3$ if and only if there is a continuous function $f^* : \beta S_d \times \beta S_d \to C$ such that $f^*(s, t) = f(s \cdot t)$ for all $s, t \in S$.*

Proof. Necessity. Pick a continuous function $g : \mathcal{AP}(S) \to C$ such that $f = g \circ \eta_3$. Let $\widetilde{\eta_3} : \beta S_d \to \mathcal{AP}(S)$ denote the continuous extension of η_3. Define $f^*(x, y) = g\big(\widetilde{\eta_3}(x) \cdot \widetilde{\eta_3}(y)\big)$. Then given $s, t \in S$ one has $g\big(\widetilde{\eta_3}(s) \cdot \widetilde{\eta_3}(t)\big) = g\big(\eta_3(s) \cdot \eta_3(t)\big) = g\big(\eta_3(s \cdot t)\big) = f(s \cdot t)$.

To see that f^* is continuous, let $x, y \in \beta S_d$ and let W be a neighborhood of $g\big(\widetilde{\eta_3}(x) \cdot \widetilde{\eta_3}(y)\big)$. Pick neighborhoods U and V of $\widetilde{\eta_3}(x)$ and $\widetilde{\eta_3}(y)$ in $\mathcal{AP}(S)$ such that $U \cdot V \subseteq g^{-1}[W]$. Then $f^*\big[\widetilde{\eta_3}^{-1}[U] \times \widetilde{\eta_3}^{-1}[V]\big] \subseteq W$.

Sufficiency. Since $\theta_3(f, C)$ holds, there is a function $g : \beta S_d/\equiv \to C$ such that $f = g \circ \pi$. By Lemma 21.10, g is continuous. $\qquad\square$

Theorem 21.28. *Let S be a semitopological semigroup and let $f \in C(S)$. The following statements are equivalent:*

(1) *The function f is almost periodic.*

(2) *There is a continuous function $f^* : \beta S_d \times \beta S_d \to \mathbb{C}$ such that for all $s, t \in S$, $f^*(s, t) = f(s \cdot t)$.*

(3) *For every $\epsilon > 0$, there is an equivalence relation $\approx$ on S with finitely many equivalence classes such that $|f(s \cdot t) - f(s' \cdot t')| < \epsilon$ whenever $s \approx s'$ and $t \approx t'$.*

Proof. The equivalence of (1) and (2) is a special case of Theorem 21.27.

(2) implies (3). Let $\epsilon > 0$ be given. Each element of $\beta S_d \times \beta S_d$ has a neighborhood of the form $U \times V$, where U and V are clopen subsets of βS_d and $|f^*(x, y) - f^*(u, v)| < \epsilon$ whenever $(x, y), (u, v) \in U \times V$. We can choose a finite family $\mathcal{F}$ of sets of this form which covers $\beta S_d \times \beta S_d$. Let

$$\mathcal{W} = \{U : U \times V \in \mathcal{F} \text{ for some } V\} \cup \{V : U \times V \in \mathcal{F} \text{ for some } U\}.$$

Define an equivalence relation $\approx$ on S by $s \approx t$ if and only if for all $W \in \mathcal{W}$, either $\{s, t\} \subseteq W$ or $\{s, t\} \cap W = \emptyset$. Then $\approx$ is an equivalence relation with finitely many equivalence classes.

Now assume that $s \approx s'$ and $t \approx t'$. Pick U and V such that $U \times V \in \mathcal{F}$ and $(s, t) \in U \times V$. Then since $s \approx s'$ and $t \approx t'$, $(s', t') \in U \times V$ and so $|f^*(s, t) - f^*(s', t')| < \epsilon$. That is, $|f(s \cdot t) - f(s' \cdot t')| < \epsilon$.

(3) implies (2). Define $f^*(x, y) = \widetilde{f}(x \cdot y)$ for all $x, y \in \beta S_d$. Then trivially $f^*(s, t) = f(s \cdot t)$ for all $s, t \in S$.

To see that f^* is continuous, let $x, y \in \beta S_d$ and let $\epsilon > 0$ be given. Pick an equivalence relation $\approx$ on S with finitely many equivalence classes such that $|f(s \cdot t) - f(s' \cdot t')| < \frac{\epsilon}{3}$ whenever $s \approx s'$ and $t \approx t'$. Let $\mathcal{R}$ be the set of $\approx$-equivalence classes and pick $A, B \in \mathcal{R}$ such that $x \in \overline{A}$ and $y \in \overline{B}$. We claim that for all $(u, v) \in \overline{A} \times \overline{B}$, one has $|f^*(x, y) - f^*(u, v)| < \epsilon$. Since $\widetilde{f}(x \cdot y) = x\text{-}\lim_{s \in S} y\text{-}\lim_{t \in S} f(s \cdot t)$, we can choose $(s, t) \in A \times B$ such that $|\widetilde{f}(x \cdot y) - f(s \cdot t)| < \frac{\epsilon}{3}$. Similarly, we can choose $(s', t') \in A \times B$ such that $|\widetilde{f}(u \cdot v) - f(s' \cdot t')| < \frac{\epsilon}{3}$. Since $s \approx s'$ and $t \approx t'$, one has $|f(s \cdot t) - f(s' \cdot t')| < \frac{\epsilon}{3}$ and so $|\widetilde{f}(x \cdot y) - \widetilde{f}(u \cdot v)| < \epsilon$. $\qquad\square$

Exercise 21.3.1. Show that $\mathcal{WAP}(\mathbb{N})\backslash\mathbb{N}$ contains a dense open set V disjoint from $(\mathcal{WAP}(\mathbb{N})\backslash\mathbb{N}) + (\mathcal{WAP}(\mathbb{N})\backslash\mathbb{N})$.

Exercise 21.3.2. Show that $\mathcal{WAP}(\mathbb{N})\backslash\mathbb{N}$ contains weak P-points.

Exercise 21.3.3. Show that $\mathcal{WAP}(\mathbb{N})$ is not an F-space. (Hint: An F-space cannot contain an infinite compact right topological group. See [182, Corollary 3.4.2].)

Exercise 21.3.4. We have seen that $K\big(\mathcal{WAP}(\mathbb{N})\big)$ can be identified with $\mathcal{SAP}(\mathbb{N})$ (by Theorem 21.20). Show that the same statement holds for $K\big(\mathcal{AP}(\mathbb{N})\big)$.

Exercise 21.3.5. Let S be a semitopological semigroup and suppose that $f \in C(S)$ is almost periodic. Show that every sequence $\langle s_n \rangle$ in S contains a subsequence $\langle s_{n_r} \rangle$ for which the sequence of functions $t \mapsto f(s_{n_r} t)$ converges uniformly on S. (Hint: Use Theorem 21.28.)

21.4 Semigroup Compactifications as Spaces of Filters

We have seen that if S is a discrete semigroup, then βS is the $\mathcal{LMC}$-compactification of S. We have also found it useful throughout this book to have a specific representation for the points of βS, namely the ultrafilters on S.

As we noted in Section 21.3, the $\mathcal{WAP}$-, $\mathcal{AP}$-, and $\mathcal{SAP}$-compactifications are all quotients of βS. The points of these compactifications correspond to closed subsets of βS. Since the closed subsets of βS correspond to filters on S by Theorem 3.20, each of these compactifications — indeed any semigroup compactification — can be viewed as a set of filters. In this section we characterize those sets of filters that are, in a natural way, semigroup compactifications of S.

Remark 21.29. *Let S be a discrete space, let X be a compact space, and let g be a continuous function from βS onto X. Then given any $x \in X$ and any neighborhood U of x, $g^{-1}[U] \cap S \in \bigcap g^{-1}[\{x\}]$.*

The first step in the characterization is to define an appropriate topology on a set of filters.

Definition 21.30. Let S be a discrete space, let $\mathfrak{R}$ be a set of filters on S, and let $A \subseteq S$.
(a) $A^{\sharp} = \{\mathcal{A} \in \mathfrak{R} : A \in \mathcal{A}\}$.
(b) The *quotient topology* on $\mathfrak{R}$ is the topology with basis $\{A^{\sharp} : A \subseteq S\}$.

Observe that if $A, B \subseteq S$, then $A^{\sharp} \cap B^{\sharp} = (A \cap B)^{\sharp}$ so that $\{A^{\sharp} : A \subseteq S\}$ does form a basis for a topology on $\mathfrak{R}$.

That the term "quotient topology" is appropriate is part of the content of the following theorem.

Recall that, given a set $\mathfrak{R}$ of sets, a choice function for $\mathfrak{R}$ is a function $f : \mathfrak{R} \to \bigcup \mathfrak{R}$ such that for all $A \in \mathfrak{R}$, $f(A) \in A$.

Theorem 21.31. *Let S be a discrete space, let X be a compact space, and let g be a continuous function from βS onto X. Let $\Re = \{\bigcap g^{-1}[\{x\}] : x \in X\}$. Then $\Re$ is a set of filters on S satisfying*

(a) *given any choice function f for $\Re$ there exists $\mathfrak{F} \in \mathcal{P}_f(\Re)$ such that $S = \bigcup_{\mathcal{A} \in \mathfrak{F}} f(\mathcal{A})$ and*

(b) *given distinct $\mathcal{A}$ and $\mathcal{B}$ in $\Re$, there exist $A \in \mathcal{A}$ and $B \in \mathcal{B}$ such that whenever $\mathcal{C} \in \Re$, either $S\backslash A \in \mathcal{C}$ or $S\backslash B \in \mathcal{C}$.*

Further, with the quotient topology on $\Re$, the function $h : X \to \Re$ defined by $h(x) = \bigcap g^{-1}[\{x\}]$ is a homeomorphism.

Proof. Given $x \in X$, $g^{-1}[\{x\}]$ is a set of ultrafilters whose intersection is therefore a filter.

Let f be a choice function for $\Re$ and suppose that for each $\mathfrak{F} \in \mathcal{P}_f(\Re)$,

$$S\backslash \bigcup_{\mathcal{A} \in \mathfrak{F}} f(\mathcal{A}) \neq \emptyset.$$

Then $\{S\backslash f(\mathcal{A}) : \mathcal{A} \in \Re\}$ has the finite intersection property so pick $p \in \beta S$ such that $\{S\backslash f(\mathcal{A}) : \mathcal{A} \in \Re\} \subseteq p$. Let $x = g(p)$ and let $\mathcal{A} = \bigcap g^{-1}[\{x\}]$. Then $f(\mathcal{A}) \in \mathcal{A} \subseteq p$ and $S\backslash f(\mathcal{A}) \in p$, a contradiction.

Now let $\mathcal{A}$ and $\mathcal{B}$ be distinct members of $\Re$ and pick x and y in X such that $\mathcal{A} = \bigcap g^{-1}[\{x\}]$ and $\mathcal{B} = \bigcap g^{-1}[\{y\}]$. Pick neighborhoods U of x and V of y such that $c\ell\, U \cap c\ell\, V = \emptyset$. Let $A = g^{-1}[U] \cap S$ and let $B = g^{-1}[V] \cap S$. Then by Remark 21.29, $A \in \mathcal{A}$ and $B \in \mathcal{B}$. Now let $\mathcal{C} \in \Re$ and pick $z \in X$ such that $\mathcal{C} = \bigcap g^{-1}[\{z\}]$. Then without loss of generality $z \notin c\ell\, U$ so if $C = g^{-1}[X\backslash c\ell\, U] \cap S$ we have by Remark 21.29 that $C \in \mathcal{C}$. Then $C \cap A = \emptyset$ so, since $\mathcal{C}$ is a filter, $S\backslash A \in \mathcal{C}$.

Finally define $h : X \to \Re$ by $h(x) = \bigcap g^{-1}[\{x\}]$. Then trivially h is one-to-one and onto $\Re$. Since X is compact, in order to show that h is a homeomorphism it suffices to show that $\Re$ is Hausdorff and h is continuous.

To see that $\Re$ is Hausdorff, let $\mathcal{A}$ and $\mathcal{B}$ be distinct members of $\Re$ and pick $A \in \mathcal{A}$ and $B \in \mathcal{B}$ as guaranteed by (b). Then $A^{\sharp}$ and $B^{\sharp}$ are disjoint neighborhoods of $\mathcal{A}$ and $\mathcal{B}$ respectively.

To see that h is continuous, let $x \in X$ and let $A \subseteq S$ such that $A^{\sharp}$ is a basic open neighborhood of $h(x)$. Let $U = X\backslash g[\overline{S\backslash A}]$. Now $g[\overline{S\backslash A}]$ is the continuous image of a compact set so is compact. Thus U is an open subset of X. We claim that $x \in U$ and $h[U] \subseteq A^{\sharp}$. For the first assertion suppose instead that we have some $p \in \overline{S\backslash A}$ such that $g(p) = x$. Then $h(x) \subseteq p$ so $A \in p$, a contradiction. Now let $y \in U$. Then by Remark 21.29, we have $g^{-1}[U] \cap S \in h(y)$. Since $g^{-1}[U] \cap S \subseteq A$ we have $A \in h(y)$ so $h(y) \in A^{\sharp}$ as required. $\qquad\square$

The following theorem tells us that the description of quotients provided by Theorem 21.31 is enough to characterize them.

Theorem 21.32. *Let S be a discrete space and let $\Re$ be a set of filters on S satisfying*

(a) *given any choice function f for $\mathfrak{R}$ there exists $\mathfrak{F} \in \mathscr{P}_f(\mathfrak{R})$ such that $S = \bigcup_{\mathcal{A} \in \mathfrak{F}} f(\mathcal{A})$ and*

(b) *given distinct $\mathcal{A}$ and $\mathcal{B}$ in $\mathfrak{R}$, there exist $A \in \mathcal{A}$ and $B \in \mathcal{B}$ such that whenever $C \in \mathfrak{R}$, either $S \backslash A \in C$ or $S \backslash B \in C$.*

Then, with the quotient topology, $\mathfrak{R}$ is a compact Hausdorff space. Further, for each $p \in \beta S$ there is a unique $\mathcal{A} \in \mathfrak{R}$ such that $\mathcal{A} \subseteq p$ and the function $g : \beta S \to \mathfrak{R}$ defined by $g(p) \subseteq p$ is a continuous surjection.

Proof. That $\mathfrak{R}$ is Hausdorff is an immediate consequence of (b). That $\mathfrak{R}$ is compact will follow from the fact that $\mathfrak{R}$ is a continuous image of βS.

Now let $p \in \beta S$ and suppose first that for no $\mathcal{A} \in \mathfrak{R}$ is $\mathcal{A} \subseteq p$. Then for each $\mathcal{A} \in \mathfrak{R}$ pick $f(\mathcal{A}) \in \mathcal{A} \backslash p$. By (a) pick some $\mathfrak{F} \in \mathscr{P}_f(\mathfrak{R})$ such that $S = \bigcup_{\mathcal{A} \in \mathfrak{F}} f(\mathcal{A})$. Then $S \in p$ so pick some $\mathcal{A} \in \mathfrak{F}$ such that $f(\mathcal{A}) \in p$, a contradiction.

Next suppose that we have distinct $\mathcal{A}$ and $\mathcal{B}$ in $\mathfrak{R}$ with $\mathcal{A} \subseteq p$ and $\mathcal{B} \subseteq p$. Pick $A \in \mathcal{A}$ and $B \in \mathcal{B}$ as guaranteed by (b). Then $S \backslash A \in \mathcal{B} \subseteq p$ and $A \in \mathcal{A} \subseteq p$, a contradiction.

The function g is trivially onto $\mathfrak{R}$. To see that g is continuous, let $p \in \beta S$ and let $A \subseteq S$ with $g(p) \in A^\sharp$. For each $\mathcal{B} \in \mathfrak{R} \backslash \{g(p)\}$ pick by (b) some $f(\mathcal{B}) \in \mathcal{B}$ and some $D(\mathcal{B}) \in g(p)$ so that for all $C \in \mathfrak{R}$, $S \backslash f(\mathcal{B}) \in C$ or $S \backslash D(\mathcal{B}) \in C$. Let $f\big(g(p)\big) = A$. Pick by (a) some $\mathfrak{F} \in \mathscr{P}_f(\mathfrak{R})$ such that $S = \bigcup_{\mathcal{B} \in \mathfrak{F}} f(\mathcal{B})$ and let $\mathfrak{G} = \mathfrak{F} \backslash \{g(p)\}$. Then $S \backslash A \subseteq \bigcup_{\mathcal{B} \in \mathfrak{G}} f(\mathcal{B})$. Let $D = \bigcap_{\mathcal{B} \in \mathfrak{G}} D(\mathcal{B})$. Then $D \in g(p) \subseteq p$. We claim that $g[\overline{D}] \subseteq A^\sharp$. So let $q \in \overline{D}$ and suppose that $A \notin g(q)$. Pick by Corollary 3.9 some $r \in \beta S$ such that $g(q) \cup \{S \backslash A\} \subseteq r$ and pick $\mathcal{B} \in \mathfrak{G}$ such that $f(\mathcal{B}) \in r$. Then one cannot have $S \backslash f(\mathcal{B}) \in g(r) = g(q)$ so $S \backslash D(\mathcal{B}) \in g(q) \subseteq q$, a contradiction. $\square$

Now we bring the algebra of S into play. The operation $*$ on filters is a natural generalization of the operation $\cdot$ on ultrafilters.

Definition 21.33. Let S be a semigroup and let $\mathcal{A}$ and $\mathcal{B}$ be filters on S. Then

$$\mathcal{A} * \mathcal{B} = \{A \subseteq S : \{x \in S : x^{-1}A \in \mathcal{B}\} \in \mathcal{A}\}.$$

Remark 21.34. *Let S be a semigroup and let $\mathcal{A}$ and $\mathcal{B}$ be filters on S. Then $\mathcal{A} * \mathcal{B}$ is a filter on S.*

Definition 21.35. Let S be a semigroup and let $\mathfrak{R}$ be a set of filters on S. If for all $\mathcal{A}$ and $\mathcal{B}$ in $\mathfrak{R}$, there is a unique $C \in \mathfrak{R}$ such that $C \subseteq \mathcal{A} * \mathcal{B}$, then define an operation $\cdot$ on $\mathfrak{R}$ by $\mathcal{A} \cdot \mathcal{B} \in \mathfrak{R}$ and $\mathcal{A} \cdot \mathcal{B} \subseteq \mathcal{A} * \mathcal{B}$.

Theorem 21.36. *Let S be a discrete semigroup and let (φ, Y) be a semigroup compactification of S. Let $\mathfrak{R} = \{\bigcap \widetilde{\varphi}^{-1}[\{x\}] : x \in Y\}$. Then $\mathfrak{R}$ is a set of filters on S satisfying*

(a) *given any choice function f for $\mathfrak{R}$ there exists $\mathfrak{F} \in \mathscr{P}_f(\mathfrak{R})$ such that $S = \bigcup_{\mathcal{A} \in \mathfrak{F}} f(\mathcal{A})$,*

(b) *given distinct $\mathcal{A}$ and $\mathcal{B}$ in $\mathfrak{R}$, there exist $A \in \mathcal{A}$ and $B \in \mathcal{B}$ such that whenever $C \in \mathfrak{R}$, either $S \backslash A \in C$ or $S \backslash B \in C$, and*

(c) *given any $\mathcal{A}$ and $\mathcal{B}$ in $\mathfrak{R}$ there is a unique $C \in \mathfrak{R}$ such that $C \subseteq \mathcal{A} * \mathcal{B}$.*

Further, with the quotient topology and the operation $\cdot$ on $\mathfrak{R}$, the function $h : Y \to \mathfrak{R}$ defined by $h(x) = \bigcap \widetilde{\varphi}^{-1}[\{x\}]$ is an isomorphism and a homeomorphism.

Proof. Conclusions (a) and (b) follow from Theorem 21.31. To establish (c), let $\mathcal{A}, \mathcal{B} \in \mathfrak{R}$. We note that it is enough to show that there exists $C \in \mathfrak{R}$ such that $C \subseteq \mathcal{A} * \mathcal{B}$. (For then, if $\mathcal{D} \in \mathfrak{R}$ and $\mathcal{D} \neq C$ pick by (b) some $C \in C$ such that $S \backslash C \in \mathcal{D}$. Then $C \in \mathcal{A} * \mathcal{B}$ so $S \backslash C \in \mathcal{D} \backslash (\mathcal{A} * \mathcal{B})$.) Pick $x, y \in Y$ such that $\mathcal{A} = \bigcap \widetilde{\varphi}^{-1}[\{x\}]$ and $\mathcal{B} = \bigcap \widetilde{\varphi}^{-1}[\{y\}]$ and let $C = \bigcap \widetilde{\varphi}^{-1}[\{xy\}]$. To see that $C \subseteq \mathcal{A} * \mathcal{B}$, let $A \in C$. Let $U = Y \backslash \overline{\widetilde{\varphi}[S \backslash A]}$. Then U is open in Y and $xy \in U$. (For if $xy \notin U$ pick some $p \in \overline{S \backslash A}$ such that $\widetilde{\varphi}(p) = xy$. Then $A \notin p$ so $A \notin \bigcap \widetilde{\varphi}^{-1}[\{xy\}]$, a contradiction.)

Since $xy \in U$ and Y is right topological, pick a neighborhood V of x such that $Vy \subseteq U$. Then by Remark 21.29, $\varphi^{-1}[V] = \widetilde{\varphi}^{-1}[V] \cap S \in \mathcal{A}$. We claim that $\varphi^{-1}[V] \subseteq \{s \in S : s^{-1}A \in \mathcal{B}\}$ so that $A \in \mathcal{A} * \mathcal{B}$ as required. To this end, let $s \in \varphi^{-1}[V]$. Then $\varphi(s)y \in U$. Since $\lambda_{\varphi(s)}$ is continuous pick a neighborhood W of y such that $\varphi(s)W \subseteq U$. Then by Remark 21.29, $\varphi^{-1}[W] \in \mathcal{B}$ so it suffices to show that $\varphi^{-1}[W] \subseteq s^{-1}A$. Let $t \in \varphi^{-1}[W]$. Then $\varphi(st) = \varphi(s)\varphi(t) \in U$ so $st \notin S \backslash A$, i.e., $st \in A$ as required.

By Theorem 21.31, the function $h : Y \to \mathfrak{R}$ defined by $h(x) = \bigcap \widetilde{\varphi}^{-1}[\{x\}]$ is a homeomorphism from Y onto $\mathfrak{R}$. To see that it is a homomorphism, let $x, y \in Y$. We have just shown that $h(xy) = \bigcap \widetilde{\varphi}^{-1}[\{xy\}] \subseteq h(x) * h(y)$ so that $h(x) \cdot h(y) = h(xy)$. $\square$

The following theorem tells us that the description of semigroup compactifications provided by (a), (b), and a weakening of (c) of Theorem 21.36 in fact characterizes semigroup compactifications.

Theorem 21.37. *Let S be a discrete semigroup and let $\mathfrak{R}$ be a set of filters on S satisfying*

(a) *given any choice function f for $\mathfrak{R}$ there exists $\mathfrak{F} \in \mathcal{P}_f(\mathfrak{R})$ such that $S = \bigcup_{\mathcal{A} \in \mathfrak{F}} f(\mathcal{A})$,*

(b) *given distinct $\mathcal{A}$ and $\mathcal{B}$ in $\mathfrak{R}$, there exist $A \in \mathcal{A}$ and $B \in \mathcal{B}$ such that whenever $C \in \mathfrak{R}$, either $S \backslash A \in C$ or $S \backslash B \in C$, and*

(c) *given any $\mathcal{A}$ and $\mathcal{B}$ in $\mathfrak{R}$ there exists $C \in \mathfrak{R}$ such that $C \subseteq \mathcal{A} * \mathcal{B}$.*

Then in fact for any $\mathcal{A}$ and $\mathcal{B}$ in $\mathfrak{R}$ there exists a unique $C \in \mathfrak{R}$ such that $C \subseteq \mathcal{A} * \mathcal{B}$. Also, with the quotient topology and the operation $\cdot$, $\mathfrak{R}$ is a compact right topological semigroup and the function $g : \beta S \to \mathfrak{R}$ defined by $g(p) \subseteq p$ is a continuous homomorphism from βS onto $\mathfrak{R}$ with $g[S] \subseteq \Lambda(\mathfrak{R})$ and $g[S]$ dense in $\mathfrak{R}$. In particular, $(g_{|S}, \mathfrak{R})$ is a semigroup compactification of S.

Proof. As in the proof of Theorem 21.36 we see that statements (b) and (c) imply that for any $\mathcal{A}$ and $\mathcal{B}$ in $\mathfrak{R}$ there exists a unique $C \in \mathfrak{R}$ such that $C \subseteq \mathcal{A} * \mathcal{B}$, so the operation $\cdot$ is well defined. By Theorem 21.32, with the quotient topology $\mathfrak{R}$ is a compact Hausdorff space and g is a (well defined) continuous function from βS onto $\mathfrak{R}$.

We show next that g is a homomorphism. To this end, let $p, q \in \beta S$. To see that $g(p \cdot q) = g(p) \cdot g(q)$, we show that $g(p) \cdot g(q) \subseteq p \cdot q$. Let $A \in g(p) \cdot g(q)$. Then $A \in g(p) * g(q)$ so $B = \{s \in S : s^{-1}A \in g(q)\} \in g(p) \subseteq p$ and if $s \in B$, then $s^{-1}A \in g(q) \subseteq q$ so $\{s \in S : s^{-1}A \in q\} \in p$. That is, $A \in p \cdot q$ as required.

Since g is a homomorphism onto $\mathfrak{R}$, we have immediately that the operation $\cdot$ on $\mathfrak{R}$ is associative. By Exercise 2.2.2, $\mathfrak{R}$ is a right topological semigroup.

Let $s \in S$. To see that $\lambda_{g(s)}$ is continuous, note that since g is a homomorphism $g \circ \lambda_s = \lambda_{g(s)} \circ g$. Let U be open in $\mathfrak{R}$ and let $V = \lambda_{g(s)}^{-1}[U]$. Then $g^{-1}[V] = (g \circ \lambda_s)^{-1}[U]$ is open in βS. Then $V = \mathfrak{R} \backslash g[\beta S \backslash g^{-1}[V]]$ is open in $\mathfrak{R}$.

Since S is dense in βS and g is continuous, $g[S]$ is dense in $\mathfrak{R}$.   □

21.5 Uniform Compactifications

Every uniform space X has a compactification $(\phi, \gamma_u X)$ defined by its uniform structure. For many uniform spaces X which are also semigroups, $(\phi, \gamma_u X)$ is a semigroup compactification of X. It has the property that the map $(a, x) \mapsto \phi(a)x$ is a continuous map from $X \times \gamma_u X$ to $\gamma_u X$, and $\gamma_u X$ is the maximal semigroup compactification of X for which this holds.

This is true for all discrete semigroups and for all topological groups. If X is a discrete semigroup, then $\gamma_u X \approx \beta X$ and so the material in this section generalizes our definition of βX as a semigroup.

We first remind the reader of the definition of a uniform structure.

Definition 21.38. Let X be any set. If $U, V \subseteq X \times X$, we define U^{-1} and UV by $U^{-1} = \{(y, x) : (x, y) \in U\}$ and $UV = \{(x, y) : (x, z) \in V$ and $(z, y) \in U$ for some $z \in X\}$. We may use U^2 to denote UU. Δ will denote the diagonal $\{(x, x) : x \in X\}$.

A *uniform structure* (or *uniformity*) $\mathcal{U}$ on X is a filter of subsets of $X \times X$ with the following properties:

(1) $\Delta \subseteq U$ for every $U \in \mathcal{U}$;
(2) for every $U \in \mathcal{U}$, $U^{-1} \in \mathcal{U}$; and
(3) for every $U \in \mathcal{U}$, there exists $V \in \mathcal{U}$ for which $V^2 \subseteq U$.

Let $\mathcal{U}$ be a uniform structure on X and let $U \in \mathcal{U}$. For each $x \in X$, we put $U(x) = \{y \in X : (x, y) \in U\}$, and, for each $Y \subseteq X$, we put $U[Y] = \bigcup_{y \in Y} U(y)$.

$\mathcal{U}$ generates a topology on X in which a base for the neighborhoods of the point $x \in X$ are the sets of the form $U(x)$, where $U \in \mathcal{U}$. If X has this topology, $(X, \mathcal{U})$ is called a uniform space and X is called a uniformizable space.

If $(X, \mathcal{U})$ and $(Y, \mathcal{V})$ are uniform spaces, a function $f : X \to Y$ is said to be *uniformly continuous* if, for each $V \in \mathcal{V}$, there exists $U \in \mathcal{U}$ such that $(f(x_1), f(x_2)) \in V$ whenever $(x_1, x_2) \in U$.

Metric spaces and topological groups provide important examples of uniformizable spaces.

If (X, d) is a metric space, the filter which has as base the sets of the form $\{(x, y) \in X \times X : d(x, y) < r\}$, where $r > 0$, is a uniform structure on X. This example includes all discrete spaces. If X is discrete, it has the trivial uniform structure $\mathcal{U} = \{U \subseteq X \times X : \Delta \subseteq U\}$.

If G is a topological group, its topology is defined by the right uniform structure which has as base the sets $\{(x, y) \in G \times G : xy^{-1} \in V\}$, where V denotes a neighborhood of the identity. In this section, we shall assume that we have assigned this uniform structure to any topological group to which we refer.

It is precisely the completely regular topological spaces which are uniformizable. Suppose that X is a space and that $C_{\mathbb{R}}(X)$ denotes the subalgebra of $C(X)$ consisting of the real-valued functions in $C(X)$. X is said to be completely regular if, for every closed subset E of X and every $x \in X \backslash E$, there is a function $f \in C_{\mathbb{R}}(X)$ for which $f(x) = 0$ and $f[E] = \{1\}$. For each $f \in C_{\mathbb{R}}(X)$ and each $\epsilon > 0$, we put $U_{f, \epsilon} = \{(x, y) \in X \times X : |f(x) - f(y)| < \epsilon\}$. The finite intersections of the sets of the form $U_{f, \epsilon}$ then provide a base for a uniform structure on X.

In particular, every compact space X is uniformizable. In fact, X has a unique uniform structure given by the filter of neighborhoods of the diagonal in $X \times X$ (see Exercise 21.5.1).

In the next lemma we establish a relation between compactifications of X and subalgebras of $C_{\mathbb{R}}(X)$.

Lemma 21.39. *Let X be any topological space and let A be a norm closed subalgebra of $C_{\mathbb{R}}(X)$ which contains the constant functions. There is a compact space Y and a continuous function $\phi : X \to Y$ with the property that $\phi[X]$ is dense in Y and $A = \{f \in C_{\mathbb{R}}(X) : f = g \circ \phi$ for some $g \in C_{\mathbb{R}}(Y)\}$. The mapping ϕ is an embedding if, for every closed subset E of X and every $x \in X \backslash E$, there exists $f \in A$ such that $f(x) = 0$ and $f[E] = \{1\}$.*

Proof. For each $f \in A$, let $I_f = \{t \in \mathbb{R} : |t| \leq \|f\|\}$. Let $C = \times_{f \in A} I_f$ and let $\phi : X \to C$ be the evaluation map defined by $\phi(x)(f) = f(x)$. Let $Y = c\ell_C \phi[X]$.

For each $f \in A$, let $\pi_f : C \to I_f$ be the projection map. If $f \in A$, we have $f = \pi_{f|Y} \circ \phi$. Conversely, let $G = \{g \in C_{\mathbb{R}}(Y) : g \circ \phi \in A\}$. By Exercise 21.5.2, G is a closed subalgebra of $C_{\mathbb{R}}(Y)$ which contains the constant functions. We claim that G also separates the points of Y. To see this, let $\vec{y}$ and $\vec{z}$ be distinct points in Y and choose $f \in A$ such that $y_f \neq z_f$. Since $\pi_{f|Y} \circ \phi = f$, we have $\pi_{f|Y} \in G$ and $\pi_{f|Y}(\vec{y}) \neq \pi_{f|Y}(\vec{z})$. So G separates the points of Y and it follows from the Stone–Weierstrass Theorem that $G = C_{\mathbb{R}}(Y)$. Thus $A = \{f \in C_{\mathbb{R}}(X) : f = g \circ \phi$ for some $g \in C_{\mathbb{R}}(Y)\}$.

Suppose now that, for every closed subset E of X and every $x \in X \backslash E$, there exists $f \in A$ such that $f(x) = 0$ and $f[E] = \{1\}$. Then ϕ is clearly injective. To see that ϕ is an embedding, let E be closed in X and let $y \in \phi[X] \backslash \phi[E]$. Pick $x \in X$ such that $y = \phi(x)$ and pick $f \in A$ such that $f(x) = 0$ and $f[E] = \{1\}$. Pick $g \in C(Y)$ such that $f = g \circ \phi$. Then $g(y) = g(\phi(x)) = 0$ and $g[\phi[E]] = \{1\}$ so $y \notin c\ell \, \phi[E]$. $\square$

The following result establishes the existence of the Stone–Čech compactification for any completely regular space. (Of course, since any subset of a compact space is completely regular, this is the greatest possible generality for such a result.)

Theorem 21.40. *Let X be a completely regular space. Then X can be embedded in a compact space βX which has the following universal property: whenever $g : X \to C$ is a continuous function from X to a compact space, there is a continuous function $\widetilde{g} : \beta X \to C$ which is an extension of g.*

Proof. Let $A = C_{\mathbb{R}}(X)$. By Lemma 21.39, X can be densely embedded in a compact space βX with the property that every function in A extends to a continuous function in $C_{\mathbb{R}}(\beta X)$. We shall regard X as being a subspace of βX.

Let C be a compact space and let F denote the set of all continuous functions from C to $[0,1]$. We define a natural embedding $i : C \to {}^F[0,1]$ by putting $i(u)(f) = f(u)$ for every $u \in C$ and every $f \in F$.

Suppose that $g : X \to C$ is continuous. Then, for every $f \in F$, $f \circ g$ has a continuous extension $(f \circ g)\widetilde{} : \beta X \to [0,1]$. We define $h : \beta X \to {}^F[0,1]$ by $h(y)(f) = (f \circ g)\widetilde{}(y)$ for every $y \in \beta X$ and every $f \in F$. We observe that $h(x) = i\big(g(x)\big)$ whenever $x \in X$. Since X is dense in βX, $h[X]$ is dense in $h[\beta X]$. So $h[\beta X] \subseteq i[C]$, because $h[X] \subseteq i[C]$ and $i[C]$ is compact. We can therefore define $\widetilde{g}$ by $\widetilde{g} = i^{-1} \circ h$. $\square$

Recall that a *topological* compactification (φ, Y) of a space X has the property that φ is an embedding of X into Y.

Theorem 21.41. *Let $(X, \mathcal{U})$ be a uniform space. There is a topological compactification $(\phi, \gamma_u X)$ of X such that it is precisely the uniformly continuous functions in $C_{\mathbb{R}}(X)$ which have continuous extensions to $\gamma_u X$. (That is, $\{f \in C_{\mathbb{R}}(X) : f = g \circ \phi \text{ for some } g \in C_{\mathbb{R}}(\gamma_u X)\} = \{f \in C_{\mathbb{R}}(X) : f \text{ is uniformly continuous}\}$.)*

Proof. It is a routine matter to prove that the uniformly continuous functions in $C_{\mathbb{R}}(X)$ are a norm closed subalgebra of $C_{\mathbb{R}}(X)$. The conclusion then follows from Exercise 21.5.3 and Lemma 21.39. $\square$

Since ϕ is an embedding, we shall regard X as being a subspace of $\gamma_u X$. The compactification $\gamma_u X$ will be called the *uniform compactification* of X.

It can be shown that all possible topological compactifications of a completely regular space X arise in this way as uniform compactifications.

We now see that $\gamma_u X$ may be very large.

Lemma 21.42. *Let $(X, \mathcal{U})$ be a uniform space. Suppose that there exist a sequence $\langle x_n \rangle_{n=1}^{\infty}$ in X and a set $U \in \mathcal{U}$ such that $x_m \notin U(x_n)$ whenever $m \neq n$. Let $f : \mathbb{N} \to X$ be defined by $f(n) = x_n$. Then $\widetilde{f} : \beta\mathbb{N} \to \gamma_u X$ is an embedding.*

Proof. We claim that, for any two disjoint subsets A and B of $\mathbb{N}$, $c\ell_{\gamma_u X} f[A] \cap c\ell_{\gamma_u X} f[B] = \emptyset$. To see this, we observe that $U[\{x_n : n \in A\}]$ and $\{x_n : n \in B\}$

are disjoint, and hence that there is a uniformly continuous function $g : X \rightarrow [0, 1]$ for which $g[\{x_n : n \in A\}] = \{0\}$ and $g[\{x_n : n \in B\}] = \{1\}$ (by Exercise 21.5.3). By Theorem 21.41, g can be extended to a continuous function $\overline{g} : \gamma_u X \rightarrow \mathbb{R}$. Since $\overline{g}[c\ell_{\gamma_u X} f[A]] = \{0\}$ and $\overline{g}[c\ell_{\gamma_u X} f[B]] = \{1\}$, we have $c\ell_{\gamma_u X} f[A] \cap c\ell_{\gamma_u X} f[B] = \emptyset$.

Now suppose that u and v are distinct elements of $\beta\mathbb{N}$. We can choose disjoint subsets A and B of $\mathbb{N}$ such that $A \in u$ and $B \in v$. Since $\widetilde{f}(u) \in c\ell_{\gamma_u X}(f[A])$ and $\widetilde{f}(v) \in c\ell_{\gamma_u X}(f[B])$, it follows that $\widetilde{f}(u) \neq \widetilde{f}(v)$. So $\widetilde{f}$ is injective and is therefore an embedding. $\square$

We now assume that X is a semigroup and that $(X, \mathcal{U})$ is a uniform space. We shall give sufficient conditions for $(\phi, \gamma_u X)$ to be a semigroup compactification of X. We observe that these conditions are satisfied in each of of the following cases:

(1) X is discrete (with the trivial uniformity);
(2) (X, d) is a metric space with a metric d satisfying $d(xz, yz) \leq d(x, y)$ and $d(zx, zy) \leq d(x, y)$ for every $x, y, z \in X$;
(3) X is a topological group.

Notice that requirement (ii) below is stronger than the assertion that ρ_a is uniformly continuous for each $a \in X$.

Theorem 21.43. *Suppose that X is a semigroup and that $(X, \mathcal{U})$ is a uniform space. Suppose that the two following conditions are satisfied:*

(i) *For each $a \in X$, $\lambda_a : X \rightarrow X$ is uniformly continuous and*
(ii) *for each $U \in \mathcal{U}$, there exists $V \in \mathcal{U}$ such that for every $(s, t) \in V$ and every $a \in X$, one has $(sa, ta) \in U$.*

Then we can define a semigroup operation on $\gamma_u X$ for which $(\phi, \gamma_u X)$ is a semigroup compactification of X.

Proof. For each $a \in X$, there is a continuous extension $\overline{\lambda}_a : \gamma_u X \rightarrow \gamma_u X$ of λ_a (by Exercise 21.5.4). We put $ay = \overline{\lambda}_a(y)$ for each $y \in \gamma_u X$.

Given $y \in \gamma_u X$, we define $r_y : X \rightarrow \gamma_u X$ by $r_y(s) = sy$. We shall show that r_y is uniformly continuous. We have observed that the (unique) uniformity on $\gamma_u X$ is generated by $\{U_{g,\epsilon} : g \in C_{\mathbb{R}}(\gamma_u X)$ and $\epsilon > 0\}$, where $U_{g,\epsilon} = \{(u, v) \in \gamma_u X \times \gamma_u X : |g(u) - g(v)| < \epsilon\}$. So let $g \in C_{\mathbb{R}}(\gamma_u X)$ and $\epsilon > 0$ be given.

Now $g_{|X}$ is uniformly continuous by Theorem 21.41 so pick $W \in \mathcal{U}$ such that $|g(s) - g(t)| < \frac{\epsilon}{2}$ whenever $(s, t) \in W$. Pick by condition (ii) some $V \in \mathcal{U}$ such that for all $a \in X$ and all $(s, t) \in V$, one has $(sa, ta) \in W$. Then, given $(s, t) \in V$, one has $|g(sy) - g(ty)| = \lim_{a \rightarrow y} |g(sa) - g(ta)| \leq \frac{\epsilon}{2}$, where a denotes an element of X, and so $\left(r_y(s), r_y(t) \right) \in U_{g,\epsilon}$.

It follows from Exercise 21.5.4 that r_y can be extended to a continuous function $\overline{r}_y : \gamma_u X \rightarrow \gamma_u X$.

We now define a binary operation on $\gamma_u X$ by putting $xy = \bar{r}_y(x)$ for every $x, y \in \gamma_u X$. We observe that, for every $a \in X$, the mapping $y \mapsto ay$ from $\gamma_u X$ to itself is continuous; and, for every $y \in \gamma_u X$, the mapping $x \mapsto xy$ from $\gamma_u X$ to itself is continuous, because these are the mappings $\bar{\lambda}_a$ and $\bar{r}_y$ respectively.

It follows that the operation defined is associative because, for every $x, y, z \in \gamma_u X$,

$$x(yz) = \lim_{s \to x} \lim_{t \to y} \lim_{u \to z} s(tu) \text{ and}$$
$$(xy)z = \lim_{s \to x} \lim_{t \to y} \lim_{u \to z} (st)u,$$

where s, t, and u denote elements of X.

So $x(yz) = (xy)z$. Thus $\gamma_u X$ is a semigroup compactification of X. $\quad\square$

We now show that the semigroup operation on $\gamma_u X$ is jointly continuous on $X \times \gamma_u X$.

Theorem 21.44. *Let $(X, \mathcal{U})$ satisfy the hypotheses of Theorem 21.43. Then the map $(s, x) \mapsto sx$ is a continuous map from $X \times \gamma_u X$ to $\gamma_u X$.*

Proof. Let $s \in X$ and $x \in \gamma_u X$. Choose any $f \in C_{\mathbb{R}}(\gamma_u X)$ and any $\epsilon > 0$.

Since $f_{|X}$ is uniformly continuous, there exists $U \in \mathcal{U}$ such that $|f(s) - f(s')| < \epsilon$ whenever $(s, s') \in U$. By condition (ii) of Theorem 21.43, there exists $V \in \mathcal{U}$ such that $(st, s't) \in U$ for every $(s, s') \in V$ and every $t \in X$. If $x' \in \gamma_u X$, then $|f(sx') - f(s'x')| = \lim_{t \to x'} |f(st) - f(s't)| \leq \epsilon$ whenever $(s, s') \in V$. Thus, if $s' \in V(s)$ and $x' \in \lambda_s^{-1}[\{y \in \gamma_u X : |f(sx) - f(y)| < \epsilon\}]$, we have $|f(sx) - f(sx')| < \epsilon$ and $|f(sx') - f(s'x')| \leq \epsilon$. So $|f(sx) - f(s'x')| < 2\epsilon$. $\quad\square$

Theorem 21.45. *Suppose that $(X, \mathcal{U})$ satisfies the conditions of Theorem 21.43. Suppose that Y is a compact right topological semigroup and that $h : X \to Y$ is a uniformly continuous homomorphism with $h[X] \subseteq \Lambda(Y)$. Then h can be extended to a continuous homomorphism $\bar{h} : \gamma_u X \to Y$.*

Proof. By Exercise 21.5.4, h can be extended to a continuous function $\bar{h} : \gamma_u X \to Y$. For every $u, v \in \gamma_u X$, we have

$$\bar{h}(uv) = \lim_{s \to u} \lim_{t \to v} h(st) = \lim_{s \to u} \lim_{t \to v} h(s)h(t) = \lim_{s \to u} h(s)\bar{h}(v) = \bar{h}(u)\bar{h}(v),$$

where s and t denote elements of X. So $\bar{h}$ is a homomorphism. $\quad\square$

We now show that, if X is a topological group, $\gamma_u X$ is maximal among the semigroup compactifications of X which have a joint continuity property.

Theorem 21.46. *Let X be a topological group and let (θ, Y) be a semigroup compactification of X. Suppose that the mapping $(x, y) \mapsto \theta(x)y$ from $X \times Y$ to Y is continuous. Then there is a continuous homomorphism $\bar{\theta} : \gamma_u X \to Y$ such that $\theta = \bar{\theta}_{|X}$.*

Proof. We shall show that θ is uniformly continuous.

Let e denote the identity of X and let $\mathcal{N}_e$ be the set of neighborhoods of e. Suppose that θ is not uniformly continuous. Then there exists an open neighborhood U of the diagonal in $Y \times Y$ such that, for every $V \in \mathcal{N}_e$, there are points $s_V, t_V \in X$ satisfying $s_V t_V^{-1} \in V$ and $\big(\theta(s_V), \theta(t_V)\big) \notin U$. The net $\langle(\theta(s_V), \theta(t_V))\rangle_{V \in \mathcal{N}_e}$ has a limit point $(y, z) \in (Y \times Y)\backslash U$. Now $s_V = e_V t_V$, where $e_V \in V$, and so $\theta(s_V) = \theta(e_V)\theta(t_V)$. Since $e_V \to e$, $\theta(e_V) \to \theta(e)$, and $\theta(e)$ is easily seen to be an identity for Y. So our continuity assumption implies that $y = z$, a contradiction.

Thus θ is uniformly continuous and therefore can be extended to a continuous homomorphism $\overline{\theta} : \gamma_u X \to Y$ (by Theorem 21.45). $\qquad\qquad\square$

Theorem 21.45 shows that $\gamma_u X$ can be identified with βX if X is a discrete semigroup with the trivial uniformity. However, there are many familiar examples in which $\gamma_u X$ and βX are different. Suppose, for example, that $X = (\mathbb{R}, +)$. We shall see in Theorem 21.47 that $\beta\mathbb{R}$ cannot be made into a semigroup compactification of $(\mathbb{R}, +)$,

It is natural, given a nondiscrete topological semigroup such as $(\mathbb{R}, +)$, to attempt to proceed as we did with a discrete semigroup in Theorem 4.1. And, indeed, one can do the first part of the extension the same way. That is, given any $s \in S$, one can define $\ell_s : S \to S \subseteq \beta S$ by $\ell_s(t) = st$. Then by Theorem 21.40, there is a continuous function $L_s : \beta S \to \beta S$ such that $L_s|S = \ell_s$ so one can define for $s \in S$ and $q \in \beta S$, $sq = L_s(q)$. As before one still has the function $r_q : S \to \beta S$ defined by $r_q(s) = sq$. However, in order to invoke Theorem 21.40 to extend r_q, one would need r_q to be continuous. We now see that this condition does not hold in many familiar semigroups, including $(\mathbb{R}, +)$.

Note that in the following theorem we do not assume that the operation on βS is associative.

Theorem 21.47. *Let S be a semigroup which is also a metric space with a metric δ for which $\delta(st, st') = \delta(t, t')$ and $\delta(ts, t's) = \delta(t, t')$ for every $s, t, t' \in S$. Suppose that S contains a sequence $\langle y_n \rangle_{n=1}^{\infty}$ such that $\delta(y_m, y_n) > 1$ whenever $m \neq n$. Let $\cdot$ be a binary operation on βS which extends the semigroup operation of S and has the property that the mapping $\xi \mapsto x \cdot \xi$ from βS to itself is continuous for every $x \in S$. Then there is a point $p \in \beta S$ such that the mapping $x \mapsto x \cdot p$ is discontinuous at every non-isolated point x of S.*

Proof. Let p be any limit point of $\langle y_n \rangle_{n=1}^{\infty}$ and let $x \in S$ be the limit of a sequence $\langle x_n \rangle_{n=1}^{\infty}$ of distinct elements of S. We may suppose that $\delta(x_m, x_n) < \frac{1}{2}$ for every $m, n \in \mathbb{N}$.

For every $m, m', n, n' \in \mathbb{N}$ with $n \neq n'$, we have

$$\delta(x_m y_n, x_{m'} y_{n'}) \geq \delta(x_m y_n, x_m y_{n'}) - \delta(x_m y_{n'}, x_{m'} y_{n'}) = \delta(y_n, y_{n'}) - \delta(x_m, x_{m'}) > \tfrac{1}{2}.$$

Let $A = \{x_m y_n : m < n$ and m is even$\}$ and $B = \{x_m y_n : m < n$ and m is odd$\}$. We claim that A and B have no points of accumulation in S. This can be seen from the fact that, for any $s \in S$, $\{t \in S : \delta(s, t) < \frac{1}{4}\}$ cannot contain two points of the form $x_m y_n$ and $x_{m'} y_{n'}$ with $n \neq n'$, and can therefore contain only a finite number of points in $A \cup B$. Thus $c\ell_S A = A$, $c\ell_S B = B$ and $c\ell_S A \cap c\ell_S B = \emptyset$.

It follows from Urysohn's Lemma that there is a continuous function $f : S \to [0, 1]$ for which $f[A] = \{0\}$ and $f[B] = \{1\}$. Let $\widetilde{f} : \beta S \to [0, 1]$ denote the continuous extension of f. Then $\widetilde{f}(x_m \cdot p) = 0$ if m is even and $\widetilde{f}(x_m \cdot p) = 1$ if m is odd. So the sequence $\langle x_m \cdot p \rangle_{m=1}^{\infty}$ cannot converge to $x \cdot p$.   □

We shall now look at some of the properties of $\gamma_u \mathbb{R}$, where $\mathbb{R}$ denotes the topological group formed by the real numbers under addition. We first note that $\beta \mathbb{Z}$ can be embedded topologically and algebraically in $\gamma_u \mathbb{R}$. By Theorem 21.45, the inclusion map $i : \mathbb{Z} \to \mathbb{R}$ can be extended to a continuous homomorphism $\overline{i} : \beta \mathbb{Z} \to \gamma_u \mathbb{R}$. We can show that $\overline{i}$ is injective and therefore an embedding, by essentially the same argument as the one used in the proof of Lemma 21.42.

We shall therefore assume that $\beta \mathbb{Z} \subseteq \gamma_u \mathbb{R}$, by identifying $\beta \mathbb{Z}$ with $\overline{i}[\beta \mathbb{Z}]$. This identifies $\beta \mathbb{Z}$ with $c\ell_{\gamma_u \mathbb{R}}(\mathbb{Z})$, because $\overline{i}[\beta \mathbb{Z}] = c\ell_{\gamma_u \mathbb{R}}(i[\mathbb{Z}])$.

The following theorem gives an expression for an element of $\gamma_u \mathbb{R}$ analogous to the expression of a real number as the sum of its fractional and integral parts.

Theorem 21.48. *Every $x \in \gamma_u \mathbb{R}$ can be expressed uniquely as $x = t + z$, where $t \in [0, 1)$ and $z \in \beta \mathbb{Z}$.*

Proof. Let $\pi : \mathbb{R} \to \mathbb{T} = \mathbb{R}/\mathbb{Z}$ denote the canonical map. Since π is uniformly continuous, π can be extended to a continuous homomorphism $\overline{\pi} : \gamma_u \mathbb{R} \to \mathbb{T}$.

We claim that $\overline{\pi}(x) = 0$ if and only if $x \in \beta \mathbb{Z}$. On the one hand, $\overline{\pi}(x) = 0$ if $x \in \beta \mathbb{Z}$, because $\pi[\mathbb{Z}] = \{0\}$. On the other hand, suppose that $x \notin \beta \mathbb{Z}$. Then there is a continuous function $f : \gamma_u \mathbb{R} \to [0, 1]$ for which $f(x) = 1$ and $f[\mathbb{Z}] = \{0\}$. Since $f_{|\mathbb{Z}}$ is uniformly continuous, there exists $\delta > 0$ such that $f\big[\bigcup_{n \in \mathbb{Z}}[n - \delta, n + \delta]\big] \subseteq [0, \frac{1}{2}]$. So $x \notin c\ell_{\gamma_u \mathbb{R}}\big(\bigcup_{n \in \mathbb{Z}}[n - \delta, n + \delta]\big)$ and therefore $x \in c\ell_{\gamma_u \mathbb{R}}\big(\bigcup_{n \in \mathbb{Z}}(n + \delta, n + 1 - \delta)\big)$ and $\overline{\pi}(x) \neq 0$, because $0 \notin c\ell_{\mathbb{T}} \pi \big[\bigcup_{n \in \mathbb{Z}}(n + \delta, n + 1 - \delta)\big]$.

Let t denote the unique number in $[0, 1)$ for which $\pi(t) = \overline{\pi}(x)$. If $z = -t + x$, then $\overline{\pi}(z) = 0$ and so $z \in \beta \mathbb{Z}$. Thus we have an expression for x of the type required.

To prove uniqueness, suppose that $x = t' + z'$, where $t' \in [0, 1)$ and $z' \in \beta \mathbb{Z}$. Then $\overline{\pi}(x) = \pi(t')$ and so $t' = t$ and therefore $z' = -t + x = z$.   □

Theorem 21.48 allows us to analyze the algebraic structure of $\gamma_u \mathbb{R}$ in terms of the algebraic structure of $\beta \mathbb{Z}$.

Corollary 21.49. *The left ideals of $\gamma_u \mathbb{R}$ have the form $\mathbb{R} + L$, where L is a left ideal of $\beta \mathbb{Z}$; and the right ideals of $\gamma_u \mathbb{R}$ have the form $\mathbb{R} + R$, where R is a right ideal of $\beta \mathbb{Z}$.*

Proof. This follows easily from Theorem 21.48 and the observation that $\mathbb{R}$ is contained in the center of $\gamma_u \mathbb{R}$.   □

We omit the proofs of the following corollaries, as they are easy consequences of Theorem 21.48.

Corollary 21.50. $K(\gamma_u \mathbb{R}) = \mathbb{R} + K(\beta \mathbb{Z})$.

Corollary 21.51. *Every idempotent of $\gamma_u \mathbb{R}$ is in $\beta \mathbb{Z}$.*

Corollary 21.52. *$\gamma_u \mathbb{R}$ has 2^c minimal left ideals and 2^c minimal right ideals, each containing 2^c idempotents.*

Proof. This follows from Corollary 21.49 and Theorem 6.9.   □

Exercise 21.5.1. Let $(X, \mathcal{U})$ be a uniform space. Show that every $U \in \mathcal{U}$ is a neighborhood of the diagonal Δ in $X \times X$. If X is compact, show that $\mathcal{U}$ is the filter of all neighborhoods of Δ in $X \times X$.

Exercise 21.5.2. Let G and Y be as in the proof of Lemma 21.39. Prove that G is a closed subalgebra of $C_{\mathbb{R}}(Y)$ which contains the constant functions.

Exercise 21.5.3. Let $(X, \mathcal{U})$ be a uniform space. Suppose that E and F are subsets of X and that $U[E] \cap F = \emptyset$ for some $U \in \mathcal{U}$. Show that there is a uniformly continuous function $f : X \to [0, 1]$ for which $f[E] = \{0\}$ and $f[F] = \{1\}$.
(Hint: Choose a sequence $\langle U_n \rangle_{n \in \omega}$ in $\mathcal{U}$ by putting $U_0 = U \cap U^{-1}$ and choosing U_n to satisfy $U_n = U_n^{-1}$ and $U_n^2 \subseteq U_{n-1}$ for every $n > 0$. Then define subsets E_r of X for each dyadic rational $r \in [0, 1]$ with the following properties:

$$E_0 = E,$$
$$E_1 = X \backslash F, \text{ and}$$
$$\text{for each } n \in \mathbb{N} \text{ and each } k \in \{0, 1, 2, \ldots, 2^n - 1\}, U_n[E_{\frac{k}{2^n}}] \subseteq E_{\frac{k+1}{2^n}}.$$

This can be done inductively by putting $E_0 = E$ and $E_1 = X \backslash F$, and then assuming that $E_{\frac{k}{2^n}}$ has been defined for every $k \in \{0, 1, 2, \ldots, 2^n\}$. If $k = 2m + 1$ for $m \in \{0, 1, \ldots, 2^n - 1\}$, $E_{\frac{k}{2^{n+1}}}$ can be defined as $U_{n+1}[E_{\frac{m}{2^n}}]$. Once the sets E_r have been constructed, define $f : X \to [0, 1]$ by putting $f(x) = 1$ if $x \in F$ and $f(x) = \inf\{r : x \in E_r\}$ otherwise.)

Exercise 21.5.4. Suppose that $(X, \mathcal{U})$ and $(Y, \mathcal{V})$ are uniform spaces and that $\theta : X \to Y$ is uniformly continuous. Show that θ has a continuous extension $\overline{\theta} : \gamma_u X \to \gamma_u Y$. (Hint: Let $A = C_{\mathbb{R}}(\gamma_u Y)$. The mapping $\phi : \gamma_u Y \to {}^A\mathbb{R}$ defined by $\phi(y)(f) = f(y)$ is then an embedding. If $f \in A$, $f_{|Y}$ is uniformly continuous and so $f \circ \theta$ is uniformly continuous and has a continuous extension $\hat{f} : \gamma_u X \to \mathbb{R}$. Let $\psi : \gamma_u X \to \phi[\gamma_u Y]$ be defined by $\psi(x)(f) = \hat{f}(x)$. For every $x \in X$, $\psi(x) = \phi(\theta(x))$.)

Exercise 21.5.5. A uniform space $(X, \mathcal{U})$ is said to be *totally bounded* if, for every $U \in \mathcal{U}$, there exists a finite subset F of X such that $X = \bigcup_{x \in F} U(x)$. If $(X, \mathcal{U})$ is not totally bounded, show that $\gamma_u X$ contains a topological copy of $\beta \mathbb{N}$. (Hint: Apply Lemma 21.42.)

Exercise 21.5.6. Let X be a topological group. Show that $f \in C_{\mathbb{R}}(X)$ is uniformly continuous if and only if the map $x \mapsto f \circ \lambda_x$ from X to $C_{\mathbb{R}}(X)$ is continuous, where the topology of $C_{\mathbb{R}}(X)$ is that defined by its norm. (For this reason, $\gamma_u X$ may be called the *left uniformly continuous compactification* of X and denoted by $\mathcal{LUC}(X)$.)

Notes

It is customary to define the $\mathscr{LMC}$, $\mathscr{WAP}$, $\mathscr{AP}$, and $\mathscr{SAP}$ compactifications only for semitopological semigroups S. Of course, if S is not semitopological, then the map η_1 cannot be an embedding (because $\eta_1[S]$ is semitopological). It was shown in [150] (a result of collaboration with P. Milnes) that there exists a completely regular semitopological semigroup S such that η_1 is neither one-to-one nor open and that there exist completely regular semigroups which are neither left nor right topological for which η_1 is one-to-one and other such semigroups for which η_1 is open as a map to $\eta_1[S]$.

In keeping with our standard practice, we have assumed that all hypothesized topological spaces are Hausdorff. However, the results of Section 21.1 remain valid if S is any semigroup with topology, without any separation axioms assumed.

We defined weakly almost periodic and almost periodic functions in terms of extendibility to the $\mathscr{WAP}$- and $\mathscr{AP}$-compactifications. It is common to define them in terms of topologies on $C(S)$. One then has that a function $f \in C(S)$ is weakly almost periodic if $c\ell\{f \circ \rho_s : s \in S\}$ is compact in the weak topology on $C(S)$ and is almost periodic provided $c\ell\{f \circ \rho_s : s \in S\}$ is compact in the norm topology on $C(S)$. For the equivalence of these characterizations with our definitions see [40].

Theorem 21.7 and Corollary 21.8 are due to J. Berglund, H. Junghenn, and P. Milnes in [39], where they credit the "main idea" to J. Baker.

The equivalence of statements (1) and (3) in Theorem 21.18 is due to A. Grothendieck [112].

Theorem 21.22 is due to W. Ruppert [217].

The algebraic properties of $\mathscr{WAP}(\mathbb{N})$ are much harder to analyze than those of $\beta(\mathbb{N})$. It is difficult to prove that $\mathscr{WAP}(\mathbb{N})$ contains more than one idempotent. T. West was the first to prove that it contains at least two [245]. It has since been shown, in the work of G. Brown and W. Moran [54], W. Ruppert [220] and B. Bordbar [51], that $\mathscr{WAP}(\mathbb{N})$ has 2^c idempotents. Whether the set of idempotents in $\mathscr{WAP}(\mathbb{N})$ is closed, was an open question for some time. It has recently been answered in the negative by B. Bordbar and J. S. Pym [53].

The results of Section 21.4 are from [37] and were obtained in collaboration with J. Berglund. In [37] a characterization of the $\mathscr{WAP}$-compactification as a space of filters was also obtained.

Theorem 21.40 is due (independently) to E. Čech [60] and M. Stone [227]. See the notes to Chapter 3 for more information about the origins of the Stone–Čech compactification.

Theorem 21.48 is due to M. Filali [90], who proved it for the more general case of $\gamma_u \mathbb{R}^n$.

Suppose that S is a semitopological semigroup and that $f \in C(S)$. It is fairly easy to prove that f is a $\mathscr{WAP}$ function if and only if $\{(f \circ \lambda_s)\widetilde{\ } : s \in S\}$ is relatively compact in $\{\widetilde{g} : g \in C(S)\}$ for the topology of pointwise convergence on $C(\beta S_d)$. It follows from [81, Theorem IV.6.14] that this is equivalent to $\{f \circ \lambda_s : s \in S\}$ being weakly relatively compact in $C(S)$.

Bibliography

[1] P. Anthony, *Ideals in the Stone–Čech compactification of noncommutative semigroups*, Ph.D. Dissertation (1994), Howard University.

[2] P. Anthony, *The smallest ideals in the two natural products on βS*, Semigroup Forum **48** (1994), 363–367.

[3] R. Arens, *The adjoint of a bilinear operation*, Proc. Amer. Math. Soc. **2** (1951), 839–848.

[4] J. Auslander, *On the proximal relation in topological dynamics*, Proc. Amer. Math. Soc. **11** (1960), 890–895.

[5] J. Auslander, *Minimal flows and their extensions*, North Holland, Amsterdam, 1988.

[6] J. Baker and R. Butcher, *The Stone–Čech compactification of a topological semigroup*, Math. Proc. Cambridge Philos. Soc. **80** (1976), 103–107.

[7] J. Baker, N. Hindman, and J. Pym, *κ-topologies for right topological semigroups*, Proc. Amer. Math. Soc. **115** (1992), 251–256.

[8] J. Baker, N. Hindman, and J. Pym, *Elements of finite order in Stone–Čech compactifications*, Proc. Edinburgh Math. Soc. **36** (1992), 49–54.

[9] J. Baker, A. Lau, and J. Pym, *Identities in Stone–Čech compactifications of semigroups*, manuscript.

[10] J. Baker and P. Milnes, *The ideal structure of the Stone–Čech compactification of a group*, Math. Proc. Cambridge Philos. Soc. **82** (1977), 401–409.

[11] B. Balcar and A. Blaszczyk, *On minimal dynamical systems on Boolean algebras*, Comment. Math. Univ. Carolin. **31** (1990), 7–11.

[12] B. Balcar and F. Franek, *Structural properties of universal minimal dynamical systems for discrete semigroups*, Trans. Amer. Math. Soc. **349** (1997), 1697–1724.

[13] B. Balcar and P. Kalášek, *Nonexistence of multiple recurrent point in the ultrafilter dynamical systems*, Bull. Polish Acad. Sci. Math. **37** (1989), 525–529.

[14] H. Bauer, *Wahrscheinlichkeitstheorie und Grundzüge der Maßtheorie*, Walter de Gruyter, Berlin, 1978.

[15] J. Baumgartner, *A short proof of Hindman's Theorem*, J. Combin. Theory Ser. A **17** (1974), 384–386.

[16] K. Berg, D. Gove and K. Haddad, *On the connection between beta compactifications and enveloping semigroups of subshifts*, Abstracts Amer. Math. Soc. **15** (1994), 320.

[17] V. Bergelson, *Sets of recurrence of $\mathbb{Z}^m$-actions and properties of sets of differences in $\mathbb{Z}^m$*, J. London Math. Soc. **31** (1985), 295–304.

[18] V. Bergelson, *A density statement, generalizing Schur's Theorem*, J. Combin. Theory Ser. A **43** (1986), 338–343.

[19] V. Bergelson, *Ergodic Ramsey Theory*, in: Logic and Combinatorics, S. Simpson (ed.), Contemp. Math. **69** (1987), 63–87.

[20] V. Bergelson, A. Blass, and N. Hindman, *Partition theorems for spaces of variable words*, Proc. London Math. Soc. **68** (1994), 449–476.

[21] V. Bergelson, W. Deuber, and N. Hindman, *Rado's Theorem for finite fields*, in: Proceedings of the Conference on Sets, Graphs, and Numbers, Budapest, 1991, Colloq. Math. Soc. János Bolyai **60** (1992), 77–88.

[22] V. Bergelson, W. Deuber, N. Hindman, and H. Lefmann, *Rado's Theorem for commutative rings*, J. Combin. Theory Ser. A **66** (1994), 68–92.

[23] V. Bergelson, H. Furstenberg, N. Hindman, and Y. Katznelson *An algebraic proof of van der Waerden's Theorem*, Enseign. Math. **35** (1989), 209–215.

[24] V. Bergelson and N. Hindman, *Density versions of two generalizations of Schur's Theorem*, J. Combin. Theory Ser. A **48** (1988), 32–38.

[25] V. Bergelson and N. Hindman, *A combinatorially large cell of a partition of $\mathbb{N}$*, J. Combin. Theory Ser. A **48** (1988), 39–52.

[26] V. Bergelson and N. Hindman, *Ultrafilters and multidimensional Ramsey theorems*, Combinatorica **9** (1989), 1–7.

[27] V. Bergelson and N. Hindman, *Nonmetrizable topological dynamics and Ramsey Theory*, Trans. Amer. Math. Soc. **320** (1990), 293–320.

[28] V. Bergelson and N. Hindman, *Ramsey Theory in non-commutative semigroups*, Trans. Amer. Math. Soc. **330** (1992), 433–446.

[29] V. Bergelson and N. Hindman, *Some topological semicommutative van der Waerden type theorems and their combinatorial consequences*, J. London Math. Soc. **45** (1992), 385–403.

[30] V. Bergelson and N. Hindman, *Additive and multiplicative Ramsey Theorems in* $\mathbb{N}$ *— some elementary results*, Combin. Probab. Comput. **2** (1993), 221–241.

[31] V. Bergelson and N. Hindman, *IP*-sets and central sets*, Combinatorica **14** (1994), 269–277.

[32] V. Bergelson and N. Hindman, *IP* sets in product spaces* in: Papers on General Topology and Applications, S. Andima et. al., eds., Annals of the New York Academy of Sciences **806** (1996), 28–41.

[33] V. Bergelson, N. Hindman, and B. Kra, *Iterated spectra of numbers — elementary, dynamical, and algebraic approaches*, Trans. Amer. Math. Soc. **348** (1996), 893–912.

[34] V. Bergelson, N. Hindman, and I. Leader, *Sets partition regular for n equations need not solve n + 1*, Proc. London Math. Soc. **73** (1996), 481–500.

[35] V. Bergelson, N. Hindman, and I. Leader, *Additive and multiplicative Ramsey Theory in the reals and the rationals*, J. Combin. Theory Ser. A, to appear.

[36] V. Bergelson, N. Hindman, and B. Weiss, *All-sums sets in* (0, 1] *— category and measure*, Mathematika **44** (1997), 61–87.

[37] J. Berglund and N. Hindman, *Filters and the weak almost periodic compactification of a discrete semigroup*, Trans. Amer. Math. Soc. **284** (1984), 1–38.

[38] J. Berglund and N. Hindman, *Sums of idempotents in* $\beta\mathbb{N}$, Semigroup Forum **44** (1992), 107–111.

[39] J. Berglund, H. Junghenn, and P. Milnes, *Compact right topological semigroups and generalizations of almost periodicity*, Lecture Notes in Math. **663**, Springer-Verlag, Berlin, 1978.

[40] J. Berglund, H. Junghenn, and P. Milnes, *Analysis on semigroups*, Wiley, N.Y., 1989.

[41] J. Berglund and K. Hofmann, *Compact semitopological semigroups and weakly almost periodic functions*, Lecture Notes in Math. **42**, Springer-Verlag, Berlin, 1967.

[42] A. Bernstein, *A new kind of compactness for topological spaces*, Fund. Math. **66** (1970), 185–193.

[43] A. Blass, *The Rudin–Keisler ordering of P-points*, Trans. Amer. Math. Soc. **179** (1973), 145–166.

[44] A. Blass, *Ultrafilters related to Hindman's finite-unions theorem and its extensions*, in: Logic and Combinatorics, S. Simpson (ed.), Contemp. Math. **69** (1987), 89–124.

[45] A. Blass, *Ultrafilters: where topological dynamics = algebra = combinatorics*, Topology Proc. **18** (1993), 33–56.

[46] A. Blass and N. Hindman, *Sums of ultrafilters and the Rudin–Keisler and Rudin–Frolík orders*, in: General Topology and Applications, R. Shortt (ed.), Lecture Notes in Pure and Appl. Math. **123** (1990), 59–70.

[47] A. Blass and N. Hindman, *On strongly summable ultrafilters and union ultrafilters*, Trans. Amer. Math. Soc. **304** (1987), 83–99.

[48] A. Blass, J. Hirst, and S. Simpson, *Logical analysis of some theorems of combinatorics and topological dynamics*, in: Logic and Combinatorics, S. Simpson (ed.), Contemp. Math. **69** (1987), 125–156.

[49] M. Blümlinger, *Lévy group action and invariant measures on $\beta\mathbb{N}$*, Trans. Amer. Math. Soc. **348** (1996), 5087–5111.

[50] D. Booth, *Ultrafilters on a countable set*, Ann. Math. Logic **2** (1970), 1–24.

[51] B. Bordbar, *Weakly almost periodic functions on $\mathbb{N}$ with a negative base*, J. London Math. Soc., to appear.

[52] B. Bordbar and J. Pym, *The weakly almost periodic compactification of a direct sum of finite groups*, Math. Proc. Cambridge Philos. Soc., to appear.

[53] B. Bordbar and J. Pym, *The set of idempotents in the weakly almost periodic compactification of the integers is not closed*, Trans. Amer. Math. Soc., to appear.

[54] G. Brown and W. Moran, *The idempotent semigroup of a compact monothetic semigroup*, Proc. Royal Irish Acad. Sect. A **72** (1972), 17–33.

[55] T. Budak, *Compactifying topologised semigroups*, Semigroup Forum **46** (1993), 128–129.

[56] T. Budak, N. Isik, and J. Pym, *Subsemigroups of Stone–Čech compactifications*, Math. Proc. Cambridge Philos. Soc. **116** (1994), 99–118.

[57] R. Butcher, *The Stone–Čech compactification of a semigroup and its algebra of measures*, Ph.D Dissertation (1975), University of Sheffield.

[58] T. Carlson, *Some unifying principles in Ramsey Theory*, Discrete Math. **68** (1988), 117–169.

[59] T. Carlson and S. Simpson, *A dual form of Ramsey's Theorem*, Adv. Math. **53** (1984), 265–290.

[60] E. Čech, *On bicompact spaces*, Ann. of Math. **38** (1937), 823–244.

[61] C. Chou, *On a geometric property of the set of invariant means on a group*, Proc. Amer. Math. Soc. **30** (1971), 296–302.

[62] G. Choquet, *Construction d'ultrafilters sur* $\mathbb{N}$, Bull. Sci. Math. **92** (1968), 41–48.

[63] J. Christensen, *Joint continuity of separately continuous functions*, Proc. Amer. Math. Soc. **82** (1981), 455–461.

[64] P. Civin and B. Yood, *The second conjugate space of a Banach algebra as an algebra*, Pacific J. Math. **11** (1961), 847–870.

[65] J. Clay, *Nearrings: geneses and applications*, Oxford Press, 1992.

[66] W. Comfort, *Ultrafilters: some old and some new results*, Bull. Amer. Math. Soc. **83** (1977), 417–455.

[67] W. Comfort, *Some recent applications of ultrafilters to topology*, in: Proceedings of the Fourth Prague Topological Symposium, 1976, J. Novak (ed.), Lecture Notes in Math. **609** (1977), 34–42.

[68] W. Comfort, *Ultrafilters: an interim report*, Surveys in General Topology, Academic Press, New York, 1980, 33–54.

[69] W. Comfort and S. Negrepontis, *The theory of ultrafilters*, Springer-Verlag, Berlin, 1974.

[70] M. Daguenet-Teissier, *Ultrafiltres à la façon de Ramsey*, Trans. Amer. Math. Soc. **250** (1979), 91–120.

[71] D. Davenport, *The algebraic properties of closed subsemigroups of ultrafilters on a discrete semigroup*, Ph.D. Dissertation (1987), Howard University.

[72] D. Davenport, *The minimal ideal of compact subsemigroups of* βS, Semigroup Forum **41** (1990), 201–213.

[73] D. Davenport and N. Hindman, *Subprincipal closed ideals in* $\beta\mathbb{N}$, Semigroup Forum **36** (1987), 223–245.

[74] D. Davenport and N. Hindman, *A proof of van Douwen's right ideal theorem*, Proc. Amer. Math. Soc. **113** (1991), 573–580.

[75] M. Day, *Amenable semigroups*, Illinois J. Math. **1** (1957), 509–544.

[76] W. Deuber, *Partitionen und lineare Gleichungssysteme*, Math. Z. **133** (1973), 109–123.

[77] W. Deuber, D. Gunderson, N. Hindman, and D. Strauss, *Independent finite sums for* K_m*-free graphs*, J. Combin. Theory Ser. A **78** (1997), 171–198.

[78] W. Deuber and N. Hindman, *Partitions and sums of* (m,p,c)*-sets*, J. Combin. Theory Ser. A **45** (1987), 300–302.

[79] W. Deuber, N. Hindman, I. Leader, and H. Lefmann, *Infinite partition regular matrices*, Combinatorica **15** (1995), 333–355.

[80] E. van Douwen, *The Čech–Stone compactification of a discrete groupoid*, Topology Appl. **39** (1991), 43–60.

[81] N. Dunford and J. Schwartz, *Linear Operators, I.*, Wiley, New York, 1958.

[82] E. Ellentuck, *A new proof that analytic sets are Ramsey*, J. Symbolic Logic **39** (1974), 163–165.

[83] R. Ellis, *A note on the continuity of the inverse*, Proc. Amer. Math. Soc. **8** (1957), 372–373.

[84] R. Ellis, *Locally compact transformation groups*, Duke Math. J. **24** (1957), 119–125.

[85] R. Ellis, *A semigroup associated with a transformation group*, Trans. Amer. Math. Soc. **94** (1960), 272–281.

[86] R. Ellis, *Lectures on topological dynamics*, Benjamin, New York, 1969.

[87] A. El-Mabhouh, J. Pym, and D. Strauss, *On the two natural products in a Stone–Čech compactification*, Semigroup Forum **48** (1994), 255–257.

[88] A. El-Mabhouh, J. Pym, and D. Strauss, *Subsemigroups of $\beta\mathbb{N}$*, Topology Appl. **60** (1994), 87–100.

[89] P. Erdős, *Problems and results on combinatorial number theory, II*, J. Indian Math. Soc. **40** (1976), 285–298.

[90] M. Filali, *The uniform compactification of a locally compact abelian group*, Proc. Cambridge. Philos. Soc. **108** (1990), 527–538.

[91] M. Filali, *Right cancellation in βS and UG*, Semigroup Forum, to appear.

[92] M. Filali, *Weak p-points and cancellation in βS*, in: Papers on General Topology and Applications, S. Andima et. al., eds., Annals of the New York Academy of Sciences **806** (1996), 130–139.

[93] M. Filali, *On the semigroup βS*, Semigroup Forum, to appear.

[94] D. Fremlin, *Consequences of Martin's Axiom*, Cambridge Univ. Press, Cambridge, 1984.

[95] Z. Frolík, *Sums of ultrafilters*, Bull. Amer. Math. Soc. **73** (1967), 87–91.

[96] Z. Frolík, *Fixed points of maps of $\beta(\mathbb{N})$*, Bull. Amer. Math. Soc. 74 (1968), 187–191.

[97] H. Furstenberg, *Ergodic behavior of diagonal measures and a theorem of Sze-merédi on arithmetic progressions*, J. Anal. Math. **31** (1977), 204–256.

[98] H. Furstenberg, *Recurrence in ergodic theory and combinatorical number theory*, Princeton University Press, Princeton, 1981.

[99] H. Furstenberg and Y. Katznelson, *Idempotents in compact semigroups and Ramsey Theory*, Israel J. Math. **68** (1989), 257–270.

[100] H. Furstenberg and B. Weiss, *Topological dynamics and combinatorial number theory*, J. Anal. Math. **34** (1978), 61–85.

[101] S. García-Ferreira, *Three orderings on $\beta(\omega)\backslash\omega$*, Topology Appl. **50** (1993), 199–216.

[102] S. García-Ferreira, *Comfort types of ultrafilters*, Proc. Amer. Math. Soc. **120** (1994), 1251–1260.

[103] S. García-Ferreira, N. Hindman and D. Strauss, *Orderings of the Stone–Čech remainder of a discrete semigroup*, Topology Appl., to appear.

[104] L. Gillman and M. Jerison, *Rings of continuous functions*, van Nostrand, Princeton, 1960.

[105] S. Glasner, *Divisibility properties and the Stone–Čech compactification*, Canad. J. Math. **32** (1980), 993–1007.

[106] S. Glasner, *On minimal actions of Polish groups*, in: Proceedings of the 8th Prague topological symposium (1996), to appear.

[107] A. Gleason, *Projective topological spaces*, Illinois J. Math. **2** (1958), 482–489.

[108] I. Glicksberg and K. de Leeuw, *Applications of almost periodic compactifications*, Acta Math. **105** (1961), 63–97.

[109] R. Graham, S. Lin, and C. Lin, *Spectra of numbers*, Math. Mag. **51** (1978), 174–176.

[110] R. Graham and B. Rothschild, *Ramsey's Theorem for n-parameter sets*, Trans. Amer. Math. Soc. **159** (1971), 257–292.

[111] R. Graham, B. Rothschild, and J. Spencer, *Ramsey Theory*, Wiley, New York, 1990.

[112] A. Grothendieck, *Critères de compacité dans les espaces fonctionels généraux*, Amer. J. Math. **74** (1952), 168–186.

[113] A. Hales and R. Jewett, *Regularity and positional games*, Trans. Amer. Math. Soc. **106** (1963), 222–229.

[114] E. Hewitt and K. Ross, *Abstract Harmonic Analysis, I*, Springer-Verlag, Berlin, 1963.

[115] E. Hewitt and K. Stromberg, *Real and Abstract Analysis*, Springer-Verlag, Berlin, 1965.

[116] D. Hilbert, *Über die Irreduzibilität ganzer rationaler Funktionen mit ganzzahligen Koeffizienten*, J. Reine Angew. Math. **110** (1892), 104–129.

[117] N. Hindman, *The existence of certain ultrafilters on* $\mathbb{N}$ *and a conjecture of Graham and Rothschild*, Proc. Amer. Math. Soc. **36** (1972), 341–346.

[118] N. Hindman, *Finite sums from sequences within cells of a partition of* $\mathbb{N}$, J. Combin. Theory Ser. A **17** (1974), 1–11.

[119] N. Hindman, *Partitions and sums of integers with repetition*, J. Combin. Theory Ser. A **27** (1979), 19–32.

[120] N. Hindman, *Partitions and sums and products of integers*, Trans. Amer. Math. Soc. **247** (1979), 227–245.

[121] N. Hindman, *Simultaneous idempotents in* $\beta\mathbb{N}\backslash\mathbb{N}$ *and finite sums and products in* $\mathbb{N}$, Proc. Amer. Math. Soc. **77** (1979), 150–154.

[122] N. Hindman, *Ultrafilters and combinatorial number theory*, in: Number Theory Carbondale 1979, M. Nathanson (ed.), Lecture Notes in Math. **751** (1979), 119–184.

[123] N. Hindman, *Partitions and sums and products – two counterexamples*, J. Combin. Theory Ser. A **29** (1980), 113–120.

[124] N. Hindman, *Sums equal to products in* $\beta\mathbb{N}$, Semigroup Forum **21** (1980), 221–255.

[125] N. Hindman, *Minimal ideals and cancellation in* $\beta\mathbb{N}$, Semigroup Forum **25** (1982), 291–310.

[126] N. Hindman, *On density, translates, and pairwise sums of integers*, J. Combin. Theory Ser. A **33** (1982), 147–157.

[127] N. Hindman, *Partitions and pairwise sums and products*, J. Combin. Theory Ser. A **37** (1984), 46–60.

[128] N. Hindman, *Ramsey's Theorem for sums, products, and arithmetic progressions*, J. Combin. Theory Ser. A **38** (1985), 82–83.

[129] N. Hindman, *The minimal ideals of a multiplicative and additive subsemigroup of* $\beta\mathbb{N}$, Semigroup Forum **32** (1985), 283–292.

[130] N. Hindman, *The ideal structure of the space of κ-uniform ultrafilters on a discrete semigroup*, Rocky Mountain J. Math. **16** (1986), 685–701.

[131] N. Hindman, *Summable ultrafilters and finite sums*, in: Logic and Combinatorics, S. Simpson (ed.), Contemp. Math. **65** (1987), 263–274.

[132] N. Hindman, *Some equivalents of the Erdős sum of reciprocals conjecture*, European J. Combin. **9** (1988), 39–47.

[133] N. Hindman, *Solving equations in $\beta\mathbb{N}$*, Annals N. Y. Acad. Sci. **552** (1989), 69–73.

[134] N. Hindman, *Ultrafilters and Ramsey Theory – an update*, in: Set Theory and its Applications, J. Steprāns and S. Watson (eds.), Lecture Notes in Math. **1401** (1989), 97–118.

[135] N. Hindman, *The semigroup $\beta\mathbb{N}$ and its applications to number theory*, in: The Analytical and Topological Theory of Semigroups — Trends and Developments, K. Hofmann, J. Lawson, and J. Pym (eds.), de Gruyter Exp. Math. **1**, Walter de Gruyter, Berlin, 1990, 347–360.

[136] N. Hindman, *Strongly summable ultrafilters on $\mathbb{N}$ and small maximal subgroups of βN*, Semigroup Forum **42** (1991), 63–75.

[137] N. Hindman, *The groups in $\beta\mathbb{N}$*, in: General Topology and Applications, S. Andima et. al. (eds.), Lecture Notes in Pure and Appl. Math. **134** (1991), 129–146.

[138] N. Hindman, *The topological-algebraic system $(\beta\mathbb{N}, +, \cdot)$*, in: Papers on General Topology and Applications, S. Andima et. al. (eds.), Annals of the New York Academy of Sciences **704** (1993), 155–163.

[139] N. Hindman, *Recent results on the algebraic structure of βS*, in: Papers on General Topology and Applications, S. Andima et. al. (eds.), Annals of the New York Academy of Sciences **767** (1995), 73–84.

[140] N. Hindman, *Algebra in βS and its applications to Ramsey Theory*, Math. Japonica **44** (1996), 581–625.

[141] N. Hindman, J. Lawson, and A. Lisan, *Separating points of $\beta\mathbb{N}$ by minimal flows*, Canadian J. Math. **46** (1994), 758–771.

[142] N. Hindman and I. Leader, *Image partition regularity of matrices*, Combin. Probab. Comput. **2** (1993), 437–463.

[143] N. Hindman and I. Leader, *The semigroup of ultrafilters near 0*, Semigroup Forum, to appear.

[144] N. Hindman and H. Lefmann, *Partition regularity of* ($\mathcal{M}$, $\mathcal{P}$, $\mathcal{C}$)-*systems*, J. Combin. Theory Ser. A **64** (1993), 1–9.

[145] N. Hindman and H. Lefmann, *Canonical partition relations for (m,p,c)-systems*, Discrete Math. **162** (1996), 151–174.

[146] N. Hindman and A. Lisan, *Does* $\mathbb{N}^*$*contain a topological and algebraic copy of* $\beta\mathbb{N}$? Topology Appl. **35** (1990), 291–297.

[147] N. Hindman and A. Lisan, *Points very close to the smallest ideal of* βS, Semigroup Forum **49** (1994), 137–141.

[148] N. Hindman, A. Maleki, and D. Strauss, *Central sets and their combinatorial characterization*, J. Combin. Theory Ser. A **74** (1996), 188–208.

[149] N. Hindman, J. van Mill, and P. Simon, *Increasing chains of ideals and orbit closures in* $\beta\mathbb{Z}$, Proc. Amer. Math. Soc. **114** (1992), 1167–1172.

[150] N. Hindman and P. Milnes, *The* $\mathcal{LMC}$-*compactification of a topologized semigroup*, Czechoslovak Math. J. **38** (1988), 103–119.

[151] N. Hindman, I. Protasov, and D. Strauss, *Strongly Summable Ultrafilters*, manuscript.

[152] N. Hindman and J. Pym, *Free groups and semigroups in* $\beta\mathbb{N}$, Semigroup Forum **30** (1984), 177–193.

[153] N. Hindman and J. Pym, *Closures of singly generated subsemigroups of* βS, Semigroup Forum **42** (1991), 147–154.

[154] N. Hindman and D. Strauss, *Cancellation in the Stone–Čech compactification of a discrete semigroup*, Proc. Edinburgh Math. Soc. **37** (1994), 379–397.

[155] N. Hindman and D. Strauss, *Nearly prime subsemigroups of* $\beta\mathbb{N}$, Semigroup Forum **51** (1995), 299–318.

[156] N. Hindman and D. Strauss, *Topological and algebraic copies of* $\mathbb{N}^*$ *in* $\mathbb{N}^*$, New York J. Math. **1** (1995), 111–119.

[157] N. Hindman and D. Strauss, *Chains of idemptotents in* $\beta\mathbb{N}$, Proc. Amer. Math. Soc. **123** (1995), 3881–3888.

[158] N. Hindman and D. Strauss, *Algebraic and topological equivalences in the Stone–Čech compactification of a discrete semigroup*, Topology Appl. **66** (1995), 185–198.

[159] N. Hindman and D. Strauss, *Prime properties of the smallest ideal of* $\beta\mathbb{N}$, Semigroup Forum **52** (1996), 357–364.

[160] N. Hindman and D. Strauss, *Compact subsemigroups of* ($\beta\mathbb{N}$, +) *containing the idempotents*, Proc. Edinburgh Math. Soc. **39** (1996), 291–307.

[161] N. Hindman and D. Strauss. *An algebraic proof of Deuber's Theorem*, Combin. Probab. Comput., to appear.

[162] N. Hindman and W. Woan, *Central sets in semigroups and partition regularity of systems of linear equations*, Mathematika **40** (1993), 169–186.

[163] K. Hofmann and P. Mostert, *Elements of compact semigroups*, Charles E. Merrill, Columbus, 1966.

[164] S. Hong-ting and Y. Hong-wei, *Nonmetrizable topological dynamic characterization of central sets*, Fund. Math., to appear.

[165] P. Johnstone, *Stone spaces*, Cambridge Univ. Press, Cambridge, 1982.

[166] R. Kadison and J. Ringrose, *Fundamentals of the theory of operator algebras (Volume 1)*, Academic Press, New York, 1983.

[167] M. Katetǒv, *A theorem on mappings*, Comment. Math. Univ. Carolin. **8** (1967), 431–433.

[168] M. Katetǒv, *Characters and types of point sets*, Fund. Math. **50** (1961), 369–380.

[169] J. Kelley, *General topology*, van Nostrand, New York, 1955.

[170] K. Kunen, *Weak P-points in* $\mathbb{N}^*$, Colloq. Math. Soc. János Bolyai **23** (1978), 741–749.

[171] K. Kunen, *Set theory: an introduction to independence proofs*, North Holland, Amsterdam, 1980.

[172] A. Lau, P. Milnes, and J. Pym, *Locally compact groups, invariant means and the centres of compactifications*, J. London Math. Soc. **56** (1997), 77–90.

[173] J. Lawson, *Joint continuity in semitopological semigroups*, Illinois Math. J. **18** (1972), 275–285.

[174] J. Lawson, *Points of continuity for semigroup actions*, Trans. Amer. Math. Soc. **284** (1984), 183–202.

[175] J. Lawson and A. Lisan, *Transitive flows: a semigroup approach*, Mathematika **38** (1991), 348–361.

[176] K. Leeb, *A full Ramsey-theorem for the Deuber-category*, Colloq. Math. Soc. János Bolyai **10** (1975), 1043–1049.

[177] A. Lisan, *The ideal structure of the space of ultrafilters on a discrete semigroup*, Ph.D. Dissertation (1988), Howard University.

468 Bibliography

[212] M. Rudin, *Partial orders on the types of* $\beta\mathbb{N}$, Trans. Amer. Math. Soc. **155** (1971), 353–362.

[213] M. Rudin and S. Shelah, *Unordered Types of Ultrafilters*, Topology Proc. **3** (1978), 199–204.

[214] W. Rudin, *Homogeneity problems in the theory of Čech compactifications*, Duke Math. J. **23** (1966), 409–419.

[215] W. Ruppert, *Rechtstopologische Halbgruppen*, J. Reine Angew. Math. **261** (1973), 123–133.

[216] W. Ruppert, *Compact semitopological semigroups: an intrinsic theory*, Lecture Notes in Math. **1079**, Springer-Verlag, Berlin, 1984.

[217] W. Ruppert, *On weakly almost periodic sets*, Semigroup Forum **32** (1985), 267–281.

[218] W. Ruppert, *In a left-topological semigroup with dense center the closure of any left ideal is an ideal*, Semigroup Forum **36** (1987), 247.

[219] W. Ruppert, *Endomorphic actions of* $\beta\mathbb{N}$ *on the torus group*, Semigroup Forum **43** (1991), 202–217.

[220] W. Ruppert, *On signed* α*-adic expansions and weakly almost periodic functions*, Proc. London Math. Soc. **63** (1991), 620–656.

[221] I. Schur, *Über die Kongruenz* $x^m + y^m = z^m$ (mod p), Jahresber. Deutsch. Math.-Verein. **25** (1916), 114–117.

[222] S. Shelah, *Proper forcing*, Springer-Verlag, Berlin, 1982.

[223] S. Shelah and M. Rudin, *Unordered types of ultrafilters*, Topology Proc. **3** (1978), 199–204.

[224] T. Skolem, *Über einige Eigenschaften der Zahlenmengen* $[\alpha n + \beta]$ *bei irrationalem* α *mit einleitenden Bemerkungen über einige kombinatorische Probleme*, Norske Vid. Selsk. Forh. **30** (1957), 42–49.

[225] G. Smith, *Partition regularity of sums of products of natural numbers*, Ph.D. Dissertation (1994), Howard University.

[226] G. Smith, *Partitions and (m and n) sums of products*, J. Combin. Theory Ser. A **72** (1995), 77–94.

[227] M. Stone, *Applications of the theory of Boolean rings to general topology*, Trans. Amer. Math. Soc. **41** (1937), 375–481.

[228] D. Strauss, $\mathbb{N}^*$*does not contain an algebraic and topological copy of* $\beta\mathbb{N}$, J. London Math. Soc. **46** (1992), 463–470.

[229] D. Strauss, *Semigroup structures on βℕ*, Semigroup Forum **41** (1992), 238–244.

[230] D. Strauss, *Ideals and commutativity in βℕ*, Topology Appl. **60** (1994), 281–293.

[231] A. Suschkewitsch, *Über die endlichen Gruppen ohne das Gesetz der eindeutigen Umkehrbarkeit*, Math. Ann. **99** (1928), 30–50.

[232] E. Szemerédi, *On sets of integers containing no k elements in arithmetic progression*, Acta. Math. **27** (1975), 199–245.

[233] A. Taylor, *A canonical partition relation for finite subsets of ω*, J. Combin. Theory Ser. A **21** (1976), 137–146.

[234] E. Terry, *Finite sums and products in Ramsey Theory*, Ph.D. Dissertation (1997), Howard University.

[235] H. Umoh, *Ideals of the Stone–Čech compactification of semigroups*, Semigroup Forum **32** (1985), 201–214.

[236] H. Umoh, *The ideal of products in βS\S*, Ph.D. Dissertation (1987), Howard University.

[237] H. Umoh, *Maximal density ideal of βS*, Math. Japonica **42** (1995), 245–247.

[238] B. van der Waerden, *Beweis einer Baudetschen Vermutung*, Nieuw Arch. Wiskunde **19** (1927), 212–216.

[239] R. Walker, *The Stone–Čech compactification*, Springer-Verlag, Berlin, 1974.

[240] A. Wallace, *A note on mobs, I*, An. Acad. Brasil. Ciênc. **24** (1952), 329–334.

[241] A. Wallace, *A note on mobs, II*, An. Acad. Brasil. Ciênc. **25** (1953), 335–336.

[242] A. Wallace, *The structure of topological semigroups*, Bull. Amer. Math. Soc. **61** (1955), 95–112.

[243] A. Wallace, *The Rees–Suschkewitsch structure theorem for compact simple semigroups*, Proc. National Acad. Sciences **42** (1956), 430–432.

[244] H. Wallman, *Lattices and topological spaces*, Ann. of Math. **39** (1938), 112–126.

[245] T. West, *Weakly monothetic semigroups of operators in Banach spaces*, Proc. Roy. Irish Acad. Sect. A **67** (1968), 27–37.

[246] E. Zelenuk, *Finite groups in βℕ are trivial* (Russian), Ukranian National Academy of Sciences Institute of Mathematics, Preprint 96.3 (1996).

[247] E. Zelenuk, *Topological groups with finite semigroups of ultrafilters* (Russian), Matematychni Studii **6** (1996), 41–52.

List of Symbols

Index

action
    measure preserving, 407
    monotone, 408
almost disjoint family 238
almost left invariant 121
almost periodic function 436, 439
alphabet 281
Anthony, P. 275
anti-homomorphism 4
anti-isomorphic 4
anti-isomorphism 4
Arens, R. 88
arithmetic progression 280
associative 3
associativity of $(\beta S, \cdot)$ 73
Auslander, J. 47, 411
Axiom of Choice 69

Baire set 361, 387
Baker, J. viii, 88, 135, 204, 221, 453
Balcar, B. 275, 411
Banach density 412
base
    filter, 48, 52
Baumgartner, J. 102
Bergelson, V. viii, 103, 221, 295, 319,
        338, 367, 368, 393, 411, 424
Berglund, J. viii, 47, 135, 453
bicyclic 22
binary expansion 107
Blass, A. viii, 103, 185, 234, 257, 393
Boolean algebra 52, 71
    homomorphism, 52
Budak, T. 135
Butcher, R. 88

c.c.c. 237

cancelability
    and $K(S)$, 30
    and $K(\beta S)$, 169
    in $\beta\mathbb{Z}$, 176
cancelable elements in βS 175
cancelable elements in $\beta\mathbb{Z}$ 175
cancelable elements of S
    cancelable in βS, 158
cancelable elements 115
cancelable
    right, 161
cancellative 7
    βS not, 158
    weakly left, 122, 323
    weakly right, 122
cardinality of U_κ 67
cardinality of βD 67
cardinality of closed sets in βD 67
cardinal 2
Carlson's Theorem 381, 386
Carlson, T. 386, 393
Čech, E. 69
center 7, 38, 80
    $(\mathbb{Z}, +)$, 107
    of $p(\beta S)p$, 129, 130
    of $p + \beta\mathbb{N} + p$, 130
    of $U_\kappa(S)$ empty, 124
    of βS equal to that of S, 125
Central Sets Theorem 279, 294
    commutative, 284
    noncommutative, 287
    satisfied by noncentral set, 289
central* sets 297
    examples, 337
centralizer
    nowhere dense in $U_\kappa(S)$, 124
central 86, 279, 284, 320